HIGHER
ALGEBRA

HIGHER ALGEBRA

S BARNARD
J M CHILD

New Academic Science
An Imprint of
New Age International (UK) Ltd.
27 Old Gloucester Street, London, WC1N 3AX, UK
www.newacademicscience.co.uk • e-mail: info@newacademicscience.co.uk

Preface

This volume is essentially a textbook of Algebra for students working for Higher School Certificate, mathematical scholarships, and examinations of similar standard; it will be followed by a second volume, entitled *Advanced Algebra*, in which the subject will be developed further, along the lines of honours courses in Universities. It is assumed that students possess a knowledge of Elementary Algebra so far as Progressions and Permutations and Combinations.

In the preparation of the present volume, the authors have aimed at producing a treatise in which the subject is developed logically, complete so far as it goes, and serving as an introduction to modern analysis. The scope, and the order of treatment, have, however, been largely determined by the consideration of questions set for mathematical scholarships and similar examinations in recent years.

In Chaps. 1, 2 and 5, starting with the positive integers, the number system is extended to include irrational and complex numbers. In Chap. 2, an irrational is regarded as defined by some rule which gives it ordinal rank with the rational numbers, so that it can be represented by an endless decimal. This is sufficient for practical purposes; and only those who intend to specialise in Pure Mathematics require Chap. 13, which contains the Theory of Irrationals, based on Dedekind's definitions.

Limits and Continuity are fully discussed in Chaps. 15, 17, 19. In Chap. 17, the symbol d/dx is introduced and the rules of differentiation are proved.

A full account of the Theory of Equations, so far as Sturm's Theorem and the numerical solution of equations, is given in Chaps. 6, 11, 12, 18, 28; so that no separate textbook on this subject is required except by advanced students.

Summation of series is fully discussed in Chap. 8, and in the chapters on the Exponential and Binomial Theorems. Consideration of the series $1^r + 2^r + \ldots n^r$ leads to a rapid way of calculating Bernoulli's Numbers.

All the usual tests for the convergence of infinite series and products are given in Chaps. 16, 20, 30. For the Binomial Theorem, in addition to Euler's proof, a next proof is given for the case of a real index, depending entirely on elementary considerations.

Chapter 9 contains all the elementary properties of determinants, including those of Symmetric and Skew-symmetric Determinants.

All the fundamental inequalities occur in Chaps. 14, 19, 20.

Chapters 1, 26, 27 deal with the Theory of Numbers. The subject is carried further than in the usual textbooks; congruences and methods of solving them, Primitive Roots and Recurring Decimals are considered.

Elementary Finite Difference Equations are introduced in Chap. 22, on account of their usefulness in reckoning.

Curve Tracing is considered in Chap. 29, in connection with Implicit Functions, and as affording the best possible exercise in 'Approximations.'

In Chaps. 24, 33, simple and recurring Continued Fractions are discussed. A new method is given for finding the period belonging to $\left(\sqrt{N}+b\right)/r$ by the G.C.M. process.

Most of the examples are from the Mathematical Tripos, Parts I and II, examinations for Oxford Senior and Junior Mathematical Scholarships, Cambridge Entrance Scholarships, Higher School Certificate, and similar examinations. A few of them have been taken from Wolstenholme's *Mathematical Problems*. Some of those in Exercises LIII, LIV, LV are from Whitworth's *Choice and Chance;* and some, in Exercise LI are from Frost's *Curve Tracing*. A large proportion of these have formed the 'stock' used by the authors for teaching purposes; and their source cannot now be definitely stated. The rest are original; and for many of the most interesting the authors are indebted to Dr. G.T. Bennett.

Among the books to which the authors have made reference are: Bromwich, *Theory of Infinite Series*; Chrystal, *Algebra*; Frost, *Curve Tracing*; Gauss, *Recherches Arithmétiques*; Hardy, *Pure Mathematics*; Salmon, *Higher Algebra;* Serret, *Algèbre Supérieure*; Tannery, *Lecons d'Algèbre et d'Analyse*; Whitworth, *Choice and Chance.*

The solution of $x^{17}-1=0$, and the construction of a regular polygon of 17 sides, are due to H.W. Richmond and G.T. Bennett.

The suggested course of reading for Higher School Certificate (for example, Group IV for Northern Universities) is as follows:

Chapters (1), 1-7, 11; (2), 11-17; (3; 4), 1-3; (5; 6), 1-10, 16, 17; (7; 8), 1-7 ; (9), 1-16 ; (10), 1-4, 6, 8, (11); 11, 1; (12), 1-8, 12, 14 ; (16), 1-15, 18, 19, 23-25, (17), 1, 2, 4 ; (18), 1-4 ; (19), 1-4, 11-13, 16-18, 21, 22 ; (21), 1-3, 6-10: (22), 1-3 ; (28), 1-6, 9-11 ; (29), 1, 2, 4 to end; (31), 1-4.

The course for scholarships, Final Ordinary, and First Year Honours at the Universities is practically the whole of the book; except that Chaps. 8, 5-7, 11, 5, 6, 17, 23, and the ' strict' proof of Rolle's Theorem may be omitted, at any rate on first reading ; while Chap. 13 need only be 'read' to obtain a general idea of its contents.

The authors are under a deep obligation to Dr. G.T. Bennett for much valuable assistance in many parts of the text, for many excellent examples, and for his criticisms on the proofs; also to Mr. W. J. Fearn, Mr. W. J. Lewis, and Mr. J. Kilgour, for leading the proofs and for many valuable suggestions

S. BARNARD
J. M. CHILD

Publisher's Note

This book has been thoroughly read by our subject experts and efforts have been made to present an error free edition. Though the overall format is not disturbed but the presentation and the rubric has been amplified. More clarity is incorporated wherever necessary and the figures are redrawn for better understanding. Suggestions for further improvement shall be received with appreciation.

PUBLISHER

Contents

CHAPTER 1

Theory of Numbers

In this section, we discuss properties peculiar to whole numbers. The word 'number' is taken to mean 'whole number', and, unless otherwise stated, 'positive whole number.'

■ 1. Division:

(1) Let b be any positive whole number, and consider the sequence

$$... -3b, -2b, -b, 0, b, 2b, 3b, ...$$

continued indefinitely both ways. Any whole number a (positive, negative or zero) is either a term of this sequence, or it lies between two consecutive terms. Thus two numbers q and r can be determined uniquely so that

$$a = bq + r \text{ and } 0 \leq r < b. \qquad ...(A)$$

To *divide* a by b is to find the numbers q and r which satisfy these conditions; q is called the *quotient* and r the *remainder.*

If $r = 0$, we say that a is *divisible* by b or is a *multiple* of b, and that b is a *divisor* or a *factor* of a. Among the divisors of a number we count the number itself and 1.

Whatever b may be, $0 = b \cdot 0 + 0$; hence zero must be regarded as divisible by every whole number.

(2) If $r = b - r'$, we have

$$a = b(q + 1) - r' \quad \text{and} \quad 0 \leq r' < b;$$

also if $r \geq \dfrac{1}{2} b$ then $r' \leq \dfrac{1}{2} b$. Hence it is always possible to find numbers Q and R such that

$$a = bQ + R \quad \text{and} \quad |R| \leq \dfrac{1}{2} b.$$

▋ EXAMPLE 1. *Every number is of one of the forms 5n, 5n ± 1, 5n ± 2.*

For if any number is divided by 5, the remainder is one of numbers 0, 1, 2, 5 – 2, 5 – 1.

▋ EXAMPLE 2. *Every square number is of one of the forms 5n, 5n ± 1.*

The square of every number is of one of the forms $(5m)^2$, $(5m \pm 1)^2$, $(5m \pm 2)^2$. If these are divided by 5, the remainders are 0, 1, 4; and, since $4 = 5 - 1$, the forms are $5n$, $5n + 1$, and $5n - 1$.

▌ **EXAMPLE 3.** *The square of every odd number is of the form $8n + 1$.*

For $(2k + 1)^2 = 4k(k + 1) + 1$, and either k or $k + 1$ must be even, so that $k(k + 1)$ is divisible by 2.

▨ **2. Theorems on Division:**

(1) *If both a and b are divisible by c, so also is $ma \pm nb$.*

(2) *If r is the remainder when a is divided by b, then cr is the remainder when ca is divided by cb.*

For if $\qquad\qquad a = bq + r \qquad$ and $\qquad 0 \le r < b,$

then $\qquad\qquad ca = (cb)q + cr \qquad$ and $\qquad 0 \le cr < cb.$

(3) *If both a and b are divisible by c, and r is the remainder when a is divided by b, then (i) r is divisible by c and (ii) r/c is the remainder when a/c is divided by b/c.*

For if $a = bq + r$ and $0 \le r < b$, then since a and b are divisible by c, so also is r, which is equal to $a - bq$. Thus we have

$$a/c = (b/c)q + r/c \text{ where } 0 \le r/c < b/c.$$

▨ **3. Theorems on the Greatest Common Divisor:**

(1) *If r is the remainder when a is divided by b, the common divisors of a and b are the same as those of b and r.*

For, since $a = bq + r$, every common divisor of b and r is a divisor of a; and, since $r = a - bq$, every common divisor of a and b is a divisor of r. This proves the statement in question.

(2) *If a and b are any two numbers, there exists a number g, and one only, such that the common divisors of a and b are the same as the divisors of g.*

For numbers (q_1, r_1), (q_2, r_2), etc., can always be found in succession, and uniquely,

so that $\qquad a = bq_1 + r_1, b = r_1q_2 + r_2, \quad r_1 = r_2q_3 + r_3,$ etc., $\qquad$ ...(A)

where $\qquad\qquad b > r_1 > r_2 > r_3 \ldots \ge 0.$

Since the number of positive integers less than b is limited, a zero remainder must occur; and, supposing that $r_{n+1} = 0$, the process terminates with

$$r_{n-2} = r_{n-1}q_n + r_n, r_{n-1} = r_nq_{n+1}.$$

Hence, by (A), the following pairs of numbers have the same common divisors:

$$(a, b) (b, r_1), (r_1, r_2), \ldots, (r_{n-1}, r_n);$$

and, since $r_{n-1} = r_nq_{n+1}$, the common divisors of r_{n-1} and r_n are the divisors of r_n. Hence the number g exists, and its value is r_n.

Since g is divisible by any common divisor of a and b, it is the *greatest common divisor* of these numbers and is known in elementary arithmetic as the greatest common measure (G.C.M.) of a and b. *

(3) *If a and b are each multiplied by any number m, or are divided by a common divisor m, then g, the greatest common divisor of a and b, is multiplied or divided by m.*

For each of the r's in the proof of (2) is multiplied or divided by m.

(4) *If a, b, c, ..., are serveral numbers, and g_1 is the greatest common divisor of a and b, g_2 that of g_1 and c, g_3 that fo g_2 and d, and so on, then the following sets of*

---

* The process, here stated algebraically, is that used in Elementary Arithmetic.

numbers have the same common divisors:

$$(a,\ b,\ c,\ d,\ ...),\ (g_1,\ b,\ c,\ d,\ ...),\ (g_2,\ c,\ d,\ ...),\ \text{etc.}$$

Hence we arrive eventually at a number g, whose divisors are the common divisors of a, b, c, ...; also g is uniquely determined. This number g is the *greatest common divisor*, or the *greatest common measure* of a, b, c, ...

■ **4. Numbers Prime to each other:** Two numbers, a and b, are said to be prime to each other when their greatest common divisor is 1, so that they have no common divisor except 1. This is often expressed by saying that a is prime to b, or that b is prime to a. The following theorems are fundamental.

(1) *If the product ab is divisible by a number m, and m is prime to one factor a, then m is a divisor of the other factor b.*

For, if a is prime to m, the greatest common divisor of a and m is 1; hence the greatest common divisor of ab and mb is b. But, by hypothesis, m is a divisor of ab, and is therefore a common divisor of ab and mb; hence m is equal to, or is a divisor of, b.

(2) *If a is prime to b, and each of these numbers is a divisor of N, then ab is a divisor of N.*

Suppose that $N = aq$, then b is a divisor of aq, and since b is prime to one factor a, it must be a divisor of the other factor q. Let $q = mb$, then $N = mab$, and ab is a divisor of N.

(3) *If a is prime to b, positive integers x, y can be found such that*

$$ax - by = \pm 1.$$

For it follows from equations (A) of Art. 3, (2), that

$$r_1 = a - bq_1,\ r_2 = -aq_2 + b\,(1 + q_1q_2),$$
$$r_3 = a\,(1 + q_2\,q_3) - b\,(q_1 + q_3 + q_1q_2q_3).$$

Continuing thus, *every remainder can be expressed in the form $\pm\,(ax - by)$, where x, y are positive integers.*

If a is prime to b, the last remainder is 1, and the theorem follows.

■ **5. Theorems on Prime Numbers:** A number which has no divisors except itself and 1 is called a *prime number*, or simply a *prime*. Numbers which are not prime are said to be *composite*.

(1) *A prime number, p, is prime to every number which is not a multiple of p.*

For, if a is any such number, q and r can be found such that $a = pq + r$ where $0 < r < p$. Now p and 1 are the only divisors of p, and as $r < p$, the only common divisor of p and r is 1, that is to say, p is prime to r and therefore to a.

(2) *If a prime p is a divisor of a product $abcd ... hk$, it is a divisor of at least one of the factors a, b, ... k.*

For if the prime p is not a divisor of a, by Theorem (1) it is prime to a, hence it is a divisor of $bcd .... k$. If, in addition, p is not a divisor of b, it must be a divisor of $cd ... k$. Continuing thus, it can be shown that if p is not a divisor of any of the numbers a, b, c, ... h, it must be a divisor of k.

■ **6. Theorems on Numbers, Prime or Composite:**

(1) *Every composite number N has at least one prime divisor.*

For, since N is not a prime, it has a divisor, n, different from N and from 1, which is *not greater than any other divisor*. Further, this divisor must be a prime; for, otherwise, it would have a divisor *less* than itself and greater than 1, and this latter would be a divisor of N. This contradicts the hypothesis that n is not greater than any other divisor.

It follows that every composite number, N, can be expressed as the product of prime factors.

For, since N has at least one prime factor, p, we have $N = pa$, where $1 < p < N$. If a is not a prime, it has at least one prime factor, q; and $a = qb$, where $1 < b < a$.

Thus, $N = pa = pqb$; and so on.

But the numbers less than N are limited; and $N > a > b > \ldots$, therefore the set N, a, b, ... must finally end in a prime. Hence, N can be expressed in the form, $N = pqr \ldots u$, where $p, q, r, \ldots, u$ are all primes, not necessarily all different.

That is to say, any composite number, N, can be expressed as
$$N = p^a . q^b . r^c \ldots u^s,$$
where p, q, r, ... , u are all different primes.

(2) *A composite number can be expressed as the product of prime factors in one way only.*

For suppose that $N = p^a q^b r^c \ldots = P^A Q^B R^C \ldots$ where p, q, r, ..., P, Q, R, ..., are primes; then, since the prime P is a divisor of the product $p^a q^b r^c \ldots$, it is a divisor of one of the factors p, q, r, ... , and is therefore equal to one of them. In the same way each of the set P, Q, R, ... is equal to one of the set p, q, r, ... , and no prime factor can occur in one of the expressions for N which does not occur in the other. Suppose then that
$$N = p^a q^b r^c \ldots = p^A q^B r^C \ldots .$$

If $a \neq A$, one of them must be the greater. Let $A > a$, and suppose that $A = a + e$, then $q^b . r^c . \ldots t^m = p^e . q^B . r^C . \ldots t^M$; but this is impossible, as the left-hand side of the equality is a number prime to p, and the right-hand side is divisible by p.

Hence, a must be equal to A; and similarly, $b = B$, ..., $m = M$; and thus the two expressions for N are identical.

The above theorem is one of the most important in the Theory of Numbers, and the following propositions are immediately deducible.

(3) *If m is prime to each of the numbers, a b, ..., k, it is prime to the product $ab \ldots k$.*

(4) *If a is prime to b, then a^n is prime to b^n, where n is any integer; and conversely.*

(5) *A number N is a prime, if it is not divisible by any prime number greater than 1 and less than, or equal to, $\sqrt{N}$.*

For, if $N = ab$, where $b \geq a$, then $N \geq a^2$; that is, $a \leq \sqrt{N}$.

(6) *The sequence of primes is endless.*

For, if p is any prime, the number $\lfloor p + 1$ is greater than p and is not divisible by p or by any smaller prime. If then $\lfloor p + 1$ is not a prime, it must have a prime divisor greater than p, and in either case a prime greater than p exists.

All the primes less than a given number N can be obtained in order by a process called the *Sieve of Eratosthenes*. The process consists in writing down in order all the numbers from 1 to $N - 1$, and erasing all multiples of the primes 2, 3, 5, 7, ... which are less than $\sqrt{N}$. Nevertheless, the problem of discovering whether a large number is prime or composite is one of great difficulty.

EXAMPLE 1. *If n is any number, prove that n(n + 1) (n + 2) is divisible by 6.*

Of the two consecutive numbers, n and $n + 1$, one is divisible by 2; and one of the three consecutive numbers, n, $n + 1$, $n + 2$, is divisible by 3. Hence the product $n (n + 1) (n + 2)$ is divisible by 2 and by 3; and, since 2 is prime to 3, $n (n + 1) (n + 2)$ is divisible by 6.

EXAMPLE 2. *Prove that $3^{2n + 1} + 2^{n + 2}$ is divisible by 7.*

We have $3^{2n+1} = 3 . 9^n = 3 (7 + 2)^n = 7k + 3 . 2^n$, by the Binomial theorem,

and
$$2^{n+2} = 4 . 2^n;$$

$$\therefore \quad 3^{2n + 1} + 2^{n + 2} = 7k + 2^n (3 + 4) = 7 (k + 2^n).$$

7. The Divisors of a Given Number N: *Let $N = p^a . q^b . r^c. ...$, where p, q, r, ..., are primes; and let n be the number, and s the sum, of the divisors of N, including 1 and N.*

Then, **(1)**
$$n = (a + 1) (b + 1) (c + 1) ....;$$

$$s = \frac{p^{a+1} - 1}{p - 1} . \frac{q^{b+1} - 1}{q - 1} . \frac{r^{c+1} - 1}{r - 1} . ... .$$

For the divisors of N are the terms in the expansion of
$$(1 + p + p^2 + ... + p^a) (1 + q + q^2 + ... + q^b) (1 + r + r^2 + ... + r^c) ... ;$$
and the expressions for n and s follow immediately.

(2) *The number of ways in which N can be expressed as the product of two factors, including N and 1, is $\frac{1}{2} (n + 1)$ or $\frac{1}{2} n$, according as N is, or is not, a perfect square.*

For, if N is a perfect square, each of the numbers, a, b, c, ..., is even, and therefore n is odd; but if N is not a perfect square, at least one of these numbers is odd, and therefore n is even.

Further, if $d_1 (= 1), d_2, d_3, ..., d_{n-2}, d_{n-1}, d_n (= N)$, are the divisors of N in ascending order, the different ways of expressing N as the product of two factors are:
$$d_1 d_n, d_2 d_{n-1}, d_3 d_{n-2}, ..., d_x d_x, \quad \text{when } n = 2x - 1,$$
and
$$d_1 d_n, d_2 d_{n-1}, d_3 d_{n-2}, ..., d_y d_{y+1}, \quad \text{when } n = 2y.$$

Hence, the number of ways is either x or y; *i.e.* either $\frac{1}{2} (n + 1)$ or $\frac{1}{2} n$, according as N is, *or* is not, a perfect square.

(3) *The number of ways in which N can be expressed as the product of two factors, which are prime to one another, is 2^{m-1}, where m is the number of different prime factors of N.*

For, such factors are the terms in the expansion of
$$(1 + p^a) (1 + q^b) (1 + r^c) ... ;$$
and their number is 2^m. Hence the number of pairs is 2^{m-1}.

For example, if $N = 2^2 . 3^3 . 5 = 540, n = (2 + 1) (3 + 1) (1 + 1) = 24,$

$$s = \frac{2^3 - 1}{2 - 1} . \frac{3^4 - 1}{3 - 1} . \frac{5^2 - 1}{5 - 1} = 1680, \text{ and } m = 3, 2^{m-1} = 4.$$

■ **8. The Symbol I [x/y]:** If a is a fraction or an irrational number, the symbol $I(a)$ will be used to denote the integral part of a. Thus if $x = qy + r$ where $0 \leq r < y$, then $I(x/y) = q$.

(1) *If $n_1, n_2, n_3, \ldots$, are any integers, and s is their sum and a is any number, then*

$$I[s/a] \geq I[n_1/a] + I[n_2/a] + I[n_3/a] + \ldots.$$

Let $n_1 = aq_1 + r_1$, $n_2 = aq_2 + r_2$, $n_3 = aq_3 + r_3$, etc.; then

$$s = a(q_1 + q_2 + q_3 + \ldots) + (r_1 + r_2 + r_3 + \ldots).$$

Hence, $\quad I[s/a] = (q_1 + q_2 + q_3 + \ldots) + I[(r_1 + r_2 + r_3 + \ldots)/a];$

and, since $q_1 = I[n_1/a]$, $q_2 = I[n_2/a], \ldots$, the result follows.

(2) *The highest power of a prime p which is contained in $\lfloor n$ is*

$$I[n/p] + I[n/p^2] + I[n/p^3] + \ldots.$$

For, of the numbers from 1 to n inclusive, there are $I[n/p]$ which are divisible by p; of these $I[n/p^2]$ are divisible by p^2 ; and so on; hence the result follows.

■ **9. Theorems:**

(1) *The product of any n consecutive integers is divisible by $\lfloor n$.*

For $(m + 1)(m + 2) \ldots (m + n)/\lfloor n = \lfloor m+n/\lfloor m \lfloor n$, and to show that the last expression is an integer it is sufficient to show that any prime p which occurs in $\lfloor m \lfloor n$ occurs to at least as high a power in $\lfloor m+n$. Thus we have to show that

$$I[(m + n)/p] + I[(m + n)/p^2] + I[(m + n)/p^3] + \ldots$$
$$\geq I[m/p] + I[m/p^2] + I[m/p^3] + \ldots$$
$$+ I[n/p] + I[n/p^2] + I[n/p^3] + \ldots.$$

Now $I[(m + n)/p] \geq I[m/p] + I[n/p]$, and the same is true if we replace p by p^2, $p^3, \ldots$, in succession: hence the result in question.

(2) *If n is a prime, C_r^n is divisible by n.*

For by the preceding $n(n - 1)(n - 2) \ldots (n - r + 1)$ is divisible by $\lfloor r$, and since n is a prime and r is supposed to be less than n, $\lfloor r$ is prime to n.

Hence, $\lfloor r$ is a divisor of $(n - 1)(n - 2) \ldots (n - r + 1)$

and $n(n - 1) \ldots (n - r + 1)/\lfloor r$ is divisible by n.

Thus *if n is a prime, all the coefficients in the expansion of $(1 + x)^n$, except the first and last, are divisible by n.*

> **Note:** The reader is supposed to be acquainted with what is said in elementary textbooks about 'permutations and combinations' and the 'binomial theorem for a positive integral index.' In what follows, P_r^n, denotes the number of permutations, and C_r^n the number of combinations, of *n things taken r at a time.*

▮ **EXAMPLE 1.** *Find the highest power of 5 contained in $\lfloor 158$.*

We have $I[158/5] = 31$, $I[158/5^2] = I[31/5] = 6$, $I[158/5^3] = I[6/5] = 1$: therefore the required power has an index $= 31 + 6 + 1 = 38$.

▮ **EXAMPLE 2.** *If n is an odd prime, the integral part of $(\sqrt{5}+2)^n - 2^{n+1}$ is divisible by 20n.*

Let $(\sqrt{5}+2)^n = N + f$ where $0 < f < 1$; and let $(\sqrt{5}+2)^n = f'$. Then, since $0 < \sqrt{5}-2 < 1$, we have also $0 < f' < 1$.

Again, since n is odd, $N + f - f' = (\sqrt{5}+2)^n - (\sqrt{5}-2)^n =$ an integer; hence, since f and f' are positive and less than 1, $f = f'$, and thus

$$N = 2 \left(C_1^n . 2 . 5^{\frac{1}{2}(n-1)} + C_3^n . 2^3 . 5^{\frac{1}{2}(n-3)} + \ldots + C_2^n . 2^{n-2}. 5 + 2^n \right).$$

Moreover, since n is a prime, C_r^n, is divisible by n, and therefore $N - 2^{n+1}$ is divisible by $20n$; which is the required result.

▨ **10. Numbers in Arithmetical Progression:**

(1) *Let a be prime to n, then if the first n terms of the arithmetical progression x, x + a, x + 2a, … , are divided by n, the remainders are the numbers 0, 1, 2, …, n – 1,* taken in a certain order.

If we suppose that two of the terms as $x + ma$, $x + m'a$ leave equal remainders, then their difference $(m - m')a$ would be divisible by n. This is impossible, for a is prime to n and $|m - m'| < n$.

Therefore the remainders are all different, and as each is less than n, they must be numbers 0, 1, 2, …, $n - 1$, taken in some order or other.

If the progression is continued beyond the n-th term, the remainders recur in the same order.

For the terms $x + ma$, $x + m'a$ leave the same remainder if $m = m' + qn$.

(2) *If a and n are not prime to one another and g is their greatest common divisor, the remainder recur in a cycle of n/g numbers.*

For let $a = ga'$, $n = gn'$, so that a' is prime to n'. The terms $x + ma$, $x + m'a$ leave the same remainder if, and only if, $a(m - m')$ is divisible by n, that is, if $a'(m - m')$ is divisible by n'. Since a' is prime to n', this can only happen when $m - m'$ is divisible by n'. Thus the first n' terms leave different remainders and, after that, they recur in order.

▨ **11. Method of Induction:** Many theorems relating to whole numbers can be proved by a process known as *mathematical induction*. In some cases this is the only method available. The method may be described as follows.

Let $f(n)$ be a function of an integral variable n. Suppose that a certain statement S, relating to $f(n)$, is true when $n = a$.

Further, suppose we can prove that, if S is true when $n = m$, it is also true when $n = m + 1$.

Then since S is true when $n = a$, it is true when $n = a + 1, a + 2, a + 3, \ldots$ in succession, that is, when $n \geq a$.

▮ **EXAMPLE 1.** *Show that $2^{2n} - 3n - 1$ is divisible by 9.*

Let $f(n) = 2^{2n} - 3n - 1$, then $f(1) = 0$ and 0 is divisible by 9, so the theorem is true for $n = 1$.

Again, $f(n + 1) - f(n) = 2^{2(n+1)} - 2^{2n} - 3 = 3(2^{2n} - 1)$.

Also $2^{2n} - 1 = (3 + 1)^n - 1 = 3k,$ where k in an integer,

$$\therefore \qquad f(n + 1) - f(n) = 9k.$$

Hence if $f(n)$ is divisible by 9, so is $f(n + 1)$; and, since $f(1)$ is so divisible, it follows in succession that $f(2), f(3), f(4)$, etc., are so divisible; that is, the theorem holds for all values of n.

Or more easily, using the method of (9) Ex. 2, $f(n) = (3 + 1)^n - 3n - 1$, etc.

■ **EXAMPLE 2.** *If n is a positive integer, prove that*

$$\frac{1}{n} + \frac{1}{n+1} + \frac{1}{n+2} + \dots + \frac{1}{2n-1} = 1 - \frac{1}{2} + \frac{1}{3} - \frac{1}{4} + \dots + \frac{1}{2n-1}.$$

If

$$u_n = \frac{1}{n} + \frac{1}{n+1} + \dots + \frac{1}{2n-1},$$

then

$$u_{n+1} = \frac{1}{n+1} + \dots + \frac{1}{2n-1} + \frac{1}{2n} + \frac{1}{2n+1};$$

$$\therefore \qquad u_{n+1} - u_n = \frac{1}{2n} + \frac{1}{2n+1} - \frac{1}{n} = -\frac{1}{2n} + \frac{1}{2n+1}.$$

Again, if

$$v_n = 1 - \frac{1}{2} + \frac{1}{3} - \dots + \frac{1}{2n-1},$$

then

$$v_{n+1} = 1 - \frac{1}{2} + \frac{1}{3} - \dots + \frac{1}{2n-1} - \frac{1}{2n} + \frac{1}{2n+1};$$

$$\therefore \qquad v_{n+1} - v_n = -\frac{1}{2n} + \frac{1}{2n+1} = u_{n+1} - u_n.$$

Hence if $u_n = v_n$, then $u_{n+1} = v_{n+1}$. Now the statement holds for $n = 1$; hence, in succession, it is true for $n = 2, 3, 4, \dots$, that is, for any value of n.

EXERCISE I

1. If q is the quotient and r the remainder when a is divided by b, show that q is the quotient when a is divided by $b + 1$, provided that $r \geq q$.

2. If a and b are prime to each other, show that
   (*i*) $a + b$ and $a - b$ have no common factor other than 2;
   (*ii*) $a^2 - ab + b^2$ and $a + b$ have no common factor other than 3.

3. If a and b are prime to each other, and n is a prime, prove that
   $$(a^n + b^n) / (a + b) \text{ and } a + b$$
   have no common factor, unless $a + b$ is a multiple of n.

4. If a is prime to b and y, and b is prime to x, then $ax + by$ is prime to ab.

5. If $X = ax + by$ and $Y = a'x + b'y$, where $ab' - a'b = 1,$ the greatest common divisor of X and Y is the same as that of x and y.

6. If p is a prime, and $p = a^2 - b^2$, then $a = \frac{1}{2}(p + 1), b = \frac{1}{2}(p - 1).$

7. Express 55 in the form $a^2 - b^2$ in two ways.

8. If x takes the values 1, 2, 3, ... , in the expressions

    (*i*) $x^2 + x + 17$,　　　　　　　　　　　　(*ii*) $2x^2 + 29$,

  (*iii*) $x^2 + x + 41$,

the resulting values of the expressions are primes, provided that in (*i*) $x < 16$, in (*ii*) $x < 29$, and in (*iii*) $x < 40$.

Verify for $x = 15, 28, 39$ repectively.

9. If $f(x)$ is a polynomial, it cannot represent primes only.

[Let $u = f(x)$ and $v = f(x + ku)$, where k is any integer. Prove that u is a factor of v.]

10. For the values 2, 3, 4, ... 10 of x, the number $2.3.5.7 + x$ is composite. Hence write down nine consecutive numbers none of which is a prime.

11. If n is any odd number, then $n(n^2 - 1)$ is divisible by 24; and if n is an odd prime greater than 3, then $n^2 - 1$ is divisible by 24.

12. Show that $2^n + 1$ or $2^n - 1$ is divisible by 3, according as n is odd or even.

13. If n is prime to 5, then $n^2 + 1$ or $n^2 - 1$ is divisible by 5, and therefore n^4 is of the form $5m + 1$.

14. If n is prime to 5, then $n^5 - n$ is divisible by 30; hence the fifth power of any number has the same right-hand digit as the number itself.

15. Show that $2^{2n} + 1$ or $2^{2n} - 1$ is divisible by 5, according as n is odd or even.

16. Show that $5^{2n} + 1$ or $5^{2n} - 1$ is divisible by 13, according as n is odd or even.

17. If $3^n - 1$ is divided by 13, show that the remainder is either 0, 2, or 8, according as n is of the form $3m$, $3m + 1$, or $3m - 1$; hence prove that $3^n - 1$ or $3^{2n} + 3^n + 1$ is divisible by 13, according as n is, or is not, a multiple of 3.

18. If $2^n + 1$ is a prime, then n must be a power of 2.

19. Show that $7^{2n} - 48n - 1$ is divisible by 2304.

20. Show that $7^{2n} + 16n - 1$ is divisible by 64.

21. Prove that $2^{2n+1} - 9n^2 + 3n - 2$ is divisible by 54.

22. Find the number of divisors of 2000, and their sum.

23. Let s be the sum of the divisors of N, excluding N itself; if $s = N$, then n is called a *perfect number.*

Show that, if $2^n - 1$ is a prime, then $2^{n-1}(2^n - 1)$ is a perfect number; and find the three least numbers given by this formula.

24. If n, r, s are the numbers, and P, R, S the products, of the divisors of N, L, M respectively, where $N = LM$, and L and M are prime to one another, prove that (*i*) $n = r.s$, (*ii*) $P = R^s . S^r$.

25. If n is the number, and P the product, of the divisors of N, prove that $P^2 = N^n$.

26. If the product of the divisors of N, excluding N itself, is equal to N, then N is the product of two primes or the cube of a prime.

27. If N has 16 divisors, it cannot have more than 4 prime factors, a, b, c, d, and it must be of one of the forms $abcd$, a^3b^3, a^3bc, a^7b, a^{15}. Hence find the smallest number having 16 divisors.

28. Find the smallest number with 24 divisors.

29. Prove by induction that $n(n+1)(n+2)...(n+r-1)$ is divisible by $\lfloor r$.

30. Find the highest power of the prime p in $\lfloor N$, when

    (*i*) $p^r - 1 < N < p^r + p$;　(*ii*) $p^r - p < N < p^r$.

31. If N is expressed as a polynomial in a prime p, with each of the coefficients less than p, and s is the sum of these coefficients, prove that the power of p contained in $\lfloor\underline{N}$ is $(N - s)/(p - 1)$.

32. Show that the index of the highest power of 2 contained in $\lfloor\underline{N}$ is $N - 1$ when N is a power of 2, and $N - r$ when N is equal to $2^r - 1$.

33. Show that $\lfloor\underline{2n-1}\big/\!\left(\lfloor\underline{n}\,\lfloor\underline{n-1}\right)$ is odd or even according as n is, or is not, a power of 2.

34. Prove that $\lfloor\underline{2n}$ is divisible by $\lfloor\underline{n}.\lfloor\underline{n+1}$.

35. Prove that, if g is the greatest common divisor of m and $n + 1$, then $g.\lfloor\underline{m+n}$ is divisible by $\lfloor\underline{m}.\lfloor\underline{n+1}$.

36. Prove that the power of 2 in $\lfloor\underline{3n}$ is greater than or equal to the power of 2 in $\lfloor\underline{n}.\lfloor\underline{n+1}.\lfloor\underline{n+2}$.

37. Prove that, if n is greater than 2, then $\lfloor\underline{3n}$ is divisible by $\lfloor\underline{n}.\lfloor\underline{n+1}.\lfloor\underline{n+2}$.

38. If a, b, c, ..., are numbers whose sum is a prime, p, then

$$\lfloor\underline{p}\big/\!\left(\lfloor\underline{a}.\lfloor\underline{b}.\lfloor\underline{c}...\right) \text{ is an integer divisible by } p.$$

39. If $u_n = (3 + \sqrt{5})^n + (3 - \sqrt{5})^n$, show that u_n is an integer, and that

$$u_{n+1} = 6u_n - 4u_{n-1}.$$

Hence prove that the integer next greater than $\left(3+\sqrt{5}\right)^n$ is divisible by 2^n.

40. The integer next greater than $\left(\sqrt{7} + \sqrt{3}\right)^{2n}$ is divisible by 2^{2n}.

41. If $2^n + 1 = xy$, prove that $x - 1$ and $y - 1$ are divisible by the same power of 2.

■ ■ ■

Rationals and Irrationals

■ **1. Rational Numbers:** Starting with the natural numbers, the number system is enlarged by

    (*i*) the introduction of fractions, in order that division may be always possible;

    (*ii*) the introduction of negative numbers and zero, so that subtraction may be always possible.

We thus obtain the *system of rational numbers* or *rationals,* consisting of positive and negative integers and fractions, with the number zero.

This system can be arranged in a definite order so as to form the *rational scale.*

With reference to any two rationals x and y, the terms 'greater than', 'equal to' and 'less than' are defined as referring to their relative positions on the rational scale.

To say that $x > y$ or $y < x$ is to say that x follows y on the scale. To say that $x = y$ is to say that x and y stand for the same number.

■ **2. The Fundamental Laws of Order are as follows:** (1) If $a = b$ then $b = a$. (2) if $a = b$ and $b = c$, then $a = c$. (3) if $a \geq b$ and $b > c$, or if $a > b$ and $b \geq c$, then $a > c$. (4) if $a > b$ then $-a < -b$.

Hence we deduce the following rules for equalities and inequalities, where it is assumed that zero is not used as a divisor:

(5) If $a = b$ then
$$a + x = b + x, \quad a - x = b - x, \quad ax = bx, \ a/x = b/x.$$

(6) If $a = b$ and $x = y$, then
$$a + x = b + y, \quad a - x = b - y, \quad ax = by, \ a/x = b/y.$$

(7) If $a > b$ then
$$a + x > b + x \quad \text{and} \quad a - x > b - x.$$

Also $ax \gtrless bx$ and $a/x \gtrless b/x$ according as x is positive or negative.

(8) If $a > b$ and $x > y$, then $a + x > b + y$, and if a and y or b and x are both positive, then $ax > by$.

■ **3. Fundamental Laws of Arithmetic:** Any two rationals can be combined by the operations of addition, subtraction, multiplication and division, the result in each case being a definite rational number, excepting that zero cannot be used as a divisor. This is what is meant when it is said that the system of rationals is *closed* for these operations.

The fundamental laws of addition and multiplication are:

    (1) $a + b = b + a,$ (2) $(a + b) + c = a + (b + c),$

(3) $ab = ba,$ (4) $(a + b)\, c = ac + bc,$ and

(5) $(ab)\, c = a\,(bc).$

The first and third constitute the *Commutative Law*, the second and fifth the *Associative Law*, the fourth the *Distributive Law*.

■ **4. Theorem of Eudoxus:*** *If a and b are any two positive rationals, an integer n exists such that nb > a.* This simply amounts to saying that an integer exists which is greater than a/b.

■ **5. Representation of Numbers by Points on a Line:** Take a straight line $X'OX$ as axis; in OX take a point 1 so that the segment $O1$ contains the unit of length. To find the point a which is to represent any positive rational a, let $a = m/n$. Divide the segment $O1$ into n equal parts and set off a length Oa, along OX, equal to m of these parts. The point which represents $-a$ is in OX', at the same distance from O as the point a.

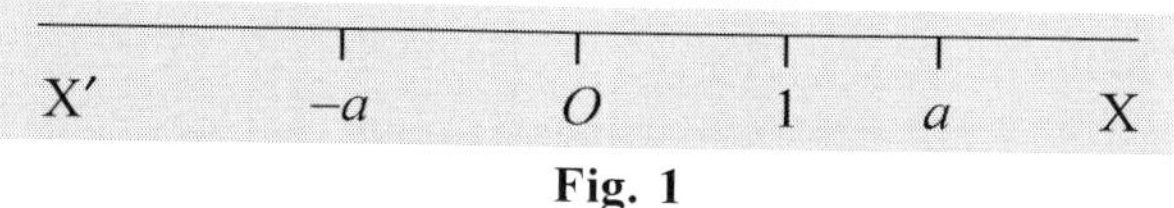

Fig. 1

Points constructed thus represent the rational numbers in the following respects:

(*i*)   For every number there is one point and one only.

(*ii*)   The points occur in the order in which the corresponding numbers stand on the rational scale.

The point X is generally taken to the right of O so that, if $a > b$, the point a is to the right of the point b.

■ **6. Absolute Values:** The *absolute* or *numerical* value of a is $+a$ or $-a$, according as a is positive or negative, and is denoted by $|a|$. Thus $|a - b| = |b - a|$. It is obvious that

$$|ab| = |a|.|b|, \quad |a + b| \le |a| + |b|, \quad |a + b| \ge |a| - |b|.$$

If a is positive, to say that $|x| < a$ is the same as saying that $-a < x < a$.

If $|x| > a$, then $x > a$ or $x < -a$.

■ **7. Large and Small Numbers:** Whether a number is regarded as large or small depends on the purpose to which numbers are applied. An error of 6 inches would be large in measuring a table, but small in setting out a mile course.

If for purposes of comparison we choose some positive number $\in$ which we regard as small, then any number x is said to be *small* if $|x| < \in$.

Any number x is said to be *large* if $|x| > N$ where N is some *previously chosen* positive number, which is regarded as large.

We say that x is large or small compared with y when x/y is large or small. If x/y is neither large nor small, x and y are said to be *large* or *small numbers of the same order*.

We use the abbreviation 'y *is* $O(x)$' to indicate that y is of the same order as x.

If $\in$ is small, numbers which are of the same order as $\in, \in^2, \in^3, \dots$ are called *small numbers of the first, second, third, ... orders* respectively.

---

*   Sometimes ascribed to Archimedes.

If N is large, numbers of the same order as N, N^2, N^3, ... are called *large numbers of the first, second, third, ... orders* respectively.

If $x - y$ is small, we say that x and y are *nearly equal*.

■ **8. Meaning of 'Tends':** To say that x *tends to zero* is to say that x varies in such a way that its numerical value becomes and remains less than any positive number that we may choose, no matter how small. This is expressed by writing $x \to 0$.

If $x - a$ tends to zero, where a is constant, we say that x *tends to a*, and we write $x \to a$.

If x tends to a and is always greater than a, we say that x tends to *a from above*, or *from the right*. This is expressed by writing $x \to a + 0$.

If x tends to a and is always less than a, we say that x tends to *a from below*, or *from the left*, and we write $x \to a - 0$.

To say that x *tends to infinity* $(x \to \infty)$ is to say that x becomes and remains greater than any positive number that we may choose, however great that number may be.

In this case we also say that $-x$ *tends to* $-\infty$. Thus, as x tends to zero from above, $1/x$ tends to ∞, and $-1/x$ tends to $-\infty$.

■ **9. Aggregate:** Any collection of numbers is called an *aggregate* or *set;* the numbers themselves are the *elements* of the set.

If the number of elements exceeds any positive integer we may choose, however great, we say that their number is *infinite,* and we have what is called an *infinite set.*

■ **10. System Everywhere Dense:** An important property of the system of rationals is that *between any two rationals a and b there are infinitely many rationals.*

For if $a < b$ and k is any positive rational, it is easily seen that

$$a < (a + kb)/(1 + k) < b.$$

This fact is sometimes expressed by saying that *the system of rationals is everywhere dense.*

■ **11. Sequence:** A succession of numbers u_1, u_2, u_3, ... u_n, ..., formed according to some definite rule, is called a *sequence,* which is generally denoted by (u_n). Such a rule defines u_n as a *function of the positive integral variable n.*

The rule may be quite arbitrary, and it is unnecessary that we should be able to express u_n in terms of n by an algebraical formula. For instance, u_n may denote the nth prime number, or the integral part of $\sqrt{n}$.

A sequence in which each term is followed by another term is called an *infinite sequence.*

■ **12. Approximate Values:** Suppose that the object of an experiment is to determine a certain number represented by A. No matter what care may be taken to ensure accuracy, it is unlikely that the *exact* value of A can be found. All that can be done in most cases is to find two numbers a and a' between which A must lie.

If we find that $a < A < a'$, we call a a *Lower Limit* and a' an *Upper Limit* to the value of A; a and a' are also called *approximate values* of A.

Definition. If a is an approximate value of A, $A - a$ is called the *(Absolute) Error,* and $(A - a)/A$ the *Relative Error,* in taking a to represent A.

If $a < A < a'$, we say that *a and a' are approximate values of A, the first* in defect *and the second* in excess *with errors numerically less than* $a' - a$.

Observe that the error $A - a$ cannot exceed $a' - a$; therefore $a' - a$ is an upper limit to the error in taking a to represent A.

Again, if a is positive, we have $1/A < 1/a$ and $A - a < a' - a$; hence
$$(A - a)/A < (a' - a)/a;$$
and $(a' - a)/a$ is an upper limit to the relative error.

We estimate the *comparative accuracy* of different experiments by comparing the *relative* errors: *the smaller the relative error, the greater is the degree of accuracy.*

In writing down decimal approximations of a number, we adopt the following scheme:

(*i*) If we write $\pi = 3.1416$ (*approx.*), we shall mean that π is some number lying between 3.14155 and 3.14165, and we say that 3.1416 is the value of π *correct to 5 significant figures, or to 4 places of decimals.*

Since $3.14155 < \pi < 3.14165$, the error in representing π by 3.1416 is numerically less than 0.00005, or $\dfrac{1}{2} \cdot \dfrac{1}{10^4}$.

(*ii*) If we write $\pi = 3.14159 \ldots$, we shall mean that the figures 3, 1, 4, ...9 are those which actually occur in the decimal representation of π, and that π is some number lying between 3.14159 and 3.14160. The error in representing π by 3.14159 is positive and is less than $1/10^5$.

■ **13. Fundamental Theorems:** In theoretical and practical work, the following theorems are of fundamental importance.

If a, a', b, b' are given numbers and A, B are any numbers such that
$$a < A < a' \quad \text{and} \quad b < B < b',$$
then (*i*) $a + b < A + B < a' + b'$; (*ii*) $a - b' < A - B < a' - b$;
and if all the letters denote positive numbers,
(*iii*) $ab < AB < a'b'$; (*iv*) $a/b' < A/B < a'/b$.

The truth of (*i*) and (*iii*) follows from Art. 2, (8).

Proof of (*ii*). Since $a < A$ and $B < b'$, $\therefore a + B < A + b'$.

Subtracting $B + b'$ from each member of the last inequality,
$(a + B) - (B + b') < (A + b') - (B + b')$; $\therefore a - b' < A - B$.

Similarly it can be shown that $A - B < a' - b$.

Proof of (*iv*). Since $a < A$ and $B < b'$, and a, A, B, b' are positive,
therefore $\qquad aB < Ab'$. $\qquad\qquad$ ...(A)

Dividing each side of (A) by Bb', which is positive,
$$\frac{aB}{Bb'} < \frac{Ab'}{Bb'} ; \quad \therefore \quad \frac{a}{b'} < \frac{A}{B} .$$

Similarly it can be shown that $A/B < a'/b$.

■ **EXAMPLE 1.** *Show that the error in representing* $\dfrac{1234567}{3456789}$ *by* $\dfrac{12}{35}$ *is less than 0.05.*

We have $\dfrac{12}{35} < \dfrac{1234567}{3456789} < \dfrac{13}{34}$. Also $\dfrac{12}{35} = 0.34...$ and $\dfrac{13}{34} = 0.38...$. Hence,

$$\frac{13}{34} - \frac{12}{35} < 0.39 - 0.34 = 0.05;$$

therefore the error is less than 0.05.

■ **EXAMPLE 2.** *If $0 < x < 1$, then $1 - 2x$ is an approximate value of $1/(1 + x)^2$ with an error in defect less than $5x^2$.*

By division we find that $\dfrac{1}{(1+x)^2} = 1 - 2x + \dfrac{3x^2 + 2x^2}{(1+x)^2} < 1 - 2x + 5x^2;$

$\therefore\ 1 - 2x < 1/(1 + x)^2 < 1 - 2x + 5x^2$.

■ **14. Non-perfect Powers:** *If A is an integer which is not a perfect n-th power, then no rational x exists such that $x^n = A$.*

For, by hypothesis, x cannot be an integer. If possible, let $x = p/q$, where p/q is a fraction in lowest terms. Then $(p/q)^n$ is a fraction in lowest terms, and so it cannot be equal to the integer A.

■ **15. Need for New Numbers:** There are many purposes for which rational numbers are inadequate; for example: (*i*) there is no rational whose square equals 7 (*ii*) if the diameter of a circle is 1 inch and we denote the length of the circumference by x inches, it can be shown that x is not a rational number.

We extend our number-system by inventing a new class of numbers called *Irrationals*.

An irrational is defined by some rule which gives it ordinal rank with the rational numbers. Such a rule is the following.

Meaning of $\sqrt[n]{A}$. We consider a positive number A which is not a perfect nth power, so that no rational exists whose nth power is A.

The symbol $\sqrt[n]{A}$ (the nth root of A) is used to denote the irrational whose place

among the numbers on the rational scale is determined by the following rule: $\sqrt[n]{A}$ *is to follow every positive rational whose n-th power is less than A, and to precede every rational whose n-th power is greater than A.* Thus if a, a' are positive rationals such that

$$a^n < A < a'^n,$$

then a, a' are *approximate values of* $\sqrt[n]{A}$ with errors, respectively in defect and excess, less than $a' - a$.

■ **16. Representation of a Number by an Endless Decimal:** Take the following instance. No rational exists whose square is 7. Hence 7 lies between two consecutive terms of the sequences $1^2, 2^2, 3^2, 4^2, ...$. Now $2^2 < 7 < 3^2$, therefore 7 lies between two consecutive terms of $(2.0)^2, (2.1)^2, (2.2)^2, ... (3.0)^2$. We find that $(2.6)^2 < 7 < (2.7)^2$, therefore 7 lies between consecutive terms of $(2.60)^2, (2.61)^2, ... (2.70)^2$, and we find that $(2.64)^2 < 7 < (2.65)^2$.

However far this process is carried on, it can be continued further; for otherwise it would be possible to find a rational whose square is 7.

The process therefore gives rise to an *endless decimal* 2.6457 ... which is such that 7 lies between the two classes of numbers.

2^2, $(2.6)^2$, $(2.64)^2$, $(2.645)^2$ ... and ...$(2.646)^2$, $(2.65)^2$, $(2.7)^2$, 3^2, and $\sqrt{7}$ lies between the two classes of numbers

2, 2.6, 2.64, 2.645 ... and ... 2.646, 2.65, 2.7, 3.

This is what is meant when we write $\sqrt{7}$ = 2.6457 ... .

Observe that $\sqrt{7}$ is related to the decimal 2.6457 ... in the same way that 2/3 is related to 0.6666 ... .

For if we say that 2/3 *is equal to* the endless decimal 0.6666 ..., this is only an abbreviation for saying that 2/3 lies between the two classes of numbers

$$0.6, 0.66, 0.666, ... \text{ and } ... 0.667, 0.67, 0.7.$$

It should further be noticed that the decimal equivalent of $\sqrt{7}$ cannot recur, for if it did, $\sqrt{7}$ would be a rational number.

By a process similar to that just described we can find the decimal representation of any root of a given number; but, of course, in practice other methods would be used.

■ **17. Real Numbers:** The rational and irrational numbers together form the system of *Real Numbers.* In Ch. 13 the rules for equalities and inequalities and the definitions of addition, subtraction, multiplication and division, already given for rationals, are extended so as to apply to all real numbers.

Theoretically it is essential to make this extension, but it is unnecessary for purposes of practical reckoning, where we replace any irrational which may occur by a sufficiently close rational approximation.

■ **EXAMPLE 1.** *Given that* $\sqrt[3]{2}$ = *1.259921 ..., prove that*

$$2 - d_n^3 < 6/10^n,$$

where d_n *is the decimal continued to n places.*

Let $d'_n = d_n + 1/10^n$, then $d_n^3 < 2 < d'^3_n$;

therefore $2 - d_n^3 < d'^3_n - d_n^3 = (d'_n - d_n)(d_n^2 + d_n d'_n + d'^2_n)$

Now $$d_n < d'_n < 1.3;$$

$\therefore$ $$d_n^2 + d_n d'_n + d'^2_n < (1.3)^2. \, 3 < 6:$$

$\therefore$ $$2 - d_n^3 < 6/10^n.$$

It follows that $2 - d_n^3 \to 0$ as $x \to \infty$.

Hence we are justified in writing $\left(\sqrt[3]{2}\right)^2$ = 2.

EXAMPLE 2. *Given that* $\sqrt{5} = 2.236...$, $\sqrt[3]{6} = 1.817...$, *find a decimal approximation to* . $\sqrt{5}/\sqrt[3]{6}$. *Find also an upper limit to the error.*

We have $\qquad 2.236/1.818 < \sqrt{5}/\sqrt[3]{6} < 2.237/1.817.$

Now $2.236/1.818 = 1.229 ...$ and $2.237/1.817 = 1.231 ...;$

$\therefore \qquad\qquad 1.229 < \sqrt{5}/\sqrt[3]{6} < 1.232.$

$\therefore$ 1.229 is an approximate value of $\sqrt{5}/\sqrt[3]{6}$ with an error in defect less than 0.003.

EXAMPLE 3. *If a is positive and less than 1, show that $1 - a/2$ is an approximate value of $\sqrt{(1-a)}$ with an error in excess less than $a^2/2$.*

Applying the square root process to $1 - a$, we have

$$N = 1 - a \left(1 - \tfrac{1}{2}a - \tfrac{1}{2}a^2\right)$$

$$
\begin{array}{r}
1 \\ \hline
\end{array}
$$

$$
2 - \tfrac{1}{2}a \, \Big|\; \begin{array}{l} -a \\ -a + \tfrac{1}{4}a^2 \end{array}
$$

$$
2 - a - \tfrac{1}{2}a^2 \, \Big|\; \begin{array}{l} -\tfrac{1}{4}a^2 \\ -a^2 + \tfrac{1}{2}a^3 + \tfrac{1}{4}a^4 \end{array} \qquad = N - \left(1 - \tfrac{1}{2}a\right)^2
$$

$$
\tfrac{3}{4}a^2 - \tfrac{1}{2}a^3 - \tfrac{1}{4}a^4 = N - \left(1 - \tfrac{1}{2}a - \tfrac{1}{2}a^2\right)^2.
$$

Now $\tfrac{3}{4}a^2 - \tfrac{1}{2}a^3 - \tfrac{1}{4}a^4 = \tfrac{1}{4}a^2 \, (3 - 2a - a^2) = \tfrac{1}{4}a^2 \, (3 + a) \, (1 - a)$; and since $0 < a < 1,$

$$\therefore \qquad\qquad a^2 \, (3 + a) \, (1 - a) > 0$$

and $\qquad \left(1 - \tfrac{1}{2}a - \tfrac{1}{2}a^2\right)^2 < 1 - a < \left(1 - \tfrac{1}{2}a\right)^2.$

▨ 18. The Function a^x.

(1) *If $a > 1$ and n is a positive integer, then*
$$a^n > 1 + n \, (a - 1).$$

For, if $a = 1 + b$, by the Binomial theorem, $a^n = (1 + b)^n > 1 + nb.$

(2) *If $a > 0$, then $a^n \gtreqless 1$ and $a^{\frac{1}{n}} \gtreqless 1$ according as $a \gtreqless 1$.*

For if $a > 1$, then $a^n > 1$. If $a < 1$, let $a = 1/b$; then $b > 1$ and $a^n = 1/b^n < 1.$

Again, if $a > 1$, then $a^{\frac{1}{n}} > 1$; for otherwise we should have $a = (a^{\frac{1}{n}})^n < 1.$

Similarly, if $a < 1$, then $a^{\frac{1}{n}} < 1.$

(3) *If $a > 0$ and x is a positive rational, then $a^x \gtrless 1$ according as $a \gtrless 1$.*

For if $x = p/q$, $a^x = \sqrt[q]{a^p} \gtrless 1$, according as $a \gtrless 1$.

(4) *If $a > 0$ and $x > y$, then $a^x \gtrless a^y$ according as $a \gtrless 1$.*

For $a^x/a^y = a^{x-y} \gtrless 1$ according as $a \gtrless 1$.

(5) *If $a > 1$ and N is any positive number, however great, a positive integer m can be found such that*

$$a^n > N \quad for \quad n \geq m.$$

For we can choose m so that $m(a - 1) > N - 1$, and then by (1),

$$a^n \geq a^m > 1 + m\,(a - 1) > N.$$

(6) *If a is positive and not equal to 1, then $a^x \to \infty$ or 0 as $x \to \infty$ according as $a \gtrless 1$.*

First let $a > 1$. Then in (5), if $x > m$, $a^x > a^m > N$, and so $a^x \to \infty$.

If $a < 1$, let $a = 1/b$. Then $b>1$, and as $x \to \infty$, $b^x \to \infty$ and $a^x = 1/b^x \to 0$.

(7) *If $a > 0$, a positive integer m can be found such that*

$$\left| a^{\frac{1}{n}} - 1 \right| < \epsilon \ \ for \ n \geq m,$$

where ϵ is any positive number, however small.

If $a > 1$, as in (5), m can be chosen so that $(1 + \epsilon)^m > a$. Hence if $n \geq m$,

$$(1 + \epsilon)^n > a \ \text{and} \ a^{\frac{1}{n}} - 1 < \epsilon.$$

If $a < 1$, assuming that $0 < \epsilon < 1$, m can be chosen so that $\left(\dfrac{1}{1-\epsilon}\right)^m > \dfrac{1}{a}$, that is,

$(1 - \epsilon)^m < a$. Hence, if $n \geq m$, $(1 - \epsilon)^n < (1 - \epsilon)^m < a$, and therefore $1 - a^{\frac{1}{n}} < \epsilon$.

If $a = 1$, then $a^{\frac{1}{n}} - 1 = 0$ for all values of n.

(8) *If a is positive and $x \to 0$, then $a^x \to 1$.*

First, let $x \to 0$ through positive values. If $a > 1$, choose m as in the first part of

(7); then if $x < 1/m$, $a^x - 1 < a^{\frac{1}{m}} - 1 < \epsilon$.

If $a < 1$, choose m as in the second part of (7); then if $x < 1/m$, $a^x > a^{\frac{1}{m}}$ and

$1 - a^x < 1 - a^{\frac{1}{m}} < \epsilon$.

Thus in both cases, $a^x - 1 \to 0$ and $a^x \to 1$. If $x \to 0$ through negative values, let $a = 1/b$ and $x = -y$. Then $a^x = 1/b^y \to 1$.

EXERCISE II

APPROXIMATE VALUES

1. If $a < A < a'$, where a, a' are known numbers, and any number c lying between a and a' is chosen as an approximate value of A, the numerical value of the error in representing A by c cannot exceed the greater of the numbers $(c - a)$, $(a' - c)$.

   Hence show that, if $\frac{1}{2}(a + a')$ is chosen as an approximate value of A, the greatest numerical value of the error is less than it would be for any other approximation.

2. If $f(x) = 2 + 3x - 4x^2 - 5x^3 + 8x^4$ and x is small, find an upper limit to the numerical value of the error in representing $f(x)$ by $2 + 3x$.

   [If e is the error, $|e| = |-4x^2 - 5x^3 + 8x^4| < |4x^2| + |5x^3| + |8x^4| < 17x^2$, when $|x| < 1$.]

3. The length of a man's step is approximately 30 in., this value being correct to the nearest inch. If the number of steps which he takes per mile is calculated on the assumption that he steps exactly 30 in., show that the error in the result may be as much as 35, but cannot exceed 36.

4. If $0.2 < x < 0.3$ and $0.6 < y < 0.7$, show that

   $$(i)\ 0.3 < y - x < 0.5; \quad (ii)\ 2 < \frac{1}{y-x} < 4; \quad (iii)\ 0.4 < \frac{x^2 + y^2}{(x+y)^2} < 0.91.$$

5. If x, y, z are all positive and $x < y < z$, show that

   $$\frac{x^2}{z} < \frac{x^2 + y^2 + z^2}{x + y + z} < \frac{z^2}{x}.$$

6. If the numerical value of the error in taking a to represent A cannot exceed α, and the numerical value of the error in taking a' to represent a cannot exceed α', then $\alpha + \alpha'$ is an upper limit of the error in taking a' for A. [*i.e.* given that $a - \alpha < A < a + \alpha$ and $a' - \alpha' < a \leq a' + \alpha'$, it is required to show that

   $$a' - (\alpha + \alpha') < A < a' + (\alpha + \alpha').]$$

7. If $a\ b$ are observed values of two numbers A and B, subject to errors (positive or negative), whose numerical values cannot exceed α and β respectively, show that an upper limit to the numerical value of the error in taking

   (*i*) $a + b$ to represent $A + B$ is $\alpha + \beta$;

   (*ii*) $a - b$ to represent $A - B$ is $\alpha + \beta$;

   (*iii*) ab to represent AB is $\alpha b + \beta(a + \alpha)$;

   (*iv*) a/b to represent A/B is $\dfrac{1}{b'}\left(\alpha + \dfrac{a}{b}\beta\right)$, where b' is any positive number less than $b - \beta$.

8. If α is small, show that $1/(1 + \alpha)^3$ is nearly equal to $1 - 3\alpha$. Also show that, if $0 < \alpha < 0.01$, then the error < 0.0007.

9. If $0 < x < 1$, show that the error in taking $\dfrac{2}{3} + \dfrac{1}{9}x$ as the value of $\dfrac{2+3x}{3+4x}$ is numerically less than $\dfrac{1}{6}x^2$.

10. If $x > 100$, the error in taking $\dfrac{3}{4}$ as the value of $\dfrac{2+3x}{3+4x}$ is less than 0.0007.

   [Divide $3x + 2$ by $4x + 3$ in *descending* powers of x.]

11. If x is small, show that:

   (i) $\dfrac{1+ax}{1+bx}$ is nearly equal to $1 + (a - b) x - b (a - b) x^2$;

   (ii) If $a = 5$, $b = 2$ and $x < 0.1$, the error in this approximation is less than 0.01.

12. By the division transformation, show that

$$\frac{1+ax}{(1+bx)(1+cx)} = 1 + (a - b - c) x + \frac{R}{(1+bx)(1+cx)} \cdot x^2,$$

   where $R = (b + c - a) (b + c + bcx) - bc$.

13. By means of Ex.12, show that if x is small, $\dfrac{1+ax}{(1+bx)(1+cx)}$ is nearly equal to

   $1 + (a - b - c) x$.

   Show also that if $a = 2$, $b = 3$, $c = 4$ and $x < 0.01$, the error in this approximation is less than 0.003.

14. The focal length (f) of a lens is given by the formula $1/f = 1/v - 1/u$. The values of u and v may be in error by as much as 2 per cent of the corresponding true values. If the true values of u and v are 20 and 13, show that the value of f as calculated from the observed values may be in error as much as 9.9 per cent of its true value.

15. The weight (w grammes) of water displaced by a solid is given by the formula $w = w_1 - w_2$, where w_1, w_2 are the weights (in grammes) of the solid (*i*) *in vacuo*, (*ii*) in water. In determining the values of w_1, w_2, the error in each case may be as much as 1 per cent of the value. If the true values of w_1, w_2 are 13 and 10, show that the value of w calculated from the observed values of w_1, w_2 may be in error by as much as 7 per cent of its true value.

   [The greatest and least possible values of w_1 are 13.13 and 12.87, those of w_2 are 10.1 and 9.9, hence those of $w_1 - w_2$ are 3.23 and 2.77.]

16. Given that the endless decimal corresponding to $\sqrt[5]{123}$ is 2.618060 ..., prove that $123 - (2.618060)^5 < 0.0003$.

   [Proceed as in Ex. 1 of Art. 17, using the identity
   $$a'^5 - a^5 = (a' - a) (a'^4 + a'^2a + a'^2a^2 + a'a^3 + a^4).]$$

17. If x and y are positive and $x > y$, then $x - y^2/(2x)$ is an approximate value of $\sqrt{(x^2 - y^2)}$ with an error in excess less than $y^4/(2x^3)$.

   [In Ex. 3 of Art. 17, put $a = y^2/x^2$.]

18. If a and b are nearly equal, show that $\dfrac{1}{2} (a + b) - \dfrac{(a-b)^2}{4(a+b)}$ is an approximate value

   of $\sqrt{ab}$ with an error in excess less than $\dfrac{(a-b)^4}{4(a+b)^3}$. [Use Ex. 17.]

19. Given that $\sqrt{14} = 3.741\ ...,\ \sqrt[3]{3} = 1.442\ ...,$ for each of the following obtain a decimal approximation with an error in defect, finding also an upper limit to the error:

(*i*) $\sqrt{14} \times \sqrt[3]{3}$; (*ii*) $\sqrt{14} / \sqrt[3]{3}$; (*iii*) $\sqrt{\sqrt{14} + \sqrt[3]{3}}$.

20. If $0 < a^3 - A < \epsilon$, then $a - \sqrt[3]{A} < \dfrac{\epsilon}{3\left(\sqrt[3]{A}\right)^2}$.

[Put $y = \sqrt[3]{A}$ in the identity $a^3 - y^3 = (a - y)\,(a^2 + ay + y^2)$.]

21. If x is small, prove that (*i*) $\sqrt[3]{1+x}$ is nearly equal to $1 + x/3$; (*ii*) if x is positive, the error in taking $1 + x/3$ to represent $\sqrt[3]{1+x}$ is negative, and is numerically less than $x^2/8$.

[Prove that if $x < 1$, then $\left(1+\dfrac{x}{3}\right)^3 - (1+x) < \dfrac{10x^2}{27}$, and use the last example.]

22. Approximate values of $A,\ B$ are $a = 3083,\ b = 8377$, each correct to the nearest digit. Using Ex. 7, show that an upper limit to the error in taking a/b for A/B is $1/10^4$.

[The error $< \dfrac{1}{8000}\ (0.5 + 0.4 \times 0.5) < \dfrac{1}{10^4}$.]

23. In the last example, suppose a/b expressed as a decimal to n places, to the nearest digit. Show that the numerical value of the error in taking this as the value of A/B is less than

$$\frac{1}{10^4} + \frac{1}{2} \cdot \frac{1}{10^n} \ .$$

Hence show that it is useless to carry on the division to more than four places of decimals.

24. If $A = 3.141592\ ...,\ B = 1.414213\ ...,$ by using Ex. 7 find how many figures must be kept in $A,\ B$ to find an approximate value of AB with an error numerically less than $1/10^3$.

[Choose $a,\ b$ so that $\alpha b + \beta\,(a + \alpha) < 1/10^3$. Now $b < 2$ and $a + \alpha < 4$, $\therefore$ we may take $\alpha,\ \beta$ such that $2\alpha + 4\beta < 1/10^3$. This is satisfied if $\alpha < \dfrac{1}{6} \cdot \dfrac{1}{10^3},\ \beta < \dfrac{1}{6} \cdot \dfrac{1}{10^3}$. We may therefore take $a = 3.1416,\ b = 1.4142$.]

■■■

Polynomials

■ **1. Definitions:** An expression of the form
$$a_0x^n + a_1x^{n-1} + a_2x^{n-2} + \ldots + a_n,$$
where $a_0, a_1, \ldots a_n$ are independent of x, is called a *polynomial in x of the n-th degree.*

A polynomial in x and y is the sum of a number of terms of the form $ax^m y^n$, where a is independent of x, y, and m, n are positive integers.

A polynomial in any number of variables is similarly defined, and is sometimes called a *rational integral function* of the variables : its *degree* is that of its highest term.

If the sum of the indices of the variables in each term of a polynomial is a constant number, the polynomial is said to be *homogeneous.* For instance,
$$a_0x^n + a_1x^{n-1}y + a_2x^{n-2}y^2 + \ldots + a_ny^n,$$
is a homogeneous polynomial in x , y of degree n.

A constant may be regarded as a polynomial of degree *zero.*

Polynomials of the first, second, third, fourth,... degrees are known as linear, quadratic, cubic, quartic, ... functions respectively.

In some parts of Algebra, a homogeneous polynomial is called a *quantic.*

A homogeneous polynomial in x, y of degree n is often written in the form
$$a_0x^n + na_1x^{n-1}y + \frac{n(n-1)}{\underline{|2}} a_2x^{n-2}y^2 + \ldots + a_ny^n,$$
where binomial coefficients are introduced for convenience. This is shortly denoted by
$$(a_0, a_1, a_2, \ldots a_n)(x, y)^n.$$
Using this notation, cubic and quartic functions of x may be denoted by
$$(a, b, c, d)(x, 1)^3 = ax^3 + 3bx^2 + 3cx + d,$$
$$(a, b, c, d, e)(x, 1)^4 = ax^4 + 4bx^3 + 6cx^2 + 4dx + e.$$

If A, B, C are polynomials and $A = BC$, then B and C are called *factors* or *divisors* of A, and A is said to be (*exactly*) *divisible* by B or by C.

■ **2. Division:** Let A and B be polynomials in x, A being of higher degree than B, or of the same degree.

To 'divide A by B' is to find an identity of the form
$$A = BQ + R,$$

where Q and R are polynomials (or in special cases constants) and R is of lower degree than B. Such division is always possible: the result is *unique* and is called the *division transformation*, Q being the *quotient* and R the *remainder.*

■ **3. Synthetic Division:** In dividing $ax^3 + bx^2 + cx + d$ by $x - h$, the reckoning may be arranged as on the left:

$$\begin{array}{lll} a+b & +c & +d \\ \underline{\quad ah \quad + ph \quad +qh} & & \\ a+p & +q & +r \end{array} \, ,$$

$$x-h \,) \, ax^3 + bx^2 + cx + d \, (\, ax^2 + px + q$$
$$\underline{ax^3 - ahx^2}$$
$$px^2 + cx$$
$$\underline{px^2 - phx}$$
$$qx + d$$
$$\underline{qx - qh}$$
$$r$$

where $p = ah + b$, $q = ph + c$, $r = qh + d$;

and then,   the quotient $= ax^2 + px + q$,

and   the remainder $= r$.

This process is known as *synthetic division.* It is justified by a comparison with the reckoning on the right.

It is evident that the method can be used to divide any polynomial in x by $x - h$. If any powers of x are missing, these must be supplied with zero coefficients.

❚ EXAMPLE 1. *Divide $3x^4 - x^3 + 2x^2 - 2x - 1$ by $x + 2$.*

We divide by $x - (-2)$ as follows :

$$\begin{array}{r|rrrrr} 1-(-2) & 3 & -1 & +2 & -2 & -1 \\ & & -6 & +14 & -32 & +68 \\ \hline & 3 & -7 & +16 & -34 & +67 \end{array}$$

$\therefore$ Quotient $= 3x^3 - 7x^2 + 16x - 34$; remainder $= 67$.

To divide a polynomial A by $ax - b$, we may use the following rule:

Find the quotient Q and the remainder R in the division of A by $x - b/a$, then Q/a and R are respectively the quotient and remainder in the division of A by $ax - b$.

For $A = \left(x - \dfrac{b}{a} \right) Q + R = (ax - b) \dfrac{Q}{a} + R.$

❚ EXAMPLE 2. *Divide $x^3 + 2x^2 - 3x - 4$ by $2x - 1$.*

Dividing by $x - \dfrac{1}{2}$, as on the right, it is seen that

$$\begin{array}{r|rrrr} 1-\frac{1}{2} & 1 & +2 & -3 & -4 \\ & & +\frac{1}{2} & +\frac{5}{4} & -\frac{7}{8} \\ \hline & 1 & +\frac{5}{2} & -\frac{7}{4} & -\frac{39}{8} \end{array}$$

the quotient $= x^2 + \dfrac{5}{2}x - \dfrac{7}{4}$

and the remainder $= -\dfrac{39}{8}$.

$\therefore$ The required quotient $= \dfrac{1}{2}\left(x^2 + \dfrac{5}{2}x - \dfrac{7}{4}\right)$; remainder $= -\dfrac{39}{8}$.

If the divisor is of the second or higher degree, the process is as follows: to divide $4x^4 + 3x^3 + x - 1$ by $x^2 - 2x + 3$, write the divisor in the form $x^2 - (2x - 3)$ and proceed thus,

$$
\begin{array}{r|rrrrr}
1-(2-3) & 4 & +3 & +0 & +1 & -1 \\
 & & -12 & -33 & -30 & (a) \\
 & & 8 & +22 & +20 & (b) \\
\hline
 & 4+11 & +10 & -12 & -31 & (c)
\end{array}
$$

The quotient is $4x^2 + 11x + 10$ and the remainder $-12x - 31$. The first term in line (*c*) is 4; 4 (2 – 3) = 8 – 12; put 8 in line (*b*) and – 12 in (*a*); 8 + 3 = 11; 11 (2 – 3) = 22 – 33; put 22 in (*b*) and – 33 in (*a*); and so on.

The reckoning will be understood if it is compared with ordinary division.

■ **EXAMPLE 3.** *Express* $f(x) \equiv 4x^4 + 3x^3 + x - 1$ *in the form*
$$a(x^2 - 2x + 3)^2 + (bx + c)(x^2 - 2x + 3) + dx + e,$$
where a, b, ... e are independent of x.

Divide $f(x)$ by $x^2 - 2x + 3$ as above. The remainder is $- 12x - 31$ and the quotient $4x^2 + 11x + 10$.

$$
\begin{array}{r|rrr}
1-(2-3) & 4+11 & +10 \\
 & & -12 \\
 & 8 \\
\hline
 & 4+19 & -2
\end{array}
$$

Divide $4x^2 + 11x + 10$ by $x^2 - 2x + 3$. The remainder is $19x - 2$ and the quotient 4. Hence
$$f(x) = 4(x^2 - 2x + 3)^2 + (19x - 2)(x^2 - 2x + 3) - 12x - 31.$$

■ **4. Applications of Synthetic Division:** The following transformations are often required.

Let $f(x)$ be a polynomial in x.

(1) *To express* $f(x)$ *as a polynomial in* $x - h$.

Suppose that
$$f(x) = a_0(x - h)^n + a_1(x - h)^{n-1} + \dots + a_{n-1}(x - h) + a_n.$$
Then a_n is the remainder when $f(x)$ is divided by $x - h$. If Q is the quotient, a_{n-1} is the remainder when Q is divided by $x - h$. If Q' is the quotient, a_{n-2} is the remainder when Q' is divided by $x - h$. Continuing thus, we can find all the coefficients.

The reckoning is arranged as in Ex. 1, overleaf.

▌**EXAMPLE 1.** *Express $2x^3 + x^2 - 5x - 3$ as a polynomial in $x - 2$.*

$$
\begin{array}{r|l}
1-2 & 2 + 1 - 5 - 3 \\
    & + 4 + 10 + 10 \\
\hline
1-2 & 2 + 5 + 5 + 7 \\
    & + 4 + 18 \\
\hline
1-2 & 2 + 9 + 23 \\
    & + 4 \\
\hline
    & 2 + 13
\end{array}
$$

$\therefore\ 2x^3 + x^2 - 5x - 3 = 2\,(x - 2)^3 + 13\,(x - 2)^2 + 23\,(x - 2) + 7.$

(2) *To expand $f(x + h)$ in powers of x,* we express $f(x)$ as a polynomial in $x - h$, and then change x into $x + h$.

▌**EXAMPLE 2.** *If $f(x) = 2x^3 + x^2 - 5x - 3$, find $f(x + 2)$.*

Proceed as in Ex. 1 and substitute $x + 2$ for x in the result, thus

$$f(x + 2) = 2x^3 + 13x^2 + 23x + 7.$$

(3) *If $f(n)$ is a polynomial in n of degree r, to express $f(n)$ in the form.*

$$a_0 + a_1 n + a_2 n\,(n + 1) + a_3 n\,(n + 1)\,(n + 2) + \ldots$$
$$+ a_r n\,(n + 1)\,(n + 2)\ldots(n + r - 1).$$

This can be done by dividing by $n,\ n + 1,\ n + 2,\ \ldots$ in succession.

▌**EXAMPLE 3.** *Express $n^4 + 3n^2 + 2$ in the form*

$a + bn + cn\,(n + 1) + dn\,(n + 1)\,(n + 2) + en\,(n + 1)\,(n + 2)\,(n + 3).$

Divide $n^4 + 3n^2 + 2$ by n, the quotient by $n + 1$, the quotient thus obtained by $n + 2$, if quotient by $n + 3$. The reckoning is shown below, and the required in thick type.

$$
\begin{array}{r|rrrrr}
1+1 & 1 & +0 & +3 & +0 & +2 \\
    &   & -1 & +1 & -4 &   \\
\hline
1+2 & 1 & -1 & +4 & -4 &   \\
    &   & -2 & +6 &   &   \\
\hline
1+3 & 1 & -3 & +10 &   &   \\
    &   & -3 &   &   &   \\
\hline
    & 1 & -6 &   &   &
\end{array}
$$

$\therefore\ n^4 + 3n^2 + 2 = 2 - 4n + 10n\,(n + 1) - 6n\,(n + 1)\,(n + 2)$
$$+ n\,(n + 1)\,(n + 2)\,(n + 3).$$

▮ **5. The Remainder Theorem:** *If $f(x)$ is a polynomial, then $f(h)$ is the remainder when $f(x)$ is divided by $x - h$.*

This follows on substituting h for x in the identity

$$f(x) = (x - h)\,Q + R,$$

where Q and R are repectively the quotient and remainder in the division of $f(x)$ by $x - h$; for R is independent of x.

Corollary: *If $f(h) = 0$, then $x - h$ is a factor of $f(x)$.*

For, in the preceding, $R = f(h) = 0$.

■ **EXAMPLE 1.** *Show that, if n is odd, x + 1 is a factor of $x^n + 1$.*

Since n is odd, $(-1)^n = -1$, hence if $x = -1$,

$$x^n + 1 = (-1)^n + 1 = -1 + 1 = 0.$$

∴　　　　　　　　$x + 1$ is a factor of $x^n + 1$.

■ **EXAMPLE 2.** *If $f(x) = 3x^4 + 2x^3 - 6x - 4$, find the value of $f(0.2)$.*

$f(0.2)$ is the remainder in the division of $f(x)$ by $x - 0.2$. Performing the division:

$$
\begin{array}{r|rrrrr}
1-0.2 & 3 & +2 & +0 & -6 & -4 \\
 & & +0.6 & +0.52 & +0.104 & -1.1792 \\
\hline
 & 3 & +2.6 & +0.52 & -5.896 & -5.1792
\end{array}
$$

Remainder $= -5.1792 = f(0.2)$.

■ **6. Theorem:** *If $f(x)$ is a polynomial which vanishes when x has the values α_1, α_3, ... α_n, no two of which are equal, then the product*

$$(x - \alpha_1)(x - \alpha_2)(x - \alpha_3) \dots (x - \alpha_n)$$

is a factor of $f(x)$.

For since $f(\alpha_1) = 0$, by the Remainder theorem $f(x) = (x - \alpha_1) \cdot f_1(x)$, where $f_1(x)$ is a polynomial. Therefore $f(\alpha_2) = (\alpha_2 - \alpha_1) f_1(\alpha_2)$.

Now $f(\alpha_2) = 0$ and $\alpha_2 - \alpha_1 \neq 0$, therefore $f_1(\alpha_2) = 0$, and consequently $x - \alpha_2$ is a factor of $f_1(x)$. Hence

$$f_1(x) = (x - \alpha_2) f_2(x). \text{ and } f(x) = (x - \alpha_1)(x - \alpha_2) f_2(x),$$

where $f_2(x)$ is a polynomial.

Proceeding in this way, we can show that $(x - \alpha_1)(x - \alpha_2) \dots (x - \alpha_n)$ is a factor of $f(x)$.

■ **7. Theorem:** *A polynomial $f(x)$ of the nth degree cannot vanish for more than n values of x unless all its coefficients are zero.*

For otherwise (by Art. 6) the product of more than n expressions of the form $x - \alpha$ would be a factor of the polynomial. That is to say, a polynomial of the nth degree would have a factor of degree higher than n, which is impossible.

If all the coefficients of $f(x)$ are zero, $f(x)$ is said to *vanish identically.* Thus $(x - a)(b - c) + (x - b)(c - a) + (x - c)(a - b)$ vanishes identically; for it is of the first degree in x, and it vanishes for the values a, b, c of x.

Corollary: *An equation of the nth degree cannot have more than n roots.*

Note: Later it will be shown that every equation of the nth degree has exactly n roots, not necessarily all different, which may be real or imaginary.

■ **8. Theorem:** *If for more than n values of x,*

$$ax^n + bx^{n-1} + \dots + hx + k = a'x^n + b'x^{n-1} + \dots + h'x + k',$$

then $a = a'$, $b = b'$, ..., $h = h'$, $k = k'$; that is to say the polynomials are identically equal.

For by the given conditions the polynomial

$$(a - a')x^n + (b - b')x^{n-1} + \dots + (h - h')x + (k - k')$$

vanishes for more than n values of x; hence, by the preceding article, it follows that

$$a = a', \; b = b', \; \dots, \; h = h', \; k = k'.$$

■ 9. Polynomials in Two or More Variables:

(1) *If $f(x, y)$ is a polynomial in x and y which vanishes for all values of x and y, then all the coefficients of $f(x, y)$ are zero.*

The following method applies in all cases. Let the polynomial be
$$u = ax^2 + bxy + cy^2 + dx + ey + f,$$
which is supposed to vanish for all values of x and y; then
$$u = ax^2 + x(by + d) + (cy^2 + ey + f),$$
and for any given value of y, the last expression is a polynomial in x, which vanishes for all values of x;

hence $\qquad\qquad a = 0, \; by + d = 0, \; cy^2 + ey + f = 0.$ $\qquad\qquad$ ... (A)

Also the relations (A) hold for all values of y;

$\therefore \qquad\qquad a = 0, \; b = 0, \; d = 0, \; c = 0, \; e = 0, \; f = 0.$

(2) *If for all values of x and y the polynomials $f(x, y)$ and $F(x, y)$ have equal values, then the coefficients of like terms in the polynomials are equal.*

This follows immediately from (1).

> **Note:** The fact that two polynomials, u, v, are identically equal is often expressed by writing $u \equiv v$.

■ 10. The Method of Undertermined Coefficients:

It is often necessary to enquire if a given function can be expressed in a certain specified form. In cases of this kind we may use the method of 'Undetermined Coefficients', which may be described as follows.

Assume that the function is expressed in the given form, where some of the coefficients are unknown. In order that the identity involved in this assumption may be true, the unknown coefficients must satisfy certain equations. If these equations have a solution, the function can be expressed in the specified form.

▌ **EXAMPLE 1.** *Show that values of a, b, c exist such that*
$$2x^2 - 5x - 1 \equiv a(x + 1)(x - 2) + b(x - 2)(x - 1) + c(x - 1)(x + 1),$$
and find these values.

The right-hand side $= a(x^2 - x - 2) + b(x^2 - 3x + 2) + c(x^2 - 1)$
$$= (a + b + c)x^2 - (a + 3b)x - (2a - 2b + c).$$

This is equal to the left-hand side for all values of x, if values for a, b, c can be found such that $a + b + c = 2$, $a + 3b = 5$, $2a - 2b - c = 1$

These equations have one solution, namely, $a = 2$, $b = 1$, $c = -1$.

In a question of this kind, *if it has been shown that an identity of the specified form exists, the values of the constants may be found by giving special values to the variables.*

Thus, assuming that an identity of the specified form exists, we can find a, b, c, by putting $x = 1, -1, 2$ in succession.

▌ **EXAMPLE 2.** *Search for factors of $S \equiv 2x^2 - 5xy - 3y^2 - x + 10y - 3$.*

Since $2x^2 - 5xy - 3y^2 = (2x + y)(x - 3y)$ we assume that
$$S = (2x + y + p)(x - 3y + q) = 2x^2 - 5xy - 3y^2 + x(p + 2q) + y(q - 3p) + pq.$$

Equating coefficients, we have the *three* conditions:

$$p + 2q = -1, \; -3p + q = 10, \; pq = -3.$$

From the first two, $p = -3$, $q = 1$, and these values *happen* to satisfy the third condition.

■ **11. Quadratic Functions of x and y:** Every homogeneous function of x, y, z of the second degree is of the form

$$ax^2 + by^2 + cz^2 + 2fyz + 2gzx + 2hxy. \qquad \text{...(A)}$$

When $z = 1$ this reduces to

$$ax^2 + 2hxy + by^2 + 2gx + 2fy + c, \qquad \text{...(B)}$$

which is the general form of a quadratic function of x and y.

From Ex. 2 of Art. 10 it appears that the cases in which an expression of the form (B) can be resolved into factors are *exceptional*. The condition that this may be possible is investigated in the next article.

■ **12. Theorem:** *The necessary and sufficient condition that the quadratic function*

$$S \equiv ax^2 + 2hxy + by^2 + 2gx + 2fy + c$$

can be expressed as the product of two linear factors is

$$\Delta \equiv abc + 2fgh - af^2 - bg^2 - ch^2 = 0.$$

Assume that $S = (px + qy + r)(p'x + q'y + r')$ and consider the equation $S = 0$; this can be written

$$ax^2 + 2x(hy + g) + (by^2 + 2fy + c) = 0,$$

giving
$$x = \frac{1}{a}\left\{-(hy + g) \pm \sqrt{(hy + g)^2 - a(by^2 + 2fy + c)}\right\} \qquad \text{... (C)}$$

These values of x must be the same as $-\dfrac{1}{p}(qy + r)$ and $-\dfrac{1}{p'}(q'y + r')$; they are therefore rational functions of y. Hence the expression under the root sign in (C) is a perfect square. Now

$$(hy + g)^2 - a(by^2 + 2fy + c) = y^2(h^2 - ab) + 2y(hg - af) + (g^2 - ac),$$

and the condition that this may be a perfect square is

$$(hg - af)^2 - (h^2 - ab)(g^2 - ac) = 0$$

or
$$a(abc + 2fgh - af^2 - bg^2 - ch^2) = 0.$$

The following cases must now be considered:

(*i*) *If a is not zero,* it follows that

$$abc + 2fgh - af^2 - bg^2 - ch^2 = 0 \qquad \text{...(D)}$$

(*ii*) *If $a = 0$* the reasoning fails; but, except when $b = 0$, we shall be led to the condition (D) by solving for y and proceeding as before.

(*iii*) *If both a and b are zero, and $h \neq 0$,* then

$$S = 2hxy + 2gx + 2fy + c$$

$$= 2x(hy + g) + \frac{2f}{h}(hy + g) + c - \frac{2fg}{h}$$

$$= 2\,(hy + g)\left(x + \frac{f}{h}\right) - \frac{\Delta}{h^2},$$

for in this case $\Delta = 2fgh - ch^2$. Hence S can be resolved into linear factors if, and only if, $\Delta = 0$.

(iv) If $a = b = h = 0$, S is not a quadratic function. Hence, in all cases, if S can be resolved into linear factors, then $\Delta = 0$.

Conversely, if $\Delta = 0$, then S can be resolved into linear factors.

This is evident in all cases, from the preceding.

> **Note:** *The expression $abc + 2fgh - af^2 - bg^2 - bg^2 - ch^2$ is generally denoted by Δ, and is called the* discriminant *of S. Observe that p, q, r, p′, q′, r′, are not necessarily all real.*

■ **EXAMPLE** : *If, in the above p, q, r, p′, q′, r′ are all real, prove that*
$$bc - f^2,\; ca - g^2,\; ab - h^2,$$
are all negative; and vice versa.

The square root of $(hy + g)^2 - a(by^2 + 2fy + c)$,

or $\qquad\qquad y^2\,(h^2 - ab) + 2y\,(hg - af) + g^2 - ac,$

is real if, and only if

$$h^2 - ab \text{ and } g^2 - ac \text{ are positive.}$$

Similarly, by solving $S = 0$ as a quadratic in y, we find that $f^2 - bc$ must be positive.

EXERCISE III

DIVISION AND FACTORS

In Exs. 1–6 find by synthetic division the quotient and remainder when the first expression is divided by the second.

1. $5x^4 + 6x + 2$; $x + 4$.
2. $x^5 - 5x^4 + 7x^2 - 1$; $x - 2$.
3. $x^4 - 4x^3 + 6x^2 - 4x + 1$; $3x - 2$.
4. $x^5 - 1$; $2x + 1$.
5. $3x^4 - 5x^3 - 11x^2 + x - 1$; $x^2 - 2x - 2$.
6. $3x^5 + 6x^4 - 2x^3 - x^2 - 2x + 4$; $x^2 + 2x - 1$.
7. If $4x^4 - (a - 1)\,x^3 + ax^2 - 6x + 1$ is divisible by $2x - 1$, find the value of a.
8. Express $x^3 + 2x^2 + x + 80$ as a polynomial in (i) $x + 5$; (ii) $x + 1$; (iii) $x + \dfrac{1}{3}$.
9. If x is given by the equation $x^3 - x^2 - 8x + 3 = 0$, find the equation giving y when (i) $y = x - 2$; (ii) $y = 3x - 1$; (iii) $y = 3x + 4$.
10. Express n^4 in the form
$$a + bn + cn\,(n + 1) + dn\,(n + 1)\,(n + 2) + en\,(n + 1)\,(n + 2)\,(n + 3).$$
11. If $f(x) = 6x^3 + 4x^2 + 3x + 2$, find the values of $f\left(-\dfrac{3}{2}\right)$, $f\left(-\dfrac{2}{3}\right)$, $f\left(\dfrac{1}{3}\right)$.

Factorise the expressions in Exs. 12, 13.

12. $x^4 - 2x^3 - 7x^2 + 8x + 12$.

13. $2x^4 - 3x^3y - 6x^2y^2 - 8xy^3 - 3y^4$.

14. Find a polynomial in x of the third degree which shall vanish when $x = 1$ and when $x = -2$, and shall have the values 4 and 28 when $x = -1$ and $x = 2$ respectively.

[The polynomial is of the form $(x - 1)(x + 2)(ax + b)$.]

15. Find a quadratic function of x which shall vanish when $x = -\dfrac{3}{7}$ and have the values 24 and 62 when $x = 3$ and $x = 4$ respectively.

[The function is of the form $(7x + 3)(ax + b)$.]

16. Find a homogeneous function of x and y of the second degree which shall vanish when $x = y$ and also when $x = 4$ and $y = 3$, and have the value 2 when $x = 2$ and $y = 1$.

[The function is of the form $(x - y)(ax + by)$.]

17. If n is odd, show that $x^m + 1$ is a factor of $x^{mn} + 1$.

18. Express $x^{15} + 1$ as the product of four factors.

[It follows from Ex. 17 that $x^3 + 1$ and $x^5 + 1$ are factors.]

19. Prove that the condition that $ax^2 + bx + c$ and $a'x^2 + b'x + c'$ may have a common linear factor is

$$(ca' - c'a)^2 = (bc' - b'c)(ab' - a'b).$$

[$x - \alpha$ will be a common factor if $a\alpha^2 + b\alpha + c = 0$ and $a'\alpha^2 + b'\alpha + c' = 0$. Eliminating α between these equations, we obtain the required condition.]

20. Find the condition that $ax^3 + bx + c$ and $a'x^3 + b'x + c'$ may have a common linear factor.

21. If n is any number, show that n^2 can be expressed in the form

$$a(n - 1)^2 + b(n - 2)^2 + c(n - 3)^2,$$

and find the values of a, b, c.

22. If p, q, r, s are four consecutive numbers, show that p^2, q^2, r^2, s^2 are connected by a linear equation, and find this equation.

[This fallows from Ex. 21.]

23. If x, y, z, w are four consecutive terms of an arithmetical progression, prove that x^2, y^2, z^2, w^2 are connected by a linear equation with constant coefficients. Find this equation.

[Write $x = p - 3q$, $y = p - q$, $z = p + q$, $w = p + 3q$, and find b, c, d such that $(p - 3q)^2 + b(p - q)^2 + c(p + q)^2 + d(p + 3q)^2 \equiv 0$.]

24. Show that constants a, b, c can be found such that

$$3x^2 + 16x + 23 \equiv a(x + 2)(x + 3) + b(x + 3)(x + 1) + c(x + 1)(x + 2)$$

and find the values of a, b, c.

In Exs. 25, 26 show that values of a, b can be found for which the first expression is divisible by the second, and find these values.

25. $x^5 + 2x^3 + ax^2 + b$; $x^3 + 1$.

[Find the remainder in the division of the first expression by the second, and make this identically zero.]

26. $ax^4 + bx^3 - 12x^2 + 21x - 5$; $2x^2 + 3x - 1$.

[Arrange in ascending powers of x before dividing.]

Resolve the expressions in Exs. 27, 28 into factors.

27. $3x^2 + xy - 4y^2 + 8x + 13y - 3.$

28. $3x^2 + 2xy - 8y^2 - 7x + 16y - 6.$

29. Show that $x^6 - 6x^5 + 24x^4 - 56x^3 + 96x^2 - 96x + 64$ is a perfect cube, and find the cube root.

30. If $x^3 + 3x^2 - 9x + c$ is the product of three factors, two of which are identical, show that c is either 5 or -27; and resolve the given expression into factors in each case.

31.   (*i*) Find the remainder in the division of
$$x^3 + 3px + q \quad \text{by} \quad x^2 - 2ax + a^2.$$
  (*ii*) If $x^3 + 3px + q$ has a factor of the form $(x - a)^2$, show that $q^2 + 4p^3 = 0$.

  (*iii*) If the equation $x^3 + 3px + q = 0$ has two equal roots, what is the relation connecting p and q?

32. If $x^4 + px^2 + qx + r$ can be expressed in the form $(x - a)^3 (x - b)$, show that
$$8p^3 = -27q^2 \text{ and } p^2 = -12r.$$

In Exs. 33–34 find the value, or the values, of λ for which the given expression can be resolved into factors, and for each value of λ resolve the expression into factors.

33. $x^2 - xy + 3y - 2 + \lambda (x^2 - y^2).$

34. $x^2 + y^2 - 6y + 4 + \lambda (x^2 - 3y + 2).$

35. If $ax^2 + 2hxy + by^2 + 2gx + 2fy + c$ is a perfect square, show that $f = \pm \sqrt{bc}, \ = g \pm \sqrt{ca}, \ h = \pm \sqrt{ab},$ provided that the negative sign does not precede an odd number of the radicals.

36. If $5x^2 + 4xy + y^2 + 24x - 10y + 24 = 0$, where x and y are real numbers, show that x must lie between $2 - \sqrt{5}$ and $2 + \sqrt{5}$ inclusive, and y between -4 and $+6$ inclusive. [Solve for x and y in succession.]

37. If $2x^2 + 4xy + y^2 - 12x - 8y + 15 = 0$, where x and y are real, show that x cannot lie between $1 \pm \sqrt{\dfrac{1}{2}}$ and y cannot lie between 1 and 3.

38. If $x^2 + xy - y^2 + 2x - y + 1 = 0$, where x and y are real, show that y cannot lie between 0 and $-\dfrac{8}{5}$, but x can have any real value whatever.

39. If $f(x)$ is a rational integral function of x and a, b are unequal, show that the remainder in the division of $f(x)$ by $(x - a)(x - b)$ is
$$[(x - a) f(b) - (x - b) f(a)]/(b - a).$$
[The remainder is a linear function of x, and may therefore be assumed to be $A(x - a) + B(x - b)$.]

40.   (*i*) If $ax^3 + bx + c$ has a factor of the form $x^2 + px + 1$, show that $a^2 - c^2 = ab$.

  (*ii*) In this case, prove that $ax^3 + bx + c$ and $cx^3 + bx^2 + a$ have a common quadratic factor.

[For (*ii*), in the identity $ax^3 + bx + c = (x^2 + px + 1)(ax + c)$ put $1/x$ for x and multiply each side by x^3.]

41.   (*i*) If $ax^5 + bx^2 + c$ has a factor of the form $x^2 + px + 1$, prove that
$$(a^2 - c^2)(a^2 - c^2 + bc) = a^2b^2.$$
  (*ii*) In this case, prove that $ax^5 + bx^2 + c$ and $cx^5 + bx^3 + a$ have a common quadratic factor.

■ **13. Expansion of Products:** To expand the product

$$(a + b)(p + q + r)(x + y + z + w), \qquad \ldots(A)$$

we multiply each term of $(a + b)$ by each term of $(p + q + r)$; we then multiply each of the poducts so formed by each term of $(x + y + z + w)$, and add the results.

Hence the expression (A) is equal to the sum of all the products which can be formed by choosing any term out of each of the factors and multiplying them together.

In general, *the product of any number of polynomials is equal to the sum of all the products which can be formed by choosing any term out of each polynomial and multiplying them together.*

■ **14. Theorem:** *If $p_1, p_2, p_3, \ldots$ denote the sums of the products of $a_1, a_2, a_3, \ldots, a_n$, taken one, two, three, $\ldots$ at a time, respectively, then*

$$(x + a_1)(x + a_2)(x + a_3) \ldots (x + a_n) = x^n + p_1 x^{n-1} + p_2 x^{n-2} + \ldots + p_n.$$

For $(x + a_1)(x + a_2)(x + a_3) \ldots (x + a_n)$ is equal to the sum of all the products which can be formed by choosing a term out of each of the factors and multiplying these terms together.

Choosing x out of each factor, we obtain the term x^n of the expansion.

Choosing x out of any $(n - 1)$ of the factors and the a out of the remaining factor, we obtain

$$x^{n-1}(a_1 + a_2 + a_3 + \ldots + a_n) \text{ or } p_1 x^{n-1}.$$

Choosing x out of any $(x - 2)$ of the factors and an a from each of the two remaining factors, we obtain

$$x^{n-2}(a_1 a_2 + a_1 a_3 + \ldots + a_2 a_3 + \ldots) \text{ or } p_2 x^{n-2}.$$

Choosing x out of any $(n - r)$ of the factors and an a from each of the r remaining factors, we obtain $p_r x^{n-r}$.

Finally, choosing an a out of each of the factors, we obtain $a_1 a_2 a_3 \ldots a_n$ or p_n;

$$\therefore (x + a_1)(x + a_2)(x + a_3) \ldots (x + a_n) = x^n + p_1 x^{n-1} + p_2 x^{x-2} + \ldots + p_n.$$

In particular, if $a_1 = a_2 = a_3 = \ldots = a_n = a$, we have

$$(x + a)^n = x^n + C_1^n x^{n-1} a + C_2^n x^{n-2} a^2 + \ldots C_r^n x^{n-r} a^r + \ldots + a^n,$$

which is the *Binomial Theorem for a positive integral index.*

Many identities can be obtained by expanding a function of x in two different ways by the Binomial theorem and equating coefficients.

■ **EXAMPLE 1.** *Use the identity $(1 + x)^{2n} = (1 + 2x + x^2)^n$ to prove that*

$$2^n + \frac{n(n-1)}{\lfloor\underline{1}\,\lfloor\underline{1}} \cdot 2^{n-2} + \frac{n(n-1)(n-2)(n-3)}{\lfloor\underline{2}\,\lfloor\underline{2}} \cdot 2^{n-4} + \ldots = \frac{\lfloor\underline{2n}}{\lfloor\underline{n}\,\lfloor\underline{n}}.$$

We have

$$(1 + x)^{2n} = (x^2 + 2x)^n + C_1^n (x^2 + 2x)^{n-1} + C_2^n (x^2 + 2x)^{n-2} + \ldots + 1.$$

The coefficient of x^n in $(1 + x)^{2n}$ is C_n^{2n},

the coefficient of x^n in $(x^2 + 2x)^n$ is 2^n,

the coefficient of x^n in $C_1^n (x^2 + 2x)^{n-1}$ is $C_1^n \cdot C_1^{n-1} \cdot 2^{n-2}$,

the coefficient of x^n in $C_2^n (x^2 + 2x)^{n-2}$ is $C_2^n \cdot C_2^{n-2} \cdot 2^{n-4}$,

and so on. The result follows by equating coefficients in the identity.

The use of the Multinomial theorem, explained in Art.16 on the opposite page, can sometimes be avoided, as in Ex. 2.

▍ **EXAMPLE 2.** *Find the coefficient of x^r in $(1 + x + x^2 + x^3)^n$.*

Here $1 + \Sigma a_r x^r \equiv (1 + x + x^2 + x^3)^n = (1 + x)^n (1 + x^2)^n$

$$= (1 + \Sigma C_s^n x^s)(1 + \Sigma C_t^n x^{st}).$$

Hence, if $r = 2m$, then $\qquad a_r = C_m^n + C_{m-1}^n C_2^n + C_{m-2}^n C_4^n + ...;$

and, if $r = 2m + 1$, $\qquad a_r = C_m^n C_1^n + C_{m-1}^n C_3^n + C_{m-2}^n C_5^n + ...$

■ **15. Expansion of $f(x + h)$ in Powers of x:** Here $f(x)$ is supposed to be a polynomial in x.

When the coefficients in $f(x)$ are known numbers, we use synthetic division, as in Art. 3.

The following general theorem is required later: the notation is explained in Art. 1 of this chapter.

$$\text{If } f(x) = a_0 x^n + na_1 x^{n-1} + \frac{n(n-1)}{\lfloor 2} a_2 x^{n-2} + ... + a_n,$$

$$\text{then } f(x + h) = a_0 x^n + nA_1 x^{n-1} + \frac{n(n-1)}{\lfloor 2} A_2 x^{n-2} + ... + A_n,$$

where $A_1 = a_0 h + a_1$, $A_2 = a_0 h^2 + 2a_1 h + a_2$, $A_3 = a_0 h^3 + 3a_1 h^2 + 3a_2 h + a_3$,

$$A_r = (a_0, a_1, a_2, ... a_r)(h, 1)^r.$$

For by the Binomial theorem,

$$f(x + h) = a_0 (x + h)^n + C_1^n a_1 (x + h)^{n-1} + C_2^n a_2 (x + h)^{n-2} + ... + a_n$$

$$= a_0 x^n + k_1 x^{n-1} + k_2 x^{n-2} + ... + k_r x^{n-r} + ... + k_n,$$

where $\qquad k_r = a_0 C_r^n h^r + a_1 C_1^n C_{r-1}^{n-1} h^{r-1} + a_2 C_2^n C_{r-2}^{n-2} h^{r-2} + ...$

to $r + 1$ terms.

Now $\qquad\qquad C_p^n C_{r-p}^{n-p} = \dfrac{\lfloor n}{\lfloor p \lfloor n-p} \cdot \dfrac{\lfloor n-p}{\lfloor r-p \lfloor n-r}$

$$= \dfrac{\lfloor n}{\lfloor r \lfloor n-r} \cdot \dfrac{\lfloor r}{\lfloor p \lfloor r-p}$$

$$= C_r^n \cdot C_p^n.$$

Thus $k_r = C_r^n (a_0 h^r + C_1^r a_1 h^{r-1} + C_2^r a_2 h^{r-2} + ... + a_r) = C_r^n A_r$,

which proves the theorem.

■ 16. Multinomial Theorem for a Positive Integral Index:

(1) *Expansion of* $(a + b + c + \ldots + k)^n$. The product

$(a + b + c + \ldots + k)(a + b + c + \ldots + k) \ldots$ to n factors is the sum of all the products which can be formed by choosing a term out of each of the n factors and multiplying these terms together.

The expansion is therefore the sum of a number of terms of the form $a^\alpha b^\beta c^\gamma \ldots k^\kappa$, where each index may have any of the values $0, 1, 2, \ldots, n$, subject to the condition.

$$\alpha + \beta + \gamma + \ldots + \kappa = n. \qquad \ldots(A)$$

Choose any positive integral or zero values for $\alpha, \beta, \gamma, \ldots, \kappa$ which satisfy this condition. To obtain the coefficient of the term $a^\alpha\, b^\beta\, c^\gamma \ldots k^\kappa$, divide the n factors into groups containing $\alpha, \beta, \gamma, \ldots, \kappa$ factors, respectively. Take a out of each factor of the first group, b out of each factor of the second group, and so on.

The number of ways in which this can be done is

$$\frac{\lfloor n}{\lfloor \alpha \lfloor \beta \lfloor \gamma \ldots \lfloor \kappa}, \qquad \ldots(B)$$

and this is the coefficient of $a^a b^\beta c^\gamma \ldots k^\kappa$ in the expansion (*i.e.* the number of times which this particular term occurs). Hence

$$(a + b + c + \ldots + k)^n = \Sigma\ \frac{\lfloor n}{\lfloor \alpha \lfloor \beta \lfloor \gamma \ldots \lfloor \kappa} \cdot a^\alpha b^\beta c^\gamma \ldots k^\kappa, \qquad \ldots(C)$$

where $\alpha, \beta, \gamma, \ldots, \kappa$ are to have all possible sets of values which can be chosen from the numbers $0, 1, 2, \ldots, n$, subject to the condition

$$\alpha + \beta + \gamma + \ldots + \kappa = n.$$

This result is called the *Multinomial Theorem* for a positive integral index.

Thus in the expansion of $(a - b - c + d)^6$ the coefficient of $a^3 b^2 c$ is

$$(-1)^3\ \frac{\lfloor 6}{\lfloor 3 \lfloor 2 \lfloor 1 \lfloor 0} = -60.$$

(2) *If there are m numbers a, b, c, … k, the greatest coefficient in the expansion of $(a + b + c + \ldots + k)^n$ is*

$$\lfloor n / \left(\lfloor q\right)^{m-r} \left(\lfloor q + 1\right)^r,$$

where q is the quotient and r the remainder when n is divided by m.

The coefficient $\lfloor n / \lfloor \alpha \lfloor \beta \lfloor \gamma \ldots \lfloor \kappa$ has its greatest value when $\lfloor \alpha \lfloor \beta \ldots \lfloor \kappa$ has its least value. Denote this value by v, then v is the least value of $\lfloor \alpha \lfloor \beta \ldots \lfloor \kappa$ if α and β alone vary and $\gamma, \ldots \kappa$ are constant.

Thus $\lfloor \alpha \lfloor \beta$ must have its least value subject to the condition $\alpha + \beta = C$, where C is a constant.

Let $u_a = \lfloor\underline{\alpha}\,\lfloor\underline{\beta} = \lfloor\underline{\alpha}\,\lfloor\underline{C-\alpha}$, then $\dfrac{u_a}{u_{a-1}} = \dfrac{\alpha}{C-\alpha+1}$ and $u_a \gtreqless u_{a-1}$ according as

$\alpha \lesseqgtr C - \alpha + 1$, *i.e,* as $\alpha \lesseqgtr \dfrac{1}{2}(C+1)$. Hence u_a has its least value when $\alpha = \beta = \dfrac{1}{2}C$ if C is even, and when $\alpha = \dfrac{1}{2}(C+1)$, $\beta = \dfrac{1}{2}(C-1)$ if C is odd.

Thus α and β must be equal or differ by 1, and the same is true for any pair of the numbers $a, b, \ldots k$.

We conclude that some (possibly all) of the set $a, b, \ldots k$ are equal to a certain number q, each of the rest being equal to $q + 1$.

Let r of the numbers be equal to $q + 1$ and $m - r$ of them equal to q, then $0 \leq r < m$ and $r(q + 1) + (m - r)q = n$, that is $qm + r = n$.

Hence q is the quotient and r the remainder when n is divided by m, which proves the theorem.

(3) *If there are m numbers a, b, c, ... k, the cofficient of x^r in the expansion of $(a + bx + cx^2 + \ldots + kx^m)^n$ is* $\displaystyle\sum \frac{\lfloor\underline{n}}{\lfloor\underline{\alpha}\,\lfloor\underline{\beta}\,\lfloor\underline{\gamma}\ldots\lfloor\underline{\kappa}} \cdot a^\alpha b^\beta c^\gamma \ldots k^\kappa$, *where $\alpha, \beta, \gamma, \ldots \kappa$ have any of the values 0,1, 2, ... n subject to the conditions*

$$\alpha + \beta + \gamma + \ldots + \kappa = n \text{ and } \beta + 2\gamma + \ldots + m\kappa = r.$$

This follows from (1) by putting $bx, cx^2, \ldots kx^m$ for $b, c, \ldots k$ in equation (C).

▌**EXAMPLE 1.** *Find the coefficient of x^4 in the expansion of*
$$(1 + 2x + 3x^2)^5.$$

The coefficient $= \displaystyle\sum \frac{\lfloor\underline{5}}{\lfloor\underline{\alpha}\,\lfloor\underline{\beta}\,\lfloor\underline{\gamma}} \cdot 1^\alpha \cdot 2^\beta \cdot 3^\gamma,$

where α, β, γ have any of the values 0, 1, 2, ... 5 such that

$$\alpha + \beta + \gamma = 5 \text{ and } \beta + 2\gamma = 4.$$

α	β	γ
1	4	0
2	2	1
3	0	2

The solutions of the second equation are (4, 0), (2, 1), (0, 2), and the possible values of α , β , γ are shown in the margin;

$$\therefore \text{ The coefficient} = \frac{\lfloor\underline{5}}{\lfloor\underline{1}\,\lfloor\underline{4}\,\lfloor\underline{0}} \cdot 2^4 \cdot 3^0 + \frac{\lfloor\underline{5}}{\lfloor\underline{2}\,\lfloor\underline{2}\,\lfloor\underline{1}} \cdot 2^2 \cdot 3 + \frac{\lfloor\underline{5}}{\lfloor\underline{3}\,\lfloor\underline{0}\,\lfloor\underline{2}} \cdot 2^0 \cdot 3^2$$

$$= 80 + 360 + 90 = 530.$$

▌**EXAMPLE 2.** *Find the greatest coefficient in the expansion of $(a + b + c)^7$.*

Here $m = 3$ and $7 = 2.\,3 + 1$, so that $q = 2, r = 1$;

$$\therefore \text{ Greatest coefficient} = \lfloor\underline{7} / (\lfloor\underline{2})^2 \cdot \lfloor\underline{3}.$$

EXERCISE IV

BINOMIAL COEFFICIENTS

If $(1 + x)^n = c_0 + c_1x + c_2x^2 + \ldots + c_nx^n$, prove the statements in Exs. 1–12.

1. $c_1 + 2c_2x + 3c_3x^2 + \ldots + nc_nx^{n-1} = n(1 + x)^{n-1}$.

2. $c_0^2 + c_1^2 + c_2^2 + \ldots + c_n^2 = \dfrac{\lfloor 2n}{\lfloor n \lfloor n}$.

3. $c_0 + 2c_1x + 3c_2x^2 + \ldots + (n + 1)\,c_nx^n = \{1 + (n + 1)\,x\}\,(1 + x)^{n-1}$.

4. $c_0 - 2c_1 + 3c_2 - \ldots + (-1)^n (n + 1)c_n = 0$.

5. $c_2 + 2c_3 + 3c_4 + \ldots + (n - 1)\,c_n = 1 + (n - 2)\,2^{n-1}$.

6. $c_0c_1 + c_1c_2 + c_2c_3 + \ldots + c_{n-1}c_n = \dfrac{\lfloor 2n}{\lfloor n+1 \lfloor n-1}$.

7. $c_0c_r + c_1c_{r+1} + c_2c_{r+2} + \ldots + c_{n-r}c_n = \dfrac{\lfloor 2n}{\lfloor n+r \lfloor n-r}$.

8. $\dfrac{c_1}{c_0} + \dfrac{2c_2}{c_1} + \dfrac{3c_3}{c_2} + \ldots + \dfrac{nc_n}{c_{n-1}} = \dfrac{1}{2}\,n\,(n + 1)$.

9. Show that $c_0^2 - c_1^2 + c_2^2 - \ldots + (-1)n\; c_n^2 = \dfrac{\lfloor n}{\left\lfloor \dfrac{1}{2}n \right. \left\lfloor \dfrac{1}{2}n \right.}$ or zero, according as n is even

or odd.

10. Show that the sum of the products of $c_0, c_1, c_2, \ldots, c_n$ taken two together is equal to

$$2^{2n-1} - \dfrac{\lfloor 2n}{2 \lfloor n \lfloor n}.$$

[The required sum $= \dfrac{1}{2}\,\{(c_0 + c_1 + c_2 + \ldots + c_n)^2 - (c_0^2 + c_1^2 + c_2^2 + \ldots + c_n^2)\}.$]

11. Using the result of Ex. 1, show that

$$c_1^2 + 2c_2^2 + 3c_3^2 + \ldots + nc_n^2$$

is equal to the coefficient of x^{n-1} in the expansion of $n(1 + x)^{2n-1}$. Hence show that

the sum of this series is $\lfloor 2n-1 / (\lfloor n-1)^2$.

12. Using the result of Ex. 3, show that

$$c_0^2 + 2c_1^2 + 3c_2^2 + \ldots + (n + 1)\,c_n^2$$

is equal to the coefficient of x^n in the expansion of

$$[1 + (n + 1)x]\,(1 + x)^{2n-1}.$$

Hence show that the sum of this series is $\dfrac{(n + 2)\lfloor 2n-1}{\lfloor n \lfloor n-1}$.

13. Prove that if $n > 1$,

$$\frac{1}{\lfloor 1 \lfloor n-1} + \frac{1}{\lfloor 2 \lfloor n-2} + \frac{1}{\lfloor 3 \lfloor n-3} + ... + \frac{1}{\lfloor n-1 \lfloor 1} = \frac{2^n - 2}{\lfloor n}.$$

14. If the terms of the series

$$(x + 3)^{n-1} + (x + 3)^{n-2} (x + 2) + (x + 3)^{n-3} (x + 2)^2 + ... + (x + 2)^{n-1}$$

are expanded and the whole is arranged in powers of x, show that the coefficient of

$$x^r \text{ is } \frac{\lfloor n}{\lfloor r \lfloor n-r} \; (3^{n-r} - 2^{n-r}).$$

15. Show that

$$\sum_{r=1}^{r=n-1} r\,(n-r)\left(C_r^n\right)^2 = n^2 \, C_n^{2n-2}.$$

[Using Ex. 11, the sum of the series is equal to the coefficient of x^n in the expansion of $n^2 (1 + x)^{2n-2}$.]

In Exs. 16 – 18 find the coefficients of the products stated:

16. $abcde$ in the expansion of $(a + b + c + d + e)^5$.

17. $a^3b^4c^5$ in the expansion of $(bc + ca + ab)^6$.

[Write $a^3b^4c^5 = (bc)^x (ca)^y (ab)^z$ and find x, y, z.]

18. a^2b^2cd in the expansion of $(ab + ac + ad + bc + bd + cd)^3$.

19. If $(1 + x + x^2)^n = c_0 + c_1x + c_2x^2 + ... + c_rx^r + ... + c_{2n}x^{2n}$, prove that

(*i*) $c_0 = c_{2n}, \; c_1 = c_{2n-1}, \; ... \; c_r = c_{2n-r} \; ...$.

(*ii*) $c_0^2 - c_1^2 + c_2^2 - ... + c_{2n}^2 = c_n$.

(*iii*) $c_0 + c_2 + c_4 + ... + c_{2n} = 1 + c_1 + c_3 + c_5 + ... + c_{2n-1}$.

(*iv*) $c_0 - nc_1 + \dfrac{n(n-1)}{\lfloor 2}\, c_2 - \dfrac{n(n-1)(n-2)}{\lfloor 3}\, c_3 + ... + (-1)^n c_n$

is equal to $\dfrac{\lfloor n}{\lfloor \frac{1}{3}n \, \lfloor \frac{2}{3}n}$ or zero, according as n is or is not a multiple of 3.

(*v*) $c_n = 1 + \dfrac{n(n-1)}{1^2} + \dfrac{n(n-1)(n-2)(n-3)}{1^2 . 2^2} + ...$, and write down the last term of

this series when n is even and when n is odd.

(*vi*) $c_0c_2 - c_1c_3 + c_2c_4 - ... + c_{2n-2}c_{2n} = c_{n+1}$.

[In (*i*) put $1/x$ for x and multiply by x^{2n}. In (*ii*) put $-x$ for x, and observe that

$$(1 + x + x^2)(1 - x + x^2) = 1 + x^2 + x^4.]$$

20. If $(1 + x + x^2 + x^3)^n = c_0 + c_1x + c_2x^2 + ... + c_{3n}x^{3n}$:

(*i*) Find the value of $c_0 + c_1 + c_2 + ... + c_{3n}$.

(*ii*) Prove that $c_0 + c_2 + c_4 + c_6 + \ldots = c_1 + c_3 + c_5 + c_7 + \ldots$.

(*iii*) Prove that $c_0 = c_{3n}$, $c_1 = c_{3n-1}$, $c_2 = c_{3n-2}$, etc.

21. In the expansion of

$$(1 + x)(1 + x + x^2)(1 + x + x^2 + x^3) \ldots (1 + x + x^2 + \ldots + x^n):$$

   (*i*) What is the term containing the highest power of x?

   (*ii*) Find the sum of the coefficients.

   (*iii*) Show that the sum of the coefficients of the even terms is equal to that of the odd terms.

   (*iv*) Show that the coefficients of the terms equidistant from the beginning and end are equal.

22. In the expansion of $(1 + x)^n (1 + y)^n (1 + z)^n$, show that the sum of the coefficients

of the terms of degree r is $\dfrac{\lfloor 3n}{\lfloor r \lfloor 3n - r}$.

23. By expanding the two sides of the identity

$$(2x + x^2)^n = \{(1 + x)^2 - 1\}^n$$

and equating the coefficients of x^n, show that

$$C_n^{2n} - C_1^n C_n^{2n-2} + C_2^n C_n^{2n-4} - \ldots = 2^n.$$

Also write down the last term of the series on the left (*i*) when n is even, (*ii*) when n is odd.

24. By expanding $\{(1 + x)^2 - 2x\}^m$ in two different ways, prove that, if $m > n$,

$$C_{2n}^{2m} - 2 \cdot C_1^m C_{2n-1}^{2m-2} + 2^2 \cdot C_2^m C_{2n-2}^{2m-4} - \ldots = C_n^m.$$

25. If $(1 + 2x + 2x^2)^n = a_0 + a_1 x + a_2 x^2 + \ldots + a_{2n} x^{2n}$,

prove that (*i*) $a_r = 2^{r-n} \left\{ C_r^{2n} + C_1^n C_r^{2n-2} + C_2^n C_r^{2n-4} + \ldots \right\}$,

the last term in the bracket being $C_{\frac{1}{2}r}^n$ or $(r + 1)\, C_{\frac{1}{2}(r+1)}^n$, according as r is even or odd.

In particular, show that

   (*ii*) $a_2 = 4n^2 - 2^{2-n} n \left(1 + 3C_1^{n-1} + 5C_2^{n-1} + \ldots\ldots \text{ to } n \text{ terms}\right)$.

   (*iii*) $a_4 = \dfrac{1}{3}\, 2^{3-n} n\, (n - 1)\, \{1.3 + 3.5\, C_1^{n-2} + 5.\, 7 C_2^{n-2} + \ldots \text{ to } (n - 1) \text{ terms}\}$.

[We have $(1 + 2x + 2x^2) = 2\left(x + \dfrac{1}{2}\right)^2 + \dfrac{1}{2}$;

$\therefore (1 + 2x + 2x^2)^n = 2^n \left(x + \dfrac{1}{2}\right)^{2n} + 2^{n-2}\, C_1^n \left(x + \dfrac{1}{2}\right)^{2n-2} + \ldots .]$

▨ **17. Highest Common Factor (H.C.F.):** We consider polynomials in a single variable x. These will be denoted by capital letters.

(1) *If $A \equiv BQ + R$, where A, B, Q, R are polynomials, the common factors of A and B are the same as those of B and R.*

For since $A = BQ + R$, every common factor of B and R is a factor of A and, since $R = A - BQ$, every common factor of A and B is a factor of R.

(2) *If B is not of higher degree than A, polynomials or constants (Q_1, R_1), (Q_2, R_2), (Q_3, R_3) ... can be found such that*

$$A = BQ_1 + R_1, \ B = R_1 Q_2 + R_2, \ R_1 = R_2 Q_3 + R_3, \ etc.,$$

where the degree of any one of the set B, R_1, R_2, ... is less than that of the preceding.

It follows from the last theorem that

    (*a*) if one of the pairs (A, B), (B, R_1), (R_1, R_2), (R_2, R_3) have a common factor, this is a common factor of every pair;

    (*b*) if one pair have no common factor, the same is true for every pair.

Moreover, the process must terminate and that in one of two ways: the last remainder, say R_n, must vanish identically, or be independent of x.

If $R_n = 0$, then $R_{n-2} = R_{n-1} Q_n$, so that R_{n-1} is a common factor of (R_{n-2}, R_{n-1}), and therefore also of (A, B). Moreover, every common factor of (A, B) is a factor of R_{n-1}, which is therefore the H.C.F. of A, B.

If R_n is independent of x, then (R_{n-2}, R_{n-1}) have no common factor, and the same is true of (A, B).

(3) The H.C.F. Process: Equations (1) of the last section are found by successive divisions, as represented below.

In practice, the reckoning is generally arranged as on the right.

$$
\begin{array}{ll}
B\,)\ A\ (Q_1 \\
\quad \underline{BQ_1} \\
\quad\ \ R_1\,)\ B\ (Q_2 \\
\qquad \underline{R_1 Q_2} \\
\qquad\ \ R_2\,)\ R_1\ (Q_3 \\
\qquad\qquad etc.
\end{array}
\qquad\qquad
\begin{array}{c|c|c|c}
Q_2 & B & A & Q_1 \\
 & R_1 Q_2 & BQ_1 & \\
\hline
 & R_2 & R_1 & Q_3
\end{array}
$$

$$etc.$$

The process may often be shortened by noticing that:

    (*i*) We may multiply or divide any of the set A, B, R_1, R_2, ... by any constant or by any polynomial which is not a factor of the preceding member of the set.

    (*ii*) If we arrive at a remainder R_n which can be completely factorised, the process need not be continued. For if we find which, if any, of these factors is a factor of R_{n-1}, we can write down all the common factors of (R_{n-1}, R_n).

    (*iii*) We may use the following theorem:

If $X = lA + mB$ and $Y = l'A + m'B$, where l, m, l', m' are constants different from zero and such that $lm' - l'm \neq 0$, then the H.C.F. of X and Y is the same as that of A and B.

For every common factor of A and B is a common factor of X and Y. Moreover,

$$A\,(lm' - l'm) = m'X - mY \text{ and } B\,(lm' - l'm) = lY - l'X,$$

therefore every common factor of X and Y is a common factor of A and B.

■ **EXAMPLE 1.** *Find the H.C.F. of $A = 3x^3 + x + 4$ and $B = 2x^3 - x^2 + 3$.*

Paying attention to the remark (*i*) and using detached coefficients, the reckoning is as follows:

```
                                      | 3 + 0 + 1 + 4
                                      |           2
                    2 - 1 + 0 + 3     | 6 + 0 + 2 + 8    3
                              3       | 6 - 3 + 0 + 9
         2 |  6 - 3 + 0 + 9           |     3 + 2 - 1    (a)
           |  6 + 4 - 2               |             7
        -7 |    - 7 + 2 + 9           | 21 + 14 - 7     - 3
           |    - 7 - 7               | 21 - 6 - 27
         9 |            9 + 9         | 20 ) 20 + 20
           |            9 + 9         |        1 +  1
```

Hence the H.C.F. $= x + 1$. *But it is unnecessary to go beyond the step marked* (*a*), which shows that $R_1 = 3x^2 + 2x - 1 = (3x - 1)(x + 1)$.

Now $x + 1$ is, and $3x - 1$ is not, a factor of B. Therefore $x + 1$ is the H.C.F. of B, R_1, and consequently that of A and B.

■ **EXAMPLE 2.** *Find the H.C.F. of A and B where*

$$A = 12x^5 + 5x^3 + 8 \text{ and } B = 8x^5 + 5x^2 + 12.$$

We have $\qquad 2A - 3B = 10x^3 - 15x^2 - 20 = 5\,(2x^3 - 3x^2 - 4),$

$$3A - 2B = 20x^5 + 15x^3 - 10x^2 = 5x^2\,(4x^3 + 3x - 2).$$

Putting $C = 2x^3 - 3x^2 - 4$ and $D = 4x^3 + 3x - 2$, it follows that the H.C.F. of A and B is the same as that of C and D.

Further, $\qquad C - 2D = 6x^2 + 3x + 6 = 3\,(2x^2 + x + 2),$

$$2C - D = 6x^3 + 3x^2 + 6x = 3x\,(2x^2 + x + 2).$$

Hence the H.C.F. of $C - 2D$ and $2C - D$ is $2x^2 + x + 2$: this is therefore the H.C.F. of C and D, and also that of A and B.

■ **18. Prime and Composite Functions:** If a polynomial has no factors except itself (and constants) it is said to be *prime*: otherwise it is said to be *composite*.

Thus $x^2 + 3x + 2$ is a composite function whose *prime factors* are $x + 1$ and $x + 2$.

Polynomials which have no common factor (except constants) are said to be *prime to each other.* Such expressions *have no* H.C.F.

Thus $2(x + 1)$ and $4(x^2 + 1)$ are prime to each other.

We can prove theorems for polynomials analogous to those of Ch. 1, Arts. 3, 4 and 5, relating to whole numbers.

Remembering the distinction between arithmetical and algebraical primeness, the reader should have no difficulty in making the necessary verbal alterations. Thus, the theorem corresponding to that of 4 (*i*) is as follows:

If A, B, M are polynomials and M is a factor of AB and prime to A, then M is a factor of B.

For if M is prime to A, these two have no common factor. Hence the H.C.F. of MB and AB is B. But M is a factor of AB, and is therefore a common factor of MB and AB. Hence M is equal to, or is a factor of, B.

Theorem: *If A and B are polynomials in x, then polynomials X, Y, prime to each other, can be found such that*

$$AX + BY = 1 \text{ or } AX + BY = G,$$

according as A is or is not prime to B, G being the H.C.F. of A and B in the latter case.

For if $Q_1, Q_2, \ldots$ are the quotients and $R_1, R_2, \ldots$ the remainders in the process of searching for the H.C.F. of A and B,

$$A = BQ_1 + R_1,\ B = R_1Q_2 + R_2,\ R_1 = R_2Q_3 + R_3, \text{ etc.}$$

Therefore we obtain in succession the following equations:

$$R_1 = A - BQ_1,$$
$$R_2 = -AQ_2 + B(1 + Q_1 Q_2),$$
$$R_3 = A(1 + Q_2Q_3) - B(Q_1 + Q_3 + Q_1Q_2Q_3).$$

Continuing thus, we can express any remainder in the form $AX + BY$, where X and Y are polynomials.

Now the last remainder is either G, the H.C.F.; or it is a constant c. Hence we can find X, Y such that

$$AX + BY = G \text{ or } AX + BY = c.$$

In either case X is prime to Y, for the first equation may be written $\dfrac{A}{G} \cdot X + \dfrac{B}{G} \cdot Y = 1$ and G is a factor of A and of B; thus in either case any common factor of X and Y would be a factor of a constant.

In the second case, the polynomials in question are X/c and Y/c.

Note: This theorem is fundamental in the Theory of Partial Fractions, which are dealt with in Ch. 7.

EXERCISE V

H.C.F. AND ITS USE

Find the H.C.F. of the functions in Exs. 1–6.

1. $3x^4 - 4x^3 + 1$, $4x^4 - 5x^3 - x^2 + x + 1$.

2. $4x^4 + 5x^2 + 7x + 2$, $16x^3 + 10x + 7$.

3. $2x^4 - 13x^2 + x + 15$, $3x^4 - 2x^3 - 17x^2 + 12x + 9$.

4. $x^5 + 5x^2 - 2$, $2x^5 - 5x^3 + 1$.

5. $2x^5 - 5x^2 + 3$, $3x^5 - 5x^3 + 2$.

6. $12x^3 + 2x^2 - 21x - 4$, $6x^3 + 7x^2 - 14x - 8$, $21x^4 - 28x^3 + 9x^2 - 16$.

In Exs. 7–9, find polynomials X and Y for which the statements are identically true.

7. $(x - 1)^2 X - (x + 1)^2 Y = 1$.

8. $(x^3 - 1) X - (x^5 - 1) Y = x - 1$.

9. $(x^3 - 3x^2 + 2x - 1) X - (x^3 - 2x^2 - 2) Y = 1$.

10. Use the H.C.F. process to obtain

$$3(x^2 + 3x + 4) - (3x^2 + 2x + 1) = 7x + 11,$$

$$x(3x^2 + 2x + 1) - (3x - 7)(x^2 + 3x + 4) = 10x + 28;$$

and hence find A, B, C, D, such that

$$\frac{2x - 3}{\left(x^2 + 3x + 4\right)\left(3x^2 + 2x + 1\right)} = \frac{Ax + B}{x^2 + 3x + 4} + \frac{Cx + D}{3x^2 + 2x + 1}.$$

[Multiply the first identity by λ, the second by μ, and add: find λ and μ such that

$$(7\lambda + 10\mu)/(11\lambda + 28\mu) = 2/(-3);$$

and thus obtain

$$(Ax + B)(3x^2 + 2x + 1) + (Cx + D)(x^2 + 3x + 4) = 2x - 3.]$$

■ ■ ■

CHAPTER 4

Symmetric and Alternating Functions, Substitutions

■ **1. Symmetric Functions:** A function which is unaltered by the interchange of any two of the variables which it contains is said to be *symmetric with regard to these variables.*

Thus $yz + zx + xy$ and $(x^2y + y^2z + z^2x)\, x^2z + y^2x + z^2y)$ are symmetric with regard to x, y, z. (In the second expression, the interchange of any two letters transforms one factor into the other.)

The interchange of any two letters, $x, y,$ is called the *transposition* (xy).

Terms of an expression which are such that one can be changed into the other by one or more transpositions are said to be *of the same type.*

Thus all the terms of $x^2y + x^2z + y^2z + y^2x + z^2x + z^2y$ are of the same type, and the expression is symmetric with regard to x, y, z.

A symmetric function which is the sum of a number of terms of the same type is often written is an abbreviated form thus: *Choose any one of the terms and place the letter Σ (sigma) before it.* For instance,

$x + y + z$ is represented by Σx and $yz + zx + xy$ by Σxy.

Again, $(x+y+z)^2 = x^2 + y^2 + z^2 + 2yz + 2zx + 2xy = \Sigma x^2 + 2\Sigma yz$.

It is obvious that

(*i*) *if a term of some particular type occurs in a symmetric function, then all the terms of this type occur.*

(*ii*) *the sum, difference, product and quotient of two symmetric functions are also symmetric functions.*

Considerations of symmetry greatly facilitate many algebraical processes, as in the following examples:

▌ **EXAMPLE 1.** *Expand*
$$(y + z - x)\ (z + x - y)\ (x + y - z).$$

This expression is symmetric, homogeneous, and of the third degree in x, y, z. We may therefore assume that $(y + z - x)\ (z + x - y)\ (x + y - z) = a.\ \Sigma x^3 + b.\ \Sigma x^2y + cxyz$, where $a, b, c,$ are independent of x, y, z.

In this assumed identity,

(*i*) put $x=1, y=0, z=0$; then $-1 = a$;

(*ii*) put $x=1, y=1, z=0$; then $0 = 2a + 2b,\ \therefore\ b=1$;

(iii) put $x=1$, $y=1$, $z=1$; then $1= 3a + 6b + c$, $\therefore c= -2$.

Hence the required product is

$$-x^3 - y^3 -z^3 + y^2z + yz^2 + z^2x + zx^2 + x^2y + xy^2 - 2xyz.$$

■ **EXAMPLE 2.** *Expand* $(a+b+c+d)$ $(ab + ac + ad + bc + bd + cd)$. *Test the result by putting* $a = b = c = d = 1$.

The product is the sum of all the products which can be obtained by multiplying any term of the first expression by any term of the second. Hence the terms in the product are of one of the types a^2b, abc.

The coefficient of a^2b in the product is 1; for this term is obtained as the product of a and ab, and in no other way.

The coefficient of abc is 3; for this term is obtained in each of the three ways a (bc), b (ac), c (ab).

Hence, the required product is $\Sigma a^2b + 3\Sigma abc$.

Test. The number of terms of the type a^2b is 12, and the number of terms of the type abc is 4; hence, if $a = b = c = d = 1$,

$$\Sigma a \cdot \Sigma ab = 4 \cdot 6 = 24 \text{ and } \Sigma a^2 b + 3\Sigma abc = 12 + 3 \cdot 4 = 24,$$

so that the test is satisfied.

■ **EXAMPLE 3.** *Factorise* $(x + y + z)^5 - x^5 - y^5 - z^5$.

Denote the given expression by E. Since $E = 0$ when $x = -y$, it follows that $x + y$ is a factor of E; similarly, $y+z$ and $z+x$ are factors. The remaining factor is a homogeneous symmetric function of x, y, z of the second degree. We therefore assume that

$$E = (y + z) (z + x) (x + y) \{a (x^2 + y^2 +z^2) + b (yz + zx + xy)\},$$

where a and b are independent of x, y, z, and proceed thus:

$$\left. \begin{array}{l} \text{put } x=1, y=1, z=0; \quad \text{then } 2a+b=15. \\ \text{put } x=1, y=1, z=1; \quad \text{then } a+b=10. \end{array} \right\} \quad \left. \begin{array}{l} \therefore a=5 \\ b=5 \end{array} \right\};$$

$\therefore \qquad E = 5 (y + z) (z + x) (x + y) (x^2 + y^2 + z^2 + yz + zx + xy).$

■ **2. Alternating Funtions:** If a function E of x, y, z, ... is transformed into $-E$ by the interchange of any two of the set x, y, z, ..., then E is called an *alternating function of x, y, z, ...* .

Such a function is $x^n (y - z) + y^n (z - x) + z^n (x - y)$; for the interchange of any two letters, say x and y, transforms it into

$$y^n (x - z) + x^n (z - y) + z^n (y - x) = -E.$$

Observe that *the product and the quotient of two alternating functions are symmetric functions.*

Thus $\{x^3 (y - z) + y^3 (z - x) + z^3 (x - y)\}/(y - z) (z - x) (x - y)$ is symmetric with regard to x, y, z.

■ **EXAMPLE :** *Factorise* $x^3 (y - z) + y^3 (z - x) + z^3 (x - y)$.

Denote the expression by E. Since $E = 0$ when $x = y$, it follows that $x - y$ is a factor. Similarly $y - z$ and $z - x$ are factors, thus

$$E = (y - z) (z - x) (x - y) \cdot F,$$

where F is symmetric, homogeneous and of the first degree in x, y, z. Hence

$$E = k (y - z) (z - x) (x - y) (x + y + z),$$

where k is independent of x, y, z. Equating the coefficients of $x^3 y$ on each side, we find that $k = -1$, thus

$$E = - (y - z) (z - x) (x - y) (x + y + z).$$

■ **3. Cyclic Expressions:** An expression is said to be *cyclic with regard to the letters* $a, b, c, d, \ldots h, k,$ *arranged in this order,* when it is unaltered by changing a into b, b into c, c into d, ..., h into k, and k into a.

This interchange of the letters is called *the cyclic substitution* $(abc...k)$.

Thus $a^2 b + b^2 c + c^2 d + d^2 a$ is cyclic with regard to a, b, c, d (in this order), for the cyclic substitution $(abcd)$ changes the first term into the second, the second into the third, ..., and the last into the first.

It is obvious that

    (*i*) *If a term of some particular type occurs in a cyclic expression, then the term which can be derived from this by the cyclic interchange, must also occur; and the coefficients of these terms must be equal.*

    (*ii*) *The sum, difference, product, and quotient, of two cyclic expressions are also cyclic.*

In writing a cyclic expression, it is unnecessary to write the whole.

Thus $x^2 (y - z) + y^2 (z - x) + z^2 (x - y)$ is often denoted by $\Sigma x^2 (y - z)$, where the meaning of Σ must not be confused with that in Art. 1.

Again, we sometimes denote $x^2 (y - z) + y^2 (z - x) + z^2 (x - y)$ by

$$x^2 (y - z) + \ldots + \ldots,$$

it being understood that the second term is to be obtained from the first and the third from the second by cyclic interchange.

The student should be familiar with the following important identities:

1. $(b - c) + (c - a) + (a - b) = 0.$

2. $a (b - c) + b (c - a) + c (a - b) = 0.$

3. $a^2 (b - c) + b^2 (c - a) + c^2 (a - b) = - (b - c) (c - a) (a - b).$

4. $bc (b - c) + ca (c - a) + ab (a - b) = - (b - c) (c - a) (a - b).$

5. $a (b^2 - c^2) + b (c^2 - a^2) + c (a^2 - b^2) = (b - c) (c - a) (a - b).$

6. $a^3 (b - c) + b^3 (c - a) + c^3 (a - b) = - (b - c) (c - a) (a - b) (a + b + c).$

7. $(a + b + c) (bc + ca + ab) = a (b^2 + c^2) + b (c^2 + a^2) + c (a^2 + b^2) + 3abc.$

8. $(b + c) (c + a) (a + b) = a (b^2 + c^2) + b (c^2 + a^2) + c (a^2 + b^2) + 2abc.$

9. $a^3 + b^3 + c^3 - 3abc = (a + b + c) (a^2 + b^2 + c^2 - bc - ca - ab).$

10. $(b - c)^2 + (c - a)^2 + (a - b)^2 = 2 (a^2 + b^2 + c^2 - bc - ca - ab).$

11. $(a + b + c) (b + c - a) (c + a - b) (a + b - c) = - a^4 - b^4 - c^4 + 2b^2 c^2 + 2c^2 a^2 + 2a^2 b^2.$

It will be proved later (pp. 91, 92) that

    (*i*) Any symmetric function of α, β, γ can be expressed in terms of $\Sigma \alpha, \Sigma \alpha \beta, \alpha \beta \gamma$.

(*ii*) Any symmetric function of α, β, γ, δ can be expressed in terms of $\Sigma\alpha$, $\Sigma\alpha\beta$, $\Sigma\alpha\beta\gamma$ and $\alpha\beta\gamma\delta$.

This mode of expression is extremely useful in factorising symmetric functions, and in proving identities.

■ **EXAMPLE :** *Factorise* $a (1 - b^2) (1 - c^2) + b (1 - c^2) (1 - a^2) + c (1 - a^2) (1 - b^2) - 4abc.$

Denoting the given expression by E, we have

$$E = a \{1 - (b^2 + c^2) + b^2c^2\} + ... + ... - 4abc$$
$$= \Sigma a - \Sigma ab^2 + abc\Sigma ab - 4abc.$$

Now $\quad \Sigma ab^2 = \Sigma a \,.\, \Sigma ab - 3abc;$

$\therefore \qquad E = \Sigma a - \Sigma a \,.\, \Sigma ab - abc + abc\Sigma ab$
$$= \Sigma a (1 - \Sigma ab) - abc (1 - \Sigma ab)$$
$$= (1 - bc - ca - ab) (a + b + c - abc).$$

■ 4. Substitutions:

(1) We consider processes by which one arrangement of a set of elements may be transformed into another.

Taking the permutations *cdba, bdac* of *a, b, c, d,* the first is changed into the second by replacing *a* by *c, b* by *a, c* by *b* and leaving *d* unaltered. This process is represented by the *operator*

$$\begin{pmatrix} abcd \\ cabd \end{pmatrix} \text{ or } \begin{pmatrix} abc \\ cab \end{pmatrix}, \text{ and we write } \begin{pmatrix} abc \\ cab \end{pmatrix} cdba = bdac.$$

Such a process and also the operator which effects it is called a *substitution*.

As previously stated, the interchange of two elements *a, b* is called *the transposition* (*ab*).

Also a substitution such as $\begin{pmatrix} abcd \\ bcda \end{pmatrix}$, in which each letter is replaced by the one immediately following it and the last by the first, is called a *cyclic substitution or cyclic,* and is denoted by (*abcd*).

If two operators are connected by the sign =, the meaning is that one is equivalent to the other, thus (*abcd*) = (*bcda*).

(2) Two or more substitutions may be applied successively. This is indicated as follows, *the order of operations being from right to left.*

Let $S = (ab)$, $T = (bc)$, then

STabcd = Sacbd = bcad, and *TSabcd = Tbacd = cabd.*

Thus $\qquad ST = \begin{pmatrix} abcd \\ bcad \end{pmatrix}, \; TS = \begin{pmatrix} abcd \\ cabd \end{pmatrix}.$

This process is called *multiplication of substitutions,* and the resulting substitution is called the *product.*

Multiplication of this kind is not necessarily commutative, but *if the substitutions have no common letter, it is commutative.*

The operation indicated by $(ab)\,(ab)$, in which (ab) is performed twice, produces no change in the order of the letters, and is called an *identical substitution.*

(3) *Any substitution is cyclic, or is the product of two or more cyclic subtitutions which have no common element.*

As an instance, consider the substitution

$$S = \begin{pmatrix} abcdefghk \\ chfbgaedk \end{pmatrix}.$$

Here a is changed to c, c to f, f to a, thus completing the cycle (acf). Also b is changed to h, h to d, d to b, making the cycle (bhd). Next, e is changed to g and g to e, giving the cycle (eg). The element k is unchanged, and we write

$$S = (acf)\,(bhd)\,(eg)\,(k) \quad \text{or} \quad S = (acf)\,(bhd)\,(eg).$$

This expression for S is unique, and the order of the factors is indifferent. Moreover, the method applies universally, for in effecting any substitution we must arrive at a stage where some letter is replaced by the first, thus completing a cycle. The same argument applies to the set of letters not contained in this cycle.

(4) *A cyclic substitutions of n elements is the product of $n - 1$ transpositions.*

For $\qquad (abc) = (ab)\,(bc),\ (abcd) = (abc)\,(cd) = (ab)\,(bc)\,(cd),$

$\qquad\qquad (abcde) = (abcd)\,(de) = (ab)\,(bc)\,(cd)\,(de)$, and so on.

We also have equalities such as

$(ae)\,(ad)\,(ac)\,(ab) = (abcde), \quad (ab)(ac)(ad)(ae)=(edcba).$

(5) *A substitution which deranges n letters and which is the product of r cycles is equivalent to $n - r$ transpositions.*

This follows at once from (3) and (4). Thus if $S = \begin{pmatrix} abcdefgh \\ chfbgaed \end{pmatrix}$,

then $\qquad\qquad S = (acf)\,(bhd)\,(eg) = (ac)\,(cf)\,(bh)\,(hd)\,(eg).$

If we introduce the product $(ab)\,(ab)$, S is unaltered and the number of transpositions is increased by 2.

Thus, if a given substitution is equivalent to j transpositions, the number j *is not unique.* We shall prove that

$$j = n - r + 2s \text{ where } s \text{ is a positive integer or zero.}$$

This is a very important theorem, and to prove it we introduce the notion of '*inversions.*'

(6) Taking the elements a, b, c, d, e, choose some arrangement, as $abcde$, and call it the *normal arrangement.*

Consider the arrangement $bdeac$. Here b precedes a, but follows it in the normal arrangement. On this account we say that the pair ba constitutes an *inversion.* Thus $bdeac$ contains five inversions, namely,

$$ba, \quad da, \quad dc, \quad ea, \quad ec.$$

Theorem 1. *If i is the number of inversions which are introduced or removed by a substitution which is equivalent to j transpositions, then i and j are both even or both odd.*

For consider the effect of a single transposition (fg).

If f, g are consecutive elements, the transposition (fg) does not alter the position of f or of g relative to the other elements. It therefore introduces or removes a single inversion due to the interchange of f, g.

If f, g are separated by n elements p, q, r, ... x, then f can be moved to the place occupied by g by $n + 1$ interchanges of consecutive elements, and then g can be moved to the place originally occupied by f by n such interchanges. (The steps are shown in the magin.)

$$\ldots fpq \ldots xg \ldots$$
$$\ldots pq \ldots xgf \ldots$$
$$\ldots gpq \ldots xf \ldots$$

Thus the transposition (fg) can be effected by $2n + 1$ interchanges of consecutive elements. Therefore *any* transposition introduces or removes an odd number of inversions, and the theorem follows.

Again, for a given substitution, i is a fixed number, and therefore whatever value j may have, it must be even or odd, according as i is even or odd. Hence the following.

Theorem 2. *If one arrangement A of a given set of elements is changed into another B by j transpositions, then j is always even or always odd.*

In other words:

The number of transpositions which are equivalent to a given substitution is not unique, but is always even or always odd.

Referring to (5), the minimum value of j is $n - r$, and so we have the theorem stated at the end of that section.

Thus substitutions may be divided into two distinct classes. We say that *a substitution is even or odd according as it is equivalent to an even or an odd number of transpositions.*

Rule. To determine the class of a substitution S we may express it as the product of cycles, and count the number of cycles with an even number of elements: then S is even or odd according as this number is even or odd.

Or we can settle the question by counting the number of inversions, but this generally takes longer.

Thus we see at once that the substitution in (5) is odd. Also it has 17 inversions.

EXERCISE VI

SYMMETRIC AND CYCLIC FUNCTIONS

1. Show that $(bc - ad)(ca - bd)(ab - cd)$ is symmetric with regard to a, b, c, d.
2. Show that the following expressions are cyclic with regard to a, b, c, d, taken in this order:

   (i) $(a - b + c - d)^2$;　(ii) $(a - b)(c - d) + (b - c)(d - a)$.

   Expand the expressions in Exs. 3, 4.
3. $(y + z - 2x)(z + x - 2y)(x + y - 2z)$.
4. $(x + y - z)^3 + (y + z - x)^3 + (z + x - y)^3 + (x + y - z)^3$.

   Prove the identities in Exs. 5–14.
5. $(\beta^2\gamma^2 + \gamma^2\alpha^2 + \alpha^2\beta^2)(\alpha + \beta + \gamma) = \Sigma\alpha^3\beta^2 + \alpha\beta\gamma\Sigma\alpha\beta$.

6. $(\alpha - \beta)(\alpha - \gamma) + (\beta - \gamma)(\beta - \alpha) + (\gamma - \alpha)(\gamma - \beta) = \Sigma\alpha^2 - \Sigma\alpha\beta$.

7. $(\beta - \gamma)(\beta + \gamma - \alpha) + (\gamma - \alpha)(\gamma + \alpha - \beta) + (\alpha - \beta)(\alpha + \beta - \gamma) = 0$.

8. $\alpha(\beta - \gamma)^2 + \beta(\gamma - \alpha)^2 + \gamma(\alpha - \beta)^2 = \Sigma\alpha^2\beta - 6\alpha\beta\gamma$.

9. $(\beta^2\gamma + \beta\gamma^2 + \gamma^2\alpha + \gamma\alpha^2 + \alpha^2\beta + \alpha\beta^2)(\alpha + \beta + \gamma) = \Sigma\alpha^3\beta + 2\Sigma\alpha^2\beta^2 + 2\alpha\beta\gamma\Sigma\alpha$.

10. $a^3(b + c) + b^3(c + a) + c^3(a + b) + abc(a + b + c) = \Sigma a^2 \cdot \Sigma ab$.

11. $(a + b - c)(a^2 + b^2 - c^2) + (b + c - a)(b^2 + c^2 - a^2) + (c + a - b)(c^2 + a^2 - b^2)$
$= 3\Sigma a^3 - \Sigma a^2 b$.

12. $(a^2 + b^2 + c^2)(x^2 + y^2 + z^2) = (ax + by + cz)^2 + (bz - cy)^2 + (cx - cz)^2 + (ay - bx)^2$.

13. $(b^2 - ca)(c^2 - ab) + (c^2 - ab)(a^2 - bc) + (a^2 - bc)(b^2 - ca) = -(bc + ca + ab)$
$(a^2 + b^2 + c^2 - bc - ca - ab)$.

14. $(a^2 - bc)(b^2 - ca)(c^2 - ab) = abc(a^3 + b^3 + c^3) - (b^3c^3 + c^3a^3 + a^3b^3)$.

15. If one of the numbers a, b, c is the geometric mean between the other two, use Ex.14 to show that $b^3c^3 + c^3a^3 + a^3b^3 = abc(a^3 + b^3 + c^3)$.

16. If the numbers x, y, z, taken in some order or other, form an arithmetical progression, use Ex. 3 to show that
$$2(x + y + z)^3 + 27xyz = 9(x + y + z)(yz + zx + xy).$$

17.   (*i*) Express $2(a - b)(a - c) + 2(b - c)(b - a) + 2(c - a)(c - b)$ as the sum of three squares.

    (*ii*) Hence show that $(b - c)(c - a) + (c - a)(a - b) + (a - b)(b - c)$ is negative for all real values of a, b, c, except when $a = b = c$.

    [Put $b - c = x, c - a = y, a - b = z$, and notice that
$$x^2 + y^2 + z^2 + 2yz + 2zx + 2xy = (x + y + z)^2 = 0.]$$

18. If $x + y + z = 0$, show that

    (*i*) $2yz = x^2 - y^2 - z^2$;

    (*ii*) $(y^2 + z^2 - x^2)(z^2 + x^2 - y^2)(x^2 + y^2 - z^2) + 8x^2y^2z^2 = 0$;

    (*iii*) $ax^2 + by^2 + cz^2 + 2fyz + 2gzx + 2hxy$ can be expressed in the form $px^2 + gy^2 + rz^2$; and find p, q, r in terms of a, b, c, f, g, h.

Prove the identities in Exs. 19–23, where $\Sigma\alpha$, $\Sigma\alpha\beta$, etc., denote symmetric functions of α, β, γ, δ: also verify each by putting $\alpha = \beta = \gamma = \delta = 1$.

19. $(\alpha + \beta + \gamma + \delta)(\alpha^2 + \beta^2 + \gamma^2 + \delta^2) = \Sigma\alpha^3 + \Sigma\alpha^2\beta$.

20. $(\alpha + \beta + \gamma + \delta)(\beta\gamma\delta + \gamma\delta\alpha + \delta\alpha\beta + \alpha\beta\gamma) = \Sigma\alpha^2\beta\gamma + 4\alpha\beta\gamma\delta$.

21. $(\beta\gamma\delta + \gamma\delta\alpha + \delta\alpha\beta + \alpha\beta\gamma)^2 = \Sigma\alpha^2\beta^2\gamma^2 + 2\alpha\beta\gamma\delta\Sigma\alpha\beta$.

22. $(\alpha\beta + \alpha\gamma + \alpha\delta + \beta\gamma + \beta\delta + \gamma\delta)^2 = \Sigma\alpha^2\beta^2 + 2\Sigma\alpha^2\beta\gamma + 6\alpha\beta\gamma\delta$.

23. $\Sigma\alpha\beta \cdot \Sigma\alpha\beta\gamma = \Sigma\alpha^2\beta^2\gamma + 3\alpha\beta\gamma\delta\Sigma\alpha$.

Simplify the expressions in Exs. 24–28.

24. $(b^{-1} + c^{-1})(b + c - a) + (c^{-1} + a^{-1})(c + a - b) + (a^{-1} + b^{-1})(a + b - c)$.

25. $\dfrac{(x-b)(x-c)}{(a-b)(a-c)} + \dfrac{(x-c)(x-a)}{(b-c)(b-a)} + \dfrac{(x-a)(x-b)}{(c-a)(c-b)}$.

26. $\dfrac{b^2+c^2-a^2}{(a-b)(a-c)} + \dfrac{c^2+a^2-b^2}{(b-c)(b-a)} + \dfrac{a^2+b^2-c^2}{(c-a)(c-b)}$.

27. $\dfrac{b-c}{1+bc}+\dfrac{c-a}{1+ca}+\dfrac{a-b}{1+ab}$.

28. $\dfrac{a(b-c)}{1+bc}+\dfrac{b(c-a)}{1+ca}+\dfrac{c(a-b)}{1+ab}$.

Factorise the expressions in Exs. 29–35.

29. $(b-c)^2 (b+c-2a) + (c-a)^2 (c+a-2b) + (a-b)^2 (a+b-2c)$.

[Put $b - c = x$, $c - a = y$, $a - b = z$, noting that $b + c - 2a = y - z$.]

30. $(b-c)(b+c-2a)^2 + (c-a)(c+a-2b)^2 + (a-b)(a+b-2c)^2$.

31. $(b-c)(b+c-a)^3 + (c-a)(c+a-b)^3 + (a-b)(a+b-c)^3$.

[Put $b + c - a = x$, $c + a - b = y$, $a + b - c = z$.]

32. $8(a+b+c)^3 - (b+c)^3 - (c+a)^3 - (a+b)^3$.

[Put $b + c = x$, $c + a = y$, $a + b = z$.]

33. $(a+b+c)^3 - (b+c-a)^3 - (c+a-b)^3 - (a+b-c)^3$.

34. $(a+b)^2 (a+c)^2 (b-c) + (b+c)^2 (b+a)^2(c-a) + (c+a)^2(c+b)^2(a-b)$.

[Put $b + c = x$, $c + a = y$, $a + b = z$.]

35. $(1-a^2)(1-b^2)(1-c^2) + (a-bc)(b-ca)(c-ab)$.

36. Express the following substitutions as the product of transpositions:

$(i)\ \begin{pmatrix} 123456 \\ 654321 \end{pmatrix};$
$\qquad (ii)\ \begin{pmatrix} 123456 \\ 246135 \end{pmatrix};$
$\qquad (iii)\ \begin{pmatrix} 123456 \\ 641235 \end{pmatrix}.$

■ ■ ■

CHAPTER 5

Complex Numbers

■ **1. Preliminary:** In order that the operation of taking the square root of a number may be always possible, we extend the number system so as to include numbers of a new class known as *imaginary* numbers.

Consider the equation

$$x^2 - 2x + 5 = 0.$$

Proceeding in the usual way, we obtain the *formal* solution,

$$x = 1 \pm \sqrt{-4}, \quad \text{or} \quad x = 1 \pm 2\sqrt{-1},$$

which at present has no meaning.

Suppose the symbol ι to possess the following properties:

   (i) *It combines with itself and with real numbers according to the laws of algebra.*

   (ii) *We may substitute (-1) for ι^2, wherever this occurs.*

The reader can verify for himself that (at any rate so far as addition, subtraction, multiplication and division are concerned) reckoning under these rules will not lead to results which are mutually inconsistent.

▌ **EXAMPLE :** *Show that if $x = 1 + 2\iota$, then $x^2 - 2x + 5 = 0$.*

We have $x^2 = 1 + 4\iota + 4\iota^2 = 1 + 4\iota - 4 = -3 + 4\iota$;

∴ $\qquad\qquad x^2 - 2x + 5 = -3 + 4\iota - 2(1 + 2\iota) + 5 = 0$.

Immediate consequences of the supposition are

   (i) $(x + \iota y) + (x' + \iota y') = (x + x') = \iota(y + y')$.

   (ii) $(x + \iota y) - (x' + \iota y') = (x - x') + \iota(y - y')$.

   (iii) $(x + \iota y)(x' + \iota y') = xx' - yy' + \iota(xy' + x'y)$.

   (iv) $\dfrac{x + \iota y}{x' + \iota y'} = \dfrac{(x + \iota y)(x' - \iota y')}{(x' + \iota y')(x' - \iota y')} = \dfrac{xx' + yy'}{x'^2 + y'^2} + \iota\dfrac{yx' - xy'}{x'^2 + y'^2}$.

This reckoning is at present meaningless, and can only be justified by an extension of the number system.

■ **2. A Complex Number** is represented by an expression of the form $x + \iota y$, where x, y are real numbers. *The sign $+$ does not indicate addition as hitherto understood, nor does the symbol ι (at present) denote a number. These things are parts of the scheme used to express numbers of a new class, and they signify that the pair of real numbers (x, y) are united to form a single complex number.*

The complex number $x + \iota 0$ is to be regarded as identical with the real number x, and we write

$$x + \iota 0 = x, \qquad \qquad \text{...(A)}$$

and in particular $\qquad 0 + \iota 0 = 0.$ $\qquad \qquad$ ...(B)

Thus the system of complex numbers includes all the real numbers, together with new numbers which are said to be *imaginary.*

For shortness we write $\qquad 0 + \iota y = \iota y,$ $\qquad \qquad$ ...(C)

and in particular $\qquad 0 + \iota 1 = \iota,$ $\qquad \qquad$ ...(D)

so that *as an abbreviation for $0 + \iota 1$, the symbol ι denotes a complex number.*

Thus *two* real numbers are required to express a single complex number. This may be compared with the fact that *two* whole numbers are required to express a fraction.

In order that expressions of the form $x + \iota y$ may be regarded as denoting *numbers,* we have to say what is meant by 'equality' and to define the fundamental operations.

▪ **3. Definition of Equality:** We say that

$$x + \iota y = x' + \iota y' \text{ if and only if } x = x' \text{ and } y = y'.$$

In applying this definition, we are said to *equate real and imaginary parts.* It should be noticed that the terms *'greater than'* and *'less than'* have no significance in connection with imaginary numbers.

▪ **4. The Fundamental Operations:** In defining these we are guided by the formal reckoning of Art. 1, and the equations (i)–(iv) are taken as *defining* addition, subtraction, multiplication and division.

▪ **5. Addition:** The sum of the complex numbers $x + \iota y$ and $x' + \iota y'$ is defined as

$$(x + x') + \iota (y + y'),$$

and we write

$$(x + \iota y) + (x' + \iota y') = (x + x') + \iota (y + y'). \qquad \text{...(E)}$$

It follows that

$$x + \iota y = (x + \iota 0) + (0 + \iota y), \qquad \qquad \text{...(F)}$$

so that $x + \iota y$ is the *sum* of the numbers denoted by x and ιy.

▪ **6. Subtraction** is defined by the equation

$$(x + \iota y) - (x' + \iota y') = (x - x') + \iota(y - y') \qquad \qquad \text{...(G)}$$

Hence *subtraction is the inverse of addition,* for $x - x' + \iota(y - y')$ is the number to which, if $x' + \iota y'$ is added, the result is $x + \iota y$.

The meaning of $- (x + \iota y)$ is defined by the equation

$$- (x + \iota y) = (0 + \iota 0) - (x + \iota y).$$

Hence $\qquad - (x + \iota y) = (0 - x) + \iota(0 - y) = (- x) + \iota(-y)$ $\qquad \qquad$ ...(H)

In particular, putting $x = 0$, we have

$$-\iota y = \iota (-y). \qquad \qquad \text{...(I)}$$

▪ **7. Multiplication** is defined by the equation

$$(x + \iota y) (x' + \iota y') = xx' - yy' + \iota (xy' + x'y) \qquad \qquad \text{...(J)}$$

It follows that multiplication is commutative.

Putting $x' = k$, $y' = 0$ in the last equation, we see that if k is a real number,

$$(x + \iota y)k = xk + \iota yk = k (x + \iota y) \qquad \qquad \text{...(K)}$$

Again, if we put $x = x' = 0$ and $y = y' = 1$, we have
$$(0 + \iota 1)^2 = -1 \quad \text{or} \quad \iota^2 = -1 \qquad \text{...(L)}$$

■ **8. Division** is the inverse of multiplication, that is to say, the equations
$$\frac{x+\iota y}{x'+\iota y'} = X + \iota Y \text{ and } (X + \iota Y)(x' + \iota y') = x + \iota y$$

mean the same thing. From the last equation we have
$$Xx' - Yy' + \iota(Xy' + Yx') = x + \iota y.$$

Equating real and imaginary parts, we have
$$Xx' - Yy' = x \quad \text{and} \quad Xy' + Yx' = y.$$

Therefore, $\quad X = \dfrac{xx' + yy'}{x'^2 + y'^2}, \quad Y = \dfrac{yx' - xy'}{x'^2 + y'^2},$

unless x' and y' are both zero. Thus *the equation defining division is*
$$\frac{x+\iota y}{x'+\iota y'} = \frac{xx'+yy'}{x'^2+y'^2} + \iota\frac{yx'-xy'}{x'^2+y'^2}, \qquad \text{...(M)}$$

excepting the case in which $x' + \iota y' = 0$.

Putting $x' = k$, $y' = 0$, we see that if k is a real number different from zero,
$$\frac{1}{k}(x + \iota y) = \frac{x}{k} + \iota\frac{y}{k}. \qquad \text{...(N)}$$

It is easy to see that addition, subtraction, multiplication and division (as defined above) are subject to the same laws as the corresponding operations for real numbers.

We have therefore justified, and attached a definite meaning to, the process of reckoning described in Art. 1.

■ **9. Zero Products:** For complex, as for real number, *if a product is zero, then one of the factors of the product is zero.* For let
$$(x + \iota y)(x' + \iota y') = 0,$$

where x, y, x, y' are real numbers. Then
$$xx' - yy' + \iota(xy' + x'y) = 0.$$

Hence, by the rule of equality,
$$xx' - yy' = 0 \qquad \text{and} \qquad xy' + x'y = 0;$$
$$\therefore \qquad x(x'^2 + y'^2) = 0 \qquad \text{and} \qquad y(x'^2 + y'^2) = 0.$$

If $x'^2 + y'^2$ is not zero, it follows that $x = 0$, $y = 0$, and therefore $x + \iota y = 0$.

If $x'^2 + y'^2 = 0$, then, since x', y' are real they must both vanish, and therefore $x' + \iota y' = 0$.

■ **10. Examples:**

▮ **EXAMPLE 1.** *Express* $(1 + 2\iota)^2/(2 + \iota)^2$ *in the form* $X + \iota Y$.

$$\frac{(1+2\iota)^2}{(2+\iota)^2} = \frac{1+4\iota-4}{4+4\iota-1} = \frac{-3+4\iota}{3+4\iota} = \frac{(-3+4\iota)(3-4\iota)}{(3+4\iota)(3-4\iota)};$$

$$\therefore \text{ The expression } = \frac{-9+16+24\iota}{9+16} = \frac{7}{25} + \iota\frac{24}{25}.$$

EXAMPLE 2. *Solve the equation* $x^3 = 1$.

The equation can be written $(x - 1)(x^2 + x + 1) = 0$, giving $x = 1$ or $x^2 + x + 1 = 0$. If $x^2 + x + 1 = 0$, we have in the usual way

$$x = \frac{1}{2}\left(-1 \pm \sqrt{-3}\right) = \frac{1}{2}\left(-1 \pm \iota\sqrt{3}\right).$$

Hence the numbers $1, \frac{1}{2}\left(-1 \pm \iota\sqrt{3}\right)$ *are called the three cube roots of unity.*

EXAMPLE 3. *Verify that* $\frac{1}{2}\left(-1 + \iota\sqrt{3}\right)$ *is a root of* $x^3 = 1$.

Denoting the given number by ω, we have

$$\omega^3 = \frac{1}{8}\left\{-1 + 3\iota\sqrt{3} - 3\left(\iota\sqrt{3}\right)^2 + \left(\iota\sqrt{3}\right)^3\right\} = \frac{1}{8}\left(-1 + 3\iota\sqrt{3} + 9 - 3\iota\sqrt{3}\right) = 1.$$

11. Geometrical Representation of Complex Numbers:

(1) *In naming line-segments and angles* the order of the letters will denote the direction of measurement. Thus AB denotes the distance travelled by a point in moving *from A to B,* and we write

$$AB + BA = 0, \text{ and } BA = -AB.$$

Again, we use '$\angle BAC$' to denote the *least* angle through which AB must turn in order that it may lie along AC.

Thus $\angle CAB = -\angle BAC$, and we shall suppose that

$$-\pi < \angle BAC \leq \pi.$$

(2) If $z = x + \iota y$, we say that the number z is *represented* by the point whose coordinates referred to rectangular axes are x, y.

When there is no risk of confusion, this will be called *the point z.*

A diagram showing points which represent complex numbers is called an *Argand diagram.*

Let (r, θ) be the polar coordinates of the point z, then

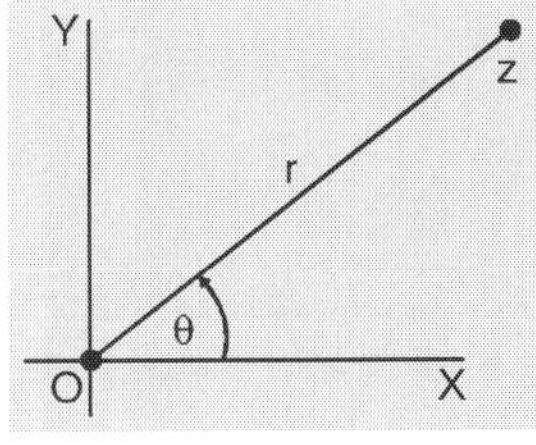

$$z = x + \iota y = r(\cos\theta + \iota\sin\theta),$$

where $x = r\cos\theta$, $y = r\sin\theta$,

Fig. 2

and hence $r = \sqrt{(x^2 + y^2)}$, $\tan\theta = y/x$.

Here r, which is essentially positive, is called the *modulus* of z and is denoted by $|z|$ or by mod z. Thus

$$\text{mod } z = |z| = r = \sqrt{(x^2 + y^2)}.$$

It follows that *if* $|z| = 0$ *then* $x = 0, y = 0,$ *and consequently* $z = 0.$

The angle θ is called the *amplitude* of z, and is denoted by am z. This angle has infinitely many values differing by multiples of 2π. The value of θ such that

$$-\pi < \theta \leq \pi$$

is called the *principal value of the amplitude.*

With the convention of § (1), it will be seen that *the principal value of* am z *is* $\angle XOz$.

Unless otherwise stated, am z will mean the principal value of the amplitude of z.

▣ 12. Addition and Subtraction:

(1) Let $z = x + \iota y$, $z' = x' + \iota y'$. Draw the parallelogram $Ozsz'$, then s *represents* $z + z'$.

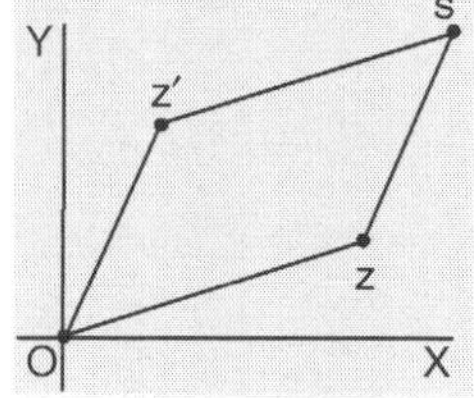

Fig. 3

For if (X, Y) are the coordinates of s,

$$\text{then} \quad X = \text{Sum of projections of } Oz, zs \text{ on } OX$$
$$= \text{Sum of projections of } Oz, Oz' \text{ on } OX$$
$$= x + x'.$$

Similarly, $Y = y + y'$, therefore s represents $z + z'$. It is to be observed that

$$|z + z'| = \text{The } length \ Os, \text{ and am } (z + z') = \angle XOs.$$

Now $Os \leq Oz + Oz'$, the sign of equality occurring if $\angle XOz = \angle XOz'$ i.e. if am $z = $ am z'. Therefore

$$|z + z'| \leq |z| + |z'|.$$

To prove this algebraically, we have to show that

$$\sqrt{\left\{(x + x')^2 + (y + y')^2\right\}} \leq \sqrt{(x^2 + y^2)} + \sqrt{(x'^2 + y'^2)},$$

where each root is to have its positive value. This will be the case if

$$(x + x')^2 + (y + y')^2 \leq x^2 + y^2 + x'^2 + y'^2 + 2\sqrt{(x^2 + y^2)(x'^2 + y'^2)},$$

i.e. if $\quad\quad xx' + yy' \leq \sqrt{(x^2 + y^2)(x'^2 + y'^2)},$

or if $\quad\quad (xx' + yy')^2 \leq (x^2 + y^2)(x'^2 + y'^2),$

or if $\quad\quad (xy' - x'y)^2 \geq 0,$ which is the case.

(2) Draw the parallelogram $Oz'zd$ (Fig. 4),

then $\quad\quad d$ *represents* $z - z'$.

For, by the last construction, the number represented by

$$d + z' = z.$$

Observe that

$$|z - z'| = \text{The } length \ z'z,$$

and $\quad\quad$ am $(z - z') = \angle XOd.$

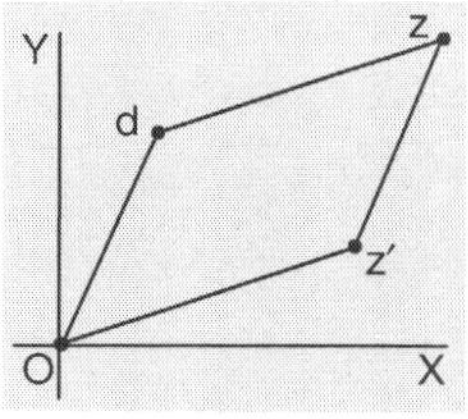

Fig. 4

(3) *If* $s_n = z_1 + z_2 + \dots + z_n$, *where* $z_1, z_2, \dots z_n$ *are given complex numbers, it is required to find the point* s_n.

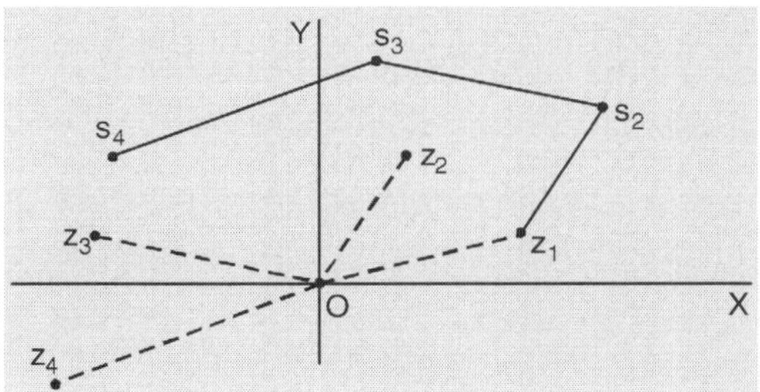

Fig. 5

Draw in succession $z_1 s_2$, $s_2 s_3$, $s_3 s_4$, ... equal to, parallel to, and in the same sense as Oz_2, Oz_3, Oz_4, ..., then s_2, s_3, ... are the points representing $z_1 + z_2$, $z_1 + z_2 + z_3$, etc.

This is a continued application of the first construction of this article.

(4) *The modulus of the sum of any number of complex numbers is less than or equal to the sum of the moduli of the numbers.*

For in the last figure, the moduli of z_1, z_2, z_3, ... are the lengths Oz_1, $z_1 s_2$, $s_2 s_3$, ..., and that of s_n is Os_n. Also

$$Os_n \le Oz_1 + z_1 s_2 + s_2 s_3 + ... + s_{n-1} s_n.$$

Therefore
$$|s_n| \le |z_1| + |z_2| + ... + |z_n|.$$

Note: The sign of equality is to be taken if, and only if, the points O, z_1, s_2 ... s_n, are in the same straight line *and occur in this order,* that is, *if* z_1, z_2, ... z_n *have the same amplitude.*

To prove this algebraically, we have by (1) of this article,
$$|s_n| \le |s_{n-1}| + |z_n| \le |s_{n-2}| + |z_{n-1}| + |z_n|,$$
and so on.

■ **13. Products and Quotients:**

(1) Let
$$z = x + \iota y = r (\cos \theta + \iota \sin \theta),$$
$$z' = x' + \iota y' = r' (\cos \theta' + \iota \sin \theta').$$

By the definition of multiplication,
$$\begin{aligned} zz' &= xx' - yy' + \iota (xy' + yx') \\ &= rr' \{\cos \theta \cos \theta' - \sin \theta \sin \theta' \\ &\quad + \iota (\cos \theta \sin \theta' + \cos \theta' \sin \theta)\}; \end{aligned}$$

$\therefore zz' + rr' \{\cos (\theta + \theta') + \iota \sin (\theta + \theta')\}$. 　　...(A)

Since division is the inverse of multiplication,

$$\frac{z}{z'} = \frac{r}{r'} \{\cos (\theta - \theta') + \iota \sin (\theta - \theta')\}. \qquad ...(B)$$

For this last is the number which when multiplied by z' produces z. Hence also

$$\frac{1}{z} = \frac{1}{r} \{\cos (-\theta) + \iota \sin (-\theta)\} = \frac{1}{r} (\cos \theta - \iota \sin \theta). \qquad ...(C)$$

If
$$z_1 = x_1 + \iota y_1 = r_1 (\cos \theta_1 + \iota \sin \theta_1),$$

with similar notation for z_2, z_3, ... z_n, by continued application of equation (A) we have

$$z_1 z_2 \ldots z_n = r_1 r_2 \ldots r_n \{\cos (\theta_1 + \theta_2 + \ldots + \theta_n) + \iota \sin (\theta_1 + \theta_2 + \ldots + \theta_n)\}. \qquad \ldots\text{(D)}$$

In (D) put $z_1 = z_2 = \ldots = z_n = z = r (\cos \theta + \iota \sin \theta)$, then

$$z^n = r^n (\cos n\theta + \iota \sin n\theta), \qquad \ldots\text{(E)}$$

where n is any positive integer.

Similarly, from (C) it follows that

$$z^{-n} = r^{-n} (\cos n\theta - \iota \sin n\theta). \qquad \ldots\text{(F)}$$

■ **EXAMPLE :** *Show that* $\tan^{-1} \dfrac{1}{5}$ *is nearly equal to* $\dfrac{\pi}{16}$.

We have $(5 + \iota) = \sqrt{26} \ (\cos \theta + \iota \sin \theta)$, where $\tan \theta = 1/5$,

and therefore $\qquad (5 + \iota)^4 = 676 (\cos 4\theta + \iota \sin 4\theta)$, by (E).

But $\qquad\qquad (5 + \iota)^4 = (24 + 10 \iota)^2 = 476 + 480 \iota$; hence we have

$\qquad\qquad \cos 4\theta = 476/676$, $\sin 4\theta = 480/676$, and $\tan 4\theta = 1$, nearly.

∴ $\qquad\qquad 4\theta = \pi/4$ approximately.

(2) Modulus and Amplitude: From (D) we see that if Z is the product of the complex numbers z_1, z_2, ... , z_n,

$$|Z| = r_1 r_2 \ldots r_n = |z_1| .| z_2| \ldots |z_n|, \qquad \ldots\text{(G)}$$

and $\qquad \text{am } Z = \theta_1 + \theta_2 + \ldots + \theta_n = \text{am } z_1 + \text{am } z_2 + \ldots + \text{am } z_n; \qquad \ldots\text{(H)}$

but note that the last equation does not give the *principal* value of am Z unless

$$- \pi < \theta_1 + \theta_2 + \ldots + \theta_n \leq \pi.$$

Again, from (B) we have

$$\left|\frac{z}{z'}\right| = \frac{r}{r'} = \frac{|z|}{|z'|}, \qquad \ldots\text{(I)}$$

and $\qquad\qquad \text{am } \dfrac{z}{z'} = \theta - \theta' = \text{am } z - \text{am } z', \qquad \ldots\text{(J)}$

although the last equation does not give the *principal* value of am z/z' unless

$$- \pi < \theta - \theta' \leq \pi.$$

It is important to notice that the amplitude of z/z' is the angle through which Oz' must be turned in order that it may lie along Oz; and that, with the convention of Art.

11, (1), the principal value of am $\dfrac{z}{z'} = \angle z'Oz$.

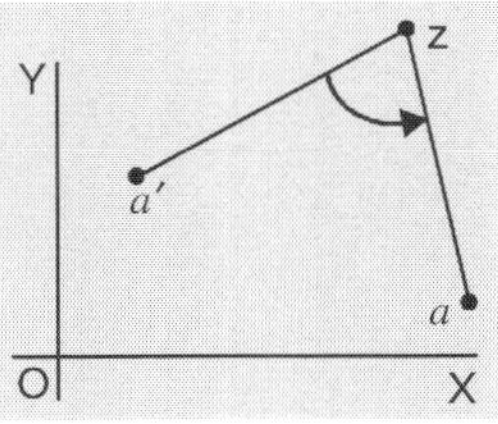

Fig. 6

Again, if z, a, a' are any numbers, $\left|\dfrac{z-a}{z-a'}\right| = \dfrac{\text{the length } az}{\text{the length } a'z}$,

and $\qquad\qquad \operatorname{am} \dfrac{z-a}{z-a'} = \angle\, a'za.$

(3) De Moivre's Theorem: From (E) and (F) it follows that *if n is a positive or negative integer*

$$(\cos\theta + \iota\sin\theta)^n = \cos n\theta + \iota\sin n\theta.$$

This, with an extension to be given later (Art. 16), is known as De Moivre's theorem.

■ **14. Conjugate Numbers:**

(1) *If $z = x + \iota y$ and $z' = x - \iota y$, then z, z' are called conjugate numbers.*

Their product is $x^2 + y^2$, and if r, θ are the polar coordinates of the point z, those of the point z' are r, $-\theta$. Hence

$$|x + \iota y| = r = \sqrt{(x^2 + y^2)} = |x - \iota y|, \quad \operatorname{am}(x + \iota y) = \theta = -\operatorname{am}(x - \iota y).$$

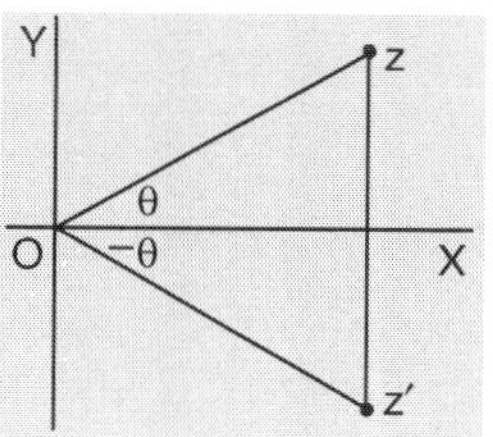

Fig. 7

(2) *If $f(z)$ is a polynomial in z with real coefficients, and $f(z) = f(x + \iota y) = X + \iota Y$, where X, Y are real, then $f(x - \iota y) = X - \iota Y$.*

We have $f(x + \iota y) = a_0 + a_1(x + \iota y) + a_2(x + \iota y)^2 + \dots$ where a_0, a_1, ... are real.

We can therefore obtain expansions of the form

$$f(x + \iota y) = \Sigma a x^m (\iota y)^n, \quad f(x - \iota y) = \Sigma a x^m (-\iota y)^n,$$

where every a is real and m, n are positive integers or zero.

Every term of $f(x + \iota y)$ in which n is even or zero is real, and occurs with the same sign in $f(x - \iota y)$.

Every term of $f(x + \iota y)$ in which n is odd is imaginary, and occurs with the opposite sign in $f(x - \iota y)$.

Hence the result follows.

(3) *If an equation with real coefficients has complex roots, then these occur in conjugate pairs.*

Let $\alpha + \iota\beta$ be a root of $f(x) = 0$, where α, β, as well as the coefficients of the polynomial $f(x)$, are real.

Let $f(\alpha + \iota\beta) = A + \iota B$, where A, B are real. Since $f(\alpha + \iota\beta) = 0$, we have $A = 0$, $B = 0$. Hence

$$f(\alpha - \iota\beta) = A - \iota B = 0.$$

Therefore, $\alpha - \iota\beta$ is also a root of $f(x) = 0$.

EXERCISE VII

COMPLEX NUMBERS

1. Find the modulus and amplitude of $\sqrt{3} + \iota, \sqrt{3} - \iota, -\sqrt{3} + \iota, -\sqrt{3} - \iota$.

2. The condition that $(a + \iota b)/(a' + \iota b')$ may be real is $ab' - a'b = 0$.

3. Find z so that $z\,(3 + \iota 4) = 2 + 3\iota$.

4. Express in the form $X + \iota Y$,

   $(i)\ \dfrac{1+\iota}{2+3\iota}$; $\qquad (ii)\ \dfrac{1}{(1-\iota)^3}$; $\qquad (iii)\ \dfrac{1}{1 - (\cos\theta + \iota\sin\theta)}$.

5. If $z = x + \iota y$, express in the form $X + \iota Y$

   $(i)\ z^n$; $\qquad\qquad (ii)\ \dfrac{1}{z}$; $\qquad\qquad (iii)\ \dfrac{z+1}{z-1}$.

6. Find the moduli of (i) $(2 + 3\iota)^2 (3 - 4\iota)^3$; (ii) $(2 + 3\iota)^2/(3 + 4\iota)^3$.

7. What is the principal value of the amplitude of $(\cos 50° + \iota \sin 50°)^6$?

8. (i) If $|z| = |z'| = 1$ and am $z = -$ am z', them $z' = \dfrac{1}{z}$.

   (ii) If $|z| = |z'|$ and am z, am z', differ by π, then $z' = -z$.

9. Let $z = x + \iota y$ where x, y are variables. Prove that
   (i) if $2|z - 1| = |z - 2|$, then $3(x^2 + y^2) = 4x$;
   (ii) if $|2z - 1| = |z - 2|$, then $x^2 + y^2 = 1$.

   Thus, in either case, the point z describes a circle.

10. Prove that $|z + z'|^2 + |z - z'|^2 = 2\,|z|^2 + 2\,|z'|^2$.

11. Prove that $|2z \cos\alpha + z^2| < 1$ $\quad$ if $\quad |z| < 0.41$.

    [Let $|z| = r$, then $|2z \cos\alpha + z^2| < 2\,|z \cos\alpha| + |z|^2 < 2r + r^2 < 1$ if $r^2 + 2r - 1 < 0$.]

12. If $z = x + \iota y$ and $Z = X + \iota Y$, show that

   (i) if $z = \dfrac{z-1}{z+1}$, then $x^2 + y^2 = \dfrac{(X-1)^2 + Y^2}{(X+1)^2 + Y^2}$;

   (ii) if $z = \dfrac{\iota(Z+1)}{Z-1}$, then $x^2 + y^2 - x = \dfrac{X^2 + Y^2 + 2X - 2Y + 1}{(X-1)^2 + Y^2}$.

13. Given that $1 + 2\iota$ is one root of the equation, $x^4 - 3x^3 + 8x^2 - 7x + 5 = 0$, find the other three roots.

14. From the formula $\cos n\theta + \iota \sin n\theta = (\cos\theta + \iota\sin\theta)^n$, where n is a positive integer, deduce that

   $$\cos n\theta = \cos^n\theta - C_2^n \cos^{n-2}\theta \sin^2\theta + C_4^n \cos^{n-4}\theta \sin^4\theta - ...,$$

   $$\sin n\theta = C_1^n \cos^{n-1}\theta \sin\theta - C_3^n \cos^{n-3}\theta \sin^3\theta + ...,$$

   $$\tan n\theta = \frac{C_1^n \tan\theta - C_3^n \tan^3\theta + ...}{1 - C_2^n \tan^2\theta + C_4^n \tan^4\theta - ...}.$$

15. By putting $z = r (\cos \theta + \iota \sin \theta)$ in the identity.

$$1 + z + z^2 + \ldots + z^{n-1} = (1 - z^n)/(1 - z),$$

prove that

$$1 + r \cos \theta + r^2 \cos 2\theta + \ldots + r^{n-1} \cos (n - 1)\theta$$

$$= \frac{1 - r \cos\theta - r^n \cos n\theta + r^{n+1} \cos (n - 1)\theta}{1 - 2r \cos \theta + r^2},$$

$$r \sin \theta + r^2 \sin 2\theta + \ldots + r^{n-1} \sin (n - 1)\theta$$

$$= \frac{r \sin \theta - r^n \sin n\theta + r^{n+1} \sin (n - 1)\theta}{1 - 2r \cos \theta + r^2}.$$

16. If $z = \cos \theta + \iota \sin \theta$ and n is a positive integer, prove that

$$z^n + z^{-n} = 2 \cos n\theta, \quad z^n - z^{-n} = 2 \iota \sin n\theta.$$

Hence show that

$$2^{n-1} \cos^n \theta = \cos n\theta + C_1^n \cos (n - 2) \theta + C_2^n \cos (n - 4) \theta + \ldots ,$$

and if n is even

$$(-1)^{\frac{n}{2}} \, 2^{n-1} \sin^n \theta = \cos n\theta - C_1^n \cos (n - 2) \theta + C_2^n \cos (n - 4) \theta - \ldots,$$

and if n is odd

$$(-1)^{\frac{n-1}{2}} \, 2^{n-1} \sin^n \theta = \sin n\theta - C_1^n \sin (n - 2) \theta + C_2^n \sin (n - 4)\theta - \ldots .$$

$[(z + z^{-1})^n = z^n + z^{-n} + C_1^n (z^{n-2} + z^{-(n-2)}) + \ldots;$ similarly for $(z - z^{-1})^n.]$

17. By the method of Ex. 16, prove that

$$16 \cos^3 \theta \sin^2 \theta = - \cos 5\theta - \cos 3\theta + 2 \cos \theta.$$

18. If $\cos \alpha + \cos \beta + \cos \gamma = 0$ and $\sin \alpha + \sin \beta + \sin \gamma = 0$, prove that

$$\cos 3\alpha + \cos 3\beta + \cos 3\gamma = 3 \cos (\alpha + \beta + \gamma)$$

and $\qquad \sin 3\alpha + \sin 3\beta + \sin 3\gamma = 3 \sin (\alpha + \beta + \gamma).$

[Put $a = \cos \alpha + \iota \sin \alpha$, $b = \cos \beta + \iota \sin \beta$, $c = \cos \gamma + \iota \sin \gamma$, then $a + b + c = 0$ and $a^3 + b^3 + c^3 = 3abc$, etc.]

19. From the identity

$$a^2 \frac{(x - b)(x - c)}{(a - b)(a - c)} + b^2 \frac{(x - c)(x - a)}{(b - c)(b - a)} + c^2 \frac{(x - a)(x - b)}{(c - a)(c - b)} = x^2$$

deduce the identities

$$\cos 2 (\theta + \alpha) \frac{\sin (\theta - \beta) \sin (\theta - \gamma)}{\sin (\alpha - \beta) \sin (\alpha - \gamma)} + \ldots + \ldots = \cos 4\theta,$$

$$\sin 2 (\theta + \alpha) \frac{\sin (\theta - \beta) \sin (\theta - \gamma)}{\sin (\alpha - \beta) \sin (\alpha - \gamma)} + \ldots + \ldots = \sin 4\theta.$$

[Put $x = \cos 2\theta + \iota \sin 2\theta$, $a = \cos 2\alpha + \iota \sin 2\alpha$, etc.]

■ 15. Roots of Complex Numbers:

(1) If n is a positive integer, any number whose nth power is equal to z is called an *nth root of z.*

Let
$$z = r(\cos \theta + \iota \sin \theta),$$

where $r > 0$ and $-\pi < \theta \le \pi$. Let $\sqrt[n]{r}$ denote the positive arithmetical nth root of r. Consider the values of

$$u = \sqrt[n]{r}\left\{\cos \frac{\theta + 2k\pi}{n} + \iota \sin \frac{q + 2k\pi}{n}\right\},$$

where k is any integer or zero. By Art. 13 we have

$$u^n = r\{\cos(\theta + 2k\pi) + \iota \sin(\theta + 2k\pi)\} = z;$$

from which it follows that u is an nth root of z.

Again, two values of u will be equal if, and only if, the corresponding angles have the same sine and cosine. That this may be the case, the angles must differ by a multiple of 2π, and the corresponding values of k must differ by a multiple of n. Therefore u has n distinct values, namely those given by

$$k = 0, 1, 2, \dots, n - 1.$$

Thus *the nth roots of* $z = r(\cos \theta + \iota \sin \theta)$ *are the values of*

$$\sqrt[n]{r}\left(\cos \frac{\theta + 2k\pi}{n} + \iota \sin \frac{\theta + 2k\pi}{n}\right) \text{ where } k = 0, 1, 2, \dots, n - 1.$$

If $-\pi < \theta \le \pi$, then $\sqrt[n]{r}\cdot\left(\cos\dfrac{\theta}{n} + \iota\sin\dfrac{\theta}{n}\right)$ is called the *principal value* of $\sqrt[n]{z}$.

(2) *To find the points representing the n values of* $\sqrt[n]{z}$ *,* we must be able to find the length represented by $\sqrt[n]{r}$ and to divide the angle θ into n equal parts.

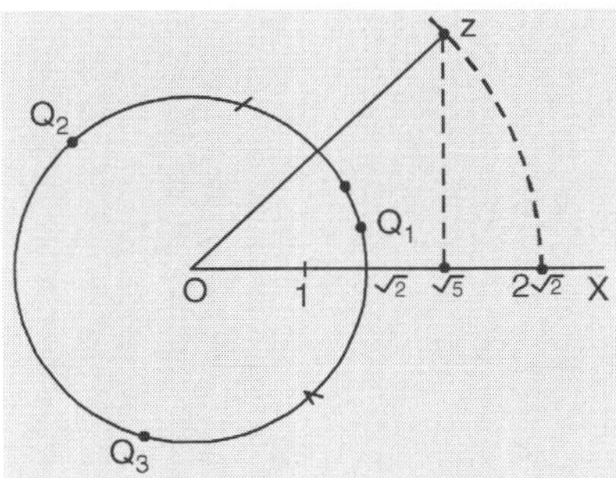

Fig. 8

■ **EXAMPLE:** *Find the points* Q_1, Q_2, Q_3 *representing the values of* $\sqrt[3]{z}$ *, where* $z = \sqrt{5} + \iota\sqrt{3}$.

Here $Oz = r = \sqrt{5 + 3} = \sqrt{8}$

and $\tan XOz = \tan \theta = \sqrt{3}/\sqrt{5}$.

Also $\sqrt[3]{r} = \sqrt{2}$, and so Q_1, Q_2, Q_3 are points on the circle with centre O and radius $\sqrt{2}$, such that

$$\angle XOQ_1 = \theta/3, \ \angle Q_1OQ_2 = \angle Q_2OQ_3 = 2\pi/3.$$

■ **16. General Form of De Moivre's Theorem:** *If x is rational, then $\cos x\theta + \iota \sin x\theta$ is one of the values of $(\cos\theta + \iota\sin\theta)^x$.*

The case in which x is a positive or negative integer has been considered in Art. 13.

Let $x = p/q$ where p, q are integers and q is positive, then since

$$(\cos\theta/q + \iota\sin\theta/q)^q = \cos\theta + \iota\sin\theta,$$

therefore

$$\cos\theta/q + \iota\sin\theta/q \text{ is a value of } (\cos\theta + \iota\sin\theta)^{\frac{1}{q}};$$

also, since p is an integer,

$$(\cos\theta/q + \iota\sin\theta/q)^p = \cos p\theta/q + \iota\sin p\theta/q;$$

therefore

$$\cos p\theta/q + \iota\sin p\theta/q \text{ is a value of } (\cos\theta + \iota\sin\theta)^{\frac{p}{q}}.$$

■ **17. The nth Roots of Unity:** In Art. 15, let $r = 1$ and $\theta = 0$, then since $\cos 0 + \iota\sin 0 = 1$, *the nth roots of unity are the values of*

$$\cos\frac{2k\pi}{n} + \iota\sin\frac{2k\pi}{n} \text{ where } k = 0, 1, 2, \dots n - 1.$$

The *principal nth root,* given by $k = 0$, is 1.

If n is even, the root -1 is given by $k = \dfrac{1}{2}n$.

Since $x^n - 1 = (x - 1)(x^{n-1} + x^{n-2} + \dots + 1)$, it follows that if α is any root, other than 1, of $x^n - 1 = 0$, then

$$\alpha^{n-1} + \alpha^{n-2} + \dots + \alpha + 1 = 0.$$

■ **18. The nth Roots of (-1):** These are the values of

$$\cos\frac{(2k+1)\pi}{n} + \iota\sin\frac{(2k+1)\pi}{n} \text{ where } k = 0, 1, 2, \dots n - 1.$$

This follows from Art. 15 by putting $r = 1$, $\theta = \pi$.

The *principal nth root of* (-1) is $\cos\dfrac{\pi}{n} + \iota\sin\dfrac{\pi}{n}$, corresponding to $k = 0$.

If n is odd, the root (-1) is given by $k = \dfrac{1}{2}(n - 1)$.

■ **19. Factors of $x^n - 1$ and $x^n + 1$:**

(1) *The n factors of $x^n - 1$ are*

$$x - \{\cos 2r\pi/n + \iota\sin 2r\pi/n\} \text{ where } r = 0, 1, 2, \dots n - 1.$$

This follows from Art. 15.

(*i*) *If n is even,* the factors $x - 1$ and $x + 1$ are given by $r = 0$, $r = n/2$. The remaining $(n - 2)$ factors can be grouped in pairs as follows:

Since $2r\,\pi/n + 2(n - r)\,\pi/n = 2\pi$, the factors corresponding to r and $n - r$ are

$$x - \left\{\cos\frac{2r\pi}{n} + \iota\sin\frac{2r\pi}{n}\right\} \text{ and } x - \left\{\cos\frac{2r\pi}{n} - \iota\sin\frac{2r\pi}{n}\right\}.$$

And the product of these is

$$\left(x - \cos\frac{2r\pi}{n}\right)^2 + \sin^2\frac{2r\pi}{n} = x^2 - 2x\cos\frac{2r\pi}{n} + 1;$$

therefore

$$x^n - 1 = (x - 1)(x + 1)\ \Pi_{r=1}^{r=\frac{1}{2}(n-2)}\left(x^2 - 2x\cos\frac{2r\pi}{n} + 1\right), \qquad \ldots(A)$$

where Π denotes the product of the factors as indicated.

(*ii*) *If n is odd,* the factor $x - 1$ is given by $r = 0$. The remaining $n - 1$ factors can be grouped in pairs, and

$$x^n - 1 = (x - 1)\ \Pi_{r=1}^{r=\frac{1}{2}(n-1)}\left(x^2 - 2x\cos\frac{2r\pi}{n} + 1\right). \qquad \ldots(B)$$

(**2**) Similarly, the n factors of $x^n + 1$ are

$$x - \{\cos(2r + 1)\pi/n + \iota\sin(2r + 1)\pi/n\}, \text{ where } r = 0, 1, 2, \ldots (n - 1).$$

(*i*) *If n is even*

$$x^n + 1 = \Pi_{r=0}^{r=\frac{1}{2}(n-2)}\left(x^2 - 2x\cos\frac{(2r+1)\pi}{n} + 1\right). \qquad \ldots(C)$$

(*ii*) *If n is odd,*

$$x^n + 1 = (x + 1)\ \Pi_{r=0}^{r=\frac{1}{2}(n-3)}\left(x^2 - 2x\cos\frac{(2r+1)\pi}{n} + 1\right). \qquad \ldots(D)$$

■ **20. The Imaginary Cube Roots of Unity:** These are the roots of $x^2 + x + 1 = 0$. Their actual values are

$$\cos\frac{2\pi}{3} \pm \iota\sin\frac{2\pi}{3} \text{ or } \tfrac{1}{2}\left(-1 \pm \iota\sqrt{3}\right).$$

Denoting either of these by ω, the other is ω^2, for $(\omega^2)^3 = (\omega^3)^2 = 1$. Moreover,

$$1 + \omega + \omega^2 = 0. \qquad \ldots(A)$$

Again, except when r is a multiple of 3, ω^r and ω^{2r} are the imaginary cube roots of 1, and so

$$1 + \omega^r + \omega^{2r} = 0. \qquad \ldots(B)$$

The following identities are important.

$$x^3 - y^3 = (x - y)(x - \omega y)(x - \omega^2 y). \qquad \ldots(C)$$
$$x^3 + y^3 = (x + y)(x + \omega y)(x + \omega^2 y). \qquad \ldots(D)$$
$$x^2 + y^2 + z^2 - yz - zx - xy = (x + \omega y + \omega^2 z)(x + \omega^2 y + \omega z). \qquad \ldots(E)$$
$$x^3 + y^3 + z^3 - 3xyz = (x + y + z)(x + \omega y + \omega^2 z)(x + \omega^2 y + \omega z). \qquad \ldots(F)$$

■ **EXAMPLE 1.** *Prove that* $(x^3 + y^3 + z^3 - 3xyz)^2 = X^3 + Y^3 + Z^3 - 3XYZ$, *where*
$$X = x^2 + 2yz,\ Y = y^2 + 2zx,\ Z = z^2 + 2xy.$$

This follows from (F) on observing that

$$(x + \omega y + \omega^2 z)^2 = X + \omega^2 Y + \omega Z, \quad (x + \omega^2 y + \omega z)^2 = X + \omega Y + \omega^2 Z$$

and

$$(x + y + z)^2 = X + Y + Z.$$

◼ **EXAMPLE 2.** *If* $(1 + x)^n = c_0 + c_1 x + c_2 x^2 + \ldots + c_n x^n$ *and*

$$S_1 = c_0 + c_3 + c_6 + \ldots, \quad S_2 = c_1 + c_4 + c_7 + \ldots, \quad S_3 = c_2 + c_5 + c_8 + \ldots,$$

each of the series being continued as far as possible, show that the values of S_1, S_2,

S_3 *are* $\dfrac{1}{3}\left(2^n + 2\cos\dfrac{r\pi}{3}\right)$, *where* $r = n$, $n - 2$, $n + 2$, *respectively.*

Putting 1, ω, ω^2 for x, in succession,

$$(1 + 1)^n = c_0 + c_1 + \ldots + c_r + \ldots + c_n, \qquad \ldots\text{(A)}$$
$$(1 + \omega)^n = c_0 + c_1\omega + \ldots + c_r\omega^r + \ldots + c_n\omega^n, \qquad \ldots\text{(B)}$$
$$(1 + \omega^2)^n = c_0 + c_1\omega^2 + \ldots + c_r\omega^{2r} + \ldots + c_n\omega^{2n}. \qquad \ldots\text{(C)}$$

If r is not a multiple of 3, $1 + \omega^r + \omega^{2r} = 0$; hence, adding (A), (B), (C) together, we have

$$3(c_0 + c_3 + c_6 + \ldots) = 2^n + (1 + \omega)^n + (1 + \omega^2)^n.$$

Now, $1 + \omega = 1 + \cos\dfrac{2\pi}{3} + \iota\sin\dfrac{2\pi}{3} = 2\cos\dfrac{\pi}{3}\left(\cos\dfrac{\pi}{3} + \iota\sin\dfrac{\pi}{3}\right),$

and, since $\cos\dfrac{\pi}{3} = \dfrac{1}{2}$, we have $1 + \omega = \cos\dfrac{\pi}{3} + \iota\sin\dfrac{\pi}{3}$;

$$\therefore \qquad\qquad (1 + \omega)^n = \cos\dfrac{n\pi}{3} + \iota\sin\dfrac{n\pi}{3}.$$

Also, $1 + \omega^2 = 1 + \cos\dfrac{2\pi}{3} - \iota\sin\dfrac{2\pi}{3}$, and $(1 + \omega^2)^n = \cos\dfrac{n\pi}{3} - \iota\sin\dfrac{n\pi}{3}$.

$$\therefore \qquad 3(c_0 + c_3 + c_6 + \ldots) = 2^n + 2\cos\dfrac{n\pi}{3}.$$

Again, multiply (B) by ω^2, (C) by ω, and add to (A); then

$$3(c_1 + c_4 + c_7 + \ldots) = 2^n + \omega^2(1 + \omega)^n + \omega(1 + \omega^2)^n.$$

Also $\qquad \omega^2(1 + \omega)^n = \left(\cos\dfrac{2\pi}{3} - \iota\sin\dfrac{2\pi}{3}\right)\left(\cos\dfrac{n\pi}{3} + \iota\sin\dfrac{n\pi}{3}\right)$

$$= \cos\dfrac{(n-2)\pi}{3} + \iota\sin\dfrac{(n-2)\pi}{3},$$

and $\qquad \omega(1 + \omega^2)^n = \cos\dfrac{(n-2)\pi}{3} - \iota\sin\dfrac{(n-2)\pi}{3}$;

$$\therefore \qquad 3(c_1 + c_4 + c_7 + \ldots) = 2^n + 2\cos\dfrac{(n-2)\pi}{3}.$$

Finally, multiply (B) by ω, (C) by ω^2, and add to (A); it will be found that

$$3(c_2 + c_5 + c_8 + \ldots) = 2^n + 2\cos\dfrac{(n+2)\pi}{3}.$$

EXERCISE VIII

COMPLEX ROOTS

1. Prove that the values of $\sqrt[4]{-1}$ are $\pm \dfrac{1}{\sqrt{2}}\,(1 \pm \iota)$.

2. Find the values of $\sqrt{(1+\iota)}$ and $\sqrt{(1-\iota)}$.

3. Use Art. 19 to show that

$$x^4 + 1 = (x^2 - \sqrt{2}x + 1)\,(x^2 + \sqrt{2}x + 1),$$

$$x^5 - 1 = (x - 1)\left(x^2 + 2x \cos \frac{2\pi}{5} + 1\right)\left(x^2 - 2x \cos \frac{2\pi}{5} + 1\right).$$

4. Prove that

$$x^5 - 1 = (x - 1)\,\{x^2 + \tfrac{1}{2}\left(\sqrt{5} + 1\right)x + 1\}\,\{x^2 - \tfrac{1}{2}\left(\sqrt{5} - 1\right)x + 1\}.$$

$[x^4 + x^3 + x^2 + x + 1 = x^2\{(x + x^{-1})^2 + (x + x^{-1}) - 1\} = x^2\,(x + x^{-1} - \alpha)\,(x + x^{-1} - \beta),$
where $\alpha,\ \beta$ are the roots of $z^2 + z - 1 = 0.]$

5. Give a geometrical construction to find the points $z_1,\ z_2$ corresponding to the values of $\sqrt{z}$.

[If $O1$ is the unit of length, Oz_1 bisects the angle XOz, and is a mean proportional to $O1$ and Oz; and z_2 is on z_1O produced, so that $Oz_2 = z_1O.]$

6. If $a,\ b$ are complex numbers, show that

$$\left|a + \sqrt{a^2 - b^2}\right| + \left|a - \sqrt{a^2 - b^2}\right| = |a + b| + |a - b|.$$

[Let $z_1 = a + \sqrt{a^2 - b^2}$, $z_2 = a - \sqrt{a^2 - b^2}$, then by Exercise VII, Ex.10,

$$|z_1|^2 + |z^2|^2 = \tfrac{1}{2}\,|z_1 + z_2|^2 + \tfrac{1}{2}\,|z_1 - z_2|^2 = 2\,|a|^2 + 2\,|a^2 - b^2|;$$

$$\therefore \quad \{|z_1| + |z_2|\}^2 = 2\{|a|^2 + |a^2 - b^2| + |b|^2\}$$

$$= \{|a + b|^2 + 2\,|a + b|\,.\,|a - b| + |a - b|^2\}$$

$$= \{|a + b| + |a - b|\}^2.]$$

7. Solve the equation

$$z^2 + 2\,(1 + 2\iota)\,z - (11 + 2\iota) = 0.$$

Verify that the sum of the roots is $-2\,(1 + 2\iota)$ and the product $-(11 + 2\iota)$.

[Put $z = x + \iota\,y$, equate real and imaginary parts to zero and solve for $x,\ y.]$

8. Prove that, with regard to the quadratic

$$z^2 + (p + \iota\,p')\,z + q + \iota\,q' = 0,$$

(*i*) if the equation has one real root, then

$$q'^2 - pp'q' + qp'^2 = 0;$$

(*ii*) if the equation has two equal roots, then

$$p^2 - p'^2 = 4q \quad \text{and} \quad pp' = 2q'.$$

[(*i*) If α is a real root, by the rule of equality $\alpha^2 + p\alpha + q = 0,\ p'\alpha + q' = 0$. Eliminate α.
(*ii*) In this case $(p + \iota\,p')^2 = 4\,(q + \iota\,q')$, etc.]

9. If $z = x + \iota y = r\,(\cos\theta + \iota \sin\theta)$, prove that

$$\sqrt{z} = \pm \frac{1}{\sqrt{2}}\left\{\sqrt{r+x} + \iota\sqrt{r-x}\right\} \quad \text{or} \quad \pm \frac{1}{\sqrt{2}}\left\{\sqrt{r+x} - \iota\sqrt{r-x}\right\},$$

according as y is positive or negative.

$$\left[\sqrt{z} = \pm\sqrt{r}\left(\cos\frac{\theta}{2} + \iota\sin\frac{\theta}{2}\right) \quad \text{and} \quad \cos\theta = \frac{x}{r};\right.$$

$$\therefore \qquad \cos\frac{\theta}{2} = \pm\sqrt{\frac{r+x}{2r}}\,,\ \sin\frac{\theta}{2} = \pm\sqrt{\frac{r-x}{2r}}\,.$$

Also if $y > 0$ then $0 < \theta < \pi$, and if $y < 0$ we have $-\pi < \theta < 0$, etc.]

10. Find the roots of $z^n = (z + 1)^n$, and show that the points which represent them are collinear.

$$\left[\text{The roots are } -\tfrac{1}{2}\left(1 + \iota\cot\frac{r\pi}{n}\right),\text{ where } r = 0, 1, 2, \dots\, n - 1.\text{ The corresponding points}\right.$$

lie on the line $x + \tfrac{1}{2} = 0.]$

11. Show that the roots of $(1 + z)^n = (1 - z)^n$ are the values of $\iota\tan\dfrac{r\pi}{n}$, where $r = 0, 1, 2, \dots\, n - 1$, but omitting $n/2$ if n is even.

12. Prove that

$$(i)\ \ x^{2n} - 2x^n\cos\theta + 1 = \Pi_{r=0}^{r=n-1}\left(x^2 - 2x\cos\frac{\theta + 2r\pi}{n} + 1\right);$$

$$(ii)\ \ x^n + x^{-n} - 2\cos\theta = \Pi_{r=0}^{r=n-1}\left(x + x^{-1} - 2\cos\frac{\theta + 2r\pi}{n}\right);$$

$$(iii)\ \ \cos n\phi - \cos n\theta = 2^{n-1}\,\Pi_{r=0}^{r=n-1}\left\{\cos\phi - \cos\left(\theta + \frac{2r\pi}{n}\right)\right\}.$$

$[(i)\ x^{2n} - 2x^n\cos\theta + 1 = \{x^n - (\cos\theta + \iota\sin\theta)\}\{x^n - (\cos\theta - \iota\sin\theta)\}$. Now use Art. 15. (iii) Put $x = \cos\phi + \iota\sin\phi$, and for θ put $n\theta$.]

13. If n is odd and not a multiple of 3, prove that $x(x + 1)(x^2 + x + 1)$ is a factor of

$$(x + 1)^n - x^n - 1.$$

[Put $x = 0, -1, \omega$, where ω is an imaginary cube root of unity.]

14. If $(1 + x + x^2)^n = a_0 + a_1 x + a_2 x^2 + \dots + a_{2n}x^{2n}$, prove that

$$a_0 + a_3 + a_6 + \dots = a_1 + a_4 + a_7 + \dots = a_2 + a_5 + a_8 + \dots$$

15. If
$$u = x + y + z + a\,(y + z - 2x),$$
$$v = x + y + z + a\,(z + x - 2y),$$
$$w = x + y + z + a\,(x + y - 2z),$$

prove that $27a^2(x^3 + y^3 + z^3 - 3xyz) = u^3 + v^3 + w^3 - 3uvw$.

16. If $(a + \omega)^{-1} + (b + \omega)^{-1} + (c + \omega)^{-1} + (d + \omega)^{-1} = 2\omega^{-1}$,

$(a + \omega')^{-1} + (b + \omega')^{-1} + (c + \omega')^{-1} + (d + \omega')^{-1} = 2\omega'^{-1}$, where ω and ω' are the imaginary cube roots of unity, prove that

$$(a + 1)^{-1} + (b + 1)^{-1} + (c + 1)^{-1} + (d + 1)^{-1} = 2.$$

[Consider the equation $(a + x)^{-1} + (b + x)^{-1} + (c + x)^{-1} + (d + x)^{-1} = 2x^{-1}$.]

17. If $z_1^2 + z_2^2 + z_3^2 - z_2 z_3 - z_3 z_1 - z_1 z_2 = 0$, prove that

$$|z_2 - z_3| = |z_3 - z_1| = |z_1 - z_2|.$$

[For $z_1 + \omega z_2 + \omega^2 z_3 = 0$, $\qquad\qquad \therefore\ z_3 - z_1 = \omega\,(z_2 - z_3)$;

$\therefore \qquad\qquad |z_3 - z_1| = |\omega| \cdot |z_2 - z_3| = |z_2 - z_3|.$]

18. If $\omega = \frac{1}{2}(-1 + \iota\sqrt{3})$, being an imaginary cube root of unity, and a, b, c are real, then

$$\sqrt{(a + \omega b + \omega^2 c)} = \pm\tfrac{1}{2}\{\sqrt{2D + L} + \iota\sqrt{2D - L}\},$$

$$\sqrt{(a + \omega^2 b + \omega c)} = \pm\tfrac{1}{2}\{\sqrt{2D + L} + \iota\sqrt{2D - L}\},$$

where $\qquad\qquad D = \sqrt{(a^2 + b^2 + c^2 - bc - ca - ab)}$,

$\qquad\qquad\quad L = 2a - b - c,$

provided that $b > c$. If $b < c$, the sign of ι must be changed.

[Use Ex. 9. For a positive number x, $\sqrt{x}$ denotes the positive square root.]

■ **21. Points representing the Product and Quotient of Two Given Numbers:**
Let $z = r\,(\cos\theta + \iota\sin\theta)$, $z' = r'\,(\cos\theta' + \iota\sin\theta')$.

(1) *Construction for the point representing the product zz'.*

Let 1 be the point on OX which represents unity, so that $O1 = 1$. Draw the triangle $Oz'P$ directly similar to the triangle $O1z$.

Then P represents the product zz'.

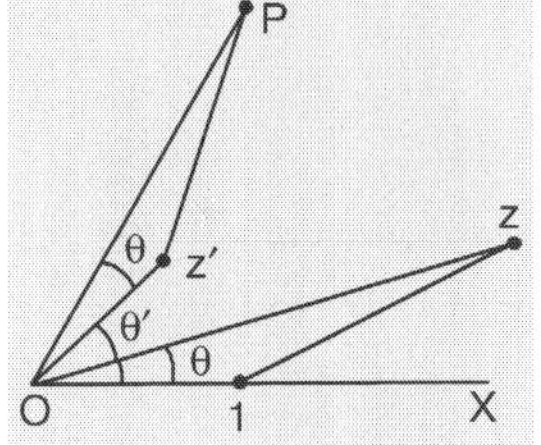

Fig. 9

For by similar triangles $\dfrac{OP}{Oz'} = \dfrac{Oz}{O1}$, that is, $\dfrac{OP}{r'} = \dfrac{r}{1}$, $\therefore OP = rr'$;

also $\angle z'OP = \angle 1Oz = \theta$, $\therefore \angle 1OP = \theta + \theta'$.

But $zz = rr'\{\cos(\theta + \theta') + \iota\sin(\theta + \theta')\}$.

Therefore P represents zz'.

(2) *Construction for the point representing the quotient z/z'.*

Draw the triangle OzQ directly similar to the triangle $Oz'1$.

Then Q represents the quotient z/z'.

For, by the last construction,

(number represented by Q). $z' = z$.

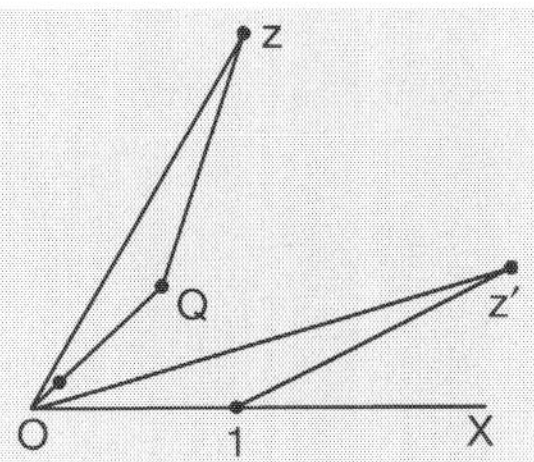

Fig. 10

(3) *If k is constant, and z varies so that*

$$\left| \frac{z-a}{z-a'} \right| = k,$$

then the point z describes a circle of which a, a′ are inverse points; unless k = 1, in which case z describes the perpendicular bisector of aa′.

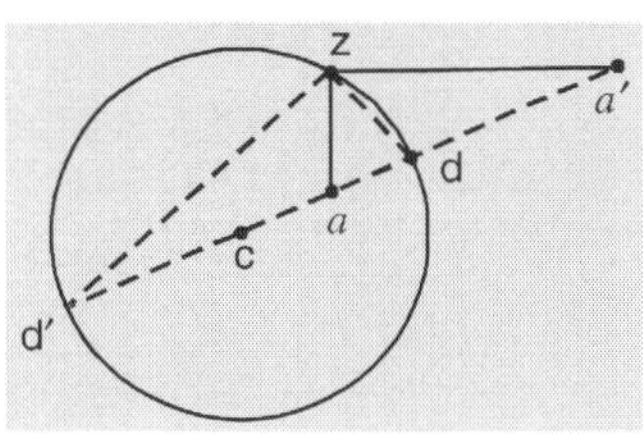

Fig. 11

For let the bisectors of $\angle$ *aza′* meet *aa′* at *d, d′*. Then, by Art. 13, (2), *az : a′z = k : 1*.

Therefore *d, d′* divide *aa′* internally and externally in the ratio *k : 1*, and are *fixed* points.

Therefore *z* describes the circle on *dd′* as diameter.

Also if *c* is the mid-point of *dd′*, we have *ca . ca′ = cd²*; hence *a, a′* are inverse points with regard to the circle.

If *k = 1*, then *az = a′z*; and *z* lies on the perpendicular bisector of *aa′*.

Conversely, if the point *z* describes a circle of which *a, a′* are inverse points, then

$$\left| \frac{z-a}{z-a'} \right| = k.$$

For, since *ca . ca′ – cd²*, then *dd′* is divided internally and externally in the same ratio, *k* to 1, say.

Hence, *az : a′z = k : 1*.*

(4) *If z varies so that*

$$\text{am } \frac{z-a}{z-a'} = \phi,$$

where ϕ is a constant angle, then the point z describes an arc of a segment of a circle on aa′, containing an angle ϕ.

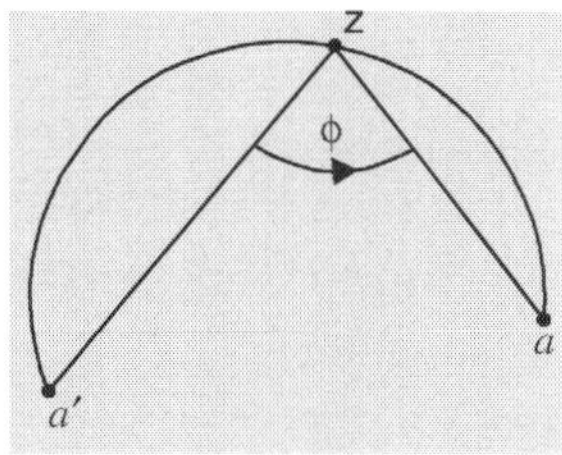

Fig. 12

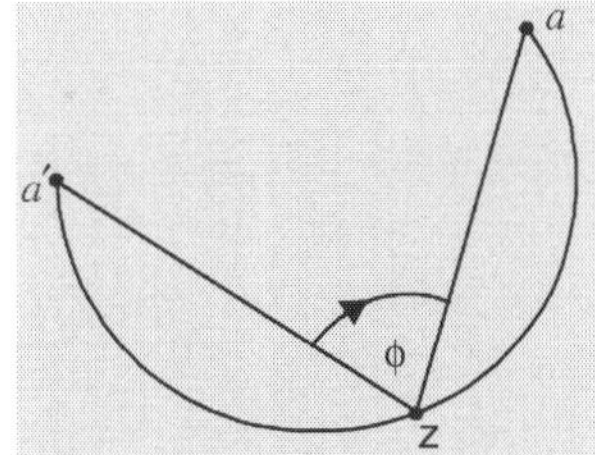

Fig. 13

For by Art. 13, $\angle a'za = \phi$. The sign of ϕ determines the side of *aa′* on which the segment lies. Thus ϕ positive in Fig. 12 and negative in Fig. 13.

* See *Elements of Geometry*, Barnard and Child, p. 316: 'Circle of Apollonius.'

■ 22. Displacements and Vectors:

(1) In connection with the geometrical representation of complex numbers, we introduce the notions of *displacement* and *directed length* or *vector*.

(2) Displacements in a given Plane: Let P, Q be two points in the plane OXY. The change of position which a point undergoes in moving from P to Q is called the *displacement* $\overline{PQ}$.

If any straight line $P'Q'$ is drawn equal to, parallel to, and in the same sense as PQ, the displacements $\overline{PQ}, \overline{P'Q'}$ are said to be *equal*.

To specify completely a displacement

$\overline{PQ}$ we must know:

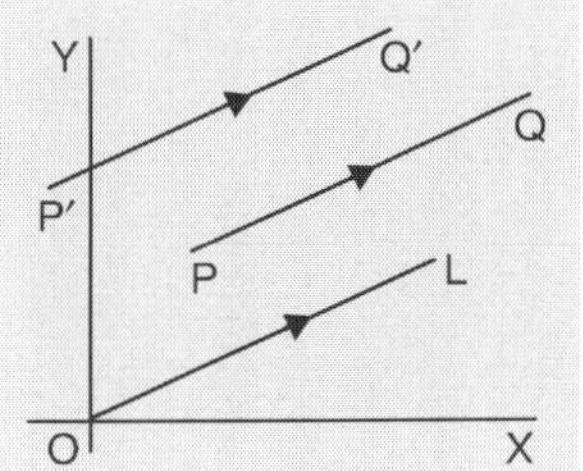

    (*i*)   its magnitude, *i.e.* the length PQ;

    (*ii*) its direction;

    (*iii*) its sense, denoted by the order of the letters, and if necessary by an arrow.

Fig. 14

If we draw OL equal and parallel to PQ and in the same sense, we say that $\angle XOL$ is the angle which PQ makes with OX.

This angle determines the direction and sense of $\overline{PQ}$.*

(3) Vectors: An expression (such as $\overline{PQ}$) used to denote a line-segment with reference to its length, direction and sense, the actual *position* of the line being indifferent, is called a *vector*.

Quantities which can be represented by lines used in this way are called *vector quantities*. Velocities and accelerations are vector quantities.

A force can be represented by a vector '*localised*' to lie in the line of action of the force.

Quantities (such as mass) which do not involve the idea of direction are called *scalar*.

(4) Connection with Complex Number: If $z = x + \iota y$ and P is the point (x, y), a one-to-one correspondence exists between the number z and any of the following:

(*i*) the point P; (*ii*) the displacement $\overline{OP}$; (*iii*) the vector (or directed length) $\overline{OP}$.

Any one of these three things may therefore be said to represent z, or to be represented by z.

■ 23. Addition of Displacements and of Vectors:

(1) Let P, Q, R be any three points. If a point moves from P to Q and then from Q to R, the resulting *change of position* is the same as if it had moved directly from P to R. We therefore define the *addition of displacements* as follows:

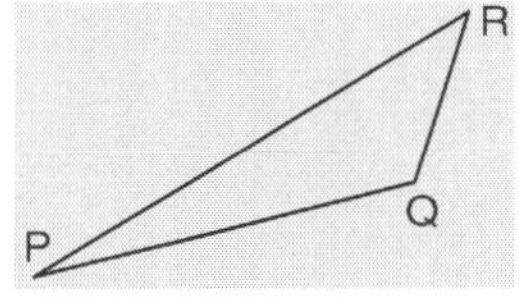

Fig. 15

---

   * Some writers use an underline instead of an overline.

The result of adding $\overline{QR}$ to $\overline{PQ}$ is defined to be $\overline{PR}$; and this is expressed by writing

$$\overline{PQ} + \overline{QR} = \overline{PR} \,. \qquad\qquad\qquad \text{...(A)}$$

This equation is also taken as defining the *addition of vectors*.

If $PQ,\ QR,\ RS$ are any three displacements or vectors (Fig. 17),

$$\overline{PQ} + \overline{QR} + \overline{RS} = \overline{PR} + \overline{RS} = \overline{PS} \,.$$

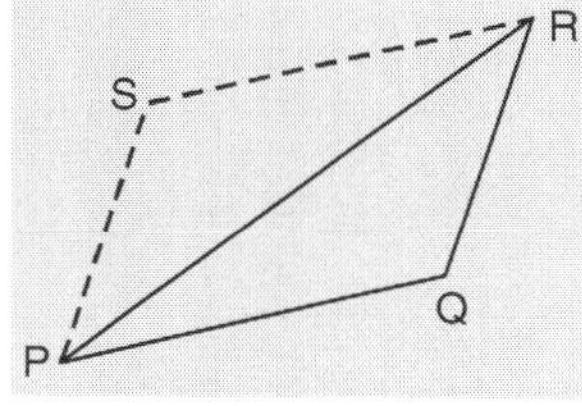
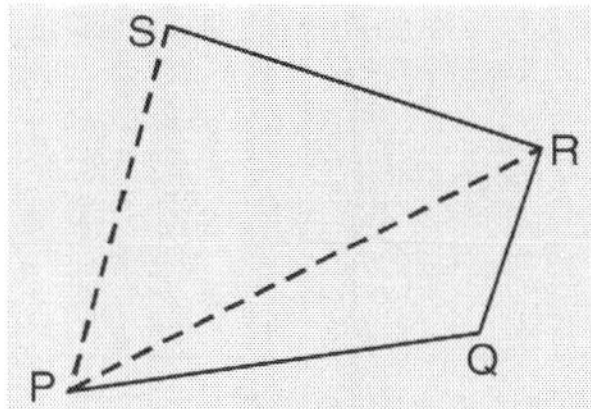

Fig. 16   Fig. 17

(2) *The Commutative and Associative Laws hold for the addition of displacements and vectors.*

(*i*) In Fig. 16, complete the parallelogram *PQRS*; then by Arts. 23, (1), and 22,

$$\overline{PQ} + \overline{QR} = \overline{PR} = \overline{PS} + \overline{SR} = \overline{QR} + \overline{PQ} \,.$$

Hence the commutative law holds, and $\overline{PR}$ is called the *sum* of $\overline{PQ}$ and $\overline{QR}$.

(*ii*) In Fig. 17,

$$(\overline{PQ} + \overline{QR}) + \overline{RS} = \overline{PR} + \overline{RS} = \overline{PS} \,,$$

$$\overline{PQ} + (\overline{QR} + \overline{RS}) = \overline{PQ} + \overline{QS} = \overline{PS} \,,$$

therefore $\qquad (\overline{PQ} + \overline{QR}) + \overline{RS} = \overline{PQ} + (\overline{QR} + \overline{RS}) \,,$

and the associative law holds.

> **Note:** In Fig. 16, $\overline{PQ} + \overline{PS} = \overline{PR}$, a fact which is expressed by saying that *displacements and vectors are added by the parallelogram law.*

■ **24. Zero and Negative Displacements and Vectors:** If after two or more displacements the moving point returns to its initial position, we say that the resulting displacement is *zero*. Thus we write

$$\overline{PQ} + \overline{QP} = 0.$$

This equation is also written in the form $-\,\overline{PQ} = \overline{QP}$, which *defines* the meaning of $-\overline{PQ}$.

These equations are also taken as defining the meaning of *zero and negative vectors*.

■ **25. Subtraction:** For displacements and vectors the meaning of $\overline{PQ} - \overline{QR}$ is defined by

$$\overline{PQ} - \overline{QR} = \overline{PQ} + (-\overline{QR}) = \overline{PQ} + \overline{RQ} .$$

Thus, $$\overline{OP} - \overline{OQ} = \overline{OP} + \overline{QO} = \overline{QO} + \overline{OP} = \overline{QP} .$$

■ **26. Definition of Multiplication by a Real Number:** To multiply a displacement or a vector $\overline{PQ}$ by a positive number k is to multiply its length by k, leaving its direction unaltered. The resulting displacement or vector is denoted by $k\,\overline{PQ}$ or by $\overline{PQ} . k$.

Further, we define $(-k)\,PQ$ by the equation

$$(-k)\,\overline{PQ} = k\,(-\overline{PQ}) = k\,\overline{QP} .$$

In particular,

$$(-1)\,\overline{PQ} = \overline{QP} .$$

So that to *multiply a vector by* (-1) *is to turn it through two right angles.*

■ **27. The Distributive Law:** *We shall prove that if k is a real number, then*

$$k\,(\overline{PQ} + \overline{QR}) = k\,\overline{PQ} + k\,\overline{QR} .$$

(i) *Let k be positive.* Along PQ set off $PQ' = kPQ$. Draw $Q'R'$ parallel to QR to meet PR in R'. By similar triangles,

$$PR' = kPR \text{ and } Q'R' = kQR;$$

$$\therefore \quad k\,(\overline{PQ} + \overline{QR}) = k\,\overline{PR} = PR' ,$$

and $k\,\overline{PQ} + k\,\overline{QR} = \overline{PQ'} + \overline{Q'R'} = PR' .$

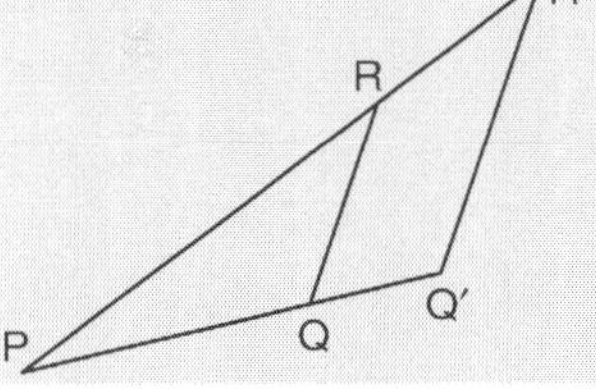

Fig. 18

Hence the theorem holds for positive numbers.

(ii) *For a negative number $(-k)$, we have*

$$(-k)\,(\overline{PQ} + \overline{QR}) = (-k)\,\overline{PR} = \overline{R'P} ,$$

and $(-k)\,\overline{PQ} + (-k)\,\overline{QR} = k\,\overline{QP} + k\,\overline{RQ} = \overline{Q'P} + \overline{R'Q'} = \overline{R'P} ;$

hence the theorem holds for negative numbers.

Thus *the distributive law holds for the multiplication of displacements and vectors by real numbers.*

Note: The diagram of Fig. 18 is drawn for a value of k greater than unity: the student should see that the same result follows from a diagram in which k is less than unity.

■ **28. Complex Numbers Represented by Vectors:** It will now be seen that, so far as addition, subtraction and multiplication by real numbers are concerned, complex numbers are subject to the same laws as the vectors which represent them. This fact is fundamental in theory and very useful in practice.

It should be noticed that if a number z is represented by a vector $\overline{AB}$, then $|z|$ is the length AB and am z is the angle which the directed line AB makes with the directed line Ox.

Theorem 1. *If C divides AB in the ratio $n : m$ and O is any point, then*

$$(m+n)\,\overline{OC} \;=\; m\,\overline{OA} + n\,\overline{OB}.$$

For

$$m\,\overline{OC} \;=\; m\,\overline{OA} + m\,\overline{AC},$$

$$n\,\overline{OC} \;=\; n\,\overline{OB} + n\,\overline{BC}.$$

Also

$$m\,\overline{AC} \;=\; n\,\overline{CB} = -\,n\,\overline{BC};$$

whence the result follows by addition.

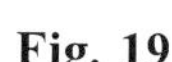

Fig. 19

Theorem 2. *If z is the point which divides the straight line joining z_1, z_2 in the ratio $n : m$, then the corresponding numbers are connected by the relation*

$$(m + n)\,z \;=\; mz_1 + nz_2.$$

This follows from Theorem 1, in accordance with the principle stated in this article.

In particular, *if z is the mid-point of z_1z_2, then $z = \frac{1}{2}\,(z_1 + z_2)$.*

■ **EXAMPLE 1.** *If a, b are complex numbers, prove geometrically that*
$$|a + b|^2 + |a - b|^2 \;=\; 2\,|a|^2 + 2\,|b|^2.$$

Let A, B be the points which represent a, b.
Bisect AB at C, then

$$\overline{OA} + \overline{OB} \;=\; 2\,\overline{OC}$$

and

$$\overline{OA} - \overline{OB} \;=\; \overline{BA} = 2\,\overline{CA}.$$

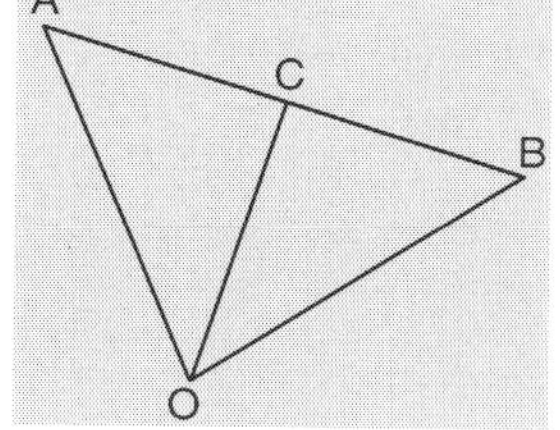

Fig. 20

Therefore $a + b$ and $a - b$ are represented by $2\,\overline{OC}$ and $2\,\overline{CA}$; hence
$$|a + b| = 2OC, \quad |a - b| = 2CA.$$
Now, since C is mid-point of base AB, $OA^2 + OB^2 = 2OC^2 + 2CA^2$;
$$\therefore \qquad 2\,|a|^2 + 2\,|b|^2 \;=\; |a + b|^2 + |a - b|^2.$$

■ **EXAMPLE 2.** *If $\overline{OA}, \overline{OB}, \overline{OC}$ are connected by the relation*

$$a\overline{OA} + b\overline{OB} + c\overline{OC} \;=\; 0, \quad \text{where} \quad a + b + c = 0,$$

then A, B, C are collinear.

[This is the converse of Theorem 1 of this article. We have

$$(a + c)\,\overline{OB} = a\,.\,\overline{OA} + c\,.\,\overline{OC}; \text{ hence } a\,.\,\overline{AB} = c\,.\,\overline{BC}.]$$

▓ **29. The Symbol ι as an Operator:*** Along two straight lines at right angles set off, consecutively, equal lengths OP, OQ, OP', OQ' in the positive direction of rotation.

---

* For the moment, the reader should forget his conception of ι as denoting a *number*.

Let the symbol ι applied to a vector denote the operation of turning it through a right angle in the positive direction of rotation.

To bring our language into conformity with that of algebra, we say that *to multiply a vector by ι is to turn it through a right angle in the positive sense.*

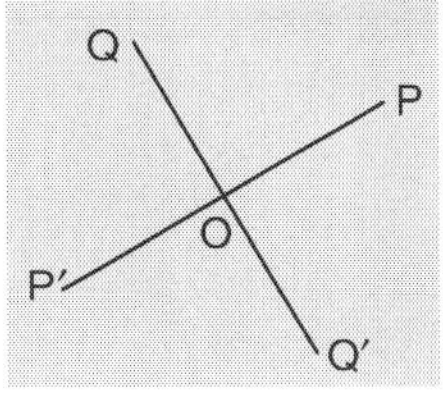

Fig. 21

Thus in Fig. 21, $\qquad \overline{OQ} = \iota\,\overline{OP}$ and $\overline{OP'} = \iota\,\overline{OQ}$.

Therefore $\qquad \overline{OP'} = \iota(\iota\,\overline{OP}) = \iota^2\,\overline{OP}$,

where $\iota^2\,\overline{OP}$ is an abbreviation for $\iota(\iota\,\overline{OP})$. Hence $\iota^2\,\overline{OP} = (-1)\,\overline{OP}$.

Thus ι^2 and -1 denote the same operation, and in this sense we write $\iota^2 = -1$.

Again, $\overline{OQ'} = \iota\,\overline{OP'} = \iota\,(-1)\,\overline{OP}$ and $\overline{OQ'} = -\,\overline{OQ} = (-1)\,\iota\,\overline{OP}$.

Either of these results is written in the form $(-\iota)\,\overline{OP} = \overline{OQ'}$, so that *to multiply a vector by $(-\iota)$ is to turn it through a right angle in the negative sense.*

Again, if $\iota^3 . \iota^2\,\overline{OP}$ is taken to mean $\iota^3\,(\iota^2\,\overline{OP})$, it is obvious that

$$\iota^3 . \iota^2\,\overline{OP} = \iota^5\,\overline{OP} = \iota^2 . \iota^3\,\overline{OP} = \overline{OQ}.$$

■ **EXAMPLE:** *If a, b are complex numbers, find numbers z, z' so that the points z, z' and a, b may be opposite corners of a square.*

Let c be the mid-point of ab, then

$$\overline{Oz} = \overline{Oc} + \overline{cz} = \overline{Oc} + \iota\,\overline{cb};$$

$\therefore \qquad Oz = \tfrac{1}{2}(Oa + Ob) + \tfrac{1}{2}\iota\,(Ob - Oa);$

$\therefore \qquad z = \tfrac{1}{2}(a + b) - \tfrac{1}{2}\iota\,(a - b).$

Similarly, $\qquad z' = \tfrac{1}{2}(a + b) + \tfrac{1}{2}\iota\,(a - b).$

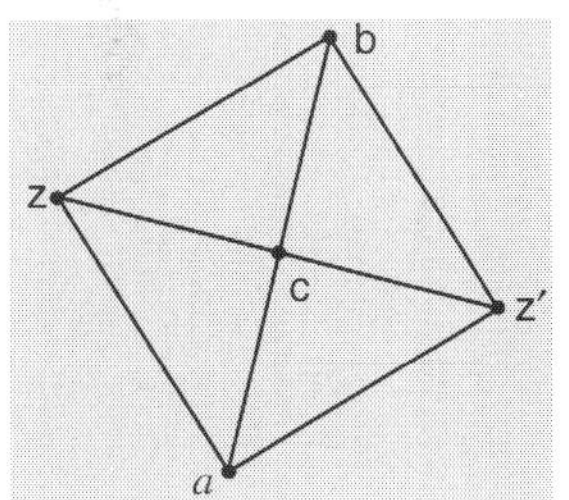

Fig. 22

■ **30. The Operator $\cos\theta + \iota\sin\theta$:** Draw two equal straight lines OP, OP' inclined at an angle θ. Draw $P'N$ perpendicular to OP. Along NP set off NQ equal to NP'. Then

$$\overline{OP'} = \overline{ON} + \overline{NP'} = \overline{ON} + \iota\,\overline{NQ}.$$

Also $\qquad \overline{ON} = \dfrac{ON}{OP}\overline{OP} = \cos\theta . \overline{OP},$

$$\overline{NQ} = \dfrac{NQ}{OP}\overline{OP} = \sin\theta . \overline{OP};$$

$\therefore \qquad \overline{OP'} = \cos\theta . \overline{OP} + \iota\sin\theta . \overline{OP},$

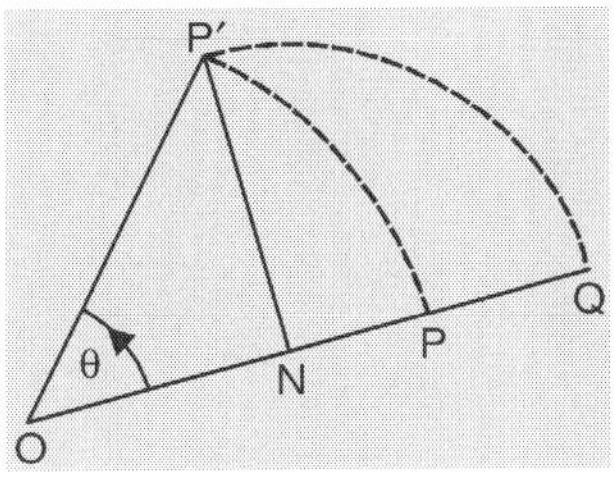

Fig. 23

which we write in the form

$$\overline{OP'} = (\cos \theta + \iota \sin \theta)\,\overline{OP},$$

and we say that *to multiply a vector by* $\cos \theta + \iota \sin \theta$ *is to turn it through the angle* θ.

■ **31. Multiplication and Division of a Vector by a Complex Number:** In accordance with Arts. 26, 29 and 30 of this chapter, we say that *to multiply a vector* $\overline{OP}$ *by the complex number r $(\cos \theta + \iota \sin \theta)$ is to multiply its length by r and turn the resulting vector through the angle* θ.

Here r is the *stretching* factor and $\cos \theta + \iota \sin \theta$ the *turning* factor. These are independent of each other, and the *order* in which they are applied is indifferent.

If $\overline{OQ}$ is the vector obtained by multiplying $\overline{OP}$ by the complex number z, we write

$$\overline{OQ} = z\overline{OP} \quad \text{and} \quad \overline{OQ}\,/\,\overline{OP} = z\,;$$

we also say that the ratio of $\overline{OQ}$ to $\overline{OP}$ is the number z.

Division is the inverse of multiplication, so that if $\overline{OQ} = z\overline{OP}$, then $\overline{OP}$ is the result of dividing $\overline{OQ}$ by z.

Therefore *to divide a vector* $\overline{OQ}$ *by z is to divide its length by r and turn the resulting vector through the angle* $(-\theta)$.

The result is the same as that obtained by multiplying $\overline{OQ}$ by $1/z$.

■ **32. Product of Complex Numbers:**

Let

$$\overline{OQ} = z'\overline{OP} \quad \text{and} \quad \overline{OR} = z\overline{OQ},$$

then we write

$$\overline{OR} = z\,(z'\,\overline{OP}) = zz'\,\overline{OP},$$

where zz' applied to $\overline{OP}$ denotes that the operators z', z are to be applied *in succession, in this order.*

Since the stretching and turning factors may be applied in any order, $\overline{OP}$ may be transformed into $\overline{OR}$ by multiplying its length by rr' and turning the resulting vector through the angle $(\theta + \theta')$.

Hence the operations denoted by

$$zz', \ z'z \ \text{and} \ rr'\{\cos (\theta + \theta') + \iota \sin (\theta + \theta')\}$$

are equivalent.

Again, if we take $(\cos \theta + \iota \sin \theta)^n$ to mean that the operation

$$(\cos \theta + \iota \sin \theta)$$

is to be applied n times, the result is the same as that given by the operator

$$\cos n\theta + \iota \sin n\theta.$$

In this sense then $\quad (\cos \theta + \iota \sin \theta)^n = \cos n\theta + \iota \sin n\theta.$

It will be seen that *complex numbers used as operators on vectors conform to the laws of algebra.*

EXERCISE IX

ARGAND DIAGRAMS : VECTORS

1. If $z = 3 + 2\iota$, $z' = 1 + \iota$, mark the points z, z' in an Argand diagram, and find by geometrical construction, the points representing

$$z + z',\ z - z',\ zz',\ z/z'.$$

2. Let z, a, b be complex numbers of which a, b are constant and z varies. If Z is given in terms of z by one of the following equations, it is required to find the point Z corresponding to a given point z. Explain the constructions indicated in the diagrams, $O1$ being the unit of length.

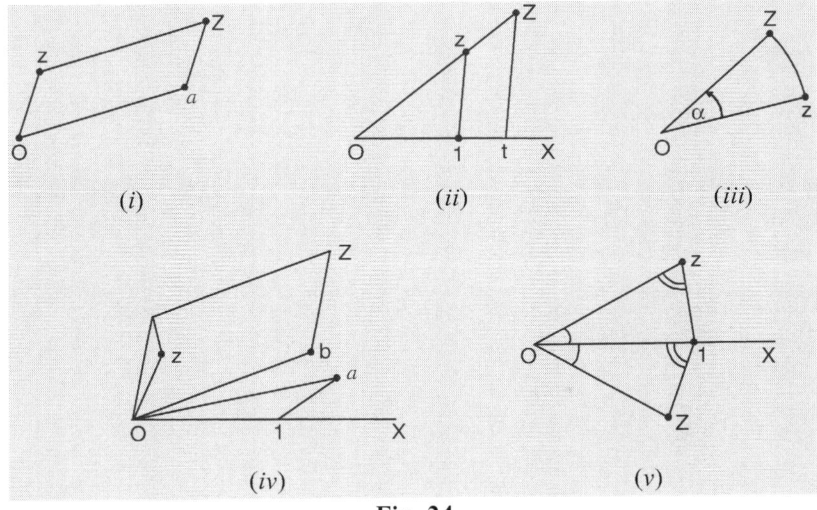

Fig. 24

 (*i*) $Z = z + a$.   (*ii*) $Z = tz$ where t is real.

(*iii*) $Z = (\cos \alpha + \iota \sin \alpha)\, z$. (*iv*) $Z = az + b$.

 (*v*) $Z = 1/z$.

3. If $Z = (1 + z)/(1 - z)$, find the point Z corresponding to a given point z, and show that if $|z| = 1$, the point Z is on the y-axis.

4. If a, b are given complex numbers and $z = a + (b - a)\, t$, find the point z corresponding to any given real value of t, and prove that as t varies from $-\infty$ to $+\infty$, z describes the entire line which passes through a, b, the segment ab corresponding to the values from 0 to 1 of t.

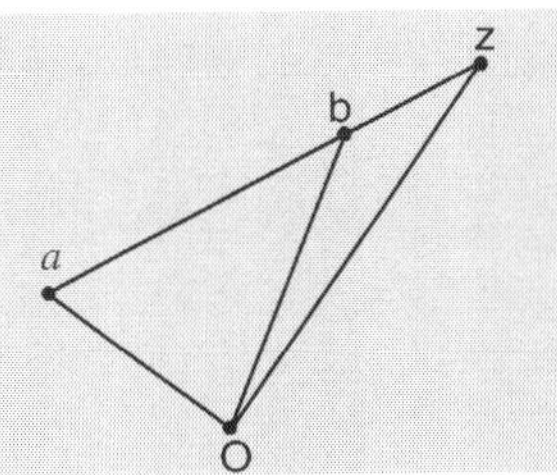

Fig. 25

[Along the line ab set off the length $az = t \cdot ab$, then $\overline{Oz} = \overline{Oa} + \overline{az} = \overline{Oa} + t \cdot \overline{ab}$;

$\therefore$ $z = a + (b - a)\, t$, etc.]

5. If $z = a\, (1 + \iota t)$ where t is a real number, prove that as t varies z describes the line through the point a perpendicular to Oa.

6. If c, a are given numbers, a being real, and,

  (*i*) if $z = c + a\,(\cos\phi + \iota\sin\phi)$,

  find the point z corresponding to a given value of ϕ;

  (*ii*) if $z = c + a\,(1 + \iota t)/(1 - \iota t)$,

  where t is real, show that as t varies from $-\infty$ to $+\infty$, the point z describes once the circle with centre c and radius a.   [Put $t = \tan\frac{1}{2}\phi$.]

7. If A, B, C, D are any four points in a plane, then

$$AD \cdot BC \le BD \cdot AC + CD \cdot AB.$$

Show that this theorem is an immediate consequence of the identity

$$(z_1 - z_4)(z_2 - z_3) + (z_2 - z_4)(z_3 - z_1) + (z_3 - z_4)(z_1 - z_2) = 0.$$

8. If G is the centroid of particles of mass m_1, m_2, m_3, ... at A_1, A_2, A_3, ... and O is any point, then $(m_1 + m_2 + m_3 + ...)\,\overline{OG} = m_1\overline{OA_1} + m_2\overline{OA_2} + m_3\overline{OA_3} + ...$.

To include cases where m_1, m_2, ... are not all of the same sign, we may suppose G to be the centre of parallel forces, proportional to m_1, m_2, acting at A_1, A_2 ... .

9. If z is the centroid of particles of mass m_1, m_2, m_3, ... at z_1, z_2, z_3, ..., then

$$(m_1 + m_2 + m_3 + ...)\,z = m_1 z_1 + m_2 z_2 + m_3 z_3 + ...$$

10. *Any three, coplanar and non-parallel vectors* $\overline{OA}, \overline{OB}, \overline{OC}$ *are connected by a relation of the form* $p\overline{OA} + q\overline{OB} + r\overline{OC} = 0$, *where* p, q, r *are real numbers.*

Moreover, $p : q : r = \Delta OBC : \Delta OCA : \Delta OAB$,

where the signs of the areas are determined by the usual rule.

Also the points A, B, C *are collinear if* $p + q + r = 0$, *and conversely.*

[For let $p : q : r = \Delta OBC : \Delta OCA : \Delta OAB$,

and let G be the centre of parallel forces, acting at A, B, C, and proportional to p, q, r, then G coincides with O, and $p\overline{OA} + q\overline{OB} + r\overline{OC} = (p + q + r)\,OG = 0$.]

11.   (*i*) *If* α, β *are non-parallel vectors and* p, q, p', q' *are real numbers, an equation of the form*

$$p\alpha + q\beta = p'\alpha + q'\beta$$

  involves the two equations $p = p'$, $q = q'$.

  (*ii*) *If* α, β, γ *are non-parallel coplanar vectors connected by the equations*

$$p\alpha + q\beta + r\gamma = 0 \text{ and } p'\alpha + q'\beta + r'\gamma = 0,$$

  where p, q, r, p', q', r' *are real, then these are one and the same equation, that is to say*

$$p/p' = q/q' = r/r'.$$

  [These theorems follow at once from Ex. 10.]

12. Any three complex numbers z_1, z_2, z_3 are connected by a relation of the form

$$pz_1 + qz_2 + rz_3 = 0,$$

where p, q, r are real numbers.

Moreover, $p : q : r = \Delta Oz_2z_3 : \Delta Oz_3z_1 : \Delta Oz_1z_2$,

the signs of the areas being determined as usual.

Also the points z_1, z_2, z_3, are collinear if $p + q + r = 0$, and conversely.

Prove this algebraically, and deduce Ex. 11.

[If $z_1 = x_1 + \iota y_1, z_2 = x_2 + \iota y_2, z_3 = x_3 + \iota y_3$, we have

$$z_1 (x_2 y_3 - x_3 y_2) + \dots + \dots = 0,$$

whence the results follow immediately.]

13. If the points A, B, C are collinear, and O is any point, then

$$\overline{OA} \,.\, BC + \overline{OB} \,.\, CA + \overline{OC} \,.\, AB = 0.$$

[In Art. 28, Theorem 1, $m : n : m + n = CB : AC : AB$.]

14. Let z_1, z_2, z_3 be complex numbers, no two of which are equal, then

(i) If the points z_1, z_2, z_3 are collinear,

$$z_1 \,|\, z_2 - z_3 \,| \pm z_2 \,|\, z_3 - z_1 \,| \pm z_3 \,|\, z_1 - z_2 \,| = 0.$$

Also, if the point z_1 lies between z_2 and z_3, the ambiguous signs are both minus.

(ii) If the above equation holds, then either z_1, z_2, z_3 are collinear, or else O is the centre of a circle which touches the sides of the triangle $z_1 z_2 z_3$.

15. If $A_1 A_2 A_3$ is an equilateral triangle, the vertices occurring in the positive direction of rotation, prove that

$$\overline{A_3 A_1} = \left(\cos \frac{2\pi}{3} + \iota \sin \frac{2\pi}{3} \right) \overline{A_2 A_3} = \omega \, \overline{A_2 A_3} ,$$

where ω is an imaginary cube root of unity.

Also, if z_1, z_2, z_3 are the numbers corresponding to A_1, A_2, A_3, prove that $z_1 + \omega z_2 + \omega^2 z_3 = 0$, and consequently

$$z_1^2 + z_2^2 + z_3^2 - z_2 z_3 - z_3 z_1 - z_1 z_2 = 0.$$

16. If $A X Y A' X' Y'$ is a regular hexagon and A, A' represent given complex numbers a, a', then the numbers represented by X, X', Y, Y' are given by

$$\frac{1}{2} (a + a') + \frac{1}{2} (a - a') (\cos \theta + \iota \sin \theta),$$

where θ has the values $\pm \dfrac{\pi}{3}, \pm \dfrac{2\pi}{3}$.

17. The triads of points A, B, C and X, Y, Z are the vertices of directly similar triangles if the corresponding complex numbers a, b, c and x, y, z are connected by the relation

$$x (b - c) + y (c - a) + z (a - b) = 0.$$

[The triangles are directly similar if $\dfrac{\overline{AB}}{\overline{AC}} = \dfrac{\overline{XY}}{\overline{XZ}}$, *i.e.* if $\dfrac{b - a}{c - a} = \dfrac{y - x}{z - x}$.]

18. If ABC is a triangle and triangles BCX, CAY, ABZ are drawn on BC, CA, AB, directly similar to one another, the centroids of XYZ and ABC coincide.

[By Ex. 17, $\dfrac{x - c}{b - c} = \dfrac{y - a}{c - a} = \dfrac{z - b}{a - b}$. Hence show that $x + y + z = a + b + c$.]

19. Equilateral triangles are drawn on the sides of a given triangle ABC, all outwards or all inwards. Prove that their centroids form an equilateral triangle.

[Let P, Q, R be the centroids of the triangles drawn outwards on BC, CA, AB. Prove that

$$\overline{QA} = \overline{CA} \cdot \frac{1}{\sqrt{3}} \,(\cos 30 + \iota \sin 30), \quad \overline{AR} = \overline{AB} \cdot \frac{1}{\sqrt{3}} \,(\cos 30 - \iota \sin 30).$$

Hence show that $\overline{QR} = \tfrac{1}{2}\overline{CB} + \dfrac{\iota}{2\sqrt{3}}\,(\overline{CA} - \overline{AB})$, and that $\overline{RP}$ has a similar value.

Hence show that $\overline{RQ} = \overline{RP}\,(\cos 60 + \iota \sin 60)$, and use Ex. 15.]

'TRANSFORMATIONS'

20. If Z, z are connected by any of the relations in Ex. 2 and if z describes a given curve s, then Z will describe a curve S.

Explain the following, where a, b are given complex numbers.

   (i) If $Z = z + a$, S can be obtained from s by a *translation*.

   (ii) If $Z = tz$ where t is real, the curves s, S are similar and similarly situated, O being the centre of similitude. In this case we say that S is a *magnification* of s.

   (iii) If $Z = (\cos \alpha + \iota \sin \alpha)\, z$, S can be obtained from s by a *rotation* about O through $\angle\, \alpha$.

   (iv) If $Z = a\,z + b$, S can be obtained from s by a *rotation, a magnification and a translation*.

   (v) If $Z = 1/z$, S is the reflection in OX of the inverse of s, O being the centre of inversion.

21. Show that each of the substitutions in Ex. 20 converts a circle c into a circle C, except that in (v), if c passes through O, then C is a straight line.

22. *Show that the substitution*

$$Z = \frac{az + b}{a'z + b'}$$

converts a circle into a circle or, in an exceptional case, into a straight line.

$\left[Z = \dfrac{a}{a'} + \dfrac{ba' - b'a}{a'} \cdot \dfrac{1}{a'z + b'} \right.$, therefore the transformation is equivalent to one or more of those in Ex. 20.]

■ ■ ■

Theory of Equations

■ **1. Roots of Equations:** Under this heading we consider equations of the type $f(x) = 0$ where $f(x)$ is a polynomial, and 'equation' will mean an equation of this kind.

The general equation of the nth degree will be written in one of the forms:

$$x^n + p_1 x^{n-1} + p_2 x^{n-2} + \ldots + p_n = 0,$$
$$a_0 x^n + a_1 x^{n-1} + a_2 x^{n-2} + \ldots + a_n = 0;$$

or,
$$a_0 x^n + n a_1 x^{n-1} + \frac{n(n-1)}{\lfloor 2} a_2 x^{n-2} + \ldots + a_n = 0,$$

where binomial coefficients are introduced. The last equation is written briefly in the form

$$(a_0, a_1, a_2, \ldots a_n)(x,\ 1)^n = 0.$$

For the present, we assume that *every equation has one root.* This is the fundamental theorem of the Theory of Equations, and will be proved in another volume.

(1) It follows that *every equation of the nth degree has exactly n roots.*

For let $f(x) \equiv x^n + p_1 x^{n-1} + \ldots + p_n$ and let α be a root ot $f(x) = 0$. By the Remainder theorem, $f(x)$ is divisible by $x - a$; we may therefore assume that

$$f(x) \equiv (x - \alpha)(x^{n-1} + q_1 x^{n-2} + \ldots + q_{n-1}) \equiv (x - \alpha).\ \phi(x).$$

Let β be a root of $\phi(x) = 0$; as before, $\phi(x)$ is divisible by $x - \beta$, and we may assume that

$$\phi(x) \equiv (x - \beta)(x^{n-2} + r_1 x^{n-3} + \ldots + r_{n-2});$$
$$\therefore \qquad f(x) \equiv (x - \alpha)(x - \beta)(x^{n-2} + r_1 x^{n-3} + \ldots + r_{n-2}).$$

Proceeding in this way, we can show that

$$f(x) \equiv (x - \alpha)(x - \beta) \ldots (x - \lambda),$$

where there are n linear factors on the right. Hence $f(x) = 0$ has n roots α, β, γ, ..., λ, and no others.

(2) Imaginary Roots: Let the coefficients of $f(x)$ be real; then, if $\alpha + \iota\beta$ is a root, so is $\alpha - \iota\beta$ (Ch. 5, 14). Therefore $f(x)$ is divisible by

$$(x - \alpha - \iota\beta)(x - \alpha + \iota\beta),$$

that is, by
$$(x - a)^2 + \beta^2.$$

Thus a polynomial in x with real coefficients can be resolved into factors which are linear or quadratic functions of x with real coefficients.

(3) Multiple Roots: If $f(x) = (x - \alpha)^r \cdot \phi(x)$ where $\phi(x)$ is not divisible by $x - \alpha$, α is called an *r-multiple root* of $f(x) = 0$.

When $r = 2$, we say that α is a *double root*.

■ **2. Relations connecting the Roots and Coefficients of an Equation:**

Theorem: *If $\alpha_1, \alpha_2, \ldots, \alpha_n$ are the roots of the equation*
$$x^n + p_1 x^{n-1} + p_2 x^{n-2} + \ldots + p_n = 0,$$
then the sums of the products of $\alpha_1, \alpha_2, \alpha_3, \ldots, \alpha_n$ taken one, two, three, ..., n at a time, are respectively equal to

$$-p_1, p_2, -p_3, \ldots, (-1)^n p_n.$$

For $x^n + p_1 x^{n-1} + p_2 x^{n-2} + \ldots + p_n = (x - \alpha_1)(x - \alpha_2) \ldots (x - \alpha_n)$
$$= x^n + \Sigma\alpha_1 \cdot x^{n-1} + \Sigma\alpha_1\alpha_2 \cdot x^{n-2} + \ldots$$
$$+ (-1)^n \alpha_1\alpha_2 \ldots \alpha_n, \quad \ldots\text{(A)}$$

and, equating coefficients, we have
$$\Sigma\alpha_1 = -p_1, \quad \Sigma\alpha_1\alpha_2 = p_2, \quad \ldots \quad \alpha_1\alpha_2 \ldots \alpha_n = (-1)^n p_n. \qquad \ldots\text{(B)}$$

Conversely, if $\alpha_1, \alpha_2, \ldots \alpha_n$ satisfy the equations (B), they are the roots of
$$x^n + p_1 x^{n-1} + p_2 x^{n-2} + \ldots + p_n = 0.$$

For under these circumstances the identity (A) holds.

It follows that the result of eliminating $\alpha_2, \alpha_3, \ldots \alpha_n$, from equations (B) is

$$\alpha_1^n + p_1\alpha_1^{n-1} + \ldots + p_n = 0.$$

▌ **EXAMPLE :** *If α, β, γ are the roots of $2x^3 + x^2 - 2x - 1 = 0$, write down the values of $\alpha + \beta + \gamma$, $\beta\gamma + \gamma\alpha + \alpha\beta$, $\alpha\beta\gamma$.*

Write the equation in the form $x^3 + \dfrac{1}{2}x^2 + (-1)x + \left(-\dfrac{1}{2}\right) = 0$;

$$\therefore \qquad \alpha + \beta + \gamma = -\frac{1}{2}, \quad \beta\gamma + \gamma\alpha + \alpha\beta = -1, \quad \alpha\beta\gamma = \frac{1}{2}.$$

■ **3. Transformation of Equations:** Let $\alpha, \beta, \gamma, \ldots$ be the roots of $f(x) = 0$, and suppose that we require the equation whose roots are $\phi(\alpha), \phi(\beta), \phi(\gamma), \ldots$ where $\phi(x)$ is a given function of x.

Let $y = \phi(x)$ and suppose that from this equation we can find x as a single-valued function of y, which we denote by $\phi^{-1}(y)$.

Transforming the equation $f(x) = 0$ by the substitution $x = \phi^{-1}(y)$, we obtain $f\{\phi^{-1}(y)\} = 0$, which is the equation required.

A case in which x is not a single-valued function of y is given in Ex. 2.

▌ **EXAMPLE 1.** *If α, β, γ are the roots of $x^3 - x - 1 = 0$, find the equation whose roots are* $\dfrac{1+\alpha}{1-\alpha}, \dfrac{1+\beta}{1-\beta}, \dfrac{1+\gamma}{1-\gamma}$. *Hence write down the value of* $\Sigma(1+\alpha)/(1-\alpha)$.

Let $y = \dfrac{1+x}{1-x}$, then $x = \dfrac{y-1}{y+1}$; and, if $x^3 - x - 1 = 0$, then y is given by

$$\left(\frac{y-1}{y+1}\right)^3 - \frac{y-1}{y+1} - 1 = 0,$$

which is equivalent to

$$y^3 + 7y^2 - y + 1 = 0.$$

This is the required equation. Hence also $\dfrac{1+\alpha}{1-\alpha} + \dfrac{1+\beta}{1-\beta} + \dfrac{1+\gamma}{1-\gamma} = -7.$

EXAMPLE 2. *If α, β, γ are the roots of $f(x) = x^3 + p_1 x^2 + p_2 x + p_3 = 0$, find the equation whose roots are α^3, β^3, γ^3.*

Let $y = x^3$, then $x = y^{1/3}$ where $y^{1/3}$ denotes *any* cube root of y. The required equation is therefore obtained by rationalising

$$f\left(y^{1/3}\right) = y + p_1 y^{2/3} + p_2 y^{1/3} + p_3 = 0;$$

for the result will be the same whichever cube root $y^{1/3}$ stands for. To rationalise the equation, let $y + p_3 = l$, $p_1 y^{2/3} = m$, $p_2 y^{1/3} = n$, then $l + m + n + 0$, and therefore $l^3 + m^3 + n^3 - 3lmn = 0$.

Hence the required equation is

$$(y + p_3)^3 + p_1^3 y^2 + p_2^3 y - 3 (y + p_3). \, p_1 y^{2/3} . \, p_2 y^{1/3} = 0,$$

which is the same as

$$y^3 + (p_1^3 - 3p_1 p_2 + 3p_3) y^2 + (p_2^3 - 3p_1 p_2 p_3 + 3p_3^2) y + p_3^2 = 0.$$

▧ **4. Special Cases:** The following transformations are often required:

Let α, β, γ, ... be the roots of $f(x) = 0$, then

(1) the equation whose roots are $-\alpha$, $-\beta$, $-\gamma$, ... is $f(-x) = 0$;

(2) the equation whose roots are $1/\alpha$, $1/\beta$, $1/\gamma$, ... is $f(1/x) = 0$;

(3) the equation whose roots are $k\alpha$, $k\beta$, $k\gamma$, ... is $f(x/k) = 0$.

This transformation is called '*multiplying the roots of $f(x) = 0$ by k.*'

(4) the equation whose roots are $\alpha - h$, $\beta - h$, $\gamma - h$ is $f(x + h) = 0$.

We find $f(x + h)$ as in Chs. 3, 4, and the transformation is called '*diminishing the roots of $f(x) = 0$ by h.*'

(5) *The second term of $f(x) \equiv n_0 x^n + \alpha_1 x^{n-1} + \alpha_2 x^{n-2} + ... + \alpha_n = 0$ can be removed by diminishing the roots by $- a_1/na_0$.*

For $f(x + h) = a_0 (x + h)^n + a_1 (x + h)^{n-1} + ...,$

so that the coefficient of x^{n-1} in $f(x + h)$ is $na_0 h + a_1$, and this is zero when $h = - a_1/na_0$.

(6) *To transform $f(x) = 0$ into an equation in which the coefficient of the first term is 1 and the other coefficients are the least possible integers,* proceed as follows:

EXAMPLE 1. *Consider the equation of*

$$72x^3 - 54x^2 + 45x - 7 = 0.$$

Write this

$$x^3 - \frac{3}{4}x^2 + \frac{5}{8}x - \frac{7}{72} = 0.$$

Putting $x = y/k$ and multiplying by k^3, the equation becomes

$$y^3 - \frac{3}{4}ky^2 + \frac{5}{8}k^2 y - \frac{7}{72}k^3 = 0 \quad \text{or} \quad y^3 - \frac{3}{2^2}ky^2 + \frac{5}{2^3}k^2 y - \frac{7}{2^3 . 3^2}k^3 = 0.$$

The least value of k which will make every coefficient an integer is 12, so that the required equation is

$$y^3 - 9y^2 + 90y - 168 = 0 \text{ where } y = 12x.$$

■ **EXAMPLE 2.** *Transform $x^3 - 6x^2 + 5x + 12 = 0$ into an eqution lacking the second term.*

We diminish the roots by $6/3 = 2$.

The reckoning on the right shows that the resulting equation is

$$x^3 - 7x + 6 = 0,$$

which is the one required.

$$
\begin{array}{r|rrrr}
1-2) & 1 & -6 & +5 & +12 \\
     &   & +2 & -8 & -6 \\
\hline
     & 1 & -4 & -3 & +6 \\
     &   & +2 & -4 &   \\
\hline
     & 1 & -2 & -7 &   \\
     &   & +2 &   &   \\
\hline
     & 1 & 0 &   &   \\
\end{array}
$$

■ **EXAMPLE 3.** *If α, β, γ are the roots of $2x^3 + 3x^2 - x - 1 = 0$, find the equation whose roots are $2\alpha + 3$, $2\beta + 3$, $2\gamma + 3$.*

$$
\begin{array}{r|rrrr}
1-(-3)\ ) & 1 & +3 & -2 & -4 \\
          &   & -3 & +0 & +6 \\
\hline
          & 1 & +0 & -2 & +2 \\
          &   & -3 & +9 &   \\
\hline
          & 1 & -3 & +7 &   \\
          &   & -3 &   &   \\
\hline
          & 1 & -6 &   &   \\
\end{array}
$$

The equation whose roots are 2α, 2β, 2γ is

$$2.\,\frac{x^3}{8} + 3.\,\frac{x^2}{4} - \frac{x}{2} - 1 = 0, \quad \text{or} \quad x^3 + 3x^2 - 2x - 4 = 0.$$

Increasing the roots of this by 3 (as in the margin, where we divide successively by $x + 3$), the required equation is

$$x^3 - 6x^2 + 7x + 2 = 0.$$

■ **EXAMPLE 4.** *If two roots α and β of $f(x) = 0$ are connected by the relation $\beta = \phi(\alpha)$ we can generally find them as follows: The equations $f(x) = 0$ and $f\{\phi(x)\} = 0$ have a common root, namely α. Therefore $x - \alpha$ is a common factor of $f(x)$ and $f\{\phi(x)\}$, and may be found by the H.C.F. process.*

If however $f(x)$ and $f\{\phi(x)\}$ and identical, then the method fails.

▮ **EXAMPLE 5.** *Solve $2x^3 + x^2 - 7x - 6 = 0$, given that the difference between two of the roots is 3.*

$$
\begin{array}{r r r r}
2 & +1 & -7 & -6 \\
  & +6 & +21 & +42 \\
\hline
  & +7 & +14 & +36 \\
  & +6 & +39 & \\
\hline
  & +13 & +53 & \\
  & +6 & & \\
\hline
  & +19 & &
\end{array}
$$

Let the roots be α, $\alpha + 3$, β. Find the equation whose roots are $\alpha - 3$, α, $\beta - 3$. The reckoning on the right shows that this equation is

$$2x^3 + 19x^2 + 53x + 36 = 0.$$

Hence, $x - \alpha$ is a common factor of

$$2x^3 + 19x^2 + 53x + 36 \text{ and } 2x^3 + x^2 - 7x - 6.$$

Find the H.C.F. of these expressions; this proves to be $x + 1$, therefore $\alpha = -1$; and hence -1 and 2 are roots.

Again, the product of the three roots is 3; and the third remaining root is $-3/2$.

▩ **5. The Cubic Equation:** We take as the standard form

$$u \equiv ax^3 + 3bx^2 + 3cx + d = 0. \qquad \ldots \text{(A)}$$

If $x = y + h$, this becomes

$$ay^3 + 3By^2 + 3Cy + D = 0,$$

where $B = ah + b$, $C = ah^2 + 2bh + c$, $D = ah^3 + 3bh^2 + 3ch + d$.

If $h = -b/a$, then $B = 0$, and the equation is

$$y^3 + \frac{3H}{a^2} y + \frac{G}{a^3} = 0, \qquad \ldots \text{(B)}$$

where $H = ac - b^2$, $G = a^2d - 3abc + 2b^3$.

Or, if we write $z = ay = ax + b$, the equation is

$$z^3 + 3Hz + G = 0. \qquad \ldots \text{(C)}$$

If α, β, γ are the roots of (A), those of (B) are $\alpha + b/a$, $\beta + b/a$, $\gamma + b/a$, and the roots of (C) are $a\alpha + b$, $a\beta + b$, $a\gamma + b$.

▩ **6. The Biquadratic Equation:** We take as the standard form

$$u \equiv ax^4 + 4bx^3 + 6cx^2 + 4dx + e = 0. \qquad \ldots \text{(A)}$$

The function u is called a *quartic.* If $x = y - b/a$ the equation becomes

$$y^4 + \frac{6H}{a^2} y^2 + \frac{4G}{a^3} y + \frac{K}{a^4} = 0, \qquad \ldots \text{(B)}$$

where $H = ac - b^2$, $G = a^2d - 3abc + 2b^3$ (as for the cubic),

and $\qquad\qquad K = a^3e - 4a^2bd + 6ab^2c - 3b^4.$

If $z = ay = ax + b$, the equation is

$$z^4 + 6Hz^2 + 4Gz + K = 0. \qquad \ldots \text{(C)}$$

If α, β, γ, δ are the roots of (A), those of (B) are $\alpha + b/a$, etc., and those of (C) are $a\alpha + b$, etc.

EXERCISE X

TRANSFORMATION OF EQUATIONS

1. If α, β, γ are the roots of $2x^3 + 3x^2 - x - 1 = 0$, find the equations whose roots are
   (*i*) $1/2\alpha$, $1/2\beta$, $1/2\gamma$; (*ii*) $\alpha - 1$, $\beta - 1$, $\gamma - 1$; (*iii*) $1/(1 - \alpha)$, $1/(1 - \beta)$, $1/(1 - \gamma)$;
   (*iv*) $\alpha + 2$, $\beta + 2$, $\gamma + 2$; (*v*) α^2, β^2, γ^2.
   [(*iii*) should be deduced from (*ii*).]

2. If α, β, γ are the roots of $8x^3 - 4x^2 + 6x - 1 = 0$, find the equations whose roots are

   (*i*) $\alpha + \dfrac{1}{2}$, $\beta + \dfrac{1}{2}$, $\gamma + \dfrac{1}{2}$; (*ii*) $2\alpha + 1$, $2\beta + 1$, $2\gamma + 1$.

   [For (*ii*) use the result of (*i*).]

3. Solve $x^4 - 2x^3 - 3x^2 + 4x - 1 = 0$, given that the product of two roots is unity.
   [If α, α^{-1} are these roots, $(x - \alpha)(x - \alpha^{-1})$ is a common factor of
   $$x^4 - 2x^3 - 3x^2 + 4x - 1 \text{ and } x^4 - 4x^3 + 3x^2 + 2x - 1.]$$

4. Solve $4x^4 - 4x^3 - 13x^2 + 9x + 9 = 0$, given that the sum of two roots is zero.
   [If α, $-\alpha$ are these roots, $(x - \alpha)(x + \alpha)$ is a common factor of
   $$4x^4 - 4x^3 - 13x^2 + 9x + 9 \text{ and } 4x^4 + 4x^3 - 13x^2 - 9x + 9.]$$

5. Solve $4x^3 - 24x^2 + 23x + 18 = 0$, given that the roots are in arithmetical progression.

   [Let the roots be $\alpha - \delta$, α, $\alpha + \delta$, $\therefore 3\alpha = \dfrac{24}{4}$.]

6. Solve $6x^3 - 11x^2 - 3x + 2 = 0$, given that the roots are in harmonical progression.
   [Solve $2y^3 - 3y^2 - 11y + 6 = 0$, whose roots are in A.P.]

7. Solve $x^4 - 8x^3 + 14x^2 + 8x - 15 = 0$, whose roots are in A.P.
   [Let the roots be $\alpha - 3\delta$, $\alpha - \delta$, $\alpha + \delta$, $\alpha + 3\delta$.]

8. Solve $2x^4 - 15x^3 + 35x^2 - 30x + 8 = 0$, whose roots are in G.P.
   [Let the roots be $\alpha\rho^{-3}$, $\alpha\rho^{-1}$, $\alpha\rho$, $\alpha\rho^3$.]

9. Solve $x^3 - 7x^2 + 36 = 0$, given that the difference between two of the roots is 5.

10. Solve $4x^4 - 4x^3 - 25x^2 + x + 6 = 0$, given that the difference between two of the roots is unity.

11. Transform into equations lacking the second term:
    (*i*) $x^3 - 6x^2 + 4x - 7 = 0$;
    (*ii*) $x^4 + 8x^3 + x^2 - x - 10 = 0$.

12. Transform into equations with integral coefficients;

    (*i*) $x^3 - \dfrac{7}{3}x^2 + \dfrac{11}{36}x - \dfrac{25}{72} = 0$;

    (*ii*) $x^4 + \dfrac{5}{6}x^3 + \dfrac{5}{12}x^2 - \dfrac{7}{150}x - \dfrac{13}{900} = 0$.

13. If n is odd, prove that

$$\tan n\theta = (-1)^{\frac{1}{2}(n-1)} \tan \theta \tan\left(\theta + \frac{\pi}{n}\right) \tan\left(\theta + \frac{2\pi}{n}\right) \dots \tan\left(\theta + \frac{(n-1)\pi}{n}\right),$$

$$n \tan n\theta = \tan \theta + \tan \left(\theta + \frac{\pi}{n}\right) + \tan \left(\theta + \frac{2\pi}{n}\right) + \ldots + \tan \left(\theta + \frac{(n-1)\pi}{n}\right).$$

[Regard the equation of Exercise VII, 14, as giving $\tan \theta$ in terms of $\tan n\theta$.]

■ **7. Character and Position of the Roots of an Equation:** We say that the *character* of the roots is known when we know how many are real and how many are imaginary.

The *position* of a real root is its position on the scale of real numbers, and is determined roughly for a non-integral root by finding two consecutive integers between which the root lies. For a complete discussion, we require Sturm's Theorem (Ch. 28), but a good deal of infomation can be derived from the elementary theorems which follow.

■ **8. Some general Theorems:** Here we suppose that

$$f(x) = x^n + p_1 x^{n-1} + p_2 x^{n-2} + \ldots + p_n,$$

where $p_1, p_2, \ldots$ are real numbers.

If $l + \iota m$ is a root of $f(x) = 0$, then $l - \iota m$ is also a root (Ch. 5, 14) and $f(x)$ has the quadratic factor, $\{x - (l + \iota m)\} \{x - (l - \iota m)\} = (x - l)^2 + m^2$.

The last expression is positive for all real values of x. Thus if $\alpha, \beta, \ldots \kappa$ are the real roots of $f(x) = 0$, these being not necessarily all different,

$$f(x) = (x - \alpha)(x - \beta) \ldots (x - \kappa) \cdot \phi(x),$$

where $\phi(x)$ is positive for all real values of x. Also this expression in factors is *unique*.

Hence, if x varies, the sign of $f(x)$ can change only when x passes through a real root of $f(x) = 0$, and we draw the following conclusions.

(1) *If x is greater than any of the roots, $f(x)$ is positive. This is also the case when all the roots are imaginary.*

Whence it follows that *for sufficiently large values of x, any polynomial has the same sign as its highest term.*

(2) Let a and b be any real numbers: than

(i) If $f(a)$ and $f(b)$ have like signs, an even number of roots of $f(x) = 0$ lie between a and b, or else there is no root between a and b.

(ii) If $f(a)$ and $f(b)$ have unlike signs, an odd number of roots of $f(x) = 0$ lie between a and b.

(3) *(i) If $f(x) = 0$ is an equation of odd degree, it has at least one real root.*

(ii) If $f(x) = 0$ is of even degree and p_n is negative, the equation has at least one positive root and at least one negative root.

For, by (1), a positive number a can be found so large that $f(a)$ is positive and $f(-a)$ has the sign of $(-1)^n$.

Hence if n is odd, $f(a)$ and $f(-a)$ have unlike signs and at least one root lies between a and $-a$. If n is even and p_n negative, $f(a)$ and $f(-a)$ are positive and $f(0)$ is negative, so that at least one root lies between 0 and a and at least one between 0 and $-a$.

■ **9. Descartes' Rule of Signs:** In what follows, every equation is supposed to have a term independent of x, so that zero roots do not occur.

In considering the signs of the terms of a polynomial taken in order from left to right, we say that a *continuation*, or a *change, of sign* occurs at any particular term according as that term has the same sign as the preceding, or the opposite sign.

Thus $x^7 - 2x^5 - 3x^4 - 4x^3 + 5x^2 - 6x + 7$ has 2 continuations, and 4 changes of sign, the continuations occurring at the terms $-3x^4$, $-4x^3$, and the changes at $-2x^5$, $+5x^2$, $-6x$, $+7$.

The equation $a_0x^n + a_1x^{n-1} + a_2x^{n-2} + ... + a_n = 0$ is said to be *complete* when no coefficient is zero. If $a_r = 0$ we say that the corresponding term is *missing.*

In a complete equation: (i) *If x is changed into $-x$, a change of sign becomes a continuation and* vice versa.

(ii) *If μ is the number of changes, and μ' the number of continuations, of sign, then $\mu + \mu' = n$, where n is the degree of the equation.*

Descartes' Rule is as follows : *The equation $f(x) = 0$ cannot have more positive roots than $f(x)$ has changes of sign, or more negative roots than $f(-x)$ has changes of sign.*

To prove the first part, we shall show that if u is any polynomial and $v = u$ $(x - a)$, where a is positive, then v, when expanded, has at least one more change of sign than u.

First suppose that no term is missing in u and consider the following instance:

Signs of terms of u,

$$+ \quad - \quad - \quad - \quad + \quad + \quad - \quad +$$
$$- \quad + \quad + \quad + \quad - \quad - \quad + \quad -$$
$$\overline{}$$

Signs of terms of v, $\quad + \quad - \quad \pm \quad \pm \quad + \quad \pm \quad - \quad + \quad -$

where $\pm$ indicates that the sign may be $+$ or $-$, or that the corresponding term is zero.

In the diagram of corresponding signs, observe that

(*i*) If the *r*th sign of u is a continuation, the *r*th sign of v is ambiguous.

(*ii*) Unlike signs precede and follow a single ambiguity or a group of ambiguities.

(*iii*) A change of sign is introduced at the end of v.

On account of (*i*) and (*ii*), v has at least as many changes of sign as u, even in the most unfavourable case in which all the ambiguities are continuations: and on account of (*iii*) v has certainly one more change of sign than u.

That no changes of sign are lost on account of any terms which may be missing from u appears on considering such instances as

$$+ \quad + \quad 0 \quad 0 \quad - \qquad\qquad - \quad + \quad 0 \quad 0 \quad -$$
$$- \quad - \quad 0 \quad 0 \quad + \qquad\qquad + \quad - \quad 0 \quad 0 \quad +$$
$$\overline{} \qquad\qquad \overline{}$$
$$+ \quad \pm \quad - \quad 0 \quad - \quad + \qquad - \quad + \quad - \quad 0 \quad - \quad +$$

Thus v has at least one more change of sign than u.

Next let $f(x) = \phi(x) . (x - \alpha)(x - \beta) ...$ where α, β, ... are the positive roots of $f(x) = 0$. If $\phi(x)$ is multiplied in succession by $x - \alpha$, $x - \beta$, ..., each multiplication introduces at least one change of sign.

Hence $f(x)$ has at least as many changes of sign as $f(x) = 0$ has positive roots.

Again, the negative roots of $f(x) = 0$ are the positive roots of $f(-x) = 0$, with their signs changed. Hence the second part of the theorem follows from the first.

▰ **10. Corollaries:** Let n be the degree of $f(x)$, and

μ the number of changes of sign in $f(x)$,

μ' the number of changes of sign in $f(-x)$,

m the number of positive roots of $f(x) = 0$,

m' the number of negative roots of $f(x) = 0$;

then (i) *if* $\mu + \mu' < n$, *the equation* $f(x) = 0$ *has at least* $n - (\mu + \mu')$ *imaginary roots;* (ii) *if all the roots of* $f(x) = 0$ *are real, then* $m = \mu$ *and* $m' = \mu'$.

For by Descartes' rule, $m \le \mu$ and $m' \le \mu'$. Hence

$$m + m' \le \mu + \mu' \le n, \qquad\qquad ...(A)$$

and the number of imaginary roots $= n - (m + m') \ge n - (\mu + \mu')$.

Again, if all the roots are real $m + m' = n$, therefore by (A) $m + m' = \mu + \mu'$. Now $m \le \mu$; and, if $m < \mu$, it follows that $m' > \mu'$, which is impossible. Therefore $m = \mu$ and $m' = \mu'$.

▌ **EXAMPLE 1.** *If* q, r, s, *are positive, show that the equation*

$$f(x) = x^4 + qx^2 + rx - s = 0$$

has one positive, one negative and two imaginary roots.

By Art. 8, (3), $f(x) = 0$ has a positive and a negative root. Also $\mu = 1$ and $\mu' = 1$, therefore these are the only real roots.

▌ **EXAMPLE 2.** *Show that* $x^5 - 2x^2 + 7 = 0$ *has at least two imaginary roots.*

Here $\mu = 2$, $\mu' = 1$; therefore $\mu + \mu' = 3$. Hence there cannot be more than three real roots, namely, two positive and one negative.

Therefore the equation has at least two imaginary roots.

▰ **11. De Gua's Rule:** *If a group of* r *consecutive terms is missing from* $f(x) = 0$, *then*

(*i*) *if* r *is even, the equation has at least* r *imaginary roots;*

(*ii*) *if* r *is odd, there are at least* $r + 1$ *or at least* $r - 1$ *imaginary roots, according as the terms which immediately precede and follow the group have like or unlike signs.*

Suppose that the r terms between hx^m and kx^{m-r-1} are missing from $f(x)$.

Let $f(x) = hx^m + kx^{m-r-1} + \phi(x)$, and $f_1(x) = \psi(x) + \phi(x)$,

where $\psi(x) = hx^m + c_1 x^{m-1} + c_2 x^{m-2} + ... c_r x^{m-r} + kx^{m-r-1}$,

none of the set, c_1, c_2, ..., c_r, being zero.

The number of changes of sign in $\psi(x)$ + the number of changes of sign in $\psi(-x)$ is $r + 1$.

Let μ = Number of changes of sign in $hx^m + kx^{m-r-1}$,

μ' = Number of changes of sign in $h(-x)^m + k(-x)^{m-r-1}$.

Thus the total numbers of changes of sign in $f(x)$ and $f(-x)$ is less than the number in $f_1(x)$ and $f_1(-x)$ by $r + 1 - (\mu + \mu')$.

Hence $f(x) = 0$ has at least $r + 1 - (\mu + \mu')$ imaginary roots.

First suppose that r is even, then $(-x)^m$ and $(-x)^{m-r-1}$ have opposite signs, and

(i) if h, k have the same signs, $\mu = 0$ and $\mu' = 1$;

(ii) if h, k have unlike signs, $\mu = 1$ and $\mu' = 0$, and in both cases

$$r + 1 - (\mu + \mu') = r.$$

If r is odd, $(-x)^m$ and $(-x)^{m-r-1}$ have the same sign, and

(i) if h, k have the same sign, $\mu = 0$, $\mu' = 0$ and $r + 1 - (\mu + \mu') = r + 1$.

(ii) if h, k have unlike signs, $\mu = 1$, $\mu' = 1$ and $r + 1 - (\mu + \mu') = r - 1$.

This completes the proof.

■ **EXAMPLE:** *If $H = ac - b^2 > 0$, the equation*

$$(a, b, c, \dots k)\,(x, 1)^n = 0$$

has at least two imaginary roots.

For by the substitution $x = y - b/a$, the equation becomes $x^n + \dfrac{nH}{a^2}\,x^{n-2} + \dots = 0$, and by De Gua's rule this equation has at least two imaginary roots if $H > 0$.

■ **12. Limits to the Roots of an Equation:**

(1) In this article, whatever is said about the roots of an equation refers to the *real* roots only. Also, the equation is supposed to be written with its first term positive.

In searching for the roots of an equation, it is advisable to begin by finding two numbers between which the real roots lie. If h, l are two such numbers and $h > l$, then h is called an *upper limit* and l a *lower limit to* the roots.

(2) Upper Limits: Any number h is an upper limit to the roots of $f(x) = 0$, provided that $f(x) > 0$ when $x \geq h$.

Method of grouping terms. The process consists in arranging the terms of the equation $f(x) = 0$ in groups, so that we can find by inspection a number h such that the sum of the terms in each group is ≥ 0 for $x \geq h$. To distribute the terms conveniently, it is often advisable to multiply $f(x)$ by some positive integer. With a little ingenuity, this method can be made to yield quite close limits, as in the examples at the end of this article. Other methods are given in the next exercise, and in Ch. 28.

(3) Lower Limits:

(i) *If h' is an upper limit to the roots of $f(-x) = 0$, then $-h'$ is a lower limit to the roots of $f(x) = 0$.*

For if α is the least root of $f(x) = 0$, then $-\alpha$ is the greatest root of $(-x) = 0$. Hence, $h' > -\alpha$, and therefore $-h' < \alpha$.

(ii) *If h'' is an upper limit to the positive roots of $f(1/x) = 0$, then $-1/h''$ is a lower limit to the positive roots of $f(x) = 0$.*

For if α' is the least positive root of $f(x) = 0$, then $1/\alpha'$ is the greatest positive root of $f(1/x) = 0$. Hence $h'' > 1/\alpha'$ and so $1/h'' < \alpha'$.

■ **EXAMPLE 1.** *Find an upper and a lower limit to the positive roots of*

$$f(x) = x^3 - 10x^2 - 11x - 100 = 0.$$

Here $6f(x) = 6x^3 - 60x^2 - 66x - 600 = 4x^2(x - 15) + x(x^2 - 66) + (x^3 - 600)$.

Thus $f(x) > 0$ for $x \geq 15$ and 15 is an upper limit to the roots.

Again, putting $x = 1/y$, the equation becomes $\phi(y) = 100y^3 + 11y^2 + 10y - 1 = 0$.
Now $\phi(y) = 5y^2(20y - 1) + (16y^2 + 10y - 1)$.

The greater root of $16y^2 + 10y - 1 = 0$ is $\dfrac{1.4\ ...}{16}$; and if y has this value, $20y - 1 > 0$.

Hence $\phi(y) > 0$ for $y \geq \dfrac{1.5}{16}$, i.e. for $x = \dfrac{1}{y} \leq \dfrac{16}{1.5}$, i.e. for $x = 10.6$.

Thus 10.6 is a lower limit to the positive roots.

EXAMPLE 2. *Find an upper and a lower limit to the roots of*
$$f(x) = 3x^4 - 61x^3 + 127x^2 + 220x - 520 = 0.$$
We have $f(x) = x^2(3x^2 - 61x + 127) + (220x - 520)$.

The greater root of $3x^2 - 61x + 127 = 0$ is $17.9\ ...$. Thus 18 is an upper limit to the roots.

Again, $f(-x) = (3x^4 - 520) + x(61x^2 + 127x - 220)$, and each group of terms is positive if $x \geq 4$. Hence -4 is a lower limit to the roots of $f(x) = 0$.

■ **13. To find the Rational Roots of an Equation:** Any such roots which may exist can be found by using the following theorem: *If $px - q$ is a factor of $ax^n + bx^{n-1} + ... + hx + k$, where $a, b, ... h, k$ are integers or zero and p, q are integers prime to one another, then p is a factor of a, and q is a factor of k.*

For, denoting the polynomial by $f(x)$, if $px - q$ is a factor of $f(x)$, then $f(q/p) = 0$, and therefore
$$aq^n + bq^{n-1}p + ... + hqp^{n-1} + kp^n = 0.$$
Hence $aq^n = p \times$ an integer, so that aq^n is divisible by p. Now p is prime to q, therefore p is prime to q^n. Hence a is divisible by p. Similarly k is divisible by q.

Hence *the rational roots of $ax^n + bx^{n-1} + ... + k = 0$ must be included among the values of $\pm q/p$, where p is prime to q, p is a factor of a and q a factor of k.*

We can test the values of $\pm q/p$ by synthetic division, or we may use the method of divisors given below.

EXAMPLE: *Search for rational roots of $f(x) = 2x^3 - 5x^2 + 5x - 3 = 0$.*

We have $f(x) = 2x^2\left(x - \dfrac{5}{2}\right) + 5\left(x - \dfrac{3}{5}\right)$, therefore $\dfrac{5}{2}$ is an upper limit of the roots. Also $f(-x) = 0$ has no positive roots, therefore $f(x) = 0$ has no negative roots.

If q/p is a root, $p = 1$ or 2, $q = 1$ or 3, and the only values of q/p which lie between 0 and $\dfrac{5}{2}$ are $1, \dfrac{1}{2}, \dfrac{3}{2}$. Testing these by synthetic division, we find that $\dfrac{3}{2}$ is a root; and it is the only rational root.

■ **14. Newton's Method of Divisors:** Let the given equation be tranformed into
$$f(x) = x^n + p_1 x^{n-1} + p_2 x^{n-2} + ... + p_n = 0, \qquad \qquad ...(A)$$
where $p_1, p_2, ...$ are the least possible integers. (See Art. 4, (6).)

By the last theorm, if α is a rational root of (A), then α is an integer and p_n is divisible by α.

Let h be any factor of p_n. If $x - h$ is a factor of $f(x)$, and we divide $f(x)$ by $x - h$, the usual reckoning is as follows:

$$
\begin{array}{l}
1-h)\quad 1+p_1\quad +p_2\ +\dots\quad +p_{n-2}\quad +p_{n-1}\quad\ +p_n\\[4pt]
\qquad\quad\ h\qquad\quad +q_1 h\qquad +q_{n-3}h\ \ +q_{n-2}h\ \ +q_{n-1}h\\[2pt]
\hline\\[-6pt]
\qquad\quad q_1\qquad\quad q_2\qquad\quad q_{n-2}\qquad\ q_{n-1}\qquad\ \ 0
\end{array}
$$

where $q_1 = p_1 + h,\ q_2 = p_2 + q_1 h,\ \dots\ p_{n-1} + q_{n-2}h = q_{n-1},\ p_n + q_{n-1}h = 0$.

Newton's method consists in performing these operations in the reverse order, thus

$$
\begin{array}{l}
p_n + p_{n-1}\qquad + p_{n-2}\ +\dots + p_2\quad + p_1\quad +1\\[4pt]
\ -q_{n-1}\qquad\ -q_{n-2}\qquad\quad\ -q_2\quad -q_1\quad -1\\[2pt]
\hline\\[-6pt]
-q_{n-2}h\quad\ -q_{n-3}h\qquad\quad -q_1 h\quad -h\qquad 0
\end{array}
$$

Here p_n is divided by h, the quotient is $-q_{n-1}$; this is added to p_{n-1}, giving $-q_{n-2}h$. Dividing this by h, we get $-q_{n-2}$, which is added to p_{n-2}, and so on.

If any q in the process is not an integer, $f(x)$ is not divisible by $x - h$ and the reckoning need not be continued.

Herein lies the advantage of Newton's method.

Further, the last number in the bottom line must be 0.

To lessen the number of trials, choose any number a, then if $x - h$ is a factor of $f(x)$, $f(a)$ is divisible by $a - h$. We generally take $a = \pm 1$.

■ **EXAMPLE:** *Find the rational roots of $f(x) = x^4 - 39x^2 + 46x - 168 = 0$.*

Since
$$f(x) = x^2 \cdot (x^2 - 39) + 2(23x - 84),$$
and $4f(-x) = 2x^2(x^2 - 78) + (x^4 - 672) + x(x^3 - 184)$,

it follows that all the real roots lie between -9 and 7.

If h is a rational root, it is an integer and $\pm h$ is a factor of $168 = 2^3 \cdot 3 \cdot 7$.

The possible values of h are therefore
$$\pm 1,\quad \pm 2,\quad \pm 3,\quad \pm 4,\quad \pm 6,\quad -7,\quad -8. \qquad\qquad \dots(A)$$

A number of these can be excluded at once, for
$$f(1) = -160 = -2^5 \cdot 5;\ f(-1) = -252 = -2^2 \cdot 3^2 \cdot 7,$$
so that ± 1 are not roots.

Also if h is a root, $1 - h$ is a factor of 160, thus we can exclude $-2, 4, -6, -8$.

Again, $-1 - h$ must be a factor of 252, and so we can exclude 4. The remaining numbers are
$$2,\quad +3,\quad -3,\quad +6,\quad -7.$$

Testing these as below by Newton's method, we find that 6 and -7 are the only rational roots.

$$
\begin{array}{llllll}
1-2)\quad -168 & +46 & -39 & +0 & +1, & \quad 1-3)\quad -168 \quad +46\\[4pt]
\qquad\quad\ \underline{-84} & \underline{-19} & \underline{-29} & & & \qquad\qquad\qquad\qquad\ \underline{-56}\\[4pt]
\qquad\quad\ -38 & -58 & -29 & & & \qquad\qquad\qquad\qquad\ -10*
\end{array}
$$

---

* The reckoning stops at these stages because 10 and 73 are not divisible by 3.

$$1+3) \quad -168 \quad +46 \quad -39 \qquad\qquad 1-6) \quad -168 \quad +46 \quad -39 \quad +0 \quad +1$$
$$\underline{\quad +56 \quad -34} \qquad\qquad\qquad\qquad \underline{\quad -28 \quad +3 \quad -6 \quad -1}$$
$$\quad 102 \quad -73^{*} \qquad\qquad\qquad\qquad \quad +18 \quad -36 \quad -6 \quad 0$$

$$1+7) \quad -168 \quad +46 \quad -39 \quad +0 \quad +1$$
$$\underline{\quad +24 \quad -10 \quad +7 \quad -1}$$
$$\quad 70 \quad -49 \quad +7 \quad 0$$

EXERCISE XI

REAL ROOTS

1. Use Descartes' rule of signs to show that:

   (*i*) If q is positive, $x^3 + qx + r = 0$ has only one real root.

   (*ii*) $x^7 - 3x^4 + 2x^3 - 1 = 0$ has at least four imaginary roots.

2. Show that the roots of $(a - x)(b - x) - h^2 = 0$ are real and are separated by a and by b.

   [Denoting the left-hand side by $f(x)$, we have
   $$f(\infty) > 0, \; f(a) = f(b) < 0, \; f(-\infty) > 0.]$$

   By grouping terms, find an upper and a lower limit to the roots of the equations in Exs. 3 - 9.

3. $x^4 - 3x^3 - 2x^2 + 7x + 3 = 0$.

4. $x^5 - 10x^4 - 5x^3 + 6x^2 - 11x - 350 = 0$.

   [For an upper limit in Ex. 4, group thus: $x^3(x^2 - 10x - 5) + (6x^2 - 11x - 350)$.

   For a lower limit, write $x = -y$ and group thus:
   $$y(y^4 - 11) + y^3(9y - 5) + (y^4 - 6y^2 + 350).]$$

5. $x^4 + 4x^3 - 11x^2 - 9x - 50 = 0$.

6. $x^4 - 2x^3 + 3x^2 - 5x + 1 = 0$.

7. $2x^3 - 11x^2 - 10x - 1 = 0$.

   [Group thus: $x(2x^2 - 11x - 11) + (x - 1).]$

8. $x^5 + x^4 - 6x^3 - 8x^2 - 15x - 10 = 0$.

   [Group thus: $x(x^4 - 6x^2 - 15) + (x^4 - 8x^2 - 10).]$

9. $x^5 - 3x^4 - 24x^3 + 95x^2 - 46x - 101 = 0$.

10. For the equation $f(x) = x^n + p_1 x^{n-1} + p_2 x^{n-2} + \dots + p_n = 0$, if the numerically greatest negative coefficient is equal to $-p$, then $p + 1$ is an upper limit to the roots.

    [Let $x > 1$, then $f(x) > 0$ if $x^n > p(x^{n-1} + x^{n-2} + \dots + 1)$, *i.e.* if $x^n > p \cdot \dfrac{x^n - 1}{x - 1}$,

    This holds if $x^n > p \cdot \dfrac{x^n}{x - 1}$, *i.e.* if $x - 1 > p.]$

11. For the equation $f(x) = x^n + p_1 x^{n-1} + p_2 x^{n-2} + \dots + p_n = 0$, if the numerically greatest negative coefficient is equal to $-p$ and the first negative coefficient is p_r, then $\sqrt[r]{p} + 1$ is an upper limit to the roots.

[Let $x > 1$, then $f(x) > 0$ if $x^n > p \, (x^{n-r} + x^{n-r-1} + \ldots + 1)$, i.e. if $x^n > p \cdot \dfrac{x^{n-r+1} - 1}{x - 1}$.

This holds if $x^n > \dfrac{p x^{n-r+1}}{x - 1}$, i.e. if $(x - 1)\, x^{r-1} > p$, which holds if $(x - 1)^r > p$, i.e. if $x > \sqrt[r]{p} + 1$.]

12. If the rule in Ex. 11 is applied to the equations in Exs. 3–7, show that the limits are (1) $4, -4$; (2) $351, -5$; (3) $9, -51$; (4) $6, -1$; (5) $7, -6$.

13. By using Ex. 11, find an upper limit to the roots of $x^7 - 100x^2 - 237 = 0$.

Find the rational roots of the equations in Exs. 14 – 19.

14. $x^3 - 9x^2 + 22x - 24 = 0$.

15. $x^3 - 5x^2 - 18x + 72 = 0$.

16. $3x^3 - 2x^2 - 6x + 4 = 0$.

17. $4x^4 - 36x^3 + 45x^2 + 54x - 81 = 0$.

18. $6x^4 - 25x^3 + 26x^2 + 4x - 8 = 0$.

19. $6x^4 + 53x^3 - 95x^2 - 25x + 42 = 0$.

▮ **15. Symmetric Functions of the Roots of an Equation:** Let $\alpha, \beta, \gamma, \delta, \varepsilon, \ldots$ be the roots of

$$x^n + p_1 x^{n-1} + p_2 x^{n-2} + \ldots + p_n = 0, \qquad \ldots\text{(A)}$$

then by Art. 2, $\qquad \Sigma\alpha = -p_1, \ \Sigma\alpha\beta = p_2, \ \Sigma\alpha\beta\gamma = -p_3,$ etc. $\qquad \ldots\text{(B)}$

It will be shown that these equations can be used to express any symmetric function of the roots in terms of the coefficients.

Functions of the type $\Sigma\alpha^a \beta^b \gamma^c \ldots$, where $a, b, c \ldots$ are positive integers, will be called *elementary* functions. These can be calculated in order by a process of multiplication, and any symmetric function of the roots can be expressed in terms of them.

▮ **EXAMPLE:** *For equation* (A) *find the values of* $\Sigma\alpha^2\beta\gamma\delta$ *and* $\Sigma\alpha^2\beta^2\gamma$.

(*i*) The product $\Sigma\alpha \,.\, \Sigma\alpha\beta\gamma\delta$ consists of terms of the types $\alpha^2\beta\gamma\delta$, $\alpha\beta\gamma\delta\varepsilon$. The first of these occurs *once* in the product. The second occurs *five* times, namely as the product of any one of $\alpha, \beta, \gamma, \delta, \varepsilon$ and a term from $\Sigma\alpha\beta\gamma\delta$. Therefore

$$\Sigma\alpha \,.\, \Sigma\alpha\beta\gamma\delta = \Sigma\alpha^2\beta\gamma\delta + 5\Sigma\alpha\beta\gamma\delta\varepsilon, \text{ and } \Sigma\alpha^2\beta\gamma\delta = -p_1 p_4 + 5 p_5.$$

(*ii*) Consider the product $\Sigma\alpha\beta \,.\, \Sigma\alpha\beta\gamma$. This consists of terms of the types $\alpha^2\beta^2\gamma$, $\alpha^2\beta\gamma\delta$, $\alpha\beta\gamma\delta\varepsilon$. The first of these occurs *once*. The term $\alpha^2\beta\gamma\delta$ occurs *three* times, namely, as each of the products $(\alpha\beta)(\alpha\gamma\delta)$, $(\alpha\gamma)(\alpha\beta\delta)$, $(\alpha\delta)(\alpha\beta\gamma)$. The term $\alpha\beta\gamma\delta\varepsilon$ occurs *ten* times, for we can select two out of the five, $\alpha, \beta, \gamma, \delta, \varepsilon$, in 10 ways.

Therefore $\Sigma\alpha\beta \,.\, \Sigma\alpha\beta\gamma = \Sigma\alpha^2\beta^2\gamma + 3\Sigma\alpha^2\beta\gamma\delta + 10\Sigma\alpha\beta\gamma\delta\varepsilon.$ $\qquad \ldots\text{(C)}$

Using the last result and equations (B), we find that

$$\Sigma\alpha^2\beta^2\gamma = 3p_1 p_4 - p_2 p_3 - 5 p_5. \qquad \ldots\text{(D)}$$

Note: These results can be tested by putting $\alpha = \beta = \gamma = \ldots = 1$. In this way (C) becomes

$$C_2^n \,.\, C_3^n = 3\, C_3^n + 12\, C_4^n + 10\, C_5^n.$$

For reference, we give the following results, indicating briefly the process of reckoning:

If α, β, γ, ... are the roots of $x^n + p_1 x^{n-1} + p_2 x^{n-2} + ... + p_n = 0$,   then

$$\Sigma\alpha^2 = (\Sigma\alpha)^2 - 2\Sigma\alpha\beta = p_1^2 - 2p_2, \quad \Sigma\alpha^2\beta = \Sigma\alpha \ . \ \Sigma\alpha\beta - 3\Sigma\alpha\beta\gamma = 3p_3 - p_1 p_2,$$

$$\Sigma\alpha^3 = \Sigma\alpha^2 . \Sigma\alpha - \Sigma\alpha^2\beta = -p_1^3 + 3p_1 p_2 - 3p_3,$$

$$\Sigma\alpha^2\beta\gamma = \Sigma\alpha \ . \ \Sigma\alpha\beta\gamma - 4\Sigma\alpha\beta\gamma\delta = p_1 p_3 - 4p_4,$$

$$\Sigma\alpha^2\beta^2 = (\Sigma.\alpha\beta)^2 - 2\Sigma\alpha^2\beta\gamma - 6\Sigma\alpha\beta\gamma\delta = p_2^2 - 2p_1 p_3 + 2p_4,$$

$$\Sigma\alpha^3\beta = \Sigma\alpha\beta.\Sigma\alpha^2 - \Sigma\alpha^2\beta\gamma = p_1^2 p_2 - 2p_2^2 - p_1 p_3 + 4p_4,$$

$$\Sigma\alpha^4 = (\Sigma\alpha^2)^2 - 2\Sigma\alpha^2\beta^2 = p_1^4 - 4p_1^2 p_2 + 4p_1 p_3 + 2p_2^2 - 4p_4.$$

Here the elementary functions of the second, third and fourth degrees in α, β, γ, ... are calculated in order, and obviously the process can be continued so as to find any function of this kind. It is important to notice that *the degree n of the equation does not occur in the reckoning.*

▨ **16. Symmetric Functions involving only the Differences of the Roots of** $f(x) = 0$:

These are unaltered if the roots are diminished by any number h. Hence they may be calculated by using the equation $f(y + h) = 0$, where h is chosen so as to remove the second term.

A useful check is provided by the theorem in Ex. 1 below. The functions in Exs. 2, 3 are important.

▌ **EXAMPLE 1.** *Let v be any symmetric function of the differences of the roots of*
$$(a_0, a_1, a_2, ...a_n)(x, 1)^n = 0,$$
then if v is expressed in terms of a_0, a_1, ...a_2, ..., the sum of the numerical coefficients is zero.

The sum of the numerical coefficients is obtained by putting $a_0 = a_1 = a_2 = ... = 1$ in v. But in this case the given equation becomes $(x + 1)^n$, every root of which is -1, and since v involves only the differences of the roots, in this case, $v = 0$.

▌ **EXAMPLE 2.** *If α, β, γ are the roots of $ax^3 + 3bx^2 + 3cx + d = 0$, find the values of*
  (i) $(\alpha^2 + 1)(\beta^2 + 1)(\gamma^2 + 1)$,
  (ii) $(\beta - \gamma)(\gamma - \alpha) + (\gamma - \alpha)(\alpha - \beta) + (\alpha - \beta)(\beta - \gamma)$.
  (i) We have $a(x - \alpha)(x - \beta)(x - \gamma) = ax^3 + 3bx^2 + 3cx + d$.
   Substituting $+\iota$ and $-\iota$ for x in this identity, we find that
   $$a(\alpha - \iota)(\beta - \iota)(\gamma - \iota) = \iota(a - 3c) + (3b - d),$$
   $$a(\alpha + \iota)(\beta + \iota)(\gamma + \iota) = -\iota(a - 3c) + (3b - d),$$
  therefore $\quad a^2(\alpha^2 + 1)(\beta^2 + 1)(\gamma^2 + 1) = (a - 3c)^2 + (3b - d)^2.$

  (ii) $\Sigma(\beta - \gamma)(\gamma - \alpha) = \Sigma\beta\gamma - \Sigma\alpha^2 = 3\Sigma\beta\gamma - (\Sigma\alpha)^2 = \dfrac{9}{a^2}(ac - b^2).$

▌ **EXAMPLE 3.** *If α, β, γ, δ are the roots of*
$$ax^4 + 4bx^3 + 6cx^2 + 4dx + c = 0, \qquad \qquad ...(A)$$
find the values of
  (i) $\Sigma(\alpha - \beta)^2$ (ii) $\Sigma(\alpha - \beta)^2 \gamma^2 \delta^2$,
  (iii) $(\beta + \gamma - \alpha - \delta)(\gamma + \alpha - \beta - \delta)(\alpha + \beta - \gamma - \delta)$.

(*i*) $\Sigma (\alpha - \beta)^2 = 3\Sigma\alpha^2 - 2\Sigma\alpha\beta = 3(\Sigma\alpha)^2 - 8\Sigma\alpha\beta = \dfrac{48}{a^2}(b^2 - ac)$.

(*ii*) The equation whose roots are $\dfrac{1}{\alpha}, \dfrac{1}{\beta}, \dfrac{1}{\gamma}, \dfrac{1}{\delta}$ is obtained by interchanging *a* and *e*, *b* and *d*, thus

$$\Sigma(\alpha - \beta)^2 \gamma^2 \delta^2 = \alpha^2\beta^2\gamma^2\delta^2. \ \Sigma\left(\frac{1}{\alpha} - \frac{1}{\beta}\right)^2$$

$$= \frac{e^2}{a^2} \cdot \frac{48}{e^2}(d^2 - ec) = \frac{48}{a^2}(d^2 - ce).$$

(*iii*) Denote the function by *v*. Since *v* is a function of the differences of α, β, γ, δ, its value may be found by using the equation

$$y^4 + \frac{6H}{a^2}y^2 + \frac{4G}{a^3}y + \frac{K}{a^4} = 0, \qquad \qquad ...(B)$$

obtained by the substitution $x = y - b/a$.

Let α', β', γ', δ' be the roots of (B). Then since $\Sigma\alpha' = 0$,

$$\beta + \gamma - \alpha - \delta = \beta' + \gamma' - \alpha' - \delta' = -2(\delta' + \alpha'),$$

and $\qquad -v/8 = (\delta' + \alpha')(\delta' + \beta')(\delta' + \gamma'),$

$$= \delta'^2 + \delta'^2(\alpha' + \beta' + \gamma') + \delta'(\beta'\gamma' + \gamma'\alpha' + \alpha'\beta') + \alpha'\beta'\gamma'$$

$$= \delta'^2 . \Sigma\alpha' + \Sigma\alpha'\beta'\gamma';$$

$\therefore \qquad v = -8\Sigma\alpha'\beta'\gamma' = 32G/a^3.$

■ **EXAMPLE 4.** *If* $\alpha + \beta + \gamma + \delta = 0$, *prove that* $\alpha^5 + \beta^5 + \gamma^5 + \delta^5 = -5\Sigma\alpha\beta.\Sigma\alpha\beta\gamma$.

Let α, β, γ, δ be the roots of $x^4 + p_3x^2 + p_3x + p_4 = 0$, and let $s_r = \Sigma\alpha^r$, then by Art. 15, $s_1 = 0$, $s_2 = -2p_2$, $s_3 = -3p_3$. Also $\alpha^5 + p_2\alpha^3 + p_3\alpha^2 + p_4\alpha = 0$, with three similar equations. Hence by addition,

$$s_5 + p_2s_3 + p_3s_2 + p_4s_1 = 0;$$

$\therefore s_5 = -p_2(-3p_3) - p_3(-2p_2) = 5p_2p_3 = -5\Sigma\alpha\beta.\Sigma\alpha\beta\gamma.$

▨ **17. Equations whose Roots are Symmetric Functions of α, β, γ :** The following are typical examples:

■ **EXAMPLE 1.** *If* α, β, γ *are the roots of* $x^3 + qx + r = 0$, *find the equation whose roots are* $(\beta - \gamma)^2$, $(\gamma - \alpha)^2$, $(\alpha - \beta)^2$.

Let $z = (\beta - \gamma)^2$, then since $\Sigma\alpha = 0$ and $\alpha\beta\gamma = -r$,

$$z = (\beta + \gamma)^2 - 4\beta\gamma = \alpha^2 + 4r/\alpha.$$

Hence the required equation can be found by eliminating α from

$$\alpha^3 + q\alpha + r = 0, \quad \alpha^3 - z\alpha + 4r = 0.$$

By subtraction, $(z + q)\alpha = 3r.$

Substituting for α in the first equation and simplifying, we have

$$(z + q)^3 + 3q(z + q)^2 + 27r^2 = 0,$$

which on expansion becomes $z^3 + 6qz^2 + 9q^2z + 4q^3 + 27r^2 = 0.$

The artifice employed in the next example is often useful.

EXAMPLE 2. (*i*) *Find the condition that the sum of two roots* α, β *of*
$$x^4 + p_1 x^3 + p_2 x^2 + p_3 x + p_4 = 0 \qquad \text{...(A)}$$
may be zero.

(*ii*) *Use the result to find the equation whose roots are the six values of* $\frac{1}{2}(\alpha + \beta)$, *where* α, β *are any two roots of*
$$ax^4 + 4bx^3 + 6cx^2 + 4dx + e = 0. \qquad \text{...(B)}$$

(*i*) Since α and $-\alpha$ are roots of (A),
$$\alpha^4 + p_1\alpha^3 + p_2\alpha^2 + p_3\alpha + p_4 = 0,$$
$$\alpha^4 - p_1\alpha^3 + p_2\alpha^2 - p_3\alpha + p_4 = 0;$$
$$\therefore \qquad \alpha^4 + p_2\alpha^2 + p_4 = 0 \quad \text{and} \quad \alpha(p_1\alpha^2 + p_3) = 0;$$

Now $\alpha \neq 0$ unless $p_4 = 0$, therefore $p_1^2 p_4 - p_1 p_2 p_3 + p_3^2 = 0$,   ...(C)
which is the required condition.

(*ii*) Let $z = \frac{1}{2}(\alpha + \beta)$, then we have $(\alpha - z) + (\beta - z) = 0$.

If then we diminish the roots of (B) by z, writing $x = y + z$, the sum of two roots of the resulting equation in y will be zero. This equation is
$$ay^4 + 4By^3 + 6Cy^2 + 4Dy + E = 0, \qquad \text{...(D)}$$
where
$$B = az + b, \quad C = az^2 + 2bz + c,$$
$$D = (a, b, c, d,)(z, 1)^3, \qquad E = (a, b, c, d, e,)(z, 1)^4.$$

For equation (D), the condition (C) gives
$$B^2 E - 6BCD + aD^2 = 0,$$
an equation of the sixth degree in z, whose roots are the values of $\frac{1}{2}(\alpha + \beta)$.

EXERCISE XII

SYMMETRIC FUNCTIONS OF ROOTS

If α, β, γ are the roots of $ax^3 + 3bx^2 + 3cx + d = 0$ and
$$H = ac - b^2, \quad G = a^2 d - 3abc + 2b^3,$$
prove that:

1. $a^3 \Sigma\alpha^3 = -3(a^2 d - 9abc + 9b^3)$.

2. $*a^3 \Sigma\beta^3\gamma^3 = 3(d^2 a - 9dcb + 9c^3)$.

3. $a^4 (\alpha\beta^2 + \beta\gamma^2 + \gamma\alpha^2)(\alpha\gamma^2 + \beta\alpha^2 + \gamma\beta^2) = a^4 (3\alpha^2\beta^2\gamma^2 + \alpha\beta\gamma\Sigma\alpha^3 + \Sigma\beta^3\gamma^3)$
$$= 9(a^2 d^2 - 6abcd + 3ac^3 + 3b^3 d).$$

4. $a^2 \{(\beta - \gamma)^2 + (\gamma - \alpha)^2 + (\alpha - \beta)^2\} = 18(b^2 - ac)$.

5. $* a^2 \{\alpha^2(\beta - \gamma)^2 + \beta^2(\gamma - \alpha)^2 + \gamma^2(\alpha - \beta)^2\} = 18(c^2 - db)$.

* Explain how to derive (2) from (1) and (5) from (4).

6. $a^2 \{\alpha(\beta - \gamma)^2 + \beta(\gamma - \alpha)^2 + \gamma(\alpha - \beta)^2\} = 9(ad - bc)$.

7. $a^4 \{(\beta - \gamma)^4 + (y - \alpha)^4 + (\alpha - \beta)^4\} = 162 H^2$.

8. $a^3 (\beta + \gamma - \alpha)(\gamma + \alpha - \beta)(\alpha + \beta - \gamma) = 18 a^2 d - 36abc + 27b^3$.

9. $a^3 (2\alpha - \beta - \gamma)(2\beta - \gamma - \alpha)(2\gamma - \alpha - \beta) = -27G$.

10.   (*i*) The condition that α, β, γ are in A.P. is $G = 0$.

  (*ii*) The condition that they are in G.P. is $ac^3 = b^3 d$.

  (*iii*) The condition that they are in H.P. is $d^2 a - 3dcb + 2c^3 = 0$.

[(*i*) If $2\alpha = \beta + \gamma$, then $\alpha = -b/a$. (*ii*) Eliminate x between $u = 0$ and $ax^3 + d = 0$. (*iii*) Put $x = 1/y$ and use condition (*i*).]

11. The condition that α, β, γ may be connected by the equation

$$(\alpha - h)(\beta - h) = (\gamma - h)^2$$

is $aC^3 = B^3 D$, where $B = ah + b$, $C = ah^2 + 2bh + c$, $D = ah^3 + 3bh^2 + 3ch + d$.

12. The equation whose roots are

$$\frac{\beta\gamma - \alpha^2}{\beta + \gamma - 2\alpha}, \ \frac{\gamma\alpha - \beta^2}{\gamma + \alpha - 2\beta}, \ \frac{\alpha\beta - \gamma^2}{\alpha + \beta - 2\gamma}$$

is $a(ax^2 + 2bx + c)^3 = (ax^3 + 3bx^2 + 3cx + d)(ax + b)^3$.

Show that this reduces to a cubic equation.   [Use Ex. 11.]

13. If α, β, γ are the roots of $x^3 + qx + r = 0$, prove that the equation whose roots are

$$\frac{\beta}{\gamma} + \frac{\gamma}{\beta}, \ \frac{\gamma}{\alpha} + \frac{\alpha}{\gamma}, \ \frac{\alpha}{\beta} + \frac{\beta}{\alpha} \ \text{is} \ r^2 (z + 1)^3 + q^3 (z + 1) + q^3 = 0.$$

[Let $z = \dfrac{\beta}{\gamma} + \dfrac{\gamma}{\beta}$. Prove that $\alpha^3 + 2q\alpha - rz = 0$. Eliminate α between this equation and

$\alpha^3 + q\alpha + r = 0$.].

14. If α, β are any two roots of $x^3 + qx + r = 0$, the equation whose roots are the six values of α/β is

$$r^2 (z^2 + z + 1)^3 + q^3 z^2 (z + 1)^2 = 0.$$

[Let $z = \dfrac{\alpha}{\beta}$; $\therefore \ \alpha = \beta z$, hence the required equation is got by eliminating β from

$$\beta^3 + q\beta + r = 0 \ \text{and} \ z^2\beta^3 + qz\beta + r = 0.]$$

15. If $a + b + c = 0$ show that

  (*i*) $a^5 + b^5 + c^5 = 5abc\,(bc + ca + ab)$;

  (*ii*) $a^7 + b^7 + c^7 = 7abc\,(bc + ca + ab)^2$.

[Let a, b, c be the roots of $x^3 + qx + r = 0$.]

16. Use the last example to show that

$$(x + a)^5 - x^5 - a^5 = 5ax\,(x + a)\,(x^2 + ax + a^2),$$
$$(x + a)^7 - x^7 - a^7 = 7ax(x + a)\,(x^2 + ax + a^2)^2.$$

17. Solve $(x + a + b)^5 - x^5 - a^5 - b^5 = 0$.

[Observe that $x + a + b + (-x) + (-a) + (-b) = 0$, and use Ex. 4, p. 94.]

18. If α, β, γ, δ are the roots of $x^4 + p_2 x^2 + p_3 x + p_4 = 0$, prove that

  (*i*) $\Sigma\alpha^4 = 2(p_2^2 - 2p_4)$,   (*ii*) $\Sigma\alpha^7 = -7p_3(p_2^2 - p_4)$.

If α, β, γ, δ are the roots of $ax^4 + 4bx^3 + 6cx^2 + 4dx + e = 0$ and $H = ac - b^2$, $G = a^2 d - 3abc + 2b^3$, prove that:

19. $a^2 \Sigma\alpha^2 = 4(4b^2 - 3ac)$.

20. $*a^2 \Sigma\alpha^2\beta^2\gamma^2 = 4(4d^2 - 3ec)$.

21. $a^2 \Sigma\alpha^2\beta = 12(ad - 2bc)$.

22. $*a^2 \Sigma\alpha\beta^2\gamma^2 = 12(eb^2 - 2dc)$.

*Note that (20) follows from (19) and (22) from (21).

23. $a^2 \Sigma(\alpha - \beta)^2 (\gamma^2 + \gamma\delta + \delta^2) = 144 (c^2 - bd)$.

24. $a^2 \Sigma(\alpha - \beta) (\alpha - \gamma) (\alpha - \delta) = -32G$.

25. If α, β, γ, ... are the roots of $x^n + p_1 x^{n-1} + p_2 x^{n-2} + ... + p_n = 0$, prove that

$$\Sigma(\alpha - \beta)^2 = (n - 1) p_1^2 - 2np_2,$$

$$\Sigma\alpha^3\beta\gamma = -p_1^2 p_3 + 2p_2 p_3 + p_1 p_4 - 5p_5,$$

$$\Sigma\alpha^3\beta^2 = -p_1 p_2^2 + 2p_1^2 p_3 + p_2 p_3 - 5p_1 p_4 + 5p_5,$$

26. Show that the squares of the roots of

$$a_0 x^n - a_1 x^{n-1} + a_2 x^{n-2} - ... + (-1)^n a_n = 0$$

are the roots of

$$b_0 x^n - b_1 x^{n-1} + b_2 x^{n-2} - ... + (-1)^n b_n = 0,$$

where

$$b_0 = a_0^2, \quad b_1 = a_1^2 - 2a_0 a_2, \quad b_2 = a_2^2 - 2a_1 a_3 + 2a_0 a_4, ...$$

$$b_r = a_r^2 - 2a_{r-1} a_{r+1} + 2a_{r-2} a_{r+2} - 2a_{r-3} a_{r+3} + ...$$

27. If the equation whose roots are the squares of the roots of the cubic

$$x^3 - ax^2 + bx - 1 = 0$$

is identical with this cubic, prove that either $a = b = 0$ or $a = b = 3$, or a, b are the roots of $z^2 + z + 2 = 0$.

■ ■ ■

Partial Fractions

■ **1. Rational Fractions:** An expression of the form P/Q, where P and Q are polynomials in x, is called a *rational fraction.* If P is of lower degree than Q, P/Q is called a *proper fraction.* If P is not of lower degree than Q, P/Q is called an *improper fraction.* By means of the division transformation, *an improper fraction can be expressed as the sum of an integral function and a proper fraction.* Thus

$$\frac{x^2+1}{x+1} = (x-1) + \frac{2}{x+1}.$$

Theorem: *If $A + \dfrac{X}{Y} \equiv A' + \dfrac{X'}{Y'}$, where the letters denote polynomials in x and X/Y, X'/Y' are proper fractions, then $A \equiv A'$ and $X/Y \equiv X'/Y'$.*

For
$$A - A' = \frac{X'}{Y'} - \frac{X}{Y} = \frac{YX' - XY'}{YY'}.$$

Also X and X' are respectively of lower degrees than Y and Y', and therefore $YX' - XY'$ is of lower degree than YY'.

Hence $A - A' \equiv 0$; for otherwise, the polynomial $A - A'$ would be identically equal to a proper fraction, which is impossible. Thus $A \equiv A'$, and consequently $X/Y \equiv X'/Y'$.

■ **2. Partial Fractions:** To resolve a given fraction into *partial fractions* is to express it as the sum of two or more simpler fractions.

Fundamental Theorem: *If C/AB is a proper fraction and the factors A, B are prime to each other, proper fractions X/A and Y/B can be found such that*

$$\frac{C}{AB} = \frac{X}{A} + \frac{Y}{B}.$$

For since A is prime to B, polynomials X', Y' can be found so that
$$BX' + AY' = 1, \qquad\qquad \text{(Ch. 3, 18.)}$$

and therefore
$$\frac{C}{AB} = \frac{CX'}{A} + \frac{CY'}{B}.$$

If CX'/A and CY'/B are not proper fractions, by division, we can find polynomials Q, Q', X, Y, such that

$$\frac{CX'}{A} = Q + \frac{X}{A} \quad \text{and} \quad \frac{CY'}{B} = Q' + \frac{Y}{B},$$

where X/A and Y/B are proper fractions, and then

$$\frac{C}{AB} = Q + Q' + \frac{X}{A} + \frac{Y}{B} = Q + Q' + \frac{BX + AY}{AB}.$$

Now X and Y are respectively of lower degrees than A and B, therefore $BX + AY$ is of lower degree than AB. Thus $(BX + AY)/AB$ is a proper fraction: so also is C/AB, therefore

$$Q + Q' \equiv 0 \quad \text{and} \quad \frac{C}{AB} = \frac{X}{A} + \frac{Y}{B}.$$

■ **3. To resolve a Proper Fraction P/Q into its simplest set of Partial Fractions:**
We shall prove that

(*i*) *To a non-repeated factor $x - a$ of Q there corresponds a fraction of the form*

$$\frac{A}{x-a}.$$

(*ii*) *To a factor $(x - b)^n$ of Q there corresponds a group of the form*

$$\frac{B_1}{x-b} + \frac{B_2}{(x-b)^2} + \frac{B_3}{(x-b)^3} + \dots + \frac{B_n}{(x-b)^n}.$$

(*iii*) *To a non-repeated quadratic factor $x^2 + px + q$ of Q there corresponds a*

fraction of the form $\dfrac{Cx + D}{x^2 + px + q}.$

(*iv*) *To a factor $(x^2 + px + q)^n$ of Q there corresponds a group of the form*

$$\frac{C_1 x + D_1}{x^2 + px + q} + \frac{C_2 x + D_2}{\left(x^2 + px + q\right)^2} + \dots + \frac{C_n x + D_n}{\left(x^2 + px + q\right)^n}.$$

Here A, B_1, B_2, … , C_1, C_2, … are independent of x.

Proof.

(*i*) Let $Q = (x - a) \, . \, B$, then since $x - a$ is a non-repeated factor, B is prime to $x - a$, and so

$$P/Q = X/(x - a) + Y/B,$$

where the fractions on the right are proper fractions. Hence X is a constant.

(*ii*) Let $Q = (x - b)^n \, . \, B$. It is assumed that $x - b$ is not a factor of B, so that B is prime to $(x - b)^n$, and consequently

$$P/Q = X/(x - b)^n + Y/B,$$

where the fractions on the right are proper fractions.

Hence X is of degree $n - 1$ at most, and it can be put in the form

$$B_1 (x - b)^{n-1} + B_2 (x - b)^{n-2} + \dots + B_n, \qquad \text{(Ch. 3, 4)}$$

where B_1, B_2, … are constants, which proves the statements in question.

(*iii*) The proof is similar to that of (*i*), but X is of the form $Cx + D$.

(*iv*) The proof is similar to that of (*ii*), but each of the set $B_1, B_2, \ldots$ is of the form $Cx + D$.

Various ways of finding the constants are explained in the following examples:

■ **EXAMPLE 1.** *Express* $\dfrac{x^4}{(x-a)(x-b)(x-c)}$ *as the sum of an integral function of* x

and three proper fractions.

We have $(x - a)(x - b)(x - c) = x^3 - x^2. \Sigma a + x\Sigma ab - abc$.

Hence it is easy to see that the quotient in the division of x^4 by $(x - a)(x - b)$ $(x - c)$ is $x + a + b + c$. We may therefore assume that

$$\frac{x^4}{(x-a)(x-b)(x-c)} = x + a + b + c + \frac{A}{x-a} + \frac{B}{x-b} + \frac{C}{x-c}. \qquad \ldots \text{(A)}$$

To find A, multiply each side of (A) by $x - a$, and put $x = a$.

$$\therefore \qquad A = \frac{a^4}{(a-b)(a-c)}.$$

The values of B and C can be found in the same way;

$$\therefore \quad \frac{x^4}{(x-a)(x-b)(x-c)} = x + a + b + c + \frac{a^4}{(a-b)(a-c)} \cdot \frac{1}{x-a} + \ldots + \ldots .$$

■ **EXAMPLE 2.** *Resolve into the simplest possible partial fractions*

(*i*) $\dfrac{x^3 - 5}{(x-1)^3 (x-2)}$; (*ii*) $\dfrac{3x^3 - 2x^2 - 1}{(x^2 + x + 1)(x^2 - x + 1)}$; (*iii*) $\dfrac{1}{(x-1)(x^2 - 1)(x^3 - 1)}$.

(*ii*) Let $\dfrac{x^3 - 5}{(x-1)^3 (x-2)} = \dfrac{a}{x-1} + \dfrac{b}{(x-1)^2} + \dfrac{c}{(x-1)^3} + \dfrac{d}{x-2}.$ $\qquad \ldots \text{(A)}$

Multiply each side by $x - 2$, and then put $x = 2$;

$$\therefore \qquad d = \frac{2^3 - 5}{(2-1)^3} = 3.$$

Now $\dfrac{x^3 - 5}{(x-1)^3 (x-2)} - \dfrac{3}{x-2} = \dfrac{-2x^2 + 5x + 1}{(x-1)^3} = \dfrac{-2(x-1)^2 + (x-1) + 4}{(x-1)^3}.$

Therefore $\dfrac{x^2 - 5}{(x-1)^3 (x-2)} = -\dfrac{2}{x-1} + \dfrac{1}{(x-1)^2} + \dfrac{4}{(x-1)^3} + \dfrac{3}{x-2}.$

Second Method. Find d as before. Next, multiply each side of (A) by $(x - 1)^3$, and then put $x = 1$;

$$\therefore \qquad c = \frac{1^3 - 5}{1 - 2} = 4.$$

Next, multiply each side of (A) by x, and let $x \to \infty$;

$\therefore \ 1 = a + d ; \quad \therefore \ a = -2.$

Finally, put $x = 0$ in (A), $\therefore \ -\dfrac{5}{2} = -a + b - c - \dfrac{d}{2}, \ \therefore \ b = 1.$

Particular attention is called to the last two steps. In this way two equations connecting the coefficients can *always* be written down. The steps are equivalent to the following: Multiply each side of (A) by $(x - 1)^3 (x - 2)$ and equate the coefficients of x^3 (the highest power of x) and the absolute terms on each side.

(*ii*) Let
$$\frac{3x^3 - 2x^2 - 1}{\left(x^2 + x + 1\right)\left(x^2 - x + 1\right)} = \frac{ax + b}{x^2 + x + 1} + \frac{cx + d}{x^2 - x + 1}. \qquad \text{...(A)}$$

Multiplying by $(x^2 + x + 1) (x^2 - x + 1)$, we find that

$3x^3 - 2x^2 - 1 = (a + c)x^3 + (-a + b + c + d) x^2 + (a - b + c + d)x + b + d,$

and, equating coefficients,

$\quad a + c = 3, \ -a + b + c + d = -2, \ a - b + c + d = 0, \ b + d = -1. \qquad \text{...(B)}$

Whence $a = 2, \ b = 1, \ c = 1, \ d = -2.$

Note that the first and last equations in (B) can be found by (*i*) multiplying (A) by x and letting $x \to \infty$, and (*ii*) putting $x = 0$. Two other equations may be quickly obtained by putting $x = 1$ and $x = -1$.

The following method is sometimes useful. From (A),

$$\frac{3x^3 - 2x^2 - 1}{x^2 - x + 1} = ax + b + \frac{cx^3 + (c + d)x^2 + (c + d)x + d}{x^2 - x + 1},$$

whence by division,

$$3x + 1 + \frac{-2x - 2}{x^2 - x + 1} = ax + b + cx + (2c + d) + \frac{(2c + 2d)x - 2c}{x^2 - x + 1}.$$

The integral functions and also the fractions are identically equal, and so

$$a + c = 3, \ b + 2c + d = 1, \ 2c + 2d = -2, \ -2c = -2,$$

giving $a = 2, \ b = 1, \ c = 1, \ d = -2.$

(*iii*) Let $\dfrac{1}{(x + 1)(x - 1)^3 (x^2 + x + 1)} = \dfrac{a}{x + 1} + \dfrac{b}{x - 1} + \dfrac{c}{(x - 1)^2} + \dfrac{d}{(x - 1)^3} + \dfrac{ex + f}{x^2 + x + 1}.$

$$\text{... (A)}$$

We first fine e and f. Multiply both sides of (A) by $x^2 + x + 1$, and then let

$$x^3 + x + 1 = 0.$$

This gives

$$\frac{1}{(x+1)(x-1)^3} = ex + f, \text{ when } x^2 + x + 1 = 0.$$

We have $x^2 = -x - 1$ and $x^3 = 1$, whence we find that

$$(x - 1)^3 = x^3 - 3x^2 + 3x - 1 = 3(x + 1) + 3x = 3(2x + 1)$$

and

$$(x + 1)(x - 1)^3 = 3(2x^2 + 3x + 1) = 3(x - 1).$$

It follows that

$$1 = 3(x - 1)(ex + f) = 3\{ex^2 - (e - f)x - f\}.$$

$$\therefore \qquad 3(f - 2e)x - 3(e + f) = 1.$$

This linear relation holds for *two* values of x (namely, the roots of $x^2 + x + 1 = 0$). It is therefore an identity, and so

$$f - 2e = 0 \text{ and } 1 = -3(e + f), \text{ giving } e = -1/9 \text{ and } f = -2/9.$$

Next, multiply (A) by $x + 1$, and then let $x = -1$;

$$\therefore \qquad a = \frac{1}{(-2)^3 \cdot 1} = -\frac{1}{8}.$$

Multiply (A) by $(x - 1)^3$, and then let $x = 1$;

$$\therefore \qquad d = \frac{1}{2 \cdot 3} = \frac{1}{6}.$$

Multiply (A) by x, and let $x \to \infty$; $\therefore 0 = a + b + e$, giving $b = \frac{1}{8} + \frac{1}{9} = \frac{17}{72}.$

In (A), let $x = 0$, $\therefore -1 = a - b + c - d + f$, giving $c = -\frac{1}{4}$;

$$\therefore \quad \frac{1}{(x-1)(x^2-1)(x^3-1)}$$

$$= -\frac{1}{8} \cdot \frac{1}{x+1} + \frac{17}{72} \cdot \frac{1}{x-1} - \frac{1}{4} \cdot \frac{1}{(x-1)^2} + \frac{1}{6} \cdot \frac{1}{(x-1)^3} - \frac{1}{9} \cdot \frac{x+2}{x^2+x+1}.$$

EXERCISE XIII

PARTIAL FRACTIONS

Resolve into partial fractions

1. $\dfrac{1}{(x-1)^2(x-2)}.$　　　　**2.** $\dfrac{3(x-6)}{x^3(x+3)}.$　　　　**3.** $\dfrac{x^2+1}{(x-2)^3(x-3)}.$

4. $\dfrac{2x^2+1}{x^3-1}.$

5. $\dfrac{(x+2)^2}{(x-2)(x^2+x+2)}.$

6. $\dfrac{x^3+x^2+1}{(x^2+2)(x^2+3)}.$

7. $\dfrac{x^5+1}{(x^2-x+1)^3}.$

8. $\dfrac{x^3+x^2+1}{(x-1)(x^3-1)}.$

9. $\dfrac{x^3-19x-15}{(x+2)^3(x^2+1)}.$

10. $\dfrac{x^4}{(x^2+1)^3(x^2+2)}.$

11. $\dfrac{x^3+x+1}{(x^2+1)^2(x^2+2)}.$

12. $\dfrac{(x^2+1)^2}{(x^3+1)^2}.$

13. $\dfrac{(x^2+1)^2}{(x^2-1)^3}.$

14. $\dfrac{1}{x^4+1}.$

15. Express $\dfrac{1-abx^2}{(1-ax)(1-bx)}$ as the sum of a constant and two proper fractions.

Hence show that if $1-abx^2$ is divided by $1-(a+b)x+abx^2$, so as to obtain a quotient of $(n+1)$ terms, this quotient is
$$1+(a+b)x+(a^2+b^2)x^2+\dots+(a^n+b^n)x^n,$$
and find the corresponding remainder.

16. Express $\dfrac{(x+a)(x+b)(x+c)}{(x-a)(x-b)(x-c)}$ as the sum of a constant and three proper fractions.

17. Express $\dfrac{x^3}{(x-a)(x-b)(x-c)}$ as the sum of a constant and three proper fractions.

18. Hence show that
$$\frac{a^3}{(a-b)(a-c)(a-d)}+\frac{b^3}{(b-c)(b-d)(b-a)}+\frac{c^3}{(c-d)(c-a)(c-b)}$$
$$+\frac{d^3}{(d-a)(d-b)(d-c)}=1.$$

19. Express $\dfrac{x^4}{(x-a)(x-b)(x-c)}$ as the sum of an intergal function of x and three proper fractions.

Hence show that
$$\frac{a^4}{(a-b)(a-c)(a-d)}+\frac{b^4}{(b-c)(b-d)(b-a)}+\frac{c^4}{(c-d)(c-a)(c-b)}$$
$$+\frac{d^4}{(d-a)(d-b)(d-c)}$$

is equal to $a+b+c+d.$

20. If $(1 + x)^n = c_0 + c_1 x + c_2 x^2 + \ldots + c_n x^n$, show that

$$\frac{\lfloor n}{x(x+1)(x+2)(x+3)\ldots(x+n)} = \frac{c_0}{x} - \frac{c_1}{x+1} + \frac{c_2}{x+2} - \ldots + (-1)n\,\frac{c_n}{x+n}\,.$$

[Assume that $\dfrac{\lfloor n}{x(x+1)(x+2)\ldots(x+n)} = \dfrac{a_0}{x} + \dfrac{a_1}{x+1} + \ldots + \dfrac{a_r}{x+r} + \ldots + \dfrac{a_n}{x+n}\,.$]

21. Hence show that

(i) $\dfrac{c_0}{1} - \dfrac{c_1}{2} + \dfrac{c_2}{3} - \ldots + (-1)^n\,\dfrac{c_n}{n+1} = \dfrac{1}{n+1}\,.$

(ii) $\dfrac{c_0}{2} - \dfrac{c_1}{3} + \dfrac{c_2}{4} - \ldots + (-1)^n\,\dfrac{c_n}{n+2} = \dfrac{1}{(n+1)(n+2)}\,.$

22. Use Ex. 21 and similar identities to prove that

(i) $\dfrac{c_0}{1.2} - \dfrac{c_1}{2.3} + \dfrac{c_2}{3.4} - \ldots + (-1)^n\,\dfrac{c_n}{(n+1)(n+2)} = \dfrac{1}{n+2}\,.$

(ii) $\dfrac{c_0}{2.3} - \dfrac{c_1}{3.4} + \dfrac{c_2}{4.5} - \ldots + (-1)^n\,\dfrac{c_n}{(n+2)(n+3)} = \dfrac{1}{(n+2)(n+3)}\,.$

(iii) $\dfrac{c_0}{1.2.3} - \dfrac{c_1}{2.3.4} + \ldots + (-1)^n\,\dfrac{c_n}{(n+1)(n+2)(n+3)} = \dfrac{1}{2(n+3)}\,.$

(iv) $\dfrac{c_0}{2.3.4} - \dfrac{c_1}{3.4.5} + \ldots + (-1)\,\dfrac{c_n}{(n+2)(n+3)(n+4)} = \dfrac{1}{2(n+3)(n+4)}\,.$

23. Prove that

$$\frac{x^{n+1}}{(x-a_1)(x-a_2)\ldots(x-a_n)} = x + a_1 + a_2 + \ldots + a_n$$

$$+ \Sigma\,\frac{a_1^{n+1}}{(a_1-a_2)\ldots(a_1-a_n)} \cdot \frac{1}{x-a_1}\,.$$

24. Use Ex. 23 to prove that

$$c_n\,(n+1)^{n+1} - c_{n-1}\,n^{n+1} + c_{n-2}\,(n-1)^{n+1} - \ldots + (-1)^n c_0\,1^{n+1} = \frac{1}{2}\lfloor n+2\,.$$

[In Ex. 23, write $n+1$ for n, and put $a_1 = 1$, $a_2 = 2$, ..., $a_n = n$, and $x = 0$.]

■ ■ ■

CHAPTER 8

Summation of Series

■ **1. Meaning of Summation:** Let u_n be a function of the positive integral variable, and let

$$s_n = u_1 + u_2 + u_3 + \dots + u_n.$$

The function s_n possesses this peculiarity—it is the sum of n terms, and these terms cannot be added up, unless the value of n is specified.

Sometimes it is possible to express s_n as the sum of p terms, *where p is a number which does not depend on n.* For example,

$$1 + 2 + 3 + \dots + n = \frac{1}{2}n^2 + \frac{1}{2}n,$$

and the number of terms on the right (namely two) is independent of n.

To sum a series to n terms or *to find the sum of the first n terms of a series* is to express s_n in the form just described; but this is not always possible.

Among the series capable of summation are arithmetic and geometric series; a harmonic series cannot be summed.

The sum of the first n terms of the series

$$u_1 + u_2 + u_3 + \dots + u_r + \dots$$

is often denoted by $\displaystyle\sum_{r=1}^{r=n} u_r$ and sometimes by Σu_n, or simply by s_n.

■ **2. Method of Differences:** *If we are able to express u_n in the form $v_n - v_{n-1}$, where v_n is some function of n, then we can sum the series to n terms.*

For, by hypothesis,

$$u_n = v_n - v_{n-1},$$
$$u_{n-1} = v_{n-1} - v_{n-2},$$
$$\dots\dots\dots\dots\dots\dots\dots,$$
$$u_2 = v_2 - v_1,$$
$$u_1 = v_1 - v_0;$$

whence by addition

$$s_n = u_1 + u_2 + u_3 + \dots + u_n = v_n - v_0.$$

▌ **EXAMPLE:** *Sum the series $1.2.3 + 2.3.4 + 3.4.5 + \dots$ to n terms.*

Here
$$u_n = n(n + 1)(n + 2)$$

$$= \frac{1}{4} n (n + 1)(n + 2)\{(n + 3) - (n - 1)\}\,*$$

$$= \frac{1}{4} n (n + 1) (n + 2) (n + 3) - \frac{1}{4} (n - 1) n (n + 1)(n + 2);$$

$$\therefore \qquad u_n = v_n - v_{n-1}, \qquad \text{where } v_n = \frac{1}{4} n (n + 1) (n + 2) (n + 3);$$

and since $v_0 = 0$, we have by the preceding,

$$s_n = v_n - v_0 = \frac{1}{4} n (n + 1) (n + 2) (n + 3).$$

■ **3. Series in which u_n is the Product of r successive terms of an A.P., beginning with the n-th:**

The last example is an instance of this type, and the same method can be applied to the general case, which is as follows:

It is required to sum to n terms the series in which

$$u_n = (a + nb) (a + \overline{n+1} . b)...(a + \overline{n+r-1} . b),$$

where a, b, r are constants.

We have

$$u_n = \frac{u_n \{(a + \overline{n+r} . b) - (a + \overline{n-1} . b)\}}{(r+1)b}$$

$$= \frac{u_n (a + \overline{n+r} . b)}{(r+1)b} - \frac{(a + \overline{n-1} . b) u_n}{(r+1)b}$$

Therefore $u_n = v_n - v_{n-1}$, where $v_n = \dfrac{u_n (a + \overline{n+r} . b)}{(r+1)b}$;

and consequently $\qquad s_n = v_n - v_0.$

Here v_0 is independent of n, hence the sum to n terms may be found by the following rule: *On the right of u_n introduce as a factor the next term of the A.P.; divide by the number of factors so increased and by the common difference of the A.P. and add a constant.*

The constant is found by putting $n = 1$, as in the next example; or by substituting 0 for n in v_n.

■ **EXAMPLE:** *Sum to n terms $1 . 3 . 5 + 3 . 5 .7 + 5 . 7 . 9 + ...$.*

Hero $u_n = (2n - 1) (2n + 1) (2n + 3)$, and applying the rule,

$$s_n = \frac{(2n-1)(2n+1)(2n+3)(2n+5)}{4.2} + C. \qquad\qquad ...(A)$$

To find the constant C, put $n = 1$ in (A), noting that $s_1 = 1. 3 . 5$;

---

* The expression in the bracket {} is formed by subtracting the term which precedes n from that which follows ($n + 2$) in the A.P. 1, 2, 3, ...

$$\therefore 1 \cdot 3 \cdot 5 = \frac{1.3.5.7}{4.2} + C, \quad \therefore C = 15 \left(1 - \frac{7}{8}\right) = \frac{15}{8} \, ;$$

$$\therefore \qquad s_n = \frac{1}{8}\{(2n - 1)(2n + 1)(2n + 3)(2n + 5) + 15\}.$$

▪ 4. Series in which u_n is a Rational Integral Function of n:

In Ch. 3, 4, (3), it has been shown that a rational integral function of n of degree r can be expressed in the form

$$a + bn + cn(n + 1) + dn(n + 1)(n + 2) + \ldots \text{ to } (r + 1) \text{ terms,}$$

where $a, b, c,\ldots$ are constants whose values may be found by synthetic division. We can therefore use the rule of the last article to sum series of this type, as in the next example.

▪ **EXAMPLE:** *Sum the series*
$$2 \cdot 3 + 3 \cdot 6 + 4 \cdot 11 + \ldots + (n + 1)(n^2 + 2).$$

Here
$$u_n = (n + 1)(n^2 + 2)$$
$$= n^3 + n^2 + 2n + 2.$$

$$\begin{array}{r} 1 + 1)1 \quad +1 \quad +2 \quad +2 \\ -1 \quad +0 \\ \hline 1 + 2)1 \quad +0 \quad +2 \\ -2 \\ \hline 1 \quad -2 \end{array}$$

Dividing *by* n, $n + 1$, $n + 2$ in succession, as on the right, the reckoning shows that
$$u_n = n(n + 1)(n + 2) - 2n(n + 1) + 2n + 2;$$

therefore
$$\Sigma u_n = \Sigma n(n + 1)(n + 2) - 2\,\Sigma n(n + 1) + 2\Sigma n + 2n$$

$$= \frac{1}{4}n(n + 1)(n + 2)(n + 3) - \frac{2}{3}n(n + 1)(n + 2) + n(n + 1) + 2n$$

$$= \frac{1}{12}n(3n^3 + 10n^2 + 21n + 38),$$

the 'constant' clearly being zero.

▪ 5. Series in which u_n is the Reciprocal of the Product of r successive terms of an A.P., beginning with the n-th. A simple instance is the following:

▪ **EXAMPLE 1.** *Sum to n terms* $\dfrac{1}{1.3.5} + \dfrac{1}{3.5.7} + \dfrac{1}{5.7.9} + \ldots$.

Here
$$u_n = \frac{1}{(2n - 1)(2n + 1)(2n + 3)} = \frac{1}{4} \cdot \frac{(2n + 3) - (2n - 1)*}{(2n - 1)(2n + 1)(2n + 3)} \, ;$$

$$\therefore \qquad u_n = \frac{1}{4} \cdot \frac{1}{(2n - 1)(2n + 1)} - \frac{1}{4} \cdot \frac{1}{(2n + 1)(2n + 3)} = v_{n-1} - v_n,$$

---

* The numerator is the difference between the first and last factors of the denominator.

where
$$v_n = \frac{1}{4} \cdot \frac{1}{(2n+1)(2n+3)} \ ;$$

$$\therefore \quad s_n = v_0 - v_n = \frac{1}{12} - \frac{1}{4} \cdot \frac{1}{(2n+1)(2n+3)} .$$

The general case is as follows: *It is required to sum to n terms the series in which* u_n *is the reciprocal of*

$$(a + nb)\left(a + \overline{n+1}.b\right)...\left(a + \overline{n+r-1}.b\right),$$

where a, b, r are constants.

Proceeding as in the last example, we have

$$u_n = \frac{1}{(r-1)b} \cdot \frac{\left(a + \overline{n+r-1}.b\right) - (a + nb)}{(a + nb)\left(a + \overline{n+1}.b\right)...\left(a + \overline{n+r-1}.b\right)} .$$

Therefore $u_n = v_{n-1} - v_n,$ and $s_n = v_0 - v_n,$ where

$$v_n = \frac{1}{(r-1)b} \cdot \frac{1}{\left(a + \overline{n+1}.b\right)\left(a + \overline{n+2}.b\right)...\left(a + \overline{n+r-1}.b\right)} .$$

Now v_0 is independent of n, hence the sum to n terms may be found by the following rule : *Remove the left-hand factor from the denominator of* u_n *; divide by the number of factors so diminished and by the common difference of the A.P. and subtract the result from a constant.*

The constant is found either by putting $n = 1$, or by substituting 0 for n in v_n. Thus, in the last example, by the rule

$$s_n = C - \frac{1}{2.2} \cdot \frac{1}{(2n+1)(2n+3)} ,$$

and putting $n = 1,\ \dfrac{1}{1.3.5} = C - \dfrac{1}{2.2} \cdot \dfrac{1}{3.5} ,\ \therefore C = \dfrac{1}{12} ,$

or
$$v_0 = \frac{1}{2.2} \cdot \frac{1}{1.3} = \frac{1}{12} .$$

▮ **EXAMPLE 2.** *Sum the series* $\dfrac{1}{1.4} + \dfrac{1}{2.5} + \dfrac{1}{3.6} + ... + \dfrac{1}{n(n+3)} .$

$$u_n = \frac{1}{n(n+3)} = \frac{n^2 + 3n + 2}{n(n+1)(n+2)(n+3)} = \frac{n(n+3)+2}{n(n+1)(n+2)(n+3)} .$$

$$\therefore \quad u_n = \frac{1}{(n+1)(n+2)} + \frac{2}{n(n+1)(n+2)(n+3)} .$$

$$\therefore \quad s_n = C - \frac{1}{n+2} - \frac{2}{3(n+1)(n+2)(n+3)} . \qquad \text{...(A)}$$

To find the constant C, put $n = 1$ in (A), observing that $s_1 = \dfrac{1}{4}$.

$$\therefore \qquad \frac{1}{1.4} = C - \frac{1}{3} - \frac{2}{3.2.3.4}, \qquad \therefore \quad C = \frac{11}{18}.$$

Alternative method, by Partial Fractions. We have

$$u_n = \frac{1}{3}\left(\frac{1}{n} - \frac{1}{n+3}\right);$$

$$\therefore \quad 3s_n = \frac{1}{1} + \frac{1}{2} + \frac{1}{3} + \ldots + \frac{1}{n} - \frac{1}{4} - \frac{1}{5} - \frac{1}{6} - \ldots - \frac{1}{n+1} - \frac{1}{n+2} - \frac{1}{n+3};$$

$$\therefore \quad s_n = \frac{1}{3}\left\{1 + \frac{1}{2} + \frac{1}{3} - \frac{1}{n+1} - \frac{1}{n+2} - \frac{1}{n+3}\right\}$$

$$= \frac{11}{18} - \frac{1}{3}\left(\frac{1}{n+1} + \frac{1}{n+2} + \frac{1}{n+3}\right);$$

which is the same result as before, in a different form.

The last example is an instance of *a series whose n-th term is*

$$P / (a + nb)\left(a + \overline{n+1}b\right)\ldots\left(a + \overline{n+r-1}b\right),$$

where P is a polynomial in n of degree m.

If $m < r - 1$, this can be summed by expressing P in the form

$$a_0 + a_1 (a + nb) + a_2 (a + nb)\left(a + \overline{n+1}b\right) + \ldots$$

If $m = r - 1$, the series cannot be summed, for to do this we should have to sum a harmonical progression.

If $m > r - 1$, the nth term consists of an integral and a fractional part which can be dealt with separately.

■ **6. Another important Type:** *If u_n is the n-th term of*

$$\frac{a}{b} + \frac{a(a+1)}{b(b+1)} + \frac{a(a+1)(a+2)}{b(b+1)(b+2)} + \ldots + \frac{a(a+1)\ldots(a+n-1)}{b(b+1)\ldots(b+n-1)},$$

then $\qquad\qquad s_n = \dfrac{1}{a-b+1}\{u_n(a+n) - a\}.$

For $\qquad\qquad u_n = u_n \cdot \dfrac{(a+n) - (b+n-1)}{a-b+1} = v_n - v_{n-1},$

where $\qquad\qquad v_n = u_n(a+n)/(a-b+1)\ (n \geq 1).$

Also $\qquad\qquad u_1 = u_1 \cdot \dfrac{(a+1) - b}{a-b+1} = v_1 - \dfrac{a}{a-b+1}.$

Therefore $\qquad s_n = v_n - \dfrac{a}{a-b+1} = \dfrac{1}{a-b+1}\{u_n(a+n) - a\}.$

■ **7. The Series** $u_0 + u_1 x + u_2 x^2 + ... + u_n x^n$ **where** u_n **is a Polynomial in** n **of Degree** m:

If $x = 1$, this series can be summed as in Art. 4. If $x \neq 1$, let

$$s = u_0 + u_1 x + u_2 x^2 + ... + u_n x^n,$$

then

$$s(1 - x) = u_0 + v_1 x + v_2 x^2 + ... + v_n x^n - u_n x^{n+1},$$

where

$$v_1 = u_1 - u_0, \ v_2 = u_2 - u_1, \ ... \ v_n = u_n - u_{n-1}.$$

Now v_n is a polynomial in n of degree $m - 1$; thus by repeatedly multiplying by $1 - x$ (m multiplications in all), we can reduce the problem of finding s to that of summing a geometric series.

It follows that $s = P/(1 - x)^{m+1}$ where P is a polynomial in x.

■ **EXAMPLE:** *Sum the series* $1^2 + 2^2 x + 3^2 x^2 + ... + n^2 x^{n-1}$.

Denoting the sum by s, we have

$$s(1 - x) = 1 + 3x + 5x^2 + ... + (2n - 1)x^{n-1} - n^2 x^n.$$

Let

$$s' = 1 + 3x + 5x^2 + ... + (2n - 1) x^{n-1},$$

then

$$s'(1 - x) = 1 + 2x + 2x^2 + ... + 2x^{n-1} - (2n - 1) x^n$$

$$= \frac{2\left(1 - x^n\right)}{1 - x} - 1 - (2n - 1)x^n;$$

$\therefore$

$$s = \frac{2\left(1 - x^n\right)}{\left(1 - x\right)^3} - \frac{1 + \left(2n - 1\right)x^n}{\left(1 - x\right)^2} - \frac{n^2 x^n}{1 - x}.$$

■ **8. The Series:** $1^r + 2^r + 3^r + ... + n^r$. We shall write

$$S_r = 1^r + 2^r + 3^r + ... + n^r = f(n),$$

and

$$\sigma = n(n + 1), \ \sigma' = 2n + 1.$$

(1) Theorem: *If* r *is a positive integer,* S_r *can be expressed as a polynomial in* n *of which the highest term is* $n^{r+1} / (r + 1)$.

For we can find $a_1, a_2, ... a_r$, independent of n, such that

$$n^r = a_1 n + a_2 n (n - 1) + ... + a_r n (n - 1) (n - 2) ... (n - r + 1), \qquad ...(A)$$

where obviously $a_r = 1$. Writing $n - 1, n - 2, ... 1$ in succession for n and adding, by Art. 3, we have

$$S_r = \frac{1}{2} a_1 (n' + 1) n + \frac{1}{3} a_2 (n + 1) n (n - 1) + ...$$

$$+ \frac{1}{r+1} a_r (n + 1) n (n - 1) ... (n - r + 1). \qquad ...(B)$$

The right-hand side, when expanded, is a polynomial in n of which the highest term is $n^{r+1} . a_r/(r + 1)$, that is, $n^{r+1}/(r + 1)$.

■ **EXAMPLE:** *Show that* $S_2 = \frac{1}{6} n (n + 1) (2n + 1), \ S_3 = \frac{1}{4} n^2 (n + 1)^2$.

We have

$$n^2 = n + n (n - 1);$$

$$\therefore \qquad S_2 = \frac{1}{2}(n+1)n + \frac{1}{3}(n+1)n(n-1) = \frac{1}{6}n(n+1)(2n+1).$$

Also
$$n^3 = n + 3n(n-1) + n(n-1)(n-2);$$

$$\therefore S_3 = \tfrac{1}{2}(n+1)n + (n+1)n(n-1) + \tfrac{1}{4}(n+1)n(n-1)(n-2) = \tfrac{1}{4}n^2(n+1)^2.$$

(2) If $S_r = b_1 n + b_2 n^2 + b_3 n^3 + \ldots + b_{r+1} n^{r+1} = f(n)$, the values of b_{r+1}, b_r, b_{r-1}, ... may be found as follows.

$$f(n) = 1^r + 2^r + \ldots + n^r \text{, and } f(n-1) = 1^r + 2^r + \ldots + (n-1)^r,$$

therefore
$$n^r = f(n) - f(n-1),$$

so that
$$n^r = b_{r+1}\{n^{r+1} - (n-1)^{r+1}\} + b_r\{n^r - (n-1)^r\} + \ldots + b_1.$$

Expanding and equating coefficients, it will be found that

$$b_{r+1} = \frac{1}{r+1}, \; b_r = \frac{1}{2}, \; b_{r-1} = \frac{r}{12}, \; b_{r-2} = 0, \; b_{r-3} = -\frac{r(r-1)(r-2)}{720},$$

and so on. If r is large, b_1, b_2, ... cannot be found conveniently in this way, and we use other methods.

(3) The values of S_1, S_3, S_5, ..., may be found in succession by considering the function $v_n = n^r(n+1)^r - (n-1)^r n^r$.

We have $v_1 + v_2 + \ldots + v_n = n^r(n+1)^r,$

and, expanding by the Binomial theorem,

$$\frac{1}{2}v_n = C_1^r n^{2r-1} + C_3^r n^{2r-3} + C_5^r n^{2r-5} + \ldots .$$

Putting $n-1$, $n-2$, ... 1 in succession for n and adding,

$$\frac{1}{2}\sigma^r = C_1^r S_{2r-1} + C_3^r S_{2r-3} + C_5^r S_{2r-5} + \ldots , \qquad \ldots\text{(A)}$$

where $\sigma = n(n+1)$. The last term is rS_{r+1} or S_r, according as r is even or odd.

When $r = 1, 2, 3, \ldots$ this formula gives

$$\frac{1}{2}\sigma = S_1$$

$$\frac{1}{2}\sigma^2 = 2S_3 \qquad\qquad S_3 = \frac{1}{4}\sigma^2$$

$$\frac{1}{2}\sigma^3 = 3S_5 + S_3 \qquad\qquad S_5 = \frac{1}{12}\sigma^2(2\sigma - 1)$$

$$\frac{1}{2}\sigma^4 = 4S_7 + 4S_5 \qquad\qquad S_7 = \frac{1}{24}\sigma^2(3\sigma^2 - 4\sigma + 2)$$

$$\frac{1}{2}\sigma^5 = 5S_9 + 10S_7 + S_5 \qquad\qquad S_9 = \frac{1}{20}\sigma^2(2\sigma^3 - 5\sigma^2 + 6\sigma - 3)$$

$$\frac{1}{2}\sigma^6 = 6S_{11} + 20S_9 + 6S_7 \qquad\qquad S_{11} = \frac{1}{24}\sigma^2(2\sigma^4 - 8\sigma^3 + 17\sigma^2 - 20\sigma + 10).$$

Thus if r is odd and >1 and $S_r = f(n)$, then $n^2 (n + 1)^2$ is a factor of $f(n)$, and the remaining factor is a rational integral function of $n(n + 1)$. Of course, it does not follow that S_r is *arithmetically* divisible by $n^2 (n + 1)^2$.

(4) The values of $S_2, S_4, S_6, \dots$, may be found as follows.

Let
$$w_n = n^r (n + 1)^r (2n + 1) - (n - 1)^r n^r (2n - 1);$$

then
$$w_1 + w_2 + \dots + w_n = n^r (n + 1)^r (2n + 1) = \sigma^r \sigma',$$

where
$$\sigma = n(n + 1) \quad \text{and} \quad a' = 2n + 1.$$

On expansion, we have

$$\frac{1}{2} w_n = (2 C_1^r + 1) n^{2r} + (2 C_3^r + C_2^r) n^{2r-2} + (2 C_5^r + C_4^r) n^{2r-4} + \dots$$

$$= (2r + 1) n^{2r} + \frac{1}{3} C_2^r (2r - 1) n^{2r-2} + \frac{1}{5} C_4^r (2r - 3) n^{2r-4} + \dots .$$

Putting $n - 1, n - 2, \dots 1$ in succession for n and adding,

$$\tfrac{1}{2} \sigma^r \sigma' = (2r + 1) S_{2r} + \tfrac{1}{3} C_2^r (2r - 1) S_{2r-2} + \tfrac{1}{5} C_4^r (2r - 3) S_{2r-4} + \dots , \quad \dots \text{(B)}$$

the last term being S_r or $(r + 2) S_{r+1}$, according as r is even or odd.

When $r = 1, 2, 3, \dots$ this formula gives

$r = 1$	$\tfrac{1}{2} \sigma\sigma' = 3S_2$	$S_2 = \tfrac{1}{6} \sigma\sigma'$
$r = 2$	$\tfrac{1}{2} \sigma^2\sigma' = 5S_4 + S_2$	$S_4 = \tfrac{1}{30} \sigma\sigma' (3\sigma - 1)$
$r = 3$	$\tfrac{1}{2} \sigma^3\sigma' = 7S_6 + 5S_4$	$S_6 = \tfrac{1}{42} \sigma\sigma' (3\sigma^2 - 3\sigma + 1)$
$r = 4$	$\tfrac{1}{2} \sigma^4\sigma' = 9S_8 + 14S_6 + S_4$	$S_8 = \tfrac{1}{90} \sigma\sigma' (5\sigma^3 - 10\sigma^2 + 9\sigma - 3)$
$r = 5$	$\tfrac{1}{2} \sigma^5\sigma' = 11S_{10} + 30S_8 + 7S_6$	$S_{10} = \tfrac{1}{66} \sigma\sigma' (3\sigma^4 - 10\sigma^3 + 17\sigma^2 - 15\sigma + 5).$

Thus if r is even and $S_r = f(n)$, then $n(n + 1)(2n + 1)$ is a factor of $f(n)$, and the remaining factor is a rational integral function of $n(n + 1)$.

It will presently appear that the formula (B) is immediately deducible from (A) of the last section.

(5) Bernoulli's Numbers: We *define* B_r as the coefficient of n in S_r, when this is expressed as a polynomial in n.

Thus $B_1 = \dfrac{1}{2}$, and $B_3, B_5, B_7, \dots$ are all zero; for $S_1 = \dfrac{1}{2} n(n + 1)$ and $S_3, S_5, S_7, \dots$ all contain the factor $n^2 (n + 1)^2$. The numbers $B_2, -B_4, B_6, -B_8, \dots$ are the famous *numbers of Bernoulli*, in terms of which a large number of functions can be expanded. For convenience we shall refer to the whole set $B_1, B_2, B_3, \dots$ as Bernoulli's numbers.

A great many formulae can be given for calculating these numbers; one of the best is the following:

If $r > 1$, then

$$(2r + 1) B_{2r} + \frac{1}{3} C_2^r (2r - 1) B_{2r-2} + \frac{1}{5} C_4^r (2r - 3) B_{2r - 4} + ... = 0, \quad ... \text{(C)}$$

the last term on the left being B_r *or* $(r + 2) B_{r+1}$, *according as r is even or odd. If* $r = 1$, *the zero on the right of* (C) *is to be replaced by* $\frac{1}{2}$.

For B_{2m} is the coefficient of n in S_{2m}. Hence by the formula (B) of Art. 8, (4), the left-hand side of (C) is the coefficient of n in the expansion of $\frac{1}{2} n^r (n + 1)^r (2n + 1)$.

Putting $r = 1, 2, 3, ... ,$ in (C),

$r = 1$	$3B_2 = \dfrac{1}{2}$	$B_2 = \dfrac{1}{6}$
$r = 2$	$5B_4 + B_2 = 0$	$B_4 = -\dfrac{1}{30}$
$r = 3$	$7B_6 + 5B_4 = 0$	$B_6 = \dfrac{1}{42}$
$r = 4$	$9B_8 + 14B_6 + B_4 = 0$	$B_8 = -\dfrac{1}{30}$
$r = 5$	$11B_{10} + 30B_8 + 7B_6 = 0$	$B_{10} = \dfrac{5}{66}$

Continuing thus, we find that

$$B_{12} = -\frac{691}{2730}, B_{14} = \frac{7}{6}, B_{16} = -\frac{3617}{510}, B_{18} = \frac{43867}{798}, B_{20} = -\frac{174611}{330}.$$

A few more of the extraordinary properties possessed by the sequence $B_1, B_2, B_3, ...$ are given below.

By the Binomial theorem,

$$(n + 1)^{r + 1} - n^{r + 1} = C_1^{r+1} n^r + C_2^{r+1} n^{r-1} + ... + C_r^{r+1} n + 1.$$

Putting $n - 1, n - 2, ... 1$ in succession for n and adding,

$$(n + 1)^{r+1} - 1 = C_1^{r+1} S_r + C_2^{r+1} S_{r-1} + ... + C_r^{r+1} S_1 + n.$$

Equating the coefficients of n on each side, we have

$$C_1^{r+1} B_r + C_2^{r+1} B_{r-1} + ... + C_r^{r+1} B_1 + 1 = r + 1. \qquad ... \text{(D)}$$

Similarly, from the equation

$$n^{r+1} - (n - 1)^{r+1} = C_1^{r+1} n^r - C_2^{r+1} n^{r-1} + ... + (-1)^r,$$

it follows that

$$n^{r+1} = C_1^{r+1} S_r - C_2^{r+1} S_{r-1} + ... + (-1)^{r-1} C_r^{r+1} S_1 + (-1)^r n,$$

and consequently

$$C_1^{r+1} B_r - C_2^{r+1} B_{r-1} + \ldots + (-1)^{r+1} C_r^{r+1} B_1 + (-1)^r = 0. \quad \ldots (E)$$

From (D) and (E) by addition and subtraction,

$$C_1^{r+1} B_r + C_3^{r+1} B_{r-2} + C_5^{r+1} B_{r-4} + \ldots = \frac{1}{2}(r+1), \qquad \ldots(F)$$

$$C_2^{r+1} B_{r-1} + C_4^{r+1} B_{r-3} + C_6^{r+1} B_{r-5} + \ldots = \frac{1}{2}(r+1), \qquad \ldots(G)$$

where each series continues till B_1, or $B_0 = 1$, appears.

Formulae (F), (G) are equivalent to equations 10′, 11′ in Chrystal's *Algebra*, Vol. II, p. 231. They may be used for calculating B_2, B_4, etc., but their use requires twice the amount of labour required by (C).

Symbolic notation. Equations (D), (E), (F), (G) can be written symbolically thus:

$$(B+1)^{r+1} - B^{r+1} = r+1, \qquad\qquad \ldots(D)$$
$$B^{r+1} - (B-1)^{r+1} = 0, \qquad\qquad \ldots(E)$$
$$(B+1)^{r+1} - (B-1)^{r+1} = (r+1), \qquad\qquad \ldots(F)$$
$$(B+1)^{r+1} + (B-1)^{r+1} - 2B^{r+1} = (r+1), \qquad\qquad \ldots(G)$$

where, after expansion, each index is to be changed into a suffix, i.e. B_m *is to be written for* B^m.

(6) Theorem: *If* $S_r = 1^r + 2^r + \ldots + n^r$ *where r is a positive integer, then*

(i) $\dfrac{d}{dn} S_r = rS_{r-1} + B_r$ *and*

(ii) $S_r = r \int S_{r-1}\, dn + nB_r$

where B_r *is independent of n and is, in fact, the Bernoullian number as defined in Art. 8, (5).*

For S_r can be expressed as a polynomial in n of degree $r+1$. Denoting this polynomial by $f(n)$, we have

$$n^r = f(n) - f(n-1).$$

This relation involves only positive integral powers of n up to n^r, and holds for all positive integral values of n. Hence it is an identity, and we may differentiate both sides, with regard to n. Therefore

$$rn^{r-1} = f'(n) - f'(n-1).$$

Writing $n-1$, $n-2$, $\ldots$ 1 in succession for n, and adding,

$$f'(n) = r\{n^{r-1} + (n-1)^{r-1} + \ldots + 2^{r-1} + 1^{r-1}\} + B_r,$$

where B_r is independent of n.

Moreover, B_r is the coefficient of n in $f(n)$, and is therefore the Bernoullian number defined in Art. 8, (5). The second statement follows from the first. No constant of integration is added, for S_r is divisible by n.

■ **EXAMPLE:** *Deduce formula (B) of Art. 8, (4), from (A) of Art. 8, (3).*

Putting $r+1$ for r in (A), we have

$$\frac{1}{2}\sigma^{r+1} = C_1^{r+1} S_{2r+1} + C_3^{r+1} + S_{2r-1} + C_5^{r+1} S_{2r-3} + \dots .$$

Differentiating with regard to n,

$$\tfrac{1}{2}(r+1)\sigma^r \sigma' = C_1^{r+1}(2r+1) S_{2r} + C_3^{r+1}(2r-1)S_{2r-2} + C_5^{r+1}(2r-3) S_{2r-4} + \dots .$$

No constant is added, for every term is divisible by n.

Dividing by $r+1$, we obtain the formula (B).

(7) Using the relation $S_r = r\int S_{r-1}\, dn + nB_r$, we can find the values of S_1, S_2, S_3 ... by successive integrations. (At any stage, the constant B_r can be found by putting $n = 1$).

Thus $S_0 = n$, $\quad S_1 = \dfrac{n^2}{2} + B_1 n,$

$$S_2 = \frac{2n^3}{2.3} + B_1\frac{2n^2}{2} + B_2 n ,$$

$$S_3 = \frac{2.3n^4}{2.3.4} + B_1\frac{2.3n^3}{2.3} + B_2\frac{3n^2}{2} + B_3 n.$$

Continuing the process, we arrive at the equation

$$\frac{S_r}{\lfloor r} = \frac{n^{r+1}}{\lfloor r+1} + B_1\frac{n^r}{\lfloor 1\lfloor r} + B_2\frac{n^{r-1}}{\lfloor 2\lfloor r-1} + B_3\frac{n^{r-2}}{\lfloor 3\lfloor r-2} + \dots + B_r\frac{n}{\lfloor r},$$

which may be written

$$(r+1) S_r = n^{r+1} + B_1\, C_1^{r+1}\, n^r + B_2\, C_2^{r+1}\, n^{r-1} + \dots + B_r\, C_r^{r+1}\, n, \qquad \dots(H)$$

and expressed symbolically, as above, in the form

$$(r+1)\, S_r = (n+B)^{r+1} - B^{r+1}. \qquad \dots(I)$$

This result is known as *Bernoulli's Theorem*.

Putting $n = 1$, we have equation (D) of Art. 8, (5).

■ **Example:** *Verify the following, either by substitution in* (H), *or, starting with*

$$S_3 = \frac{n^4}{4} + \frac{n^3}{2} + \frac{n^2}{4},$$

by successive integration.

$$S_4 = \frac{n^5}{5} + \frac{n^4}{2} + \frac{n^3}{3} - \frac{n}{30}$$

$$S_5 = \frac{n^6}{6} + \frac{n^5}{2} + \frac{5n^4}{12} - \frac{n^2}{12}$$

$$S_6 = \frac{n^7}{7} + \frac{n^6}{2} + \frac{n^5}{2} - \frac{n^3}{6} + \frac{n}{42}$$

$$S_7 = \frac{n^8}{8} + \frac{n^7}{2} + \frac{7n^6}{12} - \frac{7n^4}{24} + \frac{n^2}{12}$$

$$S_8 = \frac{n^9}{9} + \frac{n^8}{2} + \frac{2n^7}{3} - \frac{7n^5}{15} + \frac{2n^3}{9} - \frac{n}{30}$$

$$S_9 = \frac{n^{10}}{10} + \frac{n^9}{2} + \frac{3n^8}{4} - \frac{7n^6}{10} + \frac{n^4}{2} - \frac{n^2}{20}$$

$$S_{10} = \frac{n^{11}}{11} + \frac{n^{10}}{2} + \frac{5n^9}{6} - n^7 + n^5 - \frac{n^3}{2} + \frac{5n}{66}.$$

EXERCISE XIV

SUMMATION OF SERIES

A. (Arts. 3, 4.)

Sum to n terms the series in Exs. 1–9.

1. $1.2.3.4 + 2.3.4.5 + 3.4.5.6 + \dots$.
2. $2.5.8 + 5.8.11 + 8.11.14 + \dots$.
3. $1^2 + 3^2 + 5^2 + \dots$.
4. $1.5 + 2.6 + 3.7 + \dots$.
5. $1^2.2 + 2^2.3 + 3^2.4 + \dots$.
6. $1.2^2 + 2.3^2 + 3.4^2 + \dots$.
7. $1.2.4 + 2.3.5 + 3.4.6 + \dots$.
8. $1.2.3 + 4.5.6 + 7.8.9 + \dots$.
9. * $2.1 + 5.3 + 8.5 + \dots$.

Find the value of

10. $11^2 + 12^2 + 13^2 + \dots + 21^2$.
11. $11^2 - 12^2 + 13^2 - \dots - 20^2 + 21^2$.
12. $11^3 + 12^3 + 13^3 + \dots + 21^3$.
13. $20^3 - 19^3 + 18^3 - \dots + 2^3 - 1^3$.

Sum to n terms the series in Exs. 14–16.

14. $1^2.2^2 + 2^2.3^2 + 3^2.4^2 + \dots$.
15. † $1.n + 2(n - 1) + 3(n - 2) + \dots$.
16. $1.n^2 + 2(n - 1)^2 + 3(n - 2)^2 + \dots$.

---

* Here, the nth term is the product of the nth terms of the arithmetic series, 2, 5, 8, ..., and 1, 3, 5, ...

† Here $u_r = r\,(n - \overline{r - 1}) = nr - (r - 1)\,r$.

17. Show that whether n is odd or even,

(*i*) $1^2 - 2^2 + 3^2 - 4^2 + \dots$ to $(n-1)$ terms $= \frac{1}{2}(-1)^n n\,(n-1)$.

(*ii*) $1^4 - 2^4 + 3^4 - 4^4 + \dots$ to $(n-1)$ terms $= \frac{1}{2}(-1)^n n\,(n-1)\,(n^2 - n - 1)$.

18. Show that the sum of the products, taken two together, of the first n natural numbers is $\frac{1}{24} n(n^2 - 1)\,(3n + 2)$.

19. Find (*i*) the sum of the squares, (*ii*) the sum of the products taken two together of the numbers $1, 4, 7, \dots (3n - 2)$.

20. Find the number of shot which can be arranged in a pyramidal pile on a triangular base, each side of the base containing 10 shot.

21. Find the number of shot in a pile of 8 courses on a triangular base, each side of the base containing 12 shot.

22. Show that the number of shot in a pile of n courses on a square base, each side of the base containing $2n$ shot, is

$$\frac{1}{6} n\,(2n + 1)\,(7n + 1).$$

23. Find the number of shot in a pile of a rectangular base, the sides of the base containing 16 and 10 shot respectively, and the top course consisting of a single row of shot.

24. A pile of shot of n courses stands on a rectangular base, and the top course consists of a single row of x shot. Show that the number of shot in the pile is

$$\frac{1}{6} n\,(n + 1)\,(2n + 3x - 2).$$

25. Show that if n is of the form $5m + 1$ or $5m + 3$, where m is a natural number or zero, then
$1^4 + 2^4 + 3^4 + \dots + n^4$ is divisible by $1^2 + 2^2 + 3^2 + \dots + n^2$.

26. Show that if n is of the form $3m + 1$, then
$1^5 + 2^5 + 3^5 + \dots + n^5$ is divisible by $1^3 + 2^3 + 3^3 + \dots + n^3$.

B. (Arts. 5, 6, 7.)

Sum to n terms the series in Exs. 1–24.

1. $\dfrac{1}{1.2} + \dfrac{1}{2.3} + \dfrac{1}{3.4} + \dots$.

2. $\dfrac{1}{1.3} + \dfrac{1}{3.5} + \dfrac{1}{5.7} + \dots$.

3. $\dfrac{1}{1.2.3} + \dfrac{1}{2.3.4} + \dfrac{1}{3.4.5} + \dots$.

4. $\dfrac{1}{2.5.8} + \dfrac{1}{5.8.11} + \dfrac{1}{8.11.14} + \dots$.

5. $\dfrac{1}{1.2.3.4} + \dfrac{1}{2.3.4.5} + \dfrac{1}{3.4.5.6} + \dots$.

6. $\dfrac{1}{1.2.3} + \dfrac{3}{2.3.4} + \dfrac{5}{3.4.5} + \dots$.

7. $\dfrac{1}{1.5} + \dfrac{1}{3.7} + \dfrac{1}{5.9} + \dots$.

8. $\dfrac{3}{1.2.4} + \dfrac{4}{2.3.5} + \dfrac{5}{3.4.6} + \dots$.

9. $\dfrac{1}{2.4} + \dfrac{1.3}{2.4.6} + \dfrac{1.3.5}{2.4.6.8} + \dots$.

10. $\dfrac{4}{5} + \dfrac{4.7}{5.8} + \dfrac{4.7.10}{5.8.11} + \dots$.

11. $1 + 2x + 3x^2 + 4x^3 + \ ....$.

12. $1 + \dfrac{3}{2} + \dfrac{5}{2^2} + \dfrac{7}{2^3} + \ ....$.

13. $a + (a + d)\, r + (a + 2d)\, r^2 + \ ...$.

14. $1^2 + \dfrac{3^2}{2} + \dfrac{5^2}{2^2} + \dfrac{7^2}{2^3} + \ ...$.

15. $\dfrac{1}{(1 + x)(1 + 2x)} + \dfrac{1}{(1 + 2x)(1 + 3x)} + \dfrac{1}{(1 + 3x)(1 + 4x)} + \ ...$.

16. $1.\lfloor 1 + 2.\lfloor 2 + 3.\lfloor 3 + \ ...$. $\qquad\qquad$ [Here $u_n = n\,\lfloor n = \lfloor n+1 - \lfloor n\,.]$

17. $\dfrac{1}{1.3} + \dfrac{2}{1.3.5} + \dfrac{3}{1.3.5.7} + \ ...$. $\qquad$ [Here $u_n = \dfrac{1}{2}\cdot\dfrac{(2n+1)-1}{1.3.5...(2n+1)}\,.]$

18. $\dfrac{3}{1^2.2^2} + \dfrac{5}{2^2.3^2} + \dfrac{7}{3^2.4^2} + \ ...$.   [Here $u_n = \dfrac{2n+1}{n^2(n+1)^2} = \dfrac{(n+1)^2 - n^2}{n^2(n+1)^2}\,.]$

19. $\dfrac{1}{3.9.11} + \dfrac{1}{5.11.13} + \dfrac{1}{7.13.15} + \ ...$.

20. $\dfrac{1}{x+1} + \dfrac{2x}{(x+1)(x+2)} + \dfrac{3x^2}{(x+1)(x+2)(x+3)} + \ ...$.
[Observe that $nx^{n-1} = x^{n-1}(x + n) - x^n.]$

21. $1 + (1 + 2)\,x + (1 + 2 + 3)\,x^2 + (1 + 2 + 3 + 4)\,x^3 + \ ...$.
[The value of $s_n\,(1 - x)$ can be found by Ex. 11.]

22. $\dfrac{n}{1.2.3} + \dfrac{n-1}{2.3.4} + \dfrac{n-2}{3.4.5} + \ ...$.

23. $\dfrac{1}{2.3} + 2.\dfrac{2}{3.4} + 2^2\,\dfrac{3}{4.5} + \ ...$.

24. $\dfrac{1}{\lfloor 3} + \dfrac{5}{\lfloor 4} + \dfrac{11}{\lfloor 5} + ... + \dfrac{n^2 + n - 1}{\lfloor n + 2}$.

25. Prove that

$$1 + n + \dfrac{n(n+1)}{\lfloor 2} + \dfrac{n(n+1)(n+2)}{\lfloor 3} + ...\ \text{to } r \text{ terms} = \dfrac{\lfloor n+r-1}{\lfloor n\,\lfloor r-1}\,.$$

Sum to n terms

26. $\dfrac{1}{1+x} + \dfrac{2}{1+x^2} + \dfrac{4}{1+x^4} + \ ...$.

27. $\dfrac{x}{1-x^2} + \dfrac{x^2}{1-x^4} + \dfrac{x^4}{1-x^6} + \ ...$.

28. $\dfrac{a_1}{1+a_1} + \dfrac{a_2}{(1+a_1)(1+a_2)} + \dfrac{a_3}{(1+a_1)(1+a_2)(1+a_3)} + \ ...$.

C. (Art. 8.)

1. If $1^r + 2^r + 3^r + \dots + n^r = f(n)$, show that
$$f(-n) = (-1)^{r+1} f(n-1).$$
[Put $-n+1, -n+2, \dots -1$ for n in the identity $f(n) - f(n-1) = n^r$.]

2. If $S_r = b_1 n + b_2 n^2 + b_3 n^3 + \dots + b_{r+1} n^{r+1}$, show that
$$b_1 - b_2 + b_3 - b_4 + \dots = 0,$$
$$2b_2 - 3b_3, + 4b_4 - 5b_5 + \dots = 0,$$
$$3.2b_3 - 4.3b_4 + 5.4b_5 - \dots = 0.$$
[Use Art. 8, (2).]

3. Given that $S_9 = \dfrac{1}{20} \sigma^2 (2\sigma^3 - 5\sigma^2 + 6\sigma - 3)$ where $\sigma = n(n+1)$, deduce the value of S_8 by differentiation.

4. Continue the reckoning in Art. 8, (5), to find the values of B_{12} and B_{14}.

5. Show that formula (C) of Art. 8, (5), may be written symbolically in the form
$$B^r (B+1)^r (2B+1) = B^r (B-1)^r (2B-1),$$
where after expansion B_m is to be written for B^m.

6. Find the value of B_4 by using formulae (F) and (G) of Art. 8, (5).

7. Show that
$$C_1^{r+1} 2^r B_r + C_3^{r+1} 2^{r-2} B_{r-2} + C_5^{r+1} 2^{r-4} B_{r-4} + \dots = r + 1.$$
[Expand $(2n+1)^{r+1} - (2n-1)^{r+1}$.]

■■■

Determinants

■ **1. Notation:** Many complicated expressions can be easily handled if they are expressed as 'determinants.' The theory which we are about to explain arises in connection with linear equations.

If the equations $a_1x + b_1 = 0$, $a_2x + b_2 = 0$ are satisfied by the same value of x, then $a_1b_2 - a_2b_1 = 0$. The expression $a_1b_2 - a_2b_1$ is called a *determinant of the second order*, and is denoted by

$$\begin{vmatrix} a_1 & b_1 \\ a_2 & b_2 \end{vmatrix}, \text{ or by } (a_1b_2).$$

Next consider the system

$$a_1x + b_1y + c_1 = 0, \quad a_2x + b_2y + c_2 = 0, \quad a_3x + b_3y + c_3 = 0.$$

If these equations are satisfied by the same values of x and y, it is easy to show that

$$a_1(b_2c_3 - b_3c_2) + b_1(c_2a_3 - c_3a_2) + c_1(a_2b_3 - a_3b_2) = 0.$$

The expression on the left is called a *determinant of the third order,* and is denoted by

$$\begin{vmatrix} a_1 & b_1 & c_1 \\ a_2 & b_2 & c_2 \\ a_3 & b_3 & c_3 \end{vmatrix}, \text{ or by } (a_1b_2c_3).$$

This determinant in its expanded form is

$$a_1b_2c_3 - a_1b_3c_2 + a_2b_3c_1 - a_2b_1c_3 + a_3b_1c_2 - a_3b_2c_1$$

In considering the form of this expression, observe that:

(*i*) *Disregarding the signs*, the terms can be obtained by writing the letters a, b, c in their natural order and arranging the suffixes in all possible ways. The number of arrangements is $\lfloor 3$, giving six terms.

(*ii*) *The sign* of any particular term depends on *the order of the suffixes.*

As in Ch. 4, 4, (1), the interchange of two suffixes, 2, 3, for example, is called the transposition (2 3).

The term $a_1b_2c_3$, in which the suffixes occur in *their natural order,* has the sign +.

The sign of any other term is $+$ or $-$, according as the arrangement of the suffixes in that term is derived from the arrangement 1 2 3 by an even or by an odd number of transpositions. Thus, considering the second and third terms, 1 2 3 is changed to 1 3 2 by the *single* transposition (2 3). Also 1 2 3 is changed to 2 3 1 by the *two* transpositions (1 2), (1 3) performed in this order. These ideas will now be generalised.

■ **2. Definition of a Determinant:** By means of n letters $a, b, c, \ldots l$, and n suffixes 1, 2, 3, $\ldots n$, we can represent n^2 numbers thus,

$$a_1 \quad b_1 \quad c_1 \ldots l_1,$$
$$a_2 \quad b_2 \quad c_2 \ldots l_2,$$
$$\ldots\ldots\ldots\ldots\ldots\ldots$$
$$a_n \quad b_n \quad c_n \ldots l_n.$$

The determinant of the nth order, denoted by

$$\begin{vmatrix} a_1 & b_1 & c_1 \ldots l_1 \\ a_2 & b_2 & c_2 \ldots l_2 \\ \ldots\ldots\ldots\ldots\ldots \\ a_n & b_n & c_n \ldots l_n \end{vmatrix}, \text{ or by } (a_1 b_2 c_3 \ldots l_n),$$

is defined as the sum of all the products, each with the sign determined by the rule below, which can be formed by writing the letters $a, b, c, \ldots l$ in their natural order and arranging the suffixes in all possible orders.

Rule of Signs. The sign of the term $a_1 b_2 c_3 \ldots l_n$, where the suffixes occur in their natural order, is $+$. The sign of any other term is $+$ or $-$, according as the arrangement of suffixes in that term is derived from the arrangement 1 2 3 $\ldots n$ by an even or by an odd number of transpositions (that is, by an even or an odd substitution).

This rule is justified by Theorem 2 of Ch. 4, 4.

Each of the n^2 numbers $a_1, b_1, \ldots a_2, b_2 \ldots$ is called an *element* of the determinant.

The diagonal through the left-hand top corner, which contains the elements $a_1, b_2, c_3, \ldots l_n$, is called the *leading* or *principal diagonal*, and $a_1 b_2 c_3 \ldots l_n$ is called the *leading term*.

The expanded form of the determinant has $\lfloor n$ terms : half of these have the sign $+$ and half the sign.

For suppose that there are p terms of the first sort and q of the second. Since the interchange of the two suffixes transforms a positive into a negative term, $p \leq q$. Similarly $q \leq p$, and therefore $p = q$.

Every term contains one element and one only from each row, and one element and one only from each column. We can find the sign of any particular term by the rule at the end of Art. 4, Ch. 4.

▌ **EXAMPLE:** *In the expanded form of* $(a_1 b_2 c_3 d_4 e_5)$*, find the signs of*
(i) $a_2 b_4 c_5 d_1 e_3$, (ii) $a_5 b_4 c_3 d_2 e_1$.

(i) $\begin{pmatrix} 1\,2\,3\,4\,5 \\ 2\,4\,5\,1\,3 \end{pmatrix} = (1\ 2\ 4)\ (3\ 5) = (1\ 2)\ (2\ 4)\ (3\ 5), \text{ and the sign is } -.$

(The last step is unnecessary if we use the rule referred to.)

Or, we may make single transpositions, as follows,

$$2\ 4\ 5\ 1\ 3,\quad 1\ 4\ 5\ 2\ 3,\quad 1\ 2\ 5\ 4\ 3,\quad 1\ 2\ 3\ 4\ 5,$$

showing that *three* of these are required, and so the sign is –.

$$(ii)\ \begin{pmatrix} 1\ 2\ 3\ 4\ 5 \\ 5\ 4\ 3\ 2\ 1 \end{pmatrix} = (1\ 5)\ (2\ 4),\ \text{and the sign is } +\ .$$

Or if, as some textbooks advise, we count the number of 'inversions' (Ch. 4, 4) in the arrangements of the suffixes in (*i*) and (*ii*), these are 5 and 10 respectively–the first odd and the second even, giving the same results. But this takes much longer.

■ **3. Theorem:** *A determinant is unaltered by changing its rows into columns and its columns into rows.*

$$\text{Let}\quad \Delta = \begin{vmatrix} a_1 & b_1 & c_1 \\ a_2 & b_2 & c_2 \\ a_3 & b_3 & c_3 \end{vmatrix}, \qquad \Delta' = \begin{vmatrix} a_1 & a_2 & a_3 \\ b_1 & b_2 & b_3 \\ c_1 & c_2 & c_3 \end{vmatrix}.$$

The leading term of each determinant is $a_1 b_2 c_3$.

The remaining terms of Δ are derived from $a_1 b_2 c_3$ by keeping the letters in their natural order and arranging the suffixes in all possible ways, an interchange of two suffixes producing a change of sign.

The remaining terms of Δ' are derived from $a_1 b_2 c_3$ by keeping the suffixes in their natural order and arranging the letters in all possible ways, an interchange of two letters producing a change of sign.

The results in the two cases are identical. The same argument applies to a determinant of any order.

■ **4. Theorem:** *The interchange of two rows, or of two columns, changes the sign of a determinant without altering its numerical value.*

For the interchange of two rows is equivalent to the interchange of the suffixes, and the interchange of two columns is equivalent to the interchange of two letters. Hence in either case the sign of every term of the determinant is changed.

■ **5. Theorem:** *A determinant in which two rows or two columns are identical is equal to zero.*

If two rows or two columns of a determinant Δ are identical, the value of Δ is unaltered by the interchange of these rows or columns.

But this interchange transforms Δ into $-\Delta$, therefore

$$\Delta = -\Delta,\ \text{ and }\ \Delta = 0.$$

■ **6. Expansion of a Determinant in Terms of the Elements of any Row or Column:**

$$\text{Let}\quad \Delta = \begin{vmatrix} a_1 & b_1 & c_1 \\ a_2 & b_2 & c_2 \\ a_3 & b_3 & c_3 \end{vmatrix}.$$

Every term in the expansion of Δ contains one element and one only from each row, and one element and one only from each column.

Hence, Δ can be expressed in any of the six forms:

$$a_1 A_1 + \quad b_1 B_1 + \quad c_1 C_1, \quad a_1 A_1 + \quad a_2 A_2 + \quad a_3 A_3,$$
$$a_2 A_2 + \quad b_2 B_2 + \quad c_2 C_2, \quad b_1 B_1 + \quad b_2 B_2 + \quad b_3 B_3,$$
$$a_3 A_3 + \quad b_3 B_3 + \quad c_3 C_3, \quad c_1 C_1 + \quad c_2 C_2 + \quad c_3 C_3,$$

where A_1 contains no element from the row or column containing a_1, a similar remark applying to every A, B and C.

To determine A_1, every term of Δ which contains a_1 can be obtained from $a_1 b_2 c_3$ keeping the letters in the natural order, retaining 1 as the suffix of a and arranging the other suffixes in all possible orders, the interchange of two suffixes producing a change of sign.

Hence A_1 is equal to the determinant $(b_2 c_3)$, obtained by erasing the row and column in Δ which contain a_1.

To determine any other coefficient, say C_3. By the interchange of *consecutive* rows, bring the row containing c_3 to the top. By the interchange of *consecutive* columns, bring the column containing c_3 to the extreme left.

Thus we have in succession

$$\begin{vmatrix} a_3 & b_3 & c_3 \\ a_1 & b_1 & c_1 \\ a_2 & b_2 & c_2 \end{vmatrix}, \quad \begin{vmatrix} c_3 & a_3 & b_3 \\ c_1 & a_1 & b_1 \\ c_2 & a_2 & b_2 \end{vmatrix},$$

and now c_3 occupies the top left-hand corner, and the relative positions of the elements not belonging to the row and column containing c_3 is unaltered.

Every interchange of rows or of columns introduces a change of sign.

The number of interchanges is equal to the number of vertical and horizontal 'moves' (indicated by arrows) required to bring c_3 to the left-hand top corner, which in this case is 4.

Hence, C_3 is equal to the determinant obtained by erasing the row and column containing c_3, multiplied by $(-1)^4$.

Similar reasoning applies to a determinant of any order.

▌ EXAMPLE 1. *Expand the determinant* $\Delta = (a_1 b_2 c_3)$ *with reference to the elements* (*i*) *in the first row,* (*ii*) *in the second column.*

The results are

$$(i) \quad \Delta = a_1 \begin{vmatrix} b_2 & c_2 \\ b_3 & c_3 \end{vmatrix} - b_1 \begin{vmatrix} a_2 & c_2 \\ a_3 & c_3 \end{vmatrix} + c_1 \begin{vmatrix} a_2 & b_2 \\ a_3 & b_3 \end{vmatrix},$$

$$(ii) \quad \Delta = - b_1 \begin{vmatrix} a_2 & c_2 \\ a_3 & c_3 \end{vmatrix} + b_2 \begin{vmatrix} a_1 & c_1 \\ a_3 & c_3 \end{vmatrix} - b_3 \begin{vmatrix} a_1 & c_1 \\ a_2 & c_2 \end{vmatrix}.$$

■ **EXAMPLE 2.** *Prove that*

$$\begin{vmatrix} a & h & g \\ h & b & f \\ g & f & c \end{vmatrix} = abc + 2fgh - af^2 - bg^2 - ch^2.$$

Expanding with reference to the first row,

$$\Delta = a \begin{vmatrix} b & f \\ f & c \end{vmatrix} - h \begin{vmatrix} h & f \\ g & c \end{vmatrix} + g \begin{vmatrix} h & b \\ g & f \end{vmatrix}$$

$$= a(bc - f^2) - h(hc - fg) + g(hf - bg)$$
$$= abc + 2fgh - af^2 - bg^2 - ch^2.$$

■ **7. Minors and Cofactors:** For the determinant

$$\Delta = (a_1 b_2 \ldots l_n),$$

the *minor* of any element is the determinant obtained by omitting from Δ the row and column containing the element.

The *cofactor* (H_r) of an element (h_r) is the coefficient of that element in the expanded form of Δ, and is therefore equal to the corresponding minor with the proper sign prefixed.

The sign is determined as follows: Count the number of interchanges of consecutive rows and columns (*i.e.* of vertical and horizontal 'moves') required to bring the element to the left-hand top corner. The sign is + or –, according as this number is even or odd. The rule may be stated thus. *For the element h_r which is in the r-th row and the s-th column, the cofactor*

$$H_r = (-1)^{r+s}. \ (the \ minor \ of \ h_r).$$

For the number of 'moves' required to bring h_r to the left-hand top corner is $(r - 1) + (s - 1)$.

■ **EXAMPLE:** *Find the cofactors of d_2 and c_4 in the determinant $\Delta = (a_1 b_2 c_3 d_4)$.*
Omitting the row and column which contain d_2, it will be seen that the minor of

$$d_2 = (a_1 b_3 c_4).$$

Denote the cofactor in question by D_2, C_4; by the rule given above

$$D_2 = (-1)^{2+4} (a_1 b_3 c_4) = (a_1 b_3 c_4).$$

Similarly $C_4 = (-1)^{4+3} (a_1 b_2 c_3) = -(a_1 b_2 c_3)$.
The student who is acquainted with 'partial differentiation' should observe that

$$H_r = \frac{\partial \Delta}{\partial h_r}.$$

■ **8. Important Identities:** For the determinant

$$\Delta = (a_1 b_2 c_3)$$

we have a number of identities of the following types:

(*i*) $a_2 A_1 + b_2 B_1 + c_2 C_1 = 0,$ (*ii*) $b_1 A_1 + b_2 A_2 + b_3 A_3 = 0.$
where $A_1, B_1, \ldots$ are the cofactors of $a_1, b_1, \ldots$.
For $\Delta = a_1 A_1 + b_1 B_1 + c_1 C_1.$

where A_1, B_1, C_1, are independent of a_1, b_1, c_1. Hence the expression

$$a_2 A_1 + b_2 B_1 + c_2 C_1$$

is the result of substituting a_2, b_2, c_2 for a_1, b_1, c_1 in Δ, and is therefore equal to a determinant with two rows identical, therefore

$$a_2 A_1 + b_2 B_1 + c_2 C_1 = 0.$$

The second identity is obtained by putting b_1, b_2, b_3 for a_1, a_2, a_3 in the equality.

$$\Delta = a_1 A_1 + a_2 A_2 + a_3 A_3.$$

Similar identities hold for determinants of any order.

■ **9. Theorem:** *If every element in any row, or in any column, of a determinant is multiplied by a number k, then the determinant is multiplied by k.*

For every term in the expansion of a determinant Δ contains one element and one only out of each row, and one element and one only out of each column.

Thus,

$$\begin{vmatrix} ka_1 & kb_1 & kc_1 \\ a_2 & b_2 & c_2 \\ a_3 & b_3 & c_3 \end{vmatrix} = k. \begin{vmatrix} a_1 & b_1 & c_1 \\ a_2 & b_2 & c_2 \\ a_3 & b_3 & c_3 \end{vmatrix} = \begin{vmatrix} ka_1 & b_1 & c_1 \\ ka_2 & b_2 & c_2 \\ ka_3 & b_3 & c_3 \end{vmatrix}$$

Conversely, if all the elements in any row, or in any column, have a common factor k, then k is a factor of the determinant, and can be taken outside.

■ **10. Theorem:** *A determinant can be expressed as the sum of two determinants by expressing every element in any row or column as the sum of two numbers.*

For example, we shall prove that

$$\begin{vmatrix} a_1 + \alpha_1 & b_1 & c_1 \\ a_2 + \alpha_2 & b_2 & c_2 \\ a_3 + \alpha_3 & b_3 & c_3 \end{vmatrix} = \begin{vmatrix} a_1 & b_1 & c_1 \\ a_2 & b_2 & c_2 \\ a_3 & b_3 & c_3 \end{vmatrix} + \begin{vmatrix} \alpha_1 & b_1 & c_1 \\ \alpha_2 & b_2 & c_2 \\ \alpha_3 & b_3 & c_3 \end{vmatrix}$$

Denote the determinants on the right by Δ, Δ', and the one on the left by Δ''.

We have $\Delta = a_1 A_1 + a_2 A_2 + a_3 A_3$

where A_1, A_2, A_3 are independent of a_1, a_2, a_3.

Now Δ' is obtained from Δ by substituting α_1, α_2, α_3 for a_1, a_2, a_3.

And Δ'' is obtained from Δ by substituting $a_1 + \alpha_1$, $a_2 + \alpha_2$, $a_3 + \alpha_3$ for a_1, a_2, a_3.

Hence $\Delta' = \alpha_1 A_1 + \alpha_2 A_2 + \alpha_3 A_3,$

and $\Delta'' = (a_1 + \alpha_1) A_1 + (a_2 + \alpha_2)A_2 + (a_3 + \alpha_3)A_3;$

$\therefore$ $\Delta'' = \Delta + \Delta'.$

Similar reasoning applies in all cases.

■ **11. Theorem:** *A determinant is unaltered by adding to the elements of any row (or column) k times the corresponding elements of any other row (or column), where k is any given number.*

For example, we shall prove that

$$\begin{vmatrix} a_1 + kb_1 & b_1 & c_1 \\ a_2 + kb_2 & b_2 & c_2 \\ a_3 + kb_3 & b_3 & c_3 \end{vmatrix} = \begin{vmatrix} a_1 & b_1 & c_1 \\ a_2 & b_2 & c_2 \\ a_3 & b_3 & c_3 \end{vmatrix}$$

Denoting the determinant on the right by Δ and that on the left by Δ', we have

$$\Delta' = \begin{vmatrix} a_1 & b_1 & c_1 \\ a_2 & b_2 & c_2 \\ a_3 & b_3 & c_3 \end{vmatrix} + \begin{vmatrix} kb_1 & b_1 & c_1 \\ kb_2 & b_2 & c_2 \\ kb_3 & b_3 & c_3 \end{vmatrix}.$$

Now,

$$\begin{vmatrix} kb_1 & b_1 & c_1 \\ kb_2 & b_2 & c_2 \\ kb_3 & b_3 & c_3 \end{vmatrix} = k \begin{vmatrix} b_1 & b_1 & c_1 \\ b_2 & b_2 & c_2 \\ b_3 & b_3 & c_3 \end{vmatrix} = 0.$$

$$\therefore \qquad \Delta' = \Delta.$$

Note that by a second application of the above,

$$\begin{vmatrix} ha_1 + kb_1 + lc_1 & b_1 & c_1 \\ ha_2 + kb_2 + lc_2 & b_2 & c_2 \\ ha_3 + kb_3 + lc_3 & b_3 & c_3 \end{vmatrix} = h \begin{vmatrix} a_1 & b_1 & c_1 \\ a_2 & b_2 & c_2 \\ a_3 & b_3 & c_3 \end{vmatrix}.$$

Note: In simplifying determinants the following points should be carefully observed.

(*i*) When, as in the last part of Art. 11, a column (or row) is replaced by one which contains the elements of that column (or row) all multiplied by a number h, then the determinant is multiplied by h. For instance, if the second column is subtracted from the first, and the remainders are put in the *second* column, the sign of the determinant is changed.

(*ii*) When more than one column (or row) is changed by a succession of changes of this kind, *at least one column (or row) must be left unaltered*, as in Art. 13, Ex. 1.

■ **12. Theorem:** *If the elements of a determinant Δ are rational integral functions of x and two rows (or columns) become identical when x = a, then x-a is a factor of Δ.*

For Δ can be expressed as a polynomial in x, and $\Delta = 0$; when $x = a$, so that the result follows by the Remainder theorem.

Again, *if r rows become identical when a is substituted for x, then $(x - a)^{r-1}$ is a factor of Δ.*

For $r - 1$ of these rows can be replaced by rows in which the elements are the differences of corresponding elements in the original rows; hence $x - a$ is a factor of all the elements in each of these $(r - 1)$ substituted rows; and the result follows.

■ **13. Examples:** The determinant in question is denoted by Δ.

▌ **EXAMPLE 1.** *Find the value of the determinant on the right*

$$\begin{vmatrix} 6 & 8 & 7 & 11 \\ 13 & 9 & 12 & 4 \\ 7 & 10 & 3 & 5 \\ 14 & 1 & 6 & 15 \end{vmatrix}.$$

We have

$$\Delta = \begin{vmatrix} 6 & 1 & 1 & -1 \\ 13 & -3 & -1 & -22 \\ 7 & 7 & -4 & -9 \\ 14 & -5 & -8 & -13 \end{vmatrix} = \begin{vmatrix} 0 & 0 & 0 & -1 \\ 19 & -2 & -23 & -22 \\ 31 & 11 & -13 & -9 \\ 62 & 3 & 21 & -13 \end{vmatrix};$$

$$\therefore \quad \Delta = \begin{vmatrix} 19 & -2 & -23 \\ 31 & 11 & -13 \\ 62 & 3 & -21 \end{vmatrix} = \begin{vmatrix} 19 & -2 & -23 \\ 31 & 11 & -13 \\ 0 & -19 & 5 \end{vmatrix}.$$

Expanding with reference to the bottom row,

$$\Delta = 19(-19 \cdot 13 + 23 \cdot 31) + 5(19 \cdot 11 + 2 \cdot 31) = 10209.$$

[In the first step, the 3rd column is taken from the 2nd to form a new 2nd column, the 1st column is taken from the 3rd to form a new 3rd column, and twice the 1st column is taken from the 4th to form a new 4th column. In the second step, six times the 3rd column is taken from the 1st, the 3rd is subtracted from the 2nd, and the 4th added to the 3rd.

▌ **EXAMPLE 2.** *Prove that*

$$\begin{vmatrix} 1 & 1 & 1 & 1 \\ \alpha & \beta & \gamma & \delta \\ \alpha^2 & \beta^2 & \gamma^2 & \delta^2 \\ \alpha^3 & \beta^3 & \gamma^3 & \delta^3 \end{vmatrix} = -(\beta - \gamma)(\gamma - \alpha)(\alpha - \beta)(\alpha - \delta)(\beta - \delta)(\gamma - \delta).$$

If $\alpha = \beta$, two columns in Δ are identical, and consequently $\alpha - \beta$ is a factor of Δ. Similarly $\beta - \gamma$, etc., are factors. Hence

$$\Delta = k(\beta - \gamma)(\gamma - \alpha)(\alpha - \beta)(\alpha - \delta)(\beta - \delta)(\gamma - \delta).$$

Since Δ is of the sixth degree in $\alpha, \beta, ..., k$ is independent of these quantities. Now the term $\beta\gamma^2\delta^3$ occurs in Δ, and on the right-hand side the coefficient of this term is $-k$, therefore $k = -1$.

▌ **EXAMPLE 3.** *If ω is a cube root of unity, prove that $a + b\omega + c\omega^2$ is a factor of*

$$\begin{vmatrix} a & b & c \\ b & c & a \\ c & a & b \end{vmatrix}.$$

Hence show that the determinant is equal to

$$-(a^3 + b^3 + c^3 - 3abc).$$

To the first column add ω times the second and ω^2 times the third, therefore

$$\Delta = \begin{vmatrix} a + b\omega + c\omega^2 & b & c \\ b + c\omega + a\omega^2 & c & a \\ c + a\omega + b\omega^2 & a & b \end{vmatrix} = (a + b\omega + c\omega^2) \begin{vmatrix} 1 & a & c \\ \omega^2 & b & a \\ \omega & a & b \end{vmatrix};$$

$\therefore a + b\omega + c\omega^2$ is a factor of Δ, similarly $a + b\omega^2 + c\omega$ and $a + b + c$ are factors. The rest follows as in Ex. 2.

This method applies to a determinant of any order in which the second row is derived from the first, the third from the second, and so on, by a cyclic substitution.

▮ **EXAMPLE 4.** *Prove that*

$$\begin{vmatrix} (b+c)^2 & a^2 & a^2 \\ b^2 & (c+a)^2 & b^2 \\ c^2 & c^2 & (a+b)^2 \end{vmatrix} = 2abc \, (a + b + c)^3,$$

$$\Delta = \begin{vmatrix} (b+c)^2 & a^2 - (b+c)^2 & a^2 - (b+c)^2 \\ b^2 & (c+a)^2 - b^2 & 0 \\ c^2 & 0 & (a+b)^2 - c^2 \end{vmatrix} = (a+b+c)^2 \begin{vmatrix} (b+c)^2 & a-b-c & a-b-c \\ b^2 & c+a-b & 0 \\ c^2 & 0 & a+b-c \end{vmatrix}.$$

Subtracting the sum of the second and third rows from the first and taking 2 outside,

$$\Delta = 2 \, (a+b+c)^2 \begin{vmatrix} bc & -c & -b \\ b^2 & c+a-b & 0 \\ c^2 & 0 & a+b-c \end{vmatrix} = \frac{2(a+b+c)^2}{bc} \begin{vmatrix} bc & 0 & 0 \\ b^2 & bc+ba & b^2 \\ c^2 & c^2 & ac+bc \end{vmatrix}.$$

In the last step the first column is added to b times the second and to c times the third, *which is equivalent to multiplying by bc;*

$$\therefore \qquad \Delta = 2 \, (a + b + c)^2 \{bc(c + a) \, (a + b) - b^2c^2\}$$

$$\equiv 2abc \, (a + b + c)^3.$$

Alternatively.　If we substitute 0 for *a.*

$$\Delta = \begin{vmatrix} (b + c)^2 & 0 & 0 \\ b^2 & c^2 & b^2 \\ c^2 & c^2 & b^2 \end{vmatrix} = 0 \, ;$$

hence *a* is a factor; and similarly *b* and *c* are factors.

Again, if $-(b + c)$ is substituted for a,

$$\Delta = \begin{vmatrix} (-a)^2 & a^2 & a^2 \\ b^2 & (-b)^2 & b^2 \\ c^2 & c^2 & (-c)^2 \end{vmatrix} ;$$

and in this there are three identical columns; hence, by Art. 12, $(a + b + c)^2$ is a factor of Δ.

Hence, since Δ is of the sixth degree,

$$\Delta = Nabc\,(a+b+c)^3 \ ;$$

the remaining factor necessarily being of the first degree and symmetrical, since the cyclic substitution (a, b, c) gives a determinant equal to Δ.

Putting $a = 1$, $b = 1$, $c = 1$, we have

$$\begin{vmatrix} 4 & 1 & 1 \\ 1 & 4 & 1 \\ 1 & 1 & 4 \end{vmatrix} = 27\,N,$$

whence $N = 2$.

EXERCISE XV

A. SIMPLE EVALUATIONS

1. The positive terms in the expansion of $(a_1 b_2 c_3)$ are

$$a_1 b_2 c_3, \quad b_1 c_2 a_3, \quad c_1 a_2 b_3.$$

Hence the rule of Sarrus, which is as follows:

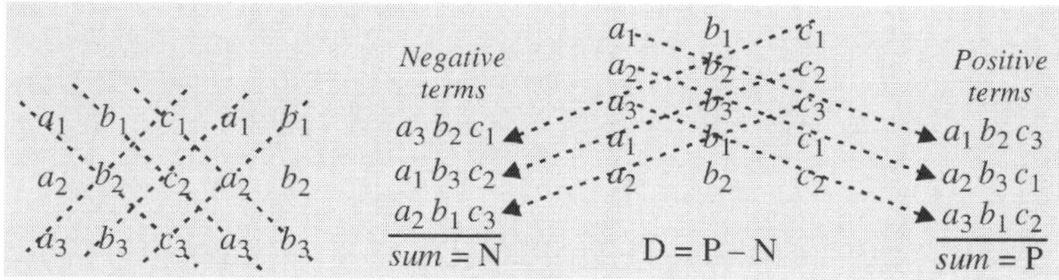

Fig. 26

The determinant $(a_1 b_2 c_3)$ is the sum of the products of the three elements lying on each of the six lines in the above diagram on the left, the sign of the product being $+$ or $-$ according as the line which determines it is drawn downwards or upwards, proceeding from left to right.

A more convenient way of using the rule in numerical cases is shown on the right.

2. Apply the rule of Ex. 1 to show that

$$\begin{vmatrix} a & h & g \\ h & b & f \\ g & f & c \end{vmatrix} = abc + 2\,fgh - af^2 - bg^2 - ch^2.$$

3. Show that in the expansion of $(a_1b_2c_3d_4)$, the following terms occur:
$$+\ b_1c_2a_3d_4,\ +\ c_1a_2b_3d_4,\ +\ a_1b_2c_3d_4,$$
$$+\ a_1d_2b_3c_4,\ +\ b_1d_2c_3a_4,\ +\ c_1d_2a_3d_4,$$
and that the other positive terms are obtained from these by the substitution (12) (34).

4. If T is any term in the expansion of $(a_1b_2c_3\ ...)$ and another term T' is derived from T by a cyclic substitution involving r letters, the order of the suffixes being unchanged, then T and T' have the same sign, or opposite signs, according as r is odd or even. [Exs. 1 and 3 are instances.]

5. In the determinant $(a_1b_2c_3...l_n)$, the sign of the term $a_nb_{n-1}c_{n-2}\ ...\ l_1$, forming the 'second diagonal' has the sign $+$ if and only if n is of the form $4k$ or $4k+1$.

6. In the expansion of $(a_1b_2c_3d_4e_5)$ show that the terms $+\ a_4b_5c_1d_2e_3, -a_3b_4c_5d_2e_1$, occur.

7. In the expansion of $(a_1b_2c_3d_4e_5f_6)$ show that the terms $+\ a_2b_4c_6d_1e_3f_5, +\ a_6b_4c_1d_2e_3f_5$ occur.

8. Expand the following:

$$(i)\ \begin{vmatrix} a & b & 0 \\ 0 & a & b \\ b & 0 & a \end{vmatrix},\ (ii)\ \begin{vmatrix} 2ab & a^2 & b^2 \\ a^2 & b^2 & 2ab \\ b^2 & 2ab & a^2 \end{vmatrix},\ (iii)\ \begin{vmatrix} 7 & 11 & 13 \\ 17 & 19 & 23 \\ 29 & 31 & 37 \end{vmatrix}.$$

9. Prove that

$$(i)\ \begin{vmatrix} 1^2 & 2^2 & 3^2 \\ 2^2 & 3^2 & 4^2 \\ 3^2 & 4^2 & 5^2 \end{vmatrix} = -8,\quad (ii)\ \begin{vmatrix} 1^2 & 2^2 & 3^2 & 4^2 \\ 2^2 & 3^2 & 4^2 & 5^2 \\ 3^2 & 4^2 & 5^2 & 6^2 \\ 4^2 & 5^2 & 6^2 & 7^2 \end{vmatrix} = 0.$$

$$(iii)\ \begin{vmatrix} 1 & a & a^2 & 0 \\ 0 & 1 & a & a^2 \\ a^2 & 0 & 1 & a \\ a & a^2 & 0 & 1 \end{vmatrix} = 1 + a^4 + a^8,\quad (iv)\ \begin{vmatrix} a & b & b & b \\ a & b & a & a \\ a & a & b & a \\ b & b & b & a \end{vmatrix} = -(a-b)^4,$$

$$(v)\ \begin{vmatrix} 1+a^2-b^2 & 2ab & -2b \\ 2ab & 1-a^2+b^2 & 2a \\ 2b & -2a & 1-a^2-b^2 \end{vmatrix} = (1+a^2+b^2)^3.$$

10. Prove that

$$\begin{vmatrix} l_1-t & m_1 & n_1 \\ l_2 & m_2-t & n_2 \\ l_3 & m_3 & n_3-t \end{vmatrix}$$

is equal to $-t^3 + t^2(l_1 + m_2 + n_3) - t(L_1 + M_2 + N_3) + (l_1m_2n_3)$, where L_1, M_2, N_3 are the cofactors of l_1, m_2, n_3 in $(l_1m_2n_3)$.

11. Prove that

$$\begin{vmatrix} x & l & m & 1 \\ \alpha & x & n & 1 \\ \alpha & \beta & x & 1 \\ \alpha & \beta & \gamma & 1 \end{vmatrix} = (x - \alpha)(x - \beta)(x - \gamma),$$

where l, m, n have any values whatever

[From the first, second and third columns respectively take α, β, γ times the fourth column.]

12. Prove that

$$\begin{vmatrix} a & b & c & d \\ -b & a & -d & c \\ -c & d & a & -b \\ -d & -c & b & a \end{vmatrix} = (a^2 + b^2 + c^2 + d^2)^2.$$

[Multiply the rows by a, $-b$, $-c$, $-d$ to form a new top row.]

13. Prove that

$$\begin{vmatrix} 1+a_1 & 1 & 1 & 1 \\ 1 & 1+a_2 & 1 & 1 \\ 1 & 1 & 1+a_3 & 1 \\ 1 & 1 & 1 & 1+a_4 \end{vmatrix} = a_1 a_2 a_3 a_4 \left\{ 1 + \frac{1}{a_1} + \frac{1}{a_2} + \frac{1}{a_3} + \frac{1}{a_4} \right\},$$

and that a similar result holds for a determinant of the nth order and of this form.

[Subtract the top row from each of the others and expand with reference to the top row.]

14. If $u = ax^4 + 4bx^3 + 6cx^2 + 4dx + e$ and $u_{11} = ax^2 + 2bx + c$, $u_{12} = bx^2 + 2cx + d$, $u_{22} = cx^2 + 2dx + e$, prove that

$$\begin{vmatrix} a & b & c & u_{11} \\ b & c & d & u_{12} \\ c & d & e & u_{22} \\ u_{11} & u_{12} & u_{22} & 0 \end{vmatrix} = -u. \begin{vmatrix} a & b & c \\ b & c & d \\ c & d & e \end{vmatrix}.$$

[From the fourth column subtract x^2 times the first column, $2x$ times the second, and the third.]

15. If $u = ax^2 + 2bxy + cy^2$, $u' = a'x^2 + 2b'xy + c'y^2$, prove that

$$\begin{vmatrix} y^2 & -xy & x^2 \\ a & b & c \\ a' & b' & c' \end{vmatrix} = \begin{vmatrix} ax+by & bx+cy \\ a'x+b'y & b'x+c'y \end{vmatrix} = -\frac{1}{y} . \begin{vmatrix} u & u' \\ ax+by & a'x+b'y \end{vmatrix}.$$

[Add x times the second column to y times the third: then add x times the first column to y times the second.]

16. Prove that

$$\begin{vmatrix} a & b & ax+by \\ b & c & bx+cy \\ ax+by & bx+cy & 0 \end{vmatrix} = -(ac - b^2)(ax^2 + 2bxy + cy^2).$$

[Multiply the first column by x, the second by y, and subtract from the third.]

17. Prove that the determinant

$$\begin{vmatrix} a-x & b & c & d \\ b & c-x & d & a \\ c & d & a-x & b \\ d & a & b & c-x \end{vmatrix}$$

is equal to $(x - a - b - c - d)(x - a + b - c + d)\{x^2 - (a - c)^2 - (b - d)^2\}$.

B. METHOD OF FACTORS

18. Prove that

$$\begin{vmatrix} x+a & b & c & d \\ a & x+b & c & d \\ a & b & x+c & d \\ a & b & c & x+d \end{vmatrix} = x^3 (x + a + b + c + d).$$

19. If $2s = a + b + c$, prove that

$$\begin{vmatrix} a^2 & (s-a)^2 & (s-a)^2 \\ (s-b)^2 & b^2 & (s-b)^2 \\ (s-c)^2 & (s-c)^2 & c^2 \end{vmatrix} = 2s^3 (s - a)(s - b)(s - c).$$

[Show that s^2, $s - a$, $s - b$, $s - c$ are factors: deduce that Ns is the remaining factor, and find N by putting $a = 2$, $b = 0$, $c = 0$.]

20. Express the following in factors:

$$(i)\ \begin{vmatrix} 1 & 1 & 1 \\ \alpha & \beta & \gamma \\ \alpha^2 & \beta^2 & \gamma^2 \end{vmatrix}, \quad (ii)\ \begin{vmatrix} 1 & 1 & 1 \\ \alpha & \beta & \gamma \\ \alpha^3 & \beta^3 & \gamma^3 \end{vmatrix}, \quad (iii)\ \begin{vmatrix} \mu\nu & \mu+\nu & 1 \\ \nu\lambda & \nu+\lambda & 1 \\ \lambda\mu & \lambda+\mu & 1 \end{vmatrix}.$$

21. Prove that

$$\begin{vmatrix} bc & bc'+b'c & b'c' \\ ca & ca'+c'a & c'a' \\ ab & ab'+a'b & a'b' \end{vmatrix} = (bc')(ca')(ab').$$

[Divide the first row by $b'c'$, the second by $c'a'$, the third by $a'b'$. Put $a/a' = \lambda$, $b/b' = \mu$, $c/c' = \nu$, and use the last example.]

22. If $\lambda = \beta\gamma + \alpha\delta$, $\mu = \gamma\alpha, + \beta\delta$, $v = \alpha\beta + \gamma\delta$, prove that

$$\begin{vmatrix} 1 & 1 & 1 \\ \lambda & \mu & v \\ \lambda^2 & \mu^2 & v^2 \end{vmatrix} = -(\beta - \gamma)(\gamma - \alpha)(\alpha - \beta)(\alpha - \delta)(\beta - \delta)(\gamma - \delta).$$

Hence show that the following determinant has the same value:

$$- \begin{vmatrix} \beta^2\gamma^2 + \alpha^2\delta^2 & \beta\gamma + \alpha\delta & 1 \\ \gamma^2\alpha^2 + \beta^2\delta^2 & \gamma\alpha + \beta\delta & 1 \\ \alpha^2\beta^2 + \gamma^2\delta^2 & \alpha\beta + \gamma\delta & 1 \end{vmatrix}.$$

23. Prove that

$$\begin{vmatrix} x^3 & x^2 & x & 1 \\ \alpha^3 & \alpha^2 & \alpha & 1 \\ \beta^3 & \beta^2 & \beta & 1 \\ \gamma^3 & \gamma^2 & \gamma & 1 \end{vmatrix} = -(\beta - \gamma)(\gamma - \alpha)(\alpha - \beta)(x - \alpha)(x - \beta)(x - \gamma).$$

By equating the coefficients of x on each side, show that

$$\begin{vmatrix} \alpha^3 & \alpha^2 & 1 \\ \beta^3 & \beta^2 & 1 \\ \gamma^3 & \gamma^2 & 1 \end{vmatrix} = -(\beta - \gamma)(\gamma - \alpha)(\alpha - \beta)(\beta\gamma + \gamma\alpha + \alpha\beta).$$

24. If ω is a root of $x^4 = 1$, show that $a + b\omega + c\omega^2 + d\omega^3$ is a factor of

$$\begin{vmatrix} a & b & c & d \\ b & c & d & a \\ c & d & a & b \\ d & a & b & c \end{vmatrix}$$

Hence show that the determinant is equal to

$$- (a + b + c + d)(a - b + c - d)\{(a - c)^2 + (b - d)^2\}.$$

■ **14. Minors of a Determinant:** The determinant obtained by omitting any number of rows and the same number of columns from a given determinant Δ is called a *minor* of Δ.

If we omit two rows and two columns, the resulting determinant is called a *second minor*, and so on.

If from Δ we omit the rows and columns which contain every element of a given minor, we get another minor which is said to be *complementary* to the first. Thus $(a_1 b_2)$ and $(c_3 d_4 e_5)$ are complementary minors of

$$(a_1 b_2 c_3 d_4 e_5)$$

By the interchange of rows and columns, any minor of Δ can be put into the left-hand top corner, and then the complementary minor can be seen at a glance.

■ **15. Expansion of a Determinant by means of its Second Minors:** Consider the determinant

$$\Delta = (a_1 b_2 c_3 d_4 e_5).$$

In the expansion of Δ, the sum of the terms which contain $a_1 b_2$ is

$$a_1 b_2 (c_3 d_4 e_5),$$

and the sum of those which contain $a_2 b_1$ is $- a_2 b_1 (c_3 d_4 e_5)$. Hence every term in the product $(a_1 b_2)(c_3 d_4 e_5)$ occurs in the expansion.

Again, every term in Δ contains one element out of the first row and one out of the second, and these elements have different suffixes.

Hence Δ is equal to the sum of all the products, each with its proper sign, of the form

$$(a_1 b_2)(c_3 d_4 e_5),$$

where the first factor is any minor formed from elements of the first two rows (or of the first two columns), and the second is the complementary minor.

The sign of a product is determined as follows. Take, for example,

$$(b_1 d_2)(a_3 c_4 e_5).$$

In the expansion of this product, the term $b_1 d_2 a_3 c_4 e_5$ occurs. Keeping the suffixes in the same order, observe that

$$\begin{pmatrix} a\,b\,c\,d\,e \\ b\,d\,a\,c\,e \end{pmatrix} = (abdc) = (ab)\,(bd)\,(dc).$$

Therefore, in the expansion of Δ, this term occurs with a negative sign, and the sign of the product is $-$.

More generally, if any two rows of a determinant Δ are

$$a_r \quad b_r \quad c_r \ldots h_r \quad \ldots k_r \quad \ldots,$$

$$a_{r'} \quad b_{r'} \quad c_{r'} \ldots h_{r'} \quad \ldots k_{r'} \quad \ldots,$$

where h_r, $h_{r'}$ belong to the sth column and k_r, $k_{r'}$ to the s'th column, then

$$\Delta = \Sigma (-1)^{r+r'+s+s'} (h_r k_{r'}) \text{ . (The complementary minor of } (h_r k_{r'})).$$

The summation is to include all the minors which can be formed from the two rows. *The sign of the product* may be determined as above, or as in Art. 7 by counting the number of 'moves' required to bring h_r to the place occupied by a_1 and $k_{r'}$ to the place occupied by b_2.

Thus, in the instance given above, the sign of $(b_1 d_2)\,(a_3 c_4 e_5)$ is given by $(-1)^{1+2+2+4} = -1$.

In a similar way we can expand a determinant in terms of the rth minors which can be formed from any r rows (or columns).

Note: In the 'double-suffix' notation of Art. 20, the rule takes the following convenient form: In $(a_{pq} a_{rs} \ldots)$ (complementary minor), the sign is $(-1)^{p+q+r+s+\ldots}$.

▮ **EXAMPLE 1.** *Expand* $\Delta = (a_1 b_2 c_3 d_4)$ *in terms of the minors of the rows containing* a_1 *and* a_2. *It is easy to verify that*

$$\Delta = (b_1 c_2)(a_3 d_4) + (c_1 a_2)(b_3 d_4) + (a_1 b_2)(c_3 d_4)$$
$$+ (a_1 d_2)(b_3 c_4) + (b_1 d_2)(c_3 a_4) + (c_1 d_2)(a_3 b_4).$$

▮ **EXAMPLE 2.** *Prove that if*

$$\Delta = \begin{vmatrix} a_1 & b_1 & c_1 & 0 & 0 & 0 \\ a_2 & b_2 & c_2 & 0 & 0 & 0 \\ a_3 & b_3 & c_3 & 0 & 0 & 0 \\ p_1 & q_1 & r_1 & l_1 & m_1 & n_1 \\ p_2 & q_2 & r_2 & l_2 & m_2 & n_2 \\ p_3 & q_3 & r_3 & l_3 & m_3 & n_3 \end{vmatrix}, \quad and \ \Delta' = \begin{vmatrix} 0 & 0 & 0 & a_1 & b_1 & c_1 \\ 0 & 0 & 0 & a_2 & b_2 & c_2 \\ 0 & 0 & 0 & a_3 & b_3 & c_3 \\ l_1 & m_1 & n_1 & p_1 & q_1 & r_1 \\ l_2 & m_2 & n_2 & p_2 & q_2 & r_2 \\ l_3 & m_3 & n_3 & p_3 & q_3 & r_3 \end{vmatrix}$$

then $\Delta = (a_1 b_2 c_3).(l_1 m_2 n_3)$, *and* $\Delta' = (-1)^3 (a_1 b_2 c_3).(l_1 m_2 n_3)$.

Considering the minors of Δ, which can be formed from the elements of the first three rows, the only one not equal to zero is $(a_1 b_2 c_3)$, and the complementary minor is $(l_1 m_2 n_3)$. Moreover, the term $a_1 b_2 c_3 l_1 m_2 n_3$ occurs in the expansion of Δ, therefore

$$\Delta = (a_1 b_2 c_3) . (l_1 m_2 n_3).$$

Again, Δ is transformed into Δ' by the interchange of the first and fourth, the second and fifth and the third and sixth columns. Whence the second result follows.

▪ **16. Product of Two Determinants of the same Order :**

As a typical case we shall prove that

$$\begin{vmatrix} a_1 & b_1 \\ a_2 & b_2 \end{vmatrix} \begin{vmatrix} l_1 & m_1 \\ l_2 & m_2 \end{vmatrix} = \begin{vmatrix} a_1 l_1 + b_1 m_1 & a_1 l_2 + b_1 m_2 \\ a_2 l_1 + b_2 m_1 & a_2 l_2 + b_2 m_2 \end{vmatrix}.$$

Denote the determinants on the left by Δ and Δ' ; then we have

$$\Delta \, \Delta' = \begin{vmatrix} a_1 & b_1 & 0 & 0 \\ a_2 & b_2 & 0 & 0 \\ -1 & 0 & l_1 & l_2 \\ 0 & -1 & m_1 & m_2 \end{vmatrix} = \begin{vmatrix} a_1 & b_1 & a_1 l_1 + b_1 m_1 & a_1 l_2 + b_1 m_2 \\ a_2 & b_2 & a_2 l_1 + b_2 m_1 & a_2 l_2 + b_2 m_2 \\ -1 & 0 & 0 & 0 \\ 0 & -1 & 0 & 0 \end{vmatrix},$$

where the second determinant is obtained from the first as follows: To the third column add l_1 times the first and m_1 times the second: to the fourth column add l_2 times the first and m_2 times the second.

Hence, as in the last example, $\Delta \, \Delta' = (-1)^2 . \begin{vmatrix} a_1 l_1 + b_1 m_1 & a_1 l_2 + b_1 m_2 \\ a_2 l_1 + b_2 m_1 & a_2 l_2 + b_2 m_2 \end{vmatrix} . \begin{vmatrix} -1 & 0 \\ 0 & -1 \end{vmatrix},$

whence the result.

The product of two determinants of the n-th order $(n > 2)$ can be expressed in a similar form, and by the same method.

Second method. It is required to show that $(a_1 b_2 c_3) . (l_1 m_2 n_3) = \Delta$, where

$$\Delta = \begin{vmatrix} a_1 l_1 + b_1 m_1 + c_1 n_1 & a_1 l_2 + b_1 m_2 + c_1 n_2 & a_1 l_3 + b_1 m_3 + c_1 n_3 \\ a_2 l_1 + b_2 m_1 + c_2 n_1 & a_2 l_2 + b_2 m_2 + c_2 n_2 & a_2 l_3 + b_2 m_3 + c_2 n_3 \\ a_3 l_1 + b_3 m_1 + c_3 n_1 & a_3 l_2 + b_3 m_2 + c_3 n_2 & a_3 l_3 + b_3 m_3 + c_3 n_3 \end{vmatrix}$$

The determinant Δ is the sum of 27 determinants, two of which are

$$\begin{vmatrix} b_1 m_1 & a_1 l_2 & c_1 n_3 \\ b_2 m_1 & a_2 l_2 & c_2 n_3 \\ b_3 m_1 & a_3 l_2 & c_3 n_3 \end{vmatrix} \text{ and } \begin{vmatrix} b_1 m_1 & b_1 m_2 & a_1 l_3 \\ b_2 m_1 & b_2 m_2 & a_2 l_3 \\ b_3 m_1 & b_3 m_2 & a_3 l_3 \end{vmatrix},$$

which are the same as $m_1 l_2 n_3 \begin{vmatrix} b_1 & a_1 & c_1 \\ b_2 & a_2 & c_2 \\ b_3 & a_3 & c_3 \end{vmatrix}$ and $m_1 m_2 l_3 \begin{vmatrix} b_1 & b_1 & a_1 \\ b_2 & b_2 & a_2 \\ b_3 & b_3 & a_3 \end{vmatrix}.$

The second of these is zero, and the first is equal to

$$m_1 l_2 n_3 \,(b_1 a_2 c_3) = -\,m_1 l_2 n_3 \,(a_1 b_2 c_3).$$

With regard to the sign, observe that $l_1 m_2 n_3$ is transformed into $m_1 l_2 n_3$ by the same number of interchanges of letters as those required to change $a_1 b_2 c_3$ into $b_1 a_2 c_3$. This number is odd, and therefore the sign is–

Of the 27 determinants, all are equal to zero except the six of the first type, in which a's are in one column, b's in another and c's in the third. Moreover, for reasons similar to the above, $(a_1 b_2 c_3)$ is a factor of each of the six, and the other factor is a term (with the proper sign) of $(l_1 m_2 n_3)$, therefore

$$\Delta = (a_1 b_2 c_3) \,.\, (l_1 m_2 n_3).$$

A similar result holds for determinants of the nth order ($n = 4, 5 \ldots$).

■ **17. Rectangular Arrays:** A number of symbols arranged in rows and columns is called a *rectangular array*.

If the number of rows is not equal to the number of columns, the array is generally enclosed by double lines, and is sometimes called a *matrix.*

(1) *Let the number of columns exceed the number of rows.*

Consider the arrays $\begin{Vmatrix} a_1 & b_1 & c_1 \\ a_2 & b_2 & c_2 \end{Vmatrix}, \begin{Vmatrix} l_1 & m_1 & n_1 \\ l_2 & m_2 & n_2 \end{Vmatrix}.$

Proceeding as in the last article, we form the determinant

$$\Delta = \begin{vmatrix} a_1 l_1 + b_1 m_1 + c_1 n_1 & a_1 l_2 + b_1 m_2 + c_1 n_2 \\ a_2 l_1 + b_2 m_1 + c_2 n_1 & a_2 l_2 + b_2 m_2 + c_2 n_2 \end{vmatrix}.$$

By reasoning similar to that in Art. 16 it will be seen that

$$\Delta = (a_1 b_2)\,(l_1 m_2) + (b_1 c_2)\,(m_1 n_2) + (c_1 a_2)\,(n_1 l_2).$$

This process is called *multiplying the arrays.*

In general, *the determinant formed thus from two such arrays is equal to the sum of the products obtained by multiplying every determinant which can be formed from one array by the corresponding determinant formed from the other array.*

(2) *Let the number of rows exceed the number of columns.*

If from the arrays $\begin{Vmatrix} a_1 & b_1 \\ a_2 & b_2 \\ a_3 & b_3 \end{Vmatrix}, \begin{Vmatrix} l_1 & m_1 \\ l_2 & m_2 \\ l_3 & m_3 \end{Vmatrix}$ we form the determinant

$$\Delta = \begin{vmatrix} a_1 l_1 + b_1 m_1 & a_1 l_2 + b_1 m_2 & a_1 l_3 + b_1 m_3 \\ a_2 l_1 + b_2 m_1 & a_2 l_2 + b_2 m_2 & a_2 l_3 + b_2 m_3 \\ a_3 l_1 + b_3 m_1 & a_3 l_2 + b_3 m_2 & a_3 l_3 + b_3 m_3 \end{vmatrix},$$ it will be seen that $\Delta = 0$.

This follows, as in Art. 16, or from the fact that $\Delta = \begin{vmatrix} a_1 & b_1 & 0 \\ a_2 & b_2 & 0 \\ a_3 & b_3 & 0 \end{vmatrix} \begin{vmatrix} l_1 & m_1 & 0 \\ l_2 & m_2 & 0 \\ l_3 & m_3 & 0 \end{vmatrix}.$

In general, *the determinant formed by multiplying two such arrays is zero.*

EXAMPLE 1. *If* $s_r = \alpha^r + \beta^r + \gamma^r + \delta^r$, *prove that*

$$\begin{vmatrix} s_0 & s_1 \\ s_1 & s_2 \end{vmatrix} = \Sigma(\alpha - \beta)^2, \quad \begin{vmatrix} s_0 & s_1 & s_2 \\ s_1 & s_2 & s_3 \\ s_2 & s_3 & s_4 \end{vmatrix} = \Sigma(\alpha - \beta)^2 (\beta - \gamma)^2 (\gamma - \alpha)^2,$$

$$\begin{vmatrix} s_0 & s_1 & s_2 & s_3 \\ s_1 & s_2 & s_3 & s_4 \\ s_2 & s_3 & s_4 & s_5 \\ s_3 & s_4 & s_5 & s_6 \end{vmatrix} = (\alpha - \beta)^2 (\beta - \gamma)^2 (\gamma - \alpha)^2 (\alpha - \delta)^2 (\beta - \delta)^2 (\gamma - \delta)^2.$$

The first two results are obtained by squaring the arrays

$$\left\| \begin{matrix} 1 & 1 & 1 & 1 \\ \alpha & \beta & \gamma & \delta \end{matrix} \right\|, \quad \left\| \begin{matrix} 1 & 1 & 1 & 1 \\ \alpha & \beta & \gamma & \delta \\ \alpha^2 & \beta^2 & \gamma^2 & \delta^2 \end{matrix} \right\|.$$

The last is got by squaring the determinant in Ex. 2 of Art. 13.

EXAMPLE 2. *Prove that*

$$\begin{vmatrix} (a_1 - b_1)^2 & (a_1 - b_2)^2 & (a_1 - b_3)^2 \\ (a_2 - b_1)^2 & (a_2 - b_2)^2 & (a_2 - b_3)^2 \\ (a_3 - b_1)^2 & (a_3 - b_2)^2 & (a_3 - b_3)^2 \end{vmatrix}$$

is equal to $2 (a_1 - a_2) (a_2 - a_3) (a_3 - a_1) (b_1 - b_2) (b_2 - b_3) (b_3 - b_1)$.

This result is obtained by forming the product

$$\begin{vmatrix} a_1^2 & a_1 & 1 \\ a_2^2 & a_2 & 1 \\ a_3^2 & a_3 & 1 \end{vmatrix} \begin{vmatrix} 1 & -2b_1 & b_1^2 \\ 1 & -2b_2 & b_2^2 \\ 1 & -2b_3 & b_3^2 \end{vmatrix}.$$

18. Reciprocal Determinants: Let $\Delta = (a_1 b_2 \dots k_n)$ and $\Delta' = (A_1 B_2 \dots K_n)$, where $A_1, B_1, \dots$ are the cofactors of $a_1, b_1, \dots$ in Δ. For reasons connected with linear transformations, $A_1, B_1, \dots$ are sometimes called *inverse* elements, and Δ' is called the *reciprocal* determinant.

(1) *If Δ is a determinant of the nth order and Δ' is the reciprocal determinant, then* $$\Delta' = \Delta^{n-1}.$$

The following method applies in all cases. Let $\Delta = (a_1 b_2 c_3)$, then using identities of the type

$$a_1 A_1 + b_1 B_1 + c_1 C_1 = \Delta,$$
$$a_1 A_2 + b_1 B_2 + c_1 C_2 = 0,$$
$$a_1 A_3 + b_1 B_3 + c_1 C_3 = 0,$$

it will be seen that $\quad \Delta\,\Delta' = \begin{vmatrix} a_1 & b_1 & c_1 \\ a_2 & b_2 & c_2 \\ a_3 & b_3 & c_3 \end{vmatrix} \cdot \begin{vmatrix} A_1 & B_1 & C_1 \\ A_2 & B_2 & C_2 \\ A_3 & B_3 & C_3 \end{vmatrix} = \begin{vmatrix} \Delta & 0 & 0 \\ 0 & \Delta & 0 \\ 0 & 0 & \Delta \end{vmatrix};$

whence it follows that $\qquad \Delta' = \Delta^2$

(2) *Any minor of Δ' of order r is equal to the complement of the corresponding minor of Δ multiplied by Δ^{r-1}, provided that $\Delta \neq 0$.*

As a typical instance, let $\Delta = (a_1 b_2 c_3 d_4)$. We shall prove that, if $\Delta \neq 0$, then $(A_1 B_2) = (c_3 d_4). \Delta$ and $(A_1 B_2 C_3) = d_4 \Delta^2$.

We have

$$\begin{vmatrix} a_1 & b_1 & c_1 & d_1 \\ a_2 & b_2 & c_2 & d_2 \\ a_3 & b_3 & c_3 & d_3 \\ a_4 & b_4 & c_4 & d_4 \end{vmatrix} \cdot \begin{vmatrix} A_1 & B_1 & C_1 & D_1 \\ A_2 & B_2 & C_2 & D_2 \\ 0 & 0 & 1 & 0 \\ 0 & 0 & 0 & 1 \end{vmatrix} = \begin{vmatrix} \Delta & 0 & c_1 & d_1 \\ 0 & \Delta & c_2 & d_2 \\ 0 & 0 & c_3 & d_3 \\ 0 & 0 & c_4 & d_4 \end{vmatrix};$$

therefore

$$\Delta\,(A_1 B_2) = \Delta^2\,(c_3 d_4) \text{ and } (A_1 B_2) = \Delta \cdot (c_3 d_4).$$

The second identity is obtained by multiplying Δ by

$$\begin{vmatrix} A_1 & B_1 & C_1 & D_1 \\ A_2 & B_2 & C_2 & D_2 \\ A_3 & B_3 & C_3 & D_3 \\ 0 & 0 & 0 & 1 \end{vmatrix}.$$

■ **19. Two Methods of Expansion:**

(1) *With reference to the elements of a row and column.*

$$\text{Let } \Delta = \begin{vmatrix} a_1 & b_1 & c_1 \\ a_2 & b_2 & c_2 \\ a_3 & b_3 & c_3 \end{vmatrix}, \quad \text{and } K = \begin{vmatrix} a_1 & b_1 & c_1 & l \\ a_2 & b_2 & c_2 & m \\ a_3 & b_3 & c_3 & n \\ l' & m' & n' & r \end{vmatrix}.$$

Let $A_1, B_1, \ldots$ be the cofactors of $a_1, b_1, \ldots$ in Δ.

In the expansion of K, the sum of the terms containing r is Δr : every other term contains one of the three l, m, n and one of the three l', m', n'.

Again, $\begin{vmatrix} a_1 & l \\ l' & r \end{vmatrix}$ and $\begin{vmatrix} b_2 & c_2 \\ b_3 & c_3 \end{vmatrix}$ are complementary minors of Δ;

hence, coefficient of ll' in $K = -$ Coefficient of $a_1 r$ in K
$$= -\text{ Coefficient of } a_1 \text{ in } \Delta = -A_1;$$

and similarly,

coefficient of mn' in $K \qquad = -$ Coefficient of $c_2 r$ in K
$$= -\text{ Coefficient of } c_2 \text{ in } \Delta = -C_2.$$

Thus it will be seen that

$$K = \Delta r - \{A_1 ll' + B_2 mm' + C_3 nn' + C_2 mn'$$
$$+ B_3 m'n + A_3 nl' + C_1 n'l + B_1 lm' + A_2 l'm\}.$$

(2) *With reference to a diagonal.*

$$\text{Let } \Delta = \begin{vmatrix} 0 & b_1 & c_1 & d_1 \\ a_2 & 0 & c_2 & d_2 \\ a_3 & b_3 & 0 & d_3 \\ a_4 & b_4 & c_4 & 0 \end{vmatrix}, \text{ and } K = \begin{vmatrix} x_1 & b_1 & c_1 & d_1 \\ a_2 & x_2 & c_2 & d_2 \\ a_3 & b_3 & x_3 & d_3 \\ a_4 & b_4 & c_4 & x_4 \end{vmatrix}.$$

Let A_1, B_2, C_3, D_4 be the cofactors of the elements in the leading diagonal of Δ, then the value of K is

$$x_1 x_2 x_3 x_4 - (x_2 x_3 d_1 a_4 + x_3 x_1 d_2 b_4 + x_1 x_2 d_3 c_4 + x_1 x_4 b_3 c_2 + x_2 x_4 c_1 a_3 + x_3 x_4 a_2 b_1)$$
$$+ x_1 A_1 + x_2 B_2 + x_3 C_3 + x_4 D_4 + \Delta.$$

Note that the expression in brackets is the sum of $x_2 x_3 d_1 a_4$, $x_1 x_4 b_3 c_2$ and terms derived from these by the cyclic transpositions (abc), (123).

The verification is left to the reader.

▪ **20. Definitions:** A determinant of the nth order is often written in the form

$$\begin{vmatrix} a_{11} & a_{12} & a_{13} \cdots a_{1n} \\ a_{21} & a_{22} & a_{23} \cdots a_{2n} \\ a_{31} & a_{32} & a_{33} \cdots a_{3n} \\ \cdots\cdots\cdots\cdots\cdots\cdots \\ a_{n1} & a_{n2} & a_{n3} \cdots a_{nn} \end{vmatrix} \equiv (a_{11}\, a_{22} \cdots a_{nn}).$$

Denoting any element by a_{rs}, the determinant is said to be *symmetric* if $a_{sr} = a_{rs}$. If $a_{sr} = -a_{rs}$, the determinant is *skew-symmetric:* it is implied that all the elements in the leading diagonal are zero.

For example, if $\qquad K = \begin{vmatrix} a & h & g & l \\ h & b & f & m \\ g & f & c & n \\ l & m & n & 0 \end{vmatrix}, \quad L = \begin{vmatrix} 0 & z & y \\ -z & 0 & x \\ -y & -x & 0 \end{vmatrix},$

the determinant K is symmetric and L is skew-symmetric. We also say that the first determinant is *bordered* by l, m, n.

■ 21. Symmetric Determinants:

(1) *If A_{rs}, A_{sr} are the cofactors of the elements a_{rs}, a_{sr} of a symmetric determinant Δ, then $A_{rs} = A_{sr}$.*

For A_{rs} is transformed into A_{sr} by changing rows into columns.

Thus if $\Delta = (a_{11}\ a_{22}\ a_{33})$,

$$A_{23} = -\begin{vmatrix} a_{11} & a_{12} \\ a_{31} & a_{32} \end{vmatrix} = -\begin{vmatrix} a_{11} & a_{31} \\ a_{12} & a_{32} \end{vmatrix} = -\begin{vmatrix} a_{11} & a_{13} \\ a_{21} & a_{23} \end{vmatrix} = A_{32}.$$

(2) For the determinants $\Delta = \begin{vmatrix} a & h & g \\ h & b & f \\ g & f & c \end{vmatrix}$, $K = \begin{vmatrix} a & h & g & l \\ h & b & f & m \\ g & f & c & n \\ l & m & n & 0 \end{vmatrix}$,

we have the following important relations:

(i) $BC - F^2 = \Delta a$, $GH - AF = \Delta f$, etc.;

(ii) $K = -(Al^2 + Bm^2 + Cn^2 + 2Fmn + 2Gnl + 2Hlm)$,

where $A, B,...$ are the cofactors of $a, b,...$ in Δ.

These follow from Arts. 18, 19.

(3) Consider the following symmetric determinants:

$$\Delta = \begin{vmatrix} a & h & g \\ h & b & f \\ g & f & c \end{vmatrix}, \quad K = \begin{vmatrix} a & h & g & l \\ h & b & f & m \\ g & f & c & n \\ l & m & n & r \end{vmatrix}.$$

Let $A, B, ...$ be the cofactors of $a, b, ...$ in Δ. We shall prove that,

if $\Delta = 0$, then $AK = -(Al + Hm + Gn)^2$

Because $\Delta = 0$, the coefficient of r in the expansion of K is zero, therefore

$$-AK = A\,(Al^2 + Bm^2 + Cn^2 + 2Fmn + 2Gnl + 2Hlm).$$

Moreover, $AB - H^2 = \Delta c$, $CA - G^2 = \Delta b$, $GH - AF = \Delta f$, etc.

Hence $AB = H^2$, $AC = G^2$, $AF = GH$, etc., and therefore

$-AK = A^2l^2 + H^2m^2 + G^2n^2 + 2GHmn + 2AGnl + 2AHlm = (Al + Hm + Gn)^2$. In the same way we can prove that

$$-BK = (Hl + Bm + Fn)^2 \quad \text{and} \quad -CK = (Gl + Fm + Cn)^2.$$

More generally, using the notation of Art. 20, *let*

$$\Delta_n = (a_{11}a_{22} \dots a_{nn}), \Delta_{n-1} = (a_{11}a_{22} \dots a_{n-1,\,n-1}),$$

the determinants being symmetric, and let A_{11}, A_{12}, ... be the cofactors of a_{11}, a_{12}, ... in Δ_{n-1}. Then, if $\Delta_{n-1} = 0$,

$$A_{11}\Delta_n = -(A_{11}a_{1n} + A_{12}a_{2n} + \dots + A_{1,\,n-1}\,a_{n-1,\,n})^2,$$

with similar values for $A_{22}\Delta_n$, $A_{33}\Delta_n$, ...

For, expanding as in Art. 19, (1), since $a_{sr} = a_{rs}$, $A_{sr} = A_{rs}$ and $\Delta_{n-1} = 0$, we have
$$\Delta_n = - (A_{11}a_{1n}^2 + A_{22}a_{2n}^2 + ... + 2A_{12}a_{1n}a_{2n} + ...).$$

Now, since $\Delta_{n-1} = 0$, every minor of the reciprocal determinant
$$(A_{11}A_{22} ... A_{n-1,\, n-1})$$
is zero. Hence, by Art. 18, (2), $A_{11}A_{pp} = A_{1p^2}$ and $A_{11}A_{pq} = A_{1p}A_{1q}$.

Using such identities, it follows that $\Delta_{11}\Delta_n$ has the value stated.

(4) *The following result is of great importance.*

Consider the symmetric determinants

$$\begin{vmatrix} a - x & h \\ h & b - x \end{vmatrix}, \quad \begin{vmatrix} a - x & h & g \\ h & b - x & f \\ g & f & c - x \end{vmatrix}, \quad \begin{vmatrix} a - x & h & g & l \\ h & b - x & f & m \\ g & f & c - x & n \\ l & m & n & d - x \end{vmatrix}.$$

Denote these by Δ_2, Δ_3, Δ_4. Let $\Delta_1 = a - x$, and suppose the sequence Δ_1, Δ_2, Δ_3 ... to be continued indefinitely by constructing determinants of the above type. Then the roots of the equations
$$\Delta_1 = 0, \quad \Delta_2 = 0, \quad \Delta_3 = 0, \quad ... \ ,$$
are all real, and the roots of any one of them are separated by those of the preceding equation.

If λ_1, λ_2 ($\lambda_1 < \lambda_2$) are the roots of $\Delta_2 = 0$, then $\lambda_1 < a < \lambda_2$, (see Exercise XI, 2).

Also, as in the last section, if $\Delta_2 = 0$, then $\Delta_1\Delta_3$ is negative. Hence if x has the values $-\infty$, λ_1, λ_2, ∞, the signs of Δ_3 are $+ - + -$.

Therefore the roots of $\Delta_3 = 0$ are real, and denoting them in ascending order by μ_1, μ_2, μ_3, we have $\mu_1 < \lambda_1 < \mu_2 < \lambda_2 < \mu_3$.

Again, as in the last section, if $\Delta_3 = 0$ then $\Delta_2\Delta_4$ is negative; hence,

if x has the values $\qquad -\infty,\ \mu_1,\ \mu_2,\ \mu_3,\ \infty,$

the signs of Δ_4 are $\qquad +, -, +, -, +;$

therefore the roots of $\Delta_4 = 0$ are real and are separated by those of $\Delta_3 = 0$; and so on without end.

■ **22. Skew-symmetric Determinants:**

(1) *For the skew-symmetric determinant $(a_{11}a_{22}, \therefore a_{nn})$, if A_{rs}, A_{sr} are the cofactors of a_{rs}, a_{sr}, then $A_{sr} = (-1)^{n-1} A_{rs}$.*

For A_{rs} is transformed into A_{sr} by changing the sign of every element and then changing rows into columns; and since A_{rs} is a determinant of $(n - 1)$ rows, the result follows.

Thus in the determinant $\qquad \begin{vmatrix} 0 & c & b \\ -c & 0 & a \\ -b & -a & 0 \end{vmatrix}$

the cofactors of c and $-c$ are $\quad -\begin{vmatrix} -c & a \\ -b & 0 \end{vmatrix}$ and $-\begin{vmatrix} c & b \\ -a & 0 \end{vmatrix},$

and one of these is $(-1)^2$ times the other.

(2) *Every skew-symmetric determinant of odd order is equal to zero.*

For if Δ is a skew-symmetric determinant of odd order n, and Δ' is the determinant obtained by changing the sign of every element in Δ,

$$\Delta' = (-1)^n \Delta = -\Delta.$$

But Δ' can also be obtained from Δ by changing rows into columns, so that $\Delta' = \Delta$: whence the result follows. For example,

$$\begin{vmatrix} 0 & c & b \\ -c & 0 & a \\ -b & -a & 0 \end{vmatrix} = (-1)^3 \begin{vmatrix} 0 & -c & -b \\ c & 0 & -a \\ b & a & 0 \end{vmatrix} = - \begin{vmatrix} 0 & c & b \\ -c & 0 & a \\ -b & -a & 0 \end{vmatrix}.$$

(3) *Every skew-symmetric determinant of even order is a perfect square.*

(*i*) Let $\Delta_n = (a_{11}a_{22} \dots a_{nn})$, $\Delta_{n-1} = (a_{11}a_{22} \dots a_{n-1, n-1})$, where n is even and the determinants are skew-symmetric. For $n = 2$,

$$\Delta_2 = \begin{vmatrix} a_{11} & a_{12} \\ a_{21} & a_{22} \end{vmatrix} = \begin{vmatrix} 0 & a_{12} \\ -a_{12} & 0 \end{vmatrix} = a_{12}^2.$$

Let $A_{11}, A_{12}, \dots$ be the cofactors of $a_{11}, a_{12}, \dots$ in Δ_{n-1}. Since n is even, by the preceding $\Delta_{n-1} = 0$ and $A_{sr} = (-1)^{n-2} A_{rs} = A_{rs}$. Remembering that $a_{sr} = -a_{rs}$, exactly as in Art. 21, (3), we can show that

$$A_{11}\Delta_n = (A_{11}a_{1n} + A_{12}a_{2n} + \dots + A_{1, n-1}\, a_{n-1, n})^2. \qquad \dots\text{(A)}$$

Hence if A_{11} is a perfect square, so also is Δ_n ; moreover, A_{11} is a skew-symmetric determinant of order $n - 2$; and it has been shown that such a determinant of order two is a perfect square, therefore the same is true for those of orders four, six, etc.

(*ii*) If n is even and $\Delta_n = \delta_n^2$, the value of δ_n* is given by the following rule.

Write down all the arrangements in pairs of the numbers 1, 2, 3, ... n. Denote any such arrangement by (hk), (lm), ... (uv), then

$$\delta_n = \Sigma(\pm\, a_{hk}a_{lm} \dots a_{uv}),$$

where the positive or negative sign is to be taken according as the arrangement hklm ... uv is derived from 1234 ...n by an even or by an odd number of transpositions. The term $a_{12}a_{34} \dots a_{n-1, n}$ is to have the sign +.

(α) The rule is suggested by the case when $n = 4$. By equation (A),

$$A_{11}\Delta_4 = (A_{11}a_{14} + A_{12}a_{24} + A_{13}a_{34})^2,$$

where A_{11}, A_{12}, A_{13} are the cofactors of a_{11}, a_{12}, a_{13} in Δ_3, so that $A_{11} = a_{23}^2$, $A_{12} = -a_{13}a_{23}$, $A_{13} = a_{12}a_{23}$. Whence we find that

$$\Delta_4 = (a_{12}a_{34} - a_{13}a_{24} + a_{14}a_{23})^2,$$

which agrees with the rule.

(β) *Assuming that the rule holds for determinants of order $n - 2$, we shall prove that it holds for those of order n.* Then, since it holds when $n = 4$, it will hold for $n = 6, 8$, etc. By equation (A),

$$\delta_n = a_{1n}\sqrt{A_{11}} + a_{2n}\frac{A_{12}}{\sqrt{A_{11}}} + \dots + a_{n-1, n}\frac{A_{1, n-1}}{\sqrt{A_{11}}}.$$

---

* Named the Pfaffian, by Cayley, after J. F. Pfaff.

Now $\Delta_{n-1} = 0$, therefore by Art. 18, (2), $A_{12}{}^2 = A_{11}A_{22}$, $A_{13}{}^2 = A_{11}A_{33}$, etc. Hence

$$\delta_n = a_{1n}\sqrt{A_{11}} \pm a_{2n}\sqrt{A_{22}} \pm \ldots \pm a_{n-1,\,n}\sqrt{A_{n-1,\,n-1}}, \qquad \ldots\text{(B)}$$

where the values of $\sqrt{A_{11}}$, $\sqrt{A_{22}}$, etc., are given by the rule. We shall show that *the signs on the right are alternately positive and negative.*

Consider the product $P = a_{1n}\sqrt{A_{11}} \cdot a_{pn}\sqrt{A_{pp}}$. The values of $\sqrt{A_{11}}$ and $\sqrt{A_{pp}}$ are obtained by the rule from the sets

$$2, 3, 4, \ldots n-1 \quad \text{and} \quad 1, 2, 3, \ldots p-1, p+1, \ldots n-1.$$

Moreover, every term in the expansion of P, with the proper sign, must occur in Δ_n.

For clearness, the element a_{rs} will be denoted by (rs), so that $(sr) = -(rs)$.

If p is even, the product P contains a term M, which is the product of

$$\underline{(1n)}\ (23)(45)\ldots(p-2,\,p-1)\ \underline{(p,\,p+1)}\ (p+2,\,p+3)\ldots(n-2,\,n-1)$$

and $(pn)\ (12)\ (34)\ldots(p-3,\,p-2)(p-1,\,p+1)\underline{(p+2,\,p+3)\ldots(n-2,\,n-1)}$.

By changing the order of the numbers in each of the $\frac{1}{2}(n-p)+1$ pairs which are underlined, it will be seen that $(-1)^{\frac{1}{2}(n-p)+1}M$ is the product of elements (rs) in which corresponding values of r, s are as below.

r	1,	2,	3,... $p-2$,	$p-1$,	p,	$p+1$,	$p+2$,	$p+3$,... $n-2$,	$n-1$,	n,
s	2,	3,	4,... $p-1$,	$p+1$,	n,	p,	$p+3$,	$p+2$,... $n-1$,	$n-2$,	1.

The numbers in the second line can be written in the same order as those in the first by $\frac{1}{2}(n+p)$ transpositions, and since

$$\tfrac{1}{2}(n-p) + 1 + \tfrac{1}{2}(n+p) = n+1, \text{ an odd number,}$$

it follows that $-M$ is a term in Δ_n, so that $-a_{pn}\sqrt{A_{pp}}$ occurs in the value of δ_n.

If p is odd, the corresponding process requires changes of order in $\frac{1}{2}(n-p+1)$ pairs, and $\frac{1}{2}(n+p-1)$ transpositions. Since the sum of these numbers is n, which is even, $+a_{pn}\sqrt{A_{pp}}$ occurs in δ_n; thus

$$\delta_n = \Sigma_{p=1}^{p=n-1}(-1)^{p-1}a_{pn}\sqrt{A_{pp}}.$$

It remains to be shown that the value of δ_n can be written down by the rule. It is obviously necessary only to consider the first term in the value of $a_{pn}\sqrt{A_{pp}}$.

This is $(pn)(12)(34)\ldots(p-1,\,p+1)(p+2,\,p+3)\ldots(n-2,\,n-1)$,

or $(pn)(12)(34)\ldots(p-2,\,p-1)(p+1,\,p+2)\ldots(n-2,\,n-1)$, according as p is even or odd. Now the arrangement

$$p, n, 1, 2, 3, \ldots p - 1, p + 1, \ldots n - 1$$

can be derived from $1, 2, 3, \ldots n$ by $p + n - 1$ transpositions. The rule asserts that the sign of the term is $+$ or $-$ according as $p + n - 1$ is even or odd, that is according as p is odd or even, which agrees with what has been proved.

EXERCISE XVI

A. MISCELLANEOUS METHODS

1. If $a + b + c = 0$, solve the equation

$$\begin{vmatrix} a-x & c & b \\ c & b-x & a \\ b & a & c-x \end{vmatrix} = 0.$$

2. Prove that

$$(b_1 c_2)(a_1 d_2) + (c_1 a_2)(b_1 d_2) + (a_1 b_2)(c_1 d_2) = 0.$$

[Use the identities

$$a_1(b_1 c_2) + b_1(c_1 a_2) + c_1(a_1 b_2) = 0, \quad a_2(b_1 c_2) + b_2(c_1 a_2) + c_2(a_1 b_2) = 0.]$$

3. If $(a_1 b_2 c_3) = 0$ and $(d_1 b_2 c_3) = 0$, prove that

$$\frac{(b_2 c_3)}{(a_2 d_3)} = \frac{(b_3 c_1)}{(a_3 d_1)} = \frac{(b_1 c_2)}{(a_1 d_2)}.$$

4. Prove that $\begin{vmatrix} b_1 + c_1 & c_1 + a_1 & a_1 + b_1 \\ b_2 + c_2 & c_2 + a_2 & a_2 + b_2 \\ b_3 + c_3 & c_3 + a_3 & a_3 + b_3 \end{vmatrix} = 2(a_1 b_2 c_3).$

5. Prove that

$$\begin{vmatrix} bc - a^2 & ca - b^2 & ab - c^2 \\ -bc + ca + ab & bc - ca + ab & bc + ca - ab \\ (a+c)(a+b) & (a+b)(b+c) & (b+c)(c+a) \end{vmatrix}$$

$$= 3(b - c)(c - a)(a - b)(a + b + c)(bc + ca + ab).$$

6. If $\begin{vmatrix} 1/(a+x) & 1/(b+x) & 1/(c+x) \\ 1/(a+y) & 1/(b+y) & 1/(c+y) \\ 1/(a+z) & 1/(b+z) & 1/(c+z) \end{vmatrix} = \dfrac{P}{Q},$

where Q is the product of the denominators, prove that

$$P = (b - c)(c - a)(a - b)(y - z)(z - x)(x - y).$$

7. Show that $\begin{vmatrix} 0 & c & b & a \\ c & 0 & a & b \\ b & a & 0 & c \\ a & b & c & 0 \end{vmatrix} = -(a + b + c)(b + c - a)(c + a - b)(a + b - c).$

8. Prove that the 'skew' determinant
$$\begin{vmatrix} x & c & b & l \\ -c & x & a & m \\ -b & -a & x & n \\ -l & -m & -n & x \end{vmatrix}$$

is equal to $x^4 + x^2 (a^2 + b^2 + c^2 + l^2 + m^2 + n^2) + (al - bm + cn)^2$.

9. Prove that
$$\begin{vmatrix} 0 & c & b & b-c \\ c & 0 & a & c-a \\ b & a & 0 & a-b \\ a & b & c & a+b+c \end{vmatrix} = 2(a + b + c)\{abc - (b - c)(c - a)(a - b)\}.$$

10. Prove that
$$\begin{vmatrix} -1 & b & c & d \\ a & -1 & c & d \\ a & b & -1 & d \\ a & b & c & -1 \end{vmatrix}$$
is equal to $(a + 1)(b + 1)(c + 1)(d + 1)$

$$\left\{1 - \frac{a}{a+1} - \frac{b}{b+1} - \frac{c}{c+1} - \frac{d}{d+1}\right\}.$$

[This follows from Ex. XV, 13, by putting $1 + a_1 = -1/a, 1 + a_2 = -1/b$, etc.; or, directly, by the same method.]

11. Prove the following:

(i)
$$\begin{vmatrix} -1 & 0 & 0 & a \\ 0 & -1 & 0 & b \\ 0 & 0 & -1 & c \\ x & y & z & -1 \end{vmatrix} = 1 - ax - by - cz.$$

(ii)
$$\begin{vmatrix} 0 & -c & b & x \\ c & 0 & -a & y \\ -b & a & 0 & z \\ x' & y' & z' & w \end{vmatrix} = -(ax + by + cz)(ax' + by' + cz').$$

(iii)
$$\begin{vmatrix} 0 & c^2 & b^2 & 1 \\ c^2 & 0 & a^2 & 1 \\ b^2 & a^2 & 0 & 1 \\ 1 & 1 & 1 & 0 \end{vmatrix} = -(a + b + c)(b + c - a)(c + a - b)(a + b - c).$$

(iv)
$$\begin{vmatrix} 0 & c^2 & b^2 & d^2 \\ c^2 & 0 & a^2 & e^2 \\ b^2 & a^2 & 0 & f^2 \\ d^2 & e^2 & f^2 & 0 \end{vmatrix} = -16s (s - ad)(s - be)(s - cf),$$

where $2s = ad + be + cf$.

12. Denoting the squares of the differences $\beta - \gamma$, $\gamma - \alpha$, $\alpha - \beta$ by l, m, n and those of $\alpha - \delta$, $\beta - \delta$, $\gamma - \delta$ by l', m', n' respectively, prove that

$$\begin{vmatrix} 0 & n & m & l' \\ -n & 0 & l & m' \\ -m & -l & 0 & n' \\ -l' & -m' & -n' & 0 \end{vmatrix} = 4\,l\,n\,l'\,n'$$

[If $\beta\gamma + \alpha\delta = \lambda$, $\gamma\alpha + \beta\delta = \mu$, $\alpha\beta + \gamma\delta = v$, then $ll' = (\mu - v)^2$, etc.]

13. If $\Delta = \begin{vmatrix} 4 & 5 & 6 & x \\ 5 & 6 & 7 & y \\ 6 & 7 & 8 & z \\ x & y & z & 0 \end{vmatrix}$, $\Delta' = \begin{vmatrix} 1^2 & 2^2 & 3^2 & 4^2 & x \\ 2^2 & 3^2 & 4^2 & 5^2 & y \\ 3^2 & 4^2 & 5^2 & 6^2 & z \\ 4^2 & 5^2 & 6^2 & 7^2 & w \\ x & y & z & w & 0 \end{vmatrix}$,

prove that
$$\Delta = (x - 2y + z)^2,$$
$$\Delta' = 8(x - 3y + 3z - w)^2.$$

14. If in a determinant Δ, when $x = a$:

(*i*) p rows become the same, then $(x - a)^{p-1}$ is a factor of Δ;

(*ii*) p rows become the same and q columns become the same, then $(x - a)^{p+q-2}$ is a factor, provided that the common elements are zero or are divisible by $(x - a)^2$.

B. PRODUCTS

15. Write down as a determinant the product

$$\begin{vmatrix} a & b & c \\ b & c & a \\ c & a & b \end{vmatrix} \cdot \begin{vmatrix} x & y & z \\ z & x & y \\ y & z & x \end{vmatrix}.$$

Hence show that the product of two expressions of the form $a^3 + b^3 + c^3 - 3abc$ is of the same form.

16. Prove that

$$\begin{vmatrix} 2bc - a^2 & c^2 & b^2 \\ c^2 & 2ca - b^2 & a^2 \\ b^2 & a^2 & 2ab - c^2 \end{vmatrix} = (a^3 + b^3 + c^3 - 3abc)^3.$$

[Consider the product $\begin{vmatrix} a & b & c \\ b & c & a \\ c & a & b \end{vmatrix} \cdot \begin{vmatrix} -a & c & b \\ -b & a & c \\ -c & b & a \end{vmatrix}$.]

17. By forming the product on the left, prove the identity on the right,

$$\begin{vmatrix} a_1 & b_1 & c_1 \\ a_2 & b_2 & c_2 \\ a_3 & b_3 & c_3 \end{vmatrix} \begin{vmatrix} c_1 & -2b_1 & a_1 \\ c_2 & -2b_2 & a_2 \\ c_3 & -2b_3 & a_3 \end{vmatrix} 2(a_1 b_2 c_3)^2 = \begin{vmatrix} 2\Delta_1 & I_{12} & I_{31} \\ I_{12} & 2\Delta_2 & I_{23} \\ I_{31} & I_{23} & 2\Delta_3 \end{vmatrix},$$

where $\Delta_1 = a_1 c_1 - b_1{}^2,\ I_{12} = a_1 c_2 + a_2 c_1 - 2b_1 b_2$, etc.

18. By forming the product

$$\begin{vmatrix} a + \iota b & c + \iota d \\ -c + \iota d & a - \iota b \end{vmatrix} \cdot \begin{vmatrix} b' + \iota a' & d' + \iota c' \\ -d' + \iota c' & b' - \iota a' \end{vmatrix},$$

prove that $\quad (a^2 + b^2 + c^2 + d^2)(a'^2 + b'^2 + c'^2 + d'^2)$

$$\equiv (aa' + bb' + cc' + dd')^2 + (l + l')^2 + (m + m')^2 + (n + n')^2,$$

where $l = (bc'),\ m = (ca'),\ n = (ab'),\ l' = (ad'),\ m' = (bd'),\ n' = (cd')$.

19. Express $(1^2 + 2^2 + 3^2 + 5^2)^2$ as the sum of four squares by taking 1, 2, 3, 5 for a, b, c, d and 2, 5, 1, 3 for a', b', c', d'.

20. By multiplying the arrays on the left, prove that the determinant Δ is zero.

$$\begin{Vmatrix} l & l' \\ m & m' \\ n & n' \end{Vmatrix}, \begin{Vmatrix} l' & l \\ m' & m \\ n' & n \end{Vmatrix};$$

$$\Delta = \begin{vmatrix} 2ll' & lm' + l'm & nl' + n'l \\ lm' + l'm & 2mm' & mn' + m'n \\ nl' + n'l & mn' + m'n & 2nn' \end{vmatrix}.$$

Hence prove that if $ax^2 + 2hxy + by^2 + 2gx + 2fy + c$ is the product of two linear factors, then

$$\begin{vmatrix} a & h & g \\ h & b & f \\ g & f & c \end{vmatrix} = 0.$$

21. Prove that

$$\begin{vmatrix} (a_1 - b_1)^2 & (a_1 - b_2)^2 & (a_1 - b_3)^2 & (a_1 - b_4)^2 \\ (a_2 - b_1)^2 & (a_2 - b_2)^2 & (a_2 - b_3)^2 & (a_2 - b_4)^2 \\ (a_3 - b_1)^2 & (a_3 - b_2)^2 & (a_3 - b_3)^2 & (a_3 - b_4)^2 \\ (a_4 - b_1)^2 & (a_4 - b_2)^2 & (a_4 - b_3)^2 & (a_4 - b_4)^2 \end{vmatrix} = 0.$$

[Multiply the two arrays of four rows $\begin{Vmatrix} a_1^2 & a_1 & 1 \\ & \text{etc.} & \end{Vmatrix}, \begin{Vmatrix} 1 & -2b_1 & b_1^2 \\ & \text{etc.} & \end{Vmatrix}$.]

C. MISCELLANEOUS THEOREMS AND EXAMPLES

22. Prove that (*i*)

$$\begin{vmatrix} x & 1 & 1 & \cdots \\ 1 & x & 1 & \cdots \\ 1 & 1 & x & \cdots \\ & & \cdots \cdots \cdots \end{vmatrix} = (x-1)^{n-1}(x+n-1),$$

the determinant being of the *n*th order.

(*ii*)

$$\begin{vmatrix} a & a+b & a+2b & \cdots & a+nb \\ a+nb & a & a+b & \cdots & a+(n-1)b \\ a+(n-1)b & a+nb & a & \cdots & a+(n-2)b \\ \cdots \cdots & \cdots \cdots & \cdots \cdots & & \cdots \cdots \\ a+b & a+2b & a+3b & \cdots & a \end{vmatrix} = (-b)^n(n+1)^n\left(a+\tfrac{1}{2}nb\right).$$

23. If $u_n =$

$$\begin{vmatrix} a_1 & 1 & 0 & 0 & \cdots & \cdots & 0 \\ -1 & a_2 & 1 & 0 & \cdots & \cdots & 0 \\ 0 & -1 & a_3 & 1 & \cdots & \cdots & 0 \\ & & & \cdots \cdots \cdots \cdots & & & \\ 0 & 0 & \cdots & \cdots & -1 & a_{n-1} & 1 \\ 0 & 0 & \cdots & \cdots & 0 & -1 & a_n \end{vmatrix},$$

prove that $u_n = a_n u_{n-1} + u_{n-2}$.

[Expand with reference to the bottom row. Such determinants are called *continuants*.]

24. If

$$\begin{vmatrix} a & b & c & d & 0 & 0 \\ a' & b' & c' & d' & 0 & 0 \\ a'' & b'' & c'' & d'' & 0 & 0 \\ 0 & a & b & 0 & c & d \\ 0 & a' & b' & 0 & c' & d' \\ 0 & a'' & b'' & 0 & c'' & d'' \end{vmatrix} = 0,$$

prove that

$$\begin{vmatrix} a & c & d \\ a' & c' & d' \\ a'' & c'' & d'' \end{vmatrix}^2 = \begin{vmatrix} a & b & d \\ a' & b' & d' \\ a'' & b'' & d'' \end{vmatrix} \times \begin{vmatrix} b & c & d \\ b' & c' & d' \\ b'' & c'' & d'' \end{vmatrix}$$

25. (*i*) By using the fact that the determinant on the right is zero, prove that

$$t\,(y-z)\,(z-x)\,(x-y) + x\,(y-z)\,(y-t)\,(z-t)$$
$$+\,y\,(z-x)\,(z-t)\,(x-t) + z(x-y)\,(x-t)\,(y-t) \equiv 0.$$

$$\begin{vmatrix} t & x & y & z \\ 1 & 1 & 1 & 1 \\ t & x & y & z \\ t^2 & x^2 & y^2 & z^2 \end{vmatrix}$$

(*ii*) Hence show that, if α, β, γ, δ are any angles,

$\sin(\beta - \gamma)\sin(\gamma - \alpha)\sin(\alpha - \beta) + \sin(\beta - \gamma)\sin(\beta - \delta)\sin(\gamma - \delta)$
$+ \sin(\gamma - \alpha)\sin(\gamma - \delta)\sin(\alpha - \delta) + \sin(\alpha - \beta)\sin(\alpha - \delta)$
$\sin(\beta - \delta) \equiv 0.$

(*iii*) Use the last result to show that if a, b, c, d and a', b', c', d' are the lengths of the sides of two quadrilaterals such that the sides a, b, c, d are parallel to a', b', c', d', respectively,

then $LMN + Lmn + Mnl + Nlm = 0$,

where $l = (bc')$, $m = (ca')$, $n = (ad')$, $L = (ad')$, $M = (bd')$, $N = (cd')$.

[(*ii*) In (*i*) put $\qquad x = \cos 2\alpha + \iota \sin 2\alpha, \qquad y = \cos 2\beta + \iota \sin 2\beta,$
$$z = \cos 2\gamma + \iota \sin 2\gamma,$$
$$t = \cos 2\delta + \iota \sin 2\delta.$$

(*iii*) Let a, b, c, d make angles α, β, γ, δ with a fixed line. By projecting the circuits a, b, c, d and a', b', c', d' on lines perpendicular to d, a, b, c, prove that

$$\frac{\sin(\alpha - \delta)}{l} = \ldots = \ldots = \frac{\sin(\beta - \gamma)}{L} = \ldots = \ldots]$$

26. Prove that the equations (*i*), (*ii*) are respectively the equation to the line joining $(x_1\, y_1)$, $(x_2 y_2)$ and the equation to the plane through $(x_1 y_1 z_1)$, $(x_2 y_2 z_2)$, $(x_3 y_3 z_3)$.

(*i*) $\begin{vmatrix} x & y & 1 \\ x_1 & y_1 & 1 \\ x_2 & y_2 & 1 \end{vmatrix} = 0$
(*ii*) $\begin{vmatrix} x & y & z & 1 \\ x_1 & y_1 & z_1 & 1 \\ x_2 & y_2 & z_2 & 1 \\ x_3 & y_3 & z_3 & 1 \end{vmatrix} = 0.$

Show also that the last equation may be written in the form

$$\begin{vmatrix} x - x_1 & y - y_1 & z - z_1 \\ x_2 - x_1 & y_2 - y_1 & z_2 - z_1 \\ x_3 - x_1 & y_3 - y_1 & z_3 - z_1 \end{vmatrix} = 0.$$

27. If A, B, C, D are four points in a plane and the lengths BC, CA, AB, AD, BD, CD are denoted by a, b, c, x, y, z, prove that these are connected by the equation

$$\begin{vmatrix} 0 & c^2 & b^2 & x^2 & 1 \\ c^2 & 0 & a^2 & y^2 & 1 \\ b^2 & a^2 & 0 & z^2 & 1 \\ x^2 & y^2 & z^2 & 0 & 1 \\ 1 & 1 & 1 & 1 & 0 \end{vmatrix} = 0.$$

[Let $(x_1, y_1,)$, (x_2, y_2), ... be the coordinates of the points. Multiply the arrays

$$\begin{Vmatrix} x_1^2 + y_1^2 & x_1 & y_1 & 1 \\ x_2^2 + y_2^2 & x_2 & y_2 & 1 \\ \hdotsfor{4} \\ 1 & 0 & 0 & 0 \end{Vmatrix}, \quad \begin{Vmatrix} 1 & -2x_1 & -2y_1 & x_1^2 + y_1^2 \\ 1 & -2x_2 & -2y_2 & x_2^2 + y_2^2 \\ \hdotsfor{4} \\ 0 & 0 & 0 & 1 \end{Vmatrix}.]$$

28. In a tetrahedron $ABCD$, the lengths BC, CA, AB, AD, BD, CD are denoted by a, b, c, d, e, f respectively; the volume is V, the radius of the circumscribing sphere is R.

Assuming equation (i), where $(x_1, y_1, z_1,)$, $(x_2, y_2, z_2,)$, ... are the coordinates of A, B ... , prove equation (ii).

$$(i)\ 6V = \begin{vmatrix} x_1 & y_1 & z_1 & 1 \\ x_2 & y_2 & z_2 & 1 \\ x_3 & y_3 & z_3 & 1 \\ x_4 & y_4 & z_4 & 1 \end{vmatrix}, \qquad (ii)\ -36R^2V^2 = \tfrac{1}{16} \begin{vmatrix} 0 & c^2 & b^2 & d^2 \\ c^2 & 0 & a^2 & e^2 \\ b^2 & a^2 & 0 & f^2 \\ d^2 & e^2 & f^2 & 0 \end{vmatrix}^*.$$

[Take the origin at the centre of the sphere. Multiply the determinants

$$6RV = \begin{vmatrix} x_1 & y_1 & z_1 & R \\ x_2 & y_2 & z_2 & R \\ \hdotsfor{4} \end{vmatrix}, \quad -6RV = \begin{vmatrix} x_1 & y_1 & z_1 & -R \\ x_2 & y_2 & z_2 & -R \\ \hdotsfor{4} \end{vmatrix},$$

and use the equations
$$x_1^2 + y_1^2 + z_1^2 = R^2, \ (x_1 - x_2)^2 + (y_1 - y_2)^2 + (z_1 - z_2)^2 = c^2, \text{ etc.}]$$

---

* Note the value of this determinant given in Ex. 11, (iv).

■ ■ ■

CHAPTER 10

Systems of Equations

■ **1. Systems of Equations:** If $u = 0$ is an equation connecting the variables x, y, ..., which is satisfied when $x = x_1$, $y = y_1$, ..., then $(x_1, y_1, ...)$ is called a *solution* of $u = 0$.

Let $u = 0$, $v = 0$,... be a system of equations connecting x, y, ... If the equations have a common solution $(x_1, y_1, ...)$, they are said to be *consistent*, and $(x_1, y_1, ...)$ is called a *solution of the system* $u = 0$, $v = 0$, ... Equations which have no common solution are said to be *inconsistent*.

Equations which are *homogeneous* in x, y,... have the common solution $(0, 0, ...)$, in which each of the variables is zero: *but they are not said to be consistent unless some other common solution exists.*

If one equation of a system is satisfied by *all* the values of x, y, ... which satisfy the others, the equations are said to be *not independent.*

To *eliminate* a variable x from two equations is to obtain from these two an equation free from x. Any result so obtained must hold if the equations are consistent.

To *solve* a system of equations is to obtain all the solutions (if any exist). This is done by a process of elimination.

Illustrations. The equations $x + y = 1$, $x + y = 2$, which represent parallel lines (or planes), are *inconsistent*. So also are the equations $u = 0$, $v = 0$, $2u + 3v = 4$.

The equations $u = 0$, $v = 0$, $lu + mv = 0$ are *not independent.*

■ **2. Equivalent Systems:**

(1) Two systems of equations are said to be *equivalent* when every solution of either system is a solution of the other. In such cases we may say that one system *is the same* as the other.

(2) *If a_1, a_2, ... are constants and $a_1 \neq 0$, the system of equations*

$$u_1 = 0, \quad u_2 = 0, \quad ... \quad u_n = 0,$$

is equivalent to the system

$$a_1 u_1 + a_2 u_2 + ... + a_n u_n = 0, \quad u_2 = 0, \quad ... \quad u_n = 0.$$

For any solution of the first system is a solution of the second and, since a_1 is a constant (not zero), any solution of the second system is a solution of the first.

(3) *If l, m, l', m' are constants such that $lm' - l'm \neq 0$, then the pair of equations $u = 0$, $v = 0$ is equivalent to the pair*

$$lu + mv = 0, \quad l'u + m'v = 0.$$

For any values of the variables x, y,... which satisfy $u = 0$, $v = 0$ also satisfy the second pair.

Moreover, any values of x, y, ... which satisfy the second pair also satisfy

$$m' (lu + mv) - m (l'u + m'v) = 0,$$
$$- l' (lu + mv) + l (l'u + m'v) = 0,$$

which are the same as

$$(lm' - l'm) u = 0, \quad (lm' - l'm) v = 0,$$

and since $lm' - l'm \neq 0$, these are the same as $u = 0$, $v = 0$.

It is left to the reader to supply proofs of the following:

(4) *The pair of equations $uv = 0$, $w = 0$ is equivalent to the two pairs*

$$u = 0, w = 0 \quad and \quad u = 0, w = 0.$$

(5) *Any solution of the system*

$$u_1 = v_1, u_2 = v_2$$

is a solution of

$$u_1 u_2 = v_1 v_2, u_2 = v_2.$$

But the second pair of equations is equivalent to the two pairs

$$u_1 = v_1, u_2 = v_2 \ and \ u_2 = 0, v_2 = 0.$$

■ **3. Systems of Linear Equations:** In considering this subject, the chief difficulty is to include all the special cases which may arise.

(1) *The equation $ax + by + c = 0$ has infinitely many solutions, unless $a = 0$, $b = 0$, $c \neq 0$, in which case there is no solution.*

A formal proof is hardly required.

(2) Consider the equations

$$u_1 \equiv a_1 x + b_1 y + c_1 = 0$$
$$u_2 \equiv a_2 x + b_2 y + c_2 = 0,$$

excluding the case in which either is of the form $0x + 0y + c = 0$.

(i) *If $(a_1 b_2) \neq 0$, there is an unique solution.*

For by Art. 2, (3), the equations are equivalent to

$$u_1 b_2 - u_2 b_1 = 0, - u_1 a_2 + u_2 a_1 = 0,$$

which are the same as

$$x (a_1 b_2) = (b_1 c_2), y (a_1 b_2) = (c_1 a_2).$$

(ii) *If $(a_1 b_2) = 0$, the equations are inconsistent unless also $(b_1 c_2) = 0$ and $(c_1 a_2) = 0$.*

For if $(a_1 b_2) = 0$, we have $u_1 b_2 - u_2 b_1 \equiv - (b_1 c_2)$, $u_1 a_2 - u_2 a_1 \equiv (c_1 a_2)$. Hence the equations cannot have a common solution unless $(b_1 c_2)$ and $(c_1 a_2)$ are both zero.

(iii) *If $(a_1 b_2)$, $(b_1 c_2)$, $(c_1 a_2)$ are all zero, the two equations are one and the same, and there are infinitely many solutions.*

For in this case

$$u_1 b_2 \equiv u_2 b_1 \ and \ u_1 a_2 \equiv u_2 a_1.$$

Hence any solution of one equation is a solution of the other, or else an a is zero *and* a b is zero. The latter alternative cannot occur.

For example, suppose that $a_1 = 0$; then by hypothesis $b_1 \neq 0$ and $a_2 b_1 = a_1 b_2 = 0$, therefore $a_2 = 0$ and $b_2 \neq 0$.

Hence the equations are one and the same.

(*iv*) *The case in which the equations are inconsistent* may be regarded from another point of view.

Suppose the coefficients of x, y to vary in such a way that $(a_1 b_2) \to 0$ while $(b_1 c_2)$ and $(c_1 a_2)$ remain finite; then $x \to \infty$, $y \to \infty$ and the fractions x/y, $- b_1/a_1$, $- b_2/a_2$ tend to equality.

We shall say that *in the limiting case when* $(a_1 b_2) = 0$, *the equations are satisfied by infinite values of* x, y *which are in the ratio* $- b_1 : a_1$ *or* $- b_2 : a_2$.

(3) *If* $a_1 x + b_1 y = 0$ *and* $a_2 x + b_2 y = 0$, *then either* $x = 0, y = 0$ *or else* $(a_1 b_2) = 0$. *Conversely, if* $(a_1 b_2) = 0$ *these equations have solutions other than* $(0, 0)$, *i.e. they are consistent.*

This is a particular case of the preceding.

The following results are required later.

■ **EXAMPLE 1.** *If* $(a_1 b_2) = 0$ *and* $(a_1 c_2) = 0$, *then* $(b_1 c_2) = 0$ *or else* $a_1 = 0, a_2 = 0$.

For a_1, a_2, satisfy the equations $a_1 b_2 - a_2 b_1 = 0$, $a_1 c_2 - a_2 c_1 = 0$.

■ **EXAMPLE 2.** *If* $(a_1 b_2) = 0$, $(b_1 c_2) = 0$ *and* $(c_1 a_2) = 0$, *where* a_1, b_1, c_1 *are not all zero and* a_2, b_2, c_2, *are not all zero, then of the pairs* (a_1, a_2), (b_1, b_2), (c_1, c_2) *there is at least one in which neither quantity is zero.*

For suppose that $a_1 \neq 0$, then if $a_2 = 0$ we should have $a_1 b_2 = 0$, $a_1 c_2 = 0$; and therefore $b_2 = 0$, c_2, $= 0$, which contradicts the hypothesis. Hence, if $a_1 \neq 0$, then $a_2 \neq 0$.

■ **EXAMPLE 3.** *Let* $\Delta = (a_1 b_2 c_3)$ *and let* A_1, B_1, ... *be the cofactors of* a_1, b_1, ... *Prove that if* $\Delta = 0$ *and* $C_3 = 0$, *then either* $C_1 = 0, C_2 = 0$ *or else* $A_3 = 0, B_3 = 0$.

For we have

$$a_1 C_1 + a_2 C_2 = 0, \quad b_1 C_1 + b_2 C_2 = 0, \quad c_1 C_1 + c_2 C_2 = 0,$$

therefore either $C_1 = 0, C_2 = 0$, or else $(b_1 c_2) = 0$, $(c_1 a_2) = 0$, i.e. $A_3, = 0, B_3 = 0$.

Or thus: Using identities of the type $(B_2 C_3) = \Delta a_1$, it will be seen that $(C_1 A_3)$, $(C_1 B_3)$, $(C_2 A_3)$, $(C_2 B_3)$ are all zero. Whence the result follows.

■ **4. Three Linear Equations:** Consider the equations

$$u_1 \equiv a_1 x + b_1 y + c_1 = 0,$$
$$u_2 \equiv a_2 x + b_2 y + c_2 = 0,$$
$$u_3 \equiv a_3 x + b_3 y + c_3 = 0.$$

Let $\Delta = (a_1 b_2 c_3)$, and let A_1, B_1, ... be the cofactors of a_1, b_1, ...

(1) *If the equations are consistent, then* $\Delta = 0$.

For on account of the identities

$$a_1 C_1 + a_2 C_2 + a_3 C_3 = 0,$$
$$b_1 C_1 + b_2 C_2 + b_3 C_3 = 0,$$
$$c_1 C_1 + c_2 C_2 + C_3 C_3 = \Delta,$$

we have $\qquad C_1 u_1 + C_2 u_2 + C_3 u_3 \equiv \Delta.$

Hence, if the equations are satisfied by the same values of x, y, we must have

$$\Delta = 0.$$

(2) *If $\Delta = 0$ the equations are consistent and have an unique solution except when every C is zero; in which case each of the equations (assumed to be distinct) is inconsistent with either of the others.*

For by the preceding, $u_1 C_1 + u_2 C_2 + u_3 C_3 \equiv 0$.

(i) *If no C is zero,* the equations are not independent.

(ii) *If* $C_1 = 0$, $C_2 \neq 0$, $C_3 \neq 0$, then $u_2 C_2 + u_3 C_3 \equiv 0$ and $u_2 = 0$, $u_3 = 0$ are the same equation.

In both of these cases, since $C_3 \neq 0$, the equations $u_1 = 0$, $u_2 = 0$ have an unique solution, which is the solution of the system.

(iii) *If* $C_1 = 0$, $C_2 = 0$, $C_3 \neq 0$, then $u_3 \equiv 0$. This case is excluded.

(iv) *If every C is zero* and the equations are distinct, each is inconsistent with either of the others. In this case we may say that *the equations are satisfied by infinite values of x, y which are in the ratio* $- b_1 : a_1$.

■ **5. The Line at Infinity:** For the sake of complete generality we say that the 'equation',

$$0x + 0y + c = 0 \quad \text{where} \quad c \neq 0,$$

represents the *line at infinity.* The *ideal* point in which a straight line $u = 0$ meets the line at infinity is called the *point at infinity* on that line.

Parallel straight lines are to be regarded as meeting on the line at infinity. For the lines $u = 0$, $u + k = 0$ intersect on the line $0x + 0y + k = 0$. These conventions enable us to include various special cases under one heading.

Illustration. If $u_1 = 0$, $u_2 = 0$, $u_3 = 0$ (as above) are the equations to three straight lines, then the necessary and sufficient condition that they may meet at a point (which may be at infinity) is $\Delta = 0$.

■ **6. Two Equations in Three Unknowns:** Consider the equations

$$u_1 \equiv a_1 x + b_1 y + c_1 z + d_1 = 0, \qquad \text{...(A)}$$
$$u_2 \equiv a_2 x + b_2 y + c_2 z + d_2 = 0, \qquad \text{...(B)}$$

excluding the case in which a_1, b_1, c_1 or a_2, b_2, c_2 are all zero.

(1) *Of the three $(a_1 b_2)$, $(b_1 c_2)$, $(c_1 a_2)$, suppose that at least one, say $(a_1 b_2)$, is not zero,* then the equations are equivalent to

$$- u_1 a_2 + u_2 a_1 = 0, \quad u_1 b_2 - u_2 b_1 = 0,$$

which are the same as
$$- (a_1 b_2)y + (c_1 a_2)z = (a_1 d_2), \qquad \text{...(C)}$$
$$- (b_1 c_2)z + (a_1 b_2)x = (b_1 d_2). \qquad \text{...(D)}$$

Corresponding to any value of z, there are definite values of x and y, provided that $(b_1 c_2)$ and $(c_1 a_2)$ are not both zero.

If $(b_1 c_2) = 0$ and $(c_1 a_2) = 0$, then x and y have definite values and z may have any value. (In fact, we have $c_1 = 0$, $c_2 = 0$.)

Thus in every case, by Art. 3, Ex. 1, there are infinitely many solutions.

(2) *Suppose that* (a_1b_2), (b_1c_2), (c_1a_2) *are all zero.* Then of the pairs (a_1, a_2), (b_1, b_2), (c_1, c_2) there is at least one, say (a_1, a_2), in which neither quantity is zero. (See Art. 3, Ex. 2.) Also we have

$$u_1a_2 - u_2a_1 \equiv - (a_1\ d_2).\qquad\qquad ...(E)$$

Thus the equations are of the form $u_1 = 0$, $u_1 + k = 0$, where $k = (a_1d_2)/a_2$, and they are inconsistent unless $k = 0$.

(3) *In particular* it follows that the equations

$$a_1x + b_1y + c_1z = 0, \quad a_2x + b_2y + c_2z = 0,$$

are always consistent (*i.e.* they have solutions other than $(0, 0, 0)$). They are equivalent to $x/(b_1c_2) = y/(c_1a_2) = z/(a_1b_2)$, provided that no denominator is zero.

If $(b_1c_2) = 0$, they are equivalent to $x = 0$, $y/(c_1a_2) = z/(a_1b_2)$.

If $(b_1c_2) = 0$ and $(c_1a_2) = 0$, then $x = 0$, $y = 0$, and z may have any value.

(4) If (α, β, γ) is any solution of (A), (B), these may be written

$$a_1(x - \alpha) + b_1\ (y - \beta) + c_1\ (z - \gamma) = 0, \quad a_2(x - \alpha) + b_2\ (y - \beta) + c_2(z - \gamma) = 0,$$

which are equivalent to

$$(x - \alpha)/(b_1c_2) = (y - \beta)/(c_1a_2) = (z - y)/(a_1b_2),\qquad\qquad ...(F)$$

provided that no denominator is zero.

If $(b_1c_2) = 0$, they are equivalent to

$$x = \alpha, \quad (y - \beta)/(c_1a_2) = (z - y)/ (a_1b_2).\qquad\qquad ...(G)$$

If $(b_1c_2) = 0$ and $(c_1a_2) = 0$, then $x = \alpha$, $y = \beta$, and z may have any value.

Note: Much labour can be saved by the following considerations:
Equations (A), (B) are unchanged by the simultaneous cyclic substitutions $(a_1b_1c_1)$, $(a_2b_2c_2)$; (xyz). If then we make these changes in any equation derived from (A) and (B), the resulting equation will also hold. For example, equation (D) can be derived from (C) in this way.

■ **7. The Equation to a Plane:**

(1) The equation $ax + by + cz + d = 0$ represents a plane unless a, b, c are all zero.

For the sake of complete generality we say that the 'equation'

$$0x + 0y + 0z + d = 0 \quad \text{where } d \neq 0$$

represents an *ideal* plane called the *plane at infinity.*

(2) Two Planes. Referring to Art. 6, let $u_1 = 0$, $u_2 = 0$ be the equations to two planes. In general these intersect in a line, and we say that *the equations to the line* are $u_1 = 0$, $u_2 = 0$. The equations may be written in one of the forms described in Art. 6, (4), where (α, β, γ) is any point on the line.

The equation to any plane through the line $u_1 = 0$, $u_2 = 0$ is of the form $u_1 + ku_2 = 0$, where k is a constant.

Parallel Planes. If (a_1b_2), (b_1c_2), (c_1a_2) are all zero, the planes $u_1 = 0$, $u_2 = 0$ do not intersect, that is to say they are parallel.

Hence the equation to any plane parallel to the plane $u_1 = 0$ is of the form $u_1 + k = 0$, where k is a constant. All this follows from Art. 6, (2).

We therefore say that *parallel planes intersect in a line on the plane at infinity.*

EXAMPLE 1: *The condition that the line whose equations are*

$$(x - \alpha)/l = (y - \beta)/m = (z - \gamma)/n \qquad \text{...(A)}$$

may be parallel to the plane

$$ax + by + cz + d = 0$$

is $la + mb + nc = 0$. *If also* $a\alpha + b\beta + c\gamma + d = 0$, *then the line is in the plane.*

If the line meets the plane at (x, y, z) and each fraction in (A) is denoted by r, then

$$x = lr + \alpha, \; y = mr + \beta, \quad z = nr + \gamma,$$

where

$$r(la + mb + nc) + a\alpha + b\beta + c\gamma + d = 0.$$

If $la + mb + nc = 0$, no value of r exists which satisfies the last equation unless $a\alpha + b\beta + c\gamma + d = 0$, in which case r may have any value. Hence the results follow.

EXAMPLE 2. *The condition that the lines whose equations are*

$$\frac{x - \alpha}{l} = \frac{y - \beta}{m} = \frac{z - \gamma}{n}, \; \frac{x - \alpha'}{l'} = \frac{y - \beta'}{m'} = \frac{z - \gamma'}{n'},$$

may meet in a point is $(\alpha - \alpha')(mn' - m'n) + (\beta - \beta')(nl' - n'l) + (\gamma - \gamma')(lm' - l'm)$
$= 0$, *and the lines are parallel if* $l/l' = m/m' = n/n'$.

If the lines meet at (x, y, z) and $(x - \alpha)/l = r$, $(x - \alpha')/l' = r'$, we have

$$lr - l'r' + (\alpha - \alpha') = 0, \; mr - m'r' + (\beta - \beta') = 0, \; nr - n'r' + (\gamma - \gamma') = 0.$$

These equations are satisfied by the same (finite) values of r, r' if

$$\begin{vmatrix} \alpha - \alpha' & l & l' \\ \beta - \beta' & m & m' \\ \gamma - \gamma' & n & n' \end{vmatrix} = 0,$$

provided that (mn'), (nl'), (lm') are not all zero: but if these three are all zero, the values of r, r' are infinite. In the latter case equations (A) are satisfied by infinite values of x, y, z and the lines are parallel.

8. Three Equations in Three Unknowns: Consider the equations

$$\left. \begin{array}{l} u_1 \equiv a_1 x + b_1 y + c_1 z + d_1 = 0 \\ u_2 \equiv a_2 x + b_2 y + c_2 z + d_2 = 0 \\ u_3 \equiv a_3 x + b_3 y + c_3 z + d_3 = 0 \end{array} \right\} . \qquad \text{...(A)}$$

Let $\Delta = (a_1 b_2 c_3)$ and let A_1, B_1, ... be the cofactors of a_1, b_1, ... in Δ.

Also let $\Delta_1 = (d_1 b_2 c_3)$, $\Delta_2 = (a_1 d_2 c_3)$, $\Delta_3 = (a_1 b_2 d_3)$.
On account of the identities

$$\left. \begin{array}{l} a_1 A_1 + a_2 A_2 + a_3 A_3 = \Delta \\ b_1 A_1 + b_2 A_2 + b_3 A_3 = 0 \\ c_1 A_1 + c_2 A_2 + c_3 A_3 = 0 \\ d_1 A_1 + d_2 A_2 + d_3 A_3 = \Delta_1 \end{array} \right\} , \qquad \text{...(B)}$$

we have

and similarly,

and

$$\left. \begin{array}{l} u_1 A_1 + u_2 A_2 + u_3 A_3 \equiv \Delta x + \Delta_1; \\ u_1 B_1 + u_2 B_2 + u_3 B_3 \equiv \Delta y + \Delta_2, \\ u_1 C_1 + u_2 C_2 + u_3 C_3 \equiv \Delta z + \Delta_3. \end{array} \right\} . \qquad \text{...(C)}$$

(1) *Suppose that* $\Delta \neq 0$. The last identities show that *if a solution exists,* it is given by

$$x = -\Delta_1/\Delta, \quad y = -\Delta_2/\Delta, \quad z = -\Delta_3/\Delta. \qquad \text{...(D)}$$

We can show by actual substitution that these values of x, y, z satisfy the equations. Or we may use the following method, in which *every step is reversible, so that verification is unnecessary.*

Because $(A_1B_2C_3) = \Delta^2 \neq 0$, therefore at least one term of $(A_1B_2C_3)$ is not zero. Suppose that $A_1B_2C_3 \neq 0$, so that no one of A_1, B_2, C_3 is zero.*

Because $A_1 \neq 0$, the equations are equivalent to

$$u_2 = 0, \quad u_3 = 0, \quad A_1u_1 + A_2u_2 + A_3u_3 = 0,$$

that is to $u_2 = 0, \; u_3 = 0, \quad \Delta x = -\Delta_1.$

In the same way, because $B_2 \neq 0$ and $C_3 \neq 0$, the equations are equivalent to each of the sets, $u_3 = 0, \; u_1 = 0, \; \Delta y = -\Delta_2; \; u_1 = 0, \; u_2 = 0, \; \Delta z = -\Delta_3.$

Hence the equations have the unique solution, $(-\Delta_1/\Delta, -\Delta_2/\Delta, -\Delta_3/\Delta)$.

(2) *If* $\Delta = 0$ *and* Δ_1, Δ_2, Δ_3 *are not all zero, the equations are inconsistent.*

For example, suppose that $\Delta_1 \neq 0$, then since

$$u_1A_1 + u_2A_2 + u_3A_3 \equiv \Delta_1, \qquad \text{...(E)}$$

it follows that the equations have no common solution.

If we suppose the coefficients to vary so that $\Delta \to 0$, and at least one of Δ_1, Δ_2, Δ_3 remains finite, then at least one of x, y, z tends to infinity. Neglecting d_1, d_2, d_3 in comparison with the large terms, we see that in general the following ratios tend to equality:

$$x : y : z, \quad A_1 : B_1 : C_1, \quad A_2 : B_2 : C_2, \quad A_3 : B_3 : C_3.$$

In the limiting case *when* $\Delta = 0$, *we say that the equations have a solution* (x, y, z) *such that at least one of x, y, z is infinite and*

$$x : y : z = A_1 : B_1 : C_1 = A_2 : B_2 : C_2 = A_3 : B_3 : C_3.$$

(3) *If* Δ, Δ_1, Δ_2, Δ_3 *are all zero, the equations are not all independent and there are infinitely many solutions; or else each of the equations (assumed to be distinct) is inconsistent with either of the others.*

For we have $u_1A_1 + u_2A_2 + u_3A_3 \equiv 0.$ ...(F)

   (i)   *If no A is zero,* the equations are not independent.

   (ii) *If* $A_1 = 0, A_2 \neq 0, A_3 \neq 0$, then $u_2A_2 + u_3A_3 \equiv 0$; and $u_2 = 0, u_3 = 0$ are the same equation.

        In both cases, since $A_3 \neq 0$, the equations $u_1 = 0, u_2 = 0$ have infinitely many solutions which are solutions of the system.

   (iii) *If* $A_1 = 0, A_2 = 0, A_3 \neq 0$, we have $u_3 \equiv 0$, which case is excluded.

---

  * It is essential to show that suffixes p, q, r exist, all different, such that none of A_p, B_q, C_r is zero. This step has been omitted in the textbooks. See Chrystal, *Algebra*, Vol. I, p. 360.

(*iv*) If, in the above, every A is replaced by the corresponding B or C, similar results follow. Therefore the only remaining case is that in which *every minor of Δ is zero*, and then every two of the equations are inconsistent; or they are one and the same.

■ **EXAMPLE 1:** *If a, b, c are unequal, solve the equations $x + y + z = 1$,*
$$ax + by + cz = d, \quad a^2x + b^2y + c^2z = d^2.$$

We have $\quad x \cdot \begin{vmatrix} 1 & 1 & 1 \\ a & b & c \\ a^2 & b^2 & c^2 \end{vmatrix} = \begin{vmatrix} 1 & 1 & 1 \\ d & b & c \\ d^2 & b^2 & c^2 \end{vmatrix}, \quad \therefore \quad x = \dfrac{(d-b)(d-c)}{(a-b)(a-c)},$

and the values of y, z are obtained by the cyclic substitution (abc), (xyz).

■ **9. Three Planes:** Referring to 8, let $u_1 = 0$, $u_2 = 0$, $u_3 = 0$ be the equations to three planes, then:

(*i*) *If $\Delta \neq 0$ they meet in the point given by the equations in Art. 8, (1).*

(*ii*) *If $\Delta = 0$ and Δ_1, Δ_2, Δ_3 are not all zero, two of the planes meet in a line which is parallel to the third plane.*

(*iii*) *If Δ, Δ_1, Δ_2, Δ_3 are all zero, the planes have a common line of intersection or they are all parallel.*

Proof of (*ii*). Suppose that $\Delta_1 \neq 0$. We have
$$d_1 A_1 + d_2 A_2 + d_3 A_3 = \Delta_1, \qquad \text{...(A)}$$
$$u_1 A_1 + u_2 A_2 + u_3 A_3 \equiv \Delta_1. \qquad \text{...(B)}$$
On account of (A), at least one A is not zero. Let $A_1 \neq 0$, then the planes $u_2 = 0$, $u_3 = 0$ are not parallel and, on account of (B), their line of intersection does not meet the plane $u_1 = 0$.

Observe that all the planes are parallel to the line
$$x/A_1 = y/B_1 = z/C_1,$$
which is parallel to the line, $u_2 = 0$, $u_3 = 0$.

The truth of (*iii*) follows from Art. 8, (3).

Hence we conclude that, *if $\Delta = 0$, the planes either meet in a point at infinity, or else they have a common line of intersection, which may be at infinity.*

■ **10. Four Equations in Three Unknowns:** Consider the equations
$$u_1 \equiv a_1x + b_1y + c_1z + d_1 = 0,$$
$$u_2 \equiv a_2x + b_2y + c_2z + d_2 = 0,$$
$$u_3 \equiv a_3x + b_3y + c_3z + d_3 = 0,$$
$$u_4 \equiv a_4x + b_4y + c_4z + d_4 = 0.$$
Let $\Delta = (a_1 b_2 c_3 d_4)$ and let A_1, B_1, ... be the cofactors of a_1, b_1 ... .

(1) *If the equations are consistent, then $\Delta = 0$.*

For, as in Art. 4, (1), we have
$$u_1 D_1 + u_2 D_2 + u_3 D_3 + u_4 D_4 \equiv \Delta.$$

(2) *If $\Delta = 0$, the equations are consistent and have an unique solution except when every D is zero.*

This follows from the identity

$$u_1D_1 + u_2D_2 + u_3D_3 + u_4D_4 \equiv 0.$$

(3) *If every D is zero, the equations are inconsistent, but they are to be regarded as having a common infinite solution, unless every minor of* Δ *is zero, in which case there are infinitely many solutions.*

This is explained with reference to four planes in Exercise XVII, 17.

The methods already described apply to *systems of linear equations connecting more than three variables,* and lead to similar results.

EXERCISE XVII

SIMULTANEOUS LINEAR EQUATIONS

1. Solve by means of determinants:

   (*i*) $x - y + z = 0$,
   $\qquad 2x + 3y - 5z = 7$,
   $\qquad 3x - 4y - 2x = -1$.

   (*ii*) $x - y + z - w = -1$,
   $\qquad 3x + 2y + 4z + 2w = 1$,
   $\qquad 2x + 4y - 5z = -1$,
   $\qquad 3y - 2z + 3w = 5$.

2. With regard to the equations

   $$4x + 7y - 14z = 10, \quad 2x + 3y - 4z = -4, \quad x + y + z = 6,$$

   show that the first can be expressed in the form

   $$l(2x + 3y - 4z) + m(x + y + z) = n,$$

   and find $l : m : n$. Hence show that the equations are inconsistent.

3. With regard to the equations

   $$4x + 7y - 14z = -24, \quad 2x + 3y - 4z = -4, \quad x + y + z = 6,$$

   show that the first can be put in the form

   $$l(2x + 3y - 4z + 4) + m(x + y + z - 6) = 0,$$

   and find $l : m$. Hence show that the equations are not independent.

4. If x, y, z are not all zero and

   $$ax + by + cz = bx + cy + az = cx + ay + bz = 0,$$

   prove that $\qquad a^3 + b^3 + c^3 + 3abc = 0$.

5. Having given the equations

   $$x = cy + bz, \quad y = az + cx, \quad z = bx + ay$$

   where x, y, z are not all zero, prove that

   $$a^2 + b^2 + c^2 + 2abc = 1$$

   and $\quad x^2/(1 - a^2) = y^2/(1 - b^2) = z^2/(1 - c^2)$.

6. Having given the equations

   $a = b \cos C + c \cos B$, $\ b = c \cos A + a \cos C$, $\ c = a \cos B + b \cos A$, deduce the following :

   $$\cos^2 A + \cos^2 B + \cos^2 C + 2 \cos A \cos B \cos C = 1,$$
   $$2bc \cos A = b^2 + c^2 - a^2, \quad a/\sin A = b/\sin B = c/\sin C.$$

7. Having given that
$$x + y + z = 1, \quad ax + by + cz = d, \quad a^2x + b^2y + c^2z = d^2,$$
prove that $a^3x + b^3y + c^3z = d^3 - (d - a)(d - b)(d - c)$
and $a^4x + b^4y + c^4z = d^4 - (d - a)(d - b)(d - c)(a + b + c + d)$.

8. If the equations
$$\frac{cy + bz}{b - c} = \frac{az + cx}{c - a} = \frac{bx + ay}{a - b} = \frac{ax + by + cz}{a + b + c}$$
are consistent, prove that either
$$a + b + c = 0 \text{ .or } (b - c)(c - a)(a - b) = abc.$$
[Put each fraction equal to k and eliminate x, y, z, k.]

9. If the three equations of Art. 8 are distinct and if $\Delta = 0$ and $\Delta_1 = 0$, show that $\Delta_2 = 0$ and $\Delta_3 = 0$ unless $b_1 : c_1 = b_2 : c_2 = b_3 : c_3$.

[Since $\Delta = 0$, $A_1 : A_2 : A_3 = B_1 : B_2 : B_3$, and since $d_1A_1 + d_2A_2 + d_3A_3 = \Delta_1 = 0$, therefore $\Delta_2 = d_1B_1 + d_2B_2 + d_3B_3 = 0$ unless A_1, A_2, A_3 are all zero.]

10. If the equations $\quad x = by + cz + du, \qquad\qquad y = cz + ax + du,$
$$z = ax + by + du, \qquad\qquad u = ax + by + cz$$
are consistent and no one of a, b, c, d is equal to -1, then
$$a/(a + 1) + b/(b + 1) + c/(c + 1) + d/(d + 1) = 1.$$
Also if $a = -1$, then one of b, c, d is equal to -1. $\hfill$ [Ex. XVI, 10.]

11. Using the notation of Art, 8, show that if $u_1 = 0$, $u_2 = 0$, $u_3 = 0$, are the equations to three straight lines, the coordinates of the vertices of the triangle formed by them are
$$\left(\frac{A_1}{C_1}, \frac{B_1}{C_1}\right), \quad \left(\frac{A_2}{C_2}, \frac{B_2}{C_2}\right), \quad \left(\frac{A_3}{C_3}, \frac{B_3}{C_3}\right),$$
and that the area of the teiangle is $\frac{1}{2}\Delta^2 / C_1C_2C_3$.

12. Prove that if $c_1 \neq 0$, the equatons $u_1 = 0$, $u_2 = 0$, $u_3 = 0$ of Art. 8 are equivalent to $u_1 = 0$, $B_3x - A_3y = (d_1c_2)$, $-B_2x + A_2y = (d_1c_3)$;
and if $\Delta \neq 0$ deduce the solution in the form already given.

13. Show that the condition that the points (x_1, y_1, z_1), (x_2, y_2, z_2), (x_3, y_3, z_3), (x_4, y_4, z_4) may be in the same plane is
$$\begin{vmatrix} x_1 & y_1 & z_1 & 1 \\ x_2 & y_2 & z_2 & 1 \\ x_3 & y_3 & z_3 & 1 \\ x_4 & y_4 & z_4 & 1 \end{vmatrix} = 0.$$

14. Using the notation of Art. 10, consider the lines whose equations are $u_1 = 0$, $u_2 = 0$, and $u_3 = 0$, $u_4 = 0$. It is required to find the equation to a plane which passes through either of these and is parallel to the other.

[We have $u_1D_1 + u_2D_2 + u_3D_3 + u_4D_4 \equiv \Delta$, therefore the required planes are $u_1D_1 + u_2D_2 = 0$ and $u_3D_3 + u_4D_4 = 0$.]

15. Consider the lines whose equations are
$$\frac{x - \alpha}{l} = \frac{y - \beta}{m} = \frac{z - \gamma}{n}, \frac{x - \alpha'}{l'} = \frac{y - \beta'}{m'} = \frac{z - \gamma'}{n'}.$$

Show that the equation to the plane which contains the first line and is parallel to the second is

$$(x - \alpha)(mn' - m'n) + (y - \beta)(nl' - n'l) + (z - \gamma)(lm' - l'm) = 0$$

Deduce the condition that the lines may meet in a point or be parallel.

[The required equation is $a(x - \alpha) + b(y - \beta) + c(z - \gamma) = 0$ where $al + bm + cn = 0$ and $al' + bm' + cn' = 0$.]

16. Using the notation of Art. 10, show that if $u_1 = 0$, $u_2 = 0$, $u_3 = 0$, $u_4 = 0$ are the equations to four planes, the vertices of the tetrahedron bounded by them are the

points $\left(\dfrac{A_1}{D_1}, \dfrac{B_1}{D_1}, \dfrac{C_1}{D_1}\right)$, $\left(\dfrac{A_2}{D_2}, \dfrac{B_2}{D_2}, \dfrac{C_2}{D_2}\right)$, etc., and that the volume of the tetrahedron is

$$\tfrac{1}{6} \Delta^3 / D_1 D_2 D_3 D_4.$$

[*It is shown in works on Coordinate Geometry that the volume of the tetrahedron whose vertices are (x_1, y_1, z_1), etc., is one-sixth of the determinant in Ex.13.*]

17. Using the notation of Art. 10, let $u_1 = 0$, $u_2 = 0$, $u_3 = 0$, $u_4 = 0$ be the equations to four planes, and suppose that $\Delta = 0$, then

(*i*) If at least one D, say D_1, is not zero, the planes meet at the point
$$(-A_1/D_1, -B_1/D_1, -c_1/D_1).$$

(*ii*) Suppose that every D is zero, then

(α) If all the minors of Δ are not zero, the four planes are parallel to a certain line, that is to say, they have a common point at infinity.

(β) If every minor of Δ is zero, the planes have a common line of intersection or they are all parallel.

For example, if $A_4 \neq 0$ and we consider the planes u_1, u_2, u_3. Since $D_4 = 0$, two of these, say u_1, u_2, meet in a line which is parallel to the third.

Also, considering the planes, u_1, u_2, u_4, since $D_3 = 0$ the plane u_4 contains or is parallel to the line $u_1 = 0$, $u_2 = 0$. Hence all the planes are parallel to the line
$$x/(b_1 c_2) = y/(c_1 a_2) = z/(a_1 b_2).$$

18. Given the equations
$$a_2 c_3 + a_3 c_2 - 2b_2 b_3 = 0, \quad a_3 c_1 + a_1 c_3 - 2b_1 b_3 = 0, \quad a_1 c_2 + a_2 c_1 - 2b_1 b_2 = 0,$$
prove that

(*i*) $\quad \dfrac{a_2 b_3 - a_3 b_2}{a_1\left(a_1 c_1 - b_1^2\right)} = \dfrac{a_3 b_1 - a_1 b_3}{a_2\left(a_2 c_2 - b_1^2\right)} = \dfrac{a_1 b_2 - a_2 b_1}{a_3\left(a_3 c_3 - b_3^2\right)}.$

(*ii*) $\quad \dfrac{a_1^2}{a_1 c_1 - b_1^2} + \dfrac{a_2^2}{a_2 c_2 - b_2^2} + \dfrac{a_3^2}{a_3 c_3 - b_3^2} = 0.$

[Prove that $\dfrac{a_1}{(a_2 b_3)} = \dfrac{-2b_1}{(c_2 a_3)} = \dfrac{c_1}{(b_2 c_3)} = \dfrac{2\left(a_1 c_1 - b_1^2\right)}{(a_1 b_2 c_3)}$.]

19. If $\quad x = \dfrac{l_1 X + m_1 Y + n_1}{l_3 X + m_3 Y + n_3}, y = \dfrac{l_2 X + m_2 Y + n_2}{l_3 X + m_3 Y + n_3}.$

prove that $\quad X = \dfrac{L_1 x + L_2 y + L_3}{N_1 x + N_2 y + N_3}, \quad Y = \dfrac{M_1 x + M_2 y + M_3}{N_1 x + N_2 y + N_3},$

where L_1, M_1, etc., are the cofactors of l_1, m_1, etc., in the determinant $(l_1 m_2 n_3)$.

20. If x, y are connected with X, Y by the equations of Ex. 19, and X, Y are given by
$$N_1 X - N_2 Y = L_1 - M_2, \quad N_2 X + N_1 Y = L_2 + M_1,$$
prove that the corresponding values of x, y are given by
$$l_3 x - m_3 y = l_1 - m_2, \qquad\qquad m_3 x + l_3 y = m_1 + l_2.$$

■ **11. Systems of Equations of any Degree:** The question of elimination is considered in detail in another volume. Here we illustrate various methods of dealing with systems of special types.

(1) *Systems which are symmetrical with regard to x, y, ...*

❙ **EXAMPLE 1.** *Find the rational solutions of*
$$x^2 + y^2 + z^2 = 50, \quad (y + z)\,(z + x)\,(x + y) = 14,$$
$$x^3 + y^3 + z^3 - 3xyz = 146.$$

If $\Sigma x = p$, $\Sigma yz = q$, $xyz = r$, the given equations are equivalent to
$$p^2 - 2q = 50. \quad pq - r = 14, \quad p^3 - 3pq = 146;$$

whence we find that $p^3 - 150p + 292 = 0$, giving $p = 2$ or $-1 \pm \sqrt{147}$.

If $p = 2$, then $q = -23$, $r = -60$ and x, y, z are the roots of
$$\theta^3 - 2\theta^2 - 23\theta + 60 = 0$$

Thus x, y, z may have the values $3, 4, -5$ taken in any order, and there is no other rational solution.

(2) *A solution may sometimes be discovered by considering the properties of a particular determinant.*

❙ **EXAMPLE 2.** *Solve*
$$x^2 - yz = a, \ y^2 - zx = b, \ z^2 - xy = c.$$
Consider the determinant
$$\Delta = \begin{vmatrix} x & z & y \\ z & y & x \\ y & x & z \end{vmatrix} = 3xyz - x^3 - y^3 - z^3.$$

The cofactors of x, y, z, in the top line, are $yz - x^2, zx - y^2, xy - z^2$;
$$\therefore \qquad (zx - y^2)\,(xy - z^2) - (yz - x^2)^2 = \Delta x.$$
Hence *if (x, y, z) is a solution,*
$$bc - a^2 = \Delta x, \text{ and similarly, } ca - b^2 = \Delta y, \ ab - c^2 = \Delta z.$$
Multiplying these by a, b, c, and adding,
$$3abc - a^3 - b^3 - c^3 = \Delta\,(ax + by + cz) = -\Delta^2.$$
Therefore the only possible solutions are given by
$$\frac{x}{a^2 - bc} = \frac{y}{b^2 - ca} = \frac{z}{c^2 - ab} = \pm \frac{1}{\sqrt{(a^3 + b^3 + c^3 - 3abc)}}.$$

It is easily verified by substitution that these are solutions unless $a^3 + b^3 + c^3 = 3abc$, in which there is no solution unless a, b, c are all zero.

(3) *A change of the variables.*

EXAMPLE 3. *Solve*
$$x(x - a) = yz, \quad y(y - b) = zx, \quad z(z - c) = xy.$$
Let $x = 1/X$, $y = 1/Y$, $z = 1/Z$, then the equations are equivalent to
$$\frac{X^2 - YZ}{a} = \frac{Y^2 - ZX}{b} = \frac{Z^2 - XY}{c} = -XYZ.$$
Therefore, as in Ex. 2,
$$X/A = Y/B = Z/C \quad \text{or} \quad Ax = By = Cz = k,$$
where $A = a^2 - bc$, $B = b^2 - ca$, $C = c^2 - ab$, and k is to be determined. Substituting in the first equation
$$\frac{k}{A}\left(\frac{k}{A} - a\right) = \frac{k^2}{BC},$$
therefore, either $\qquad k = 0$, giving the solution $(0, 0, 0)$,

or $\qquad k(BC - A^2) = aABC$, giving $k(a^3 + b^3 + c^3 - 3abc) = -ABC$;

$\therefore \qquad x = -(b^2 - ca)(c^2 - ab) / (a^3 + b^3 + c^3 - 3abc),$

with similar values for y, z.

(4) *In the case of two variables x, y the substitution $y = tx$ is sometimes useful. In equations involving x, y, z, we may write $y = tx$, $z = t'x$.*

EXAMPLE 4. *Solve*
$$ax^2 + bxy + cy^2 = lx + my, \quad a'x^2 + b'xy + c'y^2 = l'x + m'y.$$
One solution is $(0, 0)$, and if $y = tx$ the equations become
$$(a + bt + ct^2) = x(l + mt), \quad x^2(a' + b't + c't^2) = x(l' + m't).$$
Hence if $x \neq 0$, the solutions are given by $x = (l + mt)/(a + bt + ct^2)$, $y = tx$, where $\quad (a + bt + ct^2)(l' + m't) = (a' + b't + c't^2)(l + mt).$

If $x = 0$ and $y \neq 0$, the equations become $cy^2 = my$, $c'y^2 = m'y$, $\therefore cm' - c'm = 0$ and $y = -m/c$. In this one value of t is infinite.

(5) *If a solution $x = x_1$, $y = y_1$, ... can be guessed, then a suitable substitution is*
$$x = x_1 + X, \quad y = y_1 + Y, \dots .$$

EXAMPLE 5. *Solve* $(y - c)(z - b) = a^2$, $(z - a)(x - c) = b^2$, $(x - b)(y - a) = c^2$.
One solution is $(b + c, c + a, a + b)$, and if we write
$$x = b + c + X, \quad y = c + a + Y, \quad z = a + b + Z,$$
the equations become
$$YZ + a(Y+Z) = 0, \quad ZX + b(Z + X) = 0, \quad XY + c(X + Y) = 0.$$
Hence *if none of the three, X, Y, Z, is zero,* we have
$$\frac{1}{Y} + \frac{1}{Z} = -\frac{1}{a}, \quad \frac{1}{Z} + \frac{1}{X} = -\frac{1}{b}, \quad \frac{1}{X} + \frac{1}{Y} = -\frac{1}{c};$$
$$\therefore \qquad \frac{2}{X} = \frac{1}{a} - \frac{1}{b} - \frac{1}{c},$$
with similar equations. This gives the only solution, beside the obvious one.

Special cases. If $X = 0$, $Y = 0$ and $Z \neq 0$, then $a = 0$, $b = 0$ and Z may have any value.

If $X = 0$, $Y \neq 0$, $Z \neq 0$, then $b = 0$, $c = 0$ and Y, Z may have any values such that $YZ + a(Y + Z) = 0$.

■ **EXAMPLE 6.** *The eight solutions of*

$$x^2 + 2yz = a, \quad y^2 + 2zx = b, \quad z^2 + 2xy = c, \qquad \ldots(A)$$

are given by $\quad x = \dfrac{1}{3}(k - X), \quad y + z = \dfrac{1}{3}(2k + X), \quad y - z = (b - c)/X,$

where $\quad X^2 = 2a - b - c \pm 2\sqrt{(a^2 + b^2 + c^2 - bc - ca - ab)}$

and $\qquad\qquad\qquad k = \pm \sqrt{(a + b + c)}.$

By addition, $\quad (x + y + z)^2 = a + b + c, \quad \therefore\ x + y + z = k.$

From the second and third of the given equations, by subtraction,

$$(y - z)(y + z - 2x) = b - c, \quad \therefore\ (y - z)X = b - c,$$

where $\qquad\qquad\qquad X = k - 3x.$

Adding the second and third of equations (A) and subtracting the first, we easily find that

$$(y - z)^2 + x(k + X) = b + c - a;$$

$$\therefore \qquad \left(\frac{b - c}{X}\right)^2 + \frac{k^2 - X^2}{3} = b + c - a;$$

$$\therefore \qquad X^4 - 2X^2(2a - b - c) - 3(b - c)^2 = 0,$$

and solving for X^2, we obtain the values stated in the question.

Another solution. Equations (A) are equivalent to

$$(x + y + z)^2 = a + b + c = A \text{ (say)},$$
$$(x + \omega^2 y + \omega z)^2 = a + \omega b + \omega^2 c = B,$$
$$(x + \omega y + \omega^2 z)^2 = a + \omega^2 b + \omega c = C,$$

where ω is an imaginary cube root of 1.

Therefore the eight solutions are given by

$$x + y + z = \sqrt{A}, \quad x + \omega^2 y + \omega z = \sqrt{B}, \quad x + \omega y + \omega^2 z = \sqrt{C}, \qquad \ldots(B)$$

where $\sqrt{A}$ is either square root of A and so for $\sqrt{B}$, $\sqrt{C}$.

From equations (B), by addition, $\quad 3x = \sqrt{A} + X,$

where $X = \sqrt{B} + \sqrt{C}$, and therefore

$$X^2 = B + C + 2\sqrt{B}.\sqrt{C}$$
$$= 2a - b - c \pm \sqrt{(a^2 + b^2 + c^2 - bc - ca - ab)}.$$

Also, from the second and third of equations (B),

$$2x - (y + z) = X, \quad \therefore\ y + z = \frac{1}{3}(2\sqrt{A} - X),$$

and $\qquad (\omega^2 - \omega)(y - z) = \sqrt{C} - \sqrt{B} = \dfrac{C - B}{X} = \dfrac{(\omega^2 - \omega)(c - b)}{X};$

$\therefore \qquad\qquad\qquad y - z = - (b - c) / X.$

These results agree with the preceding if we change the sign of X, which is permissible.

■ **EXAMPLE 7.** *Show that, if the equations*

$$by^2 + cz^2 = 2fyz, \quad cz^2 + ax^2 = 2gzx, \quad ax^2 + by^2 = 2hxy \qquad\qquad ...(A)$$

are consistent, then $\quad abc + 2fgh - af^2 - bg^2 - ch^2 = 0.$ $\qquad\qquad\qquad ...(B)$

From equation (A), by multiplication,

$$ax^2(b^2y^4 + c^2z^4) + ... + ... + 2abcx^2y^2z^2 = 8fghx^2y^2z^2. \qquad\qquad ...(C)$$

Also $\qquad\qquad\qquad b^2y^4 + c^2z^4 = (4f^2 - 2bc)\, y^2z^2,$

with two similar equations. Substituting in (C) and dividing by $x^3y^3z^3$ (assumed not to be zero), the result (B) follows.

Special cases. Considering the original equations it will be seen that

(*i*) If $x = 0,\ y \neq 0,\ z \neq 0$, then $cz^2 = 0$, and $c = 0$;

hence the equations become, $by^2 = 2fyz,\ by^2 = 0$, and we have $b = 0$, and $f = 0$.

(*ii*) If $x = 0,\ y = 0$ and $z \neq 0$, the equations reduce to the single equation $cz^2 = 0$; and thus do not constitute a system of equations.

Thus, if the equations (A) form a system of equations with a solution other than $(0, 0, 0)$, then the condition (B) holds.

■ **EXAMPLE 8.** *Prove that, if*

$$ax^2 + by^2 + cz^2 - (b + c)\, yz - (c+a)\, zx - (a + b)\, xy = 0$$

and $\qquad\qquad\qquad x^2 + y^2 + z^2 - 2yz - 2zx - 2xy = 0,$

then, if no two of a, b, c are equal,

$$\frac{x}{(b-c)^2} = \frac{y}{(c-a)^2} = \frac{z}{(a-b)^2} = \frac{xyz}{a^2x^2 + b^2y^2 + c^2z^2 - 2bcyz - 2cazx - 2abxy}.$$

The given equations can be written

$$x\,(-ax + by + cz) + y\,(ax - by + cz) + z\,(ax + by - cz) = 0,$$
$$x\,(-x + y + z) + y\,(x - y + z) + z\,(x + y - z) = 0.$$

Multiplying these by $(x + y - z)$, $(ax + by - cz)$ respectively and subtracting, we find that

$$xy\,\{y\,(a - b) + z(c - a)\} = xy\,\{z\,(b - c) + x\,(a - b)\};$$

and two similar equations can be obtained. Hence, *assuming that none of the three, z, y, z, is zero*, we may write

$$y(a - b) + z(c - a) = k, \quad z(b - c) + x(a - b) = k, \quad x(c - a) + y\,(b - c) = k.$$

Hence, *if no two of a, b, c are equal,* $x = - \dfrac{k\,(b - c)}{(c - a)\,(a - b)},$

$\therefore \qquad\qquad \dfrac{x}{(b-c)^2} = \dfrac{y}{(c-a)^2} = \dfrac{z}{(a-b)^2} = \lambda \text{ (say)};$

whence also $\quad \lambda = \dfrac{-ax + by + cz}{-a(b-c)^2 + b(c-a)^2 + c(a-b)^2} = \dfrac{ax - by - cz}{(b+c)(c-a)(a-b)} \;;$

$\therefore \quad \Sigma ax\,(ax - by - cz) = \lambda^2\,(b-c)\,(c-a)\,(a-b)\,.\,\Sigma a(b^2 - c^2)$
$$= \lambda^2\,(b-c)^2\,(c-a)^2\,(a-b)^2 = xyz/\lambda,$$

and the result in question follows.

The consideration of special cases is left to the reader.

EXERCISE XVIII

1. If (α, β), (α', β') are the roots of
$$u = ax^2 + bx + c = 0, \quad u' = a'x^2 + b'x + c' = 0,$$
prove that
$$a^2 a'^2(\alpha - \alpha')\,(\alpha - \beta')\,(\beta - \alpha')\,(\beta - \beta') = (ac' - a'c)^2 - (ab' - a'b)\,(bc' - b'c),$$
and consequently *the necessary and sufficient condition that the equations have a common root is R = 0, where*
$$R = (ac' - a'c)^2 - (ab' - a'b)\,(bc' - b'c).$$

2. Eliminate x, y from the equations
$$ax^2 + bx + c = 0, \quad a'y^2 + b'y + c' = 0, \quad xy = z,$$
and show that the roots of the resulting equation in z are $\alpha\alpha'$, $\alpha\beta'$, $\beta\alpha'$, $\beta\beta'$, where (α, β), (α', β') are the roots of
$$ax^2 + bx + c = 0, \quad a'x^2 + b'x + c' = 0.$$
Show that this equation is
$$a^2 a'^2 z^4 - aa'bb'z^3 + (b^2 a'c' + b'^2 ac - 2aca'c')\,z^2 - bcb'c'z + c^2 c'^2 = 0.$$
Prove also that the equation whose roots are α/α', α/β', β/α', β/β' is obtained from the last equation by interchanging a' and c'.

Solve the equations in Exs. 3–28.

3. $x^3 + y^3 = 8, \quad xy + (x + y) = 2.$ [Put $x + y = \lambda$, $xy = \mu$.]

4. $x^2 y + xy^2 + x + y = 3, \quad 2(x - 1)(y - 1) = -1.$

5. $x\,(1 - y^2) = 2y, \quad y(1 - x^2) = 2x.$

6. $x + y + z = 9, \quad x^2 + y^2 + z^2 = 29, \quad x^{-1} + y^{-1} + z^{-1} = \dfrac{13}{12}.$

7. $(y + z)(z + x)(x + y) = \dfrac{45}{4}, \quad (x + 1)(y + 1)(z + 1) = 9, \quad xyz = 1.$

8. $\dfrac{x^2}{3} + \dfrac{y^2}{5} = \dfrac{9}{x} + \dfrac{25}{y} = 8\,.$

9. $yz + 2y + 4z = 2zx + 4z + x = 4xy + x + 2y = 7.$

10. $\dfrac{x^2}{y} = 16 - xy, \quad \dfrac{y^2}{x} = 9 - xy.$

11. $\dfrac{x^4 + 3}{y^2} = \dfrac{x^2 y^2 - 6}{xy} = \dfrac{y^4 + 3}{x^2}\,.$

12. $\dfrac{x^2}{a-y} + \dfrac{y^2}{a-x} = a, \qquad \dfrac{x}{a^2-y^2} + \dfrac{y}{a^2-x^2} = \dfrac{1}{a}.$

13. $(1-x)^2 + (1-y)^2 = (1-x)(1-y)(1-xy),\ ay(1-x)^2 + bx(1-y)^2 = 0.$

   [Put $x = 1 - \lambda,\ y = 1 - \mu$ and prove that $a\lambda^2 - \lambda(a-b) + a - b = 0$ and $b\mu^2 - \mu(b-a) + b - a = 0.$]

14. $y^2 + z^2 = a + y + z, \quad z^2 + x^2 = b + z + x, \quad x^2 + y^2 = c + x + y.$

15. $yz + zx + xy = a^2 - x^2 = b^2 - y^2 = c^2 - z^2.\ [(x+y)(x+z) = a^2,\ \text{etc.}]$

16. $x(y + z - x) = a^2, \quad y(z + x - y) = b^2, \quad z(x + y - z) = c^2.$

17. $x - \dfrac{1}{y} = a - \dfrac{1}{b}, \quad y - \dfrac{1}{z} = b - \dfrac{1}{c}, \quad z - \dfrac{1}{x} = c - \dfrac{1}{a}.$

   [Put $x = a + \lambda,\ y = b + \mu,\ z = c + v.$]

18. $ayz + by + cz = bzx + cz + ax = cxy + ax + by = a + b + c.$

19. $x^2 + y + z = y^2 + z + x = z^2 + x + y = 3.$

20. $\dfrac{x}{y+z+1} = \dfrac{y}{z+x} = \dfrac{z}{x+y-1} = x + y + z.$

21. $x^2 + (y - z)^2 = a^2, \quad y^2 + (z - x)^2 = b^2, \quad z^2 + (x - y)^2 = c^2.$

22. $x\,(bc - xy) = y\,(xy - ac),$
   $xy(ay + bx - xy) = abc(x + y - c).$

23. $(x - y)(x - z) = ax, \quad (y - z)(y - x) = by, \quad (z - x)(z - y) = cz.$

24. $yz(a - x) = ax^2, \quad zx(b - y) = by^2, \quad xy(c - z) = cz^2.$

25. $3(x + yz) = 4\,(y + zx) = 5\,(z + xy), \quad 8x - 9y = 1.$

26. $u + v = 2, \quad vx + vy = -1, \quad ux^2 + vy^2 = 11, \quad ux^3 + vy^3 = 5.$

27. $x + y = x' + y' = 2a, \quad \dfrac{1}{x} + \dfrac{1}{y'} = \dfrac{2}{y}, \quad \dfrac{1}{y} + \dfrac{1}{x'} = \dfrac{2}{x}.$

28. $\sqrt{x+1} + \sqrt{y-2} = \sqrt{x-1} + \sqrt{y+2} = 3.$

29. If $\dfrac{3x+1}{5x-3} = \dfrac{2y+1}{5y-2} = \dfrac{x+y+5}{8}$ prove that $y + 1 = \pm\, x$, and solve the equations.

30. If $\dfrac{x^2}{a} + \dfrac{y^2}{b} = \dfrac{a^2}{x} + \dfrac{b^2}{y} = a + b$, show that either $\dfrac{x}{a} + \dfrac{y}{b} + 1 = 0$ or $x + y = a + b.$

   [Put $x = aX,\ y = bY$, and show $(X + Y + 1)\,(aX + bY - a - b) = 0.$]

31. Eliminate $x,\ y,\ z$ from

   $\dfrac{x}{a}\,(y + z - x) = \dfrac{y}{b}\,(z + x - y) = \dfrac{z}{c}\,(x + y - z)$ and $ax + by + cz = 0.$

32. If $x^3 - y^4 = y^3 - x^4 = a^4$ and $x + y = -a$, and $x,\ y$ are unequal, prove that
   (i) $xy = a^2\,(a - 1)/(2a - 1)$,   (ii) $3a^3 - 3a^2 + 3a - 1 = 0$,   (iii) $3a^2\,(x^2 + y^2) = -\,1$,
   (iv) $3a\,(x^3 + y^3) = 2 - a$, and (v) $3a\,(x^4 + y^4 + 1) = 2 + 4a^2.$

33. If $\qquad\qquad y^2z^2 + (y + z)^2 + \lambda yz = 1 + \lambda,$
   $\qquad\qquad\qquad z^2x^2 + (z + x)^2 + \lambda zx = 1 + \lambda$ and $x \neq y,$
   then will $\qquad\quad x^2y^2 + (x + y)^2 + \lambda xy = 1 + \lambda.$

34. If x, y, z satisfy the equations

$$\frac{1}{x} + \frac{1}{y+z} = \frac{1}{a}, \quad \frac{1}{y} + \frac{1}{z+x} = \frac{1}{b}, \quad \frac{1}{z} + \frac{1}{x+y} = \frac{1}{c},$$

prove that $x(b + c - a) = y(c + a - b) = z(a + b - c)$: and hence solve the equations.

35. If $ax^2 + by^2 + cz^2 = 0$, $ayz + bzx + cxy = 0$ and $x^3 + y^3 + z^3 + \lambda xyz = 0$, prove that $a^3 + b^3 + c^3 + \lambda abc = 0$.

36. If $a + b + c = 0$ and $x + y + z = 0$, prove that

$$\Sigma a^2 x^2 - \Sigma bcyz = \tfrac{1}{4} \Sigma a^2 \cdot \Sigma x^2.$$

[Use the identities $(by + cz - ax)^2 = (bz + cy)^2$, etc.]

37. Given that

$$ax(y + z - x) + by(z + x - y) + cz(x + y - z) = 0$$

and $\qquad a^2 x^2 + b^2 y^2 + c^2 z^2 - 2bcyz - 2cazx - 2abxy = 0,$

Show that

$$\frac{x}{a(b-c)^2} = \frac{y}{b(c-a)^2} = \frac{z}{c(a-b)^2}.$$

38. Eliminate x, y, z from

$$(z + x - y)(x + y - z) = ayz,$$
$$(x + y - z)(y + z - x) = bzx,$$
$$(y + z - x)(z + x - y) = cxy,$$

showing that the result is

$$abc = (a + b + c - 4)^2.$$

39. Eliminate x, y, z from

$$x + y - z = a, \quad x^2 + y^2 - z^2 = b^2, \quad x^3 + y^3 - z^3 = c^3, \quad xyz = d^3.$$

40. Eliminate x and y from

$$x^2 + xy = a^2, \quad y^2 + xy = b^2, \quad x^2 + y^2 = c^2.$$

41. If $x^2 + y^2 + z^2 = 1$, $x + 2y + 3z = 3$, $3x + y + z = 3$, prove that $x - 8y + 5z = \pm 3$.

42. Show that the equations

$$x + y + z = 0, \quad \frac{x^2}{32} - \frac{y^2}{25} - \frac{z^2}{63} = 0, \quad 32yz - 25zx - 63xy = 0$$

are consistent, and find $x : y : z$.

43. If x, y, z are not all zero and

$$x^2 - ayz = y^2 - bzx = z^2 - cxy = \tfrac{1}{2}(x^2 + y^2 + z^2),$$

prove that $a^2 + b^2 + c^2 = 2abc + 1$ and $x^2/(1 - a^2) = y^2/(1 - b^2) = z^2/(1 - c^2)$.

44. Show that if $x = a, y = b, z = c$ is a solution of

$$px^3 + qy^3 + rz^3 = sxyz,$$

then

$$x = a(qb^3 - rc^3), \quad y = b(rc^3 - pa^3), \quad z = c(pa^3 - qb^3)$$

is another solution.

45. If $\dfrac{x}{l}\,(y + z - x) = \dfrac{y}{m}\,(z + x - y) = \dfrac{z}{n}\,(x + y - z)$ and $l + m + n = 0$, show that

$$x/l^2 = y/m^2 = z/n^2.$$

46. Eliminate x, y from $(b - x)(c - y) = a^2$, $(c - x)(a - y) = b^2$, $(a - x)(b - y) = c^2$.

47. If the four expressions

$$x^2 + y^2 + z^2 + m\,(yz + zx + xy), \qquad\qquad y^2 + z^2 + u^2 + m\,(zu + uy + yz),$$
$$z^2 + u^2 + x^2 + m\,(ux + xz + zu), \qquad\qquad u^2 + x^2 + y^2 + m\,(xy + yu + ux),$$

are equal, prove that each is equal to $\tfrac{1}{2}\,(x^2 + y^2 + z^2 + u^2)$.

48. If $x_1,\ x_2,\ \dots\ x_5$ are all different and

$$\frac{x_1^2 + x_2^2}{x_1 x_2} = \frac{x_2^2 + x_3^2}{x_2 x_3} = \frac{x_3^2 + x_4^2}{x_3 x_4} = \frac{x_4^2 + x_5^2}{x_4 x_5} = k,$$

prove that $x_1{}^2 + x_5{}^2 = (k^4 - 4k^2 + 2)\,x_1 x_5$.

49. (i) Eliminate x, y, z, t from the equations

$$ax + by + cz + dt = 0, \quad a'x + b'y + c'z + d't = 0,$$

$$\frac{a}{x} + \frac{b}{y} + \frac{c}{z} + \frac{d}{t} = 0, \quad \frac{a'}{x} + \frac{b'}{y} + \frac{c'}{z} + \frac{d'}{t} = 0,$$

showing that the result is

$$LMN + Lmn + Mnl + Nlm = 0,$$

where
$$l = (bc'),\ m = (ca'),\ n = (ab'),$$
$$L = (ad'),\ M = (bd'),\ N = (cd').$$

(ii) Use this to prove Ex. XVI, 25, (iii).

$$[(i)\ \text{Prove that}\ \ \frac{l}{(x^2 - t^2)yz} = \dots = \dots = \frac{L}{(y^2 - z^2)\,xt} = \dots = \dots,$$

the unwritten fractions being obtained by the cyclic substitutions (xyz), (lmn), (LMN). Now put x^2, y^2, z^2, t^2 for x, y, z, t in Ex. XVI, 25 (i).

(ii) Let $x = \cos\alpha + \iota\sin\alpha$, $y = \cos\beta + \iota\sin\beta$, etc., where α, β, γ, δ are the angles which the sides a, b, c, d make with a fixed line. By projecting along and perpendicular to this line, prove that the four given equations hold.]

■ ■ ■

Reciprocal and Binomial Equations

■ **1. Reciprocal Equations:** A *reciprocal equation* is an equation which possesses the following property : *If α is any root, then $1/\alpha$ is also a root.*

Let α, β, ... be the roots of

$$a_0 x^n + a_1 x^{n-1} + \dots + a_{n-1} x + a_n = 0, \qquad \dots\text{(A)}$$

then $1/\alpha$, $1/\beta$, ... are the roots of

$$a_n x^n + a_{n-1} x^{n-1} + \dots + a_1 x + a_0 = 0.$$

Hence (A) is a reciprocal equation if and only if

$$\frac{a_0}{a_n} = \frac{a_1}{a_{n-1}} = \dots = \frac{a_{n-1}}{a_1} = \frac{a_n}{a_0} .$$

Denoting each of these fractions by k, we have

$$k^2 = \frac{a_0}{a_n} \cdot \frac{a_n}{a_0} = 1,$$

so that $k = \pm 1$.

We shall say that *Reciprocal Equations are of the first or second types according as $k = +1$ or $k = -1$, that is according as $a_r = a_{n-r}$ or $a_r = -a_{n-r}$, for $r = 0, 1, 2, \dots n$.*

Theorem 1. *The solution of any reciprocal equation depends on that of a reciprocal equation of the first type and of even degree.*

Let $f(x) = a_0 x^n + a_1 x^{n-1} + \dots + a_n = 0$ be a reciprocal equation. *If this is of the first type, $x^n f\left(\dfrac{1}{x}\right) = f(x).$*

Let n be odd and equal to $2m + 1$. Then the equation may be written

$$f(x) = a_0 (x^{2m+1} + 1) + a_1 x(x^{2m-1} + 1) + \dots + a_m x^m (x + 1) = 0.$$

Therefore $f(x)$ is divisible by $x + 1$, and if $\phi(x)$ is the quotient,

$$x^{2m}\phi\left(\frac{1}{x}\right) = x^{2m}\frac{f\left(\frac{1}{x}\right)}{\frac{1}{x}+1} = x^{2m+1}\cdot\frac{f\left(\frac{1}{x}\right)}{1+x} = \frac{f(x)}{x+1} = \phi(x);$$

hence $\phi(x) = 0$ is a reciprocal equation of the first type of degree $2m$.

If $f(x) = 0$ is of the second type, $x^n f\left(\dfrac{1}{x}\right) = -f(x)$.

Let $n = 2m + 1$. Grouping the terms as before, we can show that $f(x)$ is divisible by $x - 1$, and that, if $\phi(x)$ is the quotient, then $x^{2m} \phi\left(\dfrac{1}{x}\right) = \phi(x)$; with the same result as before.

Again, if $n = 2m$, since $a_r = -a_{2m-r}$, it follows that $a_m = -a_m$, and therefore $a_m = 0$. Hence the equation may be written

$$f(x) = a_0 (x^{2m} - 1) + a_1 x (x^{2m-2} - 1) + \ldots + a_{m-1} x^{m-1} (x^2 - 1) = 0.$$

Therefore $f(x)$ is divisible by $x^2 - 1$, and if $\phi(x)$ is the quotient,

$$x^{2m-2} \phi\left(\frac{1}{x}\right) = x^{2m-2} \frac{f\left(\frac{1}{x}\right)}{\frac{1}{x^2} - 1} = -x^{2m} \frac{f\left(\frac{1}{x}\right)}{x^2 - 1} = \frac{f(x)}{x^2 - 1} = \phi(x).$$

Thus in every case the solution of $f(x) = 0$ depends on that of a reciprocal equation $\phi(x) = 0$ of the first type and of even degree.

We say that a reciprocal equation is of the *standard form* when it is of the first type and of even degree.

Theorem 2. *The solution of a reciprocal equation of the first type and of degree $2m$ depends on that of an equation of degree m.*

Let the equation be

$$a_0 x^{2m} + a_1 x^{2m-1} + \ldots + a_1 x + a_0 = 0.$$

Dividing by x^m, this may be written

$$a_0 \left(x^m + \frac{1}{x^m}\right) + a_1 \left(x^{m-1} + \frac{1}{x^{m-1}}\right) + \ldots + a_m = 0. \qquad \ldots(A)$$

Let $x + \dfrac{1}{x} = z$ and $x^r + \dfrac{1}{x^r} = u_r$, so that $u_1 = z$; then

$$u_2 = \left(x + \frac{1}{x}\right)^2 - 2 = z^2 - 2.$$

Also $\left(x + \dfrac{1}{x}\right)\left(x^{r-1} + \dfrac{1}{x^{r-1}}\right) = \left(x^r + \dfrac{1}{x^r}\right) + \left(x^{r-2} + \dfrac{1}{x^{r-2}}\right)$,

therefore $u_r = z u_{r-1} - u_{r-2}$.

Putting $r = 3, 4, \ldots$ in succession,

$$u_3 = z(z^2 - 2) - z = z^3 - 3z,$$
$$u_4 = z(z^3 - 3z) - (z^2 - 2) = z^4 - 4z^2 + 2,$$

and so on.

Thus equation (A) can be expressed as an equation of degree m in z, and if z_1 is one of its roots, the corresponding values of x are given by $x^2 - z_1 x + 1 = 0$.

The reader may verify that

$$u_5 = z^5 - 5z^3 + 5z, \qquad\qquad u_6 = z^6 - 6z^4 + 9z^2 - 2,$$
$$u_7 = z^7 - 7z^5 + 14x^3 - 7z, \qquad u_8 = z^8 - 8z^6 + 20z^4 - 16z^2 + 2.$$

EXAMPLE 1: *Solve* $6x^5 + 11x^4 - 33x^3 - 33x^2 + 11x + 6 = 0.$

One root is -1. Dividing by $x + 1$, we have $6x^4 + 5x^3 - 38x^2 + 5x + 6 = 0$. Dividing by x^2 and grouping the terms,

$$6\left(x^2 + \frac{1}{x^2}\right) + 5\left(x + \frac{1}{x}\right) - 38 = 0.$$

If $z = x + \dfrac{1}{x}$, this becomes $6(z^2 - 2) + 5z - 38 = 0$, that is $\quad 6z^2 + 5z - 50 = 0$;

$$\therefore z = \frac{5}{2} \text{ or } -\frac{10}{3}, \text{ and } x \text{ is given by } x + \frac{1}{x} = \frac{5}{2} \text{ or } -\frac{10}{3}.$$

Whence we find $x = 2, \dfrac{1}{2}, -3$ or $-\dfrac{1}{3}$. Thus the roots are $-1, 2, \dfrac{1}{2}, -3, -\dfrac{1}{3}$.

■ **2. The Binomial Equation $x^n - 1 = 0$:**

(1) *If α is a root of $x^n - 1 = 0$, so also is α^r, where r is any integer.*

For $(\alpha^r)^n = (\alpha^n)^r = 1.$

(2) *If m is prime to n, the equations $x^m - 1 = 0$ and $x^n - 1 = 0$ have no common root except 1.*

For if α is a common root, then $\alpha^{pm} = 1$ and $\alpha^{qn} = 1$, where p, q are any positive integers. Therefore $\alpha^{pm-qn} = 1$, and since m is prime to n, p and q can be found so that $pm - qn = \pm 1$. [See Ch. 1, 4, (3)]   Hence $\alpha = 1.$

(3) *If n is a prime number and α is any imaginary root of $x^n - 1 = 0$, the n roots are $1, \alpha, \alpha^2, \dots \alpha^{n-1}$.*

For every one of this set is a root and no two of them are equal. For suppose that the two α^a and α^b $(a > b)$ are equal. Then $\alpha^{a-b} = 1$, and the equations $x^n - 1 = 0$ and $x^{a-b} = 1$ have a common imaginary root. This is impossible; for n is a prime and $1 \le a - b < n$, thus n is prime to $a - b$.

(4) *If $n = pqr \dots$ where $p, q, r, \dots$ are primes or powers of primes, the roots of $x^n - 1 = 0$ are the n terms of the product*

$$(1 + \alpha + \alpha^2 + \dots + \alpha^{p-1})(1 + \beta + \beta^2 + \dots \beta^{q-1})(1 + \gamma + \gamma^2 + \dots + \gamma^{r-1}) \dots ,$$

where α is a root of $x^p - 1 = 0$, β of $x^q - 1 = 0$, γ of $x^r - 1 = 0$, etc.

Take the case of three factors p, q, r ; similar reasoning applying in all cases.

Any term of the product, for instance $\alpha^a \beta^b \gamma^c$, is a root. For

$$(\alpha^a)^n = (\alpha^p)^{aqr} = 1,$$

and similarly $(\beta^b)^n = 1$ and $(\gamma^c)^n = 1.$

If any two terms are equal, for instance, if $\alpha^a \beta^b \gamma^c = \alpha^{a'} \beta^{b'} \gamma^{c'}$, then $\beta^{b-b'} \gamma^{c-c'}$ $= \alpha^{a'-a}$, which is impossible, for $\beta^{b-b'} \gamma^{c-c'}$ is a root of $x^{qr} - 1 = 0$ and $\alpha^{a'-a}$ is a root of $x^p - 1 = 0$; and since p is prime to qr, these equations have no common root except 1.

(5) As explained in Ch. 5, 17, the roots of $x^n-1 = 0$ are

$$\cos \frac{2r\pi}{n} + \iota \sin \frac{2r\pi}{n}, \text{ where } r = 0, 1, 2, \ldots n-1.$$

The imaginary roots are therefore

$$\cos \frac{2r\pi}{n} \pm \iota \sin \frac{2r\pi}{n}, \text{ where } r = 1, 2, \ldots \frac{n-1}{2} \text{ or } \frac{n-2}{2},$$

according as n is odd or even.

If x^n-1 is divided by $x-1$ or x^2-1 according as n is odd or even and $\phi(x)$ is the quotient, then $\phi(x) = 0$ is a reciprocal equation, and if this is transformed by the substitution $z = x + \dfrac{1}{x}$, the values of z are

$$2 \cos \frac{2r\pi}{n} \text{ where } r = 1, 2, \ldots \frac{n-1}{2} \text{ or } \frac{n-2}{2}.$$

■ **3. Special Roots of $x^n-1 = 0$.**

(1) If r is less than n and prime to n, then $\cos\dfrac{2r\pi}{n} + \iota \sin \dfrac{2r\pi}{n}$ is a root of $x^n - 1 = 0$, *but it it not a root of any equation of lower degree, and of the same type.*

For, if $\left(\cos \dfrac{2r\pi}{n} + \iota \sin \dfrac{2r\pi}{n}\right)^m = 1$, where $m < n$,

then

$$\cos \frac{2rm\pi}{n} + \iota \sin \frac{2rm\pi}{n} = 1.$$

Therefore rm is divisible by n, and since n is prime to r, m must be divisible by n. This is impossible, for $m < n$.

Definition: Any root of $x^n-1 = 0$ which is not a root of an equation of the same type and of lower degree is called a *special* root of the equation.

The special roots of $x^n-1 = 0$ are $\cos \dfrac{2r\pi}{n} + \iota \sin \dfrac{2r\pi}{n}$, where $r < n$ and prime to n. Thus, $x^n-1 = 0$ has $\phi(n)$ special roots, where $\phi(n)$ is the number of numbers less than n and prime to n, including unity.

If $r < n$ and prime to n then $n - r$ is prime to n, and therefore

$$\cos \frac{2r\pi}{n} \pm \iota \sin \frac{2r\pi}{n}$$

are special roots. These can be arranged in pairs, $\left(\alpha, \dfrac{1}{\alpha}\right)$, $\left(\beta, \dfrac{1}{\beta}\right)$, etc., so that they are the roots of a reciprocal equation.

(2) *If α is a special root of $x^n - 1 = 0$, the n roots are $1, \alpha, \alpha^2, \ldots \alpha^{n-1}$.*

For every one of the set is a root. And if two of them are equal, for instance, if $\alpha^a = \alpha^b$, then $\alpha^{a-b} = 1$, so that α is a root of $x^{a-b} = 1$. This is impossible, for the last equation is of lower degree than $x^n - 1 = 0$.

(3) *If α is any special root of $x^n-1 = 0$, the complete set of special roots is α^a, α^b, α^c, ... where a, b, c, ... are the numbers less than n and prime to it, including unity.*

For if a is any number less than n and prime to it, the remainders when a, $2a$, $3a$ ... $(n–1)$ a are divided by n are the numbers 1, 2, 3, ... $n–1$, taken in some order or other (Ch. 1, 10). And since $\alpha^n = 1$, it follows that every number in the set α^a, α^{2a}, α^{3a}, ... $\alpha^{(n-1)a}$, occurs somewhere in the set α, α^2, α^3, ... α^{n-1}. Therefore α^a is a special root.

■ **4. The Equation $x^n–A = 0$.**

(1) Let $A = a\,(\cos \alpha + \iota \sin \alpha)$, where $a = |\,A\,|$, then the roots of $x^n - A = 0$ are

$$\sqrt[n]{a} \cdot \left\{ \cos \frac{\alpha + 2r\pi}{n} + \iota \sin \frac{\alpha + 2r\pi}{n} \right\}, \text{ where } r = 0, 1, 2, \dots n - 1.$$

For by De Moivre's theorem, this expression is a root for every r, and its n values are all different.

$$\text{Since } \cos \frac{\alpha + 2r\pi}{n} + \iota \sin \frac{\alpha + 2r\pi}{n} = \left(\cos \frac{\alpha}{n} + \iota \sin \frac{\alpha}{n} \right) \left(\cos \frac{2r\pi}{n} + \iota \sin \frac{2r\pi}{n} \right),$$

it follows that *the n values of $\sqrt[n]{A}$ may be found by multiplying any value of $\sqrt[n]{A}$ by the nth roots of unity.*

(2) The roots of $x^n +1 = 0$ are

$$\cos \frac{(2r+1)\pi}{n} + \iota \sin \frac{(2r+1)\pi}{n}, \text{ where } r = 0, 1, 2, \dots n - 1.$$

This is obtained by putting $a = 1$, $\alpha = \pi$ in (1).

■ **EXAMPLE 1.** *Solve $x^5 - 1 = 0$. Deduce the values of $\cos 36°$ and $\cos 72°$.*
The real root is 1. Dividing by $x - 1$,

$$x^4 + x^3 + x^2 + x + 1 = 0, \text{ that is } x^2 + \frac{1}{x^2} + x + \frac{1}{x} + 1 = 0.$$

Let $z = x + \dfrac{1}{x}$ then $z^2 + z - 1 = 0$, giving $z = \dfrac{1}{2}\,(-1 \pm \sqrt{5}\,)$.

The values of x are $\cos \dfrac{2r\pi}{5} \pm \iota \sin \dfrac{2r\pi}{5}$ $(r = 1, 2)$.

Therefore the values of z are $2 \cos \dfrac{2\pi}{5}$, and $2 \cos \dfrac{4\pi}{5}$. Now $\cos \dfrac{4\pi}{5} = - \cos \dfrac{\pi}{5}$;

$$\therefore \cos \frac{\pi}{5} = \cos 36° = \frac{\sqrt{5}+1}{4}, \text{ and } \cos \frac{2\pi}{5} = \cos 72° = \frac{\sqrt{5}-1}{4}.$$

■ **EXAMPLE 2.** *What are the special roots of $x^{15}–1= 0$. Find the equation of which these are the roots. Show that this can be reduced to an equation of the fourth degree. What are its roots?*

The numbers < 15 and prime to 15 are
$$1, 2, 4, 7, 15 - 7, 15 - 4, 15 - 2, 15 - 1.$$

The special roots are $\cos \dfrac{2r\pi}{15} \pm \iota \sin \dfrac{2r\pi}{15}$ $(r = 1, 2, 4, 7)$.

The non-special roots are the roots of $x^3 - 1 = 0$ and $x^5 - 1 = 0$.

The L.C.M. of $x^3 - 1$ and $x^5 - 1$ is $(x^2 + x + 1)(x^5 - 1)$ and

$$\frac{x^{15} - 1}{\left(x^2 + x + 1\right)\left(x^5 - 1\right)} = \frac{x^{10} + x^5 + 1}{x^2 + x + 1} = x^8 - x^7 + x^5 - x^4 + x^3 - x + 1.$$

Therefore the equation whose roots are the special roots is

$$x^8 - x^7 + x^5 - x^4 + x^3 - x + 1 = 0,$$

that is, $\left(x^4 + \dfrac{1}{x^4}\right) - \left(x^3 + \dfrac{1}{x^3}\right) + \left(x + \dfrac{1}{x}\right) - 1 = 0.$

If $z = x + \dfrac{1}{x}$, this becomes

$$(z^4 - 4z^2 + 2) - (z^3 - 3z) + z - 1 = 0,$$

that is, $z^4 - z^3 - 4z^2 + 4z + 1 = 0$,

of which the roots are $2 \cos \dfrac{2r\pi}{15}$ $(r = 1, 2, 4, 7)$.

■ **5. The Equation $x^{17} - 1 = 0$.** Gauss discovered that *the solution of this equation depends on that of four quadratic equations.* This can be proved as follows.

The imaginary roots of $x^{17} - 1 = 0$ are

$$\cos \frac{2r\pi}{17} \pm \iota \sin \frac{2r\pi}{17} , \text{ where } r = 1, 2, 3, \dots 8.$$

Let $y_r = \cos \dfrac{2r\pi}{17}$, then we have

$$y_{17-r} = y_r \text{ and } 2y_r y_s = y_{r+s}, + y_{r-s}. \qquad \dots(A)$$

Also the sum of the imaginary roots is -1, therefore

$$y_1 + y_2 + y_3 + \dots + y_8 = -\frac{1}{2}. \qquad \dots(B)$$

Equation (A) lead to an arrangement in pairs of y_1, y_2, ... y_8, which has a remarkable cyclic property. Starting with the pair $y_1 \, y_4$, we have

$$\left. \begin{array}{l} 2y_1 y_4 = y_3 + y_5, \; 2y_3 y_5 = y_2 + y_8 \\ 2y_2 y_8 = y_6 + y_7, \; 2y_6 y_7 = y_1 + y_4 \end{array} \right\} . \qquad \dots(C)$$

On trial, it will appear that no y except y_4, taken with y_1, will lead to a like result.

Let $\quad y_1 + y_4 = 2\alpha, \; y_3 + y_5 = 2\beta, \; y_2 + y_8 = 2\gamma, \; y_6 + y_7 = 2\delta \left. \right\}$

then $\qquad\qquad\qquad y_1 y_4 = \beta, \; y_3 y_5 = \gamma, \; y_2 y_8 = \delta, \; y_6 y_7 = \alpha \quad \left. \right\}$ $\qquad \dots(D)$

It follows that *any relation connecting* α, β, γ, δ, *continues to hold after the cyclic substitution* $(\alpha \, \beta \, \gamma \, \delta)$.

Such relations are the following.

By (B) $\alpha + \beta + \gamma + \delta = -\dfrac{1}{4}.$ ...(E)

Again, using equation (A),

$$\alpha\gamma = \frac{1}{4}\,(y_1 y_2 + y_1 y_8 + y_4 y_2 + y_4 y_8) = \frac{1}{8}\,(y_1 + y_3 + y_7 + y_8 + y_2 + y_6 + y_4 + y_5),$$

therefore $\alpha\gamma = -\dfrac{1}{16}.$ ...(F)

Consequently $\beta\delta = -\dfrac{1}{16}.$ ...(G)

Similarly we can show that

$$16\alpha\beta = -1 + 4\alpha - 4\beta,$$ ...(H)
$$4\alpha^2 = 1 + 2\beta + \gamma.$$ ...(I)

Finally let $\alpha + \gamma = 2\lambda, \quad \beta + \delta = 2\mu.$ ...(J)

Then by (E), $\lambda + \mu = -\dfrac{1}{8}.$

Also $4\lambda\mu = \alpha\beta + \beta\gamma + \gamma\delta + \alpha\delta.$

Hence by (H) and three equations derived from this by the cyclic substitution $(\alpha\beta\gamma\delta)$, we find that

$$\lambda\mu = -\frac{1}{16}.$$

Thus,

$$
\left.
\begin{aligned}
&y_1, y_4 \text{ are the roots of } y^2 - 2\alpha y + \beta = 0\\
&y_3, y_5 \text{ are the roots of } y^2 - 2\beta y + \gamma = 0\\
&y_2, y_8 \text{ are the roots of } y^2 - 2\gamma y + \delta = 0\\
&y_6, y_7 \text{ are the roots of } y^2 - 2\delta y + \alpha = 0\\
&\alpha, \gamma \text{ are the roots of } y^2 - 2\lambda y - \tfrac{1}{16} = 0\\
&\beta, \delta \text{ are the roots of } y^2 - 2\mu y - \tfrac{1}{16} = 0\\
&\lambda, \mu \text{ are the roots of } y^2 + \tfrac{1}{8}y - \tfrac{1}{16} = 0
\end{aligned}
\right\}.
$$ ...(K)

Since $\cos\theta$ decreases as θ increases from 0 to π, *the greater root of each quadratic is that which is stated first.*

Thus, starting with the last equation and working upwards, the values of $y_1, y_2, \ldots y_8$ can be found, and then the roots of $x^{17} - 1 = 0$ are given by $x^2 - 2y_1 x + 1 = 0$, etc.

■ **EXAMPLE 1:** *Express* $\cos\dfrac{2\pi}{17}$ *as a surd.*

By (K), $\lambda = -\dfrac{1}{16} + \dfrac{1}{16}\sqrt{17}, \quad \mu = -\dfrac{1}{16} - \dfrac{1}{16}\sqrt{17},$

and $$\alpha = \lambda + \sqrt{p},\ \gamma = \lambda - \sqrt{p},\ \beta = \mu + \sqrt{q}.$$

where $$p = \frac{1}{16^2}(34 - 2\sqrt{17}),\ q = \frac{1}{16^2}(34 + 2\sqrt{17}).$$

Also $$\cos\frac{2\pi}{17} = y_1 = \alpha + \sqrt{(\alpha^2 - \beta)};$$

and by (I), $$\alpha^2 - \beta = \frac{1}{4}(1 + \gamma - 2\beta) = \frac{1}{4}(1 + \lambda - 2\mu - \sqrt{p - 2}\sqrt{q});$$

therefore
$$\cos\frac{2\pi}{17} = \frac{1}{16}\left\{-1 + \sqrt{17} + \sqrt{(34 - 2\sqrt{17})}\right\}$$
$$+ \frac{1}{8}\sqrt{\left\{17 + 3\sqrt{17} - \sqrt{(34 - 2\sqrt{17})} - 2\sqrt{(34 + 2\sqrt{17})}\right\}}.$$

Notes: (*i*) The cosines of $\dfrac{4\pi}{17}$, $\dfrac{6\pi}{17}$, ... $\dfrac{16\pi}{17}$ can be expressed in a similar form; the surd values of the sines of these angles involve one more radical sign. (Ex. XIX, 22)

(*ii*) Any length determined by a quadratic equation can be found geometrically. It is therefore possible, *by the use of ruler and compasses only,* to effect the following construction.

■ **6. To inscribe a regular 17-sided Polygon in a Circle:** Take the radius of the circle as unit of length. Let O be the centre. Draw $x'Ox$, $y'Oy$ at right angles, cutting the circle in A, A' and B, B'. Let the vertices of a regular inscribed 17-gon be marked 0,1, 2,... 17, the point O coinciding with B. The ordinates y_1, y_2,... y_8 of the points 1, 2,... 8 are given by the equations of Art. 5. To solve these geometrically, observe that the roots of $y^2 - 2ay + b = 0$ are the ordinates of the points where the axis $x = 0$ cuts the circle $x^2 + y^2 - 2ay + b = 0$.

Thus to find the points $(0, y_1)$, $(0, y_2)$, ... $(0, y_8)$ (two of which are marked y_3, y_5 in Fig. 27), we have to draw the circles in the following table.

Circle	Centre	Ends of chord $y = 0$	Ends of diameter $x = 0$	Name of circle
$x^2 + y^2 + \frac{1}{8}y - \frac{1}{16} = 0$	$\left(0, -\frac{1}{16}\right)$	$\left(\pm\frac{1}{4}, 0\right)$	$(0, \lambda)$, $(0, \mu)$	ω
$x^2 + y^2 - 2\lambda y - \frac{1}{16} = 0$	$(0, \lambda)$	$\left(\pm\frac{1}{4}, 0\right)$	$(0, \alpha)$, $(0, \gamma)$	λ
$x^2 + y^2 - 2\mu y - \frac{1}{16} = 0$	$(0, \mu)$	$\left(\pm\frac{1}{4}, 0\right)$	$(0, \beta)$, $(0, \delta)$	μ
$x^2 + y^2 - 2\alpha y + \beta = 0$	$(0, \alpha)$	$\left(\pm\sqrt{-\beta}, 0\right)$	$(0, y_1)$, $(0, y_4)$	α

$x^2 + y^2 - 2\beta y + \gamma = 0$	$(0, \beta)$	$\left(\pm\sqrt{-\gamma}, 0\right)$	$(0, y_3), (0, y_5)$	β
$x^2 + y^2 - 2\gamma y + \delta = 0$	$(0, \gamma)$	$\left(\pm\sqrt{-\delta}, 0\right)$	$(0, y_2), (0, y_8)$	γ
$x^2 + y^2 - 2\delta y + \alpha = 0$	$(0, \delta)$	$\left(\pm\sqrt{-\alpha}, 0\right)$	$(0, y_6), (0, y_7)$	δ

The centres of the circles are marked ω, λ, μ, α, etc., in Fig. 27. The circles ω, λ, μ are easily drawn, giving the points α, β, γ, δ. To draw the last four circles notice that:

(*i*) *The circles α, β, γ, δ meet $y = 0$ at the same points as the circles on Bβ, Bγ, Bδ, Bα as diameters, respectively.*

For, since the equation to the circle on Bβ as diameter is

$$x^2 + \left(y - \frac{1+\beta}{2}\right)^2 = \left(\frac{1-\beta}{2}\right)^2,$$

which is satisfied by $x = \pm\sqrt{-\beta}$, $y = 0$, we can draw the circles β and γ.

(*ii*) *The diameters of the circles α, β, γ, δ are equal to the tangents from B to the circles β, γ, δ, α, respectively.*

For (diameter of $\odot$ α)^2 = 4 $(\alpha^2 - \beta) = (2\beta + \gamma + 1) - 4\beta$, and the square of tangent from B to $\odot$ β =1 $- 2\beta + \gamma$. Hence we can draw the circles α, δ.

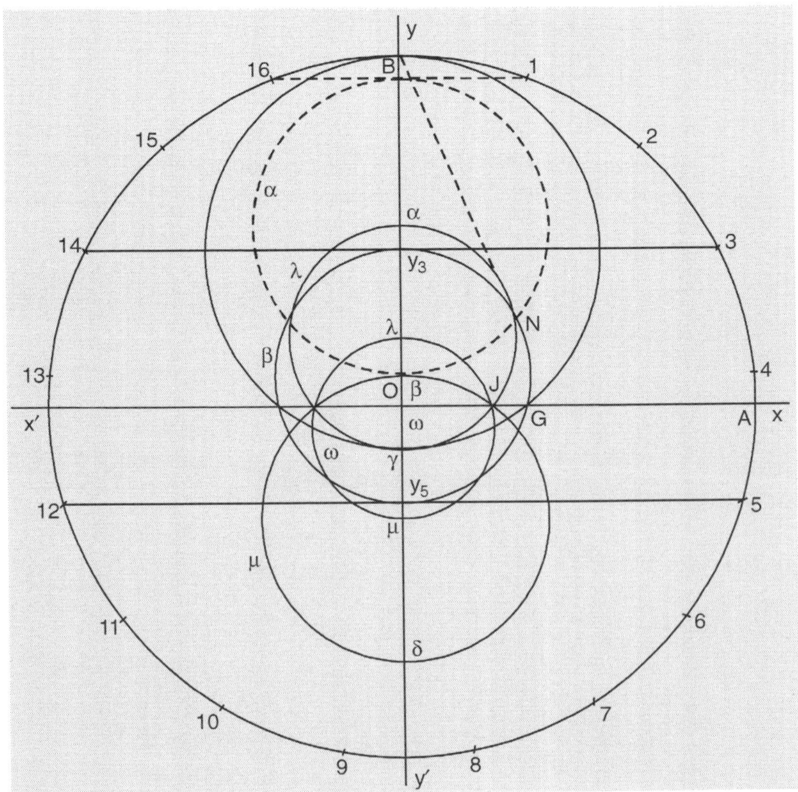

Fig. 27

Construction. Along Ox take $OJ = \dfrac{1}{4} OA$. Along Oy' take $O\omega = \dfrac{1}{4} OJ$. With centre ω and radius ωJ draw a circle cutting yOy' in λ, μ. With centre λ and radius λJ draw a circle cutting yOy' in α, γ. With centre μ and radius μJ draw a circle cutting yOy' in β, δ. Draw the circle on $B\gamma$ as diameter cutting Ox in G. With centre β and radius βG draw a circle cutting yOy' in y_3, y_5.

Through y_3, y_5 draw parallels to $x'Ox$. These lines meet the given circle in the vertices 3, 14 and 5, 12 of the required 17-gon, which can be completed by bisecting the arc 35 at 4 and setting off arcs equal to 34.

Or we may draw the circles α, γ, δ as explained above.

EXERCISE XIX

RECIPROCAL AND BINOMIAL EQUATIONS

Solve the equations in Exs. 1–9 :

1. $x^4 - 8x^3 + 17x^2 - 8x + 1 = 0$.

2. $x^4 + 6x^3 - 5x^2 + 6x + 1 = 0$.

3. $2x^4 + 3x^3 + 5x^2 + 3x + 2 = 0$.

4. $2x^4 + x^3 - 17x^2 + x + 2 = 0$.

5. $2x^5 + 5x^4 - 5x - 2 = 0$.

6. $3x^5 - 10x^4 - 3x^3 - 3x^2 - 10x + 3 = 0$.

7. $2x^6 - x^5 - 2x^3 - x + 2 = 0$.

8. $2x^7 - x^4 - 3x^4 - 3x^3 - x + 2 = 0$.

9. $x^8 + 1 + (x + 1)^8 = 2 (x^2 + x + 1)^4$.

   [Write this $x^4 + x^{-4} + (x + x^{-1} + 2)^4 = 2(x + x^{-1} + 1)^4$, and show that this reduces to $(x + x^{-1} + 1) (x + x^{-1} + 2) = 0$.]

10. If $s_n = 1 + u_1 + u_2 + \dots + u_n$ where $u_n = x^n + x^{-n}$ and $z = x + x^{-1}$, prove that

$$s_n = \frac{u_{n+1} - u_n}{z - 2} \text{ and } s_n = zs_{n-1} - s_{n-2}.$$

Hence show that

$$s_0 = 1, \; s_1 = z + 1, \; s_2 = z^2 + z - 1,$$
$$s_3 = z^3 + z^2 - 2z - 1, \; s_4 = z^4 + z^3 - 3z^2 - 2z + 1,$$
$$s_5 = z^5 + z^4 - 4z^3 - 3z^2 + 3z + 1,$$
$$s_6 = z^6 + z^5 - 5z^4 - 4z^3 + 6z^2 - 3z - 1.$$

Note that $s_n = 0$ is the equation on which the solution of $x^{2n+1} - 1 = 0$ depends.

$$\left[1 + s_n = \frac{1 - x^n}{1 - x} + \frac{1 - x^{-n}}{1 - x^{-1}} \text{ and } (1 - x)\left(1 - x^{-1}\right) = 2 - z. \right]$$

11. If α is an imaginary root of $x^7 - 1 = 0$, the equation whose roots are $\alpha + \alpha^6$, $\alpha^2 + \alpha^5$, $\alpha^3 + \alpha^4$ is $s_3 = z^3 + z^2 - 2z - 1 = 0$.

12. If n is a prime, the special roots of $x^{2n} - 1 = 0$ are the roots of $x^n + 1 = 0$.

13. The special roots of $x^9 - 1 = 0$ are the roots of $x^6 + x^3 + 1 = 0$, and their values are

$$\cos \frac{2r\pi}{9} \pm \iota \sin \frac{2r\pi}{9}, \text{ where } r = 1, 2, 4.$$

14. The special roots of $x^{12} - 1 = 0$ are the roots of $x^4 - x^2 + 1 = 0$, and their values are

$$\pm \left(\cos \frac{\pi}{6} \pm \iota \sin \frac{\pi}{6} \right), \text{ that is, } \pm \frac{1}{2} \left(\sqrt{3} \pm \iota \right).$$

[To find the equation, divide $x^{12} - 1$ by the L.C.M. of $x^2 - 1$, $x^3 - 1$, $x^4 - 1$, $x^6 - 1$, *i.e.* by the L.C.M. of $x^4 - 1$, $x^6 - 1$.]

15. If α, β, γ are the roots of $ax^3 + bx^2 + cx + d = 0$, prove that the equation whose roots

are $\alpha + \dfrac{1}{\alpha}$, $\beta + \dfrac{1}{\beta}$, $\gamma + \dfrac{1}{\gamma}$ is

$$adz^3 + (ac + bd)\, z^2 + (ab + bc + cd - 3ad)\, z + (a-c)^2 + (b-d)^2 = 0.$$

[Prove that $ad\, \pi\, \{x^2 - x\, (\alpha + \alpha^{-1}) + 1\} = \{(ax^3 + bx^2 + cx + d)\,(dx^3 + cx^2 + bx + a)\}$, and write $(x^2 + 1)/x = z$.]

16. Show that there is a value of k for which $x^6 - 15x^3 - 8x^2 + 2$ is divisible by $x^2 + kx + 1$, and find this value.

[Show that $x^6 - 15x^3 - 8x^2 + 2 = 0$ has two roots α, β such that $\alpha\beta = 1$. Then $k = -(\alpha + \beta)$.]

17. For the equation $x^4 + px^3 + qx^2 + px + 1 = 0$, prove that

    (*i*) the roots are all real if, and only if,

$$4\,(q - 2) < p^2 < \left(\frac{1}{2}q + 1 \right)^2 \quad \text{and} \quad 2q + 4 < p^2;$$

    (*ii*) the roots are all imaginary if, and only if, $p^2 < \left(\dfrac{1}{2}q + 1 \right)^2$ and at least one of the two inequalities $p^2 < 4\,(q - 2)$, $p^2 < 2q + 4$ holds;

    (*iii*) two roots are real and two imaginary if $\left(\dfrac{1}{2}q + 1 \right)^2 < p^2$.

[If $z = x + x^{-1}$, then $z^2 + pz + q - 2 = 0$. Let z_1, z_2 be the values of z, then

$$x = \frac{1}{2} \left(z_1 \pm \sqrt{z_1^2 - 4} \right) \quad \text{or} \quad \frac{1}{2} \left(z_2 \pm \sqrt{z_2^2 - 4} \right).$$

If *any* value of x is real, one and therefore both values of z must be real, and so $p^2 > 4\,(q - 2)$. The conditions in question can now be obtained by considering the equation whose roots are $z_1^2 - 4$, $z_2^2 - 4$.]

18. *If in the last example all the roots of the given equation are imaginary and the values of z are real, prove that* $|p| < 4$ *and* $q < 6$.

[For then $4\,(q - 2) < p^2 < 2q + 4$, etc.]

19. If $(x-1)^5 = a(x^5 - 1)$ and $x \neq 1$, prove that

$$x + \frac{1}{x} = \frac{4 + a \pm \sqrt{5a\,(4 + a)}}{2\,(1 - a)}.$$

Also if $0 < a > 16$, the given equation has three real and two imaginary roots, but if $a < 0$ or $a > 16$, the only real root is 1.

20. Show that, if r is any integer, the roots of

$$1 - nx - \frac{n\,(n-1)}{\lfloor 2} x^2 + \frac{n\,(n-1)\,(n-2)}{\lfloor 3} x^3 + \dots + (-1)^{\frac{1}{2}n(n+1)} x^n = 0$$

are the values of $\tan \dfrac{(4r+1)\pi}{4n}$.

21. If $y_r = \cos \dfrac{2r\pi}{5}$, show that $2y_1 y_2 = y_1 + y_2 = -\dfrac{1}{2}$,

and deduce the construction indicated in the Fig. 28 for inscribing a regular pentagon in a circle.

In the figure $OH = \dfrac{1}{2} OA$, and $OK = \dfrac{1}{2} OH$.

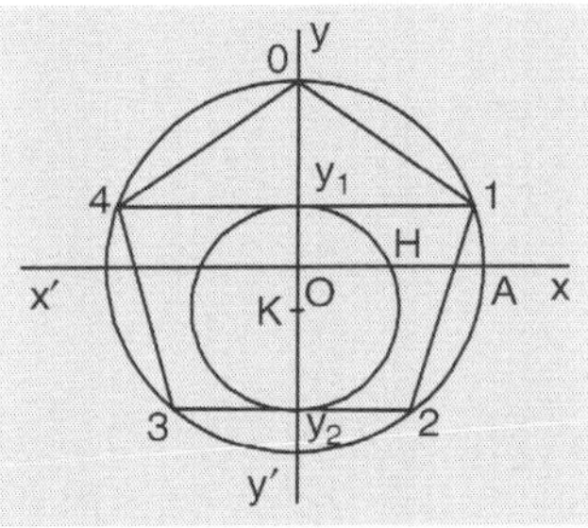

Fig. 28

22. Prove that the values of $\sin \dfrac{2\pi}{17}$ and $\cos \dfrac{\pi}{17}$ are respectively

$$\sqrt{[\tfrac{1}{32}\{17 - \sqrt{17} + \sqrt{(34 - 2\sqrt{17})}\}}$$

$$-\frac{1}{16}\sqrt{\{17 + 3\sqrt{17} + \sqrt{(34 - 2\sqrt{17})} + 2\sqrt{(34 + 2\sqrt{17})}\}}],$$

and $\dfrac{1}{16}\{1 - \sqrt{17} + \sqrt{(34 - 2\sqrt{17})}\}$

$$+\frac{1}{8}\sqrt{\{17 + 3\sqrt{17} + \sqrt{(34 - 2\sqrt{17})} + 2\sqrt{(34 + 2\sqrt{17})}\}}$$

[Referring to Art. 5, Ex. 1, $\sin \dfrac{2\pi}{17} = \sqrt{\left(\dfrac{1}{2} - \dfrac{1}{2}y_2\right)}$ and

$$\delta = \mu - \sqrt{q}, \; y_2 = \gamma + \sqrt{(\gamma^2 - \delta)} = \gamma + \frac{1}{2}\sqrt{(1 + \alpha - 2\delta)} ;$$

$$\therefore \quad y_2 = \gamma + \frac{1}{2}\sqrt{\{1 + \lambda - 2\mu + \sqrt{p + 2\sqrt{q}}\}} .]$$

▪▪▪

12

Cubic and Biquadratic Equations

■ **1. The Cubic Equation:** The standard form of the cubic equation is
$$u = ax^3 + 3bx^2 + 3cx + d = 0. \qquad \text{...(A)}$$

If $x = y - b/a$ this equation becomes

$$y^3 + \frac{3H}{a^2}\, y + \frac{G}{a^3} = 0, \qquad \text{...(B)}$$

where
$$H = ac - b^2, \quad G = a^2d - 3abc + 2b^3.$$

If $z = ay = ax + b$, the equation is
$$z^3 + 3Hz + G = 0. \qquad \text{...(C)}$$

If α, β, γ are the roots of (A), those of (B) are $\alpha + b/a$, $\beta + b/a$, $\gamma + b/a$, those of (C) are $a\alpha + b$, $a\beta + b$, $a\gamma + b$.

■ **2. Equation whose Roots are the Squares of the Differences of the Roots:** The differences of the roots are the same for equations (A) and (B). Hence the equation whose roots are

$$(\beta - \gamma)^2, \ (\gamma - \alpha)^2, \ (\alpha - \beta)^2,$$

is obtained by putting $q = 3H/a^2$, $r = G/a^3$ in Ch. 6, 17.

The resulting equation is
$$a^6z^3 + 18a^4Hz^2 + 81a^2H^2z + 27(G^2 + 4H^3) = 0, \qquad \text{...(D)}$$

whence the important equality
$$a^6(\beta - \gamma)^2 \, (\gamma - \alpha)^2 \, (\alpha - \beta)^2 = -27(G^2 + 4H^3). \qquad \text{...(E)}$$

It is easily verified that
$$G^2 + 4H^3 = a^2\Delta, \qquad \text{...(F)}$$

where
$$\Delta = a^2d^2 - 6abcd + 4ac^3 + 4b^3d - 3b^2c^2, \qquad \text{...(G)}$$

so that
$$a^4(\beta - \gamma)^2 \, (\gamma - \alpha)^2 \, (\alpha - \beta)^2 = -27\Delta. \qquad \text{...(H)}$$

The function Δ is called the *discriminant* of the cubic u, and its vanishing is the necessary and sufficient condition that the equation $u = 0$ should have two equal roots.

■ **3. Character of the Roots:** Excluding the case in which two roots are equal, and remembering that imaginary roots occur in pairs, the possibilities are

(*i*) *All the roots may be real,* and then by (H), $\Delta < 0$.

(*ii*) *One root may be real and two imaginary*; and, denoting the roots by α, $\lambda \pm \iota\mu$, we find that the product of the squares of the differences of the roots is equal to $-4\mu^2\{(\alpha - \lambda)^2 + \mu^2\}^2$, so that $\Delta > 0$.

182

Hence *if $\Delta < 0$, all the roots are real and unequal ; if $\Delta > 0$, one root is real and two are imaginary.*

Also it follows from equation (D) that $H = 0$, $G = 0$ are the necessary and sufficient conditions for *three equal roots.*

■ **4. Cardan's Solution:** If $z = ax + b$, the equation

$$ax^3 + 3bx^2 + 3cx + d = 0 \qquad \text{...(A)}$$

becomes $\qquad\qquad z^3 + 3Hz + G = 0. \qquad\qquad\qquad\text{...(B)}$

Let $\qquad\qquad z = m^{\frac{1}{3}} + n^{\frac{1}{3}}$; then $z^3 - 3\, m^{\frac{1}{3}} n^{\frac{1}{3}} z - (m + n) = 0. \qquad \text{...(C)}$

Comparing (B) and (C), $\quad m^{\frac{1}{3}} n^{\frac{1}{3}} = -H, \quad m + n = -G.$

Therefore m and n are the roots of $\quad t^2 + Gt - H^3 = 0, \qquad\qquad \text{...(D)}$

and we may take $\qquad\qquad m = \tfrac{1}{2}\left(-G + \sqrt{G^2 + 4H^3}\right).$

If Q denotes any one of the three values of

$$\sqrt[3]{\left\{\tfrac{1}{2}\left(-G + \sqrt{G^2 + 4H^3}\right)\right\}} ,$$

the three values of $m^{\frac{1}{3}}$ are Q, ωQ, $\omega^2 Q$, where ω is an imaginary cube root of unity.

Also, because $m^{\frac{1}{3}} n^{\frac{1}{3}} = -H$, the corresponding values of $n^{\frac{1}{3}}$ are
$$-H/Q, \quad -\omega^2 H/Q, \quad -\omega H/Q.$$

Hence the values of z, that is of $ax + b$, are
$$Q - H/Q, \quad \omega Q - \omega^2 H/Q, \quad \omega^2 Q, - \omega H/Q.$$

Notes: *If $G^2 + 4H^3 < 0$, all the roots of the cubic are real, but Cardan's solution gives them in an imaginary form, which is very unsuitable for practical purposes.* In this case a solution can be easily obtained as follows.*

■ **5. Trigonometrical Solution when $G^2 + 4H^3 < 0$:** Taking the equation:

$$z^3 + 3Hz + G = 0, \qquad\qquad \text{...(A)}$$

when $G^2 + 4H^3 < 0$, a solution can be obtained by using the equation

$$\cos 3\theta = 4\cos^3\theta - 3 \cos \theta.$$

If $\cos 3\theta$ is known, $\cos \theta$ is determined by

$$\cos^3\theta - \tfrac{3}{4} \cos \theta - \tfrac{1}{4} \cos 3\theta = 0. \qquad\qquad \text{...(B)}$$

Let $z = q \cos \theta$, so that

$$\cos^3 \theta + \frac{3H}{q^2} \cos\theta + \frac{G}{q^3} = 0. \qquad\qquad \text{...(C)}$$

Equations (B) and (C) are identical if

$$q = 2\sqrt{-H} \quad \text{and} \quad \cos 3\theta = -\frac{4G}{q^3} = -\frac{G}{2\sqrt{-H^3}}. \qquad\qquad \text{...(D)}$$

---

* This is sometimes called the *irreducible case.*

Now $G^2 < -4H^3$, and so a real value of θ can be found to satisfy the last equation. If α is any such value, the roots of (A) are

$$q \cos \alpha, \quad q \cos\left(\frac{2\pi}{3} + \alpha\right), \quad q \cos\left(\frac{2\pi}{3} - \alpha\right).$$

■ **EXAMPLE :** *If all the roots of $ax^3 + 3bx^2 + 3cx + d = 0$ are real, show that the equation can be reduced to*

$$t^3 - t + \mu = 0, \text{ where } \quad 27\mu^2 < 4,$$

by a substitution of the form $x = p + qt$, where p and q are real.

If $z = ax + b$, the equation becomes $z^3 - hz + G = 0$,

where $h = -3H = 3(b^2 - ac)$. Since all the roots are real, $G^2 + 4H^3 < 0$, $\therefore H < 0$ and $h > 0$.

Writing $z = \sqrt{h.t}$, the equation becomes $t^3 - t + \mu = 0$ where $\mu - G/\sqrt{h^3}$, and the substitution is

$$x = -\frac{b}{a} + \frac{\sqrt{h}}{a}.t.$$

> **Note:** The roots of the cubic can be expressed as convergent series by using this reduction. (See Ch. 29, 3.)

■ **6. Two Important Functions of the Roots:**

Let $\qquad\qquad L = \alpha + \omega\beta + \omega^2\gamma, \quad M = \alpha + \omega^2\beta + \omega\gamma,$

where ω is an imaginary cube root of unity, and observe that

$$\omega^3 = 1, \quad 1 + \omega + \omega^2 = 0, \quad \omega - \omega^2 = \pm\iota\sqrt{3}.$$

(1) *The interchange of any two of the letters α, β, γ transforms either of the functions L^3, M^3 into the other.*

Thus the transposition (α, β) changes L^3 into

$$(\beta + \omega\alpha + \omega^2\gamma)^3 = (\omega\alpha + \omega^3\beta + \omega^2\gamma)^3 = \omega^3 M^3 = M^3.$$

It is the existence of functions possessing this property which makes the solution of a cubic equation depend on that of a quadratic.

(2) *L and M are functions of the differences of α, β, γ.* For since

$$1 + \omega + \omega^2 = 0,$$

they are unaltered by writing $\alpha + h$, $\beta + h$, $\gamma + h$ for α, β, γ.

(3) We have the following equalities:

$$L + M = 2\alpha - \beta - \gamma, \qquad L - M = (\omega - \omega^2)(\beta - \gamma),$$
$$\omega^2 L + \omega M = 2\beta - \gamma - \alpha, \qquad \omega^2 L - \omega M = (\omega - \omega^2)(\gamma - \alpha),$$
$$\omega L + \omega^2 M = 2\gamma - \alpha - \beta, \qquad \omega L - \omega^2 M = (\omega - \omega^2)(\alpha - \beta).$$

Also, by Ex. XII, 9,

$$(2\alpha - \beta - \gamma)(2\beta - \gamma - \alpha)(2\gamma - \alpha - \beta) = -27G/a^3.$$

Therefore $\qquad\qquad L^3 + M^3 = -27G/a^3, \qquad\qquad\qquad ...(A)$

$$L^3 - M^3 = - 3(\omega - \omega^2)(\beta - \gamma)(\gamma - \alpha)(\alpha - \beta). \quad ...(B)$$

Moreover,

$$LM = \Sigma\alpha^2 - \Sigma\beta\gamma = - 9H/a^2. \qquad ...(C)$$

(4) From equations (A) and (C) it follows that $\left(\tfrac{1}{3}aL\right)^3$, $\left(\tfrac{1}{3}aM\right)^3$ *are the roots of*

$$t^2 + Gt - H^3 = 0,$$

which is the auxiliary quadratic in Cardan's solution.

7. The Cubic as the Sum of Two Cubes:

If $u \equiv ax^3 + 3bx^2 + 3cx + d$, then constants A, B, λ, μ can be found such that

$$u \equiv A(x - \lambda)^3 + B(x - \mu)^3, \qquad ...(A)$$

provided that u has no square factor. Also λ, μ are the roots of

$$H \equiv (ac - b^2)x^2 + (ad - bc)x + (bd - c^2) = 0.$$

By expanding and equating coefficients it will be seen that the identity (A) holds

if
$$\left.\begin{array}{ll} A + B = a & \lambda A + \mu B = - b \\ \lambda^2 A + \mu^2 B = c & \lambda^3 A + \mu^3 B = - d \end{array}\right\} . \qquad ...(B)$$

Eliminating A, B from the first three and from the last three of these equations,

$$a(\lambda^2\mu - \lambda\mu^2) + b(\lambda^2 - \mu^2) + c(\lambda - \mu) = 0,$$
$$b(\lambda^3\mu^2 - \lambda^2\mu^3) + c(\lambda^3\mu - \lambda\mu^3) + d(\lambda^2\mu - \lambda\mu^2) = 0.$$

Assume that neither λ nor μ is zero and that $\lambda \neq \mu$, and divide the first of these equations by $\lambda - \mu$ and the second by $\lambda\mu(\lambda - \mu)$, then

$$a\lambda\mu + b(\lambda + \mu) + c = 0, \quad \text{and} \quad b\lambda\mu + c(\lambda + \mu) + d = 0.$$

Eliminating μ,

$$(a\lambda + b)(c\lambda + d) - (b\lambda + c)^2 = 0.$$

Therefore λ and, by symmetry, μ are the roots of

$$(ax + b)(cx + d) - (bx + c)^2 = 0,$$

which is the same as

$$H \equiv (ac - b^2)x^2 + (ad - bc)x + (bd - c^2) = 0. \quad ...(C)$$

Also A, B are given by $\quad A + B = a, \quad \lambda A + \mu B = -b.$

If $\lambda = 0$, then by equation (B),

$$bd - c^2 = 0 \text{ and } \mu = -\frac{c}{b} = -\frac{d}{c} = -\frac{ad - bc}{ac - b^2},$$

so that also in this case λ and μ are the roots of $H = 0$.

If $\lambda = \mu$, the reasoning fails. This case will arise if, and only if, $u = 0$ has two equal roots. For using equation (G) of Art. 2, we find that

$$(ad - bc)^2 - 4(ac - b^2)(bd - c^2) = \Delta.$$

Hence, if $u = 0$ has two equal roots, so also has $H = 0$ and *vice versa*.

■ **EXAMPLE :** *Use the preceding to solve the equation* $2x^3 + 3x^2 - 21x + 19 = 0$.

Let $2x^3 + 3x^2 - 21x + 19 = A(x - \lambda)^3 + B(x - \mu)^3$.   ...(A)

Then λ, μ are the roots of $(2x + 1)(-7x + 19) - (x - 7)^2 = 0$, giving $\lambda = 1$, $\mu = 2$; and, equating the coefficients of x^3 and x^2 in (A),

$$A + B = 2, \quad A + 2B = -1, \quad \text{so that } A = 5, \ B = -3.$$

Hence the given equation may be written $5(x - 1)^3 = 3(x - 2)^3$, and the roots are given by $\sqrt[3]{5}(x - 1) = k \cdot \sqrt[3]{3(x - 2)}$,   where $k = 1$, ω, ω^2.

Since $k^3 = 1$,
$$x = \frac{5^{\frac{1}{3}} - 2k3^{\frac{1}{3}}}{5^{\frac{1}{3}} - k3^{\frac{1}{3}}} = \frac{\left(5^{\frac{1}{3}} - 2k3^{\frac{1}{3}}\right)\left(5^{\frac{2}{3}} - k5^{\frac{1}{3}}.3^{\frac{1}{3}} + k^2 3^{\frac{2}{3}}\right)}{2}.$$

Hence the roots are

$$\tfrac{1}{2}\left(-1 - \sqrt[3]{75} - \sqrt[3]{45}\right), \qquad \tfrac{1}{2}\left(-1 - \omega\sqrt[3]{75} - \omega^2\sqrt[3]{45}\right), \quad \tfrac{1}{2}\left(-1 - \omega^2\sqrt[3]{75} - \omega\sqrt[3]{45}\right).$$

■ **8. The Hessian:**

(1) The function H which occurs in the last article is called the *Hessian*, and is of great importance in the theory of the cubic. The roots of $H = 0$ will be denoted by λ, μ, and we shall write

$$H = h_0 x^2 + h_1 x + h_2,$$

where
$$h_0 = H = ac - b^2, \quad h_1 = ad - bc, \quad h_2 = bd - c^2,$$

so that
$$h_1{}^2 - 4h_0 h_2 = \Delta. \qquad \text{...(A)}$$

(2) Using the identities of Exercise XII, 4, 5, 6, we find that
$$-18H = a^2\{(x - \alpha)^2(\beta - \gamma)^2 + (x - \beta)^2(\gamma - \alpha)^2 + (x - \gamma)^2(\alpha - \beta)^2\}. \qquad \text{...(B)}$$

(3) We can find the values of λ, μ in terms of α, β, γ as follows. Let

$$L = \alpha + \omega\beta + \omega^2\gamma, \qquad\qquad M = \alpha + \omega^2\beta + \omega\gamma,$$
$$L' = \beta\gamma + \omega\gamma\alpha + \omega^2\alpha\beta, \qquad\qquad M' = \beta\gamma + \omega^2\gamma\alpha + \omega\alpha\beta.$$

Then, if $P = (x - \alpha)(\beta - \gamma)$, $Q = (x - \beta)(\gamma - \alpha)$, $R = (x - \gamma)(\alpha - \beta)$, we have
$$P + Q + R = 0, \quad \text{therefore } \Sigma P^2 = -2\Sigma QR$$

and $$\Sigma P^2 = \tfrac{2}{3}(\Sigma P^2 - \Sigma QR) = \tfrac{2}{3}(P + \omega Q + \omega^2 R)(P + \omega^2 R + \omega R).$$

It follows from (B) that one root λ of $H = 0$ is given by
$$(\lambda - \alpha)(\beta - \gamma) + \omega(\lambda - \beta)(\gamma - \alpha) + \omega^2(\lambda - \gamma)(\alpha - \beta) = 0,$$

that is, $$\lambda(\omega^2 - \omega)L + (\omega^2 - \omega)L' = 0,$$

so that $\lambda = -L'/L$ and similarly $\mu = -M'/M$.   ...(C)

(4) It is easy to verify that the substitution
$$t = H(ax + b)$$

reduces the equation $h_0 x^2 + h_1 x + h_2 = 0$ to $t^2 + Gt - H^3 = 0$; which is the auxiliary quadratic in Cardan's solution.

Thus the roots of the auxiliary quadratic are
$$H(a\lambda + b), \ H(a\mu + b).$$

EXERCISE XX

THE CUBIC

1. Find to five places of decimals the real root of

   (*i*) $x^3 + 29x - 97 = 0$,   (*ii*) $x^3 + 6x^2 + 27x - 26 = 0$,   (*iii*) $x^3 - 3x^2 + 5x - 43 = 0$.

2. Find to five places of decimals the real roots of

   (*i*) $x^3 - 3x + 1 = 0$,   (*ii*) $2x^3 + 11x^2 - 10x + 1 = 0$.

3. If α, β, γ are the roots of $x^3 - 3qx + r = 0$ and $k = \pm \sqrt{3\left(4q^3 - r^2\right)}$, show that

   (*i*) $(\beta - \gamma)(\gamma - \alpha)(\alpha - \beta) = 3k$.

   (*ii*) $(\beta - \gamma)(\gamma - \alpha) + (\gamma - \alpha)(\alpha - \beta) + (\alpha - \beta)(\beta - \gamma) = -9q$.

   (*iii*) The equation whose roots are $\beta - \gamma, \gamma - \alpha, \alpha - \beta$, is
   $$x^3 - 9qx + 3k = 0.$$

   (*iv*) $\alpha^2\beta + \beta^2\gamma + \gamma^2\alpha = \frac{3}{2}(r + k)$, $\alpha^2\gamma + \beta^2\alpha + \gamma^2\beta = \frac{3}{2}(r - k)$.

   (*v*) $\alpha^3\beta + \beta^3\gamma + \gamma^3\alpha = \alpha^3\gamma + \beta^3\alpha + \gamma^3\beta = -9q^2$.

   (*vi*) If the roots are all real and $\alpha > \beta > \gamma$, the difference between any two of the roots
   cannot exceed $2\sqrt{(3q)}$, and the difference between the greatest and least must
   exceed $3\sqrt{q}$.

4. If α, β, γ are the roots of $ax^3 + 3bx^2 + 3cx + d = 0$,

   (*i*) the equation whose roots are $\beta - \gamma, \gamma - \alpha, \alpha - \beta$ is

   $$a^3x^3 + 9Hax \pm \sqrt{-27\left(G^2 + 4H^3\right)} = 0;$$

   (*ii*) the equation whose roots are $\alpha^2\beta + \beta^2\gamma + \gamma^2\alpha, \alpha^2\gamma + \beta^2\alpha + \gamma^2\beta$ is
   $$a^4x^2 - 3a^2(ad - 3bc)x + 9(a^2d^2 - 6abcd + 3ac^3 + 3b^3d) = 0.$$

   *In the following examples the letters L, M, L', M', h_0, h_1, h_2 have the meanings
   assigned in Art. 8, and unless otherwise stated α, β, γ are the roots of*
   $$u \equiv ax^3 + 3bx^2 + 3cx + d = 0.$$

5. Show that $L^2 + \dfrac{3b}{a}M = 3M'$, and $M^2 + \dfrac{3b}{a}L = 3L'$.

6. If $u = A(x - \lambda)^3 + B(x - \mu)^3$ and α, β, γ are the roots of $u = 0$, assume that

   $$A^{\frac{1}{3}}(\alpha - \lambda) + B^{\frac{1}{3}}(\alpha - \mu) = \omega A^{\frac{1}{3}}(\beta - \lambda) + B^{\frac{1}{3}}(\beta - \mu) = \omega^2 A^{\frac{1}{3}}(\gamma - \lambda)$$
   $$+ B^{\frac{1}{3}}(\gamma - \mu) = 0. \qquad \qquad \text{...(A)}$$

   Hence prove that $A/B = -L^3/M^3$, $\lambda = -L'/L$, $\mu = -M'/M$.

   [Multiply equations (A) by $(1,1,1)$, $(1, \omega, \omega^2)$, $(1, \omega^2, \omega)$, add and use Ex. 5.]

7. If $k = \pm \sqrt{\left(-\frac{1}{3}\Delta\right)}$, prove that

   $$\alpha^2\beta + \beta^2\gamma + \gamma^2\alpha + 3\frac{d}{a} = \frac{9}{2a^2}(h_1 + k), \text{ and } \alpha^2\gamma + \beta^2\alpha + \gamma^2\beta + 3\frac{d}{a} = \frac{9}{2a^2}(h_1 - k).$$

8. Find the values of l, m, n such that
$$\beta\gamma + l\beta + m\gamma + n = 0, \quad \gamma\alpha + l\gamma + m\alpha + n = 0, \quad \alpha\beta + l\alpha + m\beta + n = 0;$$

proving that $2h_0 l = h_1 + k$, $2h_0 m = h_1 - k$, $h_0 n = h_2$, where $k = \pm \sqrt{\left(-\tfrac{1}{3}\Delta\right)}$.

That is to say, *any two roots* (α, β) *of the cubic equation* $u = 0$ *are connected by the* 'homographic' *relation*
$$2h_0\alpha\beta + (h_1 + k)\,\alpha + (h_1 - k)\beta + 2h_2 = 0.$$

9. Except when $\Delta = 0$, there are two substitutions of the form
$$x = (ly + m)/(l'y + m'),$$
which will transform the cubic equation $u = 0$ into itself, and these are
$$x = -\frac{(h_1 + k)\,y + 2h_2}{2h_0 y + h_1 - k},$$
where $3k^2 = 4h_0 h_2 - h_1^2$ and h_0, h_1, h_2 are the coefficients of the Hessian.
Explain why this fails if $\Delta = 0$.

[This is merely another way of stating the last part of Ex. 8.]

10. Find substitutions of the form $x = (ly + m) / (l'y + m')$ which will transform the equation $x^3 - 3x^2 + 3 = 0$ into itself, showing that these are
$$x = (2y - 3) / (y - 1) \quad \text{and} \quad x = (y - 3) / (y - 2).$$

11. Find substitutions of the form $x = (ly + m) / (l'y + m')$ which will transform the equation $x^3 + 3x^2 - 6x + 1 = 0$ into itself, showing that these are
$$x = (y - 1) / y \quad \text{and} \quad x = 1/(1 - y).$$

12. Show that any pair of roots α, β of the equation $x^3 - 3x^2 + 3x - 2 = 0$ are connected by a relation of the form $p\alpha + q\beta + r = 0$ where p, q, r are the same numbers whatever pair is chosen, proving that the relation is
$$\left(\sqrt{3} - \imath\right)\alpha + \left(\sqrt{3} + \imath\right)\beta - 2\sqrt{3} = 0.$$

13. If α, β, γ are the roots of $x^3 - 3qx + r = 0$, find l, m, n such that
$$l\alpha^2 + m\alpha + n = \beta, \quad l\beta^2 + m\beta + n = \gamma, \quad l\gamma^2 + m\gamma + n = \alpha;$$
proving that, if α, β, γ are real and $\alpha > \beta > \gamma$, then
$$\frac{l}{q} = \frac{2m}{r - k} = \frac{n}{-2q^2} = \frac{1}{k},$$

where
$$k = \sqrt{\tfrac{1}{3}\left(4q^3 - r^2\right)}.$$

Hence show that with these values of l, m, n the substitution
$$y = lx^2 + mx + n$$
transforms the equation $x^3 - 3qx + r = 0$ into itself.

14. If α, β, γ are the roots of $x^2 - 21x + 35 = 0$, show that $\alpha^2 + 2\alpha - 14$ is equal to either β or γ. (In other words, the substitution $y = x^2 + 2x - 14$ transforms the equation into itself.)

15. If α, β, γ are the roots of the cubic $u = 0$ and
$$\alpha^2 + p\alpha + q = k, \quad \beta^2 + p\beta + q = \omega k, \quad \gamma^2 + p\gamma + q = \omega^2 k,$$

prove that $\quad p = \dfrac{3b}{a} + \lambda, \quad q = \dfrac{2c}{a} + \lambda\dfrac{b}{a}$, where λ is a root of the Hessian $H = 0$.

Further, if $\lambda = - L'/L$, $\mu = - M'/M$, then $\quad 3k = M(\lambda - \mu) = -\dfrac{a^2 M}{27 H}\,(L^3 - M^3)$.

[Multiply by $(1,1,1)$, $(1, \omega, \omega^2)$, $(1, \omega^2, \omega)$, and add.]

16. Prove that the equation $u = 0$ is transformed into $y^3 = k^3$ by the substitution*
$$y = x^2 + px + q,$$

if $\quad p = \dfrac{3b}{a} + \lambda, \quad q = \dfrac{2c}{a} + \lambda\dfrac{b}{a},$

where λ is a root of $H = 0$ and k is determined as in Ex. 15.

17. Find substitutions of the form $y = x^2 + px + q$ which will transform
$$u \equiv x^3 + 3x^2 + 15x + 13 = 0$$
into the form $y^3 = k^3$, giving the value of k in each case.

[λ, μ are the roots of $x^2 + 2x - 3 = 0$ and $\left(\tfrac{1}{3}L\right)^3$, $\left(\tfrac{1}{3}M\right)^3$ are the roots of
$$t^2 + Gt - H^3 = 0,$$
i.e. of $t^2 = 4^3$. If we take $\lambda = 1$, $\mu = -3$, we must take $L = -6$, $M = 6$. It is thus found that each of the substitutions, $y = x^2 + 4x + 11$ and $y = x^2 + 7$, will reduce $u = 0$ to $y^3 = 8^3$.]

18. In Ex. 17, if $y = x^2 + 4x + 11$, we may have $y = 8$, giving $x^2 + 4x + 3 = 0$.

Explain why it is not to be expected that both roots of this equation satisfy $u = 0$.

■ **9. The Standard Form of the Biquadratic Equation is**
$$u = ax^4 + 4bx^3 + 6cx^2 + 4dx + e = 0. \qquad \text{...(A)}$$
If $x = y - b/a$, this becomes

$$ay^4 + \frac{6H}{a}y^2 + \frac{4G}{a^2}y + \frac{K}{a^3} = 0, \qquad \text{...(B)}$$

where $\quad H = ac - b^2$, $\quad G = a^2 d - 3abc + 2b^3$, $\quad K = a^3 e - 4a^2 bd + 6ab^2 c - 3b^4$.

If $z = ay = ax + b$, the equation becomes
$$z^4 + 6Hz^2 + 4Gz + K = 0. \qquad \text{...(C)}$$

If α, β, γ, δ are the roots of (A), those of (B) are $\alpha + b/a$, $\beta + b/a$, $\gamma + b/a$, $\delta + b/a$, and those of (C) are $a\alpha, + b$, $a\beta + b$, $a\gamma + b$, $a\delta + b$.

■ **10. Some Important Functions of the Roots:**

Let
$$\lambda = \beta\gamma + \alpha\delta, \quad \mu = \gamma\alpha + \beta\delta, \quad v = \alpha\beta + \gamma\delta.$$

(1) Any rearrangement of the letters α, β, γ, δ transforms any one of the functions λ, μ, v into itself or into one of the other two.

It will be found that the existence of functions having this property makes the solution of (A) depend on that of a cubic equation.

(2) The cyclic substitution $(\alpha\beta\gamma)$ changes λ to μ, μ to v, v to λ.

---

* This is a case of '*Tschirnhausen's Transformation.*' (See Art. 17)

We have
$$\left.\begin{array}{l}(\beta-\gamma)(\alpha-\delta)=\nu-\mu \\ (\gamma-\alpha)(\beta-\delta)=\lambda-\nu \\ (\alpha-\beta)(\gamma-\delta)=\mu-\lambda\end{array}\right\}, \qquad \dots(D)$$

Hence if two of α, β, γ, δ are equal, then two of λ, μ, ν are equal, and *vice versa.*

Also if three of α, β, γ, δ are equal, then $\lambda = \mu = \nu$, and *vice versa.*

(3) *The functions* λ, μ, ν *are the roots of*
$$a^3y^3 - 6a^2cy^2 + 4a(4bd - ae)\,y - 8(2ad^2 + 2eb^2 - 3ace) = 0. \qquad \dots(E)$$

For
$$\Sigma\lambda = \Sigma\alpha\beta = 6c/a,$$
$$\Sigma\mu\nu = \Sigma\alpha^2\beta\gamma = \Sigma\alpha \,.\, \Sigma\alpha\beta\gamma - 4\alpha\beta\gamma\delta = 4\,(4bd - ae)/a^2,$$
$$\lambda\mu\nu = \Sigma\alpha^2\beta^2\gamma^2 + \alpha\beta\gamma\delta\Sigma\alpha^2$$
$$= \frac{4}{a^2}\left(4d^2 - 3ce\right) + \frac{4e}{a^3}\left(4b^2 - 3ac\right)$$
$$= \frac{8}{a^3}\left(2ad^2 + 2eb^2 - 3ace\right).$$

The second term of this equation can be removed by the substitution
$$y = (4t + 2c)/a, \qquad \dots(F)$$
and the equation becomes
$$4t^3 - It + J = 0, \qquad \dots(G)$$
where
$$I = ae - 4bd + 3c^2, \qquad \dots(H)$$
$$J = ace + 2bcd - ad^2 - c^3 - eb^2. \qquad \dots(I)$$

Expressed as a determinant
$$J = \begin{vmatrix} a & b & c \\ b & c & d \\ c & d & e \end{vmatrix}. \qquad \dots(J)$$

It will be shown that the solution of (A) depends on equation (G), which is called *the reducing cubic.*

The functions I and J are of great importance in the theory of the biquadratic.

(4) Roots of the Reducing Cubic: Let t_1, t_2, t_3 be the roots of equation (G) corresponding to λ, μ, ν respectively. Then by (F)

$$\frac{4t_1}{a} = \lambda - \frac{2c}{a} = \lambda - \tfrac{1}{3}\left(\lambda + \mu + \nu\right) = \tfrac{1}{3}\left\{\lambda - \mu - \left(\nu - \lambda\right)\right\}; \qquad \dots(K)$$

whence, by (A), $\quad t_1 = \tfrac{1}{12}\,a\{(\gamma - \alpha)\,(\beta - \delta) - (\alpha - \beta)\,(\gamma - \delta)\}, \qquad \dots(L)$

with similar values for t_2, t_3.

If α, β, γ, δ *are all real, or if they are all imaginary, then* t_1, t_2, t_3 *are all real, and conversely.*

If two of the roots α, β, γ, δ *are real and two are imaginary, then only one of the three* t_1, t_2, t_3 *is real, and conversely.*

For λ, μ, ν, and therefore also t_1, t_2, t_3, are all real if α, β, γ, δ are all real, or if α, β, γ, δ are of the forms $l \pm \iota m$, $l' \pm \iota m'$.

Also if α, β are real and $\gamma = l + \iota m$, $\delta = l - \iota m$, then ν is real and λ, μ are imaginary. Moreover, one of these cases must arise, and so the converse statements are true.

(5) The Functions I, J: Since t_1, t_2, t_3, are functions of the *differences* of α, β, γ, δ, *so also are I, J.*

It follows that I and J are the same for equations (A) and (B), and therefore in equations (H) and (I) we may put

$$b = 0, \quad c = H/a, \quad d = G/a^2, \quad e = K/a^3,$$

giving
$$a^2 I = K + 3H^2 \quad \text{and} \quad a^3 J = HK - G^2 - H^3,$$

whence the important identities
$$K = a^2 I - 3H^2, \quad G^2 + 4H^3 = a^2(HI - a J). \qquad \text{...(M)}$$

Another important equality is
$$(\beta - \gamma)^2 (\alpha - \delta)^2 + (\gamma - \alpha)^2 (\beta - \delta)^2 + (\alpha - \beta)^2 (\gamma - \delta)^2 = 24I/a^2. \qquad \text{...(N)}$$

For
$$\Sigma (\beta - \gamma)^2 (\alpha - \delta)^2 = \Sigma (\nu - \mu)^2 = 2(\Sigma\lambda^2 - \Sigma\mu\nu),$$

and
$$\Sigma\lambda^2 - \Sigma\mu\nu = (\Sigma\lambda)^2 - 3\Sigma\mu\nu$$

$$= \left(\frac{6c}{a}\right)^2 - 3\frac{4(4bd - ae)}{a^2} = \frac{12}{a^2}\left(ae - 4bd + 3c^2\right) = 12I/a^2.$$

(6) The Discriminant: By equations (D),
$$a^6(\beta - \gamma)^2 (\gamma - \alpha)^2 (\alpha - \beta)^2 (\alpha - \delta)^2 (\beta - \delta)^2 (\gamma - \delta)^2$$
$$= a^6(\mu - \nu)^2 (\nu - \lambda)^2 (\lambda - \mu)^2$$
$$= 4^6 (t_2 - t_3)^2 (t_3 - t_1)^2 (t_1 - t_2)^2 \text{ by equation (K)}$$
$$= 256\Delta \text{ by Art. 2,} \qquad \text{...(O)}$$

where
$$\Delta = I^3 - 27 J^2. \qquad \text{...(P)}$$

The function Δ is called the *discriminant* of the quartic u, and its vanishing is the necessary and sufficient condition that the equation $u = 0$ may have two equal roots.

■ **11. Character of the Roots:** If two roots are equal, then $\Delta = 0$. Excluding this case and remembering that if there are any imaginary roots, then these occur in pairs, the possible cases are:

(1) *All the roots may be real*, and then $\Delta > 0$.

(2) *Two roots may be real and two imaginary.* Denoting the roots by α, β, $l \pm \iota m$, it is easily shown that the product of the squares of the differences of the roots is $- 4m^2 (\alpha - \beta)^2 \{(\alpha - l)^2 + m^2\}^2 \{(\beta - l)^2 + m^2\}^2$, and so $\Delta < 0$.

(3) *All the roots may be imaginary.* Denoting them by $l \pm \iota m$, $l' \pm \iota m'$, we find that the product of the squares of the differences of the roots is $16m^2 m'^2 \{(l - l')^2 + (m + m')^2\}^2 \{(l + l')^2 + (m - m')^2\}^2$, and so $\Delta > 0$.

Hence **(4)** *If $\Delta < 0$, two roots are real and two imaginary.*

(5) If $\Delta > 0$, *the roots are all real or all imaginary.*

For the criterion distinguishing the two cases of (5), see Art. 15.

■ **12. Ferrari's Solution of the Biquadratic:** Writing the equation
$$u = ax^4 + 4bx^3 + 6cx^2 + 4dx + e = 0, \qquad \text{...(A)}$$

we assume that

$$au = (ax^2 + 2bx + s)^2 - (2mx + n)^2. \qquad \text{...(B)}$$

Expanding and equating coefficients, we have

$$2m^2 = as + 2b^2 - 3ac, \quad mn = bs - ad, \quad n^2 = s^2 - ae. \qquad \text{...(C)}$$

Eliminating m, n,

$$(s^2 - ae)(as + 2b^2 - 3ac) = 2(bs - ad)^2,$$

which reduces to

$$s^3 - 3cs^2 + (4bd - ae)s + (3ace - 2ad^2 - 2eb^2) = 0. \qquad \text{...(D)}$$

The second term can be removed by the substitution

$$s = 2t + c \qquad \text{...(E)}$$

and equation (D) becomes

$$4t^3 - It + J = 0, \qquad \text{...(F)}$$

which is the 'reducing cubic' (See Art. 10,). Equation (C) become

$$\left. \begin{aligned} m^2 &= at + b^2 - ac = at - H \\ mn &= 2bt + bc - ad \\ n^2 &= (2t + c)^2 - ae \end{aligned} \right\} . \qquad \text{...(G)}$$

Thus, if t_1 is a root of (F) *and*

$$m_1 = \sqrt{\left(at_1 + b^2 - ac\right)}, \quad n_1 = (2bt_1 + bc - ad)/m_1,$$

the equation $u = 0$ can be put into the form

$$(ax^2 + 2bx + c + 2t_1)^2 - (2m_1x + n_1)^2 = 0,$$

and its roots are the roots of the quadratics

$$ax^2 + 2bx + c + 2t_1 = \pm (2m_1x + n_1).$$

It should be noticed that the three roots of (F) correspond to the three ways of expressing u as the product of two quadratic factors.

■ **Example :** *Solve $u = x^4 + 3x^3 + x^2 - 2 = 0$.*

Let
$$u = (x^2 + px + s)^2 - (mx + n)^2.$$

Expanding and equating coefficients, we have

$$p = \tfrac{3}{2}, \quad m^2 = 2s + \tfrac{5}{4}, \quad mn = \tfrac{3}{2}s, \quad n^2 = s^2 + 2;$$

$$\therefore \qquad 4s^3 - 2s^2 + 8s + 5 = 0.$$

The last equation is satisfied if $s = -\tfrac{1}{2}$, and then $m^2 = -1 + \tfrac{5}{4}$, $mn = -\tfrac{3}{4}$. Thus

we may take
$$s = -\tfrac{1}{2}, \quad m = \tfrac{1}{2}, \quad n = -\tfrac{3}{2},$$

and
$$\begin{aligned} u &= \{x^2 + (p + m)x + s + n\}\{x^2 + (p - m)x + s - n\} \\ &= (x^2 + 2x - 2)(x^2 + x + 1). \end{aligned}$$

Therefore the roots are $-1 \pm \sqrt{3}$, ω, ω^2, where ω is an imaginary cube root of 1.

■ **13. Deductions from Ferrari's Solution:**

Let β, γ be the roots of $ax^2 + 2bx + c + 2t_1 + (2m_1x + n_1) = 0$, ...(H) then α, δ are the roots of $ax^2 + 2bx + c + 2t_1 - (2m_1x + n_1) = 0.$ ...(I)

Further, let (γ, α), (α, β) be the roots of the equations obtained from (H) by changing the suffix 1 into 2 and 3 respectively.

Now
$$\alpha\beta\gamma = c + 2t_1 + n_1, \qquad a\alpha\delta = c + 2t_1 - n_1,$$
$$a(\beta + \gamma) = -2(b + m_1), \quad a(\alpha + \delta) = -2(b - m_1),$$

therefore
$$a\,(\beta\gamma + \alpha\delta) = 2c + 4t_1,$$

which agrees with equation (F) of Art. 10.

Also
$$a(\beta\gamma - \alpha\delta) = 2n_1, \quad a(\beta + \gamma - \alpha - \delta) = -4m_1, \qquad ...(J)$$

with similar equations given by the cyclic substitutions (α, β, γ) and $(1, 2, 3)$ of the suffixes.

Hence (*i*) *the values of m are*

$$-\tfrac{1}{4}a(\beta + \gamma - \alpha - \delta), \; -\tfrac{1}{4}a(\gamma + \alpha - \beta - \delta), \; -\tfrac{1}{4}a(\alpha + \beta - \gamma - \delta).$$

In connection with these functions it should be noted that by Ch. 6, 16, Ex. 3, (*iii*), p. 92,

$$(\beta + \gamma - \alpha - \delta)\,(\gamma + \alpha - \beta - \delta)\,(\alpha + \beta - \gamma - \delta) = 32G/a^3, \qquad ...(K)$$

Also, by (F) and (G), *the equation whose roots are*

$$\tfrac{1}{16}a^2(\beta + \gamma - \alpha - \delta)^2, \; \tfrac{1}{16}a^2(\gamma + \alpha - \beta - \delta)^2, \; \tfrac{1}{16}a^2(\alpha + \beta - \gamma - \delta)^2$$

is
$$4(y + H)^3 - a^2I\,(y + H) + a^3J = 0. \qquad ...(L)$$

This is an important equation, obtained in another way in Art. 15.

(*ii*) *The values of n are* $\tfrac{1}{2}a(\beta\gamma - \alpha\delta), \; \tfrac{1}{2}a(\gamma\alpha - \beta\delta), \; \tfrac{1}{2}a(\alpha\beta - \gamma\delta),$

(*iii*) The values of $-\dfrac{n_1}{2m_1}, \; -\dfrac{n_2}{2m_2}, \; -\dfrac{n_3}{2m_3}$ are

$$\frac{\beta\gamma - \alpha\delta}{\beta + \gamma - \alpha - \delta}, \; \frac{\gamma\alpha - \beta\delta}{\gamma + \alpha - \beta - \delta}, \; \frac{\alpha\beta - \gamma\delta}{\alpha + \beta - \gamma - \delta},$$

and from (F), (G) it follows that *these functions are the values of z found by eliminating t from*

$$z = -\frac{1}{2}\cdot\frac{2bt + bc - ad}{at + b^2 - ac} \quad \text{and} \quad 4t^3 - It + J = 0. \qquad ...(M)$$

■ **14. Biquadratic as the Product of Quadratic Factors:**

Descartes' method.   Let $u = ax^4 + 4bx^3 + 6cx^2 + 4dx + e$, and assume that
$$u = a\,(x^2 + 2lx + m)\,(x^2 + 2l'x + m').$$

Expanding and equating coefficients, we have

$$l + l' = 2\,\frac{b}{a}, \quad m + m' = 6\,\frac{c}{a} - 4ll', \quad lm' + l'm = 2\,\frac{d}{a}, \quad mm' = \frac{e}{a}.$$

Now,
$$\begin{vmatrix} 1 & 1 & 0 \\ l & l' & 0 \\ m & m' & 0 \end{vmatrix} \begin{vmatrix} 1 & 1 & 0 \\ l' & l & 0 \\ m' & m & 0 \end{vmatrix} = \begin{vmatrix} 2 & l + l' & m + m' \\ l + l' & 2ll' & lm' + l'm \\ m + m' & lm' + l'm & 2mm' \end{vmatrix} = 0.$$

Substituting the above values for $l + l'$, $m + m'$, etc., and multiplying each row by $a/2$, we have

$$\begin{vmatrix} a & b & 3c - 2all' \\ b & all' & d \\ 3c - 2all' & d & e \end{vmatrix} = 0.$$

If we write

$$t = c - all',$$

the equation becomes

$$\begin{vmatrix} a & b & c + 2t \\ b & c - t & d \\ c + 2t & d & e \end{vmatrix} = 0,$$

which when expanded is

$$4t^3 - It + J = 0.$$

Corresponding to any root t_1 of this equation, we can now find values of l, m, l', m' which will satisfy the conditions.

Again, if the equation is given in the form $u \equiv x^4 + px^2 + qx + r = 0$, we proceed as in Art. 15, using l instead of $2l$, to obtain the cubic in the form $l^2(l^2 + p)^2 - 4rl^2 - q^2 = 0$, which reduces to an equation lacking the second term on substituting $z - 2p/3$ for l^2.

> **Note:** In numerical work, unless this cubic has rational roots, the solution becomes very laborious by the methods described in this chapter; the student can convince himself of this by verifying the steps of Cardan's method for solving the equation, $z^3 - 11z - 15 = 0$, which is obtained as above for the biquadratic, $x^4 - 3x^2 - x + 2 = 0$.* Later, by one of the methods described in Ch.27, it will be found that one root can be found approximately with very little trouble. This is all that is required, since it is only necessary to have one way of breaking up u into quadratic factors.

■ **15. Four Real Roots:** In Art. 11, it has been shown that $\Delta > 0$ is a *necessary* condition that the four roots should be real; also, by De Gua's rule (Ch. 6, 11, Ex. 1), if $H > 0$, at least two of the roots must be imaginary.

Hence, $\Delta > 0$, $H < 0$ are *necessary* conditions that all the roots should be real; but these two alone are not *sufficient*.

The complete set of conditions may be found by using Sturm's theorem, given later. This, the usual course adopted in textbooks, involves rather tedious reckoning; the conditions can be found much more easily by the use of either Ferrari's or Descartes' solution of the biquadratic.

Theorem. *The necessary and sufficient conditions that the roots of*

$$u \equiv ax^4 + 4bx^3 + 6cx^2 + 4dx + e = 0$$

are all real is (i) $\Delta > 0$, $H < 0$ *and* $12H^2 > a^2 I$.

---

* Thus, $u^3 + v^3 = 15$, $uv = 11/3$, $u^3 - v^3 = \pm 5.27397$, $u = 2.16423$, $v = 1.6943$, $z = u + v = 3.85845$, from which $u \equiv (x^3 + 2.42042x + 1.63580)(x^3 - 2.42042x + 1.22265) = 0$.

or their equivalents   (*ii*) $\Delta > 0$, $H < 0$ *and* $2HI > 3aJ$.

Let $z = ax + b$; then by Art. 9, (C), and Art. 10, (M), z is given by
$$v \equiv z^4 + 6Hz^2 + 4Gz + K = 0, \text{ where } K = a^2I - 3H^2.$$

Now, $v = 0$ has the same number of real roots as $u = 0$.

For $v = 0$, using the method of Descartes, and supposing that
$$v \equiv (z^2 + 2lz + m)(z^2 - 2lz + m'),$$
we get, by equating coefficients,
$$-4l^2 + m + m' = 6H, \quad -l(m - m') = 2G, \quad mm' = K.$$

Eliminating m and m' from these equations, $(2l^3 + 3Hl)^2 - G^2 = l^2K$,

or $$4l^6 + 12Hl^4 + (9H^2 - K)l^2 - G^2 = 0.$$

Hence, since $K = a^2I - 3H^2$, the values of l^2 are the roots of
$$4y^3 + 12Hy^2 + (12H^2 - a^2I)y - G^2 = 0; \qquad \qquad ...(A)$$
which, with the substitution, $y + H = at$, reduces to $4t^3 - It + J = 0$.

Now the three values of l^2 correspond to the three ways of expressing v as the product of two quadratic factors. Hence,

(*i*)   If the roots of $v = 0$ are all real, the three values of l^2 must be all real and positive, *i.e.* the three roots of (A) are all positive.

(*ii*) If the roots of $v = 0$ are all imaginary, then, since their sum is zero, we may denote them by $\lambda \pm \iota\mu$, $-\lambda \pm \iota\mu'$, and thus the values of l^2 are λ^2, $-\frac{1}{4}(\mu + \mu')^2$, $-\frac{1}{4}(\mu - \mu')^2$, *i.e.* the roots of (A) are all real, but only one is positive; and conversely.

(*iii*) If two of the roots of $v = 0$ are real and two imaginary, we may denote them by λ, μ, $\lambda' \pm \iota\mu'$, where $\lambda + \mu + 2\lambda' = 0$; and then the values of l^2, *i.e.* the roots of (A), are λ'^2, $\frac{1}{4}(\lambda + \lambda' \pm \iota\mu')^2$, of which one is real and the other two imaginary; and conversely.

It follows that the roots of $v = 0$, and therefore those of $u = 0$, are all real, if, and only if, the roots of (A) are all real and positive.

Now, the roots of (A) are real if those of $4t^3 - It + J = 0$ are real; and this is the case if, and only if, $\Delta = I^3 - 27J^2 > 0$; also, the roots of (A) are, by Descartes' rule (Ch. 6, 9, 10), all real and positive, if, and only if, H is negative and $12H^2 - a^2I$ is positive; which proves the first part of the theorem.

To prove the equivalence of the second set of conditions, suppose first that
$$\Delta > 0, H < 0 \text{ and } 12H^2 > a^2I.$$

Then, since $\Delta = I^3 - 27J^2 > 0$, we have $I > 0$; hence, since $H < 0$, it follows that
$$HI < 0.$$

Also, $12H^2I^2 > a^2I^3 > 27a^2J^2$, so that $4H^2I^2 > 9a^2J^2$, *i.e.* $\left| 2HI \right| > \left| 3aJ \right|$; and, since HI is negative, therefore $2HI < 3aJ$.

Next, suppose that $\Delta > 0$, $H < 0$,   and $2HI < 3aJ$.

First observe that, if p and q are any real numbers, positive or negative, $p^2 + pq + q^2 = (p + \frac{1}{2}q)^2 + \frac{3}{4}q^2 > 0$; hence, since $p^3 - q^3 = (p - q)(p^2 + pq + q^2)$, it follows that $p^3 \gtrless q^3$ according as $p \gtrless q$.*

Then, since $2HI < 3aJ$, and $\Delta > 0$; we have $8H^3I^3 < 27a^3J^3 < a^3I^3J$; but I is positive, and therefore $8H^3 < a^3J$. Now, $G^2 + 4H^3 = a^2(HI - aJ)$; hence, by addition, $\qquad G^2 + 12H^3 < a^2HI$; therefore $12H^3 < a^2HI$, and dividing by H, which is negative, $12H^2 > a^2I$.

Thus, the two sets of conditions are equivalent.

Alternatively, using Ferrari's method, the equation whose roots are

$$\tfrac{1}{16}a^2(\beta + \gamma - \alpha - \delta)^2, \ \tfrac{1}{16}a^2(\gamma + \alpha - \beta - \delta)^2, \ \tfrac{1}{16}a^2(\alpha + \beta - \gamma - \delta)^2$$

is $4(y + H)^3 - a^2I(y + H) + a^3J = 0$,

which, when expanded, is equation (A) on p. 195.

There are three possibilities.

(*i*) All the roots of $u = 0$ may be real; in that case the roots of (A) are *all three real and positive.*

(*ii*) All the roots of $u = 0$ may be imaginary; in that case we may take

$$\alpha = \lambda + \iota\mu, \ \ \beta = l - \iota\mu, \ \ \gamma = \lambda' + \iota\mu', \ \ \delta = \lambda' - \iota\mu';$$

and the roots of (A) are $-\tfrac{1}{4}a^2(\mu - \mu')^2, \ \tfrac{1}{4}a^2(\mu + \mu')^2, \ \tfrac{1}{4}a^2(\lambda - \lambda')^2$, which are all three *real*, but *one only* is *positive.*

(*iii*) Two of the roots of $u = 0$ may be real and two imaginary; we may take α, β, to be the real roots and $\gamma = \lambda + \iota\mu, \ \delta = \lambda - \iota\mu$; and then

$$\tfrac{1}{16}a^2(\beta - \alpha + 2\iota\mu)^2, \ \tfrac{1}{16}a^2(\alpha - \beta + 2\iota\mu)^2, \ \tfrac{1}{16}a^2(\alpha + \beta - 2\lambda)^2,$$

are the roots of (A), of which *one only is real.*

The proof proceeds as before.

Another method of solving the biquadratic, due to *Euler*, is given in Exercise XXI, 17.

■ **16. Transformation into the Reciprocal Form:**

The substitution $x = py + q$ transforms the equation $u = 0$ into

$$ap^4y^4 + 4Bp^3y^3 + 6Cp^2y^2 + 4Dpy + E = 0, \qquad\qquad ...(A)$$

where $\qquad\qquad\qquad B = aq + b, \ \ C = aq^2 + 2bq + c,$

$$D = (a, b, c, d)(q, 1)^3 \quad E = (a, b, c, d, e)(q, 1)^4.$$

---

* In general, if n is *odd*, $p^n - q^n = (p - q)(p^{n-1} + p^{n-2} + q + ... + q^{n-1})$, where the last factor is the product of pairs of complex conjugate factors, and is therefore positive; hence $p^n \gtrless q^n$ according as $p \gtrless q$.

But, if n is even, $p^n - q^n = (p - q)(p + q)$ (a positive factor), hence $p^n \gtrless q^n$ as $|p| \gtrless |q|$, and not as $p \gtrless q$.

This will be a reciprocal equation if $ap^4 = E$, $Bp^3 = Dp$,

that is, if q, p are given by $aD^2 = B^2E$, $p^2 = D/B$.   ...(B)

Suppose that the conditions (B) are satisfied, and that the roots of (A) are y_1, y_2, y_3, y_4 where $y_2y_3 = y_1y_4 = 1$. Let α, β, γ, δ be the corresponding roots of $u = 0$, so that

$$\alpha = py_1 + q, \ \beta = py_2 + q, \ \gamma = py_3 + q, \ \delta = py_4 + q.$$

It follows that

$$p^2 = (\beta - q)(\gamma - q) = (\alpha - q)(\delta - q), \qquad ...(C)$$

and therefore

$$q = \frac{\beta\gamma - \alpha\delta}{\beta + \gamma - \alpha - \delta}, \quad p^2 = \frac{(\beta - \alpha)(\beta - \delta)(\gamma - \alpha)(\gamma - \delta)}{(\beta + \gamma - \alpha - \delta)^2}. \qquad ...(D)$$

Hence by Art. 13, (*iii*), *the values of q are the roots of the equation in z found by eliminating t from*

$$z = -\frac{1}{2} \cdot \frac{2bt + bc - ad}{at + b^2 - ac} \ \text{ and } \ 4t^3 - It + J = 0. \qquad ...(E)$$

Note: The points corresponding to q and $q \pm p$, as given by (C), are respectively the centre C and the foci F, F' of the involution determined by the points α, β, γ, δ, in which (β, γ) and (α, δ) are corresponding pairs of points. That is to say, the line segments are such that $C\beta . C\gamma = C\alpha . C\delta = CF^2 = F'C^2$.

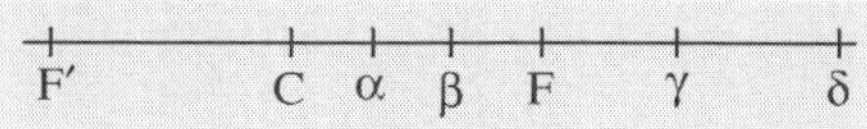

Fig. 29

EXAMPLE : *Find a substitution of the form $x = py + q$, which will change the equation*

$$x^4 + x^3 - 2x - 1 = 0 \qquad ...(A)$$

into the reciprocal form. Use this to solve the equation.

Here $\qquad a = 1, \quad b = \tfrac{1}{4}, \quad c = 0, \quad d = -\tfrac{1}{2} \ e = -1;$

$\therefore \qquad I = -1 + \tfrac{1}{2} = -\tfrac{1}{2}, \quad J = -\tfrac{1}{4} + \tfrac{1}{16} = -\tfrac{3}{16}.$

The equation $4t^3 - It + J = 0$ is $4t^3 + \tfrac{1}{2}t - \tfrac{3}{16} = 0$, that is $(4t)^3 + 2(4t) - 3 = 0$.

One solution is $t = \tfrac{1}{4}$, and by (E) of the text, the corresponding value of q is

$$q = -\frac{1}{2} \cdot \frac{\tfrac{1}{8} + \tfrac{1}{2}}{\tfrac{1}{4} + \tfrac{1}{16}} = -1.$$

and by (B) of the text, the corresponding value of p^2 is

$$p^2 = \frac{aq^3 + 3bq^2 + 3cq + d}{aq + b} = \frac{-1 + \tfrac{3}{4} - \tfrac{1}{2}}{-1 + \tfrac{1}{4}} = 1.$$

Taking $p = 1$, one substitution of the kind required is $x = y - 1$.

Substituting in (A), the equation becomes $y^4 - 3y^3 + 3y^2 - 3y + 1 = 0$.

Dividing by y^2 and putting $z = y + y^{-1}$, we find that

$$z^2 - 3z + 1 = 0, \quad \text{giving} \quad z = \tfrac{1}{2}(3 \pm \sqrt{5}).$$

Also, $y^2 - yz + 1 = 0$, $\quad \therefore y = \tfrac{1}{2}\left(z \pm \sqrt{z^2 - 4}\right)$, and $x = \tfrac{1}{2}\left(z - 2 \pm \sqrt{z^2 - 4}\right)$;

$$\therefore x = \tfrac{1}{4}\left(-1 + \sqrt{5} \pm \sqrt{-2 + 6\sqrt{5}}\right), \quad \text{or} \quad \tfrac{1}{4}\left(-1 - \sqrt{5} \pm \sqrt{-2 - 6\sqrt{5}}\right).$$

■ **17. Tschirnhausen's Transformation:** If we eliminate x between

$$f(x) = a_0 x^n + a_1 x^{n-1} + a_2 x^{n-2} + \ldots + a_n = 0$$

and

$$y = x^r + p_1 x^{r-1} + p_2 x^{r-2} + \ldots + p_r \ (r < n),$$

we shall obtain an equation of the form

$$b_0 y^n + b_1 y^{n-1} + b_2 y^{n-2} + \ldots + b_n = 0;$$

for a single value of y corresponds to each of the n values of x. Thus *theoretically, in general* $p_1, p_2, \ldots p_r$ can be chosen so that r of the coefficients $b_1, b_2, \ldots b_n$ are zero. This process is called *Tschirnhausen's Transformation:* it has been applied to the cubic in Exercise XX, 16.

Case of the biquadratic. Let $\alpha, \beta, \gamma, \delta$ be the roots of

$$u \equiv ax^4 + 4bx^3 + 6cx^2 + 4dx + e = 0.$$

We shall prove that *in general three sets of values of p and q can be found such that if x is eliminated between*

$$u = 0 \quad \text{and} \quad y = x^2 + px + q,$$

the resulting equation is of the form

$$y^4 + fy^2 + g = 0. \qquad \qquad \text{...(A)}$$

Further, the values of p and q are given by

$$ap = 4b + 2az, \quad aq = 3c + 2bz, \qquad \qquad \text{...(B)}$$

where z is one of the three

$$\frac{\beta\gamma - \alpha\delta}{\beta + \gamma - \alpha - \delta}, \quad \frac{\gamma\alpha - \beta\delta}{\gamma + \alpha - \beta - \delta}, \quad \frac{\alpha\beta - \gamma\delta}{\alpha + \beta - \gamma - \delta}.$$

One value of y corresponds to each of the four values of x ; the resulting equation in y is therefore of the fourth degree. Let p, q be chosen so that the coefficients of y and y^3 are zero, and let $\pm y_1, \pm y_2$ be the values of y. Then we may take

$$\left. \begin{array}{ll} \alpha^2 + p\alpha + q = y_1, & \delta^2 + p\delta + q = -y_1 \\ \beta^2 + p\beta + q = y_2, & \gamma^2 + p\gamma + q = -y_2 \end{array} \right\}, \qquad \text{...(C)}$$

whence by addition and subtraction,

$$\beta^2 + \gamma^2 - \alpha^2 - \delta^2 + p(\beta + \gamma - \alpha - \delta) = 0, \qquad \text{...(D)}$$

and

$$s_2 + ps_1 + 4q = 0, \qquad \qquad \text{...(E)}$$

where $s_1 = \Sigma\alpha$, $s_2 = \Sigma\alpha^2$. Hence we have, from (D),

$$-p = \frac{\beta^2 + \gamma^2 - \alpha^2 - \delta^2}{\beta + \gamma - \alpha - \delta} = \frac{(\beta + \gamma)^2 - (\alpha + \delta)^2 - 2(\beta\gamma - \alpha\delta)}{\beta + \gamma - \alpha - \delta}.$$

Now, $\qquad (\beta + \gamma)^2 - (\alpha + \delta)^2 = s_1(\beta + \gamma - \alpha - \delta);$

therefore $-p = s_1 - 2z,$ where $z = \dfrac{\beta\gamma - \alpha\delta}{\beta + \gamma - \alpha - \delta}.$

Hence $\qquad\qquad p = \dfrac{4b}{a} + 2z.$ $\hfill ...(F)$

Substituting in (E), we find that

$$\left(\frac{4b}{a}\right)^2 - 2 \cdot \frac{6c}{a} - \frac{4b}{a}\left(\frac{4b}{a} + 2z\right) + 4q = 0,$$

giving $\qquad\qquad q = \dfrac{3c}{a} + \dfrac{2bz}{a}.$ $\hfill ...(G)$

Moreover, if p, q have these values, it is obvious that the equation in y is of the form (A).

As will be seen in the next example, *each value of y gives one and only one value of x which satisfies the equation $u = 0$.*

▮ **EXAMPLE :** *Find a substitution of the form $y = x^2 + px + q$ which will reduce the equation*

$$x^4 + 12x - 5 = 0 \hfill ...(A)$$

to the form $y^4 + fy^2 + g = 0$. Find the values of f, g, and use the result to solve the equation.

Here $a = 1, b = 0, c = 0, d = 3, e = -5$, so that $I = -5, J = -9$; and using Art.13. (*iii*), we have

$$z = \frac{3}{2t} \quad \text{where } 4t^3 + 5t - 9 = 0.$$

One value of t is 1, giving $z = \dfrac{3}{2}$, and then $p = 3, q = 0$. One such substitution is therefore

$$y = x^2 + 3x. \hfill ...(B)$$

To eliminate x from (A) and (B) we have*

$$y(x^2 - 3x) = x^4 - 9x^2 = -9x^2 - 12x + 5, \text{ from (A)};$$

$\therefore \qquad\qquad (y + 9)x^2 - 3(y - 4)x - 5 = 0. \hfill ...(C)$

But $\qquad\qquad x^2 + 43x - y = 0;$

$\therefore \qquad\qquad \dfrac{x^2}{y^2 - 4y + 5} = \dfrac{3x}{y^2 + 9y - 5} = \dfrac{1}{2y + 5};$

$\therefore \qquad\qquad (y^2 + 9y - 5)^2 = 9(y^2 - 4y + 5)(2y + 5),$

which reduces to

$$y^4 + 98y^2 - 200 = 0.$$

The values of y are $\pm \sqrt{2}$, $\pm 10\,\iota$, and the corresponding values of x are given by

$$x = \frac{1}{3}(y^2 + 9y - 5)/(2y + 5).$$

Hence we find that the roots of (A) are

$$-1 \pm \sqrt{2}, \; 1 \pm 2\iota.$$

EXERCISE XXI

THE BIQUADRATIC

Unless otherwise stated α, β, γ, δ *are the roots of*

$$u = ax^4 + 4bx^3 + 6cx^2 + 4dx + e = 0.$$

1. If $\beta + \gamma = \alpha + \delta$, show that $a^2d + 2b^3 - 3abc = 0$.

2. If $(\beta + \gamma)\,\alpha\delta = (\alpha + \delta)\,\beta\gamma$, show that $e^2b + 2d^3 - 3edc = 0$.
    [Deduce from Ex. 1.]

3. If $\beta\gamma = \alpha\delta$, show that $ad^2 = b^2e$.

4. If

$$L = \beta\gamma + \alpha\delta + \omega\,(\gamma\alpha + \beta\delta) + \omega^2\,(\alpha\beta + \gamma\delta),$$
$$M = \beta\gamma + \alpha\delta + \omega^2\,(\gamma\alpha + \beta\delta) + \omega\,(\alpha\beta + \gamma\delta),$$

show that the interchange of any two of four α, β, γ, δ changes L^3 into M^3 and M^3 into L^3.

5. Show that the equation whose roots are

$$\frac{\beta\gamma - \alpha\delta}{\beta + \gamma - \alpha - \delta}, \quad \frac{\gamma\alpha - \beta\delta}{\gamma + \alpha - \beta - \delta}, \quad \frac{\alpha\beta - \gamma\delta}{\alpha + \beta - \gamma - \delta}$$

is
$$aD^2 = B^2E$$

where $B = az + b$, $D = (a, b, c, d)(z, 1)^3$, $E = (a, b, c, d, e)(z, 1)^4$.
[Diminish the roots of $u = 0$ by z and use Ex. 3.]

6. If α, β, γ, δ are the roots of $u \equiv x^4 + qx^2 + rx + s = 0$, show that the equation whose roots are

$$\beta\gamma + \alpha\delta, \; \gamma\alpha + \beta\delta, \; \alpha\beta + \gamma\delta$$

is
$$z^3 - qz^2 - 4sz + 4qs - r^2 = 0,$$

and that if z_1 is a root of this equation, then

$$u \equiv \left(x^2 - \frac{rx}{\lambda_1 - \lambda_2} + \lambda_1\right)\left(x^2 + \frac{rx}{\lambda_1 - \lambda_2} + \lambda_2\right),$$

where λ_1, λ_2 are the roots of

$$\lambda^2 - \lambda z_1 + s = 0.$$

---

* In the case of the general equation, $u = 0$ and the substitution, $y = x^2 + px + q$, the method for obtaining the second quadratic (C) is

$$y[ax^2 + (4b - ap)x] = (x^2 + px + q)[ax^2 + (4b - ap)x] = ax^4 + 4bx^3 + mx^2 + nx,$$

and thus, $y[ax^2 + (4b - ap)x] + 6cx^2 + 4dx + e = mx^3 + nx.$

7. Show that the equation whose roots are

$$(\beta - \gamma)(\alpha - \delta), \ (\gamma - \alpha)(\beta - \delta), \ (\alpha - \beta)(\gamma - \delta)$$

is
$$a^3 z^3 - 12aIz \pm 16\sqrt{\Delta} = 0.$$

[Use equations (D), (G) of Art. 10, and Exercise XX, 3, (*iii*).]

8. Solve $x^4 - 2x^3 + a(2x - 1) = 0$ by putting it in the form

$$(x^2 - x + s)^2 = (2s + 1)x^2 - 2(a + s)x + a + s^2,$$

and choosing s so that the righ-hand side is a perfect square. Hence show that the equation has two and only two real roots unless $a = 1$ or 0.

Solve by Ferrari's method:

9. $x^4 + 12x - 5 = 0.$

10. $x^4 + x^3 - 2x - 1 = 0.$

11. $x^4 - 3x^2 - 4x - 3 = 0.$

12. $x^4 - 4x^3 + 5x + 2 = 0.$

13. $x^4 + 12x^3 + 54x^2 + 96x + 40 = 0.$

14. Express $x^4 - 4x^3 + 7x^2 - 6x + 3$ as the product of quadratic factors in three different ways.

15. If $\alpha, \beta, \gamma, \delta$ are the roots of $x^4 + 3x^3 + x^2 - 2 = 0$, prove that the equation whose roots are

$$(\beta + \gamma - \alpha - \delta)^2, \ (\gamma + \alpha - \beta - \delta)^2, \ (\alpha + \beta - \gamma - \delta)^2$$

is
$$z^3 - 19z^2 + 243z - 225 = 0.$$

[Use equation (L) of Art. 13.]

16. Show that the equation whose roots are $\ \dfrac{1}{16} a^2 \{\beta\gamma(\alpha + \delta) - \alpha\delta(\beta + \gamma)\}^2$

and two similar expressions is $\ \ 4(x + L)^3 - e^2 I(x + L) + e^3 J = 0$

where $L = ce - d^2$.

[This follows from equation (L) of Art. 13 by changing $\alpha, \beta, \gamma, \delta$ into their reciprocals.]

17. ***Euler's Solution.*** If $\sqrt{l}, \sqrt{m}, \sqrt{n}$ denote either of the square roots of l, m, n and $z = \sqrt{l} + \sqrt{m} + \sqrt{n}$, prove that

(*i*) $z^4 - 2\Sigma l.\ z^2 - 8\sqrt{l}\ \sqrt{m}\ \sqrt{n}.\ z + \Sigma l^2 - 2\Sigma mn = 0.$

(*ii*) If this equation is identical with $\ \ z^4 + 6Hz^2 + 4Gz + K = 0,$

then l, m, n are the values of s given by $4s^3 + 12Hs^2 + (9H^2 - K)s - G^2 = 0,$

which is the same as $\ \ 4(s + H)^3 - a^2 I(s + H) + a^3 J = 0.$

(*iii*) The complete solution of the biquadratic $u = 0$ is given by

$$a\alpha + b = -\sqrt{l} + \sqrt{m} + \sqrt{n}, \ a\beta + b = \sqrt{l} - \sqrt{m} + \sqrt{n},$$

$$a\gamma + b = \sqrt{l} + \sqrt{m} - \sqrt{n}, \ a\delta + b = -\sqrt{l} - \sqrt{m} - \sqrt{n},$$

the square roots being chosen so that $\sqrt{l}.\sqrt{m}.\sqrt{n}$ has the same sign as G.

Observe that the values of l, m, n are $\frac{1}{16} a^2 (\beta + \gamma - \alpha - \delta)^2$, etc., see equations (L) and (K) of Art. 13.

18. Find a substitution of the form $x = py + q$ which will transform the equation
$$x^4 - 14x^2 - 40x - 11 = 0$$
into the reciprocal form. Show that one such substitution is $x = 2y - 1$. Use this to solve the equation.

19. In Ex. 1 of Art. 17, if each value of y which satisfies the equation
$$y^4 + 98y^2 - 200 = 0$$
is substituted for y in $y = x^2 + 3x$, we get eight values of x, four of which satisfy
$$x^4 + 12x - 5 = 0.$$
Without solving any equations, show that the other four values of x are the roots of
$$x^4 + 12x^3 + 54x^2 + 96x + 40 = 0. \qquad\qquad (Cf.\ Ex.\ 13)$$
[The left-hand side is
$$\{(x^2 + 3x)^4 + 12\,(x^2 + 3x) - 5\} \div (x^4 + 12x - 5).]$$

20. Find a substitution of the form $y = x^2 + px + q$ which will reduce the equation $x^4 + x^3 - 2x - 1 = 0$ to the form $y^4 + fy^2 + g = 0$.

Show that one such substitution is $y = x^2 - x - \frac{1}{2}$, this reducing the equation to $(2y)^4 + 26\,(2y)^2 - 11 = 0$. Use this to solve the equation. $\qquad (Cf.\ Ex.\ 10)$

21. Find a substitution of the form $y = x^2 + px + q$ which will reduce the equation
$$2x^4 + 8x^3 + 4x + 13 = 0$$
to the form $y^4 + fy^2 + g = 0$, showing that one such substitution is
$$y = x^2 + 3x - 1,$$
this reducing the equation to
$$4\left(\frac{y}{3}\right)^4 + 4\left(\frac{y}{3}\right)^2 - 9 = 0.$$

■ ■ ■

Theory of Irrationals

1. Sections of the System of Rational Numbers: Suppose the rational numbers to be represented by points in a straight line, and let the line be cut at any point P.

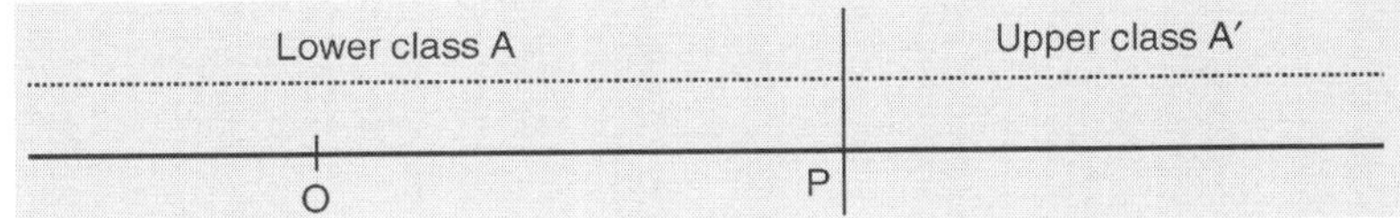

Fig. 30

The system of rationals is thus divided into two parts, which will be called *the lower class A* and *the upper class A'*. The class A contains all the rationals to the left of P, and the class A' contains all those to the right of P. Thus, any number in A is less than any number in A'. Two distinct cases arise, according as the point P does, or does not, represent a rational number.

First case. Suppose P to represent some rational, 3 for example. The class A contains every rational less than 3, and the class A' contains every rational greater than 3. *The number 3 may be assigned to either class.* If 3 belongs to A, it is the greatest number of this class, and there is no least number in A'.

For if a' is supposed to be the least number in A', rationals exist which are less than a' and greater than 3, and which therefore belong to the class A'.

Similarly if 3 is assigned to A', it is the least number of this class, and the class A has no greatest number.

Thus *corresponding to any rational,* the classification is such that *either the class A has a greatest number or the class A' has a least number.*

Second case. Suppose that the point P does not represent a rational number. For example, let OP be equal to the side of a square whose area is 7 square units.

Classify the rationals according to the following rule:

The lower class A is to contain all the negative numbers, zero and every positive number whose square is less than 7.

The upper class A' is to contain every positive number whose square is greater than 7.

The classification is such that

(*i*) *Every rational is included in one or other of two classes A, A'.* For no rational exists whose square is equal to 7.

(*ii*) *Any number in A is less than any number in A'.* For if a is any positive number in A and a' is any number in A', we have $a^2 < 7 < a'^2$, and so $a < a'$.

(*iii*) *The class A has no greatest number and the class A' has no least number.* For, suppose that a is the greatest number of A, then $a^2 < 7$, and we can find a rational b such that $b > a$ and $b^2 < 7$. To do this, we have to find b (greater than a), so that $b^2 - a^2 < 7 - a^2$. This will be the case if

$$b - a < \frac{7 - a^2}{b + a} < \frac{7 - a^2}{2a}.$$

Thus, b belongs to the class A, and a is not the greatest number in A. Similarly it can be shown that A' has no least number.

This classification is said to *define the irrational number* which we denote by $\sqrt{7}$, and we assign it a place on the scale with rationals as follows : $\sqrt{7}$ is to follow or to precede any positive rational a, according as $a^2 < 7$ or $a^2 > 7$.

■ **2. Dedekind's Definition:** Suppose that a certain rule enables us to divide the whole system of rationals into two classes, a lower class A and an upper class A', so that any number in A is less than any number in A'. Two cases arise:

(*i*) If A has a greatest number or if A' has a least number, the classification defines a *rational number*, namely the greatest number in A or the least number in A'.

(*ii*) If there is no greatest number in A and no least number in A', the classification defines an *irrational number* which is to follow all the numbers in A and to precede all those in A'.

Any rational or irrational number is called a *real number*, and the aggregate of rationals and irrationals is known as the *system of real numbers.*

The classification just described will be called the *classification* or the *section* (A, A'), and the real number defined by it may be known as the number (A, A').

The *real number zero* is defined by the section (A, A'), where A contains all the negative rationals and A' contains all the positive rationals.

If A contains some positive rationals, the section (A, A') defines a *positive real number.*

If A' contains some negative rationals, (A, A') defines a *negative real number.*

We are not justified in regarding a '*real number*' as a 'number' until we have given fresh definitions of equality, inequality and the fundamental operations of arithmetic. Moreover, these definitions must be in agreement with those already given for the system of rationals.

■ **3. Equality and Inequality:** Two real numbers are said to be *equal* when they are defined by the same classification of rationals.

If α and β are real numbers, and some of the rationals which follow α precede β, we say that α is *less than* β and that β is *greater than* α.

If α and β are unequal real numbers, infinitely many rationals lie between them.

For let $\alpha < \beta$, where α and β are defined by the classifications (A, A'), (B, B') respectively. Since these are different classifications, at least one rational r lies between α and β. Then r follows α and precedes β, therefore r belongs to each of the classes A' and B. If r is the *only* rational between α and β, then r is the least

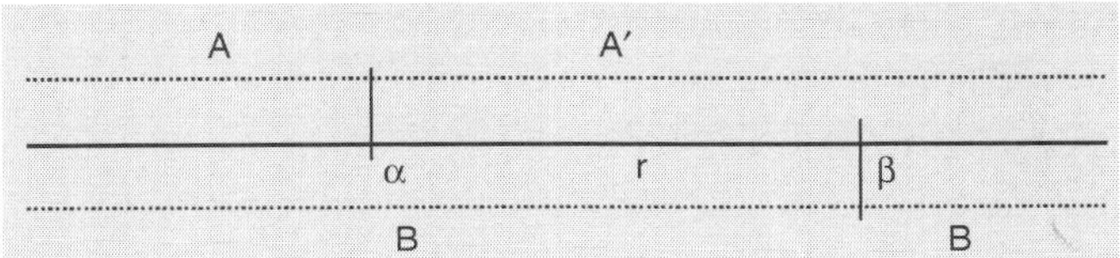

Fig. 31

number in A' and the greatest in B. Hence, by Art. 2, (i), each of the classifications (A, A'), (B, B') defines the same number r, and we should have $\alpha = r = \beta$, which is impossible, for $\alpha < \beta$. Thus at least two, and therefore infinitely many, rationals lie between α and β. (See Ch. 2, 10.)

■ **4. Theorem:** *Let A and A' be two classes of rationals such that*

   (i) *Each class contains at least one number.*

  (ii) *Any number in A is less than any number in A'.*

 (iii) *Two numbers a and a' can be chosen from A and A' respectively, so that*

$$a' - a < \varepsilon,$$

where ε is any positive number we may choose, however small.

Then there is one and only one real number α such that $a \leq \alpha \leq a'$ where a is any number in A and a' is any number in A'.

Proof: Divide the system of rationals into two classes according to the following rule: The lower class is to contain every rational less than any number a' in A'. The upper class is to contain every rational greater than any number a in A.

It will be proved that not more than one rational can escape classification. For if two rationals l and l' ($l < l'$) so escape, then for every a and a'

$$a < l < l' < a'.$$

This contradicts the hypothesis, for we can choose a and a' so that

$$a' - a < l' - l,$$

so that either a or a' lies between l and l'.

If there is one rational which escapes classification, then it is to be assigned to either the upper or the lower class. Thus, by Dedekind's definition, the classification defines a real number α which satisfies the conditions stated above.

This theorem may also be stated as follows:

If (a_n) and (a_n') are sequences of rationals such that

  (i) $a_1 \leq a_2 \leq a_3 ... \leq a_n ... \leq a_n' ... \leq a_3' \leq a_2' \leq a_1',$

and (ii) *it is possible to find n such that*

$$a_n' - a_n < \varepsilon,$$

where ε is any positive number we may choose, however small, then one real number α, and only one, exists such that $a_n \leq \alpha \leq a_n'$ for every n.

■ **5. Endless Decimals:** Using the ordinary notation, let

$$a_0 . a_1 a_2 ... a_n ...$$

be an endless decimal, where the successive figures are formed according to some definite rule.

Let $d_n = a_0 \cdot a_1 a_2 ... a_n$ and $d_n' = d_n + 1/10^n$.

The sequences (d_n) and (d_n') obviously satisfy the conditions of the last theorem, and therefore a real number δ is defined by

$$d_0 \leq d_1 \leq d_2 \; ... \leq \delta \; ... \leq d_2' \leq d_1' \leq d_0'.$$

We say that the decimal *represents* or *is equal to* this real number δ.

Conversely, *any real number δ can be represented by a decimal.* For, using the same notation as in the preceding, if δ is known, then for any suffix n, we can find d_n so that $d_n \leq \delta \leq d_n'$. That is to say, we can calculate to any number of figures a decimal which represents δ.

Of course, a terminating decimal may be regarded as an endless decimal in which, after a certain stage, all the figures are zeros.

Since any rational can be represented by a decimal which terminates or recurs, it follows that *a non-recurring endless decimal represents an irrational number, and conversely.*

■ **6. The Fundamental Operations of Arithmetic:** In what follows, a Greek letter denotes a *real* number, a small Roman letter represents a *rational*. Real numbers will be defined by using the theorem of Art. 4, and when we say that a real number α is defined by $a \leq \alpha \leq a'$, it is presumed that a, a' are any numbers in classes A, A' of rationals which satisfy the conditions of the theorem.

(1) *Addition.* If α, β are real numbers defined by

$$a \leq \alpha \leq a' \text{ and } b \leq \beta \leq b',$$

then $\alpha + \beta$ is difined by $a + b \leq \alpha + \beta \leq a' + b'$.

This classification defines a real number, for

  (*i*) There is at least one $a + b$ and one $a' + b'$.

 (*ii*) Any $a + b \leq$ any $a' + b'$.

(*iii*) We can choose a, a', b, b' so that

$$a' - a < \frac{1}{2}\varepsilon \text{ and } b' - b < \frac{1}{2}\varepsilon,$$

where ε is any positive number, however small, and then

$$(a' + b') - (a + b) = (a' - a) + (b' - b) < \varepsilon.$$

We define $\alpha + \beta + \gamma$ as meaning $(\alpha + \beta) + \gamma$, and it is easy to show that the commutative and associative laws hold good.

■ **EXAMPLE 1.** *Show that $\alpha + \beta = \beta + \alpha$.*

It is obvious that $\alpha + \beta$ and $\beta + \alpha$ are defined by the same classification of rationals.

■ **EXAMPLE 2.** *Prove that $\alpha + 0 = \alpha$.*

Zero is defined by $b < 0 < b'$, where b is any negative rational and b' is any positive rational.

Let α be defined by $a < \alpha < a'$, then $\alpha + 0$ is defined by

$$a + b < \alpha + 0 < a' + b',$$

and, since every b is negative and every b' positive,

$$a + b < \alpha < a' + b'.$$

Thus $\alpha + 0$ and α are defined by the same classification of rationals and are equal.

EXAMPLE 3. *Prove that* $\alpha + (-\alpha) = 0$.

If α is defined by $a < \alpha < a'$, then $\alpha + (-\alpha)$ is defined by

$$a - a' < \alpha + (-\alpha) < a' - a.$$

Now every $a <$ any a', therefore $a - a' < 0 < a' - a$.

Hence $\alpha + (-\alpha)$ and 0 are defined by the same classification, and are equal.

(2) *Subtraction.* We define $\alpha - \beta$ as $\alpha + (-\beta)$. Whence it follows that $(\alpha - \beta) + \beta = \alpha$, for

$$\{\alpha + (-\beta)\} + \beta = \alpha + \{\beta + (-\beta)\} = \alpha + 0 = \alpha.$$

(3) *Multiplication.* If α, β are positive real numbers defined by

$$a \leq \alpha \leq a' \quad \text{and} \quad b \leq \beta \leq b',$$

then $\alpha\beta$ is defined by $ab \leq \alpha\beta \leq a'b'$. These conditions define a real number, for

(*i*) there is at least one ab and one $a'b'$,

(*ii*) every $ab \leq$ any $a'b'$,

(*iii*) we can choose a, b, a', b' so that $a'b' - ab < \varepsilon$, where ε is any assigned positive number. For $a'b' - ab = a'(b' - b) + b(a' - a)$.

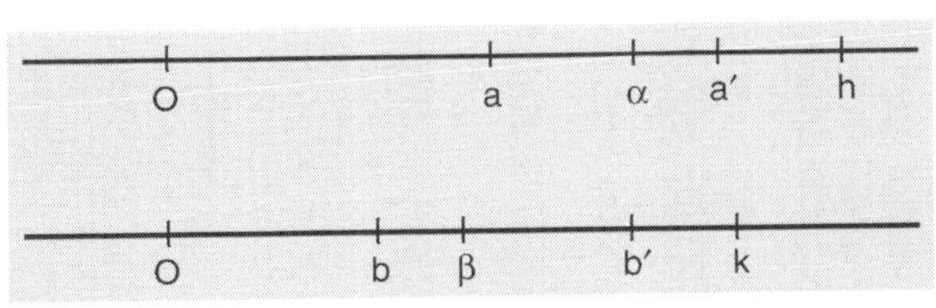

Fig. 32

Choose rationals h, k, so that $h > \alpha$ and $k > \beta$, then if $a' < h$,

$$a'b' - ab < h(b' - b) + k(a' - a).$$

We can now choose a, b, a', b', so that

$$b' - b < \varepsilon/2h \quad \text{and} \quad a' - a < \varepsilon/2k,$$

and then $a'b' - ab < \varepsilon$.

For zero and negative factors, the definitions are

$$\alpha \cdot 0 = 0 = 0 \cdot \alpha,$$

$$\alpha(-\beta) = -\alpha\beta = \beta(-\alpha) \quad \text{and} \quad (-\alpha)(-\beta) = \alpha\beta.$$

Further, $\alpha\beta\gamma$ is taken to mean $(\alpha\beta)\gamma$. It is easy now to show that the commutative, distributive and associative laws hold good.

EXAMPLE 4. *Show that* $\sqrt{7} \cdot \sqrt{7} = 7$.

$\sqrt{7}$ is defined by $a < \sqrt{7} < a'$, where a, a' are any positive rationals such that

$$a^2 < 7 < a'^2.$$

Hence $\sqrt{7} \cdot \sqrt{7}$ is defined by $a^2 < 7 < a'^2$, and this classification also defines the number 7.

(4) *Division.* If α is a positive real number defined by $a \leq \alpha \leq a'$, then $1/\alpha$ is defined by $1/a' \leq 1/\alpha \leq 1/a$.

These conditions define a real number, for

(*i*) there is at least one $1/a'$ and one $1/a$.

(*ii*) every $1/a' \leq$ any $1/a$,

(*iii*) we can choose a, a' so that $1/a - 1/a' < \varepsilon$, where ε is any assigned positive number.

To prove this, we have to show that a, a' can be chosen so that

$$a' - a < aa'\varepsilon.$$

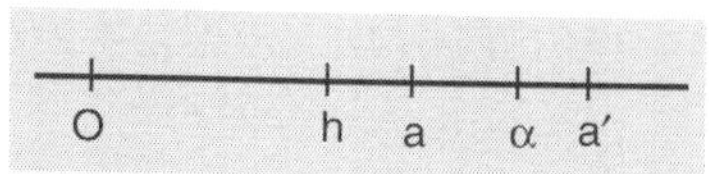

Fig. 33

First choose h so that $0 < h < \alpha$. Next choose a, a' so that $a > h$ and $a' - a < h^2\varepsilon$. Since $h < a < a'$, we have $h^2 < aa'$ and $a' - a < aa'\varepsilon$.

Note: This reasoning depends on the existence of positive rationals less than α. Consequently *no meaning is assigned to 1/0.*

Further definitions are $1/(-\alpha) = -1/\alpha$ and $\alpha/\beta = \alpha \cdot 1/\beta$: whence it follows that $(\alpha/\beta) \cdot \beta = \alpha$; for $(\alpha \cdot 1/\beta) \cdot \beta = \alpha (1/\beta \cdot \beta) = \alpha$.

■ **EXAMPLE 5.** *If α, β are positive real numbers defined by $a < \alpha < a'$ and $b < \beta < b'$, then α/β is defined by $a/b' < \alpha/\beta < a'/b$.*

This follows from the definition of $1/\beta$ and multiplication.

■ **7. Powers and Roots:** If n is a positive integer and α a real number, the nth power of α (written α^n) is defined by $\alpha^n = \alpha \cdot \alpha \cdot \alpha \dots$ to n factors. If m, n are positive integers, it follows that $\alpha^m \cdot \alpha^n = \alpha^{m+n}$ and $(\alpha^m)^n = \alpha^{mn}$.

That these formulae may hold for zero and negative values of m and n, we must have $\alpha^0 = 1$ and $\alpha^{-n} = 1/\alpha^n$.

Theorem 1. (*i*) *If α is a positive real number, n a positive integer and x, x' positive rationals such that $x^n < \alpha < x'^n$, then positive rationals x_1, x_1' exist such that*

$$x^n < x_1^n < \alpha < x_1'^n < x'^n.$$

For if $x < x_1 < x'$, then

$$x_1^n - x^n = (x_1 - x)(x_1^{n-1} + x_1^{n-2} x + \dots + x^{n-1}) < (x_1 - x) \cdot nx'^{n-1}.$$

We can choose $x_1 > x$, so that $x_1 - x < (\alpha - x^n)/nx'^{n-1}$; and then

$$x_1^n - x^n < \alpha - x^n, \quad \text{so that } x_1^n < \alpha.$$

The existence of a number x_1' such that $\alpha < x_1'^n < x'^n$ can be proved in a similar way.

(*ii*) *If ε is any positive number, however small, then $x_1'^n - x_1^n < \varepsilon$, provided that $x_1' - x_1 < \varepsilon/nx^{m-1}$.*

For as in the preceding, $x_1'^n - x_1^n < (x_1' - x_1) \cdot nx'^{n-1}$.

Theorem 2. *If α is a positive real number, there is one positive real number ξ, and one only, whose nth power is equal to α.*

If $\alpha = x^n$ where x is a positive rational, there is only one such number x, for the nth powers of unequal positive rationals are unequal.

If no such x exists, divide the system of rationals into a lower class A and an upper class A', according to the following rule:

The class A is to contain every negative number, zero, and every positive number x such that $x^n < \alpha$.

The class A' is to contain every positive number x' such that $x'^m > \alpha$.

Then

    (i) no rational escapes classification; for no rational x exists such that $x^n = \alpha$,

    (ii) every $x <$ every x', for $x^n < \alpha < x'^n$,

    (iii) there is no greatest x and no least x'. This follows from the last theorem. Therefore the classification defines a real number ξ.

Again, for every x and every x',

$$x^n < \alpha < x'^n \text{ and } x^n < \xi^n < x'^n.$$

Also every rational is an x or an x', therefore these may be chosen so that $x' - x$ is as small as we like. Hence, as in Theorem 1, they may be chosen so that $x'^m - x^n < \varepsilon$, where ε is any positive number, however small.

Thus both α and ξ^n are defined by the same classification of rationals, a classification which satisfies the conditions of the theorem of Art. 4 and therefore $\xi^n = \alpha$.

Definitions: If α is a positive real number, the *principal nth root* of α is defined as the positive real number whose nth power is equal to α. This is written $\sqrt[n]{\alpha}$, and is generally called the *nth root of α*.

Thus, $$\left(\sqrt[n]{\alpha}\right)^n = \alpha.$$

It follows that

$$\sqrt[n]{\alpha} \cdot \sqrt[n]{\beta} = \sqrt[n]{(\alpha\beta)} \text{ and } \sqrt[m]{\left(\sqrt[n]{\alpha}\right)} = \sqrt[mn]{\alpha} = \sqrt[n]{\left(\sqrt[m]{\alpha}\right)}.$$

If n is odd, we have

$$\left(-\sqrt[n]{\alpha}\right)^n = (-1)^n \left(\sqrt[n]{\alpha}\right)^n = -\alpha.$$

When n is an odd integer, we therefore define the *principal nth root,* or simply the *nth root,* of $-\alpha$ as $-\sqrt[n]{\alpha}$; thus

$$\sqrt[n]{(-\alpha)} = -\sqrt[n]{\alpha}.$$

If n is even, the *n-th* power of every real number is positive, and therefore no real number exists which is the nth root of a negative number.

Indices which are Rational Fractions: If p, q are positive integers,

$$a^{p/q} \text{ is defined as } \sqrt[q]{a^p} \text{ or } \left(\sqrt[q]{a}\right)^p.$$

This definition assigns no meaning to a^x when x is *irrational*. This case is considered in Art. 9.

■ **8. Surds:** If a is not a perfect nth power, $\sqrt[n]{a}$ is called a *surd* of the nth order.

Two surds of the same order are *like* surds when their quotient is rational: otherwise they are *unlike* surds.

The following theorems and examples depend on the fact that a rational number cannot be equal to an irrational.

Theorem 1. *If $x + \sqrt{y} = a + \sqrt{b}$ where x, y, a, b are rationals, then $x = a$ and $y = b$, or else y and b are the squares of rationals.*

For suppose that $x \neq a$ and let $x = a + z$, then $z + \sqrt{y} = \sqrt{b}$, and by squaring,

$$2z\sqrt{y} = b - y - z^2.$$

Therefore $\sqrt{y}$ is rational, and from the given equation $\sqrt{b}$ is rational. On the other hand, if $x = a$, then $y = b$.

■ **EXAMPLE 1.** *If $a + b\sqrt{p} + c\sqrt{q} = 0$ where a, b, c are rationals and $\sqrt{p}$, $\sqrt{q}$ are unlike surds, then $a = 0$, $b = 0$, $c = 0$.*

By transposing and squaring we can show that

$$2ab\sqrt{p} = c^2q - a^2 - b^2p.$$

If $ab \neq 0$, the left-hand side would be irrational and the right-hand side would be rational; which is impossible.

Therefore $ab = 0$, and consequently $a = 0$ or $b = 0$.

If $a = 0$, then $b\sqrt{p} + c\sqrt{q} = 0$; and, if $b \neq 0$, then $\sqrt{p}/\sqrt{q} = -c/b$; so that $\sqrt{p}$ and $\sqrt{q}$ would be like surds; which is not the case.

Therefore $b = 0$ and $c = 0$.

If $b = 0$ then $a + c\sqrt{q} = 0$, and therefore $a = 0$ and $c = 0$.

■ **EXAMPLE 2.** *If $a + b\sqrt[3]{p} + c\sqrt[3]{p^2} = 0$ where a, b, c, p are rationals and p is not a perfect cube, then a, b, c are all zero.*

Multiplying the given equation by $\sqrt[3]{p}$ we have

$$cp + a\sqrt[3]{p} + b\sqrt[3]{p^2} = 0,$$

and eliminating the terms containing $\sqrt[3]{p^2}$,

$$(b^2 - ac)\sqrt[3]{p} = c^2p - ab.$$

Since $\sqrt[3]{p}$ is irrational, it follows that

$$b^2 = ac \text{ and } c^2p = ab,$$

therefore $\qquad c^4p^2 = a^2b^2 = a^3c.$

If $c \neq 0$ we should have $p^2 = \left(\dfrac{a}{c}\right)^3$, so that $\sqrt[3]{p^2}$ would be rational, which is not the case.

Hence $c = 0$, and therefore also $a = 0$ and $b = 0$.

Theorem 2. *Suppose that p, q are rationals and that q is not a perfect square, then if* $p + \sqrt{q}$ *is a root of* $f(x) = 0$ *where* $f(x)$ *is a polynomial with rational coefficients,* $p - \sqrt{q}$ *is also a root of* $f(x) = 0$.

For we have $\{x - (p + \sqrt{q})\}\{x - (p - \sqrt{q})\} = (x - p)^2 - q$, and if $f(x)$ is divided by $(x - p)^2 - q$, we obtain an identity of the form

$$f(x) = Q\{(x - p)^2 - q\} + Rx + S,$$

where R, S are rationals. Putting $x = p + \sqrt{q}$, we have

$$0 = R(p + \sqrt{q}) + S.$$

Since $\sqrt{q}$ is irrational it follows that $Rp + S = 0$ and $R = 0$. Consequently $S = 0$ and $f(x) = Q\{(x - p)^2 - q\}$, therefore $p - \sqrt{q}$ is a root of $f(x) = 0$.

9. Irrational Indices: The theorems of Art. 7 hold when a is a positive *real* number, the indices being *rationals,* and lead to the following:

Definition. Let a be a real number greater than unity, and let ξ be an irrational defined by $x < \xi < x'$ where x, x' belong to classes of rationals satisfying the conditions stated in Art. 4. Then a^ξ is defined by

$$a^x < a^\xi < a^{x'}.$$

These conditions define a real number, for

    (*i*) there is at least one a^x and one $a^{x'}$,

    (*ii*) every a^x < every $a^{x'}$,

    (*iii*) we can choose x and x' so that $a^{x'} - a^x < \varepsilon$ where ε is any assigned positive number, however small.

To prove this, choose a number $h > \xi$, then $a^x < a^h$ and

$$a^{x'} - a^x = a^x \{a^{x'-x} - 1) < a^h (a^{x'-x} - 1).$$

Next, by Ch. 2, 18, (7), we can find a positive integer n such that

$$a^{1/n} - 1 < \varepsilon / a^h.$$

Finally, choose x, x' so that $x' - x < 1/n$, and then

$$a^{x'} - a^x < a^h \left(\frac{1}{a^{\frac{1}{n}} - 1} \right) < \varepsilon.$$

> **Note:** The real positive value of a^ξ, defined in this way, is called its *principal value,* to distinguish it from other values which will be found later. At present a^ξ will stand for its principal value.

If $0 < a < 1$, a^ξ is defined by the classification $a^{x'} < a^\xi < a^x$. Or it may be defined as $(1/b)^\xi$, where $b = 1/a$. We define 1^ξ as 1. At present, we are not able to assign a meaning to a^ξ when a is negative.

It is easy to show that if $a > 0$, the index laws

$$a^x \cdot a^y = a^{x+y} \quad \text{and} \quad (a^x)^y = a^{xy}$$

hold for all real values of x and y.

■ 10. Logarithms:

Theorem. *If a and N are positive real numbers and $a \neq 1$, there exists a single real number ξ such that $a^\xi = N$.*

Proof. Suppose that no rational x exists such that $a^x = N$ and first let $a > 1$. Then, by Art. 7, rationals x, x' exist such that $a^x < N < a^{x'}$.

Divide the system of rationals into a lower class A and an upper class A' by the following rule:

The class A is to contain every rational x such that $a^x < N$.

The class A' is to contain every rational x' such that $a^{x'} > N$.

Then

(*i*) no rational escapes classification; for it is assumed that no rational x exists such that $a^x = N$.

(*ii*) every $x <$ every x'; for $a^x < a^{x'}$,

(*iii*) there is no greatest x; for if x_1 is supposed to be the greatest, as in the last article, we can find $x_2 > x_1$, so that $a^{x_2} - a^{x_1} < N - a^{x_1}$, *i.e.* so that $a^{x_2} < N$. Similarly it can be shown that there is no least x'.

The classification therefore defines a real number ξ.

Again, for every x and every x',

$$a^x < N < a^{x'} \quad \text{and} \quad a^x < a^\xi < a^{x'};$$

and since every rational is an x or an x', these may be chosen so that $x' - x$ is as small as we like. Hence, as in Art. 9, they may be chosen so that $a^{x'} - a^x < \varepsilon$, where ε is any positive number, however small.

Thus both N and a^ξ are defined by the same classification of rationals, a classification which satisfies the conditions of Art. 4, and so $a^\xi = N$.

If $a < 1$, we find ξ so that $(1/a)^\xi = 1/N$, and then $a^\xi = N$.

Definition. If any positive real number a is chosen as ' base' and N is any positive real number, the number ξ given by the relation $a^\xi = N$ is called the *logarithm* of N to the base a, and we write $\xi = \log_a N$.

■ 11. Definitions:
We say that (*i*) the set of real numbers (x) which lie between a and b forms the (*open*) *interval* (a, b) or the *range* $a < x < b$.

(*ii*) The real values of x such that $a \leq x \leq b$ form the *closed interval* (a, b) or the *range* $a \leq x \leq b$.

(*iii*) If $a < x \leq b$, the real numbers x form an interval (a, b) *open at a and closed at b, or the range* $(a < x \leq b)$.

'Interval' generally means 'open interval.'

'Any number x in the interval (a, b) is often called a *point* in the interval.

(*iv*) If x_1, x_2 are any numbers in the interval (a, b) and $f(x)$ is a function of x such that $f(x_1) \leq f(x_2)$ when $x_1 < x_2$, we say that $f(x)$ *increases steadily* through the interval, unless the sign is always that of equality, in which case $f(x)$ is constant in the interval.

If $f(x_1) < f(x_2)$ when $x_1 < x_2$, then $f(x)$ is said to *increase steadily in the stricter sense.*

If $f(x_1) \geq f(x_2)$ when $x_1 < x_2$, $f(x)$ *decreases steadily;* and if $f(x_1) > f(x_2)$ when $x_1 < x_2$, $f(x)$ *decreases steadily in the stricter sense.*

(v) If a is a real number, the '*neighbourhood of a*' is the set of real numbers in the interval $(a - \varepsilon, a + \varepsilon)$; where ε is as small as we like.

■ 12. Sections of the System of Real Numbers:

(1) Dedekind's Theorem: *If the system of real numbers is divided into two classes A and A' such that*

(i) *each class contains at least one number,*

(ii) *every real number belongs to one or other of the classes,*

(iii) *any number in A is less than any number in A', then there is a single real number α, such that all numbers less than α belong to A and all numbers greater than α belong to A'. The number α may be regarded as belonging to either of the classes.*

Proof. Consider the *rationals* in A and A'. These form two classes A_1 and A_1', which form a section of the system of rationals and define a real number α. Two cases arise.

(i) If α is rational, it is the greatest number in A_1 or the least in A_1'. If α is the greatest number in A_1, it is also the greatest number in A. For suppose that there is a number β in A greater than α. Between α and β there are rationals greater than α which therefore belong to A_1' and also to A'. Hence β must belong to A', which is not the case.

Similarly if α is the least number in A_1', it is also the least in A'.

(ii) If α is irrational, it is greater than all the numbers in A_1 and less than all those in A_1'. Also, α must belong to A or to A', and just as in the preceding, we can show that it is the greatest number in A or the least in A'.

The theorem just proved is of great importance. It shows that *any section of the system of real numbers defines a real number.* Thus the consideration of sections of this kind leads to no further generalisation of our idea of number. All this is sometimes expressed by saying that *the system of real numbers is closed,* or *the aggregate of real numbers is perfect.*

On the other hand, the rational numbers do not form a closed system, or in other words, the aggregate of rational numbers is not perfect, for a section of the system of rationals does not always define a rational.

(2) *If the numbers in the interval (a, b) are divided into two classes A and A', as in* (1), *the section determines a real number α.*

For if we take with A all the real numbers x such that $x \leq a$ and with A' all the real numbers x' such that $x' \geq b$, we obtain a section of the system of real numbers.

■ 13. The Continuum:
It follows from Art. 1 that if we take any point O in a straight line as origin and any length we may choose as unit of length, then to any point P in the line there corresponds a real number p.

By choosing suitable axioms regarding the characteristics of a straight line we can ensure the truth of the converse of this statement, namely, that to every real number p there corresponds a point P in the line. Thus the correspondence between the points of a straight line and the real numbers is complete.

If p is the real number corresponding to the point P of the line, we say that p is the measure of the length of OP. Hence the system of real numbers is adequate for the measurement of the *length* of a straight line. This is an instance of what is known as a *continuous magnitude*.

The aggregate of real numbers is called the *arithmetic continuum*, and the aggregate of the points of a straight line is called the *linear continuum*.

RATIO AND PROPORTION

> **Note:** *In Arts. 14–16 capital letters will denote concrete magnitudes and not numbers; small letters will denote positive integers.*

■ **14. Equality and Inequality of Concrete Magnitudes:** In considering how a magnitude or a quantity of any kind is to be measured, at the outset we must *define* what is meant by saying that 'A is equal to B', 'A is greater than or less than B', where A and B are quantities of the kind. The exact meaning of these statements depends on *the particular kind* of quantity we are considering. For instance, let A and B denote segments of straight lines; if A can be made to coincide with B, we say $A = B$; if A can be made to coincide with a part of B, we say $A < B$ and $B > A$.

The student should reflect on the meaning of equal angles, equal velocities, equal forces, equal quantities of heat, etc.

■ **15. Ratio:** We assume that a magnitude B can be divided into any number (n) of equal parts. Denoting any one of these parts by C, B is said to contain C (exactly) n times, and C is called the nth part of B. This is expressed shortly by writing $B = nC$ or $C = \dfrac{1}{n} B$. If A contains the nth part of B exactly m times, we denote this by writing $A = \dfrac{m}{n} B$.

Definition. If A and B are magnitudes of the same kind, and if integers a and b exist such that A contains the bth part of B exactly a times $\left(\text{that is, if } A = \dfrac{a}{b} B \right)$, the magnitudes A and B are said to be *commensurable*, and the *ratio of A to B* (written $A : B$) is defined as the number a/b.

If no integers a and b exist such that $A = \dfrac{a}{b} B$, the magnitudes A and B are said to be *incommensurable*.

Axiom of Archimedes. If A and B are magnitudes of the same kind, no matter how small A may be compared with B, an integer m can always be found such that $mA > B$ or $\dfrac{1}{m} B < A$.

16. Ratio of Two Incommensurables: If A and B are incommensurable magnitudes, the ratio of A to B (written $A : B$) is defined as the irrational number determined by the following classification of the entire system of rationals:

The Lower Class is to contain every rational a/b for which $bA > aB$,

The Upper Class is to contain every rational a'/b' for which $b'A < a'B$.

This classification defines an irrational number; because

(1) every rational falls into one or other of the two classes; for, by hypothesis, there is no rational a/b for which $bA = aB$:

(2) every a/b is less than any a'/b'; for $\dfrac{a}{b} B < A < \dfrac{a'}{b'} B$:

(3) there is no greatest a/b; for suppose a_1/b_1 to be the greatest, then by hypothesis $b_1A > a_1B$.

Now by the axiom of Archimedes, a multiple of a magnitude (however small the magnitude may be) can be found which exceeds any magnitude of the same kind, however great. Therefore an integer r can be found such that $r\,(b_1\,A - a_1B) > a_1B$, and therefore $rb_1\,A > (r + 1)\,a_1\,B$. Let $rb_1 = b_2$ and $(r + 1)\,a_1 = a_2$, then $b_2A > a_2B$ and $a_2/b_2 = a_1\,(r + 1)/rb_1 > a_1/b_1$, so that a_1/b_1 is not the greatest a/b. Similarly it can be shown that there is no least a'/b'. Hence the classification defines an irrational number.

Theorem. If A, B, C, D are four magnitudes of the same kind, the ratio of A to B is equal to that of C to D if $mA > =$ or $< nB$ according as $mC > =$ or $< nD$, for all positive integral values of m and n.*

Proof. If rationals m, n exist such that $mA = nB$, then by hypothesis $mC = nD$ and $A/B = C/D = n/m$. If no such rationals m, n exist, then A, B and also C, D are incommensurables, and the ratios A/B, C/D are irrationals defined by the same classification and are therefore equal.

If A, B, C, D are magnitudes of the same kind and $A : B = C : D$, then A, B, C, D are said to be *in proportion*.

EXERCISE XXII

IRRATIONALS

1. Show that no rational exists whose nth power is equal to a/b, where a/b is a positive fraction in its lowest terms unless a and b are perfect nth powers.

[Suppose that $\left(\dfrac{x}{y}\right)^{n} = \dfrac{a}{b}$ where x/y is a positive fraction in its lowest terms.

Then $bx^n = ay^n$. Since y is prime to x, y^n must be a divisor of b. Let $b = ky^n$, where k is a whole number. Then $a = kx^n$. Hence $k = 1$, for a is prime to b. Therefore $a = x^n$ and $b = y^n$, that is to say, a and b are perfect nth powers.]

2. If $\sqrt{x} + \sqrt{y} = \sqrt{a} + \sqrt{b}$, then either $x = a$, $y = b$ or $x = b$, $y = a$, or $\sqrt{x}$, $\sqrt{y}$, $\sqrt{a}$, $\sqrt{b}$ are all rational or all like surds.

[Squaring, $x + y = a + b$, and $\sqrt{xy} = \sqrt{ab}$, unless $\sqrt{xy}$, $\sqrt{ab}$ are rational.]

---

* This is Euclid's definition of the equality of two ratios.

3. Find the condition that rationals x, y may exist such that

$$\sqrt{a + \sqrt{b}} = \sqrt{x} + \sqrt{y},$$

where a, b are given rationals and b is not a perfect square. Also find x and y in terms of a and b.

[Proceed as in Ex. 2, and show that $x = \dfrac{1}{2}\left(a \pm \sqrt{a^2 - b}\right)$.]

4. If $a + b\sqrt{p} + c\sqrt{q} + d\sqrt{pq} = 0$ where $\sqrt{p}$, $\sqrt{q}$ are unlike surds, then a, b, c, d are all zero.

[We have $a + b\sqrt{p} + \left(c + d\sqrt{p}\right)\sqrt{q} = 0$.

$$\therefore \left(a + b\sqrt{p}\right)\left(c - d\sqrt{p}\right) + (c^2 - d^2 p)\sqrt{q} = 0.$$

Hence by Art. 8, Ex. 1,

$$ac = bdp, \quad bc = ad, \quad c^2 = dp,$$

and consequently, if $c \neq 0$ and $a \neq 0$, then $p = c^2$.]

5. If $x + \sqrt[3]{y} = a + \sqrt[3]{b}$ where x, y, a, b are rational and y, b are not perfect cubes, then $x = a$ and $y = b$.

[Put $z = x - a$. Show that $z^2 + y - b + 3z\sqrt[3]{by} = 0$; $\therefore \sqrt[3]{y} = m\sqrt[3]{b^2}$ where m is rational. Now use Art. 8, Ex. 2.]

6. If $a\sqrt[3]{p^2} + b\sqrt[3]{pq} + c\sqrt[3]{q^2} = 0$ and neither of p^2q, pq^2 is a perfect cube, show that a, b, c are all zero.

[Show that $c^3q^2 - a^3p^2 - b^3pq = 3ab\left(ap\sqrt[3]{pq^2} + bq\sqrt[3]{p^2q}\right)$, and that the expression in brackets is not zero, since p/q is not a cube.]

7. Show that $\sqrt{2}$ and $\sqrt{3}$ are cubic functions of $\sqrt{2} + \sqrt{3}$ with rational coefficients, and that $\sqrt{2} - \sqrt{6} + 3$ is the ratio of two linear functions of $\sqrt{2} + \sqrt{3}$.

[For the first part, let $y = \sqrt{2} + \sqrt{3}$; find y^2, and eliminate $\sqrt{3}$ and $\sqrt{2}$ in turn : for the second part, find the value of the product $\left(\sqrt{2} - \sqrt{6} + 3\right)\left(\sqrt{2} + \sqrt{3} + x\right)$, and show that this is equal to $y + 5$ when $x = 1$.]

8. Find the equation of lowest degree with rational coefficients of which one root is

    (*i*) $\sqrt{2} + \sqrt{3}$;   (*ii*) $\sqrt[3]{2} + 3\sqrt[3]{4}$;   (*iii*) $\sqrt[3]{2} + \sqrt[3]{3}$.

    [(*i*) If $x = \sqrt{2} + \sqrt{3}$, $x^2 - 1 = 2\sqrt{2x}$ etc.;

    (*ii*) if $y = \sqrt[3]{2} + 3\sqrt[3]{4}$, $y^3 = 110 + 18\left(\sqrt[3]{2} + 3\sqrt[3]{4}\right)$;

    (*iii*) if $z = \sqrt[3]{2} + \sqrt[3]{3}$, $z^3 = 5 + 3\sqrt[3]{6} \cdot z$, and $(z^3 - 5)^3 = 162\,z^3$.]

9. If all the solutions of

$$ax^2 + 2hxy + by^2 = 1, \quad a'x^2 + 2h'xy + b'y^2 = 1$$

are rational and a, h, b, a', h', b' are rational, then $(h - h')^2 - (a - a')(b - b')$ and $(ab' - a'b)^2 - 4(ah' - a'h)(hb' - h'b)$ are the squares of rationals.

[Let $(\alpha_1, \beta_1), (\alpha_2, \beta_2), (-\alpha_1, -\beta_1) (-\alpha_2, -\beta_2)$ be the solutions. Prove that $\dfrac{\alpha_1}{\beta_1}, \dfrac{\alpha_2}{\beta_2}$ are the roots of

$$(a - a')\, t^2 + 2(h - h')\, t + b - b' = 0.$$

Also, show that $\alpha_1\beta_1, \alpha_2\beta_2$ are the roots of

$$z^2\{(ab')^2 - 4(ah')(hb')\} + 2z\{(a - a')(hb') - (b - b')(ah')\} + (a - a')(b - b') = 0,$$

where $(ab') = ab' - a'b$.   Hence show that

$$\alpha_1{}^2\, \alpha_2{}^2 = \frac{(b - b')^2}{(ab')^2 - 4(ah')(hb')},$$

whence the second result follows.]

10. If a, b, x, y are rationals such that

$$(ay - bx)^2 + 4(a - x)(b - y) = 0,$$

prove that either $x = a, y = b$, or $1 - ab$ and $1 - xy$ are the squares of rationals.

[Put $x - a = X, y - b = Y; \quad \therefore (aY - bX)^2 + 4XY = 0.$

If X and Y are not zero, solve for $\dfrac{X}{Y}$, and show that $1 - ab$ is a perfect square.

Next observe that the given equation is unaltered by the interchange of x and a, y and b.]

■ ■ ■

CHAPTER 14

Inequalities

In this chapter, various methods of dealing with inequalities are explained. Many of the results are of fundamental importance.

■ **1. Weierstrass' Inequalities:** *If a_1, a_2,... a_n are positive numbers less than 1 whose sum is denoted by s_n, then*

$$1 - s_n < (1 - a_1)(1 - a_2) \dots (1 - a_n) < 1/(1 + s_n),$$
$$1 + s_n < (1 + a_1)(1 + a_2) \dots (1 + a_n) < 1/(1 - s_n),$$

where, in the last inequality, it is supposed that $s_n < 1$.

For
$$(1 - a_1)(1 - a_2) = 1 - (a_1 + a_2) + a_1 a_2 > 1 - (a_1 + a_2),$$
$$(1 - a_1)(1 - a_2)(1 - a_3) > \{1 - (a_1 + a_2)\}(1 - a_3) > 1 - (a_1 + a_2 + a_3),$$

and continuing thus, we can prove that

$$(1 - a_1)(1 - a_2) \dots (1 - a_n) > 1 - s_n.$$

In the same way

$$(1 + a_1)(1 + a_2) \dots (1 + a_n) > 1 + s_n.$$

Again, $1 - a_1 < 1/(1 + a_1)$, and $1 + a_1 < 1/(1 - a_1)$, for $0 < a_1 < 1$. Using these and similar inequalities, we have

$$(1 - a_1)(1 - a_2) \dots (1 - a_n) < 1/(1 + a_1)(1 + a_2) \dots (1 + a_n) < 1/(1 + s_n),$$

and if $s_n < 1$,

$$(1 + a_1)(1 + a_2) \dots (1 + a_n) < 1/(1 - a_1)(1 - a_2) \dots (1 - a_n) < 1/(1 - s_n).$$

■ **2. Many Inequalities Depend on the Fact that the Square of a Real Number is Positive:**

▌ **EXAMPLE 1.** *If $a > 0$ and $b > 0$, then $\frac{1}{2}(a + b) > \sqrt{ab}$.*

For
$$\left(\sqrt{a} - \sqrt{b}\right)^2 > 0.$$

▌ **EXAMPLE 2.** *If a, b, c are positive and not all equal, then*
$$(a + b + c)(bc + ca + ab) > 9abc.$$

For $(a + b + c)(bc + ca + ab) - 9abc$
$$= a(b^2 + c^2) + b(c^2 + a^2) + c(a^2 + b^2) - 6abc$$
$$= a(b - c)^2 + b(c - a)^2 + c(a - b)^2 > 0.$$

■ **3. A special arrangement of terms or factors is sometimes useful:**

▌ **EXAMPLE 1.** *Show that $\lfloor n > n^{\frac{n}{2}}$.*

We have $$(\lfloor n)^2 = 1 \cdot n \cdot 2\,(n-1) \cdot 3\,(n-2) \ldots r\,(n-r+1) \ldots n \cdot 1,$$

and $r\,(n-r+1) > n$ if $r^2 - r\,(n+1) + n < 0$; that is if $(r-1)\,(r-n) < 0$ or if $1 < r < n$.

Therefore $\quad (\lfloor n)^2 > n^n$.

■ **EXAMPLE 2.** *If a, b, c are positive numbers any two of which are together greater than the third, then*

$$1/(b+c-a) + 1/(c+a-b) + 1/(a+b-c) > 1/a + 1/b + 1/c.$$

Now, since

$$a\,(c+a-b)\,(a+b-c) > 0, \text{ we have } 1/(c+a-b) + 1/(a+b-c) > 2/a,$$

if $\quad a(a+b-c+c+a-b) > 2\,(c+a-b)\,(a+b-c)$ or $a^2 > a^2 - (b-c)^2$; which is the case. Similarly, $1/(a+b-c) + 1/(b+c-a)$, $1/(b+c-a) + 1/(c+a-b)$, are greater than $2/b$, $2/c$ respectively; and the result is obtained by addition.

■ **4.** *If a, b, x are positive, then $(a+x)/(b+x) \gtrless a/b$, according as $a \lessgtr b$.*

For since $b\,(b+x)$ is positive, $(a+x)/(b+x) \gtrless a/b$

according as $ab + bx \gtrless ab + ax$, that is, according as $ax \lessgtr bx$ or $a \lessgtr b$.

■ **EXAMPLE :** ***Prove that*** $\quad \dfrac{1}{2\sqrt{(n+1)}} < \dfrac{1}{2} \cdot \dfrac{3}{4} \cdot \dfrac{5}{6} \ldots \dfrac{2n-1}{2n} < \dfrac{1}{\sqrt{(2n+1)}}.$

Let $\quad u_n = \dfrac{1}{2} \cdot \dfrac{3}{4} \cdot \dfrac{5}{6} \ldots \dfrac{(2n-1)}{2n}$, then $\quad u_n < \dfrac{2}{3} \cdot \dfrac{4}{5} \ldots \dfrac{2n}{2n+1}$; $\therefore u_n^2 < \dfrac{1}{2n+1}$.

Also $\quad (2n+1)u_n = \dfrac{3}{2} \cdot \dfrac{5}{4} \ldots \dfrac{2n+1}{2n} > \dfrac{4}{3} \cdot \dfrac{6}{5} \ldots \dfrac{2n+2}{2n+1}$; $\therefore (2n+1)^2 u_n^2 > n+1.$

Therefore $\quad u_n < \dfrac{1}{\sqrt{2n+1}}$ and $u_n > \dfrac{\sqrt{(n+1)}}{2n+1} > \dfrac{1}{2\sqrt{(n+1)}}.$

■ **5.** *If both sides of an inequality are symmetric functions of a, b, c,... h, k, there is no loss of generality in assuming that $a \geq b \geq c \ldots \geq h \geq k$.*

■ **EXAMPLE :** *If a, b, c are all positive and $n \geq 0$ or $n \leq -1$, then*

$$a^n\,(a-b)\,(a-c) + b^n\,(b-a)\,(b-c) + c^n\,(c-a)\,(c-b) \geq 0.$$

First let $n > 0$. On account of symmetry, we may assume that $a \geq b \geq c$.

Hence,

$a^n \geq b^n \geq c^n$ and $a^n\,(a-b)\,(a-c) \geq b^n\,(a-b)\,(b-c)$, also $c^n\,(c-a)\,(c-b) \geq 0$;

therefore $\quad a^n\,(a-b)\,(a-c) + c^n(c-a)\,(c-b) \geq b^n\,(a-b)\,(b-c),$

whence the result in question.

If $n = 0$, $\quad \Sigma a^n\,(a-b)\,(a-c) = \Sigma\,(a-b)\,(a-c) = \Sigma a^2 - \Sigma bc$

and $\quad \Sigma a^2 - \Sigma bc = \tfrac{1}{2} \Sigma\,(a-b)^2 > 0.$

If $n \leq -1$, let $n = -m - 1$, so that $m \geq 0$. Let $a = 1/\alpha$, $b = 1/\beta$, $c = 1/\gamma$,

then it is easily seen that $\Sigma a^n (a - b)(a - c) = \dfrac{1}{\alpha\beta\gamma} \Sigma \alpha^m (\alpha - \beta)(\alpha - \gamma) \geq 0$ by the preceding.

■ **6.** *If* $a_1, a_2, \ldots a_n$ *and* $b_1, b_2, \ldots b_n$ *are two sets of real numbers, then*
$$(a_1 b_1 + a_2 b_2 + \ldots + a_n b_n)^2 \leq (a_1^2 + a_2^2 + \ldots + a_n^2)(b_1^2 + b_2^2 + \ldots + b_n^2),$$
the sign of equality occurring only when $a_1/b_1 = a_2/b_2 = \ldots = a_n/b_n$.

For let $A = \Sigma a_r^2$, $B = \Sigma a_r b_r$, $C = \Sigma b_r^2$, then for all values of λ,
$$\Sigma (a_r + \lambda b_r)^2 = A + 2\lambda B + \lambda^2 C.$$

Therefore $A + 2\lambda B + \lambda^2 C \geq 0$. If the sign of equality occurs, we must have $a_r + \lambda b_r = 0$ for every r, and so
$$-\lambda = a_1/b_1 = a_2/b_2 = \ldots = a_n/b_n.$$

Hence the roots of $A + 2\lambda B + \lambda^2 C = 0$ are equal and $B^2 = AC$.

Otherwise $A + 2\lambda B + \lambda^2 C > 0$ for all values of λ, and therefore $B^2 < AC$.

The theorem also follows from the identity
$$\Sigma a_r^2 \cdot \Sigma b_r^2 = (\Sigma a_r b_r)^2 + S,$$

where S is the sum of the $\frac{1}{2} n (n - 1)$ squares of the form $(a_1 b_2 - a_2 b_1)^2$.

■ **7.** *If* $a_1 \, a_2, \ldots a_n$ *are n positive numbers, not all equal to one another, then*
$$n\Sigma a_r^{x + y} \gtrless \Sigma a_r^x \cdot \Sigma a_r^y,$$
according as x and y have the same or opposite signs.

First suppose that x and y have the same sign. Let r, s be any two of the numbers $1, 2, \ldots n$, then $a_r^x - a_s^x$ and $a_r^y - a_s^y$ are both positive, both negative or zero. Hence $(a_r^x - a_s^x)(a_r^y - a_s^y) \geq 0$, and consequently
$$a_r^{x + y} + a_s^{x + y} \geq a_r^x a_s^y + a_s^x a_r^y.$$

There are $\frac{1}{2} n (n - 1)$ relations of this sort, *not all* equalities. Also $a_r^{x + y}$ occurs in $n - 1$ of them and $a_r^x a_s^y$ in only one.

Hence by addition
$$(n - 1) \Sigma a_r^{x + y} > \Sigma a_r^x a_s^y \text{ where } s \neq r.$$

Therefore $\quad n\Sigma a_r^{x + y} > \Sigma a_r^{x + y} + \Sigma a_r^x a_s^y = \Sigma a_r^x \cdot \Sigma a_r^y.$

If x and y have opposite signs, then $(a_r^x - a_s^x)(a_r^y - a_s^y) \leq 0$, and $<$ must be substituted for $>$ in the preceding .

■ **8.** The inequality in Art. 7 can be written
$$\frac{\Sigma a_r^{x + y}}{n} \gtrless \frac{\Sigma a_r^x}{n} \cdot \frac{\Sigma a_r^y}{n}.$$

Hence it follows that *if* $a_1, a_2, \ldots a_n$ *are n positive numbers, not all equal to one another, and x, y, z, … are all positive or all negative, then*
$$\frac{\Sigma a_r^{x + y + z + \ldots}}{n} > \frac{\Sigma a_r^x}{n} \cdot \frac{\Sigma a_r^y}{n} \cdot \frac{\Sigma a_r^z}{n} \ldots$$

■ **9.** *If a is any positive number except* 1, *and x, y are positive rationals,* *

then $$(a^x - 1)/x > (a^y - 1)/y \text{ if } x > y.$$

First suppose that $x = p$, $y = q$ where p, q are positive integers. If $p > 1$,
we have
$$(a^p - 1)/p - (a^{p-1} - 1)/(p - 1)$$
$$= \{p(a^p - a^{p-1}) - (a^p - 1)\}/p(p - 1)$$
$$= (a - 1)\{pa^{p-1} - (a^{p-1} + a^{p-2} + \ldots + 1)\}/p(p - 1).$$

Now, the expression in the large brackets can be written
$$\{(a^{p-1} - a^{p-2}) + (a^{p-1} - a^{p-3}) + \ldots + (a^{p-1} - 1)\}$$

and each term of this is positive or negative according as $a \gtrless 1$.

Hence, in either case, $(a^p - 1)/p - (a^{p-1} - 1)/(p - 1) > 0$,

that is, $(a^p - 1)/p$ decreases as p decreases, and therefore
$$(a^p - 1)/p > (a^q - 1)/q \text{ if } p > q.$$

If one or both of x, y are fractions, we may take $x = p/d$, $y = q/d$ where p, q, d are
positive integers and $p > q$. We have to show that

$$\frac{a^{p/d} - 1}{p/d} > \frac{a^{q/d} - 1}{q/d}, \text{ that is } \frac{b^p - 1}{p} > \frac{b^q - 1}{q},$$

where $b = a^{\frac{1}{d}}$. This follows from the preceding, for $b > 0$ and $\neq 1$.

Note: It has been shown that if $a > 0$ and $\neq 1$, as x passes through positive rational values,
$(a^x - 1)/x$ *increases with x.* Putting $1/a$ for a, it follows that $(a^{-x} - 1)/x$ increases,
and consequently $(1 - a^{-x})/x$ *decreases as x increases.* Further, putting $1/x$ for x,
we conclude that $x(a^{\frac{1}{x}} - 1)$ *decreases and* $x(1 - a^{-\frac{1}{x}})$ *increases as x increases.*

▌ **EXAMPLE :** *If $a > 1$ and $n \geq 1$, then $n(a^{\frac{1}{n}} - 1) > 1 - \frac{1}{a}$.*

If $n > 1$, $n(a^{\frac{1}{n}} - 1) = a^{\frac{1}{n}} \cdot n(1 - a^{-\frac{1}{n}}) > a^{\frac{1}{n}}(1 - a^{-1}) > 1 - a^{-1}$.

If $n = 1$, the inequality becomes $a - 1 > 1 - \frac{1}{a}$, or $a(a - 1) > a - 1$, i.e.
$(a - 1)^2 > 0$.

■ **10.** *If a and b are positive and unequal and x is any rational number except* 1, *then*
$$xa^{x-1}(a - b) > a^x - b^x > xb^{x-1}(a - b),$$
unless $0 < x < 1$, in which case $xa^{x-1}(a - b) < a^x - b^x < xb^{x-1}(a - b)$.

Proof. First observe that, in either of the cases, the truth of one of the inequalities
involves that of the other. Thus, if for all positive and unequal values of a and b, $xa^{x-1}(a - b) > a^x - b^x$, we may interchange a and b, so that $xb^{x-1}(b - a) > b^x - a^x$, and
therefore $a^x - b^x > xb^{x-1}(a - b)$.

Thus it is only necessary to prove one inequality in each case.

Let $a/b = k$, then

---

* It can be shown that this is true when x, y are positive real numbers. (See Ex. XXIII,
37–39.)

(*i*) if x is positive $a^x - b^x \gtrless xb^{x-1}(a - b)$ according as $k^x - 1 \gtrless x(k - 1)$, or according as $(k^x - 1)/x \gtrless (k - 1)/1$, that is according as $x \gtrless 1$; (See Art. 9)

(*ii*) if x is negative and equal to $-y$, then $xa^{x-1}(a - b) \gtrless a^x - b^x$ according as $-yk^{-y-1}(k - 1) \gtrless k^{-y} - 1$, that is according as $-y(k - 1) \gtrless k - k^{y+1}$, or according as $k^{y+1} - 1 \gtrless k(y + 1) - y - 1$, or according as

$$(k^{y+1} - 1)/(y + 1) \gtrless (k - 1)/1;$$

that is (by Art. 9) according as $y + 1 \gtrless 1$, or $y \gtrless 0$.   Now $y > 0$, and so

$$xa^{x-1}(a - b) > a^x - b^x.$$

This completes the proof of the theorem.

■ **11.** When $b = 1$, the second inequalities in Art. 10 become

$$a^x - 1 > x(a - 1) \text{ if } x < 0 \text{ or } > 1,$$
$$a^x - 1 < x(a - 1) \text{ if } 0 < x < 1.$$

Writing n for x and $1 + x$ for a, it follows that, *if n is any rational except* 1, *and $x > -1$ and $\neq 0$, then*

$$(1 + x)^n > 1 + nx \text{ if } n < 0 \text{ or } > 1,$$
$$(1 + x)^n < 1 + nx \text{ if } 0 < n < 1.$$

Or again, changing the sign of x, *if $x < 1$ and $\neq 0$,*

$$(1 - x)^n > 1 - nx \text{ if } n < 0 \text{ or } > 1,$$
$$(1 - x)^n < 1 - nx \text{ if } 0 < n < 1.$$

■ **EXAMPLE 1.** *If m, n are positive and $m < n$, it is possible to find positive values of x such that*

$$(1 + x)^m < 1 + nx.$$

For this inequality holds if $x > 0$ and

$$(1 + x)^{-m} > 1/(1 + nx).$$

Now $(1 + x)^{-m} > 1 - mx$, hence the inequality holds if $1 - mx > 1/(1 + nx)$, or if $(1 + nx)(1 - mx) > 1$, or if $x\{(n - m) - mnx\} > 0$, or if $0 < x < (n - m)/mn$.

■ **EXAMPLE 2.** *If $s_n = a_1 + a_2 + \dots + a_n$ where $a_1, a_2, \dots a_n$ are all positive, then*

$$1 + s_n + \frac{s_n^2}{\lfloor 2} + \dots + \frac{s_n^2}{\lfloor n} - (1 + a_1)(1 + a_2)\dots(1 + a_n) \geq 0.$$

Denote the left-hand side by u_n, then if $n > 1$,

$$u_n - u_{n-1} > \sum_{r=1}^{r=n} \frac{s_n^r - s_{n-1}^r}{\lfloor r} - a_n(1 + a_1)(1 + a_2)\dots(1 + a_{n-1}).$$

Now $s_n^r - s_{n-1}^r > ra_n s_{n-1}^{r-1}$, when $r > 1$ (Art. 10), and when $r = 1$, this becomes an equality, therefore

$$\frac{u_n - u_{n-1}}{a_n} > \sum_{r=1}^{r=n} \frac{s_{n-1}^{r-1}}{\lfloor r-1} - (1 + a_1)(1 + a_2)\dots(1 + a_{n-1}) > u_{n-1};$$

therefore $\quad u_n - u_{n-1} > a_n u_{n-1}$ and $u_n > (1 + a_n) u_{n-1} > u_{n-1}.$

Now $u_1 = 0$, therefore $u_n > 0$ for $n = 2, 3, 4$, etc.

■ **12. Arithmetic and Geometric Means:** If a, b, c,... k are n numbers and

$$A = \frac{1}{n}(a + b + c +... + k) \quad \text{and} \quad G = \sqrt[n]{(abc...k)},$$

then A is called the *arithmetic mean* and G the *geometric mean* of a, b, c,... k.

■ **13. Theorem:** *The arithmetic mean of n positive numbers, which are not all equal to one another, is greater than their geometric mean.*

For let A be the arithmetic mean and G the geometric mean of the n positive numbers

$$a, b, c, d,... k. \tag{...(A)}$$

Suppose that $a \geq$ any other number of the set and that $b \leq$ any other number, then

$$na > a + b + c +... + k > nb \quad \text{and} \quad a > A > b.$$

Let $b' = a + b - A$ so that $b' > 0$, and consider the set

$$A, b', c, d, ... k, \tag{...(B)}$$

obtained by substituting A, b' for a, b in (A). We have

$$Ab' - ab = A(a + b - A) - ab = (a - A)(A - b) > 0.$$

Thus $A + b' = a + b$ and $Ab' > ab$. Hence the numbers of the set (B) have the same arithmetic mean as those of (A), but a greater geometric mean.

Again, if the numbers in (B) are not all equal, then A is neither the greatest nor the least of the set. By repeating the process we can therefore obtain a set of n positive numbers containing two A's, with the same arithmetic mean as before, but a greater geometric mean.

Continuing thus, we shall finally arrive at a set of n numbers each equal to A of which the geometric mean is greater than G, and since A is the geometric mean of these equal numbers, it follows that $A > G$.

Thus for the numbers 10, 1, 2, 7, the successive sets are

$$10, 1, 2, 7; \quad 5, 6, 2, 7; \quad 5, 6, 4, 5; \quad 5, 5, 5, 5.$$

The arithmetic mean of the numbers in each set is 5, the geometric mean increasing from left to right.

Alternative proof. Taking the set (A), we have

$$ab \leq \{\tfrac{1}{2}(a + b)\}^2, \quad cd \leq \{\tfrac{1}{2}(c + d)\}^2,$$

$$abcd \leq \{\tfrac{1}{2}(a + b). \tfrac{1}{2}(c + d)\}^2 \leq \{\tfrac{1}{4}(a + b + c + d)\}^4.$$

Proceeding in this way, we can show that if n is a power of 2, then

$$abc ... k \leq \left\{\frac{1}{n}(a + b + ... k)\right\}^n,$$

so that $G \leq A$.

If n is not a power of 2, consider the set a, b, c,... k, A, A, A,... , where A occurs r times and $n + r$ is a power of 2. By the preceding,

$$abc ... k.A^r \leq \left\{\frac{a + b + c + ... k + rA}{n + r}\right\}^{n+r}$$

The right-hand side is equal to A^{n+r}, and therefore $G^n A^r \leq A^{n+r}$, so that $G \leq A$, the sign of equality only occurring when all the numbers a, b, ... k are equal.

■ **EXAMPLE 1.** *Show that* $n^n > 1 . 3 . 5 ... (2n - 1)$.

We have
$$\{1 + 3 + 5 + ... + (2n - 1)\}/n > \sqrt[n]{1.3.5...(2n-1)}.$$

Now
$$1 + 3 + 5 + ... + 2n - 1 = \tfrac{1}{2} n \{1 + (2n - 1)\} = n^2;$$

$$\therefore \quad n > \sqrt[n]{1.3.5...(2n-1)} \quad \text{and} \quad n^n > 1 . 3 . 5 ... (2n - 1).$$

■ **EXAMPLE 2.** *If* a_1, a_2, a_3, ... a_n *are positive and* $(n - 1) s = a_1 + a_2 + ... + a_n$, *then*
$$a_1 a_2 a_3 ... a_n \geq (n - 1)^n (s - a_1) (s - a_2) ... (s - a_n).$$
For $(s - a_2) + (s - a_3) + ... + (s - a_n) = a_1$,
therefore $a_1/(n - 1) \geq \{(s - a_2)(s - a_3) ... (s - a_n)\}^{1/(n-1)}$.

There are n inequalities (or equalities) of this kind, and the result follows by multiplication.

■ **14. Theorem:** *If a, b, ... k and x, y, ... w are two sets of n positive numbers, those in the second set being rational, then*

$$\left(\frac{ax + by + ... + kw}{x + y + ... + w} \right)^{x+y+...+w} \geq a^x b^y ... k^w.$$

For in Art. 13 we may take $x = x' / g$, $y = y' / g$, ... $w = w' / g$, where x', y', ... w' and g are positive integers, and then it is sufficient to prove that

$$\frac{ax' + by' + ... kw'}{x' + y' + ... + w'} \geq (a^{x'} b^{y'} ... k^{w'})^{1/(x'+y'+...+w')}.$$

Now the left-hand side is the arithmetic mean and the right-hand side is the geometric mean of a, a,... b, b,... k, k ... , where a occurs x' times, b occurs y' times,... and k occurs w' times, so that the result follows by the last theorem.

■ **15. Theorem:** *If a, b, c,... k are n positive numbers which are not all equal to one another, and m is any rational except 0 or 1, then*

$$\frac{a^m + b^m + c^m + ... + k^m}{n} \gtrless \left(\frac{a + b + c + ... + k}{n} \right)^m,$$

according as m does not or does lie between 0 and 1.

For let
$$\frac{a}{x} = \frac{b}{y} = \frac{c}{z} = ... = \frac{k}{w} = \frac{a + b + c + ... + k}{n},$$

then
$$x + y + z + ... + w = n,$$

and
$$\frac{1}{n} (a^m + b^m + c^m + ... + k^m) \gtrless \left[\frac{1}{n} (a + b + c + ... + k) \right]^m,$$

according as
$$x^m + y^m + z^m + ... + w^m \gtrless n.$$

If m does not lie between 0 and 1, by Art. 11,

$$x^m - 1 > m (x - 1), \quad y^m - 1 > m (y - 1), \quad \dots \quad w^m - 1 > m(w - 1).$$

Hence $\quad x^m + y^m + \dots + w^m - n > m (x + y + \dots + w - n) = 0,$

and therefore $\quad\quad\quad\quad\quad\quad\quad\quad x^m + y^m + \dots + w^m > n.$

If $0 < m < 1$, by Art. 11, the sign $>$ must be replaced by $<$ in all these inequalities, so that $x^m + y^m + \dots + w^n < n.$

This completes the proof of the theorem.

■ **16. Theorem:** *If a, b, $\dots k$ and x, y, $\dots w$ are two sets of n-rationals and those of the first set are not all equal to one another, then, m being any rational,*

$$\frac{xa^m + yb^m + \dots + wk^m}{x + y + \dots + w} \gtrless \left(\frac{xa + yb + \dots + wk}{x + y + \dots + w} \right)^m,$$

according as m does not or does lie between 0 and 1.

This depends on Art. 15 in the same way that Art. 14 depends on Art. 13.

■ **17. Application to Questions of Maxima and Minima Values:** In the theorems of Arts. 13–15 the inequalities become equalities when all the numbers a, b, c,$\dots$ k are equal. Hence we draw the following conclusions:

Suppose that x, y, z, $\dots$ w are n positive variables and that c is a constant, then:

(1) If $x+y+z+ \dots +w = c$, the value of $xyz \dots w$ is greatest when

$$x = y = \dots = w = c/n,$$

so that the greatest value of $xyz \dots w$ is $(c/n)^n$.

(2) If $xyz \dots w = c$, the value of $x + y + \dots w$ is least when $x = y = \dots = w$,

so that the least value of $x + y + \dots + w$ is $nc^{\frac{1}{n}}$.

Again, if m is any rational except 0 or 1, from Art. 3 we conclude that:

(3) If $x + y + z + \dots + w = c$ then, according as m does not or does lie between 0 and 1, the least or the greatest value of $x^m + y^m + \dots + w^m$ occurs when $x = y = \dots = w$, the value in question being $n^{1-m}.c^m$.

(4) If $x^m + y^m + \dots + w^m = c$, then according as m does not or does lie between 0 and 1, the greatest or the least value of $x + y + \dots + w$ occurs when $x = y = \dots = w$, the value in question being $n^{1-1/m}.c^{1/m}$.

▌ **Example 1.** *Find the greatest value of xyz for positive values of x, y, z, subject to the condition* $\quad\quad yz + zx + xy = 12.$

Since $yz + zx + xy = 12$, the value of $(yz) (zx) (xy)$ is greatest when $yz = zx = xy$, that is when $\quad x = y = z = 2.$

Hence the greatest value of $(yz) (zx) (xy)$ is 2^6, and the greatest value of xyz is 8.

▌ **Example 2.** *If the sum of the sides of a triangle is given, prove that the area is greatest when the triangle is equilateral.*

Let a, b, c be the sides of the triangle, and let $a + b + c = 2s$.

If Δ is the area, then $\Delta^2 = s(s - a) (s - b) (s - c)$.

Now $(s - a) + (s - b) + (s - c) = s = a$ constant,

Hence the value of $(s - a) (s - b) (s - c)$ is greatest when $s - a = s - b = s - c$.

Therefore the value of Δ is greatest when $a = b = c$.

■ **EXAMPLE 3.** *Find the least value of 3x + 4y for positive values of x and y, subject to the condition $x^2y^3 = 6$.*

Since $x^2y^3 = 6$, if λ, μ are any constants, we have $(\lambda x)(\lambda x)(\mu y)(\mu y)(\mu y) = 3\lambda^2\mu^3$.

Therefore $\lambda x + \lambda x + \mu y + \mu y + \mu y$ is least when $\lambda x = \mu y = (6\lambda^2\mu^3)^{\frac{1}{5}}$.

Hence the least value of $2\lambda x + 3\mu y$ is $5(6\lambda^2\mu^3)^{\frac{1}{5}}$.

Putting $2\lambda = 3$ and $3\mu = 4$, it follows that the least value of $3x + 4y$ is

$$5\left(6.\frac{3^2}{2^2}.\frac{4^3}{3^3}\right)^{\frac{1}{5}} = 10.$$

■ **EXAMPLE 4.** *Find the least value of $x^{-1} + y^{-1} + z^{-1}$ for positive values of x, y, z which satisfy the condition x + y + z = c.*

In Art. 17, (3), putting $m = -1$, $n = 3$, since m does not lie between 0 and 1, the least value of $x^{-1} + y^{-1} + z^{-1}$ is 3^2c^{-1} or $9/c$.

EXERCISE XXIII

*In the inequalities marked with an *, all the numbers concerned are supposed to be positive.*

1. If $a_1, a_2, \dots a_n$ and $b_1, b_2, \dots b_n$ are two sets of numbers and all those in the second set are positive, then $(a_1 + a_2 + \dots + a_n) / (b_1 + b_2 + \dots + b_n)$ lies between the greatest and least of $a_1/b_1, a_2/b_2, \dots, a_n/b_n$.

2. If a, b, c are any real numbers, show that
$$(b + c - a)^2 + (c + a - b)^2 + (a + b - c)^2 \geq bc + ca + ab.$$

3. If $l^2 + m^2 + n^2 = 1$ and $l'^2 + m'^2 + n'^2 = 1$, then $ll' + mm' + nn' \leq 1$.
$$[\Sigma(mn' - m'n)^2 + (ll' + mm' + nn')^2 = \Sigma l^2\, \Sigma l'^2.]$$

4. Show that $a(x + y)^2 + bz^2 + cz(x - y)$ cannot be positive for all real values of the quantities involved.

5. If any two of a, b, c are together greater than the third and $x + y + z = 0$, show that
$$a^2yz + b^2zx + c^2xy \leq 0.$$

6. If $a > 0$ and $x > y$, show that $a^x + a^{-x} > a^y + a^{-y}$.

7. * If a, b, c, d are in harmonical progression, then $a + d > b + c$.
   [Let $p - 3q, p - q, p + q, p + 3q$ be the reciprocals of a, b, c, d.]

8. * If $x \neq y$, then $(ax^n + by^n) / (ax^{n-1} + by^{n-1})$ increases as n increases through integral values.

9. If a, b, c, d are all > 1, then $8(abcd + 1) > (a + 1)(b + 1)(c + 1)(d + 1)$.
   [$2(ab + 1) > (a + 1)(b + 1)$, etc.]

10. * $\dfrac{b^2 + c^2}{b + c} + \dfrac{c^2 + a^2}{c + a} + \dfrac{a^2 + b^2}{a + b} \geq a + b + c$.

   [For, $(b^2 + c^2) / (b + c) \geq \frac{1}{2}(b + c)$, etc.]

11. If $x > 0$ or < -1, then $x + x^2 + x^3 + ... + x^{2n} \leq n(1 + x^{2n+1})$.

    [Show that $x^{r+1} + x^{2n-r} \leq 1 + x^{2n+1}$.]

12. * (i) $a(a - b)(a - c) + b(b - c)(b - a) + c(c - a)(c - b) \geq 0$,

    (ii) $a^3 + b^3 + c^3 + 3abc \geq a^2(b + c) + b^2(c + a) + c^2(a + b)$.

    [The second inequality is merely another form of the first.]

13. If $a, b, c ... k$ are n positive numbers, then

    (i) $\Sigma a . \Sigma 1/a \geq n^2$; (ii) $\Sigma a^3 . \Sigma 1/a \geq n \Sigma a^2$.

14. If a, b, c are positive and not all equal, then

$$6abc < a^2(b + c) + b^2(c + a) + c^2(a + b) < 2(a^3 + b^3 + c^3).$$

    [$3\Sigma a^3 > \Sigma a . \Sigma a^2$ and $\Sigma a(b - c)^2 > 0$.]

15. If $a \neq b$ and x, y are any two positive numbers such that $x + y = 1$, then

$$a^x b^y < ax + by.$$

$$\left[\text{If } a > b, \text{ then } \left(\frac{a}{b} \right)^x - 1 < x \left(\frac{a}{b} - 1 \right). \right]$$

16. If $n > 0$ and $x > 1$, then

$$\frac{x^{n+1} - x^{-(n+1)}}{x^n - x^{-n}} > \frac{n+1}{nx}. \quad \text{[Use Art. 9.]}$$

17. If $p > 1$ and r is any rational except zero, then

$$(p + 1)^{r+1} - p^{r+1} \gtrless (r + 1) p^r \gtrless p^{r+1} - (p - 1)^{r+1},$$

    where the upper or lower signs are to be taken according as r does not, or does, lie between 0 and -1.

    Hence show that $(1^r + 2^2 + 3^r + ... + n^r) / n^{r+1}$ lies between $1/(r + 1)$ and

$$\{(1 + 1/n)^{r+1} - 1/n^{r+1}\} / (r + 1).$$

    [In Art. 10, let $x = r + 1$. First put $a = p + 1$, $b = p$, and next $a = p$, $b = p - 1$. Finally let $p = 1, 2, 3, ... n$ in succession, and add.]

18. * $\dfrac{a_1}{a_2} + \dfrac{a_2}{a_3} + \dfrac{a_3}{a_4} + ... + \dfrac{a_{n-1}}{a_n} + \dfrac{a_n}{a_1} \geq n$.

19. * $a^2 b + b^2 c + c^2 a \geq 3abc$.

20. $(n + 1)^n > 2^n \lfloor n$.

21. If $x > 1$ and n is a positive integer, $(x^n - 1)/(x - 1) > n x^{\frac{1}{2}(n-1)}$.

22. If any two of a, b, c are together greater than the third, then

$$(a + b + c)^3 > 27(b + c - a)(c + a - b)(a + b - c).$$

23. If all the factors are positive, then

    (i) $abc \geq \Pi(b + c - a)$;   (ii) $abcd \geq \Pi(b + c + d - 2a)$.

24. The greatest value of $xyz(d - ax - by - cz)$ is $d^4/4^4 abc$, provided that all the factors are positive, and a, b, c, d are given positive numbers.

25. If a, b, c are positive rationals and x, y, z are positive variables such that $x + y + z$ is constant, show that $x^a y^b z^c$ has its greatest value when $x/a = y/b = z/c$.

26. If x, y, z are positive and $x + y + z = 1$, then $8xyz < (1 - x)(1 - y)(1 - z) < \frac{8}{27}$.

[Observe that $\Sigma(1 - x) = 2$.]

27. If $s_n = 1 + \dfrac{1}{2} + \dfrac{1}{3} \dots + \dfrac{1}{n}$ and $n > 2$, show that

$$n\,(n + 1)^{\frac{1}{n}} - n < s_n < n - (n - 1)\,n^{-\frac{1}{n-1}}.$$

[Consider the sequences $\dfrac{1}{2}, \dfrac{2}{3}, \dfrac{3}{4}, \dots \dfrac{n-1}{n}$ and $\dfrac{2}{1}, \dfrac{3}{2}, \dfrac{4}{3}, \dots \dfrac{n+1}{n}$.]

28. Let $a_1, a_2, \dots a_n$ be a sequence of positive numbers and let A, G, H respectively be the arithmetic, geometric and harmonic means of a_1, and a_n, then

(*i*) If $a_1, a_2, \dots a_n$ is an arithmetical progression, $G^n < a_1 a_2, \dots a_n < A^n$.

(*ii*) If $a_1, a_2, \dots a_n$ is a harmonical progression,

$$nH < a_1 + a_2 + \dots + a_n < nA \quad \text{and} \quad H^n < a_1 a_2 \dots a_n < G^n.$$

29. If $x_1 x_2 x_3 \dots x_n = a^n$ where a is a constant, then

(*i*) The sum of the products of every r of the x's $> C_r^n\, a^r$.

(*ii*) The least value of $(x_1 + k)(x_2 + k) \dots (x_n + k)$ is $(a + k)^n$.

30. If a is any positive number except 1, and x, y, z are rationals no two of which are equal and which may be positive or negative, then

$$a^x (y - z) + a^y (z - x) + a^z (x - y) > 0.$$

[First let x, y, z be integers and let $x > y > z$ (Art. 5). Consider the A.M. and the G.M. of the set of numbers $a^x, a^x, a^x, \dots, a^z, a^z, a^z, \dots$ where a^x occurs $(y - z)$ times and a^z occurs $(x - y)$ times.

If x, y, z are fractions, denote them by $p/d, q/d, r/d$ where p, q, r, d are integers and d is positive. This case can then be at once deduced from the preceding.]

31. Deduce the theorem of Art. 9 from Ex. 30.

32. If $a_1, a_2, \dots, a_n$ are positive and $p > q$ then

$$(a_1^q + a_2^q + \dots + a_n^q) < n^{p-q} (a_1^p + a_2^p + \dots + a_n^p).$$

[In Art. 15 put $m = \dfrac{q}{p}$, $a = a_1^p$, etc.]

33. * $\dfrac{3}{b+c+d} + \dfrac{3}{c+d+a} + \dfrac{3}{d+a+b} + \dfrac{3}{a+b+c} \geq \dfrac{16}{a+b+c+d}$.

[In Art. 15, put $m = -1$, $n = 4$, $\frac{1}{3}(b + c + d)$ for a, etc.

34. *Use Art. 16 to show that

$$\dfrac{a^{-1}b + b^{-1}c + c^{-1}a}{a+b+c} \geq \dfrac{a+b+c}{bc+ca+ab}.$$

35. * $3\,(a^2 b + b^2 c + c^2 a)(ab^2 + bc^2 + ca^2) \geq abc\,(a + b + c)^3$.   [Use Ex. 34]

36. If $a_1, a_2 ..., a_n$ and $b_1, b_2 ..., b_n$ are two sets of positive numbers arranged in descending order, then

$$(a_1 + a_2 + ... + a_n)(b_1 + b_2 + ... + b_n) \leq n (a_1 b_1 + a_2 b_2 + ... + a_n b_n).$$

[Let $u_n = n\Sigma a_r b_r - \Sigma a_r . \Sigma b_r$. Use the relation $(a_r - a_n)(b_r - b_n) \geq 0$ to show that $u_n > u_{n-1}$.]

37. *If $a > 1$ and p, q an positive integers, show that*

$$\frac{a^p - 1}{p} - \frac{a^q - 1}{q} > \frac{p - q}{2} (a - 1)^2 \text{ when } p > q.$$

[Let $u_p = (a^p - 1)/p - (a^{p-1} - 1) / (p - 1)$, then by Art. 9,

$$u_p = \frac{(a-1)^2}{p(p-1)} \{a^{p-2} + a^{p-3}(a + 1) + a^{p-4}(a^2 + a + 1) + ...+ (a^{p-2} + a^{p-3} + ... + 1)\};$$

$$\therefore u_p > \frac{(a-1)^2}{p(p-1)} \{1 + 2 + 3 + ... + (p + 1)\}, \therefore u_p > \frac{1}{2}(a - 1)^2$$

and $\dfrac{a^p - 1}{p} - \dfrac{a^q - 1}{q} = u_p + u_{p-1} + ... + u_{q+1} > \dfrac{p - q}{2}(a - 1)^2.$]

38. *If $a > 1$ and x, y are positive rationals, then*

$$\frac{a^x - 1}{x} - \frac{a^y - 1}{y} > \frac{x - y}{2} \cdot \left(\frac{a - 1}{2}\right)^2 \text{ if } x > y.$$

[Put $x = p/d, y = q/d$ where p, q, d are positive integers. Let $a^{1/d} = b$; then, using Ex. 37, we find that

$$\frac{a^x - 1}{x} - \frac{a^y - 1}{y} > d.\frac{p - q}{2}.(b - 1)^2 = \frac{x - y}{2} \{d(a^{\frac{1}{d}} - 1)\}^2,$$

and by Art. 9, Ex. 1, $\quad d(a^{\frac{1}{d}} - 1) > 1 - a^{-1}$.]

39. *If $a > 1$ and x, y are positive real numbers, then*

$$(a^x - 1) / x > (a^y - 1) /y \text{ if } x > y.$$

[Let x', y' be rational approximations to x, y such that $x' > x, y' < y$, and therefore $x' - y' > x - y$. Let $f(x) = (a^x - 1) / x$, and suppose that

$$f(x) = f(x') + \varepsilon, f(y) = f(y') + \varepsilon'.$$

then $\quad f(x) - f(y) = f(x') - f(y') + \varepsilon - \varepsilon' > \frac{1}{2}(x' - y')(1 - a^{-1})^2 + \varepsilon - \varepsilon';$

$$\therefore \quad f(x) - f(y) > \frac{1}{2}(x - y)(1 - a^{-1})^2 + \varepsilon - \varepsilon'.$$

We can choose x', y' so that $|\varepsilon - \varepsilon'| < \frac{1}{2}(x - y)(1 - a^{-1})^2$, and so $f(x) - f(y) > 0$, when $x > y$. See Ch. 13, 7.]

■ ■ ■

CHAPTER 15

Sequences and Limits

■ **1. Limit of a Function of the Positive Integral Variable n:** When n is large, $1 + \dfrac{1}{n}$ is nearly equal to 1; further, by making n large enough, we can make $1 + \dfrac{1}{n}$ as nearly equal to 1 as we like.

This is roughly what is meant when we say that 1 is the limit of $1 + \dfrac{1}{n}$ as n tends to infinity. In this explanation, the meaning of 'large enough', 'as nearly equal to 1 as we like', is by no means clear.

If u_n is a function of the positive integral variable n, the 'limit of u_n as n tends to infinity' is usually defined as follows:

Definitions:

(1) If corresponding to any positive number ε that we may choose, no matter how small, there is a positive integer m such that for every integer n greater than or equal to m

$$|u_n - l| < \varepsilon,$$

where l is a fixed number, then l is called the *limit of u_n as n tends to infinity.*

This is expressed by writing

$$\lim_{n \to \infty} u_n = l, \text{ or simply } \lim u_n = l.$$

We also say that u_n *tends to l as n tends to infinity*; and express this by

$$u_n \to l \text{ as } n \to \infty.$$

(2) If, corresponding to any positive number M that we may choose, no matter how great, there is a positive integer m, such that for every integer n greater than or equal to m,

$$u_n > M,$$

then u_n is said to *tend to infinity as n tends to infinity,* and we write

$$u_n \to \infty \text{ or } \lim u_n = \infty.$$

If $u_n \to \infty$ and $v_n = -u_n$, we say that v_n *tends to minus infinity* $(-\infty)$.

If $u_n \to l$ we say that the sequence (u_n) is *convergent*, and that it converges to the limit l. If $u_n \to \infty$ or $-\infty$, the sequence (u_n) is said to be *divergent*.

If u_n does not tend to a limit, nor to $+\infty$, nor to $-\infty$, it is said to *oscillate* and the sequence (u_n) is called *oscillatory*.

If u_n oscillates and a positive number M can be found such that, for every positive integer n,

$$|u_n| < M,$$

then u_n is said to *oscillate finitely*. Otherwise, u_n *oscillates infinitely*.

▮ **EXAMPLE 1.** *How does the function a^n behave as $n \to \infty$?*

Referring to Ch. 2, 18, (5), it will be seen that if $a > 1$, $a^n \to \infty$; if $a = 1$, $a^n = 1$ for every n; if $|a| < 1$, $a^n \to 0$; if $a = -1$, $a^n = \pm 1$ and a^n oscillates finitely; if $a < -1$, $a^n \to \infty$ or $- \infty$ according as n is even or odd, and so a^n oscillates infinitely.

It is often useful to represent the first few terms of a sequence (u_n) graphically. This is done by plotting points whose coordinates are $(1, u_1), (2, u_2), (3, u_3), \ldots$. It is usual to join points representing consecutive terms.

▮ **EXAMPLE 2.** *If* $u_n = (-1)^n \left(1 + \dfrac{1}{n} \right)$, *represent graphically the first few terms of the sequence (u_n), and examine the character of the sequence as $n \to \infty$.*

As $n \to \infty$, $u_{2n} \to 1$ and $u_{2n+1} \to -1$.

Thus u_n oscillates finitely.

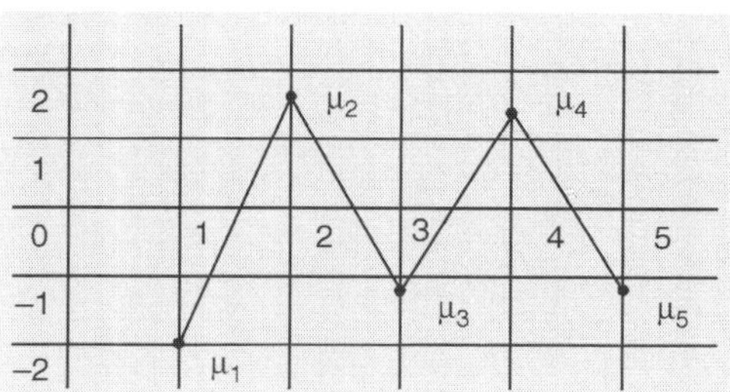

Fig. 34

▨ **2. Fundamental Theorems on Limits:**

(1) *If $u_n \to l$ and $u_n \to l'$, then*

(i) $u_n + v_n \to l + l'$,   *(ii)* $u_n - v_n \to l - l'$,

(iii) $u_n v_n \to ll'$,   *(iv)* $1/u_n \to 1/l$, unless $l = 0$,

(v) $u_n / v_n \to l/l'$, *unless* $l' = 0$.

Proof. Choose a positive number ε, no matter how small.

(i) and *(ii)*. By the definition of a limit we can choose m_1 and m_2 so that

$$|u_n - l| < \frac{1}{2}\varepsilon \text{ for } n \geq m_1$$

and

$$|v_n - l'| < \frac{1}{2}\varepsilon \text{ for } n \geq m_2.$$

If we denote the greater of m_1, m_2 by m, then each of these inequalities will hold for $n \geq m$, and for such values of n we have

$$|u_n + v_n - (l + l')| \leq |u_n - l| + |v_n - l'| < \varepsilon$$

and

$$|u_n - v_n - (l - l')| \leq |u_n - l| + |v_n - l'| < \varepsilon.$$

Hence, by the definition of a limit,

$$u_n + v_n \rightarrow l + l' \text{ and } u_n - v_n \rightarrow l - l'.$$

(*iii*) Let $u_n = l + \alpha$ and $v_n = l' + \beta$. Then

$$u_n v_n - ll' = \beta l + \alpha (l' + \beta).$$

Choose any positive number L greater than $|l' + \beta|$, then

$$| u_n v_n - ll' | < | \beta l | + | \alpha | . L.$$

By the definition of a limit we can choose m so that

$$|\alpha| = |u_n - l| < \frac{1}{2} \frac{\varepsilon}{L} \text{ and } | \beta | = |v_n - l'| < \frac{1}{2} \frac{1}{|l|},$$

provided only that $n \geq m$. Hence also $| u_n v_n - ll'| < \varepsilon$.

Therefore $$u_n v_n \rightarrow ll'.$$

(*iv*) Let $u_n = l + \alpha$, then

$$\frac{1}{u_n} - \frac{1}{l} = - \frac{\alpha}{l (l + \alpha)}.$$

If l is not zero we can choose a positive number L less than $|l + \alpha|$, and then

$$\left| \frac{1}{u_n} - \frac{1}{l} \right| < \frac{|\alpha|}{|l| L}.$$

Now by the definition of a limit we can choose m so that

$$|\alpha| = |u_n - l| < | l | . L . \varepsilon,$$

provided only that $n \geq m$.   Hence also

$$\left| \frac{1}{u_n} - \frac{1}{l} \right| < \varepsilon, \text{ and therefore } \frac{1}{u_n} \rightarrow \frac{1}{l}.$$

(*v*) This follows from (*iii*) and (*iv*), for

$$\frac{u_n}{v_n} = u_n . \frac{1}{v_n} \rightarrow l . \frac{1}{l'}, \text{ when } l' \neq 0.$$

(2) *If* $u_n \rightarrow l$ *and* $v_n - u_n \rightarrow 0$, *then* $v_n \rightarrow l$.

For $v_n - l = (v_n - u_n) + (u_n - l)$, therefore $| v_n - l | \leq | v_n - u_n | + |u_n - l|$.
Now we can choose m so that for $n \geq m$ both

$$|u_n - l| < \frac{1}{2} \varepsilon \text{ and } |v_n - u_n| < \frac{1}{2} \varepsilon.$$

Hence for such values of n,

$$|v_n - l| < \varepsilon, \text{ and therefore } v_n \rightarrow l.$$

■ **3. Monotone Sequences:** If $u_{n+1} \geq u_n$ for all values of n, u_n is said to *increase steadily*, and (u_n) is called a *monotone increasing sequences*.

If $u_{n+1} \leq u_n$ for all values of n, u_n is said to *decrease steadily*, and (u_n) is called a *monotone decreasing sequence*.

Theorem 1. *If every term of a monotone increasing sequence* (u_n) *is less than a fixed number k, then the sequence converges to a limit equal to, or less than, k.*

Proof. Divide the system of rational numbers into two classes as follows:

The *lower class* is to contain every rational a such that, for some value of n (and therefore for all greater values), $u_n \geq a$.

The *upper class* is to contain every rational a' such that $u_n < a'$ for all values of n.

No rational escapes classification, and every a is less than any a'. Hence the classification defines a real number A.

We shall prove that the sequence converges to A as a limit.

First observe that u_n cannot exceed A. For if $u_n > A$, u_n would exceed every rational which lies between u_n and A. This is impossible, for such a rational would belong to the upper class.

Next, choose a positive number ε, no matter how small, and let a be a rational such that

$$A - \varepsilon < a < A.$$

Then a belongs to the lower class, and is exceeded by some term u_m of the sequence and by all succeeding terms;

therefore $\qquad\qquad\qquad 0 < A - u_n < \varepsilon$ if $n \geq m.$

Hence $\qquad\qquad\qquad\qquad \lim u_n = A.$

Again $k \geq A$, for otherwise k would be exceeded by some term of the sequence, thus

$$\lim u_n \leq k.$$

It follows that if u_n increases steadily as $n \to \infty$, then u_n tends to a limit or to $+ \infty$.

For if u_n does not tend to $+ \infty$, a positive number k exists such that $u_n < k$ for all values of n.

▓ **Theorem 2.** *If every term of a monotone decreasing sequence (u_n) is greater than a fixed number k, then the sequence converges to a limit equal to or greater than k.*

This can be proved by an argument similar to the preceding, or thus: Let $v_n = -u_n$, then (v_n) is an increasing sequence of which every term is less than $-k$. Therefore (v_n) converges to a limit less than $-k$. Hence (u_n) converges to a limit greater than k.

Hence also *if u_n decreases steadily as $n \to \infty$, then u_n tends to a limit or to $-\infty$.*

▮ **EXAMPLE :** *If $u_n = n\left(\dfrac{1}{x^{\frac{1}{n}}} - 1\right)$ where $x > 0$, show that u_n tends to a limit as $n \to \infty$.*

If $x = 1$, $u_n = 0$ for every n, and so $\lim u_n = 0$.

If $x > 1$, $u_n > 0$, and by XIV, 9, u_n decreases as n increases. Hence u_n tends to a limit, which we shall denote by $f(x)$, as $n \to \infty$.

If $0 < x < 1$, let $x = 1/y$, then $y > 1$ and

$$u_n = n\left(y^{-\frac{1}{n}} - 1\right) = -y^{-\frac{1}{n}} \cdot n\left(y^{\frac{1}{n}} - 1\right) \to -f(y).$$

▓ **4. Theorems:**

(1) *If for all values of n,[*] u_n is positive and $u_{n+1} > k u_n$ where k is a constant greater than unity, then $u_n \to \infty$.*

---

[*] Or, if $n \geq m$ where m is a fixed number, *i.e.* after a certain stage.

For $u_n > k u_{n-1} > k^2 u_{n-2} \ldots > k^{n-1} u_1$. Now $k > 1$, therefore $u_n \to \infty$.

(2) *If for all values of n, u_n is positive and $u_{n+1} < k u_n$ where k is a positive constant less than unity, then $u_n \to 0$.*

The proof is similar to that of Theorem (1).

(3) *If u_n is positive and $\lim u_{n+1}/u_n = l$, then $u_n \to \infty$ if $l > 1$, and $u_n \to 0$ if $l < 1$.*

For if $l > 1$ we can choose k so that $l > k > 1$, and then by the definition of a limit we can find m such that $u_{n+1}/u_n > k$ for $n \geq m$. Hence by Theorem (1), $u_n \to \infty$.

If $l < 1$, we can choose k so that $l < k < 1$, and then we can find m so that $u_{n+1}/u_n < k$, for $n \geq m$. Hence, by Theorem (2), $u_n \to 0$.

■ **EXAMPLE 1:** *If p is a given integer, positive or negative, show that*

(i) $\dfrac{x^n}{n^p} \to \infty$ *if* $x > 1$;　(ii) $\dfrac{x^n}{n^p} \to 0$ *if* $|x| < 1$.

If $u_n = \dfrac{x^n}{n^p}$ then $\dfrac{u_{n+1}}{u_n} = \left(\dfrac{n}{n+1}\right)^p \cdot x \to x$ *as* $n \to \infty$,

and the results follow from Theorem (3).

■ **5. Exponential Inequalities and Limits:**

(1) *If n is a positive integer,* $\left(1+\dfrac{1}{n}\right)^n < 3$.

For, by the Binomial theorem,

$$\left(1+\frac{1}{n}\right)^n = 1 + 1 + \frac{1}{\lfloor 2} \left(1-\frac{1}{n}\right) + \frac{1}{\lfloor 3} \left(1-\frac{1}{n}\right)\left(1-\frac{2}{n}\right) + \ldots \text{ to } n + 1 \text{ terms}$$

$$< 1 + 1 + \frac{1}{\lfloor 2} + \frac{1}{\lfloor 3} + \ldots + \frac{1}{\lfloor n}$$

$$< 1 + 1 + \frac{1}{2} + \frac{1}{2^2} + \ldots + \frac{1}{2^{n-1}} < + \frac{1}{1-\dfrac{1}{2}} < 3.$$

(2) *If m and n are positive integers and m > n, then*

$$\left(1+\frac{1}{m}\right)^m > \left(1+\frac{1}{n}\right)^n \text{ and } \left(1-\frac{1}{m}\right)^{-m} < \left(1-\frac{1}{n}\right)^{-n}.$$

For since $m/n > 1$, by XIV, 11,

$$\left(1+\frac{1}{m}\right)^{\frac{m}{n}} > 1 + \frac{m}{n} \cdot \frac{1}{m}, \text{ and therefore } \left(1+\frac{1}{m}\right)^m > \left(1+\frac{1}{n}\right)^n.$$

Moreover, $0 < n/m < 1$, therefore by XIV, 11, $\left(1-\dfrac{1}{n}\right)^{\frac{n}{m}} < 1 - \dfrac{n}{m} \cdot \dfrac{1}{n}$.

Hence $\left(1-\dfrac{1}{n}\right)^n < \left(1-\dfrac{1}{m}\right)^m$ and $\left(1-\dfrac{1}{m}\right)^{-m} < \left(1-\dfrac{1}{n}\right)^{-n}$.

(3) The number e. *If* $u_n = \left(1 + \dfrac{1}{n}\right)^n$ *and* $u_n' = \left(1 - \dfrac{1}{n}\right)^{-n}$ *, then as* $n \to \infty$ *through positive integral values,* u_n *tends to a finite limit, so also does* u_n'*; and the limit is the same for both. This limit is denoted by e, and is one of the most important numbers in mathematics.*

For (i) it has been shown that u_n increases with n and that $u_n < 3$ for every n. Therefore $u_n \to e$ as $n \to \infty$, where e is a fixed number less than 3.

(ii) Again, $u_n' - u_n = u_n \left(\dfrac{u_n'}{u_n} - 1\right) = u_n \left\{\left(1 - \dfrac{1}{n^2}\right)^{-n} - 1\right\}.$

Now if $n > 1$, $1 < \left(1 - \dfrac{1}{n^2}\right)^{-n} < 1 + n \cdot \dfrac{1}{n^2}$ (by XIV, 11),

therefore $\qquad\qquad\qquad 0 < u_n' - u_n < \dfrac{u_n}{n} < \dfrac{3}{n} \cdot$ Now $\dfrac{3}{n} \to 0,$

therefore $\qquad\qquad\qquad u_n' - u_n \to 0$ and $u_n' \to e.$

It should be noticed that

$$u_n < e < u_n'.$$

It can be shown that $e = 2.7182818284 \dots$.

EXERCISE XXIV

1. State how the following functions behave as $n \to \infty$:

    (i) $2 + (-1)^n$ (ii) $n + (-1)^n$ (iii) $\dfrac{1}{n} + 2\,(-1)^n$

    (iv) $2 + (-1)^n \dfrac{1}{n}$ (v) $n^2 + (-1)^n\, n.$

2. Prove that $\dfrac{x^n}{\lfloor n} \to 0$ as $n \to \infty$.

3. Represent graphically the first few terms of the following sequences, and say how each behaves as $n \to \infty$.

    (i) $\dfrac{1}{2}, \dfrac{2}{3}, \dots \dfrac{n}{n+1}, \dots$.

    (ii) $-1 + 1,\ 1 + \dfrac{1}{2},\ \dots (-1)^n + \dfrac{1}{n}, \dots$.

4. If $|x| < 1$, then $\dfrac{n\,(n-1)\,(n-2)\dots(n-r+1)}{\lfloor r}\, x^n \to 0$ as $r \to \infty$.

5. Find $\lim\limits_{n \to \infty} \dfrac{x^n + n}{x^n - n}$. [The limit is 1, if $|x| > 1$, and -1 otherwise.]

6. If $u_n = \dfrac{1}{n+1} + \dfrac{1}{n+2} + \dfrac{1}{n+3} + \ldots + \dfrac{1}{2n}$, prove that $\dfrac{1}{2} < u_n < 1$, and that u_n tends to a limit as $n \to \infty$.

7. Show that $\dfrac{1}{2}n < \dfrac{1}{2} + \dfrac{1}{3} + \dfrac{1}{4} + \ldots + \dfrac{1}{2n} < n.$

Hence show that $1 + \dfrac{1}{2} + \dfrac{1}{3} + \ldots + \dfrac{1}{n} \to \infty$ as $n \to \infty.$

[Group the terms thus : $\dfrac{1}{2} + \left(\dfrac{1}{3} + \dfrac{1}{4}\right) + \left(\dfrac{1}{5} + \dfrac{1}{6} + \dfrac{1}{7} + \dfrac{1}{8}\right) + \ldots .]$

8. If $\lim\limits_{n \to \infty} n\left(x^{\frac{1}{n}} - 1\right) = f(x)$, prove that if $x > 0$ and $y > 0$,

$$(i)\ f(1) = 0;\quad (ii)\ f\left(\dfrac{1}{x}\right) = -f(x);\quad (iii)\ f(xy) = f(x) + f(y).$$

[See Art. 3, Ex. 1. For the last part we have $f(xy) = \lim n\left(x^{\frac{1}{n}} y^{\frac{1}{n}} - 1\right).$

Now $x^{\frac{1}{n}} y^{\frac{1}{n}} - 1 = y^{\frac{1}{n}}\left(x^{\frac{1}{n}} - 1\right) + \left(y^{\frac{1}{n}} - 1\right)$, and $\lim y^{\frac{1}{n}} = 1;$

$\therefore\ f(xy) = \lim n\left(x^{\frac{1}{n}} - 1\right) + \lim n\ (y^{\frac{1}{n}} - 1) = f(x) + f(y).]$

9. If $u_n = \dfrac{1 \cdot 3 \cdot 5 \ldots (2n-1)}{2 \cdot 4 \cdot 6 \ldots 2n}$, $v_n = \dfrac{3 \cdot 5 \cdot 7 \ldots (2n+1)}{2 \cdot 4 \cdot 6 \ldots 2n}$, prove that

$$u_n \to 0,\ v_n \to \infty \text{ and } \dfrac{1}{2} < u_n v_n < 1$$

[If $y > 0$, then $\dfrac{x}{y} \gtrless \dfrac{x+1}{y+1}$, according as $x \gtrless y$, hence

$$u_n < \dfrac{2}{3} \cdot \dfrac{4}{5} \ldots \dfrac{2n}{2n+1},\ \therefore u_n^2 < \dfrac{1}{2n+1} \text{ and } u_n \to 0;$$

$$v_n > \dfrac{4}{3} \cdot \dfrac{6}{5} \ldots \dfrac{2n+2}{2n+1},\ \therefore v_n^2 > \dfrac{2n+2}{2} \text{ and } v_n \to \infty;$$

$$u_n v_n < \dfrac{2}{3} \cdot \dfrac{4}{5} \ldots \dfrac{2n}{2n+1} \cdot v_n = 1 \text{ and } v_n u_n > \dfrac{4}{3} \cdot \dfrac{6}{5} \ldots \dfrac{2n+2}{2n+1}\ u_n = \dfrac{1}{2} \cdot \dfrac{2n+2}{2n+1} > \dfrac{1}{2}.]$$

10. If $w_n = \dfrac{1 \cdot 3 \cdot 5 \ldots (2n-1)}{2 \cdot 4 \cdot 6 \ldots 2n} \cdot \sqrt{2n+1}$, prove that w_n tends to a limit which lies between $1/\sqrt{2}$ and 1 as $n \to \infty.$

[Show that $w_{n+1} < w_n$, then by Ex. 9, $w_n{}^2 = u_n v_n$ and $\dfrac{1}{\sqrt{2}} < w_n < 1$. Wallis has shown that this limit $= \sqrt{\dfrac{2}{\pi}}$.]

11. If $n \geq 3$, prove that $\sqrt[n+1]{(n+1)} < \sqrt[n]{n}$. Hence show that $\sqrt[n]{n} \to 1$ as $n \to \infty$.

[For the first part use the inequality $\left(1 + \dfrac{1}{n}\right)^n < 3$. Thus $\sqrt[n]{n}$ decreases steadily; also $\sqrt[n]{n} > 1$, therefore $\sqrt[n]{n} \to l$ where $l \geq 1$. If $l > 1$, for $n > 3$ we should have $\sqrt[n]{n} > l$ and $l^n/n < 1$. But $l^n/n \to \infty$ when $l > 1$ (XV, 4, Ex. 1).]

12. If r is any rational except zero and $u_n = (1^r + 2^r + 3^r + \ldots + n^r)/n^{r+1}$, show that

    (i) $(r + 1)\, u_n$ lies between 1 and $\left(1 + \dfrac{1}{n}\right)^{r+1} - \dfrac{1}{n^{r+1}}$.

    (ii)   If $r + 1 > 0$, $\displaystyle\lim_{n \to \infty} u_n = 1/(r + 1)$.

[In Exercise XXIII, 17, put 1, 2, 3, ... , n in succession for p, and add. The necessity for the condition $r + 1 > 0$ in (ii) should be explained.]

13. Show that $\displaystyle\lim_{n \to \infty} \{(a + 1)^r + (a + 2)^r + \ldots + (a + n)^r\}/n^{r+1} = 1/(r + 1)$.

■ **6. General Principle of Convergence:** The necessary and sufficient condition for the convergence of the sequence (u_n) is as follows.

Corresponding to any positive number ε that we may choose, no matter how small, it must be possible to find m so that, for every positive integer p,

$$| u_{m+p} - u_m | < \varepsilon. \qquad\qquad \text{...(A)}$$

That is to say, the difference between u_m and any subsequent term must be numerically less than ε, so that if the terms of the sequence are represented graphically, all the terms beginning with u_m lie in a strip of width 2ε.

This statement, called the *general principle of convergence*, is of the greatest importance, and may be proved as follows.

(i) *The condition is necessary.* For if $u_n \to l$, we can find m so that

$$| u_m - l | < \tfrac{1}{2}\,\varepsilon \quad \text{and} \quad | u_{m+p} - l | < \tfrac{1}{2}\,\varepsilon$$

for every positive integer p. Now $u_{m+p} - u_m = (u_{m+p} - l) + (l - u_m)$.

Hence $| u_{m+p} - u_m | \leq | u_{m+p} - l | + | u_m - l |$, and $| u_{m+p} - u_m| < \varepsilon$.

(ii) *The condition is sufficient.* Represent the terms $u_1, u_2, \ldots$ by the points $u_1, u_2, \ldots$ on the x-axis. The condition asserts that m can be found so that all the points,

$$u_{m+1},\ u_{m+2},\ u_{m+3},\ \ldots ,$$

lie on a segment of length 2ε with its middle point at u_m.

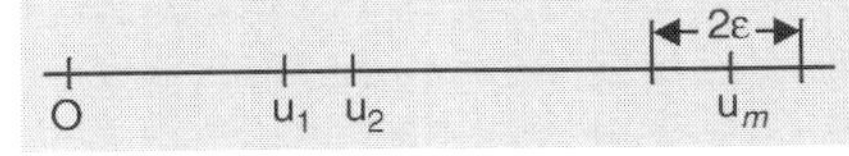

Fig. 35

By choosing ε small enough, we can make this segment as short as we like. Hence it *seems reasonable* to conclude that as $n \to \infty$, the point u_n tends to a limiting position somewhere near to u_m. If the student is satisfied with this reasoning, he may omit the proof on the next page.

Given that, for every positive ε, no matter how small, there exists a positive integer m such that

$$| u_n - u_m | < \varepsilon$$

for every integer n greater than m, it is required to prove that the sequence (u_n) is convergent.

Proof. Using the given condition, we can determine a positive integer m_1 so that, for every integer $m_2 > m_1$, u_{m_2} is within the interval

$$(u_{m_1} - \varepsilon, \; u_{m_1} + \varepsilon).$$

Denote this interval by (a_1, b_1).

Again, we can determine m_2 so that, for every integer $m_3 > m_2$, u_{m_3} is within the interval $(u_{m_2} - \frac{1}{2}\varepsilon, \; u_{m_2} + \frac{1}{2}\varepsilon)$. Now u_{m_3} is within (a_1, b_1), for $m_3 > m_2$. Thus the whole or part of $(u_{m_2} - \frac{1}{2}\varepsilon, \; u_{m_2} + \frac{1}{2}\varepsilon)$ is within (a_1, b_1).

Denote the common part of these intervals by (a_2, b_2).

This is of length $\leq \varepsilon$, lies entirely within (a_1, b_2), and contains u_n for $n > m_2$. The figure illustrates the case in which the intervals overlap.

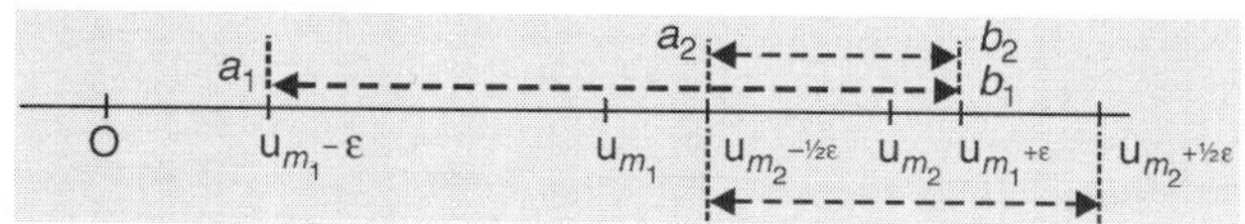

Fig. 36

By repeating the process, we can find a succession of intervals (a_1, b_1), (a_2, b_2), ... (a_r, b_r) ... , such that (*i*) each contains all that follow;

(*ii*) for sufficiently large values of n, u_n is within each of the intervals;

(*iii*) $b_n - a_n \leq \dfrac{1}{2^{n-2}} \varepsilon \to 0$ as $n \to \infty$.

Hence by Ch. 13, 4, there is one real number l and one only, such that $a_n < l < b_n$ for every n, and this is the limit of the sequence (u_n).

Note: The necessary and sufficient condition for the convergence of the sequence (u_n) is usually stated (less simply) as follows.

Corresponding to any positive number ε that we may choose, no matter how small, it must be possible to find m so that

$$| u_{n+p} - u_n | < \varepsilon \qquad \qquad \text{...(B)}$$

if $n \geq m$, for every positive integer p.

The condition (B), though apparently more general than (A), is equivalent to it. To prove this, we have only to show that if (A) is possible, then (B) is possible when $n > m$.

By the condition (A), we can find m so that

$$|u_{m+p} - u_m| < \frac{1}{2}\,\varepsilon,$$

and then all the terms u_{m+1}, u_{m+2}, u_{m+3}, ... lie in the interval

$$(u_m - \frac{1}{2}\,\varepsilon,\ u_m + \frac{1}{2}\,\varepsilon).$$

Therefore the difference between any two terms which follow u_m is less than ε. Hence we can find m so that

$$|u_{n+p} - u_n| < \varepsilon \text{ if } n > m.$$

Therefore the conditions (A) and (B) are one and the same.

■ **7. Bounds of a Sequence:** If there exists a number M such that, for all values of n,

$$u_n \leq M,$$

the sequence (u_n), and also the function u_n, are said to be *bounded above* (or *on the right*).

If there exists a number N such that, for all values of n,

$$u_n \geq N,$$

then (u_n), and also u_n, are said to be *bounded below* (or *on the left*).

A sequence (u_n), or a function u_n, which is bounded above and below, is said to be *bounded*.

■ **EXAMPLES :**

1. Let $u_n = (-1)^n \left(1 + \dfrac{1}{n}\right)$.

Here (u_n) has a *greatest* term, namely $u_2 = \dfrac{3}{2}$, ∴ it has also a *least* term, namely $u_1 = -2$. These are called *the upper and lower bounds* of (u_n), or of u_n (Fig. 37).

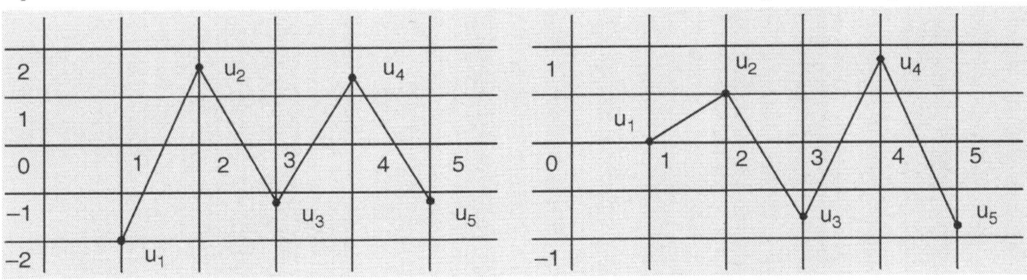

Fig. 37**Fig. 38**

2. *Let* $u_n = (-1)^n \left(1 - \dfrac{1}{n}\right)$.

Although there is no greatest term, we say that 1 is the *upper bound* of (u_n), or of u_n, for the following reasons:

(*i*) No term of (u_n) exceeds 1.

(*ii*) Infinitely many terms exceed any number less than 1, for $u_{2n} \to 1$.

For similar reasons, we say that -1 is the *lower bound* of (u_n) of u_n (Fig. 38).

Theorem. *If u_n is bounded above there exists a number h such that*

(i) *no term of (u_n) exceeds h;*

(ii) *at least one term of (u_n) exceeds any number less than h.*

This number h is called the *upper bound* of the sequence (u_n) or of the function u_n.

Proof. Divide the system of rationals into two classes.

The lower class is to contain every rational a such that for some value of n

$$u_n \geq a.$$

The upper class is to contain every rational a' such that for all values of n

$$u_n < a'.$$

Since u_n is bounded above, both classes exist. No rational escapes classification, and every a is less than any a'. Hence the classification defines a real number h.

Moreover, no term of (u_n) exceeds h, for in that case it would exceed a rational a' greater than h. This is impossible, for a' belongs to the upper class.

Also if ε is any positive number, no matter how small, we can find a rational a such that

$$h - \varepsilon < a < h.$$

Now a belongs to the lower class and is exceeded by some term of (u_n). Hence at least one term of (u_n) exceeds $h - \varepsilon$.

Hence the number h satisfies the conditions stated above.

This theorem is often stated as follows: *If u_n is bounded above, then it has an upper bound.*

In a similar way we can show that: *If u_n is bounded below there exists a number l such that*

(i) *no term of (u_n) is less than l;*

(ii) *at least one term of (u_n) is less than any number greater than l.*

The number l is called the *lower bound* of the sequence (u_n) or of the function u_n.

Note: *If (u_n) has no greatest term, infinitely many terms of the sequence lie between h and $h - \varepsilon$.*

For every term is less than h, and at least one term u_m is greater than $h - \varepsilon$. Since there is no greatest term, infinitely many terms are greater than u_m, and therefore also greater than $h - \varepsilon$.

Similarly, *if (u_n) has no least term, infinitely many terms of the sequence lie between l and $l + \varepsilon$.*

Terms selected from $u_1, u_2, u_3, \ldots$ according to some definite rule are said to form a *subsequence*.

Example : *Explain how to select a subsequence which will converge to the upper bound h of (u_n).*

If $u_1 \neq h$, let a_2 be the first term of (u_n) greater than u_1. If $a_2 \neq h$, let a_3 be the first term of (u_n) greater than a_2. Continuing in this way we form the required subsequence $v_1, a_2, a_3, \ldots$. The formal proof is left to the student.

▮ 8. Upper and Lower Limits (of Indetermination) of a Sequence:

(1) Let (u_n) be a bounded sequence and let $h_1, h_2, h_3, \ldots$ and $l_1, l_2, l_3, \ldots$ be the upper and lower bounds of the sequences $u_1, u_2, u_3, \ldots$, $u_2, u_3, u_4 \ldots$, $u_3, u_4, u_5 \ldots$, etc., respectively.

If $h_1 = u_1$, then u_1 is not less than any term of $u_2, u_3, u_4, \ldots$; therefore $h_1 \geq h_2$. If $h_1 \neq u_1$, then h_1 is the upper bound of $u_2, u_3, u_4, \ldots$, and so $h_1 = h_2$. Thus, in any case, $h_1 \geq h_2$. Similarly $h_2 \geq h_3$, $h_3 \geq h_4$, and so on.

Hence $h_1, h_2, h_3, \ldots$ is a steadily decreasing sequence and every term is greater than l_1. Therefore this sequence tends to a limit H greater than or equal to l_1.

Similarly $l_1, l_2, l_3, \ldots$ is an increasing sequence and every term is less than h_1. Therefore this sequence tends to a limit L less than or equal to h_1.

The numbers H and L are called the *upper and lower limits* (*of indetermination*) of the sequence (u_n).

It is usual to write $\lim u_n = H$ and $\underline{\lim} \, u_n = L$.

(2) Examples:

(*i*) *Let* $u_n = (-1)^n \left(1 - \dfrac{1}{n}\right)$. *Here* $h_1 = h_2 = h_3 = \ldots = H$. *(Fig. 39)*

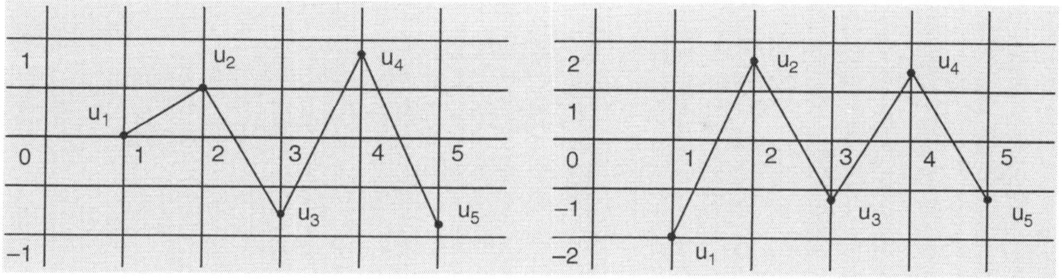

Fig. 39   Fig. 40

(*ii*) *Let* $u_n = (-1)^n \left(1 + \dfrac{1}{n}\right)$.

Here $h_1 = h_3 = u_2$; $h_3 = h_4 = u_4$; etc. $\therefore H = \lim u_{2n} = 1$. (Fig. 40)

Also $l_1 = u_1$; $l_2 = l_3 = u_3$; etc. $\therefore L = \lim u_{2n+1} = -1$.

(3) It is easy to see that, if $h_n > h_{n+1}$, then $h_n = u_n$ and $h_1, h_2, \ldots h_{n-1}$ are all terms of $u_1, u_2, \ldots , u_{n-1}$. As $n \to \infty$ two cases arise:

(*i*) After a certain stage, all the h's are equal, that is to say, m exists so that
$$h_m = h_{m+1} = h_{m+2} = \ldots = H.$$

In this case, (u_n) has no greatest term. None of the terms from u_n onwards can exceed H, but infinitely many terms exceed $H - \varepsilon$.

(*ii*) There is no such stage as that described in (*i*). That is, no matter how great m may be, we can find n, greater than m, so that $h_n > h_{n+1}$. Therefore every h is a term of (u_n), and in all cases H is such that, after a certain stage, every term of (u_n) is less than $H + \varepsilon$.

Hence *in all cases, infinitely many terms of* (u_n) *are greater than* $H - \varepsilon$; *also, after a certain stage all the terms of* (u_n) *are less than* $H + \varepsilon$, *in other words, only a finite number of terms exceed* $H + \varepsilon$.

In a similar way we can show that *infinitely many terms of* u_n *are less than* $L + \varepsilon$ *and, after a certain stage, all the terms are greater than* $L - \varepsilon$; *in other words, only a finite number of terms are less than* $L - \varepsilon$.

It follows that, *if* $H = L = l$, *then the sequence converges to* l *as a limit.*

For, after a certain stage, all the terms of (u_n) lie between $L - \varepsilon$ and $H + \varepsilon$, that is between $l - \varepsilon$ and $l + \varepsilon$.

(4) *General Principle of Convergence.* We can at once deduce a proof that the condition in Art. 6 is *sufficient* for convergence.

Given that m can be found so that $|u_n - u_m| < \varepsilon$ *for* $n > m$, *it is required to prove that* (u_n) *is convergent.*

First observe that u_n cannot tend to $+ \infty$ or to $- \infty$. Therefore (u_n) has upper and lower limits H and L. We can therefore choose m and n $(n > m)$ so that

$$|H - u_n| < \frac{1}{3} \varepsilon, \; |u_m - L| < \frac{1}{3} \varepsilon, \; |u_n - u_m| < \frac{1}{3} \varepsilon; \text{ and since } H - L = H - u_n + u_m$$
$$- L + u_n - u_m,$$
$$|H - L| \leq |H - u_n| + |u_n - L| + |u_n - u_m| < \varepsilon.$$

That is to say, the difference between the *fixed* numbers H and L is less than any positive number ε, however small. Therefore $H = L$ and (u_n) is convergent.

■ **9. Theorems:**

(1) *If* (u_n) *is an increasing sequence and if, after a certain stage,*
$$u_n - u_{n-1} < k (u_{n-1} - u_{n-2}),$$
where k *is a positive constant less than unity, then the sequence is convergent. But if* $u_n - u_n - 1 \geq u_{n-1} - u_{n-2}$, *the sequence is divergent.*

In the first case, suppose that the conditions hold for $n \geq s + 2$,

Then if $m \geq s$, $u_{m+1} - u_m < k (u_m - u_{m-1}) \dots < k^{m-s} (u_{s+1} - u_s)$,

therefore $\qquad u_{m+1} - u_m < k^m c$ where $c = (u_{s+1} - u_s)/k^s$.

Also $\qquad\qquad\qquad\qquad u_{m+2} - u_{m+1} < k^{m+1} c,$

$$\dotsb\dotsb\dotsb\dotsb\dotsb\dotsb$$

$$u_{m+n} - u_{m+n-1} < k^{m+n-1} c.$$

Hence by addition $\quad u_{m+n} - u_m < k^m c (1 + k + k^2 + \dots k^{n-1}) < k^m c/(1 - k).$

Choose any positive number ε, no matter how small; then, since $k < 1$, we can find m so that $k^m < \varepsilon (1 - k)/c$, and since $u_{m+n} > u_m$, it follows that

$$|u_{m+n} - u_m| < \varepsilon \text{ for every } n.$$

Therefore the sequence (u_n) converges.

In the second case, if $n \geq s$, $u_{n+1} - u_n \geq u_{s+1} - u_s$.

Now $u_{s+1} - u_s$ is a fixed positive quantity, therefore obviously $u_n \to \infty$.

(2) *If* $u_n > 0$ *for all values of n and* $u_{n+1}/u_n \to l > 0$ *where l is a fixed number, then*

$$u_n^{\frac{1}{n}} \to l.$$

Because $l > 0$, we can choose a positive number $\varepsilon < l$ and as small as we like. Then we can find m so that

$$l - \varepsilon < u_{m+1}/u_m < l + \varepsilon,$$
$$l - \varepsilon < u_{m+2}/u_{m+1} < l + \varepsilon,$$
$$\cdots\cdots\cdots\cdots\cdots\cdots\cdots$$
$$l - \varepsilon < u_n/u_{n-1} < l + \varepsilon,$$

and by multiplication,

$$(l - \varepsilon)^{n-m} < u_n/u_m < (l + \varepsilon)^{n-m}.$$

Since $l - \varepsilon > 0$, it follows that

$$(l - \varepsilon)^n < \frac{u_n}{u_m}\,(l - \varepsilon)^m < \frac{u_n}{u_m}\cdot l^m, \text{ and } (l + \varepsilon)^n > \frac{u_n}{u_m}\,(l + \varepsilon)^m > \frac{u_n}{u_m}\cdot l^m;$$

therefore

$$(l - \varepsilon)^n < \frac{u_n}{u_m}\cdot l^m < (l + \varepsilon)^n.$$

and

$$\left(\frac{u_m}{l_m}\right)^{\frac{1}{n}}\cdot (l - \varepsilon) < u_n^{\frac{1}{n}} < \left(\frac{u_m}{l_m}\right)^{\frac{1}{n}} (l + \varepsilon).$$

Now u_m/l^m is a fixed number, therefore $(u_m/l^m)^{\frac{1}{n}} \to 1$ and $u_n^{\frac{1}{n}} \to l$.

Note: *The converse of this theorem is not true.* That is to say, if $u_n^{\frac{1}{n}} \to l$, it does *not* follow that $u_{n+1}/u_n \to l$. For consider as a special case the sequence $1, 2, 1, 2, \ldots$ in which the terms are alternately 1 and 2. Here $u_n^{\frac{1}{n}} \to 1$, but u_{n+1}/u_n does not tend to a limit.

(3) *If $s_n = u_1 + u_2 + \ldots + u_n$ and $\lim\limits_{n \to \infty} u_n = l$, then $\lim\limits_{n \to \infty} \dfrac{1}{n} s_n = l$.*

We can find m so that for any positive ε,

$$l - \varepsilon < u_{m+1} < l + \varepsilon,$$
$$l - \varepsilon < u_{m+2} < l + \varepsilon,$$
$$\cdots\cdots\cdots\cdots\cdots\cdots\cdots$$
$$l - \varepsilon < u_n < l + \varepsilon.$$

Hence by addition,

$$(n-m)\,(l - \varepsilon) < s_n - s_m < (n-m)\,(l + \varepsilon) \text{ for } n > m$$

and

$$\left(1 - \frac{m}{n}\right)(l - \varepsilon) < \frac{s_n - s_m}{n} < \left(1 - \frac{m}{n}\right)(l + \varepsilon). \qquad \ldots\text{(A)}$$

Now let m and n tend to ∞ in such a way that $n/m \to \infty$. (For example, we may suppose that $m \to \infty$ and $n > m^2$.)

We shall show that under these circumstances $s_m/n \to 0$.

Since $u_n \to l$ we can find a fixed number k such that $|u_r| < k$ for every r, and then

$$|s_m| \le |u_1| + |u_2| + \ldots + |u_m| < mk,$$

so that $\dfrac{1}{n}\,|\,s_m\,| < \dfrac{m}{n}\,k$ and $s_m/n \to 0$.

Thus in the inequality (A), m/n and s_m/n tend to zero and ε can be taken as small as we like, therefore $s_n/n \to l$.

(4) *If* $P_n = a_n b_1 + a_{n-1} b_2 + a_{n-2} b_3 + \ldots + a_1 b_n$ *and* $\lim\limits_{n \to \infty} a_n = a, = \lim\limits_{n \to \infty} b_n = b,$

then $\lim\limits_{n \to \infty} \dfrac{1}{n}\, P_n = ab.$

Let $a_n = a + \alpha_n$ and $b_n = b + \beta_n$, so that $\alpha_n \to 0$, $\beta_n \to 0$, then

$$\frac{1}{n}\, P_n = ab + \frac{a}{n}\,(\beta_1 + \beta_2 + \ldots + \beta_n) + \frac{b}{n}\,(\alpha_1 + \alpha_2 + \ldots + \alpha_n) + Q_n,$$

where $$Q_n = \frac{1}{n}\,(\alpha_n \beta_1 + \alpha_{n-1}\beta_2 + \ldots + \alpha_1 \beta_n).$$

Now by (3), $\dfrac{1}{n}\,(\alpha_1 + \alpha_2 + \ldots + \alpha_n) \to 0$ and $\dfrac{1}{n}\,(\beta_1 + \beta_2 + \ldots + \beta_n) \to 0,$

and we shall prove that $Q_n \to 0$.

We have $$|\,Q_n\,| \le \frac{1}{n}\,\sum_{r=1}^{r=n} |\,\alpha_r \beta_{n-r+1}\,|,$$

and since $\alpha_n \to 0$ and $\beta_n \to 0$ we can find a fixed number k such that for every s

$$|\,\alpha_s\,| < k \text{ and } |\,\beta_s\,| < k.$$

We can also find m so that

$$|\,\alpha_s\,| < \varepsilon/k \text{ and } |\,\beta_s\,| < \varepsilon/k \text{ for } s > m.$$

Hence $|\,\alpha_r \beta_{n-r+1}\,| < k\,|\,\beta_{n-r+1}\,| < \varepsilon \text{ for } n-r+1 > m,$ ...(A)

and $|\,\alpha_r \beta_{n-r+1}\,| < k\,|\,\alpha_r\,| < \varepsilon \text{ for } r > m.$ ...(B)

If $n > 2m$, the inequality (A) will hold for $r < m + 1$, *i.e.* for $r \le m$, so that, with this proviso, $|\,\alpha_r \beta_{n-r+1}\,| < \varepsilon$ for $1 \le r \le n$.

Therefore $|\,Q_n\,| < \varepsilon$: consequently $Q_n \to 0$ and $\dfrac{1}{n}\, P_n \to ab.$

■ **10. Complex Sequences:**

(1) If $z_n = x_n + \iota y_n$ where x_n and y_n are functions of a positive integral variable n, the sequence (z_n) is said to be *complex*.

If corresponding to any positive ε, however small, there is a positive integer m such that $|\,z_n - z\,| < \varepsilon$ for $n \ge m,$
where z is a fixed number, then z is called the *limit of* z_n as $n \to \infty$, and we write

$$\lim\limits_{n \to \infty} z_n = z \text{ or } \lim z_n = z \text{ or } z_n \to z.$$

Under these circumstances the sequence (z_n) is said to converge to the limit z.

In order that $\lim z_n = z = x + \iota y$ *it is necessary and sufficient that*

$$\lim x_n = x \text{ and } \lim y_n = y.$$

For $$|\,z_n - z\,| = \sqrt{\{(x_n - x)^2 + (y_n - y)^2\}}$$ ...(A)

Therefore $|z_n - z| \ge |\,x_n - x\,|$ and $|\,z_n - z\,| \ge |\,y_n - y\,|.$

First suppose that $z_n \to z$. By definition,

$$| z_n - z | \to 0, \text{ therefore } | x_n - x | \text{ and } | y_n - y |$$

tend to zero, so that $x_n \to x$ and $y_n \to y$.

Again, if $x_n \to x$ *and* $y_n \to y$, from equation (A) it follows that

$$| z_n - z | \to 0, \text{ therefore } z_n \to z.$$

All this is evident geometrically, for $| z_n - z |$ is the distance between the points z_n and z.

(2) The theorems of Ch. 15, 2, regarding the limits of the sum, difference, product and quotient of two functions of n, continue to hold when the functions are complex.

(3) *Limit of* z^n *as* $n \to \infty$ *when z is complex. If* $| z | < 1$, *then* $z^n \to 0$. *Otherwise* z^n *does not tend to a limit except when* $z = 1$, *and then* $\lim z^n = 1$.

For let $z = r (\cos \theta + \iota \sin \theta)$, then $z^n = r^n (\cos n \theta + \iota \sin n \theta)$, and

   (i) if $| z | = r < 1$, then $r^n \to 0$ and $z^n \to 0$;

   (ii) if $z = 1$ then $z^n = 1$, and if $| z | > 1$ then $r^n \to \infty$ and z^n does not tend to a limit.

■ **11. General Principle of Convergence:** The necessary and sufficient condition for the convergence of a complex sequence (z_n) is as follows:

Corresponding to any positive ε, *however small, it must be possible to find m so that* $| z_{m+p} - z_m | < \varepsilon$ *for every positive integer p.*

Proof. *If* (z_n) *is convergent,* then (x_n) and (y_n) are convergent sequences; and by Art. 6, we can therefore choose m so that for every positive integer p,

$$| x_{m+p} - x_m | < \frac{1}{2}\varepsilon \text{ and } | y_{m+p} - y_m | < \frac{1}{2}\varepsilon.$$

Now
$$|z_{m+p} - z_m| \le |x_{m+p} - x_m | + | y_{m+p} - y_m |,$$
therefore
$$| z_{m+p} - z_m | < \varepsilon.$$

Hence the condition is *necessary.*

Next suppose that $|z_{m+p} - z_m| < \varepsilon$, then $| x_{m+p} - x_m | \le |z_{m+p} - z_m| < \varepsilon.$

Therefore (x_m) is convergent (Art. 6), and similarly it can be shown that (y_n) is convergent. Hence (z_n) is convergent, and the condition is *sufficient.*

> **Note:** As in Art. 6, the condition may be stated thus. *Corresponding to any positive* ε, *however small, it must be possible to find m so that*
> $$| z_{n+p} - z_n | < \varepsilon,$$
> *if* $n \ge m$, *for every positive integer p.*

EXERCISE XXV

1. If u_n is a decreasing function of n, then, after a certain stage, all the terms of u_n have the same sign.

    [Either u_n tends to a limit or to $- \infty$. First let $u_n \to l$. If $l \ge 0$, u_n is positive for all values of n, for u_n tends to l *from above.* If $l < 0$, we can choose m so that
$$l < u_n < 0 \text{ for } n \ge m.$$
    Therefore the terms of (u_n) beginning with (u_m) are negative.

    If $u_n \to - \infty$, we can find m so that $u_n < - k$ for $n \ge m.$]

2. If u_n is an increasing function, the statement in Ex. 1 is true.

3. Show that the positive proper fractions can be arranged in a definite order, so as to form a sequence. Represent the first few terms graphically; indicate terms which form subsequences converging to the upper and lower bounds as a limit.

   [The sequence, omitting such fractions as $\dfrac{2}{4}$, is

   $$\frac{1}{2}, \frac{1}{3}, \frac{2}{3}, \frac{1}{4}, \frac{3}{4}, \frac{1}{5}, \frac{2}{5}, \frac{3}{5}, \frac{4}{5}, \frac{1}{6}, \frac{5}{6} \ldots$$

   The required subsequences are indicated by the dotted lines.]

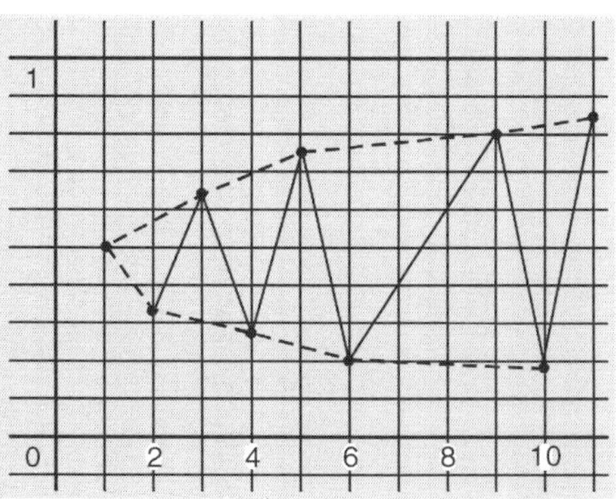

Fig. 41

4. If $u_n > 0$ for all values of n and $u_{n+1}/u_n \to l$, then $(nu_n)^{\frac{1}{n}} \to l$.

   [See Ex. XXIV, 11, and Art. 9, (2).]

5. Prove that $\lim \dfrac{1}{n}\left(1 + \dfrac{1}{2} + \dfrac{1}{3} + \ldots + \dfrac{1}{n}\right) = 0$. [Art. 9, (3)]

6. If $u_n = \dfrac{(n+1)^n}{\lfloor n}$, show that $\dfrac{u_n}{u_{n-1}} \to e$, and consequently $(n+1)/(\lfloor n)^{\frac{1}{n}} \to e$.

7. If $u_n = \dfrac{1}{1.n} + \dfrac{1}{2(n-1)} + \dfrac{1}{3(n-2)} + \ldots + \dfrac{1}{n.1}$, Prove that

   $$u_n = \frac{2}{n+1}\left(1 + \frac{1}{2} + \frac{1}{3} + \ldots + \frac{1}{n}\right) \to 0.$$

   [We have $(n+1)u_n = \left(1 + \dfrac{1}{n}\right) + \left(\dfrac{1}{2} + \dfrac{1}{n-1}\right) + \left(\dfrac{1}{3} + \dfrac{1}{n-2}\right) + \cdots.$]

8. If $u_n = \dfrac{1}{1^p.n^p} + \dfrac{1}{2^p.(n-1)^p} + \dfrac{1}{3^p(n-2)^p} + \ldots + \dfrac{1}{n^p.1^p}$, where $0 < p < \dfrac{1}{2}$, prove that $u_n \to \infty$.

[Show that $r(n - r + 1) \le \dfrac{1}{4}(n + 1)^2$ for $r = 1, 2, 3, \ldots n$. Hence show that $u_n > n$. $4^p/(n + 1)^{2p} \to \infty$.]

9. If $k > 0$ and $\alpha, -\beta$ are the positive and negative roots of $x^2 - x - k = 0$ prove that

   (i) if $u_n = \sqrt{(k + u_{n-1})}$ and $u_1 > 0$, then $u_n \to \alpha$;

   (ii) if $v_n = \sqrt{(k - v_{n-1})}$ and $v_1 < k$, then $v_n \to \beta$.

Illustrate geometrically.

[(i) Here $u_n^2 = k + u_{n-1}$, $u_{n-1}^2 = k + u_{n-2}$,

$\therefore u_n^2 - u_{n-1}^2 = u_{n-1} - u_{n-2}$;

hence $u_n \gtrless u_{n-1}$, according as $u_{n-1} \gtrless u_{n-2}$; therefore (u_n) is monotone.

First suppose that $u_1 > \alpha$, then $u_1^2 > u_1 + k$ for $u_1^2 - u_1 - k = (u_1 - \alpha)(u_1 + \beta)$,

$\therefore u_2^2 = u_1 + k < u_1^2$, $\therefore u_2 < u_1$ and (u_n) is a decreasing sequence.

Again, $u_2^2 = u_1 + k > u_2 + k$, $\therefore u_2 > \alpha$. Similarly $u_n > \alpha$ for $n = 3, 4\ldots$. Hence $u_n \to l \ge \alpha$. Thus $u_n - u_{n-1} \to 0$ and it follows that

$$(u_n - \alpha)(u_n + \beta) = (u_n^2 - u_n - k) - (u_n^2 - u_{n-1} - k) \to 0, \therefore u_n \to \alpha.$$

The case in which $u_1 < \alpha$ can be treated in the same way.

(ii) Draw graphs of $y = x$ and $y = \sqrt{(k + x)}$, meeting at A. In Fig. 42, we have

$$ON_r = N_r p_r = u_r, \text{ and } N_r P_r = \sqrt{k + u_r}$$
$$= u_{r+1},$$

and thus $P_r p_r = u_r - u_{r+1}$.

The figure illustrates the case in which

$$u_1 = ON_1 = N_1 P_1 > \alpha,$$

and shows that (u_n) is a decreasing sequence.

Thus $N_r P_r > N_r p_r$ for every r, and so N_r is always to the right of N. Thus $u_n \to l \ge \alpha$ and $u_n - u_{n+1} = P_n p_n \to 0$.

Hence N_r tends to N as a limiting position and $u_n \to \alpha$.]

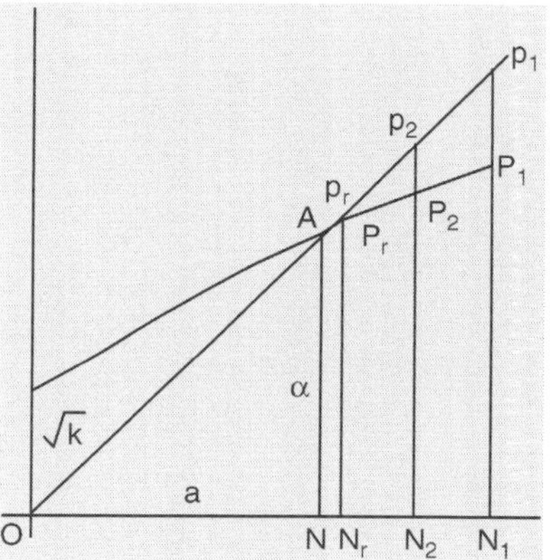

Fig. 42

10. If $u_n = f(u_{n-1})$ where $f(x)$ is such that (i) $f(x) > 0$ if $x > 0$, (ii) $f(x)$ increases with x, and (iii) the equation $x = f(x)$ has a single positive root α, prove that $u_n \to \alpha$. [It is advisable to draw a figure.]

11. Show that as the number of roots tends to infinity, the expressions

$$\sqrt{7 + \sqrt{7 + \sqrt{7 +}}} \ \ldots, \ -\sqrt{7 - \sqrt{7 - \sqrt{7 -}}} \ \ldots$$

tend to the roots of $x^2 - x - 7 = 0$ as limits.

12. If $x = \sqrt{7 - \sqrt{7 + y}}$, prove that

$$y = (x^2 - 7)^2 - 7.$$

Draw a graph of the last equation, and prove that if

$$u_n = \sqrt{7 - \sqrt{7 + \sqrt{7 - \sqrt{7 + \ldots}}}} \quad \text{to } 2n \text{ roots}$$

then $\qquad u_1, u_3, u_5 \ldots$ is a decreasing sequence,

$$u_2, u_4, u_6 \ldots \text{ is an increasing sequence,}$$

and that both sequences converge to 2 as a limit.

Prove also that

$$\sqrt{7 + \sqrt{7 - \sqrt{7 + \sqrt{7 - \ldots}}}} \quad \text{to } \infty = 3.$$

13. If u_1, v_1 are given unequal numbers and

$$u_n = \frac{1}{2} (u_{n-1} + v_{n-1}), \quad v_n = \sqrt{(u_{n-1} v_{n-1})} \quad \text{for } n \geq 2,$$

prove that (*i*) u_n decreases, and v_n increases as n increases, (*ii*) Each of the functions u_n and v_n tends to a limit, which is the same for both. This limit is called the *arithmetico-geometric mean* of u_1 and v_1, (*Gauss*).

$$\left[u_n - v_n = \frac{1}{2} \left(\sqrt{u_{n-1}} - \sqrt{v_{n-1}} \right)^2 > 0, \quad \therefore \quad u_n > v_n \text{ for every } n; \right.$$

$$\therefore \qquad u_n = \frac{1}{2} (u_{n-1} + v_{n-1}) < u_{n-1} \text{ and } v_n = \sqrt{(u_{n-1} v_{n-1})} > v_{n-1}.$$

Thus $v_1 < v_2 \ldots < v_n \ldots < u_n \ldots < u_2 < u_1$.

Hence v_n tends to a limit, and so does v_n.

Also $\lim v_n = \lim v_{n-1} = \lim (2u_n - u_{n-1}) = \lim u_n.]$

■ ■ ■

CHAPTER 16

Convergence of Series

■ **1. Definitions:** Let u_n be a function of n which has a definite value for all positive integral values of n. An expression of the form

$$u_1 + u_2 + u_3 + \dots + u_n + \dots ,$$

in which every term is followed by another term, is called an *infinite series*.

This series will be denoted by $\sum_{n=1}^{\infty} u_n$, or by Σu_n, and the sum of its first n terms by s_n.

The sum of the p terms immediately following the nth term is denoted by $R_{n,p}$, so that

$$R_{n,p} = u_{n+1} + u_{n+2} + \dots + u_{n+p} = s_{n+p} - s_n.$$

As n tends to infinity, there are four distinct possibilities: s_n may tend to a finite limit, to infinity or to minus infinity, or it may do none of these things.

If s_n tends to a finite limit s, the series is said to be *convergent* and s is called its *sum to infinity.* Thus s is defined by

$$\lim_{n \to \infty} s_n = s, \quad \text{or briefly,} \quad \lim s_n = s.$$

This is also expressed by writing

$$u_1 + u_2 + u_3 + \dots \text{ to } \infty = s \quad \text{or} \quad \sum_1^{\infty} u_n = s, \quad \text{or} \quad \Sigma u_n = s.$$

If s_n tends to ∞ or to $-\infty$, the series is said to be *divergent.*

If s_n tends to no limit, whether finite or infinite, the series is said to *oscillate.* We say that the series oscillates *finitely* or *infinitely*, according as s_n oscillates between finite limits, or between $+\infty$ and $-\infty$.

Series which diverge or oscillate are often said to be *non-convergent.*

The sum (to infinity) of a convergent series is essentially a *limit*, and is not a sum in the sense described in the definition of addition. We are therefore not justified in assuming that the sum of an infinite series is unaltered by changing the order of the terms or by the introduction or removal of brackets.

In fact, changes of this kind *may* alter the sum, or they may transform a convergent series into one which diverges or oscillates.

Thus the series $(1 - 1) + (1 - 1) + (1 - 1) + \dots$ is convergent and its sum is zero, but the series $1 - 1 + 1 - 1 + \dots$ oscillates.

■ **2. The Geometric Series:** *The series $1 + x + x^2 + ... + x^{n-1} + ...$ converges if $|x|$ < 1 and its sum is $1/(1-x)$. It diverges if $x \geq 1$, oscillates finitely if $x = -1$, and oscillates infinitely if $x < -1$.*

For $s_n = (1 - x^n)/(1 - x)$ if $x \neq 1$. If $|x| < 1$, $x^n \to 0$ and $s_n \to 1/(1-x)$. If $x \geq 1$, $s_n \to \infty$. If $x = -1$, $s_n = 1$ or 0, according as n is odd or even. If $x < -1$, $s_n \to \infty$ or $-\infty$, according as n is odd or even.

■ **3. Theorems:**

(1) *If m is a given number, the series*

$$u_1 + u_2 + u_3 + ... \quad and \quad u_{m+1} + u_{m+2} + u_{m+3} + ...$$

both converge, both diverge or both oscillate.

For let $\quad s_n = u_1 + u_2 + ... + u_n$, $t_n = u_{m+1} + u_{m+2} + ... + u_{m+n}$.

Then $t_n = s_{m+n} - s_m$ and s_m is a fixed number, hence

(*i*) if s_{m+n} tends to a finite limit or to $\pm \infty$, so also will t_n;

(*ii*) if s_{m+n} tends to no limit, whether finite or infinite, neither will t_n.

(2) *If k is a fixed number and Σu_n converges to a sum s, then $\Sigma k u_n$ is convergent and its sum is ks. Also if the first series diverges or oscillates, so does the second, unless k = 0.*

For if $s_n \to s$, then $ks_n \to ks$. If $k \neq 0$ and $s_n \to \pm \infty$, so does ks_n. If s_n tends to no fixed limit, finite or infinite, neither does ks_n.

■ **4. Addition and Subtraction of Series:** *If the series*

$$u_1 + u_2 + ... \quad and \quad v_1 + v_2 + ...$$

converge and their sums are s and t respectively, then

(*i*) $(u_1 + v_1) + (u_2 + v_2) + ...$ *converges, and its sum is $s + t$;*

(*ii*) $(u_1 - v_1) + (u_2 - v_2) + ...$ *converges, and its sum is $s - t$.*

(*i*) For lim $\{(u_1 + v_1) + (u_2 + v_2) + ... + (u_n + v_n)\}$

$$= \lim \{(u_1 + u_2 + ... + u_n) + (v_1 + v_2 + ... + v_n)\}$$

$$= \lim (u_1 + u_2 + ... + u_n) + \lim (v_1 + v_2 + ... + v_n) = s + t.$$

Note: In the proof, all the series in brackets are *finite*.

(*ii*) The proof is similar to the preceding.

SERIES OF POSITIVE TERMS

We first consider series in which *all the terms are positive*. Under this heading we include series in which all the terms are positive *beginning with some particular term*.

■ **5. Introduction and Removal of Brackets:** Let Σu_n be a series of positive terms, and let the terms be arranged in groups without altering their order. Denote the sum of the terms in the nth group by v_n, then

(*i*) *If Σu_n converges to a sum s, so does Σv_n.*

For let $\quad s_n = u_1 + u_2 + ... + u_n$,

$$t_n = v_1 + v_2 + ... + v_n.$$

Then for every n we can find m and p, so that

$$s_m \leq t_n < s_{m+p,}$$

where m tends to infinity with n.

Now $\lim s_m = \lim s_{m+p} = s$, therefore $\lim t_n = s$.

(*ii*) *If Σv_n converges to a sum s, so does Σu_n.*

For corresponding to every n we can find m, so that
$$t_m \le s_n < t_{m+1}.$$
But $\lim t_m = \lim t_{m+1} = s$, therefore $\lim s_n = s$.

■ **6. Changing the Order of Terms:** *If the terms of a convergent series of positive terms are rearranged, the series remains convergent and its sum is unaltered.*

Proof. Let Σu_n be convergent, and let the terms be rearranged in any manner. Denote the new series by Σv_n, so that every u is a v and every v is a u. Let
$$s_n = u_1 + u_2 + \dots + u_n,$$
$$t_n = v_1 + v_2 + \dots + v_n.$$

Suppose that the first n terms of Σu_n are among the first m terms of Σv_n. Further, suppose that the first m terms of Σv_n are among the first $(n + p)$ terms of Σu_n.

Thus $$n \le m \le n + p,$$
and m tends to infinity with n.

Also $$s_n \le t_m \le s_{n+p}.$$
Now $\lim s_n = \lim s_{n+p} = s$, therefore $\lim t_m = s$.

Hence Σv_n is convergent and has the same sum as Σu_n.

> **Note:** The argument fails for such a derangement as
> $$u_1 + u_3 + u_5 + \dots + u_2 + u_4 + u_6 + \dots ,$$
> where Σu_n is broken up into two (or any finite number of) infinite series.
>
> For here we cannot find m so that the first n terms of Σu_n occur among the first m terms of Σv_n. Thus u_2 does not occur till infinitely many of the terms u_1, u_3, u_5, ... have been placed.
>
> *The rule for the addition of series applies* in cases of this kind, but fails to apply when Σu_n is broken up into *infinitely many* infinite series.
>
> Derangements of the last sort are considered in another volume.

■ **7. Theorem:** *A series of positive terms cannot oscillate; and if s_n is always less than some fixed number k, the series converges and its sum is less than k.*

For s_n increases with n, and therefore tends to a finite limit or to ∞ (Ch. 15, 3). Hence the series cannot oscillate. Further, if $s_n < k$ for every n, then $s_n \to s$, where s is a fixed number less than k (Ch. 15, 3).

■ **8. Comparison Tests:**

(1) *If Σu_n and Σa_n are two series of positive terms and Σa_n is known to be convergent, then Σu_n will be convergent if*

(*i*) * $u_n \le a_n$ *for all values of n*;

(*ii*) * *or if u_n/a_n is always less than some fixed positive number k*;

(*iii*) *or if u_n/a_n tends to a finite limit.*

Proof. (*i*) If $a_1 + a_2 + \dots$ to $\infty = t$, we have
$$u_1 + u_2 + \dots + u_n \le a_1 + a_2 + \dots a_n < t \ ;$$
and since t is a fixed number, it follows that Σu_n is convergent.

(*ii*) For all values of n, we have $u_n < ka_n$. Now Σka_n is convergent, therefore Σu_n is convergent.

(*iii*) This follows at once from (*ii*), for if $\lim u_n/a_n$ is finite, we can find a positive constant k, such that $u_n/a_n < k$ for all values of n.

(2) *If Σu_n and Σa_n are two series of positive terms and Σa_n is known to be divergent, then Σu_n will be divergent if*

(*i*) * $u_n \geq a_n$ *for all values of* n;

(*ii*) * *or if u_n/a_n is always greater than some fixed positive number* k;

(*iii*) *or if u_n/a_n tends to a limit greater than zero.*

Proof. (*i*) Let N be a positive number, no matter how great. Since Σa_n is divergent, m can be found such that

$$a_1 + a_2 + ... + a_n > N \quad \text{provided that } n > m;$$

$$\therefore \qquad u_1 + u_2 + ... + u_n > N \text{ provided that } n > m,$$

hence Σu_n is divergent.

(*ii*) For all values of n, $u_n > ka_n$. Now Σa_n is divergent; therefore Σka_n is divergent, therefore Σu_n is divergent.

(*iii*) This follows at once from (*ii*), for if $\lim u_n/a_n$ is finite, a positive constant k can be found such that $u_n/a_n > k$ for all values of n.

In order to apply these tests, we require certain series which are known to be convergent or divergent. The geometric series and the series of the next article are those which are most generally useful as *test series*.

■ **9. Theorem:** *The series* $\dfrac{1}{1^p} + \dfrac{1}{2^p} + \dfrac{1}{3^p} + ... + \dfrac{1}{n^p} + ...$ *is convergent if $p > 1$ and divergent if $p \leq 1$.*

Denote the given series by S. Since S is a series of positive terms, its convergence or divergence is not affected by grouping the terms in brackets in any way we please.

First let $p > 1$. Group the terms as follows:

$$1 + \left(\frac{1}{2^p} + \frac{1}{3^p}\right) + \left(\frac{1}{4^p} + \frac{1}{5^p} + \frac{1}{6^p} + \frac{1}{7^p}\right) + ..., \qquad ...(A)$$

where the first terms in the brackets are $1/2^p, 1/2^{2p}, 1/2^{3p}$, etc., and the brackets contain 2, 4, 8, ... terms, respectively.

Each term in (A) after the first is less than the corresponding term in

$$1 + \left(\frac{1}{2^p} + \frac{1}{2^p}\right) \left(\frac{1}{4^p} + \frac{1}{4^p} + \frac{1}{4^p} + \frac{1}{4^p}\right) + ...,$$

which is the same as $1 + \dfrac{2}{2^p} + \dfrac{4}{4^p} + ...$, or $1 + \dfrac{1}{2^{p-1}} + \dfrac{1}{2^{2(p-1)}} + \qquad ...(B)$

Now (B) is a geometric series whose common ratio $1/2^{p-1}$ is less than unity, for $p > 1$. Hence (B) is convergent, and therefore S is convergent.

Next, if $p = 1$, the series S becomes $1 + \frac{1}{2} + \frac{1}{3} + \frac{1}{4} + \dots$. Group the terms thus:

$$1 + \frac{1}{2} + \left(\frac{1}{3} + \frac{1}{4}\right) + \left(\frac{1}{5} + \frac{1}{6} + \frac{1}{7} + \frac{1}{8}\right) + \dots , \qquad \dots(C)$$

where the last terms in the brackets are $1/2^2$, $1/2^3$, $1/2^4$, etc., and the brackets contain 2, 4, 8, ... terms, respectively.

Each term of (C) is greater than the corresponding term of

$$1 + \frac{1}{2} + \left(\frac{1}{4} + \frac{1}{4}\right) + \left(\frac{1}{8} + \frac{1}{8} + \frac{1}{8} + \frac{1}{8}\right) + \dots,$$

which is the same as $\quad 1 + \frac{1}{2} + \frac{2}{4} + \frac{4}{8} + \dots,\quad$ or $1 + \frac{1}{2} + \frac{1}{2} + \frac{1}{2} + \dots$. $\qquad \dots(D)$

Now (D) is divergent, and therefore S is divergent.

Lastly, if $p < 1$, each term of S after the first is greater than the corresponding term

of $1 + \frac{1}{2} + \frac{1}{3} + \frac{1}{4} + \dots$; but the last series is divergent, therefore S is divergent.

■ **10. Examples on the Comparison Tests:**

■ **EXAMPLE 1.** *Is the series,* $\dfrac{1}{1 \cdot 2 \cdot 3} + \dfrac{3}{2.3.4} + \dfrac{5}{3.4.5} + \dots$ *convergent or divergent?*

Here $u_n = \dfrac{2n - 1}{n(n + 1)(n + 2)}$; let $a_n = \dfrac{1}{n^2}$*. Then we have

$$\lim \frac{u_n}{a_n} = \lim \frac{n^2(2n - 1)}{n(n + 1)(n + 2)} = 2.$$

Now Σa_n is convergent; therefore Σu_n is convergent.

■ **EXAMPLE 2.** *Test for convergence or divergence the series*

$$\frac{x}{1+x} + \frac{x}{1+x^2} + \frac{x^2}{1+x^3} + \frac{x^{n-1}}{1+x^n} + \dots, \text{ where } x > 0.$$

If $x < 1$, $x^n \to 0$ and $u_n/x^{n-1} \to 1$. Now Σx^{n-1} is convergent, therefore Σu_n is convergent.

If $x > 1$, $x^{-n} \to 0$ and $u_n = \dfrac{1}{x} \cdot \dfrac{x^n}{1+x^n} = \dfrac{1}{x} \cdot \dfrac{1}{x^{-n}+1} \to \dfrac{1}{x}$. Hence Σu_n is divergent.

If $x = 1$, $u_n = \frac{1}{2}$, and the series is divergent.

---

* The nth term a_n of a suitable test series is found thus: Keeping only the highest powers of n in the numerator and denominator of u_n, we obtain $2n/n^3$ or $2/n^2$. Disregarding the numerical factor 2, we take $1/n^2$ for a_n.

TESTS DERIVED FROM THE GEOMETRIC SERIES

■ **11. D'Alembert's Test:** *A series Σu_n of positive terms is convergent if $u_{n+1}/u_n < k < 1$, where k is a fixed number and n may have any value.*

The series is divergent if $u_{n+1}/u_n \geq 1$ for all values of n.

For if $u_{n+1} / u_n < k < 1$ for every n, then

$$\frac{u_n}{u_1} = \frac{u_n}{u_{n-1}} \cdot \frac{u_{n-1}}{u_{n-2}} \cdots \frac{u_3}{u_2} \cdot \frac{u_2}{u_1} < k^{n-1}.$$

Thus $u_n < u_1 \cdot k^{n-1}$ and $\cdot \Sigma u_1 \cdot k^{n-1}$ is convergent, therefore Σu_n is convergent.

Again, if $u_{n+1}/u_n \geq 1$ for all values of n, then

$$u_n \geq u_{n-1} \geq u_{n-2} \ldots \geq u_1,$$

therefore $u_1 + u_2 + \ldots + u_n \geq nu_1 \to \infty$. Hence Σu_n is divergent.

In particular, *if u_{n+1} / u_n tends to a limit l, then Σu_n is convergent when $l < 1$ and divergent when $l > 1$.*

For if $l < 1$, choose k so that $l < k < 1$. Then by the definition of a limit, $u_{n+1} / u_n < k$ for $n > m$, where m is a fixed number. Hence

$$u_{m+1} + u_{m+2} + \ldots$$

is convergent, and therefore Σu_n is convergent.

Again, if $l > 1$, then m can be found so that $u_{n+1} > u_n$ for $n > m$. Hence $u_{m+1} + u_{m+2} + \ldots$ is divergent, and therefore Σu_n is divergent.

Notes: (*i*) It is sufficient that the conditions should hold *for all values of n greater than some fixed value.*

(*ii*) *Nothing is said as to the case in which $\lim u_{n+1}/u_n = 1$. In this case the series* may be convergent or it may be divergent, and we say that the *test fails.*

For example, consider the series

$$\frac{1}{1} + \frac{1}{2} + \frac{1}{3} + \frac{1}{4} + \ldots \quad \text{and} \quad \frac{1}{1^2} + \frac{1}{2^2} + \frac{1}{3^2} + \frac{1}{4^2} + \ldots .$$

The first is divergent, the second is convergent, and in both we have $\lim u_{n+1} / u_n = 1$. If, however, u_{n+1} / u_n tends to 1 as a limit *from above,* so that always $u_{n+1} > u_n$, the series is divergent. The series $1 + 2 + 3 + 4 + \ldots$ is an instance of this.

■ **12. Cauchy's Test:** *A series Σu_n of positive terms is convergent if $u_n^{\frac{1}{n}} < k < 1$ where k is a fixed number and n can have any value.*

The series is divergent if $u_n^{\frac{1}{n}} \geq 1$ for infinitely many values of n.

For if $u_n^{\frac{1}{n}} < k < 1$, then $u_n < k^n$, and since $k < 1$ the series Σk^n is convergent, and consequently Σu_n is convergent.

If $u_n^{\frac{1}{n}} \geq 1$, and consequently $u_n \geq 1$, for infinitely many values of n, then Σu_n is obviously divergent.

In particular, *if $u_n^{\frac{1}{n}}$ tends to a limit l, then Σu_n is convergent when $l < 1$ and divergent when $l > 1$.*

The proof is similar to that in the last article.

It is sufficient that the conditions should hold for $n > m$, where m is a fixed number. Also nothing is said as to the case in which $\lim u_n^{\frac{1}{n}} = 1$, but if the limit is approached *from above*, Σu_n is divergent.

■ **13. Comparing these tests,** that of D'Alembert is more generally *useful* than Cauchy's. For in most of the series with which we are concerned, u_{n+1}/u_n is a simpler function than u_n.

On the other hand, Cauchy's test is more *general* than D'Alembert's. That this is the case will be seen by noting that

(*i*) Cauchy's condition for divergence is wider than D'Alembert's. For in D'Alembert's test the condition has to be satisfied by *all* values of n greater than a fixed value, and this is not the case in Cauchy's.

(*ii*) It has been shown in Ch. 15, 9, (2), that if $u_{n+1}/u_n \to l$ then $u_n^{\frac{1}{n}} \to l$.

But if $u_n^{\frac{1}{n}} \to l$, it does *not* follow that u_{n+1}/u_n tends to any limit.

So we may expect that Cauchy's test will sometimes succeed when D'Alembert's fails, as in Ex. 3 below:

■ **EXAMPLE 1.** *Show that the series* $x + \dfrac{x^2}{2} + \dfrac{x^3}{3} + \dfrac{x^4}{4} + ...$ *is convergent if $0 < x < 1$ and divergent if $x \geq 1$.*

Here $u_n = \dfrac{x^n}{n}, \quad u_{n+1} = \dfrac{x^{n+1}}{n+1}; \quad \therefore \lim \dfrac{u_{n+1}}{u_n} = \lim \dfrac{n}{n+1} x = x.$

Hence, if $0 < x < 1$, the series is convergent, and if $x > 1$, it is divergent.

If $x = 1$, D'Alembert's test fails, but in this case the series becomes

$$1 + \tfrac{1}{2} + \tfrac{1}{3} + \tfrac{1}{4} + ...,$$

and is divergent.

■ **EXAMPLE 2.** *Show that the series $1 + x + x^2 / \lfloor 2 + x^3 / \lfloor 3 + ...$ is convergent for all positive values of x.*

Here $u_n = \dfrac{x^{n-1}}{\lfloor n-1}; \quad u_{n+1} = \dfrac{x^n}{\lfloor n}; \quad \therefore \lim \dfrac{u_{n+1}}{u_n} = \lim \dfrac{x}{n} = 0.$

Hence the series is convergent.

■ **EXAMPLE 3.** *Test for convergence the series* $a + b + a^2 + b^2 + a^3 + b^3 + \dots$.

Here $\dfrac{u_{2n}}{u_{2n-1}} = \left(\dfrac{b}{a}\right)^n$, $\dfrac{u_{2n+1}}{u_{2n}} = a.\left(\dfrac{a}{b}\right)^n$,

and D'Alembert's test is inapplicable. Using Cauchy's test, since $u_n^{\frac{1}{n}} \to a^{\frac{1}{2}}$ or $b^{\frac{1}{2}}$ according as n is odd or even, the series is convergent if $0 < a < 1$ and $0 < b < 1$, and divergent if $a \geq 1$, or if $b \geq 1$.

EXERCISE XXVI

1. State the character of the series

   (*i*) $\left(1 - \frac{1}{2}\right) + \left(1 - \frac{3}{4}\right) + \left(1 - \frac{7}{8}\right) + \dots,$ (*ii*) $1 - \frac{1}{2} + 1 - \frac{3}{4} + 1 - \frac{7}{8} + \dots,$

   showing that for the second series $\lim s_{2n} = 1$ and $\lim s_{2n+1} = 2$.

2. Show that the series $x + x^{-1} + x^2 + x^{-2} + x^3 + x^{-3} + \dots$ is never convergent.

3. Show that, if $a_1 + a_2 + a_3 + \dots$ is convergent, so also is

   $$\frac{1}{2}(a_1 + a_2) + \frac{1}{2}(a_2 + a_3) + \frac{1}{2}(a_3 + a_4) + \dots.$$

   Prove also that the converse is true if $a_n > 0$, for $n >$ some fixed number m.
   [Consider the sum to n terms of the second series.]

4. Criticise the following:

   $$2\left(\frac{1}{2} + \frac{1}{4} + \frac{1}{6} + \frac{1}{8} + \dots\right) = 1 + \frac{1}{2} + \frac{1}{3} + \frac{1}{4} + \dots = \left(1 + \frac{1}{3} + \frac{1}{5} + \dots\right)$$

   $$+ \left(\frac{1}{2} + \frac{1}{4} + \frac{1}{6} + \dots\right).$$

   $$\therefore \qquad \frac{1}{2} + \frac{1}{4} + \frac{1}{6} + \dots = 1 + \frac{1}{3} + \frac{1}{5} + \dots,$$

   which is absurd, for $\dfrac{1}{2} < 1$, $\dfrac{1}{4} < \dfrac{1}{3}$, and so on.

5. If n is a positive integer, show that

   (*i*) $\dfrac{1}{2} < \dfrac{1}{n+1} + \dfrac{1}{n+2} + \dfrac{1}{n+3} + \dots + \dfrac{1}{2n} < 1.$

   (*ii*) $\dfrac{1}{2}n < \dfrac{1}{2} + \dfrac{1}{3} + \dfrac{1}{4} + \dfrac{1}{5} + \dots + \dfrac{1}{2^p} < n.$

   (*iii*) Given that $2^{20} > 10^6 + 1$, show that the sum of the first million terms of the series $\frac{1}{2} + \frac{1}{3} + \frac{1}{4} + \frac{1}{5} + \dots$ is less than 20.

6. Show that

   (*i*) $\dfrac{1}{(m+1)(m+2)} + \dfrac{1}{(m+2)(m+3)} + \dfrac{1}{(m+3)(m+4)} + \dots$ to $\infty = \dfrac{1}{m+1}.$

(*ii*) The sum to infinity of the series $\dfrac{1}{(m+1)^2} + \dfrac{1}{(m+2)^2} + \dfrac{1}{(m+3)^2} + \ldots$

lies between $\dfrac{1}{m+1}$ and $\dfrac{1}{m+1} + \dfrac{1}{(m+1)^2}$.

(*iii*) Hence show that, in order to find the sum to infinity of the series

$\dfrac{1}{2^2} + \dfrac{1}{3^2} + \dfrac{1}{4^2} + \ldots$, with an error less than 0.001, at least 1000 terms must be taken.

Determine whether the series in Exx. 7–14 are convergent or divergent.

7. $1 + \dfrac{1}{3} + \dfrac{1}{5} + \dfrac{1}{7} + \ldots$

8. $1 + \dfrac{1}{\sqrt{2}} + \dfrac{1}{\sqrt{3}} + \dfrac{1}{\sqrt{4}} + \ldots$

9. $\dfrac{1}{a+b} + \dfrac{1}{a+2b} + \dfrac{1}{a+3b} + \ldots$

10. $\sqrt{\dfrac{1}{2^3}} + \sqrt{\dfrac{2}{3^3}} + \sqrt{\dfrac{3}{4^3}} + \ldots$

11. $\dfrac{1\cdot 2}{3^2\cdot 4^2} + \dfrac{3.4}{5^2\cdot 6^2} + \dfrac{5.6}{7^2\cdot 8^2} + \ldots$

12. $\dfrac{1}{\lfloor 1} + \dfrac{3}{\lfloor 2} + \dfrac{5}{\lfloor 3} + \dfrac{7}{\lfloor 4} + \ldots,$

13. $\dfrac{1}{1+2} + \dfrac{2}{1+2^2} + \dfrac{3}{1+2^3} + \ldots$

14. $\dfrac{1}{1+2^{-1}} + \dfrac{2}{1+2^{-2}} + \dfrac{3}{1+2^{-3}} + \ldots$

Test for convergence or divergence the series whose *n*th terms are

15. $\dfrac{\sqrt{n}}{n^2+1}$.

16. $\dfrac{n}{(a+nb)^2}$.

17. $\sqrt{\dfrac{2^n-1}{3^n-1}}$.

18. $\dfrac{n^p}{(n+1)^q}$.

19. $\dfrac{n^3+a}{2^n+a}$.

20. $\dfrac{n^3}{\lfloor n-1}$.

21. $\sqrt{n^2+1} - n$

22. $\sqrt{n^3+1} - \sqrt{n^3}$.

For what positive values of x are the series in Exs. 23-31 convergent, and for what values are they divergent?

23. $1 + 2x + 3x^2 + 4x^3 + \ldots$.

24. $1 + x/2^2 + x^2/3^2 + x^3/4^2 + \ldots$.

25. $\dfrac{1}{1\cdot 2\cdot 3} + \dfrac{x}{4\cdot 5\cdot 6} + \dfrac{x^2}{7\cdot 8\cdot 9} + \ldots$.

26. $\dfrac{x}{a+\sqrt{1}} + \dfrac{x^2}{a+\sqrt{2}} + \dfrac{x^3}{a+\sqrt{3}} + \ldots$

27. $\dfrac{1}{2^{a+x}} + \dfrac{2^a}{3^{a+x}} + \dfrac{3^a}{4^{a+x}} + \ldots$

28. $\dfrac{x}{x+1} + \dfrac{x^2}{x+2} + \dfrac{x^3}{x+3} + \ldots$

29. $\sum \dfrac{a^n}{a^n + x^n}$.

30. $\sum \dfrac{1}{x^n + x^{-n}}$.

31. $\sum \dfrac{\sqrt{n}}{\sqrt{n^n + 1}} \, x^n$.

32. Explain why D'Alembert's test cannot be stated thus: A series of positive terms is convergent if the ratio of each term to the preceding is less than unity.

Exs. 33–37 refer to series of positive terms.

33. If Σu_n is convergent and $\dfrac{v_{n+1}}{v_n} \leq \dfrac{u_{n+1}}{u_n}$ for $n \geq m$, then Σv_n is convergent.

34. If Σu_n is divergent and $\dfrac{v_{n+1}}{v_n} \geq \dfrac{u_{n+1}}{u_n}$, then Σv_n is divergent.

35. If Σu_n is convergent, so is $\Sigma u_n{}^2$.
[After a certain stage $u_n < 1$, for $u_n \to 0$; $\therefore u_n{}^2 < u_n$.]

36. If Σu_n is convergent, so is $\Sigma u_n/(1 + u_n)$.

37. If $\Sigma u_n{}^2$ is convergent, so is $\Sigma u_n/n$. [For $u_n/n \leq \tfrac{1}{2}(u_n{}^2 + 1/n^2)$.]

Find the sum to infinity of the series in Exs. 38-41.

38. $\dfrac{1}{1 \cdot 3 \cdot 5} + \dfrac{1}{3 \cdot 5 \cdot 7} + \dfrac{1}{5 \cdot 7 \cdot 9} + \dots$.

39. $\dfrac{1}{1 \cdot 4} + \dfrac{1}{2 \cdot 5} + \dfrac{1}{3 \cdot 6} \dots$.

40. $\dfrac{2}{1 \cdot 4 \cdot 5} + \dfrac{3}{2 \cdot 5 \cdot 6} + \dfrac{4}{3 \cdot 6 \cdot 7} + \dots$.

41. $\dfrac{1}{1 \cdot 3 \cdot 7} + \dfrac{2}{3 \cdot 5 \cdot 9} + \dfrac{3}{5 \cdot 7 \cdot 11} + \dots$.

TERMS ALTERNATELY POSITIVE AND NEGATIVE

■ **14. Theorem:** *The series $u_1 - u_2 + u_3 - \dots$, in which the terms are alternately positive and negative, is convergent if each term is numerically less than the preceding and $\lim u_n = 0$.*

Proof. We have

$$s_{2n} = (u_1 - u_2) + (u_3 - u_4) + \dots + (u_{2n-1} - u_{2n}), \qquad \dots(A)$$

and this can be written in the form

$$s_{2n} = u_1 - (u_2 - u_3) - (u_4 - u_5) - \dots - u_{2n}. \qquad \dots(B)$$

Now $u_1 - u_2, u_2 - u_3, u_3 - u_4, \dots$ are all positive, for

$$u_1 > u_2 > u_3 > \dots .$$

From (A) it therefore follows that s_{2n} is positive and increases with n, and from (B) it is seen that s_{2n} is always less than the fixed number u_1.

Hence, s_{2n} tends to a limit which is less than u_1 (15, 3).

Again, $s_{2n+1} = s_{2n} + u_{2n+1}$ and $\lim u_{2n+1} = 0$; therefore s_{2n+1} tends to the same limit as s_{2n}.

The series is therefore convergent, its sum is positive and less than u_1.

■ EXAMPLE : *Show that the series* $1 - \frac{1}{2} + \frac{1}{3} - \frac{1}{4} + ...$ *is convergent.*

Here the terms are alternately positive and negative, each term is numerically less than the preceding and $\lim 1/n = 0$; hence the series is convergent.

SERIES WITH TERMS EITHER POSITIVE OR NEGATIVE

■ 15. Absolutely Convergent Series: The series Σu_n, containing positive and negative terms, is said to be *absolutely convergent* if the series $\Sigma \, | \, u_n \, |$ is convergent.

If Σu_n is convergent and $\Sigma \, | \, u_n \, |$ divergent, then Σu_n is said to be *conditionally convergent* or *semi-convergent*.

Thus the series $1 - \frac{1}{2} + \frac{1}{4} - \frac{1}{8} + ...$ is absolutely convergent, because the series

$1 + \frac{1}{2} + \frac{1}{4} + \frac{1}{8} + ...$ is convergent. But, $1 - \frac{1}{2} + \frac{1}{3} - \frac{1}{4} + ...$ is conditionally convergent,

for it is convergent and $1 + \frac{1}{2} + \frac{1}{3} + \frac{1}{4} + ...$ is divergent.

Theorem: *An absolutely convergent series is convergent.*

Let Σu_n be absolutely convergent; then by definition $\Sigma \, | \, u_n \, |$ is convergent.

Now $u_n + | \, u_n \, | = 2u_n$ or 0, according as u_n is positive or negative.

Therefore every term of the series $\Sigma(u_n + | \, u_n \, |)$ is ≥ 0 and is $\leq$ the corresponding term of the convergent series $\Sigma 2 \, | \, u_n \, |$.

Hence $\Sigma(u_n + | \, u_n \, |)$ is convergent and consequently, by Art. 4, Σu_n is convergent.

Note: If the reader fails to see the point in the statement of this theorem, he should note that, if we say that Σu_n is absolutely convergent, we assert the convergence of *another* series, $\Sigma \, | \, u_n \, |$, and *not* that of Σu_n itself.

From Arts. 11 and 12, it follows that *the series Σu_n is absolutely convergent if, after a certain stage,*

$$\left| \frac{u_{n+1}}{u_n} \right| < k \text{ or if } | \, u_n \, |^{\frac{1}{n}} < k,$$

where k is a fixed positive number less than unity.

■ 16. General Condition for Convergence: Since s_n is a function of the positive integral variable n, by Ch. 15, 6, (B), *the necessary and sufficient condition for the convergence of Σu_n is as follows:*

Corresponding to any positive number ε that we may choose, however small, it must be possible to find m so that, if $n \geq m$, for every positive integer p,

$$| s_{n+p} - s_n | < \varepsilon,$$

that is, $| R_{n,p} | = | u_{n+1} + u_{n+2} + ... + u_{n+p}| < \varepsilon.$

Thus if Σu_n is convergent and p is any integer,

$$R_{n,p} = u_{n+1} + u_{n+2} + ... + u_{n+p} \to 0 \quad \text{as} \quad n \to \infty \qquad \qquad ...(A)$$

and in particular $u_n \to 0.$

But *the condition* (A) *is not sufficient to ensure the convergence of* Σu_n *unless p is allowed to tend to* ∞ *in any way whatever.*

■ **EXAMPLE :** *Apply the general condition to the series*

$$1 + \frac{1}{2} + \frac{1}{3} + ... + \frac{1}{n} + ...$$

Here $\qquad\qquad R_{n, p} = \dfrac{1}{n+1} + \dfrac{1}{n+2} + ... + \dfrac{1}{n+p} > \dfrac{p}{n+p}.$

For any *fixed* value of p, $p/(n + p) \to 0$ as $n \to \infty$. But if $p=n^2$, then $p/(n + p) \to$ 1; hence, $R_{n,p}$ cannot tend to zero, and so the series is not convergent.

■ **17. Pringsheim's Theorem:** *If* Σu_n *is a convergent series of positive decreasing terms, then* $nu_n \to 0.$

For if p is any positive integer

$$pu_{m+p} < u_{m+1} + u_{m+2} + ... + u_{m+p} \to 0 \text{ as } m \to \infty \,.$$

Let $p = m$, therefore $mu_{2m} \to 0$ and $2mu_{2m} \to 0$; *i.e.* $nu_n \to 0$, when $n = 2m$.

Let $p = m + 1$, therefore $(m + 1)\, u_{2m+1} \to 0$; and

$(2m + 1)\, u_{2m+1} = 2(m + 1)\, u_{2m+1} - u_{2m+1} \to 0$; *i.e.* $nu_n \to 0$, when $n = 2m +1$.

This proves the theorem, whether n is odd or even.

■ **EXAMPLE :** *Use this theorem to show that the series* $\Sigma \dfrac{1}{n}$ *is divergent.*

Here $nu_n = n \cdot \dfrac{1}{n} = 1$, and does not tend to zero. Hence $\Sigma \dfrac{1}{n}$ cannot be convergent; and, since it is a series of positive terms, it must be divergent.

■ **18. Introduction and Removal of Brackets:** Let the terms of Σu_n be arranged in groups without altering their order. Denote the sum of the terms of the nth group by v_n, then

(1) *If* Σu_n *is convergent, so also is* Σv_n, *and these series have the same sum.*

For let $\qquad s_n = u_1 + u_2 + ... + u_n,\ \ t_n = v_1 + v_2 + ... + v_n.$

Then for any value of n we can find m so that either

$$s_n = t_m \text{ or } s_n = t_m + (u_{m+1} + u_{m+2} + ... u_{m+p}),$$

where the terms in the bracket are included in the group $v_{m+1}.$

If Σu_n is convergent,

$$R_{n,p} = u_{n+1} + u_{n+2} + ... u_{n+p} \to 0 \text{ as } n \to \infty,$$

also $\qquad\qquad m \to \infty$, as $n \to \infty$, therefore $s_n - t_m \to 0.$

Hence Σv_n converges to the same sum as $\Sigma u_n.$

(2) *If Σv_n is convergent, Σu_n is not necessarily so.*

For, given that Σv_n is convergent, we cannot conclude that $R_{n,p} \to 0$ for all values of p.

(3) Denote $|u_n|$ by u_n', and suppose that the terms of $\Sigma u_n'$ are grouped as in § (1). Let v_n' be the sum of the terms of the nth group. Then, *if $\Sigma v_n'$ is convergent,* so also is u_n' (Art. 5); and, as $n \to \infty$,

$$|R_{n,p}| = |u_{n+1} + u_{n+2} + \dots + u_{n+p}| \le u'_{n+1} + u'_{n+2} + \dots + u'_{n+p} \to 0.$$

Therefore, as in § (1), *Σv_n converges to the same sum as Σu_n.*

▨ **19. Rearrangement of Terms:** The following example shows that a rearrangement of the terms of a *semi-convergent* series may alter its sum, and a rearrangement of the signs may change the series into one which is divergent.

▮ **EXAMPLE :** *Let s be the sum of the semi-convergent series*

$$1 - \frac{1}{2} + \frac{1}{3} - \frac{1}{4} + \dots. \qquad \qquad \dots\text{(A)}$$

With regard to the series

$$1 + \frac{1}{2} - \frac{1}{4} - \frac{1}{3} - \frac{1}{6} - \frac{1}{8} + \frac{1}{5} - \dots, \qquad \qquad \dots\text{(B)}$$

$$1 + \frac{1}{2} - \frac{1}{3} + \frac{1}{4} + \frac{1}{5} - \frac{1}{6} + \frac{1}{7} + \dots, \qquad \qquad \dots\text{(C)}$$

obtained by rearranging the terms and the signs of (A) *respectively, prove that* (B)

is convergent with a sum $\frac{1}{2}s$ and (C) *is divergent.*

(i) Let $s_n = 1 - \dfrac{1}{2} + \dfrac{1}{3} - \dfrac{1}{4} + \dots$to n terms, $\sigma_n = 1 - \dfrac{1}{2} - \dfrac{1}{4} + \dfrac{1}{3} - \dfrac{1}{6} - \dfrac{1}{8} + \dots$ to n terms;

then $\quad \sigma_{3n} = \left(1 - \dfrac{1}{2} - \dfrac{1}{4}\right) + \left(\dfrac{1}{3} - \dfrac{1}{6} - \dfrac{1}{8}\right) + \left(\dfrac{1}{5} - \dfrac{1}{10} - \dfrac{1}{12}\right) + \dots$ to n terms.

that is, $\quad \sigma_{3n} = \left(\dfrac{1}{2} - \dfrac{1}{4}\right) + \left(\dfrac{1}{6} - \dfrac{1}{8}\right) + \left(\dfrac{1}{10} - \dfrac{1}{12}\right) + \dots$ to n terms

$$= \frac{1}{2}\left(1 - \frac{1}{2} + \frac{1}{3} - \frac{1}{4} + \dots \text{ to } 2n \text{ terms}\right) = \frac{1}{2} s_{2n}.$$

$\therefore \lim \sigma_{3n} = \dfrac{1}{2} \lim s_{2n} = \dfrac{1}{2} s.$ $\qquad$ Also $\lim \sigma_{3n+1} = \lim \sigma_{3n+2} = \lim \sigma_{3n}.$

Hence, the second series converges and its sum is $\frac{1}{2} s$.

(ii) Let $\quad t_n = 1 + \dfrac{1}{2} - \dfrac{1}{3} + \dfrac{1}{4} + \dfrac{1}{5} - \dfrac{1}{6} + \dfrac{1}{7} + \dots$ to n terms,

then $\quad t_{3n} > \left(\dfrac{1}{3} + \dfrac{1}{3} - \dfrac{1}{3}\right) + \left(\dfrac{1}{6} + \dfrac{1}{6} - \dfrac{1}{6}\right) + \left(\dfrac{1}{9} + \dfrac{1}{9} - \dfrac{1}{9}\right) + \dots$ to n terms;

$$\therefore \qquad t_{3n} > \frac{1}{3}\left(1 + \frac{1}{2} + \frac{1}{3} + \dots + \frac{1}{n}\right).$$

Now the series $1 + \frac{1}{2} + \frac{1}{3} + \dots$ is divergent; therefore t_{3n} and consequently t_{3n+1} and t_{3n+2} all tend to ∞. Hence the series (C) is divergent.

Theorem: *If the terms of an absolutely convergent series are rearranged, the series remains convergent and its sum is unaltered.*

Denote the new series by Σv_n, so that every v is a u and every u is a v.

We have $u_n + |u_n| = 2u_n$ or 0, according as u_n is positive or negative. And since $\Sigma |u_n|$ is a convergent series of positive terms, so also is $\Sigma (u_n + |u_n|)$, for its terms are $\leq$ the corresponding terms of $\Sigma 2 |u_n|$.

Let $\Sigma |u_n| = s$ and $\Sigma (u_n + |u_n|) = s'$, then $\Sigma u_n = s' - s$, and since $\Sigma |u_n|$ and $\Sigma (u_n + |u_n|)$ are series of positive terms, their sums are unaltered by any rearrangement of terms (Art. 6), therefore

$$\Sigma |v_n| = s, \quad \Sigma(v_n + |v_n|) = s', \quad \text{and} \quad \Sigma v_n = s' - s = \Sigma u_n.$$

■ **EXAMPLE :** *Find a range of values of x for which the series*

$$1 + (x - x^2) + (x - x^2)^2 + (x - x^2)^3 + \dots \qquad \dots(A)$$

can be arranged as a convergent series of ascending powers of x.

Expanding the terms of (A) and removing brackets, we have the series

$$1 + x - x^2 + x^2 - 2x^3 + x^4 + x^3 - 3x^4 + \dots \qquad \dots(B)$$

Now, if $|x - x^2| < 1$, the series (A) is convergent; but this condition *does not justify* either the removal of brackets or a derangement of the terms of (B).

However, by Arts. 15, 18(1), 19, each of these steps are permissible *if* (B) *is absolutely convergent*; and, by Art. 5, this will be the case if

$$1 + (x' + x'^2) + (x' + x'^2)^2 + \dots,$$

where $x' = |x|$, is convergent; that is, if $x' + x'^2 < 1$, giving

$$x' < \tfrac{1}{2}\left(\sqrt{5} - 1\right), \text{ or } -\tfrac{1}{2}\left(\sqrt{5} - 1\right) < x < +\tfrac{1}{2}\left(\sqrt{5} - 1\right);$$

which is the required range.

> **Note:** What has been said does not preclude the possibility that the derangement can be made over a *wider range.*

■ **20. Approximations:** In searching for the sum of an infinite series, we generally have to be content with an approximate value. Consider the series

$$u_1 + u_2 + u_3 \dots + u_n + \dots, \qquad \dots(A)$$
$$u_{n+1} + u_{n+2} + u_{n+3} + \dots \qquad \dots(B)$$

If (A) is convergent, so also is (B), and its sum, denoted by R_n, is called the *remainder after n terms* of the series (A). Thus we have

$$s = s_n + R_n,$$

and R_n is the error in taking s_n as an approximate value of s. It should be noticed that $\lim R_n = 0$, for $\lim s_n = s$.

Limit of Error. To find an upper limit to the numerical value of the error in representing s by s_n is to find a number which $|R_n|$ cannot exceed. First consider a convergent series

$$u_1 - u_2 + u_3 - u_4 + \ldots + (-1)^n u_n + \ldots ,$$

with alternately positive and negative terms. Here

$$R_n = (-1)^{n+1} \{u_{n+1} - u_{n+2} + u_{n+3} - \ldots\}.$$

Now the sum of the series in brackets is positive and less than u_{n+1} (Art. 14); therefore $|R_n| < u_{n+1}$. Hence *the error in taking s_n as an approximate value of s is numerically less than u_{n+1}.*

For example, if $0 < x < 1$, the error in taking $x - x^2/2$ as the sum of the series $x - x^2/2 + x^3/3 - x^4/4 + \ldots$ is less than $x^3/3$.

Next, take the case of a series in which all the terms are positive or in which, *after a certain stage,* the terms are all of the same sign. This is more difficult to deal with; we can often proceed as follows.

■ **Example :** *Consider the series* $x + \dfrac{x^2}{2} + \dfrac{x^3}{3} + \ldots + \dfrac{x^n}{n} + \ldots$ *where $0 < x < 1$. The series is convergent, and*

$$R_n = \frac{x^{n+1}}{n+1} + \frac{x^{n+2}}{n+2} + \ldots < \frac{x^{n+1}}{n+1}(1 + x + x^2 + \ldots) < \frac{x^{n+1}}{(n+1)(1-x)}.$$

Thus the error in taking s_n as the value of s is numerically less than $x^{n+1} / (n+1)(1-x)$.

■ **21. Rapidity of Convergence or Divergence:** For a convergent series, the smaller the value of R_n for a given value of n, *the more rapidly is the series said to converge.*

For example, the smaller the value of r, the more rapid is the convergence of the geometric series $a + ar + ar^2 + \ldots$.

Again, the series $1/2^2 + 1/3^2 + 1/4^2 + \ldots$ is convergent, but in order to find its sum to three places of decimals, at least 1000 terms must be taken [Ex. XXVI, 6, (*iii*)]. We therefore say that the series *converges slowly.*

Next, consider the series $1 + \frac{1}{2} + \frac{1}{3} + \frac{1}{4} + \ldots$; this is divergent, but the sum of a million terms is less than 20 [see Ex. XXVI, 5, (*iii*)]. The series is therefore said to *diverge slowly. In practice,* it is very important to arrange the reckoning so that we have to deal with rapidly converging series only.

For instance, the series, $\quad s = 1 - \frac{1}{2} + \frac{1}{3} - \frac{1}{4} + \ldots;\quad$ converges very slowly.

By grouping the terms, we find that

$$s_n = \left(1 - \frac{1}{2}\right) + \left(\frac{1}{3} - \frac{1}{4}\right) + \ldots = \frac{1}{2} + \frac{1}{12} + \ldots + \frac{1}{(2n-1)\,2n}.$$

This series converges more rapidly: further, in Ch. 19, 19, Ex. 1, we find that $s = 2\left[\frac{1}{3} + \frac{1}{3}\left(\frac{1}{3}\right)^3 + \frac{1}{5}\left(\frac{1}{3}\right)^5 + \ldots\right]$; and this series converges more rapidly than a geometrical series whose common ratio is $1/9$.

■ **22. Series of Complex Terms: (1)** Let $z_n = x_n + \iota y_n$, and suppose that the series Σx_n and Σy_n converge to sums σ and τ, respectively. Let s_n, σ_n, τ_n be the sums to n terms of Σz_n, Σx_n, Σy_n respectively; then, as in Ch. 15, 10,

$$\lim s_n = \lim (\sigma_n + \iota \tau_n) = \sigma + \iota \tau,$$

and we say that Σz_n converges to the sum $\sigma + \iota \tau$.

Thus Σz_n *is convergent if* Σx_n *and* Σy_n *are convergent.*

(2) By Ch. 15, 11, *the necessary and sufficient condition for the convergence of* Σz_n *is as follows: Corresponding to any positive number* ε, *however small, it must be possible to find m so that, if* $n \geq m$, *for every positive integer p,*

$$| s_{n+p} - s_n | < \varepsilon, \quad \text{that is to say,} \quad | R_{n,p} | = | z_{n+1} + z_{n+2} + \ldots + z_{n+p} | < \varepsilon.$$

(3) Definition of Absolute Convergence: If Σz_n is a series of complex terms and $\Sigma | z_n |$ is convergent, then Σz_n is said to be *absolutely convergent.*

As for real series, *an absolutely convergent series of complex terms is convergent.*

For $| z_n | = \sqrt{(x_n{}^2 + y_n{}^2)}$, therefore $| x_n | \leq | z_n |$ and $| y_n | \leq | z_n |$. If then $\Sigma | z_n |$ is convergent, so are $\Sigma | x_n |$ and $\Sigma | y_n |$.

Consequently Σx_n and Σy_n are convergent, and so is Σz_n.

(4) It is important to notice that *if* Σz_n *is absolutely convergent, so are* Σx_n *and* Σy_n, *and conversely.*

The first part of this has just been proved. For the second part, we have

$$| z_n | = \sqrt{(x_n{}^2 + y_n{}^2)} \leq | x_n | + | y_n |.$$

Hence if $\Sigma | x_n |$ and $\Sigma | y_n |$ are convergent, so is $\Sigma | z_n |$.

(5) The theorems of Arts. 18 and 19, about the introduction and removal of brackets and the derangement of terms, hold for absolutely convergent complex series. This follows immediately from (3) and (4).

■ **23. The Geometric Series:** *If z is complex, the series*

$$1 + z + z^2 + \ldots + z^{n-1} + \ldots$$

is convergent if and only if $| z | < 1$, *and when convergent, its sum is* $1/(1 - z)$.

For $\qquad s_n = 1 + z + z^2 + \ldots + z^{n-1} = (1 - z^n) / (1 - z).$

If $z = 1$, the series is divergent; and if $| z | > 1$, z^n does not converge to a limit. If $| z | < 1$, then, by Ch. 15, 10, (3), $z^n \to 0$ and $s_n \to 1/ (1 - z)$, so that the series is convergent and its sum is $1/(1 - z)$.

Putting $z = r (\cos \theta + \iota \sin \theta)$ where $0 < r < 1$, we have

$$1 + r (\cos \theta + \iota \sin \theta) + r^2 (\cos 2\theta + \iota \sin 2\theta) + \ldots \text{ to } \infty$$

$$= \frac{1}{1 - r (\cos \theta + \iota \sin \theta)} = \frac{1 - r \cos \theta + \iota r \sin \theta}{1 - 2r \cos \theta + r^2}.$$

Equating real and imaginary parts,

$$1 + r \cos \theta + r^2 \cos 2\theta + \ldots + r^n \cos n\theta + \ldots = \frac{1 - r \cos \theta}{1 - 2r \cos \theta + r^2},$$

$$r \sin \theta + r^2 \sin 2\theta + \dots + r^n \sin n\theta + \dots = \frac{r \sin \theta}{1 - 2r \cos \theta + r^2}.$$

Changing θ into $\pi + \theta$, it will be seen that these results are also true when $-1 < r < 0$.   Thus they hold for $-1 < r < 1$.

EXAMPLE : *Show on an Argand diagram the points representing the values of s_1, s_2, ... s_6 for the series*

$$1 + z + z^2 + \dots \text{ when } z = \frac{1}{2}$$

$$\left(\cos \tfrac{\pi}{3} + \iota \sin \tfrac{\pi}{3} \right).$$

It will be seen that the point s_n rapidly approaches a limiting position near s_6.

Here $|z| = \tfrac{1}{2} < 1$, and the series is absolutely convergent.

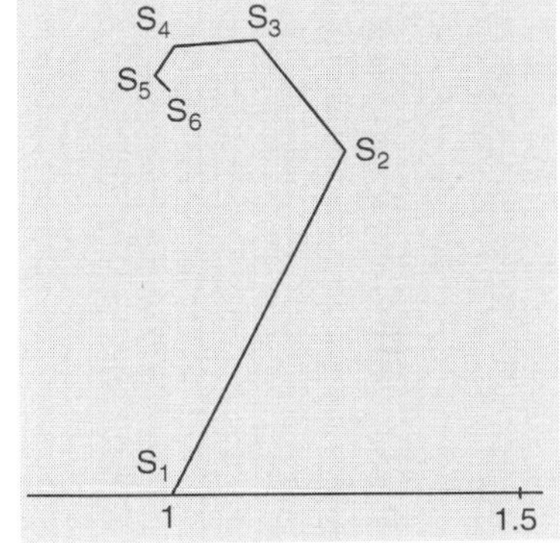

Fig. 43

24. We shall frequently require the following theorem:

If $\dfrac{u_n}{u_{n+1}} = 1 + \dfrac{a_n}{n}$ *where* a_n *tends to a positive number* a, *then* $u_n \to 0$.

For since $a > 0$, we can choose k so that $a > k > 0$, and then we can find m so that $a_n > k$ for $n \geq m$. Therefore, it follows that

$$\frac{u_m}{u_n} = \frac{u_m}{u_{m+1}} \cdot \frac{u_{m+1}}{u_{m+2}} \dots \frac{u_{n-1}}{u_n} > \left(1 + \frac{k}{m}\right)\left(1 + \frac{k}{m+1}\right) \dots \left(1 + \frac{k}{n-1}\right)$$

$$> 1 + k\left(\frac{1}{m} + \frac{1}{m+1} + \dots + \frac{1}{n-1}\right).$$

Now the series $\dfrac{1}{m} + \dfrac{1}{m+1} + \dfrac{1}{m+2} + \dots$ is divergent, therefore

$$u_m/u_n \to \infty \text{ and } u_n \to 0.$$

25. The Binomial Series: For any given value of n, the series

$$1 + nx + \frac{n(n-1)}{\underline{\,2\,}} x^2 + \dots + \frac{n(n-1)(n-2)\dots(n-r+1)}{\underline{\,r\,}} x^r + \dots$$

is called the *Binomial Series*. When n is a positive integer, the series terminates and its sum is $(1 + x)^n$. In all other cases, it is an infinite series, which we denote by

$$u_0 + u_1 + u_2 + \dots + u_r + \dots .$$

(1) Ultimate Signs of the Terms. *From the beginning, or after a certain stage, the terms are alternately positive and negative, or they all have the same sign according as $x \gtrless 0$.*

For $$u_{r+1}/u_r = (n - r) x/(r + 1).$$

Hence if m is the least of the numbers 0, 1, 2,... which is greater than n, the terms u_m, u_{m+1}, u_{m+2}, ... are alternately positive and negative, or all have the same sign according as $x > 0$.

(2) Convergence, etc. We shall prove that *the series is absolutely convergent if* $|x| < 1$; *and when* $x = 1$, *the series converges if* $n > -1$, *oscillates finitely if* $n = -1$ *and oscillates infinitely if* $n < -1$.

Proof. Denote the series by $u_0 + u_1 + u_2 + \dots$, then

$$\frac{u_{r+1}}{u_r} = \frac{n-r}{r+1} x \to -x, \text{ as } r \to \infty \text{ ; therefore } \lim_{x \to \infty} \left| \frac{u_{r+1}}{u_r} \right| = |x|,$$

and the series is absolutely convergent if $|x| < 1$.

Next let $x = 1$. If $r > n$, u_{r+1} and u_r have opposite signs, so that, after a certain stage, the terms are alternately positive and negative. Also

$$\left| \frac{u_r}{u_{r+1}} \right| = \frac{r+1}{r-n} = 1 + \frac{a_r}{r} \text{ where } a_r = (n+1)/\left(1 - \frac{n}{r}\right).$$

(i) If $n > -1$, as $r \to \infty$, $a \to n + 1 > 0$. Hence, by Art. 24, $u_r \to 0$.

Also $|u_{r+1}| < |u_r|$. Hence, by Art. 14, the series is convergent.

(ii) If $n = -1$ the series is $1 - 1 + 1 - 1 + \dots$ and oscillates finitely.

(iii) If $n < -1$, let $n + 1 = -m$ so that $m > 0$, then

$$|u_r| = (1 + m)\left(1 + \frac{m}{2}\right) \dots \left(1 + \frac{m}{r}\right) > 1 + m\left(1 + \frac{1}{2} + \dots + \frac{1}{r}\right).$$

Hence $|u_r| \to \infty$ and the series oscillates infinitely.

The case in which $x = -1$ is considered in Ch. 20, 6.

EXERCISE XXVII

RANGE OF CONVERGENCE

Show that the series in Exs. 1–5 are convergent.

1. $1 - \dfrac{1}{3} + \dfrac{1}{5} - \dfrac{1}{7} + \dots$

2. $1 - \dfrac{1}{2^2} + \dfrac{1}{3^2} - \dfrac{1}{4^2} + \dots$

3. $\dfrac{1}{a+1} - \dfrac{1}{a+2} + \dfrac{1}{a+3} - \dfrac{1}{a+4} + \dots$

4. $1 - \dfrac{1}{2} + \dfrac{1 \cdot 3}{2 \cdot 4} - \dfrac{1 \cdot 3 \cdot 5}{2 \cdot 4 \cdot 6} + \dots$

5. $1 + \dfrac{1}{2^2} - \dfrac{1}{3^2} + \dfrac{1}{4^2} + \dfrac{1}{5^2} - \dfrac{1}{6^2} + \dots$

6. Prove that the series $\dfrac{x}{1+x} - \dfrac{x^2}{1+x^2} + \dfrac{x^3}{1+x^3} - \dots$ is convergent if $0 < x < 1$.

7. Show that the series $1 + x + x^2/\lfloor 2 + x^3/\lfloor 3 + \ldots$ is convergent for all values of x, real or complex.

8. Show that $\dfrac{1}{a} + \dfrac{1}{a+1} - \dfrac{1}{a+2} + \dfrac{1}{a+3} + \dfrac{1}{a+4} - \dfrac{1}{a+5} + \ldots$ is a divergent series.

9. Show that the series

$$\frac{\mu}{2} + \frac{3}{\lfloor 2}\left(\frac{\mu}{2}\right)^2 + \frac{4.6}{\lfloor 3}\left(\frac{\mu}{2}\right)^3 + \ldots + \frac{(n+1)(n+3)(n+5)\ldots(3n-3)}{\lfloor n}\left(\frac{\mu}{2}\right)^n + \ldots$$

is convergent if $\mu^2 < 4/27$.

[Consider separately the sum of the odd and the sum of the even terms.]

10. (*i*) Find a range of values of x for which the series

$$1 + \left(\tfrac{5}{2}x - x^2\right) + \left(\tfrac{5}{2}x - x^2\right)^2 + \ldots$$

is convergent.

(*ii*) Find a range for which the series can be arranged as a convergent series of ascending powers of x.

11. If $0 < x \le 0.4$, show that the series
$$1 + (2x \cos\theta - x^2) + (2x \cos\theta - x^2)^2 + \ldots$$
can be arranged as a convergent series of ascending powers of x.

[It is enough to show that with the given condition $|\, 2x \cos\theta \,| + x^2 < 1$.]

12. Let the terms of Σu_n be arranged in groups, without altering their order, and denote the sum of the terms of the nth group by v_n.

Given that Σv_n is convergent, prove that Σu_n converges to the same sum as Σv_n in the following cases:

(*i*) If the number of terms in every bracket is finite and $u_n \to 0$.

(*ii*) If all the terms in each bracket have the same sign.

[See Art. 18. In case (*i*), $R_{n,p} \to 0$. In case (*ii*), $v_n \to 0$; hence, every term in v_n tends to zero, and case (*i*) applies.]

13. Show that the series $1 + z + z^2 + \ldots$ is absolutely convergent when $z = 1/(1 + \iota)$ and find its sum.

14. If $z_n = \dfrac{(-1)^n}{n} + \dfrac{1}{n^2}\iota$, show that Σz_n is convergent, but not absolutely convergent.

Prove the results in Exs. 15–17, where R_n is the remainder after n terms in the given series.

15. For the series $1 + \dfrac{1}{\lfloor 1} + \dfrac{1}{\lfloor 2} + \dfrac{1}{\lfloor 3} + \ldots,$ $R_n < \dfrac{n+1}{n \lfloor n}.$

16. For the series $1 - \dfrac{x}{\lfloor 1} + \dfrac{x^2}{\lfloor 2} - \dfrac{x^3}{\lfloor 3} + \ldots,$ $|\,R_n\,| < \dfrac{x^n}{\lfloor n}$ provided that $0 < x < 1$.

17. For the series $1 + \dfrac{x}{\lfloor 1} + \dfrac{x^2}{\lfloor 2} + \dfrac{x^3}{\lfloor 3} + \ldots,$ $R_n < \dfrac{x^n}{\lfloor n} \cdot \dfrac{n+1}{n+1-x}$ provided that $0 < x < n + 1$.

Prove the results stated in Exs. 18–23:

18. $\sum_{1}^{\infty} \dfrac{1}{4n^2 - 1} = \dfrac{1}{2}$.

19. $\sum_{1}^{\infty} \dfrac{n-1}{\lfloor n} = 1$

20. If $|x| > 1$, $\dfrac{1}{1+x} + \dfrac{2}{1+x^2} + \dfrac{4}{1-x^4} + \ldots$ to $\infty = \dfrac{1}{x-1}$.

21. $\dfrac{x}{1-x^2} + \dfrac{x^2}{1-x^4} + \dfrac{x^4}{1-x^8} + \ldots$ to $\infty = \dfrac{x}{1-x}$ or $\dfrac{1}{1-x}$, according as $|x| \lessgtr 1$.

22. If $b - 1 > a > 0$, then

$$1 + \frac{a}{b} + \frac{a(a+1)}{b(b+1)} + \frac{a(a+1)(a+2)}{b(b+1)(b+2)} + \ldots \text{ to } \infty = \frac{b-1}{b-a-1}.$$

23. If $b - 2 > a > 0$, then

$$\frac{a}{b} + 2\frac{a(a+1)}{b(b+1)} + 3\frac{a(a+1)(a+2)}{b(b+1)(b+2)} + \ldots \text{ to } \infty = \frac{a(b-1)}{(b-a-1)(b-a-2)}.$$

24. Show that the sum to n terms of the series

$$\cos \theta + \cos 2\theta + \cos 3\theta + \ldots$$

is

$$\tfrac{1}{2} \{\sin (n + \tfrac{1}{2} \theta) \operatorname{cosec} \tfrac{1}{2} \theta - 1\}.$$

Hence show that the series oscillates finitely except when θ is an even multiple of π, in which case the series diverges.

$$[2 \cos r\theta \sin \frac{\theta}{2} = \sin (r + \tfrac{1}{2} \theta) - \sin (r - \tfrac{1}{2} \theta).]$$

25. Show that the series $\Sigma \sin n\theta$ oscillates finitely except when θ is a multiple of π, in which case it converges to zero.

26. Show that the series $z + \tfrac{1}{2} z^2 + \tfrac{1}{3} z^3 + \ldots$ is convergent, but not absolutely convergent,

when $z = \cos \dfrac{\pi}{4} + \iota \sin \dfrac{\pi}{4}$. Illustrate the slowness of the convergence plotting the

points $s_1, s_2, \ldots s_7$ on an Argand diagram.

■ ■ ■

CHAPTER 17

Continuous Variable

■ **1. Definitions:** We say that *x varies continuously from a to b* when it passes *once* through all real values between *a* and *b*. Under these circumstances *x* is called a continuous real variable.

Limit of f(x). Consider the function $f(x) = \{(1 + x)^3 - 1\}/x$.

When $x = 0$, $f(x)$ has no definite value, for it then takes the form $0/0$. In other words the function $f(x)$ is *undefined* for the value zero of *x*. For every value of *x* except zero, we have $f(x) = (3x + 3x^2 + x^3)/x = 3 + 3x + x^2$.

Thus, *if x is nearly equal to zero, f(x) is nearly equal to 3: further, by making x small enough, we can make the difference between f(x) and 3 as small as we please.* This is *roughly* what is meant by saying that 3 *is the limit of f(x) as x tends to zero.*

The matter may be put precisely as follows: Choose any positive number ε, no matter how small. If $|x| < 1$, then

$$|f(x) - 3| = |3x + x^2| < 4|x|,$$

therefore $\qquad |f(x) - 3| < \varepsilon$ provided that $|x| < \dfrac{1}{4}\varepsilon$.

Thus, *as x tends to zero, $|f(x) - 3|$ becomes and remains less than any positive number ε that we may choose, no matter how small.* This is *precisely* what is meant by saying that 3 is the limit of $f(x)$ as *x* tends to zero.

It should be observed that $f(x)$ never *attains* the value 3. The usual definitions are as follows:

(1) *If, corresponding to any positive number ε which we may choose, no matter how small, there exists a number η such that*

$$|f(x) - l| < \varepsilon \text{ provided only that } |x - a| < \eta,$$

where l and a are fixed numbers, then l is called the limit of f(x) as x tends to a.

This is expressed by writing $\lim\limits_{x \to a} f(x) = l$.

Or we may say that $f(x) \to l$ as $x \to a$, where it is understood *that x may approach a from either side, i.e. x may tend to a + 0 or a - 0.*

Observe that nothing is said as to the value of $f(x)$ when $x = a$. Such a value may or may not exist. In any case, we are not concerned with it.

(2) *If, corresponding to any positive number ε, no matter how small, there exists a positive number* **N** *such that*

$$|f(x) - l| < \varepsilon, \quad provided\ only\ that\ \ x > N,$$

where l is a fixed number, then l is called the limit of f (x) as x tends to infinity.

This is expressed by writing $\lim\limits_{x \to \infty} f(x) = l$ or $f(x) \to l$ as $x \to \infty$.

(3) *If to any positive number M, no matter how great, there corresponds a positive number N such that*

$$f(x) > M,\ provided\ only\ that\ \ x > N,$$

we say that f (x) tends to infinity with x, and we write $f(x) \to \infty$ as $x \to \infty$ or (less correctly) $\lim\limits_{x \to \infty} f(x) = \infty$.

The reader should be able to frame for himself corresponding definitions for the meaning of '$f(x)$ tends to minus infinity as $x \to \infty$.'

■ **2. Fundamental Theorems:** Let u, v be functions of x which tend to the limits l, l', as x tends to a or as x tends to infinity, then

    (*i*) $\lim (u + v) = l + l'$,   (*ii*) $\lim (u - v) = l - l'$,   (*iii*) $\lim uv = ll'$,

  (*iv*) $\lim 1/u = 1/l$ unless $l = 0$,   (*v*) $\lim u/v = l/l'$ unless $l' = 0$. .

Apart from some verbal changes which the reader can easily make for himself, the proofs are the same as those in Ch. 15, 2.

■ **3. Two Theorems on Polynomials:** What is said in this article holds good, no matter whether the variable x and the coefficients are real or complex?

Let $f(x) = a_0 + a_1 x + a_2 x^2 + \ldots + a_n x^n$, then

(1) *If ε is any positive number, no matter how small, we can find a positive number η so that*

$$|f(x) - a_0| < \varepsilon\ provided\ only\ that\ |x| < \eta.$$

Proof. For shortness, let $|x| = x'$ and let k be the greatest of the numbers

$$|a_1|,\ \ |a_2|, \ldots, |a_n|\ ;$$

then $|f(x) - a_0| \le |a_1| x' + |a_2| x'^2 + \ldots + |a_n| x'^n$

$$< kx'(1 + x' + x'^2 + \ldots + x'^n) < kx'/(1 - x').$$

If $x' < 1$, then $kx'/(1 - x') < \varepsilon$, provided that

$$kx' < \varepsilon(1 - x'),\ that\ is\ if\ \ x' < \varepsilon/(k + \varepsilon).$$

Therefore $|f(x) - a_0| < \varepsilon,\ if\ |x| < \varepsilon/(k + \varepsilon)$;

and so $\varepsilon/(k + \varepsilon)$ is a suitable value of η.

(2) *If M is a positive number, no matter how great, we can find a positive number m so that*

$$|f(x)| > M,\ provided\ only\ that\ |x| \ge m.$$

The letters having the same meaning as in (1), we have

$$|f(x)| = x'^n.\ \left| a_n + a_{n-1} \cdot \frac{1}{x} + a_{n-2} \cdot \frac{1}{x^2} + \ldots + a_0 \cdot \frac{1}{x^n} \right|$$

$$\geq x'^n \cdot \left\{ \mid a_n \mid - \mid a_{n-1} \cdot \frac{1}{x} + a_{n-2} \cdot \frac{1}{x^2} + \ldots + a_0 \cdot \frac{1}{x^n} \mid \right\}.$$

Choose any number ε between 0 and $\mid a_n \mid$. Then by the last theorem

$$\mid a_{n-1} \cdot \frac{1}{x} + a_{n-2} \cdot \frac{1}{x^2} + \ldots + a_0 \cdot \frac{1}{x^n} \mid < \varepsilon$$

if $\dfrac{1}{x'} < \dfrac{\varepsilon}{k+\varepsilon}$, that is if $x' > \dfrac{k+\varepsilon}{\varepsilon}$.

For such values of x we have $\quad \mid f(x) \mid > x'^n \{ \mid a_n \mid - \varepsilon \}.$

In particular, put $\varepsilon = \dfrac{1}{2} \mid a_n \mid$, then

$$\mid f(x) \mid > \frac{1}{2} \mid a_n \mid x'^n \quad \text{if} \quad x' > \frac{2k+\mid a_n \mid}{\mid a_n \mid}.$$

Hence if m is the greater of the two numbers

$$\sqrt[n]{\frac{2M}{\mid a_n \mid}} \quad \text{and} \quad \frac{2k+\mid a_n \mid}{\mid a_n \mid},$$

then $$f(x) > M \text{ if } \mid x \mid > m.$$

From Theorem (1) it follows that $\lim\limits_{x \to 0} f(x) = a_0.$*

Again, if $x = y + a$, then $y \to 0$ as $x \to a$ and $f(a)$ is the term independent of y in the polynomial $f(y + a)$. Hence

$$\lim_{x \to a} f(x) = \lim_{y \to 0} f(y + a) = f(a).$$

■ **EXAMPLE :** *If $f(x) = \dfrac{2x^2 + x - 3}{3x^2 + 2x - 5}$, find lim f(x); (i) as $x \to 0$, (ii) as $x \to \infty$,*

(iii) as $x \to 1$.

(i) $\lim\limits_{x \to 0} f(x) = \dfrac{\lim (2x^2 + x - 3)}{\lim (3x^2 + 2x - 5)} = \dfrac{-3}{-5} = \dfrac{3}{5}.$

(ii) Let $x = 1/y$, then $y \to 0$ as $x \to \infty$ and

$$\lim_{x \to \infty} f(x) = \lim_{y \to 0} \frac{3y^2 - y - 2}{5y^2 - 2y - 3} = \frac{\lim (3y^2 - y - 2)}{\lim(5y^2 - 2y - 3)} = \frac{2}{3}.$$

(iii) Let $x = 1 + y$, then $y \to 0$ as $x \to 1$ and

$$\lim_{x \to 1} f(x) = \lim_{y \to 0} \frac{2y^2 + 5y}{3y^2 + 8y} = \lim_{y \to 0} \frac{2y + 5}{3y + 8} = \frac{5}{8}.$$

■ **4. Continuous and Discontinuous Functions:** We confine ourselves to one-valued functions of the real variable x.

---

* The reader must understand that this is *not* a roundabout way of saying that $f(x) = a_0$, when $x = 0$.

We may say that $f(x)$ is a continuous function of x in the interval (a, b) if the curve whose equation is $y = f(x)$ is continuous between the points where $x = a$ and $x = b$.

This supposes that we know what is meant by a 'continuous curve.' A representation of the curve in Fig. 44 can be drawn without allowing the pencil to leave the paper, and we say that it is *continuous*. In Fig. 45, the curve cannot be so drawn near the points where $x = x_1$ and $x = x_2$, and we say that the curve is *discontinuous* at these points.

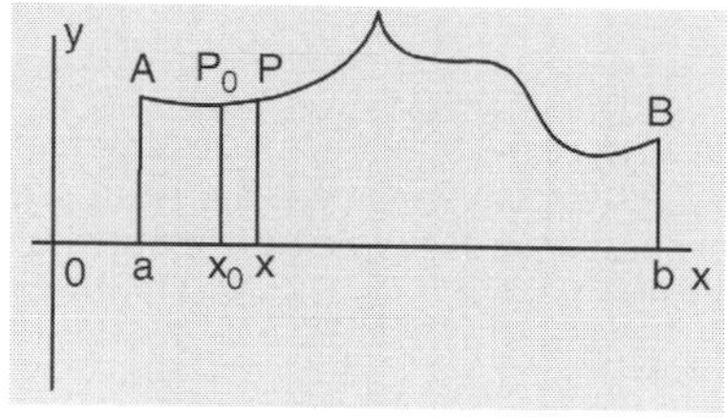

Fig. 44

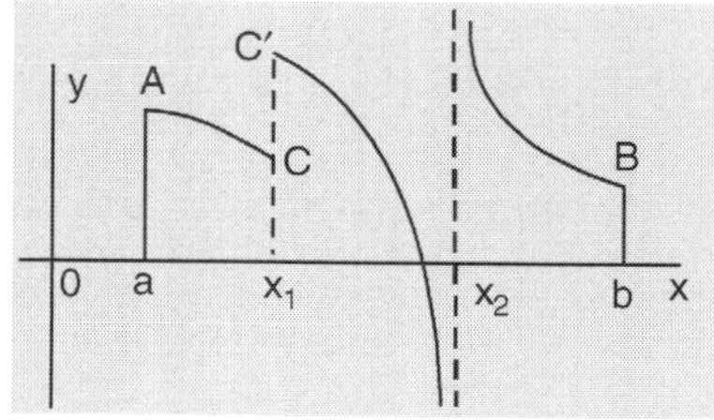

Fig. 45

We choose the following as the distinctive feature of a continuous curve : *As the point P (x, y) passes along the curve, any small change in the value of x is accompanied by a small change (or by no change at all) in the value of y.*

(This is true everywhere in Fig. 44, but is not true near the points where $x = x_1$ and $x = x_2$ in Fig. 45.)

More precisely thus : let $P_0\,(x_0, y_0)$ be any point on the curve and $P\,(x, y)$ any other point on it *on either side of P_0*. Choose a positive number ε, no matter how small. Then if $|y - y_0| < \varepsilon$ for all sufficiently small values of $|x - x_0|$, the curve is said to be *continuous at P_0*.

Further, we say that a curve is *continuous* if it is continuous at every point on it.

Thus the curve AB is continuous in Fig. 44, but in Fig. 45 it is discontinuous at the points where $x = x_1$ and $x = x_2$.

Definitions:

(1) *The single-valued function $f(x)$ is continuous at $x = x_0$ (or at the point x_0, or for x_0), if for any positive number ε that we may choose, however small, a number η exists such that $|f(x) - f(x_0)| < \varepsilon$, provided that $|x - x_0| < \eta$.*

This is equivalent to the following: $f(x)$ *is continuous at* x_0 *if* $f(x) \to f(x_0)$ *as* $x \to x_0$, *where x may tend to* x_0 *either from the left or from the right.*

It is important that the student should fully realise all that is implied in the definition just given. If $f(x)$ is continuous at x_0, then

    (*i*) $f(x)$ has a *single definite* value when $x = x_0$;

    (*ii*) if h tends to zero through *positive* values, then $f(x_0 + h) \to$ a finite number L_1 and $f(x_0 - h)$, a finite number L_2;

    (*iii*) and we must have $L_1 = f(x_0) = L_2$.

EXAMPLE : *A simple case, in which $f(x_0)$, L_1, L_2 all exist, but are not all equal, is given by Goursat.*

If $f(x) = x^2 + \dfrac{x^2}{1+x^2} + \ldots + \dfrac{x^2}{(1+x^2)^n} + \ldots$ to ∞, then, when $x = 0$, every term of $f(x)$ is zero, and so $f(x) = 0$; but, *for all values of x except zero, $f(x) = 1 + x^2$.*

Thus, $f(0) = 0$, and $\displaystyle\lim_{x \to 0} f(x) = 1$, so that $f(x)$ is discontinuous at $x = 0$.

(2) *The function $f(x)$ is continuous in any interval if it is continuous at every point of the interval.*

If the interval (a, b) is *closed* $(a < b)$, so that the end points are included, $f(x)$ *is continuous at a if $f(x) \to f(a)$ as $x \to a + 0$ and at b if $f(x) \to f(b)$ as $x \to b - 0$.*

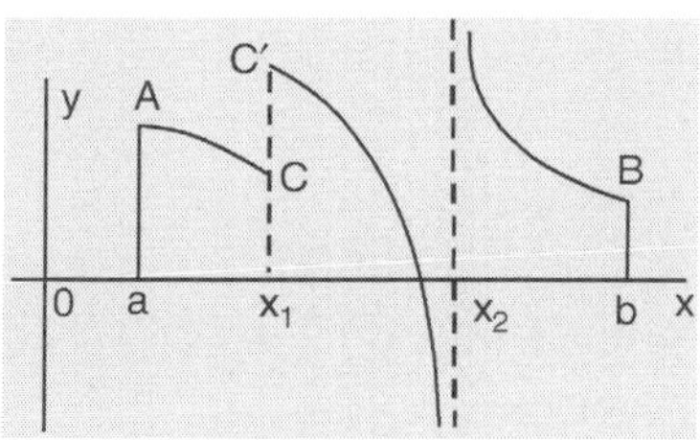

Fig. 45 (*a*)

Types of Discontinuities. (*i*) In Fig. 45 (*a*) let y_1, y_1' be the ordinates of C, C', then $y \to y_1$ as $x \to x_1$ from the left, *i.e.* as $x \to x_1 - 0$ and $y \to y_2$ as $x \to x_1$ from the right, *i.e.* as $x \to x_1 + 0$. This is expressed by writing

$$\lim_{x \to x_1 - 0} y = y_1 \text{ and } \lim_{x \to x_1 + 0} y = y_2.$$

(*ii*) At the point where $x = x_2$ in Fig. 45(*a*), it is supposed that y is infinite. Thus y is a discontinuous function of x at $x = x_1$, and at $x = x_2$. For example, $1/x$ and $1/x^2$ are discontinuous at $x = 0$.

(*iii*) It may happen that $f(x)$ has real values only on one side of the point $x = x_0$. In such a case the curve $y = f(x)$ stops abruptly and $f(x)$ is discontinuous at x_0.

Taking (as usual) $\sqrt{x}$ to denote the positive square root of the number x, we see that $\sqrt{x_0 - x}$ is discontinuous at x_0.

■ **5. Theorems on Continuous Functions:**

(1) It follows from the definition that *if $f(x)$ and $\phi(x)$ are continuous at $x = x_0$, so also are the sum, difference and product of these functions.*

The same is true for the quotient $f(x)/\phi(x)$, except when $\phi(x_0) = 0$.

(2) *Function of a Function.* If $y = \phi(u)$ where $u = f(x)$, then y is called a *function of a function of x*, and is denoted by $\phi\{f(x)\}$.

Theorem. *If $f(x)$ tends to a finite limit l as $x \to a$, and $\phi(x)$ is continuous at $x = l$, then*

$$\lim_{x \to a} \phi\{f(x)\} = \phi\{\lim_{x \to a} f(x)\}.$$

For if $\lim f(x) = l$, then $f(x) = l + \eta$, where $\eta \to 0$ as $x \to a$; also $\phi(x)$ is continuous at $x = l$, therefore

$$\lim_{x \to a} \phi\{f(x)\} = \lim_{\eta \to 0} \phi(l + \eta) = \phi(l) = \phi\{\lim_{x \to a} f(x)\}.$$

◼ **6. Continuity of Rational Functions:** If $f(x)$ is a polynomial, then, by Art. 3, $\lim_{x \to a} f(x) = f(a)$, where a is any real number.

Therefore *the polynomial is continuous for all values of x.*

Hence also *any rational function of x (i.e. a polynomial or the quotient of two polynomials) is continuous for all values of x except for such as make the denominator vanish.*

◼ **7. The Function x^n:** *If n is rational, x^n is continuous for all values of x for which x^n has definite real values.*

(*i*) *Suppose that $x > 0$.* If x_0 is any positive number, we can choose a, b so that $0 < a < x_0 < b$.

Let x be any number other than x_0 in the interval (a, b); then by Ch. 14, 10, $x^n - x_0{}^n$ lies between $nx^{n-1}(x - x_0)$ and $nx_0{}^{n-1}(x - x_0)$.

Now both x^{n-1} and $x_0{}^{n-1}$ lie between a^{n-1} and b^{n-1}, hence if k is the greater of the last two numbers, $|x^n - x_0{}^n| < k \cdot |n| \cdot |x - x_0|$.

If then ε is any assigned positive number, however small,

$$|x^n - x_0{}^n| < \varepsilon, \quad \text{provided that} \quad |x - x_0| < \frac{\varepsilon}{k|n|}.$$

Hence x^n is continuous at $x = x_0$.

(*ii*) *If $x < 0$,* let $x = -y$, so that $x^n = (-1)^n y^n$ where it is supposed that x^n has a real value. Now y^n is continuous for positive values of y, therefore x^n is continuous for negative values of x.

(*iii*) *At the point $x = 0$.* If n is positive, $\lim_{x \to 0} x^n = 0$. If n is negative, $x^n \to \infty$ or to $-\infty$, according as $x \to +0$ or to -0. Hence x^n is continuous at $x = 0$ for positive, but not for negative values of n.

◼ **8. Fundamental Theorems:**

(**1**) *If $f(x)$ is continuous at $x = a$ and $f(a) \neq 0$, then $f(x)$ has the same sign as $f(a)$ for all values of x in the neighbourhood of a; that is to say, if $a - \eta < x < a + \eta$ where η is arbitrarily small.*

For since $f(a) \neq 0$, we can choose ε so that $f(a) - \varepsilon$ and $f(a) + \varepsilon$ have the same sign as $f(a)$: and since $f(x)$ is continuous at $x = a$, we can find η so that

$$f(a) - \varepsilon < f(x) < f(a) + \varepsilon, \quad \text{provided that} \quad |x - a| < \eta.$$

For such values of x, $f(x)$ has the same sign as $f(a)$.

(**2**) *If $f(x)$ is continuous for the range $a \leq x \leq b$, then as x passes from the value a to the value b, $f(x)$ assumes at least once every value between $f(a)$ and $f(b)$.*

From the graphical point of view the truth of this statement is obvious.

Let A, B be the points on the curve $y = f(x)$ whose abscissae are a and b, and let k be any number between $f(a)$ and $f(b)$.

We assume that any straight line which passes between the points A, B cuts the curve at least once.

It follows that the line $y = k$ cuts the curve at least once. Hence as x varies from a to b, $f(x)$ takes the value k at least once.

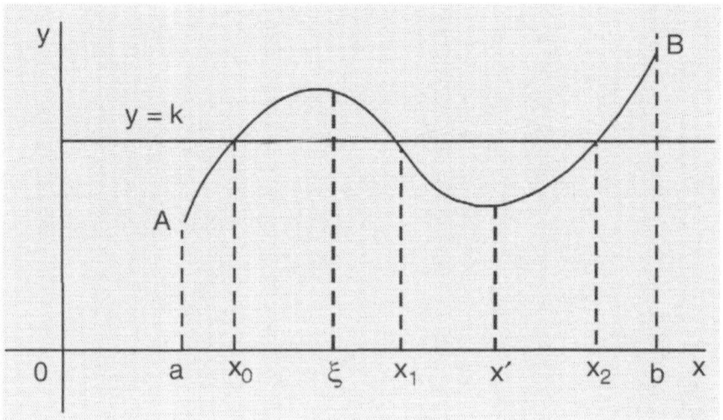

Fig. 46

Proof. Suppose that $a < b$ and $f(a) < f(b)$. Let k be any number between $f(a)$ and $f(b)$. Because $f(x)$ is continuous at $x = a$ and $f(a) < k$, there are values of x in the interval (a, b) for which $f(x) < k$: and because $f(x)$ is continuous at $x = b$ and $f(b) > k$, there are values of x in (a, b) for which $f(x) > k$.

Divide the real numbers in the interval (a, b) into two classes as follows:

The lower class A is to contain every number x such that $f(\xi) < k$ where ξ is any number in the range $a < \xi \leq x$.

The upper class A′ is to contain every number x' not included in A.

(Thus in Fig. 46, class A consists of the range $(a < x < x_0)$. The range $(x_1 \leq x' \leq x_2)$ belongs to class A', for although $f(x') < k$, yet $f(\xi) > k$ for some values of ξ in the range $a < \xi \leq x'$.)

It has been shown that both classes exist.

Also every number in the interval (a, b) is included, and every x is less than any x'.

Therefore the classification defines a real number x_0, which is either the greatest number in A or the least number in A'.

It remains to prove that $f(x_0) = k$.

If $f(x_0) - k < 0$, by reason of the continuity of $f(x)$ at x_0, we can find ε so that $f(x) - k < 0$ if $x_0 - \varepsilon \leq x \leq x_0 + \varepsilon$. Thus $f(\xi) < k$ when $\xi \leq x_0 + \varepsilon$. Hence $x_0 + \varepsilon$ belongs to class A, which is impossible, as it is greater than x_0.

If $f(x_0) - k > 0$, we can find ε so that $f(x) - k > 0$ if $x_0 - \varepsilon \leq x \leq x_0 + \varepsilon$. Thus $x_0 - \varepsilon$ belongs to class A', which is impossible, for it is less than x_0.

Hence it follows that $f(x_0) = k$. It is easy to modify the proof to suit the case when $f(a) > f(b)$.

Note: If $f(x) = k$ for more than one value of x in the interval (a, b), then x_0, determined as above, is the least of these values.

■ **9. Derivatives:** If $f(x)$ is a function of x such that

$$\{f(x+h) - f(x)\}/h$$

tends to a limit as h tends to zero, this limit is called the *derivative of $f(x)$, and is denoted by $f'(x)$.* Thus $f'(x)$ is defined by the equation

$$f'(x) = \lim_{h \to 0} \frac{f(x+h) - f(x)}{h}.$$

The function $f'(x)$ is also called the *differential coefficient* of $f(x)$.

It is possible that for some values of x, this limit does not exist. If x_0 is such a value, *$f(x)$ has no derivative at $x = x_0$. This is certainly the case when $f(x)$ is discontinuous at $x = x_0$.* For in order that $\{f(x_0 + h) - f(x_0)\}/h$ may tend to a limit as $h \to 0$, $f(x_0 + h)$ must tend to $f(x_0)$.

▌ **EXAMPLE :** *If $f(x) = x^n$, where n is a positive integer, prove that $f'(x) = nx^{n-1}$.*

We have
$$f'(x) = \lim_{h \to 0} \frac{(x+h)^n - x^n}{h}$$

$$= \lim_{h \to 0} \left(nx^{n-1} + \frac{n(n-1)}{\lfloor 2} hx^{n-2} + \ldots + h^{n-1}\right) = nx^{n-1}.$$

■ **10. Tangent to a Curve:** Let P be any point on a curve and let Q be a point which is supposed to move along the curve, so as to approach P from either side. It is supposed that the curve is continuous near P, so that Q may be as near to P as we like.

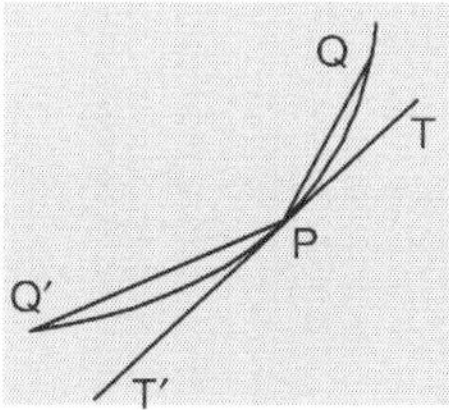

Fig. 47

If a straight line $T'PT$ exists such that, as Q approaches P from either side, the angle which PQ makes with $T'PT$ becomes and (as Q continues to approach P) remains less than any angle we may choose, however small, then $T'PT$ is called the *tangent* to the curve at P.

This is sometimes expressed by saying that *the tangent to a curve at P is the limiting position of the chord PQ as Q tends to coincidence with P.*

Notice that (*i*) *Q is not supposed to reach P, for if Q were to coincide with P, PQ would cease to be a chord.*

(*ii*) The curve in Fig. 48 is supposed to represent the path of a ball which is thrown from A, strikes the ground at B and proceeds to C. Just as the ball hits the ground, it is moving in the direction TB and, just after, in the direction BT'. We may, if we choose, say that the curve has *two* tangents at B, namely BT and BT'. Strictly speaking, according to the definition, the curve has no tangent at B.

This is an instance of *a continuous curve which has no definite tangent at one point.*

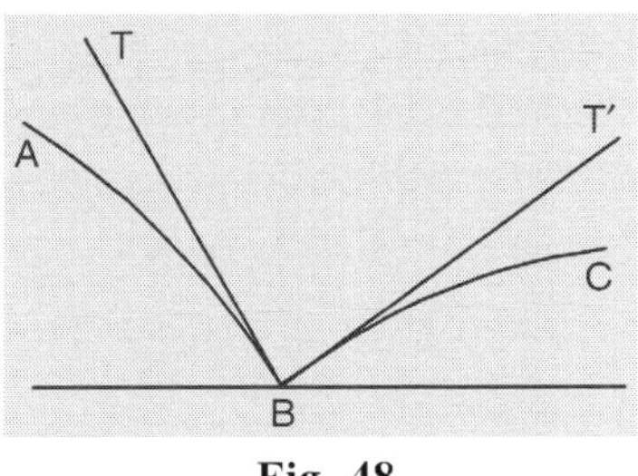

Fig. 48

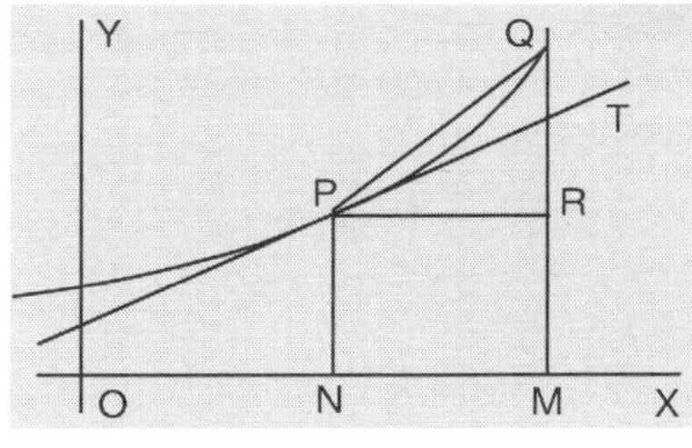

Fig. 49

■ **11. Gradient of Tangent:** Let (x, y) be the coordinates of a point P on the curve whose equation is $y = f(x)$. We shall suppose that the curve has a definite tangent PT at P, and that this tangent is not parallel to the y-axis, as in Fig. 49.

Let Q be the point on the curve whose coordinates are $(x + h, y + k)$.

Draw the ordinates PN, QM, and draw PR parallel to OX to meet QM in R. Then

$$PR = OM - ON = (x + h) - x = h,$$
$$RQ = MQ - NP = (y + k) - y = k,$$

and since P and Q are on the curve, we have

$$y = f(x) \text{ and } y + k = f(x + h); \therefore k = f(x + h) - f(x);$$

$$\therefore \text{ Gradient of chord } PQ = \frac{k}{h} = \frac{f(x+h) - f(x)}{h}.$$

Now, as Q tends to coincidence with P, PQ tends to the limiting position PT and h tends to zero;

$$\therefore \text{ Gradient of } PT = \lim_{h \to 0} \frac{f(x+h) - f(x)}{h} = f'(x).$$

Hence $f'(x)$ *is the gradient of the tangent to the curve* $y = f(x)$ *at the point* (x, y).

If PT is nearly parallel to OY, $f'(x)$ will be large, and if P tends to coincide with a point where the tangent is parallel to the y-axis, we may expect that $f'(x)$ will tend to infinity.

■ EXAMPLE : *Find the equation of the tangent to the curve, whose equation is* $y = x^3$, *at the point whose abscissa is a.*

If $f(x) = x^3$, we have $f'(x) = 3x^2$; $\therefore$ The gradient of the tangent at the point (a, a^3) is $3a^2$, and the equation to the tangent is

$$y - a^3 = 3a^2 (x - a) \text{ or } y = 3a^2x - 2a^3.$$

■ **12. Notation of the Differential Calculus:** In the differential calculus the derivative of $f(x)$ is given a different name and a different notation is used.

Any number nearly equal to x is denoted by $x + \delta x$; and, if $y = f(x)$, the value of y corresponding to $x + \delta x$ is denoted by $y + \delta y$; thus we have

$$f'(x) = \lim_{\delta x \to 0} \frac{\delta y}{\delta x}.$$

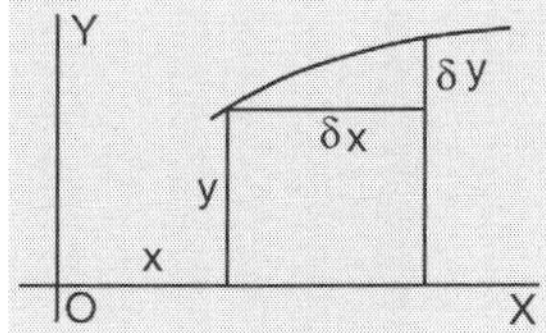

Fig. 50

This limit is denoted by $\dfrac{dy}{dx}$, and is called *the differential coefficient of y* (*with*

regard to x). In finding $\dfrac{dy}{dx}$ we are said to *differentiate y with regard to x*, and the process is called *differentiation*.

We often write $\dfrac{df(x)}{dx}$ in the form $\dfrac{d}{dx} f(x)$, where $\dfrac{d}{dx}$ denotes the *operation of*

differentiating f(x) : it is also written shortly as $\dfrac{df}{dx}$.

■ **13. Rules of Differentiation:**

(1) If u, v are functions of x, then

(i) $\dfrac{d}{dx}(u+v) = \dfrac{du}{dx} + \dfrac{dv}{dx},$ $\qquad\qquad$ (ii) $\dfrac{d}{dx}(u-v) = \dfrac{du}{dx} - \dfrac{dv}{dx},$

(iii) $\dfrac{d}{dx}(uv) = u\dfrac{dv}{dx} + v\dfrac{du}{dx},$ $\qquad\qquad$ (iv) $\dfrac{d}{dx}\left(\dfrac{1}{u}\right) = -\dfrac{1}{u^2}\dfrac{du}{dx},$

(v) $\dfrac{d}{dx}\left(\dfrac{u}{v}\right) = \dfrac{1}{v^2}\left(v\dfrac{du}{dx} - u\dfrac{dv}{dx}\right).$

It is assumed that $\dfrac{du}{dx}$ and $\dfrac{dv}{dx}$ have definite values, also that in (iv) $u \neq 0$ and in (v) $v \neq 0$.

Proof. Let δu, δv, $\delta(uv)$, etc., be the increments in u, v, uv, etc., corresponding to an increment δx in x.

Because $\dfrac{du}{dx}$, $\dfrac{dv}{dx}$ are not infinite, δu and δv tend to zero as $\delta x \to 0$.

(i) and (ii). These are left to the student.

(iii) $\delta(uv) = (u + \delta u)(v + \delta v) - uv = u\,\delta v + (v + \delta v)\,\delta u;$

$$\therefore \qquad \frac{d}{dx}(uv) = \lim\left(u\frac{\delta v}{\delta x}\right) + \lim(v + \delta v)\frac{\delta u}{\delta x}$$

$$= u\lim\frac{\delta v}{\delta x} + \lim(v + \delta v)\lim\frac{\delta u}{\delta x}$$

$$= u\frac{dv}{dx} + v\frac{du}{dx}.$$

(iv) $\delta\left(\dfrac{1}{u}\right) = \dfrac{1}{u + \delta u} - \dfrac{1}{u} = -\dfrac{\delta u}{u(u + \delta u)};$

$$\therefore \frac{d}{dx}\left(\frac{1}{u}\right) = -\lim\frac{\dfrac{\delta u}{\delta x}}{u\,(u + \delta u)} = -\frac{\lim\dfrac{du}{dx}}{u\lim(u + \delta u)} = -\frac{1}{u^2}\frac{du}{dx}.$$

(v) $\dfrac{d}{dx}\left(\dfrac{u}{v}\right)=\dfrac{d}{dx}\left(\dfrac{1}{v}.u\right)=\dfrac{1}{v}\dfrac{du}{dx}+u\dfrac{d}{dx}\left(\dfrac{1}{v}\right);$

$\therefore \dfrac{d}{dx}\left(\dfrac{u}{v}\right)=\dfrac{1}{v}\dfrac{du}{dx}-\dfrac{u}{v^2}\dfrac{dv}{dx}=\dfrac{1}{v^2}\left(v\dfrac{du}{dx}-u\dfrac{dv}{dx}\right).$

■ **EXAMPLE 1.** *Find* $\dfrac{dy}{dx}$, *(i) if* $y=\dfrac{2x-1}{3x+1}$; *(ii)* $y=\dfrac{x^2+3x}{x-1}$.

We have $\qquad\qquad$ (i) $\dfrac{dy}{dx}=\dfrac{(3x+1)2-(2x-1)3}{(3x+1)^2}=\dfrac{5}{(3x+1)^2}.$

$\qquad$ (ii) $y=x^2+x+4+\dfrac{4}{x-1}$; $\therefore \dfrac{dy}{dx}=2x+1-\dfrac{4}{(x-1)^2}.$

■ **EXAMPLE 2.** *Show that the rules for the differentiation of a product and a quotient can be written in the forms*

(i) $\dfrac{1}{p}\dfrac{dp}{dx}=\dfrac{1}{u}\dfrac{du}{dx}+\dfrac{1}{v}\dfrac{dv}{dx}$ *and* (ii) $\dfrac{1}{q}\dfrac{dq}{dx}=\dfrac{1}{u}\dfrac{du}{dx}-\dfrac{1}{v}\dfrac{dv}{dx},$

where $p = uv$ *and* $q = u/v.$

(iii) Deduce that if $z = (u_1 u_2 \dots u_n) / (v_1 v_2 \dots v_m)$, *where* $u_1, u_2, \dots, v_1, v_2, \dots$ *are functions of x ; then*

$$\frac{1}{z}\frac{dz}{dx}=\sum_{r=1}^{r=n}\left(\frac{1}{u_r}\cdot\frac{du_r}{dx}\right)-\sum_{s=1}^{s=m}\left(\frac{1}{v_s}\frac{dv_s}{dx}\right).$$

(i) If $p = uv$, then $\dfrac{dp}{dx}=v\dfrac{du}{dx}+u\dfrac{dv}{dx}$; and, if each side is divided by p ($= uv$), the result follows.

(ii) This follows in a similar manner.

(iii) Let $U = u_1 u_2 \dots u_n$, $V = v_1 v_2 \dots v_m$; then

$$z = U/V \text{ and } \frac{1}{z}\frac{dz}{dx}=\frac{1}{U}\frac{dU}{dx}-\frac{1}{V}\frac{dV}{dx}:$$

and by repeated application of the rule for a product,

$$\frac{1}{U}\frac{dU}{dx}=\frac{1}{u_1}\frac{du_1}{dx}+\frac{1}{u_2 u_3\dots u_n}\frac{d(u_2 u_3\dots u_n)}{dx}$$

$$=\frac{1}{u_1}\frac{du_1}{dx}+\frac{1}{u_2}\frac{du_2}{dx}+\frac{1}{u_3 u_4\dots u_n}\frac{d(u_3 u_4\dots u_n)}{dx}=\sum_{r=1}^{r=n}\left(\frac{1}{u_r}\cdot\frac{du_r}{dx}\right).$$

Similary $\dfrac{1}{V}\dfrac{dV}{dx}=\displaystyle\sum_{s=1}^{s=m}\left(\dfrac{1}{v_s}\dfrac{dv_s}{dx}\right)$; and the result follows.

◼ **EXAMPLE 3.** *Prove that, if y is a function of x, and $\dfrac{dy}{dx}$ exists, then for all rational values of n,*

$$\frac{d}{dx}(y^n) = ny^{n-1}\frac{dy}{dx}:$$

Let $w = y^n$, and (*i*) let *n* be a positive integer; then $w = y \cdot y \cdot y \cdot \ldots y$ (*n* factors);

$$\therefore \frac{1}{w}\frac{dw}{dx} = \frac{1}{y}\frac{dy}{dx} + \frac{1}{y}\frac{dy}{dx} + \ldots + \frac{1}{y}\frac{dy}{dx} \ (n \text{ terms}) \ = \frac{n}{y}\frac{dy}{dx}.$$

(*ii*) Let $n = -m$, where *m* is a positive integer; then $w = y^{-m}$,

$$\therefore \qquad wy^m = 1 \text{ and } \frac{1}{w}\frac{dw}{dx} + \frac{1}{y^m}\frac{dy^m}{dx} = 0.$$

Hence, by (*i*), $\dfrac{1}{w}\dfrac{dw}{dx} + \dfrac{m}{y}\dfrac{dy}{dx} = 0$ or $\dfrac{1}{w}\dfrac{dw}{dx} = \dfrac{n}{y}\dfrac{dy}{dx}.$

(*iii*) Let $n = p/q$ where *p* and *q* are integers; then $w = y^{\frac{p}{q}}$ and $w^q = y^p = z$, say.

Hence, by (*i*) or (*ii*), $\dfrac{q}{w}\dfrac{dw}{dx} = \dfrac{1}{z}\dfrac{dz}{dx} = \dfrac{p}{y}\dfrac{dy}{dx},$ and $\dfrac{1}{w}\dfrac{dw}{dx} = \dfrac{n}{y}\dfrac{dy}{dx}.$

That is, in all cases $\dfrac{1}{w}\dfrac{dw}{dx} = \dfrac{n}{y}\dfrac{dy}{dx};$ and multiplying each side by w ($= y^n$),

we have $$\frac{dw}{dx} = \frac{d}{dx}(y^n) = ny^{n-1} \cdot \frac{dy}{dx}.$$

Note: It should be carefully observed that there is always a factor $\dfrac{dy}{dx}$, unless *y*.

◼ **14. Function of a Function:**

(1) *If $u = f(x)$ is a single-valued function of x, continuous for $a \leq x \leq b$, and $y = \phi(u)$ is a single-valued function of u, continuous for the corresponding range of values of u, then y is a continuous function of x for $a \leq x \leq b$.*

Let x_0 be any value of *x* in the given range, and u_0, y_0 the corresponding values of *u* and *y*.

Because $y = \phi(u)$ is continuous at u_0, for any assigned positive number ε, we can find a number η such that

$$|y - y_0| < \varepsilon, \text{ provided that } |u - u_0| < \eta,$$

and because $f(x)$ is continuous at x_0, we can find η' so that

$$|u - u_0| < \eta, \text{ and consequently } |y - y_0| < \varepsilon \text{ if } |x - x_0| < \eta'.$$

This proves the theorem.

(2) *If the derivatives $f'(x)$ and $\phi'(u)$ exist for $a < x < b$ and for corresponding values of u, and if $F(x) = \phi(u) = \phi\{f(x)\}$, then,*

$$F'(x) = \phi'(u) \cdot f'(x), \quad \text{that is,} \quad \frac{dy}{dx} = \frac{dy}{du} \cdot \frac{du}{dx}.$$

Let $\delta u,\ \delta y$ be the increments u and y corresponding to an increment δx in x, then if $\delta u \neq 0$,

$$\frac{\delta y}{\delta x} = \frac{\delta y}{\delta u} \cdot \frac{\delta u}{\delta x}. \qquad \qquad \text{...(A)}$$

As $\delta x \to 0$, $\dfrac{\delta u}{\delta x} \to \dfrac{du}{dx}$ and since $\dfrac{du}{dx}$ is finite, $\delta u \to 0$

and therefore $\dfrac{\delta y}{\delta u} \to \dfrac{dy}{du}$.

Consequently,

$$\frac{dy}{dx} = \lim \frac{\delta y}{\delta u} \cdot \lim \frac{\delta u}{\delta x} = \frac{dy}{du} \cdot \frac{du}{dx}. \qquad \qquad \text{....(B)}$$

It may happen that $\delta u = 0$ for some value of δx, and then equation (A) does not hold.

In this case we must have $\delta y = 0$, for otherwise $\dfrac{dy}{du}$ would not be finite:

hence $\dfrac{dy}{dx} = 0$. Also $\dfrac{du}{dx} = 0$, for $\delta u = 0$. Thus the equation

$$\frac{dy}{dx} = \frac{dy}{du} \cdot \frac{du}{dx}$$

is true also in this case.

▮ **EXAMPLE 1.** *Find $\dfrac{dy}{dx}$, (i) when $y = (ax + b)^n$, (ii) when $y = \dfrac{1}{a + bx^2}$.*

(*i*) Let $u = ax + b$, then $y = u^n$, and $\dfrac{dy}{dx} = \dfrac{dy}{du} \cdot \dfrac{du}{dx} = nu^{n-1} \cdot a = na(ax + b)^{n-1}$.

(*ii*) Let $u = a + bx^2$, then $y = \dfrac{1}{u}$, and $\dfrac{dy}{dx} = \left(-\dfrac{1}{u^2}\right) \cdot 2bx = -\dfrac{2bx}{(a + bx^2)^2}$.

▮ **EXAMPLE 2.** *If $\dfrac{x^2}{a^2} + \dfrac{y^2}{b^2} = 1$, then $\dfrac{dy}{dx} = -\dfrac{b^2 x}{a^2 y}$.*

We regard y as a function of x, and differentiate both sides of the given equation with regard to x, thus $\dfrac{2x}{a^2} + \dfrac{2y}{b^2}\dfrac{dy}{dx} = 0.$

■ **15. Derivative of x^n.** *If $y = x^n$ where n is any rational, then*

$$\frac{dy}{dx} = nx^{n-1}.$$

Proof. If x is positive, $x + h$ is positive for all sufficiently small values of h, and by Ch. 14, 10, $(x + h)^n - x^n$ lies between $nh(x+h)^{n-1}$ and nhx^{n-1}. Also x^{n-1} is a continuous function of x. Therefore

$$\frac{dy}{dx} = \lim_{h \to 0} \frac{(x+h)^n - x^n}{h} = nx^{n-1}.$$

If x is negative, let $x = -u$, then $y = (-1)^n u^n$ and

$$\frac{dy}{dx} = \frac{dy}{du} \cdot \frac{du}{dx} = (-1)^n \cdot nu^{n-1} \cdot (-1) = nx^{n-1}.$$

*Or,** having proved that the derivative of x^n is nx^{n-1} when n is a positive integer, we can prove that this is also true when $n = p/q$ as follows.

Let $y = x^{\frac{p}{q}}$, therefore $y^q = x^p$. Now $\dfrac{d}{dx} y^q = qy^{q-1} \dfrac{dy}{dx}$,

therefore $qy^{q-1} \dfrac{dy}{dx} = px^{p-1}$ and $\dfrac{dy}{dx} = \dfrac{p}{q} \dfrac{x^{p-1}}{y^{q-1}} = \dfrac{p}{q} \dfrac{x^{p-1}}{x^{p/q(q-1)}} = \dfrac{p}{q} x^{\frac{p}{q}-1}$

■ **16. Meaning of the Sign of $f'(x)$:** *If $f'(x_0) > 0$, then for all values of x in the neighbourhood of x_0, $f(x) \gtrless f(x_0)$ according as $x \gtrless x_0$.*

For $\{f(x_0 + h) - f(x_0)\}/h$ tends to a positive limit as $h \to 0$. Hence for sufficiently small values of h, $f(x_0 + h) - f(x_0)$ has the same sign as h.

Similarly *if $f'(x_0) < 0$, for every x in the neighbourhood of x_0,*

$$f(x) \gtrless f(x_0) \ as \ x \lessgtr x_0.$$

Note: If $f'(x) > 0$ only for the *single* value x_0 of x, it does *not* follow that $f(x)$ increases steadily as x increases through x_0.

For if $x_1 < x_2 < x_0$ where x_1, x_2 are in the neighbourhood of x_0, it has only been proved that $f(x_1) < f(x_0)$ and $f(x_2) < f(x_0)$, and we cannot conclude that $f(x_1) < f(x_2)$.

■ **17. Complex Functions of a Real Variable x:** If $y = f(x) + \iota\phi(x)$, then y is called a *complex function* of x and its derivative is $f'(x) + \iota\phi'(x)$. The rules of Art. 13 obviously hold for such functions.

■ **18. Higher Derivatives:** If the derivative of $f'(x)$ exists, it is called the second derivative or the second differential coefficient of $f(x)$.

If $y = f(x)$, this is written in any of the forms

$$f''(x), \ \frac{d^2y}{dx^2}, \ \left(\frac{d}{dx}\right)^2 f(x).$$

---

* See also Art.13 Ex.3

So by n differentiations (when possible) we obtain the nth derivative or the nth differential coefficient of $f(x)$, which is written in any of the forms

$$f^{(n)}(x), \quad \frac{d^n y}{dx^n}, \quad \left(\frac{d}{dx}\right)^n f(x).$$

The following are important instances.

(i) If $y = x^n$, $\dfrac{dy}{dx} = nx^{n-1}$, $\dfrac{d^2 y}{dx^2} = n(n-1)x^{n-2}$,

$$\frac{d^r y}{dx^r} = n(n-1)(n-2) \dots (n-r+1)x^{n-r}.$$

(ii) If $f(x) = a_0 x^n + na_1 x^{n-1} + \dfrac{n(n-1)}{\underline{|2}} a_2 x^{n-2} + \dots + a_n$,

then $f'(x) = n\left\{ a_0 x^{n-1} + (n-1)a_1 x^{n-2} + \dfrac{(n-1)(n-2)}{\underline{|2}} a_2 x^{n-3} + \dots + a_{n-1} \right\}.$

Or, using the notation of Ch. 3, 1, if $f(x) = (a_0, a_1, a_2, \dots a_n)(x, 1)^n$,

then $\qquad\qquad f'(x) = n(a_0, a_1, a_2, \dots a_{n-1})(x, 1)^{n-1}$,

and by successive differentiation

$$f''(x) = n(n-1)(a_0, a_1, \dots a_{n-2})(x, 1)^{n-2},$$

....

$$f^{(n-1)}(x) = n(n-1) \dots 3 \cdot 2 \cdot (a_0 x + a_1),$$

$$f^{(n)}(x) = \underline{|n}\; a_0.$$

■ **19. Maxima and Minima:** We say that $f(x)$ has a *maximum value* at $x = x_0$ when $f(x_0) > f(x)$ for all values of x in the neighbourhood of x_0. In other words, if a number ε exists such that $f(x_0) > f(x)$ when

$$x_0 - \varepsilon < x < x_0 + \varepsilon,$$

then $f(x_0)$ is a maximum value of $f(x)$. If in the preceding, the sign $>$ is replaced by $<$, then $f(x_0)$ is a *minimum* value of (fx).

Maxima and minima values of a function are sometimes called *stationary* or *turning* values.

A *necessary* condition that $f(x_0)$ may be a maximum or a minimum value of $f(x)$ is that $f'(x_0) = 0$.

This follows from Art. 16; but the condition is not *sufficient*.

If $f(x_0)$ is a maximum value, then for sufficiently small values of h, $f(x_0 + h) < f(x_0)$. Hence by Art. 16, as x increases through x_0, the sign of $f'(x)$ changes from $+$ to $-$.

Suppose now that $f''(x_0)$ exists and is not zero. Then since $f'(x)$ decreases as x increases through x_0, $f''(x_0)$ *must be negative*.

Similarly if $f(x_0)$ is a minimum value, as x increases through x_0, the sign of $f'(x)$ must change from $-$ to $+$, and if $f''(x_0)$ is not zero, it must be *positive*.

If the sign of $f'(x)$ does not change as x passes through x_0, $f(x)$ is neither a maximum nor a minimum. The case in which $f''(x_0) = 0$ will be considered later. In this case $f(x_0)$ may be a maximum, a minimum or neither of these.

To illustrate this geometrically, in Fig. 51, the tangents at A, B, C to the curve $y = f(x)$ are parallel to OX.

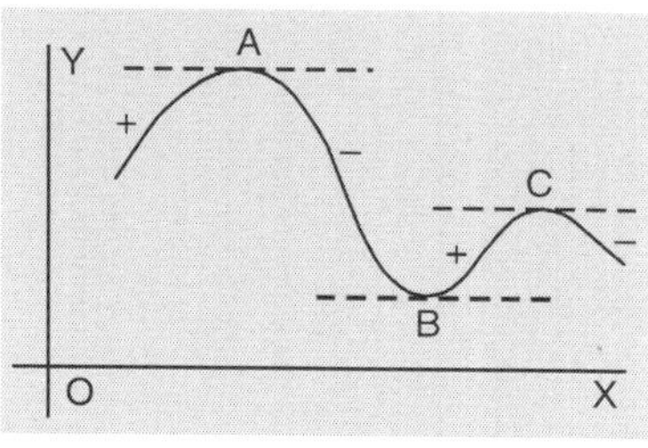

Fig. 51   **Fig. 52**

The sign of $f'(x)$ for different parts of the curve is as indicated, and $f(x)$ has maximum values at A, C and a minimum value at B.

Figure 52 represents the graph of $y = x^3 + 1$. Here $\dfrac{dy}{dx} = 3x^2$, so that $\dfrac{dy}{dx} = 0$ when $x = 0$, but as x passes through O, the sign of $\dfrac{dy}{dx}$ does not change. Thus y is neither a maximum nor a minimum when $x = 0$.

■ **EXAMPLE :** *Search for maxima and minima values of*
$$f(x) = 2x^3 - 3x^2 - 12x + 6.$$
Here $\qquad\qquad f'(x) = 6x^2 - 6x - 12 = 6\,(x + 1)(x - 2).$

Thus $\qquad\qquad\qquad f'(x) = 0$ if $x = -1$ or 2.

As x increases through -1, the sign of $f'(x)$ changes from $+$ to $-$; and as x increases through 2, the sign of $f'(x)$ changes from $-$ to $+$.

Thus $f(-1)$ is a maximum and $f(2)$ a minimum value.

Or thus, $f''(x) = 12x - 6 = 6(2x - 1)$,

so that $f''(-1) < 0$ and $f''(2) > 0$, leading to the same results as before.

■ **20. Points of Inflexion:** In Fig. 53 APB is supposed to be part of a curve represented by $y = f(x)$.

If a point moves along the curve from A to B, the gradient of the tangent increases along the arc AP and then decreases.

Thus $\dfrac{dy}{dx}$ has a maximum value at P, and therefore $\dfrac{d^2y}{dx^2} = 0$ at P.

If Q is a point on the curve near P, the secant QP cuts the curve at another point Q' near P, and we may regard the tangent TPT' as the limiting position of the line QPQ', when the three points Q, P, and Q' are very close together.

Definition. A point (x, y) on a curve at which $\dfrac{dy}{dx}$ is a maximum or a minimum is called a *point of inflexion.*

At such a point, $\dfrac{d^2 y}{dx^2} = 0$, and the curve crosses the tangent.

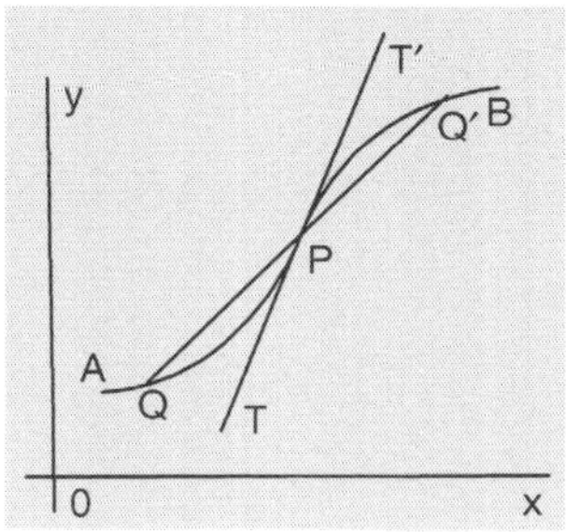

Fig. 53

EXERCISE XXVIII

Find the limits of the functions in Exs. 1–3 (*i*) as x tends to zero; (*ii*) as x tends to ∞.

1. $\dfrac{1 + 2x + 3x^2}{2 + 3x + 4x^2}$.

2. $\dfrac{2x^2 - 3x^3 + 4x^4}{3x + 4x^2 - 5x^3}$.

3. $\dfrac{(1 + x)^2 - (1 - x)^2}{(1 + x)^3 - (1 - x)^3}$.

Find the values of

4. $\displaystyle\lim_{x \to 1} \dfrac{3x^2 - 7x + 4}{2x^2 - 5x + 3}$.

5. $\displaystyle\lim_{x \to 2} \dfrac{x^3 - x^2 - x - 2}{x^3 - 3x^2 + 3x - 2}$.

6. $\displaystyle\lim_{x \to a} \dfrac{x^3 - a^3}{x^4 + 2x^2 a^2 + a^4}$.

7. $\displaystyle\lim_{h \to 0} \dfrac{(x + h)^{3/2} - x^{3/2}}{h}$.

8. $\displaystyle\lim_{x \to 0} \dfrac{\sqrt{1 + x} - \sqrt{1 - x}}{x}$.

9. Prove that (*i*) $\displaystyle\lim_{x \to 0} \cos x = 1$; (*ii*) $\displaystyle\lim_{x \to 0} \dfrac{\sin x}{x} = 1$.

10. Prove that $\sin x$ and $\cos x$ are continuous functions of x.

$$[\sin (x + h) - \sin x = 2 \sin \tfrac{1}{2} h \cos (x + \tfrac{1}{2} h).]$$

11. For what values of x is tan x discontinuous ?

12. Prove that $\dfrac{d}{dx}$ sin x = cos x, $\dfrac{d}{dx}$ cos x = $-$ sin x, $\dfrac{d}{dx}$ tan x = sec^2 x.

$$\left[\frac{\sin (x + h) - \sin x}{h} = \frac{2\sin \frac{1}{2} h}{h} \cdot \cos \left(x + \frac{1}{2} h\right) \to \cos x \text{ as } h \to 0. \right]$$

13. If $f(x) = (x-a)^m (x - b)^n$, prove that

$$f'(x) = (x - a)^{m-1} (x - b)^{n-1} [(m + n) x - mb - na.].$$

Hence, if m, n are positive, $f'(x)$ vanishes for a value of x between a and b.

14. Prove the following:

(*i*) $\left(\dfrac{d}{dx}\right)^r (ax + b)^n = n (n - 1) \ldots (n - r + 1) a^r (ax + b)^{n-r}$.

(*ii*) $\left(\dfrac{d}{dx}\right)^r \dfrac{1}{(x - a)^n} = (-1)^r \dfrac{n (n + 1) \ldots (n + r - 1)}{(x - a)^{n + r}}$.

(*iii*) $\left(\dfrac{d}{dx}\right)^n \dfrac{1}{1 - x^2} = \dfrac{1}{2} \lfloor n \left\{ \dfrac{1}{(1 - x)^{n+1}} + (-1)^n \cdot \dfrac{1}{(1 + x)^{n+1}} \right\}$.

15. Find the derivatives of

(*i*) $\dfrac{ax^2 + 2bx + c}{a'x^2 + 2b'x + c'}$.

(*ii*) $\sqrt{\left(\dfrac{1 - x}{1 + x}\right)}$.

(*iii*) $\dfrac{x}{\sqrt{(a^2 + x^2)^3}}$.

(*iv*) $\sin^n x$.

(*v*) $\cos^n x$.

(*vi*) $\tan^n x$.

16. Find $\dfrac{dy}{dx}$ if (*i*) $ax^2 + 2hxy + by^2 + 2gx + 2fy + c = 0$, (*ii*) if $x^3 + y^3 = 3axy$.

17. Prove that $\dfrac{d^2 x}{dy^2} = - \dfrac{d^2 y}{dx^2} \Big/ \left(\dfrac{dy}{dx}\right)^3$.

18. If $y = A \cos \mu x + B \sin \mu x$, then $\dfrac{d^2 y}{dx^2} = - \mu^2 y$.

19. If $y = \cos x + \iota \sin x$, then $\dfrac{dy}{dx} = \iota y$.

20. Prove that if $\theta = \tan^{-1} a/x$, then

$$\left(\frac{d}{dx}\right)^n \frac{a}{x^2 + a^2} = (-1)^n \lfloor n \cdot \frac{\sin (n + 1) \theta}{(x^2 + a^2)^{-(n+1)/2}},$$

$$\left(\frac{d}{dx}\right)^{n}\frac{x}{x^{2}+a^{2}} = (-1)^{n}\lfloor n \cdot \frac{\cos(n+1)\theta}{(x^{2}+a^{2})^{-(n+1)/2}}.$$

[Consider the values of $1/(x - \iota a) \pm 1/(x + \iota a)$.]

21. Find the equation to the tangent at the origin to the curve whose equation is $y = ax + bx^{2} + cx^{3}$.

22. If P, Q, Q' are points on the parabola $y = px^{2} + qx + r$ whose abscissae are $a, a-h, a + h$ respectively, prove that the tangent at P is parallel to the chord QQ'.

23. Prove that the function $2x^{3} - 3x^{2} + 6x - 5$ always increases with x.

24. Sketch roughly the graph of $y = x(x - 1)^{2}$, and find the greatest value of $x(x - 1)^{2}$ for values of x between 0 and 1.

25. (*i*) If $f(x) = (x - 1)^{3}(x - 2)$, for what values of x does $f'(x)$ vanish?

(*ii*) Sketch roughly the graph of $y = f(x)$ from $x = 0$ to $x = 2$, and show that one of the values of x found in (*i*) gives a minimum value of $f(x)$ and that the other gives neither a maximum nor a minimum.

26. Prove that $x = -b/2a$ gives a maximum or a minimum value of $ax^{2} + bx + c$ according as $a \lessgtr 0$.

27. (*i*) When does the function $3x^{4} - 4x^{3} - 36x^{2} - 1$ increase with x?

(*ii*) Find the turning values of the function, and state the character of each.

28. Show that the turning values of $(x - 1)(x - 2)(x - 3)$ are approximately ± 0.38.

29. Four equal squares are cut off from the corners of a rectangular sheet of tin, 8 inches long and 5 inches wide. The rest of the sheet is bent so as to form an open box on a rectangular base. Find the volume of the box of greatest capacity which can be formed in this way.

30. The regulations for Parcel Post require that the sum of the length and girth of a parcel must not exceed 6 feet. Prove that the right circular cylinder of greatest volume which can be sent is 2 feet long and 4 feet in girth.

31. Show that the height of the right circular cylinder of greatest volume which can be cut out from a spherical ball of radius r is $2r/\sqrt{3}$.

■ **21. Inverse Functions:** Let $f(x)$ be a single-valued function of x, continuous for the range $a \le x \le b$. Further, suppose that throughout the range, either $f(x)$ increases with x, or decreases as x increases. We shall prove that

(1) *The equation $y = f(x)$ determines x as a single-valued continuous function of y.*

This function is denoted by $f^{-1}(y)$, so that the statements $y = f(x)$ and $x = f^{-1}(y)$ mean the same thing. Also the function f^{-1} is called the *inverse* of f.

Proof. Since $f(x)$ is continuous, as x varies continuously from a to b, $f(x)$ takes every value from $f(a)$ to $f(b)$ at least once. Also $f(x)$ takes every value once only, for $f(x_{1}) = f(x_{2})$ only when $x_{1} = x_{2}$. Thus, corresponding to any value y_{0} of y between $f(a)$ and $f(b)$, there is a single value x_{0} of x such that $y_{0} = f(x_{0})$.

Again, $f^{-1}(y)$ is continuous. For, taking the case in which y increases with x, let y_{0} be any value of y between $f(a)$ and $f(b)$ and x_{0} the corresponding value of x, so that

$$a < x_{0} < b.$$

For sufficiently small values of ε, $x_0 - \varepsilon$ and $x_0 + \varepsilon$ are within the interval (a, b). Let $y_0 - k$ and $y_0 + k'$ be the corresponding values of y, and let η be the smaller of the two k, k'.

As y varies from $y_0 - \eta$ to $y_0 + \eta$, x lies between $x_0 - \varepsilon$ and $x_0 + \varepsilon$, for x increases with y.

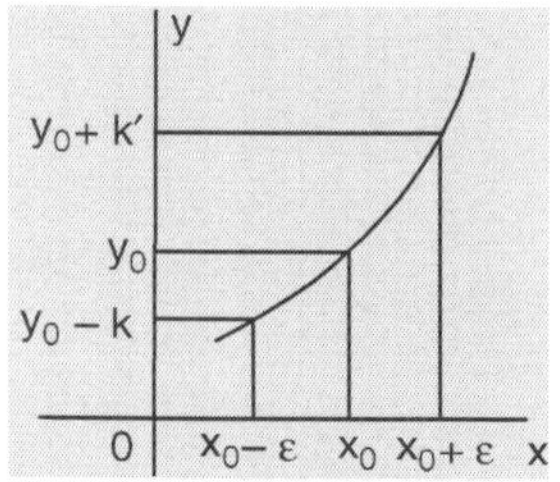

Fig. 54

Hence $| x - x_0 | < \varepsilon$, provided that $| y - y_0 | < \eta$; and therefore $x = f^{-1} (y)$ is continuous at y_0.

If y decreases as x increases, we take $y_0 + k$ and $y_0 - k'$ as the values of y corresponding to $x_0 - \varepsilon$ and $x_0 + \varepsilon$. To show that $f^{-1} (y)$ is continuous at $y = f(a)$ and $y = f(b)$, we consider the variation of y for the ranges

$$a < x < a + \varepsilon \text{ and } b - \varepsilon < x < b.$$

(2) If $\dfrac{dy}{dx}$ exists and is never zero for the range $a < x < b$, then $\dfrac{dx}{dy} = 1 / \dfrac{dy}{dx}$.

For if δx and δy are corresponding increments of x and y, then, since $\dfrac{dy}{dx}$ is finite, $\delta y \to 0$ as $\delta x \to 0$, and *vice versa*. Therefore

$$\frac{dx}{dy} = \lim \frac{\delta x}{\delta y} = \lim 1 \Big/ \frac{\delta y}{\delta x} = 1 \Big/ \lim \frac{\delta y}{\delta x} = 1 \Big/ \frac{dy}{dx}.$$

■ **22. The Inverse Circular Functions:** The numerically smallest value of y which has the same sign as x and is such that $\sin y = x$ is called the *principal value of* $\sin^{-1} x$. A similar definition applies to $\tan^{-1} x$.

The smallest positive value of y such that $\cos y = x$ is called the *principal value of* $\cos^{-1} x$.

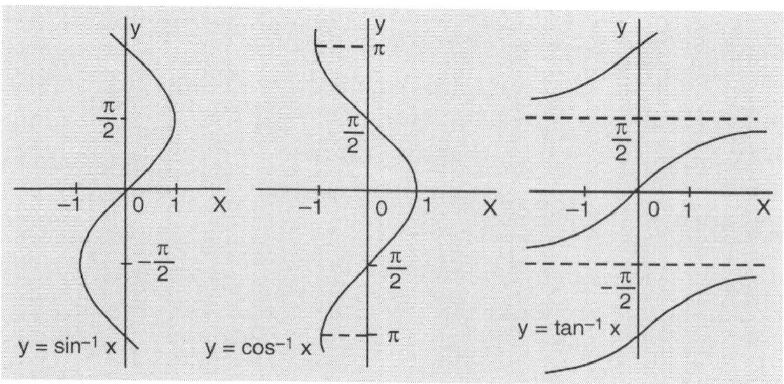

Fig. 55

Thus the principal values of $\sin^{-1} x$ and $\tan^{-1} x$ lie between $-\dfrac{\pi}{2}$ and $\dfrac{\pi}{2}$, and that of $\cos^{-1} x$ lies between 0 and π.

Unless otherwise stated, $\sin^{-1} x$, $\cos^{-1}x$, and $\tan^{-1}x$ will be used to represent the principal values of the functions; for these, the derivatives of $\sin^{-1} x$ and $\cos^{-1}x$ are positive and negative respectively, as can be seen from Fig. 55.

■ **EXAMPLE :** *Show that* $\dfrac{d}{dx} \tan^{-1} x = \dfrac{1}{1 + x^2}$.

Let $y = \tan^{-1}x$, then $x = \tan y$ and $\dfrac{dx}{dy} = \sec^2 y = 1 + x^2$; $\therefore \dfrac{dy}{dx} = \dfrac{1}{1 + x^2}$.

■ **23. Bounds of a Function:** Let $f(x)$ be a function which has a definite value for every value of x in the interval (a, b).

If a number M exists such that $f(x) \leq M$ for every x in (a, b), then $f(x)$ is said to be *bounded above*.

If a number N exists such that $f(x) \geq N$ for every x in (a, b), $f(x)$ is said to be *bounded below*.

A function which is bounded above and below is said to be *bounded*.

Suppose that $f(x)$ is bounded above in (a, b).

Let S denote the aggregate of the values of $f(x)$ as x varies continuously from a to b.

By applying Dedekind's theorem (Ch. 13, 12) to the set S, we can show that *a number h exists such that*

(*i*) *no value of f(x) exceeds h*;

(*ii*) *at least one value of $f(x)$ exceeds any number less than h.*

This number h is called the *upper bound* of $f(x)$.

Similarly, *if $f(x)$ is bounded below, a number l exists such that*

(*i*) *no value of $f(x)$ is less than l;*

(*ii*) *at least one value of f(x) is less than any number greater than l.*

This number l is called the *lower bound of $f(x)$.*

■ **EXAMPLE :** *Consider the function* $f(x) = \lim\limits_{n \to \infty} \dfrac{nx}{1 + nx^2}$. *If $x = 0$, $f(x) = 0$.*

If $x \neq 0$. $f(x) = 1/x$. Thus $f(x)$ has a definite value for every value of x. But $f(x)$ is not bounded in any interval including zero, for by making x small enough, we can make $1/x$ exceed any number we may choose.

Theorem 1. *If $f(x)$ is continuous in the closed interval (a, b), it is bounded in (a, b).*

Suppose that $f(x)$ is not bounded in (a, b). Because $f(x)$ is continuous at $x = a$, for sufficiently small values of $x–a$, $f(x)$ lies between $f(a) - \varepsilon$ and $f(a) + \varepsilon$, where ε is any small positive number. Hence there are values of x in (a, b) such that $f(x)$ is bounded in (a, x) and (if our supposition is possible) not bounded in (x, b).

Divide the real numbers in (a, b) into a lower class A and an upper class A' as follows: The number x is to be placed in A or in A' according as $f(x)$ is or is not bounded in (a, x).

According to the supposition, both classes exist and every number in A is less than every number in A'. The classification therefore defines a real number α, which is the greatest number in A or the least in A'.

Now $f(x)$ is continuous at $x = \alpha$, therefore for sufficiently small values of $x - \alpha$, $f(x)$ lies between $f(\alpha - \varepsilon)$ and $f(\alpha + \varepsilon)$, where ε is any small positive number. Hence $f(x)$ is bounded in $(a, \alpha + \varepsilon)$, and so $\alpha + \varepsilon$ belongs to A, which is impossible, for $\alpha + \varepsilon > \alpha$. This proves the theorem.

Theorem 2. * *If $f(x)$ is continuous in (a, b) and h, l are its upper and lower bounds, $f(x)$ takes the values h and l at least once as x varies continuously from a to b.*

For if ε is an assigned positive number, however small, there is at least one value of x in (a, b) for which $f(x) > h - \varepsilon$, so that

$$h - f(x) < \varepsilon \text{ and } 1/\{h - f(x)\} > 1/\varepsilon.$$

Hence $1/\{h - f(x)\}$ is not bounded. But if $h - f(x)$ does not vanish for some value of x in the interval, $1/\{h - f(x)\}$ is continuous in (a, b).

Therefore $h = f(x)$ for some value of x in (a, b).

Similarly there is a value of x for which $f(x) = l$.

■ **24. Rolle's Theorem:** *Suppose that $f(x)$ is continuous in the closed interval (a, b) and has a derivative $f'(x)$ for every x such that $a < x < b$.*

If $f(a) = 0$ and $f(b) = 0$, then $f'(x) = 0$ for at least one value of x between a and b. **

This theorem is of the greatest importance. A strict proof (as given on the next page) depends on the rather difficult considerations of the last article, but from the graphical point of view its truth is obvious.

Let the curve $y = f(x)$ cut the x-axis at A, B. Suppose that it is continuous from A to B, and that at every point it has a definite and finite gradient. The theorem asserts that *there is at least one point on the curve between A and B where the tangent is parallel to the x-axis.*

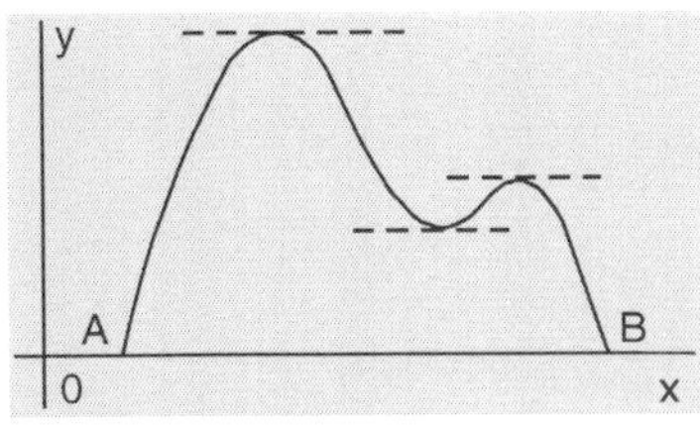

Fig. 56

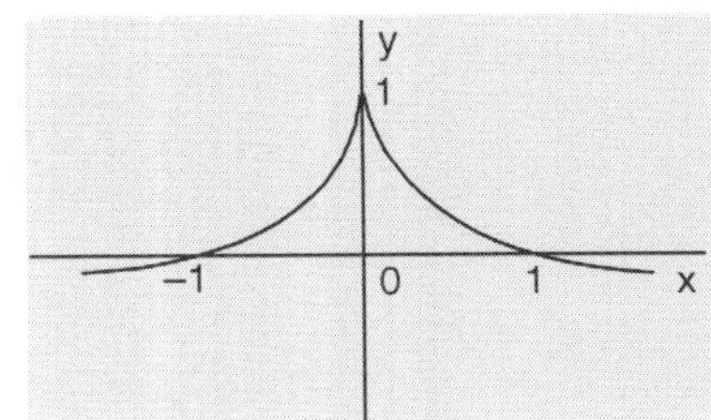

Fig. 57

Figure 57 represents the curve $y = 1 - x^{\frac{2}{3}}$. Here $f'(x) = -\infty$ at $x = 0$ and $f'(x)$ is not zero for any value of x between -1 and $+1$.

Proof. Because $f(x)$ is continuous in (a, b), it is bounded in (a, b) and attains at least once its upper bound h and its lower bound l.

Either (i) $f(x) = 0$ for all values of x in (a, b), in which case $f'(x) = 0$ for every x; or (ii) some values of $f(x)$ are positive or negative.

---

* This proof is taken from Hardy's *Pure Mathematics*.

** If $f'(x)$ exists for $a < x < b$, $f(x)$ must be continuous in the open interval (a, b), but not necessarily in the *closed* interval.

If positive values of $f(x)$ exist, h is positive and unequal to $f(a)$ or $f(b)$. Moreover, $f'(h) = 0$, for otherwise there would be values of x in the neighbourhood of h for which $f(x) > h$, which is impossible.

If there are negative values of $f(x)$, then l is negative, unequal to $f(a)$ or $f(b)$ and $f'(l) = 0$. For if $f'(l) \neq 0$, there would be values of x near l for which $f(x) < l$, which is impossible. This proves the theorem.

A simpler proof, for the case in which $f(x)$ is a polynomial, is given in the next chapter.

■ **25. The Mean-Value Theorem:** *If $f(x)$, is continuous for $a \leq x \leq b$ and $f'(x)$ exists for $a < x < b$, then*

$$f(b) - f(a) = (b - a) f'(\xi), \qquad \qquad ...(A)$$

where ξ is some number between a and b.

Proof. Let

$$\phi(x) = f(b) - f(x) - \frac{b - x}{b - a} \{f(b) - f(a)\},$$

then

$$\phi'(x) = -f'(x) + \frac{1}{b - a} \{f(b) - f(a)\}.$$

Also $\phi(a) = 0$ and $\phi(b) = 0$, hence by Rolle's theorem $\phi'(x) = 0$ for some value ξ of x between a and b, which proves the result in question.

Geometrically. Let A, B be the points on the curve $y = f(x)$ where $x = a, b$ respectively. The theorem asserts that there is a point C on the curve between A and B where the tangent is parallel to AB. If $P(x, y)$ is any point on the curve, and the ordinate NP (or NP produced) meets AB in Q, it is easily shown that

$$\phi(x) = NQ - NP = PQ.$$

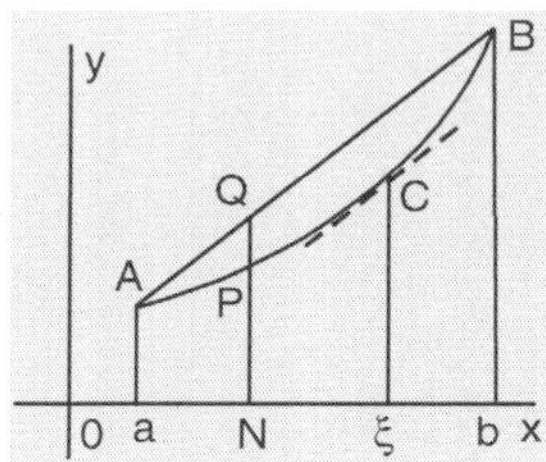

Fig. 58

Corollary. *If, throughout any given interval,*

(i) *$f'(x)$ is always positive, then $f(x)$ increases with x;*

(ii) *if $f'(x)$ is always negative, $f(x)$ decreases as x increases;*

(iii) *if $f'(x)$ is always zero, then $f(x)$ is constant throughout the interval.*

All this follows from equation (A), where it is supposed that a, b are any two values of x in the interval.

Note: It has not been assumed that $f'(x)$ is continuous.

■ **26. Integration:** Suppose that $f(x)$ is a given function of x, and that we try to find a function $\phi(x)$ such that $\phi'(x) = f(x)$. Such a function is called an *integral of $f(x)$* (*with regard to x*), and is denoted by $\int f(x)\,dx$.

If $\phi(x)$ and $\psi(x)$ are any two functions such that $\phi'(x) = \psi'(x)$, then $\phi(x)$ and $\psi(x)$ differ by a constant.

For
$$\frac{d}{dx}\{\phi(x) - \psi(x)\} = \phi'(x) - \psi'(x) = 0;$$

therefore, by Art. 25, Cor. (*iii*), $\phi(x) - \psi(x) =$ a constant C.

Hence if *any* value of $\phi(x)$ is denoted by $\int f(x)\,dx$, the *general* value is $\int f(x)\,dx + C$. Thus $\dfrac{d}{dx}\phi(x) = f(x)$, and $\phi(x) = \int f(x)\,dx + C$, are two equations which mean the same thing.

■ **EXAMPLE 1.** *Since $\dfrac{d}{dx}x^{n+1} = (n+1)x^n$, therefore $\int x^n dx = \dfrac{x^{n+1}}{n+1} + C.$*

Note: The symbol $\int \ldots dx$ is to be regarded as denoting the *operation* of integration, just as $\dfrac{d}{dx}$ denotes that of differentiation. These are *inverse* operations.

'Integration' can be applied to find the *area of a curve* as in the next example.

■ **EXAMPLE 2.** *If PN is the ordinate of any point $P(x, y)$ on the curve represented by $y = kx^n$ where $n > 0$, prove that the area ONP bounded by ON, NP and the arc OP is equal to $1/(n+1)$ of the rectangle contained by ON, NP.*

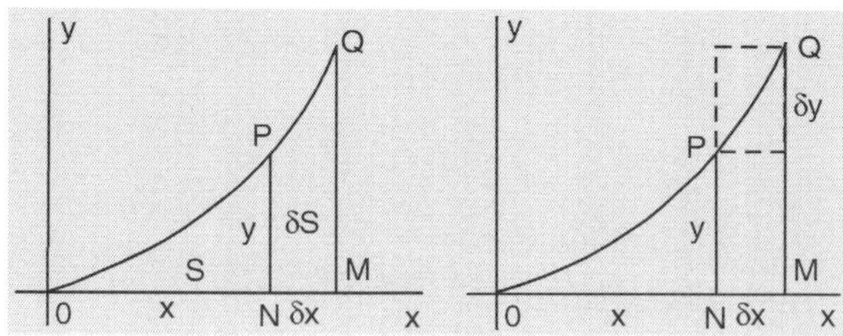

Fig. 59

Denote the area by S and let x, y be the coordinates of P. Let $\delta y,\ \delta S$ be the increments in y, S corresponding to the increment δx in x. If Q is the point $(x + \delta x, y + \delta y)$ and QM its ordinate, then rect. $PM < \delta S < $ rect. QN, i.e. $y\delta x < \delta S < (y + \delta y)\,\delta x$;

therefore
$$y < \frac{\delta S}{\delta x} < y + \delta y;\ \text{and since}\ \delta y \to 0,\ \therefore\ y = \frac{dS}{dx}.$$

Hence
$$S = \int y\, dx = \int kx^n\, dx = \frac{kx^{n+1}}{n+1} + C, \text{ where } C \text{ is a constant.}$$

Now $S = 0$ when $x = 0$; $\therefore C = 0$ and $S = kx^{n+1}/(n+1) = xy/(n+1)$.

■ **27. Taylor's Theorem, also known as the General Mean Value Theorem:**

If $f(x)$ is continuous in the closed interval (a, b) and the first n derivatives $f'(x)$, $f''(x) ...f^{(n)}(x)$ exist for all values of x between a and b, then

$$f(b) = f(a) + (b-a)f'(a) + \frac{1}{\lfloor 2} (b-a)^2 f''(a)$$

$$+ \frac{1}{\lfloor n-1} (b-a)^{n-1} f^{(n-1)}(a) + \frac{1}{\lfloor n} (b-a)^n f^{(n)}(\xi), \qquad ...(A)$$

where ξ is some number between a and b.

The proof is similar to that of Art. 25.   Let
$$F(x) = f(b) - f(x) - (b-x)f'(x)$$

$$- \frac{1}{\lfloor 2} (b-x)^2 f''(x) - ... - \frac{1}{\lfloor n-1} (b-x)^{n-1} f^{(n-1)}(x),$$

and
$$\phi(x) = F(x) - \left(\frac{b-x}{b-a}\right)^n F(a),$$

then
$$F'(x) = - \frac{1}{\lfloor n-1} (b-x)^{n-1} f^{(n)}(x),$$

and
$$\phi'(x) = - \frac{1}{\lfloor n-1} (b-x)^{n-1} f^{(n)}(x) + n \frac{(b-x)^{n-1}}{(b-a)^n} F(a)$$

$$= \frac{n(b-x)^{n-1}}{(b-a)^n} \left\{ F(a) - \frac{1}{\lfloor n} (b-a)^n f^{(n)}(x) \right\}.$$

Now $\phi(a) = 0$ and $\phi(b) = 0$, therefore by Rolle's theorem $\phi'(x) = 0$ for some value ξ of x between a and b, hence

$$F(a) = \frac{1}{\lfloor n} (b-a)^n f^{(n)}(\xi),$$

which is equivalent to equation (A).

Let $b = a + h$, then equation (A) may be written

$$f(a+h) = f(a) + hf'(a) + \frac{1}{\lfloor 2} h^2 f''(a) + ... + \frac{1}{\lfloor n-1} h^{n-1} f^{(n-1)}(a) + R_n, \qquad ...(B)$$

where
$$R_n = \frac{1}{\lfloor n} h^n f^{(n)}(a + \theta h) \text{ and } 0 < \theta < 1.$$

For any number between a and $a + h$ may be written in the form $a + \theta h$ where $0 < \theta < 1$. In general, *the value of θ depends on n.*

Taylor's Series. *Suppose that $f(x)$ is continuous in the closed interval $(a - c, a + c)$, and the n derivatives $f'(x), f''(x), \ldots f^{(n)}(x)$ exist when $a - c < x < a + c$ and that $|h| < c$. Then equation (B) holds.*

Further suppose that $R_n = \dfrac{1}{\lfloor n} h^n f^{(n)}(a + \theta h) \to 0$ *as* $n \to \infty$,

that $f(a + h) = f(a) + hf'(a) + \dfrac{1}{\lfloor 2} h^2 f''(a) + \ldots + \dfrac{1}{\lfloor n} h^n f^{(n)}(a) + \ldots$ *to* ∞.   ...(C)

For if s_n is the sum to n terms of the series,

$$f(a + h) - s_n = R_n \to 0 \text{ so that } s_n \to f(a + h).$$

The series in (C) is called *Taylor's Series* and R_n is *Lagrange's* form of the remainder after n terms. When $a = 0$, equation (A) becomes

$$f(h) = f(0) + hf'(0) + \dfrac{1}{\lfloor 2} h^2 f''(0) + \ldots \text{ to } \infty. \qquad \ldots(D)$$

This expansion is known as *Maclaurin's Series.*

THE COMPLEX VARIABLE

■ **28. The Quantity $x + \iota y$,** which we denote by z, is called the complex variable. Here x and y are real variables, independent of one another and each capable of assuming any values from $-\infty$ to $+\infty$.

As z passes from a value z_0 to a value z_1, we suppose the variation to be *continuous.* By this we mean that the point z is to move along a continuous curve from the point z_0 to the point z_1.

Thus the variation of z is continuous if the variations of x and y are continuous.

■ **29. Function of a Complex Variable z:** If $Z = X + \iota Y$, where X and Y are functions of x and y, we say that Z is a *complex function of the real variables x and y.*

If z is known, so also are x and y. Consequently X, Y and Z are known. Hence it would be perfectly reasonable to call Z a function of z.

But it is usual to restrict the meaning of the term 'function of z' as follows.

We use the equation $Z = f(z)$ to indicate that Z is the result of definite operations performed on z, the variables x and y not occurring explicitly, and we say that Z is a function of z.

Thus $2x + 3\iota y$ is a function of z in the first sense, but not in the restricted sense.

For if z is known, so is $2x + 3\iota y$; but $2x + 3\iota y$ cannot be expressed in terms of z alone, *i.e.* without the explicit use of x or y.

It follows that if $\qquad Z = X + \iota Y = f(z) = f(x + \iota y)$,

one series of operations determines X and Y in terms of x and y.

▪ 30. Definitions Relating to Limits and Continuity:

(1) To say that z *tends to zero* $(z \to 0)$ is to say that z varies in such a way that $|z|$ becomes and remains less than any positive ε we may choose, however small.

Thus, after a certain stage, the point z (on an Argand diagram) is within a circle with centre O and radius ε, which may be as small as we like.

To say that z tends to a $(z \to a)$, where a is a fixed number, is to say that $z - a \to 0$. After a certain stage, the point z is within a circle with centre a and radius ε, which may be as small as we like.

(2) We say that $f(z)$ tends to a limit l as z tends to a if, for any positive ε that we may choose, however small, η can be found so that

$$|f(z) - l| < \varepsilon, \text{ provided only that } |z - a| < \eta.$$

In other words, as z approaches a *by any path whatever* the distance of the point $f(z)$ from the fixed point l becomes and remains less than any length ε that we may choose, however small.

(3) The function $f(z)$ is continuous for $z = z_0$ or at (the point) z_0 if for any positive ε that we may choose, however small, we can find η so that

$$|f(z) - f(z_0)| < \varepsilon, \text{ provided only that } |z - z_0| < \eta.$$

This is the same as saying that as z tends to z_0 along any path whatever, $f(z)$ tends to $f(z_0)$ as a limit.

It is assumed that $f(z)$ has a definite value at every point in the neighbourhood of the point z_0.

(4) The function $f(z)$ is continuous in a region A if it is continuous at every point of the region.

Hence if z describes a continuous curve in the region of continuity of the function $f(z)$, the point $f(z)$ will also describe a continuous curve.

▪ 31. Continuity of Rational Functions of z:
All that has been said in Arts. 2 and 3, with regard to the real variable, holds for the complex variable. In particular:

(*i*) $\lim\limits_{z \to 0} (a_0 + a_1 z + a_2 z^2 + ... + a_n z^n) = a_0.$

(*ii*)   A polynomial in z is continuous for all values of z.

(*iii*)   A rational function of z is continuous for all values of z except for those which make the denominator zero.

EXERCISE XXIX

1. Prove that   (*i*) $\sin x - x$ always decreases with x.

(*ii*) $\tan x - x$ increases with x in the intervals

$$\left(-\tfrac{1}{2}\pi, \tfrac{1}{2}\pi\right), \left(\tfrac{1}{2}\pi, \tfrac{3}{2}\pi\right), ...$$

(*iii*) $\dfrac{\sin x}{x}$ decreases as x increases from 0 to $\tfrac{1}{2}\pi$.

2. If $a,\ ax + b,\ ax^2 + 2bx + c,\ ax^3 + 3bx^2 + 3cx + d$ and $ax^4 + 4bx^3 + bcx^2 + 4dx + e$ are denoted by u_0, u_1, u_2, u_3, respectively, show that, if $U = u_0^2 u_3 - 3u_0 u_1 u_2 + 2u_1^3,$

$V = u_0u_4 - 4u_1u_3 + 3u_2{}^2$, then $\dfrac{dU}{dx} = 0$ and $\dfrac{dV}{dx} = 0$, so that U and V are independent of x.

3. If $\quad u_n = \cos x - \left(1 - \dfrac{x^2}{\underline{|2}} + \dfrac{x^4}{\underline{|4}} - \ldots + (-1)^n \dfrac{x^{2n}}{\underline{|2n}}\right)$

and $\quad v_n = \sin x - \left(x - \dfrac{x^3}{\underline{|3}} + \dfrac{x^5}{\underline{|5}} - \ldots + (-1)^n \dfrac{x^{2n+1}}{\underline{|2n+1}}\right)$,

prove that $\qquad \dfrac{du_n}{dx} = -v_{n-1}, \quad \dfrac{dv_n}{dx} = u_n.$

Hence show that u_n and v_n are positive or negative according as n is odd or even.
[If $x = 0$, then $u_n = 0$ and $v_n = 0$; also $u_0 = \cos x - 1 < 0$, $\quad v_0 = \sin x - x < 0$.

Now $\qquad \dfrac{du_1}{dx} = -v_0 > 0, \therefore u_1$ is an increasing function, $\therefore u_1 > 0$;

and $\qquad \dfrac{dv_1}{dx} = u_1 > 0, \therefore v_1$ is an increasing function, $\therefore v_1 > 0$;

and so on, in succession for $(u_2, v_2), (u_3, v_3),\ldots .$]

4. Prove that for all values of x

$\quad (i) \ \cos x = 1 - \dfrac{x^2}{\underline{|2}} + \dfrac{x^4}{\underline{|4}} - \ldots$ to ∞;

$\quad (ii) \ \sin x = x - \dfrac{x^3}{\underline{|3}} + \dfrac{x^5}{\underline{|5}} - \ldots$ to ∞.

$[(i)$ Let t_r be the rth term of the series and s_r the sum to r terms. Then by Ex. 3, for every x, $s_{2n} < \cos x < s_{2n+1}$ and $s_{2n+1} - s_{2n} = t_{2n+1} \to 0$ as $n \to \infty$. Therefore $s_r \to \cos x$ as $r \to \infty$.]

5. Use Taylor's theorem to prove that

$\quad (i) \ \sin (x + h) = \sin x + h \cos x - \dfrac{h^2}{\underline{|2}} \sin x - \dfrac{h^3}{\underline{|3}} \cos x + \ldots;$

$\quad (ii) \ \cos (x + h) = \cos x - h \sin x - \dfrac{h^2}{\underline{|2}} \cos x + \dfrac{h^3}{\underline{|3}} \sin x + \ldots .$

$[(i)$ If $f(x) = \sin x, \ |R_n| < \dfrac{1}{\underline{|n}} \cdot |h^n| \to 0$ as $n \to \infty.]$

6. Deduce the expansions of $\sin x$ and $\cos x$ in Ex. 4 from Taylor's theorem.

7. Prove, by means of Art. 22, Ex. 1, that, if $0 < x \leq 1$, then

$$\tan^{-1} x = x - \dfrac{x^3}{3} + \dfrac{x^5}{5} - \dfrac{x^7}{7} + \text{ to } \infty.$$

[Let $S_r = x - \dfrac{x^3}{3} + \dfrac{x^5}{5} - \dots + (-1)^{r-1} \dfrac{x^{2r-1}}{2r-1}$, and $f(x) = \tan^{-1} x - S_r$; then

$f(0) = 0$, and $f'(x) = 1/(1 + x^2) - (1 - x^2 + x^4 - \dots + (-1)^{r-1} x^{2r-2}$, so that

$$(1 + x^2)\, f'(x) = (-1)^r x^{2r},$$

which is positive or negative according as r is even or odd.

Hence, $\tan^{-1} x$ lies between S_r and S_{r+1}, and $|\, S_r - S_{r+1} \,| \to 0$, as $r \to \infty$, so long as x is positive and not greater than unity.]

8. If $f(x)$ is a function whose first derivative is $1/x$, show that, if $0 < x \le 1$,

$$f(1 + x) = x - \frac{x^2}{2} + \frac{x^3}{3} - \dots \text{ to } \infty.$$

[Proceed as in Ex. 7.]

9. If $0 < \theta \le \pi/6$, prove that $\tan \theta$ lies between $\theta + \frac{1}{3}\theta^3$ and $\theta + \frac{8}{9}\theta^3$.

[For the second inequality, let $f(\theta) = \theta + a\theta^3 - \tan \theta$; show that $f(0), f'(0), f''(0)$, are all zero, and that $f'''(\theta) \ge 0$ throughout the interval $0 < \theta \le \pi/6$ if $9a \ge 8$; hence, in succession, $f''(\theta), f'(\theta), f(\theta)$, are all positive.]

The Binomial Theorem

10. If n is a positive integer, show that the expansion of $(a + h)^n$ by the Binomial theorem is given by Taylor's series.

11. If n is any rational number and $-1 < x < 1$, (i) show that

$$(1 + x)^n = 1 + nx + \frac{n(n-1)}{\underline{|2}}\, x^2 + \dots + \frac{n(n-1)\dots(n-r+2)}{\underline{|r-1}}\, x^{r-1} + R_r$$

where $R_r = \dfrac{n(n-1)\dots(n-r+1)}{\underline{|r}}\, x^r\, (1 + \theta x)^{n-r}$ and $0 < \theta < 1$.

(ii) If $0 < x < 1$, prove that $R_r \to 0$ as $r \to \infty$, so that in this case

$$(1 + x)^n = 1 + nx + \frac{n(n-1)}{\underline{|2}}\, x^2 + \dots \text{ to } \infty. \qquad \dots(A)$$

[For $r > n$, $(1 + \theta x)^{n-r} < 1$ and $\dfrac{n(n-1)\dots(n-r+1)}{\underline{|r}}\, x^r \to 0$ as $r \to \infty$.]

(iii) If $-1 < x < 0$, explain why we are unable to say that $R_r \to 0$.

[Here $(1 + \theta x)^{n-r} > 1$ for $r > n$, and so far as we can tell, R_r may not tend to zero. To deduce the expansion (A) in this case, another form (named after *Cauchy*) of the remainder after n terms of Taylor's series must be used. See any treatise on the 'Calculus.']

■ ■ ■

CHAPTER 18

Theory of Equations, Polynomials, Rational Fractions

■ **1. Multiple Roots:** *If α is an r-multiple root of $f(x) = 0$, where $f(x)$ is a polynomial, then $(x - \alpha)^{r-1}$ is a common factor of $f(x)$ and $f'(x)$.*

For $f(x) = (x - \alpha)^r \cdot \phi(x)$ where $\phi(x)$ is a polynomial and
$$f'(x) = r(x - \alpha)^{r-1} \cdot \phi(x) + (x - \alpha)^r \cdot \phi'(x).$$

■ **2. Rolle's Theorem for Polynomials:** *If $f(x)$ is a polynomial, at least one real root of $f'(x) = 0$ lies between any two real roots of $f(x) = 0$.*

Proof. Let α, β be consecutive real roots of $f(x) = 0$, so that
$$f(x) = (x - \alpha)^m (x - \beta)^n \cdot \phi(x)$$
where m, n are positive integers, and $\phi(x)$ is not divisible by $x - \alpha$ or $x - \beta$. Then $\phi(x)$ is of invariable sign for $\alpha \le x \le \beta$ and
$$f'(x) = (x - \alpha)^{m-1} (x - \beta)^{n-1} \cdot \psi(x),$$

where
$$\psi(x) = \{m(x - \beta) + n(x - \alpha)\} \cdot \phi(x) + (x - \alpha)(x - \beta) \phi'(x).$$

Now $\qquad \psi(\alpha) = m(\alpha - \beta) \phi(\alpha)$ and $\psi(\beta) = n(\beta - \alpha) \phi(\beta)$.

Therefore $\psi(\alpha)$ and $\psi(\beta)$ have unlike signs; moreover, $\psi(x)$ is everywhere continuous.

Hence $\psi(x)$ and consequently $f'(x)$ is equal to zero for some value of x between α and β.

■ **3. Deductions from Rolle's Theorem:**

(1) *If all the roots of $f(x) = 0$ are real, so also are those of $f'(x) = 0$, and the roots of the latter equation separate those of the former.*

For if $f(x)$ is of degree n, $f'(x)$ is of degree $n - 1$, and a root of $f'(x) = 0$ exists in each of the $n - 1$ intervals between the n roots of $f(x) = 0$.

(2) *If all the roots of $f(x) = 0$ are real, so also are those of $f'(x) = 0$, $f''(x) = 0$, $f'''(x) = 0$, ... , and the roots of any one of these equations separate those of the preceding equation.*

This follows from (1).

(3) *Not more than one root of $f(x) = 0$ can*

    (i) lie between two consecutive roots of $f'(x) = 0$, or

    (ii) be less than the least of these, or

    (iii) be greater than the greatest of them.

For let β_1, β_2, ... β_r be the real roots of $f'(x) = 0$, any of which may be multiple roots, and suppose that $\beta_1 < \beta_2 < ... < \beta_r$. Let α_1, α_2 be real roots of $f(x) = 0$.

If $\alpha_1 = \alpha_2$, then α_1 is one of the set β_1, $\beta_2 = \beta_r$. (Art. 1.)

If $\alpha_1 \neq \alpha_2$, by Rolle's theorem, a root of $f'(x)$ lies between α_1 and α_2.

Hence (*i*) if $\beta_1 < \alpha_1 < \alpha_2 < \beta_2$, then β_1 and β_2 cannot be consecutive roots; (*ii*) if $\alpha_1 < \alpha_2 < \beta_1$, then β_1 cannot be the least root of $f'(x) = 0$;

(*iii*) if $\beta_r < \alpha_1 < \alpha_2$, then β_r cannot be the greatest.

Thus, *not more than one root of $f(x) = 0$ can lie in any one of the open intervals*
$$(-\infty, \beta_1), (\beta_1, \beta_2),... (\beta_{r-1}, \beta_r), (\beta_r, \infty).$$

(4) *If $f'(x) = 0$ has r real roots, then $f(x) = 0$ cannot have more than $r + 1$ real roots.*

If $f(x) = 0$ has no multiple roots, none of the roots of $f'(x) = 0$ is a root of $f(x) = 0$, and the theorem follows from (3).

If $f(x) = 0$ has an *m*-multiple root, we regard this as the limiting case in which *m* roots tend to equality. Thus the theorem is true in all cases.

(5) *If $f^{(r)}(x)$ is any derivative of $f(x)$ and the equation $f^{(r)}(x) = 0$ has imaginary roots, then $f(x) = 0$ has at least as many.*

It follows from (4) that $f(x) = 0$ has at least as many imaginary roots as $f'(x) = 0$.

(6) If all the real roots β_1, β_2, ... of $f'(x) = 0$ are known, we can find the number of real roots of $f(x) = 0$ by considering the signs of $f(\beta_1)$, $f(\beta_2)$, ... A single root of $f(x) = 0$, or no root, lies between β_1 and β_2, according as $f(\beta_1)$ and $f(\beta_2)$ have opposite signs, or the same sign.

■ **EXAMPLE 1.** *Find the character of the roots of $f(x) \equiv 3x^4 - 8x^3 - 6x^2 + 24x + 7 = 0$.*
Here $f'(x) = 12 (x^2 - 1) (x - 2)$. The roots of $f'(x) = 0$ are -1, 1, 2.

x	$-\infty$	-1	1	2	∞
$f(x)$	$+$	$-$	$+$	$+$	$+$

and $f(x) = 0$ has a real root < -1, one between -1 and 1, and two imaginary roots.

■ **EXAMPLE 2.** *If $b^2 - ac < 0$, then the equation $f(x) = (a, b, c,... k) (x, 1)^n = 0$ has at least two imaginary roots.* See Ch. 6, 11, Ex. 1.

For $\qquad\qquad f^{(n-2)}(x) = n(n-1) ... 3. (ax^2 + 2bx + c)$

and the equation $ax^2 + 2bx + c = 0$ has two imaginary roots.

■ **4. Sums of Powers of the Roots of an Equation—Newton's Theorem:**

Let α, β, ... κ be the roots of
$$f(x) \equiv x^n + p_1 x^{n-1} + p_2 x^{n-2} +...+ p_n = 0, \qquad\qquad ...(A)$$
and let $s_r = \alpha^r + \beta^r +... + \kappa^r$ $(r = 0, 1, 2,...)$, so that $s_0 = n$, then

(*i*) $s_r + p_1 s_{r-1} + p_2 s_{r-2} + ... + s_1 p_{r-1} + r p_r = 0$ $(r < n)$,

(*ii*) $s_r + p_1 s_{r-1} + p_2 s_{r-2} + ... + p_n s_{r-n} = 0$ $(r \geq n)$.

Proof. (*i*) We have identically $f(x) = (x - \alpha) (x - \beta) ... (x - \kappa)$, $\qquad ...(B)$

and, assuming the rule for differentiating a product (Ex. 2, p. 270),

$$f'(x) = \frac{f(x)}{x-\alpha} + \frac{f(x)}{x-\beta} +...+ \frac{f(x)}{x-\kappa}. \qquad\qquad ...(C)$$

By synthetic division,

$$f(x)/(x - \alpha) = x^{n-1} + A_1 x^{n-2} + A_2 x^{n-3} + \ldots + A_{n-1},$$

where the coefficients are given by

$A_1 = \alpha + p_1, A_2 = \alpha^2 + p_1\alpha + p_2, A_3 = \alpha^3 + p_1\alpha^2 + p_2\alpha + p_3$, and so on, in succession; and finally, $A_{n-1} = \alpha^{n-1} + p_1\alpha^{n-2} + \ldots + p_{n-1}$.

If the other quotients in (C) are expressed in a similar way, it will be seen that

$$f'(x) = nx^{n-1} + Q_1 x^{n-2} + Q_2 x^{n-3} + \ldots + Q_n, \qquad \ldots\text{(D)}$$

where $Q_1 = s_1 + np_1$, $Q_2 = s_2 + p_1 s_1 + np_2$, $Q_3 = s_3 + p_1 s_2 + p_2 s_1 + np_3$, and so on; and $Q_{n-1} = s_{n-1} + p_1 s_{n-2} + \ldots + p_{n-2} s_1 + np_{n-1}$.

Now $f'(x) = nx^{n-1} + (n - 1)p_1 x^{n-2} + (n - 2)p_2 x^{n-3} + \ldots + p_{n-1}$; and comparing this with (D), $Q_1 = (n - 1)p_1$, $Q_2 = (n - 2)p_2$, ..., $Q_n = p_{n-1}$.

Therefore

$$s_1 + p_1 = 0,$$
$$s_2 + p_1 s_1 + 2p_2 = 0,$$
$$s_3 + p_1 s_2 + p_2 s_1 + 3p_3 = 0,$$

$$\cdots\cdots\cdots\cdots\cdots\cdots\cdots\cdots\cdots\cdots\cdots$$

$$s_{n-1} + p_1 s_{n-2} + \ldots + (n - 1)p_{n-1} = 0. \qquad \ldots\text{(E)}$$

(*ii*) *Let* $r \geq n$, then multiplying each side of $f(\alpha) = 0$ by α^{r-n} we have

$$\alpha^r + p_1\alpha^{r-1} + p_2\alpha^{r-2} + \ldots + p_n\alpha^{r-n} = 0,$$

and similar equations hold for each root.

Hence the second result follows by addition.

■ **Example** : *For equation* (A) *of the last article, find the value of* $\Sigma (\alpha - \beta)^r$, *where r is an even number, in terms of* $s_0, s_1, s_2, \ldots$.

If $c_1, c_2, \ldots$ are the coefficients of $x, x^2, \ldots$ in the expansion of $(1 + x)^r$, then

$$(\alpha - x)^r = \alpha^r - c_1 x\alpha^{r-1} + c_2 x^2\alpha^{r-2} - \ldots + x^r.$$

In this equation, substitute $\alpha, \beta, \gamma, \ldots \kappa$ in succession for x and add, therefore,

$$(\alpha - \alpha)^r + (\alpha - \beta)^r + (\alpha - \gamma)^r + \ldots + (\alpha - \kappa)^r$$
$$= s_0\alpha^r - c_1 s_1\alpha^{r-1} + c_2 s_2\alpha^{r-2} - \ldots + s^r.$$

Similar equations can be deduced from the expansions of $(\beta - x)^r$, $(\gamma - x)^r$, etc.: hence, by addition, we have

$$2\Sigma (\alpha - \beta)^r = s_0 s_r - c_1 s_1 s_{r-1} + c_2 s_2 s_{r-2} - \ldots + s_r s_0.$$

■ **5. Order and Weight of Symmetric Functions:** Let $\alpha, \beta, \gamma \ldots$ be the roots of

$$x_1^n + p_1 x^{n-1} + p_2 x^{n-2} + \ldots + p_n = 0. \qquad \ldots\text{(A)}$$

Consider any function of the coefficients; for example, let

$$u = p_1^2 p_2 - 2p_2^2 - p_1 p_3 + 4p_4.$$

Let u be expressed in terms of the roots by means of the equalities

$$p_1 = - \Sigma \alpha, p_2 = \Sigma \alpha\beta, p_3 = - \Sigma \alpha\beta\gamma, \ldots .$$

Consider *the highest power of any root* α *which occurs in this expression for u*, and denote it by α^s.

Each of $p_1, p_2, \ldots$ involves α to the first degree only, therefore α^s arises from the highest term $p_1^2 p_2$ and *from this alone* : also the index s is the degree of this term, namely 3.

The *order* of a symmetric function of the roots of an equation is defined as the index of the highest power of any root which occurs in the function.

Thus the *order* of $\Sigma\alpha^3\beta$ is 3, and that of $\Sigma\alpha\beta\gamma$ is 1.

From what has been said it will be clear that *the order of a symmetric function of the roots is the degree of the function when expressed in terms of the coefficients p_1, p_2, ... p_n.*

If u is a symmetric function of order s of the roots of

$$a_0x^n + a_1x^{n-1} + ... + a_n = 0, \qquad\qquad ...(B)$$

then $a_0^s u$ is a homogeneous function of a_0, a_1, a_2, ... of degree s.

For if the equation is written in the form (A), u is a function of p_1, p_2, ... in which the highest term is of degree s and $p_1 = a_1/a_0$, $p_2 = a_2/a_0$, etc.

The *weight* of a symmetric function (u) of the roots α, β, ... is its *degree in α, β, ...* .

Thus the *weight* of $\Sigma\,\alpha^3\beta$ is 4, and that of $\Sigma\,\alpha\beta\gamma$ is 3.

Theorem: *If a symmetric function (u) of the roots of equation (A) is expressed in terms of the coefficients p_1, p_2, ... the sum of the suffixes of every term, when written at full length,* is equal to the weight (w) of u.*

Proof. Denote any term in u by $p_1{}^h p_2{}^k p_3{}^l$ ... ; and in equation (A) write $x = y / \lambda$. The equation becomes $y^n + \lambda p_1 y^{n-1} + \lambda^2 p_2 y^{n-2} + ... + \lambda^n p_n ... 0$, and the roots of this are $\lambda\alpha$, $\lambda\beta$, $\lambda\gamma$, ...

If then we multiply every root by λ, the effect of this is

   (*i*)   to multiply u by λ^w ;

   (*ii*)   to multiply p_1, p_2, p_3 by λ, λ^2, λ^3, ... respectively;

   (*iii*)   to multiply $p_1{}^h p_2{}^k p_3{}^l$ ... by $\lambda^{h+2k+3l+\,\cdots}$;

$\therefore \qquad\qquad w = h + 2k + 3l +... ,$

which proves the theorem.

■ **EXAMPLE :** *Write down the literal part of $\Sigma\alpha^2\beta^2$, expressed as a function of the coefficients.*

$\Sigma\alpha^2\beta^2$ is of order 2 and weight 4. The only terms of weight 4 are such as contain $P_1{}^4$, $p_1{}^2 p_2$, $P_1 P_3$, $P_2{}^2$, p_4. The first two of these are of too high an order.

Thus we have $\qquad\qquad \Sigma\alpha^2\beta^2 = ap_1 p_3 + bp_2{}^2 + cp_4,$

where a, b, c are independent of p_1, p_2, p_3.

■ **6. Partial Derivatives:** Suppose that $u \equiv f(x, y, z, ...)$ is a function of the independent variables $x, y, z,...$. The operation of differentiating u with respect to x, on the supposition that *x alone varies,* is denoted by the symbol $\dfrac{\partial}{\partial x}$, and $\dfrac{\partial u}{\partial x}$ is called the first *partial* derivative of u with respect to x. Sometimes the ordinary symbol $\dfrac{d}{dx}$ is used in the same sense, the context implying that the differentiation is partial.

---

* Thus the term $p_1{}^2 p_2{}^3$ is written $P_1 P_1 P_2 P_2 P_2$.

We also write

$$\frac{\partial}{\partial x}\left(\frac{\partial u}{\partial y}\right) = \frac{\partial^2 u}{\partial x \partial y} \text{ and } \frac{\partial}{\partial y}\left(\frac{\partial u}{\partial x}\right) = \frac{\partial^2 u}{\partial y \partial x},$$

with similar notation for higher derivatives.

For example, if $u = x^m y^n$,

$$\frac{\partial}{\partial x}\left(\frac{\partial u}{\partial y}\right) = \frac{\partial}{\partial x}(nx^m y^{n-1}) = mnx^{m-1} y^{n-1},$$

$$\frac{\partial}{\partial y}\left(\frac{\partial u}{\partial x}\right) = \frac{\partial}{\partial y}(mx^{m-1} y^n) = mnx^{m-1} y^{n-1}.$$

Therefore $\qquad \dfrac{\partial^2 u}{\partial x \partial y} = \dfrac{\partial^2 u}{\partial y \partial x}.$ $\hspace{2cm}$ ...(A)

It follows that *equation* (A) *holds when u is a polynomial in x, y.*

It is proved in books on the 'Calculus' that the commutative property represented by (A) holds when the functions in question are continuous. A strict proof is difficult, and here we are only concerned with polynomials.

Again, we often regard $h\dfrac{\partial}{\partial x} + k\dfrac{\partial}{\partial y}$ as an operator, and we write

$$h\frac{\partial u}{\partial x} + k\frac{\partial u}{\partial y} = \left(h\frac{\partial}{\partial x} + k\frac{\partial}{\partial y}\right) u = Du,$$

$$D^2 u = D\,(Du) = D\left(h\frac{\partial u}{\partial x} + k\frac{\partial u}{\partial y}\right) = h\left(h\frac{\partial^2 u}{\partial x^2} + k\frac{\partial^2 u}{\partial x \partial y}\right) + k\left(h\frac{\partial^2 u}{\partial x \partial y} + k\frac{\partial^2 u}{\partial y^2}\right)$$

$$= h^2\frac{\partial^2 u}{\partial x^2} + 2hk\frac{\partial^2 u}{\partial x \partial y} + k^2\frac{\partial^2 u}{\partial y^2},$$

if h, k are constants; a result which we may express as follows:

$$\left(h\frac{\partial}{\partial x} + k\frac{\partial}{\partial y}\right)^2 u = \left\{h^2\left(\frac{\partial}{\partial x}\right)^2 + 2hk\frac{\partial}{\partial x}\frac{\partial}{\partial y} + k^2\left(\frac{\partial}{\partial y}\right)^2\right\} u.$$

The value of $D^n u$ can be written down in the same way by the Binomial theorem.

■ **7. Taylor's Theorem for Polynomials:**

(1) *If f(x) is a polynomial of degree n in x, then*

$$f(x + h) = f(x) + hf'(x) + \frac{h^2}{\lfloor 2}f''(x) + ... + \frac{h^n}{\lfloor n}f^{(n)}(x). \hspace{1.5cm} ...(A)$$

Proof. This follows from Ch. 17, or thus: we may assume that

$$f(x + h) = a_0 + ha_1 + \frac{h^2 a_2}{\lfloor 2} + ... + \frac{h^n}{\lfloor n}a_n$$

where $a_0, a_1, ... a_n$ are independent of h, and if $X = x + h$, we have

$$\frac{d}{dh}f(x + h) = \frac{dX}{dh} \cdot \frac{d}{dX}f(X) = f'(x + h).$$

Hence, by successive differentiation with regard to h,

$$f'(x + h) = a_1 + ha_2 + \frac{h^2}{\lfloor 2} a_3 + \ldots + \frac{h^{n-1}}{\lfloor n-1} a_n,$$

$$f''(x + h) = a_2 + ha_3 + \frac{h^2}{\lfloor 2} a_4 + \ldots + \frac{h^{n-2}}{\lfloor n-2} a_n,$$

$$\cdots\cdots\cdots ,$$

$$f^{(n)}(x + h) = a_n.$$

The result follows by putting $h = 0$ in these equations; for we find that

$$a_1 = f'(x), \ a_2 = f''(x), \ \ldots \ , \ a_n = f^{(n)}(x).$$

(2) *If $u \equiv f(x, y)$ is a polynomial of degree n in x, y, then*

$$f(x + h, y + k) = u + Du + \frac{1}{\lfloor 2} D^2 u + \ldots + \frac{1}{\lfloor n} D^n u. \qquad \ldots(B)*$$

For we may assume that

$$f(x + h, y + k) = a_0 + (b_0 h + b_1 k) + \Sigma_{m=2}^{n} \ \frac{1}{\lfloor m} \ (p_0, p_1, \ldots p_m)(h, k)^m, \qquad \ldots(C)$$

where a_0, b_0, $\ldots$ are independent of h, k.

Putting $h = 0$, $k = 0$ we have $a_0 = u$; and differentiating $m{-}r$ times with regard to h and r times with regard to k, we find that

$$\left(\frac{\partial}{\partial h}\right)^{m-r} \left(\frac{\partial}{\partial k}\right)^{r} f(x + h, y + k) = p_r + \text{terms involving } h, k. \qquad \ldots(D)$$

Then, if $X = x + h$, $Y = y + k$, the left-hand side of (D) is equal to

$$\left(\frac{\partial}{\partial X}\right)^{m-r} \left(\frac{\partial}{\partial y}\right)^{r} f(X, Y) = f_{m-r, r}(X, Y), \text{ say, } i.e., f_{m-r, r}(x + h, y + k);$$

and putting $h = 0$, $k = 0$, we have $p_r = f_{m-r, r}(x, y) = \dfrac{\partial^m u}{\partial x^{m-r} \partial y^r}$, which proves the theorem.

(3) *If $u \equiv f(x, y, z, \ldots)$ is a polynomial of degree n in x, y, z, $\ldots$ then*

$$f(x + h, y + k, z + l, \ldots) = u + Du + \frac{1}{\lfloor 2} D^2 u + \ldots + \frac{1}{\lfloor n} D^n u, \qquad \ldots(E)$$

where D denotes the operator $h \dfrac{\partial}{\partial x} + k \dfrac{\partial}{\partial y} + l \dfrac{\partial}{\partial z} + \ldots$.

The proof is similar to the above, using the Multinomial theorem.

---

*See end of Art. 6.

■ **8. Euler's Theorem:**

(1) *Let u be a polynomial homogeneous in x, y and of degree n, then*

$$x\frac{\partial u}{\partial x} + y\frac{\partial u}{\partial y} = nu, \quad x^2\frac{\partial^2 u}{\partial x^2} + 2xy\frac{\partial^2 u}{\partial x \partial y} + y^2\frac{\partial^2 u}{\partial y^2} = n(n-1)u,$$

and, in general, $\left(x\dfrac{\partial}{\partial x} + y\dfrac{\partial}{\partial y}\right)^r u = n(n-1)(n-2)\ldots(n-r+1)u.$

Proof. Let $u = f(x, y)$, then by Taylor's theorem,

$$f(x + \lambda x, y + \lambda y)$$

$$= u + \lambda\left(x\frac{\partial u}{\partial x} + y\frac{\partial u}{\partial y}\right) + \frac{\lambda^2}{\lfloor 2}\left(x^2\frac{\partial^2 u}{\partial x^2} + 2xy\frac{\partial^2 u}{\partial x \partial y} + y^2\frac{\partial^2 u}{\partial y^2}\right) + \ldots \qquad \ldots(A)$$

and since u is homogeneous and of degree n in $x, y,$

$$f(x + \lambda x, y + \lambda y) = (1 + \lambda)^n u. \qquad \ldots(B)$$

The coefficients of the powers of λ in the expressions on the right of (A) and (B) must be equal, whence the results follow.

(2) *If $u = f(x, y, z, \ldots)$ is a polynomial homogeneous in x, y, z, … and of degree n, then*

$$x\frac{\partial u}{\partial x} + y\frac{\partial u}{\partial y} + z\frac{\partial u}{\partial z} + \ldots = nu,$$

$$D^r u = n(n-1)(n-2)\ldots(n-r+1)u,$$

where D^r stands for $\left(x\dfrac{\partial}{\partial x} + y\dfrac{\partial}{\partial y} + z\dfrac{\partial}{\partial z} + \ldots\right)^r$ *in its expanded form.*

Putting $h = \lambda x,\ k = \lambda y,\ l = \lambda z,\ \ldots$ in equation (E) of Art. 7, the proof is similar to the preceding.

■ **9. Partial Fractions:** Let $f(x)/\phi(x)$ be a rational proper fraction which is to be expressed in partial fractions. *If $x-a$ is a non-repeated factor of $\phi(x)$, the fraction corresponding to $x - a$ is*

$$\frac{f(a)}{\phi'(a)} \cdot \frac{1}{x - a}.$$

For let $\phi(x) = (x-a)\psi(x)$ and $\dfrac{f(x)}{\phi(x)} = \dfrac{A}{x - a} + \dfrac{F(x)}{\psi(x)}.$

From the second equation

$$f(x) = A\psi(x) + (x - a)F(x),$$

and from the first, by differentiation,

$$\phi'(x) = (x - a)\psi'(x) + \psi(x).$$

Putting $x = a$ and observing that $\psi(a) \neq 0$, we have

$$f(a) = A\psi(a) \text{ and } \phi'(a) = \psi(a), \text{ therefore } A = f(a)/\phi'(a).$$

EXAMPLE : *Express $x^{m-1}/(x^n-1)$ in partial fractions, where m, n are positive integers and $m \le n$.*

If $x-a$ is any factor of x^n-1, then $a^n = 1$ and

$$\frac{x^{m-1}}{x^n - 1} = \Sigma \frac{a^{m-1}}{na^n - 1} \cdot \frac{1}{x - a} = \frac{1}{n} \Sigma \frac{a^m}{x - a} .$$

The imaginary roots of $x^n - 1 = 0$ are

$$\cos \frac{2r\pi}{n} \pm \iota \sin \frac{2r\pi}{n} \text{ where } r = 1, 2, \dots \frac{1}{2} (n - 1) \text{ or } \frac{1}{2} (n - 2),$$

according as n is odd or even.

Let $a = \cos \dfrac{2r\pi}{n} + \iota \sin \dfrac{2r\pi}{u}$, then $a^{-1} = \cos \dfrac{2r\pi}{n} - \iota \sin \dfrac{2r\pi}{n}$;

and denote the sum of the fractions corresponding to $x-a$ and $x - a^{-1}$ by u_r, then by the preceding,

$$u_r = \frac{1}{n} \left(\frac{a^m}{x - a} + \frac{a^{-m}}{x - a^{-1}} \right) = \frac{1}{n} \frac{x (a^m + a^{-m}) - (a^{m-1} + a^{-(m-1)})}{x^2 - x (a + a^{-1}) + 1} ;$$

$$\therefore \quad u_r = \frac{2}{n} \cdot \frac{x \cos rm\alpha - \cos r (m - 1)\alpha}{x^2 - 2x \cos r \alpha + 1} \text{ where } \alpha = \frac{2\pi}{n} .$$

Hence, according as n is odd, or even, $x^{m-1}/(x^n - 1)$ is equal to

$$\frac{1}{n} \cdot \frac{1}{x - 1} + \sum_{r=1}^{r = \frac{1}{2}(n - 1)} u_r , \text{ or } \frac{1}{n} \cdot \frac{1}{x - 1} + \frac{(-1)^m}{n} \cdot \frac{1}{x + 1} + \sum_{r=1}^{r = \frac{1}{2}(n - 2)} u_r .$$

■ **10. Theorem:** The following theorem is often useful:

If α, β, ... are the roots of $u \equiv (a_0, a_1, \dots a_n)(x, 1)^n = 0$ and

$$v \equiv \phi (a_0, a_1, \dots a_n)$$

is a symmetric function of α, β, ..., involving only differences of the roots, then v satisfies the equation

$$a_0 \frac{\partial v}{\partial a_1} + 2a_1 \frac{\partial v}{\partial a_2} + \dots + na_{n-1} \frac{\partial v}{\partial a_n} = 0. \qquad \dots(A)$$

For the substitution $x = X + h$ transforms $u = 0$ into

$$U \equiv (a_0, A_1, A_2, \dots A_n)(X, 1)^n,$$

where $A_1 = a_1 + a_0 h$, $A_2 = a_2 + 2a_1 h + a_0 h^2$, etc., and since the roots of $U = 0$ are $\alpha - h$, $\beta - h$, ... , we have

$$v = \phi (a_0, a_1, \dots a_n) = \phi (a_0, A_1, \dots A_n). \qquad \dots(B)$$

If the right-hand side of (B) is expanded in powers of h, every coefficient must vanish, and by the extension of Taylor's theorem the coefficient of h is the same as that in

$$a_0 h \frac{\partial v}{\partial a_1} + (2a_1 h + a_0 h^2) \frac{\partial v}{\partial a_2} + (3a_2 h + 3a_1 h^2 + a_0 h^3) \frac{\partial v}{\partial a_3} + \dots;$$

$$\therefore \quad a_0 \frac{\partial v}{\partial a_1} + 2a_1 \frac{\partial v}{\partial a_2} + 3a_2 \frac{\partial v}{\partial a_3} + \dots = 0.$$

■ **Example :** *For the biquadratic* $(a_0, a_1, a_2, a_3, a_4)(x, 1)^4 = 0$, *find in terms of the coefficients the value of*

$$v = (\beta + \gamma - \alpha - \delta)(\gamma + \alpha - \beta - \delta)(\alpha + \beta - \gamma - \delta).$$

This function is of order 3 and weight 3. We may therefore assume that

$$a_0^3 v = p a_0^2 a_3 + q a_0 a_1 a_2 + r a_1^3,$$

where p, q, r are independent of $a_0, a_1, \ldots$. And since v is a function of the differences of $\alpha, \beta, \ldots$, we have

$$a_0 \frac{\partial v}{\partial a_1} + 2a_1 \frac{\partial v}{\partial a_2} + 3a_2 \frac{\partial v}{\partial a_3} + 4a_3 \frac{\partial v}{\partial a_4} = 0;$$

$\therefore \qquad a_0(q a_0 a_2 + 3r a_1^2) + 2q_1 q a_0 a_1 + 3a_2 p a_0^2 = 0;$

$\therefore \qquad a_0 a_1^2 (3r + 2q) + a_0^2 a_2 (3p + q) = 0.$

This is true for all values of a_0, a_1, a_2, therefore

$$3r + 2q = 0, \ 3p + q = 0, \text{ giving } q = -3p, \ r = 2p;$$

$\therefore \qquad a_0^3 v = p(a_0^2 a_3 - 3a_0 a_1 a_2 + 2a_1^3).$

The value of p may be found by giving special values to $\alpha, \beta, \gamma, \delta$. Thus by taking $\alpha = 0, \beta = 1, \gamma = -1, \delta = 4$, we find that $p = 32$, which agrees with Ex. 3 of VI, 16.

EXERCISE XXX

1. The equation $f(x) \equiv x^3 + 3qx + r = 0$ has two equal roots if $r^2 + 4q^3 = 0$. Prove this (*i*) by applying the H.C.F. process to $f(x)$ and $f'(x)$; (*ii*) by eliminating x between $f(x) = 0$ and $f'(x) = 0$.

2. The condition that the equation $ax^3 + 3bx^2 + 3cx + d = 0$ may have two equal roots is $(ad - bc)^2 - 4(ac - b^2)(bd - c^2) = 0$.

    Show that this may be obtained by eliminating x from
    $$ax^2 + 2bx + c = 0 \quad \text{and} \quad bx^2 + 2cx + d = 0.$$

3. Find the values of a for which $ax^3 - 9x^2 + 12x - 5 = 0$ has equal roots, and solve the equation in one case.

4. The equation $x^n - qx^{n-m} + r = 0$ has two equal roots if
    $$\left\{ \frac{q}{n}(n - m) \right\}^n = \left\{ \frac{r}{m}(n - m) \right\}^m.$$

5. All the roots of $x^n - \dfrac{n \cdot n - 1}{\lfloor 2}\, x^{n-2} + \dfrac{n(n-1)(n-2)(n-3)}{\lfloor 4}\, x^{n-4} - \ldots = 0$ are real.

    [Use Ex. VII, 14.]

6. Show that the equation $x^3 - 7x + 7 = 0$ has two roots between 1 and 2 and one between -3 and -4.

7. If the equation $x^n + p_1 x^{n-1} + p_2 x^{n-2} + \ldots + p_n = 0$ has three roots, each equal to α, show that α is a root of
    $$n^2 x^{n-1} + (n-1)^2 P_1 x^{n-2} + (n-2)^2 p_2 x^{n-3} + \ldots + p_{n-1} = 0.$$

8. Show that the equation $x^4 - 4x^3 + 10x^2 + 7x - 5 = 0$ has one positive, one negative and two imaginary roots.

9. The equation $3x^4 + 8x^3 - 6x^2 - 24x + r = 0$ has four real roots if
    $$-13 < r < -8,$$

two real roots if $-8 < r < 19$, and no real root if $r > 19$.

[The roots of $f'(x) = 0$ are $-2, -1, 1$.]

10. The equation $f(x) = (x-a)^3 + (x-b)^3 + (x-c)^3 = 0$ has one real and two imaginary roots.

11. The equation $x^5 - 5ax + 4b = 0$ has three real roots or only one, according as $a^5 \gtrless b^4$.

12. The equation $x^5 + 5ax^3 + b = 0$ has one and only one real root in the following cases: (*i*) if $a > 0$; (*ii*) if $a < 0$ and $b^2 + 108\,a^5 > 0$.

13. The equation $x^5 + 5ax^2 + b = 0$ has one and only one real root in the following cases: (*i*) if a, b have the same sign; (*ii*) if a, b have opposite signs and $b\,(b^3 + 108a^5) > 0$.

[Deduce from Ex. 12 by putting $1/x$ for x.]

14. Show that if $3b^2 < a^2$ there are real values of x, y, z which satisfy

$$x + y + z = a, \quad yz + zx + xy = b^2,$$

and that these values are all positive if $a > 0$ and $a^2 < 4b^2$.

[x, y, z are the roots of an equation of the form $\theta^3 - a\theta^2 + b^2\theta + k = 0$ and if

$\theta = \phi + a/3$, this reduces to one of the form $\phi^3 + \phi\,(b^2 - \dfrac{1}{3}\,a^2) + k' = 0$.]

15. If $\lambda_1, \lambda_2, \lambda_3$ are the values of λ given by the equation

$$\frac{x^2}{a^2 - \lambda} + \frac{y^2}{b^2 - \lambda} + \frac{z^2}{c^2 - \lambda} = 1,$$

prove that these are all real; also, if a, b, c and $\lambda_1, \lambda_2, \lambda_3$ are in descending order, then

$$a^2 > \lambda_1 > b^2 > \lambda_2 > c^2 > \lambda_3$$

and

$$x^2 = \frac{(a^2 - \lambda_1)(a^2 - \lambda_2)(a^2 - \lambda_3)}{(a^2 - b^2)(a^2 - c^2)}.$$

16. If $\alpha, \beta, \gamma, \ldots$ are the roots of the equation

$$x^n + \frac{x^{n-1}}{\underline{|1}} + \frac{x^{n-2}}{\underline{|2}} + \ldots + \frac{1}{\underline{|n}} = 0,$$

show that $\Sigma_\alpha^r = 0$ for $r = 2, 3, 4, \ldots n$.

17. If $f(x) = 0$ is a cubic equation whose roots are α, β, γ, and α is the harmonic mean of the roots of $f'(x) = 0$, prove that $\alpha^2 = \beta\gamma$.

18. The sum of the ninth powers of the roots of $x^3 + 3x + 9 = 0$ is zero.

19. If s_n is the sum of the nth powers of the roots of $x^4 - x^2 + 1 = 0$, show that

$$s_{2n-1} = 0, \quad s_{2n} = 4\cos\frac{n\pi}{3}.$$

20. If $\alpha, \beta, \ldots$ are the roots of $(a_0, a_1, a_2, \ldots a_n)(t, 1)^n = 0$, show that the equation whose roots are $\alpha y - x, \beta y - x$, etc., is $(a_0, u_1, u_2, \ldots u_n)(t, 1)^n = 0$, where $u_1 = a_0 x + a_1 y$, $u_2 = a_0 x^2 + 2a_1 xy + a_2 y^2$, etc.

21. If m and n are positive integers and $m \le n$, prove that

$$\frac{x^{m-1}}{x^n + 1} = \frac{(-1)^{n-m}}{n\,(x+1)} + \frac{2}{n}\sum_{r=0}^{r=\frac{1}{2}(n-3)} u_r \quad \text{or} \quad \frac{2}{n}\sum_{r=0}^{r=\frac{1}{2}(n-2)} u_r,$$

according as n is odd or even, where

$$u_r = -\frac{x\cos(2r+1)\,m\alpha - \cos(2r+1)\,(m-1)\alpha}{x^2 - 2x\cos(2r+1)\,\alpha + 1}\quad\text{and}\quad \alpha = \frac{\pi}{n}.$$

22. Let k be given by the equation

$$\Delta' = \begin{vmatrix} a-k & h & g \\ h & b-k & f \\ g & f & c-k \end{vmatrix} = 0.$$

If this equation has a repeated root k_1, prove that

$$k_1 = a - gh/f = b - hf/g = c - fg/h,$$

provided that none of f, g, h is zero. If one of these (g) is zero, so also is another (h), and then

$$k_1 = a \quad\text{and}\quad (a-b)(a-c) = f^2.$$

Under these circumstances, show that

$$ax^2 + by^2 + cz^2 + 2fyz + 2gzx + 2hxy - k_1\,(x^2 + y^2 + z^2).$$

is a perfect square.

[Denoting the cofactors of the elements of Δ' by A', F', etc., the equations $A' = 0$, $B' = 0$, $C' = 0$ are quadratics in k, and the roots of any one of them separate those of $\Delta' = 0$. Hence if $k = k_1$, then A', B', C' and consequently F', G', H' are all zero. See Ch. 9, 21, (4).]

■ ■ ■

CHAPTER 19

Exponential and Logarithmic Functions and Series

■ **1. Continuity of a^x and $\log_a x$:** If $a > 0$, the function a^x is continuous for all real values of x. This follows from Ch. 13, 9.

Hence the inverse function $\log_a x$ is continuous for positive values of x. (Ch. 17, 21.) The reader is reminded that $\log x$ stands for $\log_e x$.

■ **2. Exponential Inequalities and Limits:** (Continued from Ch. 15, 5)

(1) *As $x \to \infty$ through real values,*

$$\left(1+\frac{1}{x}\right)^x \to e \ \text{ and } \ \left(1-\frac{1}{x}\right)^{-x} \to e \,.$$

For let $m < x < m + 1$, where m is a positive integer, then

$$1+\frac{1}{m+1} < 1+\frac{1}{x} < 1+\frac{1}{m};$$

therefore

$$\left(1+\frac{1}{m+1}\right)^m < \left(1+\frac{1}{x}\right)^x < \left(1+\frac{1}{m}\right)^{m+1}$$

If $x \to \infty$, then $m \to \infty$ and by Ch. 15, 5, (3),

$$\left(1+\frac{1}{m+1}\right)^m = \left(1+\frac{1}{m+1}\right)^{m+1} \div \left(1+\frac{1}{m+1}\right) \to e,$$

$$\left(1+\frac{1}{m}\right)^{m+1} = \left(1+\frac{1}{m}\right)^m \cdot \left(1+\frac{1}{m}\right) \to e;$$

therefore

$$\left(1+\frac{1}{x}\right)^x \to e.$$

Next let $x = y + 1$, then as $x \to \infty$, $y \to \infty$ and

$$\left(1-\frac{1}{x}\right)^{-x} = \left(\frac{y}{y+1}\right)^{-(y+1)} = \left(1+\frac{1}{y}\right)^y \cdot \left(1+\frac{1}{y}\right) \to e\,.$$

It follows that if $x \to \infty$ or $-\infty$,

$$\lim \left(1 + \frac{1}{x}\right)^x = e,$$

and consequently
$$\lim_{x \to 0} (1 + x)^{\frac{1}{x}} = e.$$

(2) *If z is any real number,*

$$\lim_{x \to \infty} \left(1 + \frac{z}{x}\right)^x = e^z.$$

This is obviously true if $z = 0$. If $z \neq 0$, let $x = yz$, then as $x \to \infty$, $y \to \infty$ or $-\infty$, according as $z \gtrless 0$. In either case, by the preceding,

$$\left(1 + \frac{z}{x}\right)^x = \left(1 + \frac{1}{y}\right)^{yz} = \left\{\left(1 + \frac{1}{y}\right)^y\right\}^z \to e^z.$$

(3) *If $a > 0$. then,*

$$\lim_{x \to 0} \frac{a^x - 1}{x} = \log a.$$

This is obviously true if $a = 1$. If $a \neq 1$, let $a^x = 1 + y$, then as $x \to 0$, $y \to 0$ and

$$\lim_{x \to 0} \frac{x}{a^x - 1} = \lim_{y \to 0} \frac{\log_a (1 + y)}{y} = \lim_{y \to 0} \log_a (1 + y)^{\frac{1}{y}}.$$

Now by Ch. 17, 5, (2),

$$\lim_{y \to 0} \log_a (1 + y)^{\frac{1}{y}} = \log_a \{\lim_{y \to 0} (1 + y)^{\frac{1}{y}}\} = \log_a e = \frac{1}{\log a} ;$$

therefore
$$\lim_{x \to 0} \frac{a^x - 1}{x} = \log a.$$

Here x may tend to zero through positive or negative values, so that *each of the functions* $\dfrac{a^x - 1}{x}$ *and* $\dfrac{1 - a^{-x}}{x}$ *tends to log a as a limit as $x \to 0$.*

It has been shown in Ch. 14, 9, that the first of these functions increases and the second decreases as x increases from zero. Hence

if $\qquad x > 0,\ then \qquad \dfrac{1 - a^{-x}}{x} < \log a < \dfrac{a^x - 1}{x}.$

■ **3. The Exponential Theorem:** *For all real values of x,*

$$e^x = 1 + x + \frac{x^2}{\underline{|2}} + \frac{x^3}{\underline{|3}} + \dots + \frac{x^n}{\underline{|n}} + \dots . \qquad \dots(A)$$

Proof. The series (A) is convergent for all values of x, and its sum is a function of x which we shall denote by $E(x)$. If n is a positive integer,

$$\left(1+\frac{x}{n}\right)^n = 1 + n \cdot \frac{x}{n} + \frac{n(n-1)}{\lfloor 2} \left(\frac{x}{n}\right)^2 + \frac{n(n-1)(n-2)}{\lfloor 3} \left(\frac{x}{n}\right)^3 + \dots \text{ to } n+1 \text{ terms}$$

$$= 1 + x + \frac{x^2}{\lfloor 2} p_1 + \frac{x^3}{\lfloor 3} p_2 + \dots + \frac{x^n}{\lfloor n} p_{n-1},$$

where $$p_r = \left(1-\frac{1}{n}\right)\left(1-\frac{2}{n}\right)\left(1-\frac{3}{n}\right)\dots\left(1-\frac{r}{n}\right).$$

Now $p_r < 1$, and since $1/n, 2/n, \dots (n-1)/n$ are positive numbers less than unity, for the values 2, 3, ..., $(n-1)$ of r we have, by Ch. 14, 1,

$$p_r > 1 - \frac{1}{n}(1 + 2 + 3 + \dots + r) = 1 - \frac{r(r+1)}{2n};$$

$$\therefore \qquad 0 < 1 - p_r < r(r+1)/2n.$$

Let s_n be the sum to n terms of (A), and let $|x| = x_1$; then

$$\left| s_{n+1} - \left(1+\frac{x}{n}\right)^n \right| \leq \frac{x_1^2}{\lfloor 2}(1-p_1) + \frac{x_1^3}{\lfloor 3}(1-p_2) + \dots + \frac{x_1^n}{\lfloor n}(1-p_{n-1})$$

$$< \frac{x_1^2}{2n}\left\{\frac{1\cdot 2}{\lfloor 2} + \frac{2\cdot 3}{\lfloor 3} x_1 + \frac{3\cdot 4}{\lfloor 4} x_1^2 + \dots + \frac{(n-1)n}{\lfloor n} x_1^{n-2}\right\}$$

$$< \frac{x_1^2}{2n}\left\{1 + x_1 + \frac{x_1^2}{\lfloor 2} + \frac{x_1^3}{\lfloor 3} + \dots \text{ to } \infty\right\} < \frac{x_1^2}{2n} \cdot E(x_1).$$

Now $E(x_1)$ is finite for all values of x_1, therefore

$$\lim_{n\to\infty} s_{n+1} = \lim_{n\to\infty} \left(1+\frac{x}{n}\right)^n = e^x;$$

but $$\lim_{n\to\infty} s_{n+1} = E(x) ; \therefore e^x = E(x).$$

This result is called the *Exponential Theorem*.

In particular, when $x = 1$ we have

$$e = 1 + 1 + \frac{1}{\lfloor 2} + \frac{1}{\lfloor 3} + \dots + \frac{1}{\lfloor n} + \dots,$$

from which it can be shown that $e = 2.7182818284 \dots$.

■ **EXAMPLE :** *Show that e is irrational.*

Suppose that $e = p/q$, where p and q are integers; then

$$\frac{p}{q} = 1 + 1 + \frac{1}{\lfloor 2} + \dots + \frac{1}{\lfloor q} + R, \qquad\qquad \dots\text{(A)}$$

where $$R = \frac{1}{\lfloor q+1}\left\{1 + \frac{1}{q+2} + \frac{1}{(q+2)(q+3)} + ...\right\} < \frac{1}{\lfloor q+1} \cdot \frac{1}{1 - \dfrac{1}{q+2}} ;$$

$$\therefore \qquad R \lfloor q < \frac{q+2}{(q+1)^2} = \frac{q+2}{q(q+2)+1} < 1 . \qquad\qquad ...(B)$$

If the equality (A) holds, we should have $R \lfloor q$ = an integer, which is impossible on account of (B). Hence e cannot be a rational number.

■ **4.** *If $a > 0$ and x is any real number.*

$$a^x = e^{x \log a} = E\,(x \log a), \qquad\qquad ...(A)$$

that is to say

$$a^x = 1 + x \log a + \frac{x^2}{\lfloor 2} (\log a)^2 + \frac{x^3}{\lfloor 3} (\log a)^3 + ... \qquad ...(B)$$

■ **5. An Irrational Index:** *When $a > 0$ and x is irrational, we take the equation $a^x = E(x \log a)$ as defining the meaning of a^x.*

This agrees with what has been said in Ch. 13, 9, for it will be shown later that the sum of the series denoted by $E(x)$ is a continuous function.

■ **6. Derivatives of a^x, $\log x$ and x^n:**

(1) *Suppose that $a > 0$, then by Art. 2, (3),*

$$\frac{d}{dx} a^x = \lim_{h \to 0} \frac{a^{x+h} - a^x}{h} = \lim_{h \to 0} a^x \cdot \frac{a^h - 1}{h} = a^x \log a.$$

In particular, $$\frac{d}{dx} e^x = e^x.$$

Note: It should be observed that, if $a < 0$, a^x has not been defined for all real values of x.

(2) If $y = \log x$, then $x = e^y$ and $\dfrac{dx}{dy} = e^y = x,$

therefore $\dfrac{dy}{dx} = \dfrac{1}{x}$; that is, $\dfrac{d}{dx} \log x = \dfrac{1}{x}.$

(3) It has been shown that when n is rational, $\dfrac{d}{dx} x^n = nx^{n-1}$. To prove that this is true for all real values of n, we proceed thus:

$$\frac{d}{dx} x^n = \frac{d}{dx} e^{n \log x} = \frac{n}{x} e^{n \log x} = \frac{n}{x} x^n = nx^{n-1}.$$

Note: Since it has been shown in (1) that, for the curve $y = a^x$, the gradient dy/dx is $y \log a$, it follows that, if the tangent at a point P on the graph meets OX in T, and PN is the ordinate of P, then $TN = 1/\log a$: *i.e.* the *subtangent is constant.* This fact gives a rapid construction for the graph of $y = a^x$, with considerable accuracy.

For instance, for $y = e^x$, where $TN = 1$, take a sheet of ruled exercise paper and draw a line perpendicular to the ruling as the axis of x; and mark one of the rulings as the axis of y. On this take OA, equal to, say, 4 intervals, as the unit; and on OX, to the left of O, take $Oa = OA$. Draw $A\,a$ until it meets the first ruling to the left of OY at B. Without removing the pencil-point from B, rotate the ruler until it passes through b, the foot of the ruling next to the left of a, and draw bB until it meets the

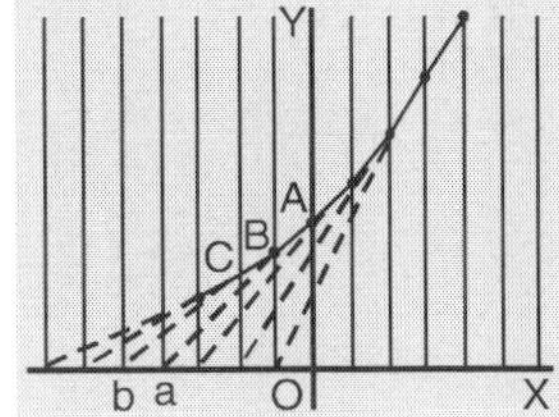

Fig. 60

ruling next to the left of A; and so on. Find points on the curve to the right of OY in the same way, and replace the broken line by a 'fair curve'.

If, in Fig. 60, OX and OY are interchanged, the curve is then the graph of $y = \log x$.

Also, if the paper is turned through $90°$ in its own plane, and then turned over, the curve as seen through the back of the paper is the graph of $y = \log x$, with the axes in the usual position.

■ **7. Inequalities and Limits (continued):**

(1) *For all real values of x other than zero $e^x > 1 + x$.*

For let $f(x) = e^x - 1 - x$, then $f(0) = 0$ and $f'(x) = e^x - 1$,

so that $f'(x) \gtrless 0$ according as $x \gtrless 0$. Hence, if x increases or decreases from zero, in either case $f(x)$ increases from zero, and is therefore positive.

(2) If $x > -1$ *and* $\neq 0$, *then* $x > \log (1 + x)$.

This follows from (1) by taking logarithms.

(3) If $x > 0$, *then* $\dfrac{1}{2} x^2/ (1 + x) < x - \log (1 + x) < \dfrac{1}{2} x^2$,

and if $-1 < x < 0$, *then* $\dfrac{1}{2} x^2 < x - \log (1 + x) < \dfrac{1}{2} x^2/(1 + x)$.

For let $f(x) = x - \log (1 + x) - \dfrac{1}{2}x^2$, then

$$f'(x) = 1 - \frac{1}{1 + x} - x = - \frac{x^2}{1 + x} < 0.$$

Hence $f(x)$ decreases as x increases from -1, and since $f(0) = 0$, it follows that $f(x) \gtrless 0$ according as $x \gtrless 0$.

Again, if $\phi (x) = x - \log (1 + x) - \dfrac{1}{2} \dfrac{x^2}{1 + x}$, then

$$\phi' (x) = 1 - \frac{1}{1 + x} - \frac{x}{1 + x} + \frac{1}{2} \frac{x^2}{(1 + x)^2} = \frac{1}{2} \left(\frac{x}{1 + x} \right)^2 > 0.$$

Hence $\phi (x)$ increases as x increases from -1, and since $\phi (0) = 0$, it follows that $\phi (x) \gtrless 0$ according as $x \gtrless 0$.

This proves all the inequalities in question.

(4) *If $x > -1$ and $\neq 0$ and l, h are respectively the smaller and the greater of the numbers 1 and $1 + x$, then*

$$\frac{1}{2}\frac{x^2}{h} < x - \log(1+x) < \frac{1}{2}\frac{x^2}{l}.$$

This is merely another way of stating the inequalities in (3).

(5) *For any positive value of n; (i)* $\lim\limits_{x\to\infty}\dfrac{\log x}{x^n} = 0$, *(ii)* $\lim\limits_{x\to 0}(x^n\log x) = 0$.

(*i*) For all positive values of m we have by Art. 2, (3),

$$\frac{\log x}{x^n} < \frac{1}{x^n}\frac{x^m - 1}{m} < \frac{1}{mx^{n-m}}.$$

Choose m so that $0 < m < n$, then as $x \to \infty$,

$$x^{n-m} \to \infty \text{ and } (\log x)/x^n \to 0.$$

(*ii*) Put $x = 1/y$, then as $x \to 0$, $y \to \infty$ and

$$x^n \log x = -(\log y)/y^n \to 0.$$

(6) *For all values of r,* $\lim\limits_{x\to\infty} x^r/e^x = 0$.

This is obvious when $r < 0$. If $r > 0$, let $e^x = y$, then, putting $\dfrac{1}{r} = n$,

$$\frac{x^r}{e^x} = \frac{(\log y)^r}{y} = \left(\frac{\log y}{y^n}\right)^r.$$

Now as $x \to \infty$, $y \to \infty$; and, by (5), since $n = \dfrac{1}{r} > 0$, $(\log y)/y^n \to 0$.

Hence it follows that $x^r/e^x \to 0$.

■ **8. The Manner in which e^x and $\log x$ tend to Infinity:** Let n be any positive number.

First suppose that $x \to \infty$, then $e^x \to \infty$ and $\log x \to \infty$.

Now, by Art. 7, (6), $x^n/e^x \to 0$, therefore *e^x tends to infinity faster than x^n, however great n may be.*

Again, by Art. 7, (5), $(\log x)/x^n \to 0$, therefore *$\log x$ tends to infinity more slowly than x^n, no matter how small n may be.*

If $x \to 0$, then $-\log x \to \infty$, and by Art. 7, (5), $(\log x)/x^{-n} \to 0$; hence *$-\log x$ tends to ∞ more slowly than $1/x^n$, no matter how small n may be.*

■ **9. Theorem:** *If $u_n = 1 + \dfrac{1}{2} + \dfrac{1}{3} + \ldots + \dfrac{1}{n} - \log n$, then as $n \to \infty$, $u_n \to \gamma$ where γ is a fixed number lying between 0 and 1.*

Proof. By Art. 7, (2),

$$u_n - u_{n-1} = \frac{1}{n} - \log\frac{n}{n-1} = \frac{1}{n} + \log\left(1 - \frac{1}{n}\right) < 0,$$

so that u_n decreases as n increases. Again, by Art. 7, (2),

$$\log \left(1+\frac{1}{n}\right) < \frac{1}{n} < -\log \left(1-\frac{1}{n}\right), \text{ that is, } \log \frac{n+1}{n} < \frac{1}{n} < \log \frac{n}{n-1}.$$

Substituting $n-1$, $n-2$, ... 2, in succession for n, we have

$$\log \frac{n}{n-1} < \frac{1}{n-1} < \log \frac{n-1}{n-2}, \quad ... \quad , \log \frac{3}{2} < \frac{1}{2} < \log \frac{2}{1}.$$

Also $\log \dfrac{2}{1} < 1$; and by addition, it follows that

$$\log (n + 1) < 1 + \frac{1}{2} + \frac{1}{3} + ... + \frac{1}{n} < 1 + \log n; \text{ i.e., } \log (n+1) - \log n < u_n < 1.$$

But $\log (n + 1) - \log n > 0$, hence $0 < u_n < 1$; and, from above, $u_n < u_{n-1}$.

Thus u_n is a decreasing function which is always > 0. Consequently $u_n \to \gamma$ where γ is a fixed number > 0; and, since $u_n < 1$, $\gamma < 1$.

Note: The number γ is known as *Euler's Constant*. It can be shown that $\gamma = 0.577215...$.
It is often convenient to state the theorem as follows:

$$1 + \frac{1}{2} + \frac{1}{3} + ... + \frac{1}{n} = \gamma + \log n + \varepsilon_n, \text{ where } \varepsilon_n \to 0 \text{ as } n \to \infty.$$

■ **10. Series for log 2:** Let s_r be the sum to r terms of the series,

$$1 - \frac{1}{2} + \frac{1}{3} - ... + (-1)^{n-1} \frac{1}{n} + ... \text{ to } \infty.$$

then $S_{2n} = \left(1 + \dfrac{1}{2} + \dfrac{1}{3} + ... + \dfrac{1}{2n}\right) - 2 \left(\dfrac{1}{2} + \dfrac{1}{4} + \dfrac{1}{6} + ... + \dfrac{1}{2n}\right)$

$$= \left(1 + \frac{1}{2} + \frac{1}{3} + ... + \frac{1}{2n}\right) - \left(1 + \frac{1}{2} + \frac{1}{3} + ... + \frac{1}{n}\right)$$

$$= (\gamma + \log 2n + \varepsilon_{2n}) - (\gamma + \log n + \varepsilon_n) = \log 2 + \varepsilon_{2n} - \varepsilon_n.$$

Therefore $s_{2n} \to \log 2$. Also $s_{2n+1} - s_{2n} \to 0$, hence $s_{2n+1} \to \log 2$.

Thus, $\log 2 = 1 - \dfrac{1}{2} + \dfrac{1}{3} - ... + (-1)^{n-1} \dfrac{1}{n} + ... \text{ to } \infty.$

■ **11. The Exponential Function:**

(1) For all values of z, real or complex, the series $\quad 1 + z + \dfrac{z^2}{\lfloor 2} + \dfrac{z^3}{\lfloor 3} + ...$ is

convergent. Its sum is called the *Exponential Function,* and is denoted by $E(z)$, or by exp z.

(2) Theorem. *If $z = x + \iota y$ where x and y are real,*

$$E(z) = \lim_{n \to \infty} \left(1 + \frac{z}{n}\right)^n = e^x (\cos y + \iota \sin y).$$

Proof. It has been shown in Art. 3 that the first of these equalities holds when z is real, and exactly the same proof holds when z is complex.

Let $1 + \dfrac{z}{n} = 1 + \dfrac{x}{n} + \iota \dfrac{y}{n} = p\,(\cos \phi + \iota \sin \phi)$, where $\rho > 0$ and $-\pi < \phi \leq \pi$.

For sufficiently large values of n, $1 + \dfrac{x}{n} > 0$; hence, since $\rho > 0$, and $\rho \cos \phi = 1 + \dfrac{x}{n}$, we have $\cos \phi > 0$, and $-\tfrac{1}{2}\pi < \phi < \tfrac{1}{2}\pi$.

Also $\tan \phi = y/(n + x) \to 0$, therefore $\phi \to 0$, and $\phi/\tan \phi \to 1$; hence $\lim n\phi = \lim (n \tan \phi) = \lim \{ny/(n + x)\} = y$.

Now $\rho^2 = \left(1 + \dfrac{x}{n}\right)^2 + \dfrac{y^2}{n^2} = \left(1 + \dfrac{x}{n}\right)^2 \sec^2 \phi$; therefore $\rho^n = \left(1 + \dfrac{x}{n}\right)^2 \sec^n \phi$; and,

since $\lim \sec^n \phi = \lim \left(1 + \tan^2 \phi\right)^{\frac{1}{2}n} = \lim \left(1 + \dfrac{y^2}{n^2}\right)^{\frac{1}{2}n} = \lim \left(ev^2\right)^{\frac{1}{2n}} = 1,$

We have $\lim \rho^n = e^x$; hence, $E(z) = \lim \rho^n (\cos n\phi + \iota \sin n\phi)$
and $\qquad\qquad E(z) = e^x (\cos y + \iota \sin y).$

(3) It follows that $E(z)$ *cannot vanish for any value of z, for neither e^x nor $\cos y + \iota \sin y$ can vanish.*

(4) *The function $E(z)$ is periodic, and its period is $2\iota\pi$.*

For if k is any integer, positive or negative, the values of $\cos y$ and $\sin y$ are unaltered by adding $2k\pi$ to y. Therefore $E(z)$ is unaltered by adding $2\iota k\pi$ to z, thus $E(z + 2\iota k\pi) = E(z)$.

■ **12. A Complex Index:** When z is real, e^z is a multiple-valued function, except when z is an integer. The real positive value of e^z is called its principal value, and it has been shown that

$$E(z) = \text{The principal value of } e^z.$$

We take this equation as defining the principal value of e^z when z is complex.

This implies that for imaginary (as for real) values of z, e^z is to be regarded as a multiple-valued function.

For the present, we take e^z as denoting its principal value defined by $E(z)$.

Thus if $z = x + \iota y$, and $z' = x' + \iota y'$, we write

$$e^z = E(z) = e^x (\cos y + \iota \sin y), \text{ and } e^{z'} = e^{x'} (\cos y' + \iota \sin y');$$

thus, $\qquad e^z \cdot e^{z'} = e^x \cdot e^{x'} (\cos y + \iota \sin y)(\cos y' + \iota \sin y')$
$$= e^{x+x'}\{\cos (y + y') + \iota \sin (y + y')\} = e^{z+z'}$$

Hence the same index laws hold for imaginary as for real indices.

Again, if a is real and positive and z is real, $a^z = e^{z \log a} = E(z \log a)$.

In agreement with this, *if $a > 0$, we define the principal value of a^z as $E(z \log a)$ and, at present, we take a^z as representing this value.*

13. Series for Sine and Cosine: For all real values of x and y, it has been shown that $E(x + \iota y) = e^x (\cos y + \iota \sin y)$. When $x = 0$, this gives

$$\cos y + \iota \sin y = E(\iota y)$$

$$= 1 + \iota y + \frac{\iota^2 y^2}{\lfloor 2} + \frac{\iota^3 y^3}{\lfloor 3} + \dots$$

$$= \left(1 - \frac{y^2}{\lfloor 2} + \frac{y}{\lfloor 4} - \dots\right) + \iota \left(y - \frac{y^3}{\lfloor 3} + \frac{y^5}{\lfloor 5} - \dots\right).$$

Equating real and imaginary parts, we have, for all real values of y,

$$\cos y = 1 - \frac{y^2}{\lfloor 2} + \frac{y^4}{\lfloor 4} - \dots, \quad \sin y = y - \frac{y^3}{\lfloor 3} + \frac{y^5}{\lfloor 5} - \dots .$$

14. Exponential Values of Sine and Cosine: Using e^z to mean exactly the same as $E(z)$, by the preceding,

$$\cos y + \iota \sin y = E(\iota y) = e^{\iota y}, \cos y - \iota \sin y = E(-\iota y) = e^{-\iota y};$$

therefore $\cos y = \dfrac{1}{2} (e^{\iota y} + e^{-\iota y})$, $\sin y = \dfrac{1}{2\iota} (e^{\iota y} - e^{-\iota y})$.

15. Series which can be summed by the Exponential Theorem:

(1) *The series* $\Sigma u_n. \dfrac{x^n}{\lfloor n}$, *where* u_n *is a polynomial in n.*

We can find $a_0, a_1, \dots a_r$, independent of n, so that

$$u_n = a_0 + a_1 n + a_2 n (n-1) + \dots + a_r n (n-1)\dots (n-r+1),$$

and then

$$\sum_{n=0}^{\infty} u_n \cdot \frac{x^n}{\lfloor n} = a_0 \sum_{n=0}^{\infty} \frac{x^n}{\lfloor n} + a_1 x \sum_{n=1}^{\infty} \frac{x^{n-1}}{\lfloor n-1} + a_2 x^2$$

$$\sum_{n=2}^{\infty} \frac{x^{n-2}}{\lfloor n-2} + \dots + a_r x^r \sum_{n=r}^{\infty} \frac{x^{n-r}}{\lfloor n-r}$$

$$= (a_0 + a_1 x + a_2 x^2 + \dots + a_r x^r) e^x.$$

EXAMPLE 1: *Find the value of* $\Sigma(n^3 + 5)/\lfloor n$.

Here $n^3 + 5 = n (n-1) (n-2) + 3n(n-1) + n + 5$; hence, putting $a_0 = 5$, $a_1 = 1$, $a_2 = 3$, $a_3 = 1$, and $x = 1$ in the result above, we have

$$\sum_{n=0}^{\infty} (n^3 + 5) = (5 + 1 + 3 + 1)e = 10e.$$

(2) *The series* $\Sigma u_n x^n/(n + a)(n + b) \dots (n + k)\lfloor n$ *where* u_n *is a polynomial in n and* $a, b, \dots k$ *are unequal positive integers.*

Series of this kind can be summed as in the following example.

■ **EXAMPLE 2:** *Find the value of*

$$\sum_{n=1}^{n=\infty} \frac{n^2+1}{n+2} \cdot \frac{x^n}{\lfloor n}.$$

Denote the sum by S, then

$$S = \sum_{n=1}^{n=\infty} \frac{(n^2+1)(n+1)}{\lfloor n+2} \cdot x^n.$$

Now $(n^2 + 1)(n + 1) = n^3 + n^2 + n + 1$

$$= a + b(n+2) + c(n+2)(n+1) + d(n+2)(n+1)n,$$

where a, b, c, d can be found by division, their values being $a = -5$, $b = 5$, $c = -2$, $d = 1$. Thus

$$S = -5 \sum_{1}^{\infty} \frac{x^n}{\lfloor n+2} + 5 \sum_{1}^{\infty} \frac{x^n}{\lfloor n+1} - 2 \sum_{1}^{\infty} \frac{x^n}{\lfloor n} + \sum \frac{x^n}{\lfloor n-1}.$$

Now

$$\sum_{1}^{\infty} \frac{x^n}{\lfloor n+2} = \frac{1}{x^2}\left(e^x - 1 - x - \frac{x^2}{2}\right),$$

$$\sum_{1}^{\infty} \frac{x^n}{\lfloor n+1} = \frac{1}{x}(e^x - 1 - x), \quad \sum_{1}^{\infty} \frac{x^n}{\lfloor n} = e^x - 1,$$

$$\sum_{1}^{\infty} \frac{x^n}{\lfloor n-1} = xe^x$$

$$\therefore \qquad S = \frac{1}{2x^2}(10 - x^2) - \frac{e^x}{x^2}(5 - 5x + 2x^2 - x^3).$$

EXERCISE XXXI

1. Show that $\displaystyle\lim_{x \to \infty}\left(1 - \frac{k}{x^2}\right)^x = 1.$

Find the sum to infinity of the series in Exs. 2–9:

2. $1 + \dfrac{2}{\lfloor 1} + \dfrac{3}{\lfloor 2} + \dfrac{4}{\lfloor 3} + \dots$

3. $\dfrac{3}{\lfloor 2} + \dfrac{5}{\lfloor 3} + \dfrac{7}{\lfloor 4} + \dfrac{9}{\lfloor 5} + \dots$

4. $\dfrac{3}{\lfloor 2} + \dfrac{5}{\lfloor 4} + \dfrac{7}{\lfloor 6} + \dfrac{9}{\lfloor 8} + \dots$

5. $1 + \dfrac{1+2}{\lfloor 2} + \dfrac{1+2+3}{\lfloor 3} + \dots$

6. $\dfrac{1}{2} - \dfrac{1}{3\lfloor 1} + \dfrac{1}{4\lfloor 2} - \dfrac{1}{5\lfloor 3} + \dots$

7. $1^3 + \dfrac{2^3}{\lfloor 1} + \dfrac{3^3}{\lfloor 2} + \dfrac{4^3}{\lfloor 3} + \dots$

8. $1^3 - \dfrac{2^3}{\lfloor 1} + \dfrac{3^3}{\lfloor 2} - \dfrac{4^3}{\lfloor 3} + ...$

9. $1 + \dfrac{x}{\lfloor 2} + \dfrac{2x^2}{\lfloor 3} + \dfrac{3x^3}{\lfloor 4} + ...$

Prove that:

10. $\displaystyle\sum_{1}^{\infty} \dfrac{n-1}{(n+2)\lfloor n} x^n = \dfrac{1}{x^2} \{(x^2 - 3x + 3)\, e^x + \dfrac{1}{2} x^2 - 3\}.$

11. $\displaystyle\sum_{1}^{\infty} \dfrac{2n-1}{(n+3)\lfloor n} = \dfrac{1}{3}(43 - 15e).$

12. $\displaystyle\sum_{0}^{\infty} \dfrac{(n+1)^3}{\lfloor n} x^n = (1 + 7x + 6x^2 + x^3)\, e^x.$

13. If $e^{ax} \cos bx = u_0 + u_1 x + ... + u_n x^n + ...,$ then $u_n = (a^2 + b^2)^{\frac{n}{2}} \dfrac{\cos n\theta}{\lfloor n},$

where $\theta = \tan^{-1} \dfrac{b}{a}.$

■ **16. Theorem:** *If* $-1 < x \leq 1,$ *then*

$$\log (1 + x) = x - \dfrac{x^2}{2} + \dfrac{x^3}{3} - ... + (-1)^{n-1} \dfrac{x^n}{n} + ...$$

Proof. (*i*) *Suppose that* $0 < x < 1.$ Let s_n be the sum to n terms of the series, then

$$\dfrac{ds_n}{dx} = 1 - x + x^2 - ... + (-1)^{n-1} x^{n-1} = \dfrac{1 - (-x)^n}{1 + x}.$$

Let $y_n = \log (1 + x) - s_n;$ then $y_n = 0$ when $x = 0,$ and

$$\dfrac{dy_n}{dx} = \dfrac{1}{1+x} - \dfrac{1 - (-x)^n}{1+x} = \dfrac{(-x)^n}{1+x}.$$

Therefore $y_n \gtrless 0,$ according as n is even or odd.

Hence $\log (1 + x)$ lies between s_n and s_{n+1} for every $n.$

But $s_{n+1} - s_n = (-1)^n x^{n+1} /(n+1) \to 0$ as $n \to \infty,$

therefore $s_n \to \log (1 + x)$ as $n \to \infty,$

which proves this case of the theorem. (*Cf.* Ex. XXIX, 8.)

(*ii*) The case in which $x = 1$ has been dealt with in Art. 10.

(*iii*) *When* $x < 0,$ changing the sign of $x,$ it has to be shown that

if $\;0 < x < 1,$ *then* $- \log (1 - x) = x + \dfrac{x^2}{2} + \dfrac{x^3}{3} + ... + \dfrac{x^n}{n} + ...$

Let $y_n = - \log (1 - x) - s_n,$ where s_n is the sum to n terms of this series; then

$y_n = 0$ when $x = 0,$ and $\dfrac{dy_n}{dx} = \dfrac{1}{1-x} - \dfrac{1 - x^n}{1-x} = \dfrac{x^n}{1-x} > 0;$ therefore $y_n > 0$ for every $n.$

Let $z_n = y_n - \dfrac{x^{n+1}}{n+1} \cdot \dfrac{1}{1-x}$; then $z_n = 0$ when $x = 0$, and

$$\frac{dz_n}{dx} = \frac{x^n}{1-x}\left\{\frac{x^n}{1-x} + \frac{1}{n+1}\cdot\frac{x^{n+1}}{(1-x)^2}\right\} = -\frac{1}{n+1}\cdot\frac{x^{n+1}}{(1-x)^2} < 0;$$

therefore $z_n < 0$. Hence $0 < y_n < \dfrac{1}{n+1}\cdot\dfrac{x^{n+1}}{1-x}$; and, since $0 < x < 1$, the last expression

$\to 0$ as $n \to \infty$; thus $y_n \to 0$ and $s_n \to -\log(1-x)$, which proves the theorem in this case.

■ **17. Logarithmic Series:** If $-1 < x < 1$, we have

$$\log(1+x) = x - \frac{x^2}{2} + \frac{x^3}{3} - \ldots + (-1)^{n-1}\frac{x^n}{n} + \ldots \qquad \ldots(A)$$

Changing the sign of x,

$$\log(1-x) = -\left(x + \frac{x^2}{2} + \frac{x^3}{3} + \ldots + \frac{x^n}{n} + \ldots\right), \qquad \ldots(B)$$

and since $\log\dfrac{1+x}{1-x} = \log(1+x) - \log(1-x)$, it follows that

$$\log\frac{1+x}{1-x} = 2\left(x + \frac{x^3}{3} + \frac{x^5}{5} + \ldots\right). \qquad \ldots(C)$$

Writing $\dfrac{1+x}{1-x} = \dfrac{n+1}{n}$, so that $x = \dfrac{1}{2n+1}$, we have

$$\log(n+1) - \log n = 2\left(\frac{1}{2n+1} + \frac{1}{3}\cdot\frac{1}{(2n+1)^3} + \frac{1}{5}\cdot\frac{1}{(2n+1)^5} + \ldots\right). \qquad \ldots(D)$$

■ **EXAMPLE :** *If $0 < x < 1$ and $s_n = x + \dfrac{x^3}{3} + \dfrac{x^5}{5} + \ldots$ to n terms, show that the error*

in taking $2s_n$ as the value of $\log\dfrac{1+x}{1-x}$ it less than $\dfrac{2x^{2n+1}}{2n+1}\cdot\dfrac{1}{1-x^2}$.

If R_n is the remainder after n terms of the series,

$$R_n = \frac{x^{2n+1}}{2n+1} + \frac{x^{2n+3}}{2n+3} + \frac{x^{2n+5}}{2n+5} + \ldots < \frac{x^{2n+1}}{2n+1}(1 + x^2 + x^4 + \ldots);$$

$\therefore$ The error $= 2R_n < \dfrac{2x^{2n+1}}{2n+1}\cdot\dfrac{1}{1-x^2}$.

18. Series which can be summed by the Logarithmic Series:

EXAMPLE 1. *If k is a positive integer and $|x| < 1$, then*

$$\sum_{n=1}^{n=\infty} \frac{x^n}{n+k} = -\frac{1}{x^k}\left\{x + \frac{x^2}{2} + \dots + \frac{x^k}{k} + \log(1-x)\right\}.$$

EXAMPLE 2. *If $|x| < 1$, find the value of $\displaystyle\sum_{n=1}^{n=\infty} \frac{n^3 + n^2 + 1}{n(n+2)} x^n$.*

The series is convergent when $|x| < 1$. Let S be the sum, then since

$$\frac{n^3 + n^2 + 1}{n(n+2)} = n - 1 + \frac{1}{2}\cdot\frac{1}{n} + \frac{3}{2}\cdot\frac{1}{n+2},$$

we have

$$S = \sum_1^\infty (n-1)x^n + \frac{1}{2}\sum_1^\infty \frac{x^n}{n} + \frac{3}{2}\frac{x^n}{n+2};$$

$$\therefore \quad S = \frac{x^2}{(1-x)^2} - \frac{1}{2}\log(1-x) - \frac{3}{2x^2}\left\{x + \frac{x^2}{2} + \log(1-x)\right\}.$$

The method of Ex. 2 can be applied to sum any series of which the general term is

$$Px^n/(n+a)(n+b)\dots(n+k),$$

where P is a polynomial in n and $a, b, \dots k$ are unequal positive integers.

19. Calculation of Napierian Logarithms:

The number e is chosen as the base of the system of logarithms used in theoretical work. Such logarithms are called *Napierian*, after Napier, the inventor of logarithms. In theoretical work, $\log N$ means $\log_e N$; just as, in practical reckoning, $\log N$ means $\log_{10} N$.

The method of applying the equations of Art. 17 to the calculation of Napierian logarithms is exhibited in the following example:

EXAMPLE 1. *Calculate $\log_e 2$ to seven places of decimals.*

Let $\dfrac{1+x}{1-x} = 2$; $\therefore x = \dfrac{1}{3}$; $\therefore \log_e 2 = 2\left\{\dfrac{1}{3} + \dfrac{1}{3}\cdot\dfrac{1}{3^3} + \dfrac{1}{5}\cdot\dfrac{1}{3^5} + \dots\right\}$.

Carrying the reckoning to nine places, we have

$1/3$	$=$	0.333	333	333	$1/3$	$=$	0.333	333	333

$$
\begin{array}{llll @{\qquad} llll}
1/3 & = 0.333 & 333 & 333 & 1/3 & = 0.333 & 333 & 333 \\
1/3^3 & = 0.037 & 037 & 037 & 1/(3\cdot 3^3) & = 0.012 & 345 & 679 \\
1/3^5 & = 0.004 & 115 & 226 & 1/(5\cdot 3^5) & = 0.000 & 823 & 045 \\
1/3^7 & = 0.000 & 457 & 247 & 1/(7\cdot 3^7) & = 0.000 & 065 & 321 \\
1/3^9 & = 0.000 & 050 & 805 & 1/(9\cdot 3^9) & = 0.000 & 005 & 645 \\
1/3^{11} & = 0.000 & 005 & 645 & 1/(11\cdot 3^{11}) & = 0.000 & 000 & 513 \\
1/3^{13} & = 0.000 & 000 & 627 & 1/(13\cdot 3^{13}) & = 0.000 & 000 & 048 \\
1/3^{15} & = 0.000 & 000 & 070 & 1/(15\cdot 3^{15}) & = 0.000 & 000 & 005 \\
& & & & & = 0.346 & 573 & 589 \\
& & & & & & & 2 \\
\hline
& & & & & 0.693 & 147 & 178
\end{array}
$$

$$\therefore \quad \log_e 2 = 0.693147178 \text{ nearly.}$$

The error in the value obtained for $\log_e 2$ is due to two causes:

(*i*) We have taken $2S_8$ for $\log_e 2$. By Ex. 1 of Art. 17, the error on this account is less than

$$\frac{2\left(\frac{1}{3}\right)^{17}}{17} \cdot \frac{1}{1-\frac{1}{9}} < 0.000\ 000\ 002.$$

(*ii*) The error in the *calculated* value of $2S_8$ is numerically less than
$$2 \times 8 \times 0.000\ 000\ 000\ 5 = 0.000\ 000\ 008.$$

Adding the results of (*i*) and (*ii*), we see that $0.693\ 147\ 178$ is an approximate value of $\log_e 2$ with an error numerically less than $0.000\ 000\ 01$.

$\therefore \qquad\qquad 0.693\ 147\ 168 < \log_e 2 < 0.693\ 147\ 188.$

Hence, to seven figures, $\log_e 2 = 0.6931472$.

Having found $\log_e 2$, the Napierian logarithms of the natural numbers can be found in succession by using formula of (D) Art. 17.

The logarithms of successive prime numbers may be calculated by the following method, due to Borda, in which the series employed converge very rapidly. Using the identities

$$x^3 - 3x + 2 = (x-1)^2\,(x+2) \quad\text{and}\quad x^3 - 3x - 2 = (x+1)^2\,(x-2)$$

it will be seen that

$$\frac{(x-1)^2\,(x+2)}{(x+1)^2\,(x-2)} = \frac{x^3 - 3x + 2}{x^3 - 3x - 2} = \frac{1 + 2/(x^3 - 3x)}{1 - 2/(x^3 - 3x)};$$

$$\therefore \qquad \log(x-1) + \frac{1}{2}\log(x+2) - \log(x+1) - \frac{1}{2}\log(x-2)$$

$$= \frac{2}{x^3 - 3x} + \frac{1}{3}\left(\frac{2}{x^3 - 3x}\right)^3 + \frac{1}{5}\left(\frac{2}{x^3 - 3x}\right)^5 + \ \dots\ .$$

Putting $x = 5, 6, 7, 8$, it will be found that

$$\log 2 - \frac{3}{2}\log 3 + \frac{1}{2}\log 7 = 0.0181838220 \ \dots\ ,$$

$$\frac{1}{2}\log 2 + \log 5 - \log 7 = 0.0101013536\dots\ ,$$

$$-2\log 2 + 2\log 3 - \frac{1}{2}\log 5 = 0.0062112599 \ \dots\ ,$$

$$-\frac{5}{2}\log 3 + \frac{1}{2}\log 5 + \log 7 = 0.0040983836 \ \dots\ .$$

Hence it can be shown that
$$\log 2 = 0.693147180 \ \dots\ \log 3 = 1.098612288 \ \dots\ ,$$
$$\log 5 = 1.609437912 \ \dots\ \log 7 = 1.945910149 \ \dots\ .$$

The logarithms of successive primes 11,13, etc., can now be obtained by putting $x + 2 = 11$, 13, etc., in the formula.

The method of the example which follows, in which logarithms to a base 10 are calculated without the use of the logarithmic series, is of historical interest, being one of the methods given by Briggs.

■ **EXAMPLE 2.** *Obtain the logarithm of 7 to the base 10, by the use of square-roots only, to 5 decimal places.*

In the reckoning below, when a square-root has been obtained which is greater than 7, the square-root of the product of it and the last square-root obtained that was less than 7 is found next; and *vice versa*. Thus we have

$$\log \sqrt{10} \; = \; \log 3.162277 \ldots = \log A = 0.5,$$

$$\log \sqrt{10A} \; = \; \log 5.623413 \ldots = \log B = 0.75,$$

$$\log \sqrt{10B} \; = \; \log 7.498942 \ldots = \log C = 0.875.$$

$$\log \sqrt{BC} \; = \; \log 6.493816 \ldots = \log D = 0.8125,$$

$$\log \sqrt{CD} \; = \; \log 6.978305 \ldots = \log E = 0.84375,$$

$$\log \sqrt{CE} \; = \; \log 7.233941 \ldots = \log F = 0.859375,$$

and so on; until we arrive at

$$\log 7.000032 \; = \; 0.845100\ldots ,$$
$$\log 6.999968 \; = \; 0.845096 \ldots;$$

thus, $\log_{10} 7 = 0.84510$ to five places, and probably is equal to 0.845098 to six places.

■ **20. Calculation of Common Logarithms:** We have

$$\log_{10} N = \log_e N / \log_e 10.$$

The expression $1/(\log_e 10)$ is called the *modulus* of the common system of logarithms and is denoted by μ. It can be shown that, to fifteen places of decimals,

$$\mu = 0.434294481903251.$$

Thus $\log_{10} N = \mu \log_e N$, *where μ has the above value.*

■ **21. The Hyperbolic Functions:**

(1) The functions known as the hyperbolic cosine, sine and tangent of x, and denoted by $\cosh x$, $\sinh x$ and $\tanh x$, are defined by

$$\cosh x = \frac{e^x + e^{-x}}{2}, \; \sinh x = \frac{e^x - e^{-x}}{2}, \; \tanh x = \frac{e^x - e^{-x}}{e^x + e^{-x}}.$$

The reciprocals of $\cosh x$, $\sinh x$, $\tanh x$ are called the hyperbolic secant, cosecant and cotangent of x, and we write

$$\operatorname{sech} x = \frac{1}{\cosh x}, \; \operatorname{cosech} x = \frac{1}{\sinh x}, \; \coth x = \frac{1}{\tanh x}$$

(2) Hence the following equations:

$$\cosh^2 x - \sinh^2 x = 1, \; 1 - \tanh^2 x = \operatorname{sech}^2 x,$$

$$\frac{d}{dx} \cosh x = \sinh x, \; \frac{d}{dx} \sinh x = \cosh x, \; \frac{d}{dx} \tanh x = \operatorname{sech}^2 x.$$

Also, cosh $(-x)$ = cosh x, sinh $(-x)$ = $-$ sinh x, tanh $(-x)$ = $-$ tanh x, so that cosh x is an *even* function, and sinh x, tanh x are odd functions of x. Graphs of these functions are shown below: rough sketches of the first two can be constructed quickly from the curves $y = e^x$, $y = e^{-x}$ (see p. 302), by observing that, in Fig. 61, C is the middle point of AB, and $ED = BC$. The ordinate of $y = $ tanh x is equal to the ratio ED/EC; this can be found numerically or by a geometrical construction.

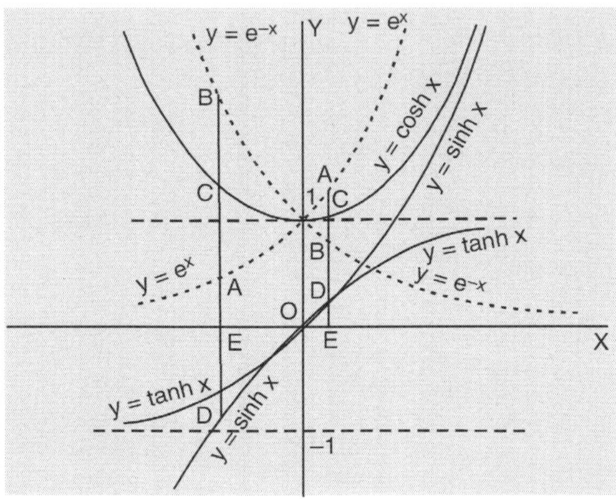

Fig. 61

(3) *Addition Formulae.* We have

$$2 \cosh x \cosh y = \frac{1}{2} (e^x + e^{-x})(e^y + e^{-y})$$

$$= \frac{1}{2} (e^{x+y} + e^{-(x+y)} + e^{x-y} + e^{-(x-y)})$$

$$= \cosh (x + y) + \cosh (x - y).$$

Similarly,

$$2 \sinh x \sinh y = \cosh (x + y) - \cosh (x - y),$$
$$2 \sinh x \cosh y = \sinh (x + y) + \sinh (x - y),$$
$$2 \cosh x \sinh y = \sinh (x + y) - \sinh (x - y).$$

Therefore

$$\cosh (x \pm y) = \cosh x \cosh y \pm \sinh x \sinh y,$$
$$\sinh (x \pm y) = \sinh x \cosh y \pm \cosh x \sinh y,$$

and consequently

$$\tanh (x \pm y) = \frac{\tanh x \pm \tanh y}{1 \pm \tanh x \tanh y}.$$

where in each equation the upper or lower sign is to be taken throughout.

(4) Using the exponential values of sin x and cos x (Art. 14), we have

$$\cos x = \frac{e^{\iota x} + e^{-\iota x}}{2} = \cosh \iota x,$$

$$\sin x = \frac{e^{\iota x} - e^{-\iota x}}{2\iota} = \frac{1}{\iota} \sinh \iota x,$$

$$\tan x = \frac{1}{\iota} \tanh \iota x,$$

and $\cosh x = \cos \iota x,\ \sinh x = \iota \sin \iota x,\ \tanh x = \iota \tan \iota x.$

■ **22. Inverse Hyperbolic Functions:** Let $y = \cosh x$, then

$$e^x + e^{-x} = 2y \text{ and } e^{2x} - 2e^x y + 1 = 0,$$

therefore $\quad e^x = y \pm \sqrt{(y^2 - 1)}$ and $x = \log \left\{ y \pm \sqrt{(y^2 - 1)} \right\}.$

Thus if $y > 1$, there are two real values of x, equal in magnitude and of opposite signs, which correspond to any given value of y.

The *positive* value is denoted by $\cosh^{-1} y$, so that

$$\cosh^{-1} y = \log \left\{ y + \sqrt{(y^2 - 1)} \right\}.$$

Next, let $y = \sinh x$, then $e^{2x} - 2e^x y - 1 = 0$, and since $e^x > 0$ for real values of x,

we have $e^x = y + \sqrt{(y^2 + 1)}$, so that $x = \log \left\{ y + \sqrt{(y^2 + 1)} \right\}$, and we write

$$\sinh^{-1} y = \log \left\{ y + \sqrt{y^2 + 1} \right\}.$$

Again, if $\qquad y = \tanh x = \dfrac{e^x - e^{-x}}{e^x + e^{-x}},$

then $\qquad e^{2x} = \dfrac{1 + y}{1 - y}$ and $x = \dfrac{1}{2} \log \dfrac{1 + y}{1 - y}.$

Hence we write

$$\tanh^{-1} y = \frac{1}{2} \log \frac{1 + y}{1 - y}.$$

Thus $\cosh^{-1} y$ is a single-valued and continuous function of y (see Ch. 17, 21) for all values of $y > 1$; $\sinh^{-1} y$ is a single-valued and continuous function of y for all values of y; $\tanh^{-1} y$ is a single-valued and continuous function of y over the range $-1 < y < 1$.

EXERCISE XXXII

The notation of Art. 9 is used in Exs. 2 and 3.

1. Show that $\dfrac{1}{n+1} + \dfrac{1}{n+2} + \dfrac{1}{n+3} + \dots + \dfrac{1}{2n} \to \log 2$ as $n \to \infty.$

2. Prove that $1 + \frac{1}{3} + \frac{1}{5} + \ldots + \frac{1}{2n-1} = \log 2 + \frac{1}{2}\gamma + \frac{1}{2}\log n + \varepsilon_{2n} - \varepsilon_n$, so that

$$1 + \frac{1}{3} + \frac{1}{5} + \ldots + \frac{1}{2n-1} - \frac{1}{2}\log n \to \log 2 + \frac{1}{2}\gamma.$$

3. Let the series $1 - \frac{1}{2} + \frac{1}{3} - \frac{1}{4} + \ldots$ be deranged as follows. Write down the first p positive terms followed by the first q negative terms, then the next p positive terms followed by the next q negative terms, and so on.

Prove that the series so deranged is convergent, and that its sum is

$$\log 2 + \frac{1}{2}\log p/q.$$

[Let σ_r be the sum of the first r groups of p positive followed by q negative terms, then

$$\sigma_r = \left(1 + \frac{1}{3} + \ldots + \frac{1}{2pr-1}\right) - \frac{1}{2}\left(1 + \frac{1}{2} + \frac{1}{3} + \ldots + \frac{1}{rq}\right)$$

$$= (\log 2 + \frac{1}{2}\gamma + \frac{1}{2}\log pr + \varepsilon_{2pr} - \varepsilon_{pr}) - \frac{1}{2}(\gamma + \log qr + \varepsilon_{pr}) \to \log 2 + \frac{1}{2}\log p/q.$$

Also, if s_n is the sum to n terms of the deranged series, r can be found so that all these terms are contained in σ_r, but not in σ_{r-1}. Thus $\sigma_r - s_n$ is the sum of a finite number of terms each of which $\to 0$. Also $r \to \infty$ as $n \to \infty$, and consequently

$$s_n \to \log 2 + \frac{1}{2}\log p/q.]$$

4. Use the series for $\log (1 + x)$ to show that:

(i) if $0 < x < 1$, then $0 < x - \log (1 + x) < \frac{1}{2}x^2$;

(ii) if $-1 < x < 0$, then $0 < x - \log (1 + x) < \frac{1}{2}x^2 /(1 + x)$.

[For (ii) let $x = -y$, then $0 < y < 1$ and

$$- \log (1 - y) = y + \frac{1}{2}y^2 + \frac{1}{3}y^3 + \ldots < y + \frac{1}{2}y^2 (1 + y + y^2 + \ldots), \text{ etc.}]$$

5. (i) Show that $\log 50 - \log 49 = \mu \left\{ \frac{1}{50} + \frac{1}{2} \cdot \frac{1}{50^2} + \frac{1}{3} \cdot \frac{1}{50^3} + \ldots \right\}.$

(ii) Hence calculate $\log_{10} 7$ to five places of decimals, given that
$$\log_{10} 2 = 0.301030, \ \mu = 0.434294.$$

6. (i) Show that $\log_{10} 1 \cdot 1 = 2\mu \left\{ \frac{1}{21} + \frac{1}{3} \cdot \frac{1}{21^3} + \frac{1}{5} \cdot \frac{1}{21^5} + \ldots \right\}.$

(ii) Given $\mu = 0.434294$, find $\log_{10} 11$ to five places of decimals.

7. (i) Show that $\log_{10} \frac{13}{12} = 2\mu \left\{ \frac{1}{25} + \frac{1}{3} \cdot \frac{1}{25^3} + \frac{1}{5} \cdot \frac{1}{25^5} + \ldots \right\}.$

(ii) Given that $\log_{10} 2 = 0.301030$, $\log_{10} 3 = 0.477121$, $\mu = 0.434294$, find $\log_{10} 13$ to five places of decimals.

8. (*i*) Show that $\log_e (1 + x + x^2) = \log_e (1 - x^3) - \log_e (1 - x)$.

(*ii*) Expand $\log_e (1 + x + x^2)$ to six terms, and show that the coefficient of x^n in the expansion is $-2/n$ or $1/n$ according as n is or is not a multiple of 3.

9. Find the coefficients of x^{4m} and x^{4m+1} in the expansion of
$$\log_e (1 + x + x^2 + x^3).$$

Sum to infinity the series in Exs. 10–14.

10. $\dfrac{1}{1\cdot 2} + \dfrac{1}{3\cdot 4} + \dfrac{1}{5\cdot 6} + \dots.$

11. $\dfrac{1}{1\cdot 2} - \dfrac{1}{2\cdot 3} + \dfrac{1}{3\cdot 4} - \dots.$

12. $\dfrac{1}{1\cdot 3} + \dfrac{1}{2\cdot 5} + \dfrac{1}{3\cdot 7} + \dfrac{1}{4\cdot 9} + \dots.$

13. $\dfrac{1}{1\cdot 2} + \dfrac{x}{3\cdot 4} + \dfrac{x^2}{5\cdot 6} + \dfrac{x^3}{7\cdot 8} + \dots.$

14. $1 + \dfrac{x^2}{3} + \dfrac{x^4}{15} + \dots + \dfrac{x^{2n}}{4n^2 - 1} + \dots.$

15. In the expansion of $(1 + bx + cx^2) \log_e (1 + x)$,

(*i*) Find the coefficients of x^3 and x^4, and determine the values of b and c so that each of these coefficients may vanish.

(*ii*) Giving to b and c the values so determined, expand the given expression as far as the term containing x^5.

(*iii*) Hence find an expression of the form $\dfrac{x + ax^2}{1 + bx + cx^2}$ which will give the best possible approximate values of $\log_e (1 + x)$ when x is small.

16. (*i*) Prove that $(6 + 6x + x^2) \log_e (1 + x) - (6x + 3x^2)$
$$= x^5 \left\{ \frac{1\cdot 2}{3\cdot 4\cdot 5} - \frac{2\cdot 3}{4\cdot 5\cdot 6} x + \frac{3\cdot 4}{5\cdot 6\cdot 7} x^2 - \dots \right\},$$

(*ii*) Denoting the series in brackets by $u_1 - u_2 + u_3 - \dots$, prove that
$$\frac{u_n}{u_{n-1}} = \frac{(n+1)^2}{(n-1)(n+4)} x.$$

(*iii*) If $0 < x < \dfrac{2}{3}$, show that $u_n < u_{n-1}$.

(*iv*) Hence show that *if* $0 < x < \dfrac{2}{3}$, *then* $\dfrac{6x + 3x^2}{6 + 6x + x^2}$ *is an approximate value of* $\log_e$ *(1 + x), with an error in defect less than* $x^5/180$.

17. Use Ex. 16 (*iv*) to show that $\dfrac{759}{32003}$ is an approximate value of $\log_e 1.024$, with an error in defect less than $5/10^{11}$. Hence find $\log_{10} 1.024$ and $\log_{10} 2$, each to ten places of decimals.

18. If $a - b = h$, where h is small, show that

(*i*) $\dfrac{a-b}{2}\left(\dfrac{1}{a}+\dfrac{1}{b}\right)=\dfrac{h}{a}+\dfrac{h^2}{2a^2}+\dfrac{h^3}{2a^3}+\dots.$

(*ii*) $2\left(\dfrac{a-b}{a+b}\right)+\dfrac{2}{3}\left(\dfrac{a-b}{a+b}\right)^3=\dfrac{h}{a}+\dfrac{h^2}{2a^2}+\dfrac{h^3}{3a^3}+\dots.$

(*iii*)* Hence show that, if a is nearly equal to b,

$$\log_e \frac{a}{b}=\frac{a-b}{2}\left(\frac{1}{a}+\frac{1}{b}\right),\ \textit{nearly,}$$

the error being approximately equal to $(a - b)^3/6a^3$.

Sum the series in Exs. 19, 20:

19. $\displaystyle\sum_{n=1}^{n=\infty}\frac{n^2+1}{n(n+2)}x^n.$

20. $\displaystyle\sum_{n=1}^{n=\infty}\frac{(n+1)^3}{n(n+3)}x^n.$

21. If s_n is the sum to n terms of the series

$$\frac{1}{1\cdot2\cdot3}+\frac{1}{5\cdot6\cdot7}+\frac{1}{9\cdot10\cdot11}+\frac{1}{13\cdot14\cdot15}+\dots,$$

prove that

$$2s_n=\frac{1}{2n+1}+\frac{1}{2n+3}+\frac{1}{2n+5}+\dots+\frac{1}{4n-1}.$$

Hence show that the sum to infinity of the series is $\frac{1}{4}\log 2$.

22. If s_n is the sum to n terms of the series

$$\frac{1}{2\cdot3\cdot4}+\frac{1}{5\cdot6\cdot7}+\frac{1}{8\cdot9\cdot10}+\dots+\frac{1}{(3n-1)\,3n\,(3n+1)}+\dots,$$

prove that $2s_n=\dfrac{1}{n+1}+\dfrac{1}{n+2}+\dfrac{1}{n+3}+\dots+\dfrac{1}{3n+1}-1.$

Hence show that the sum to infinity of the series is $\frac{1}{2}(\log 3-1)$.

23. Prove that (*i*) $\dfrac{1}{1\cdot2\cdot3}+\dfrac{1}{3\cdot4\cdot5}+\dfrac{1}{5\cdot6\cdot7}+\dots$ to $\infty=\log 2-\dfrac{1}{2}.$

(*ii*) $\dfrac{1}{1\cdot2\cdot3\cdot4}+\dfrac{1}{3\cdot4\cdot5\cdot6}+\dfrac{1}{5\cdot6\cdot7\cdot8}+\dots$ to $\infty=\dfrac{2}{3}\log 2-\dfrac{5}{12}.$

---

* This is Napier's formula.

24. By expanding $\log (1 + x + x^2)$ in various ways show that

(i) $1 - n + \dfrac{n(n-3)}{\underline{|2}} - \dfrac{n(n-4)(n-5)}{\underline{|3}} + ... = (-1)^n . 2$ or $(-1)^{n-1}$ according as n is of the form $3r$ or $3r \pm 1$.

(ii) $1 - \dfrac{n^2}{\underline{|2}} + \dfrac{n^2(n^2-1^2)}{\underline{|4}} - \dfrac{n^2(n^2-1^2)(n^2-2^2)}{\underline{|6}} + ... = (-1)^n$ or $\left(-\dfrac{1}{2}\right)^{n-1}$ according as n is of the form $3r$ or $3r \pm 1$.

(iii) $1 - \dfrac{n(n+1)}{\underline{|3}} + \dfrac{(n-1)n(n+1)(n+2)}{\underline{|5}} - \dfrac{(n-2)(n-1)n(n+1)(n+2)(n+3)}{\underline{|7}}$

$+ ... = \dfrac{(-1)^n}{2n+1}$ or $\dfrac{(-2)^{n-1}}{2n+1}$ according as n is of the form $3r$ or $3r \pm 1$.

(iv) $C_1^n - \dfrac{1}{2} C_3^{n+1} + \dfrac{1}{3} C_5^{n+2} + \dfrac{1}{4} C_7^{n+3} + ... = 0, \dfrac{1}{n}, \dfrac{3}{n}$ or $\dfrac{4}{n}$ according as n is of the form $6r, 6r \pm 1, 6r \pm 2, 6r - 3$.

[These are obtained from the identities:

(i) $\log (1 - x^3) - \log (1 - x) = \log \{1 + x(1 + x)\}$;

(ii) and (iii) $\log (1 - x^3) - \log (1 - x) - \log (1 + x^2) = \log \left\{1 + \dfrac{x}{1 + x^2}\right\}$;

(iv) $\log (1 - x^3) - 2 \log (1 - x^2) + \log (1 - x) = \log \left\{1 - \dfrac{x}{(1 + x)^2}\right\}.]$

25. Use the identities

$x^4 - 25x^2 + 144 = (x + 3)(x - 3)(x + 4)(x - 4)$, $x^4 - 25x^2 = x^2(x + 5)(x - 5)$ to show that

$\log (x + 3) + \log (x - 3) + \log (x + 4) + \log (x - 4) - 2 \log x - \log (x + 5) - \log (x - 5)$

$$= 2 \left\{\dfrac{72}{x^4 - 25x^2 + 72} + \dfrac{1}{3}\left(\dfrac{72}{x^4 - 25x^2 + 72}\right)^3 + ...\right\}.$$

This formula is useful in finding the logarithms of large primes.

■ ■ ■

CHAPTER 20

Convergence

SERIES OF POSITIVE TERMS (Continued)

■ **1. Cauchy's Condensation Test:** *If $f(n)$ is a (positive*) decreasing function of n, and a is any positive integer > 1, then the two series, $\Sigma f(n)$ and $\Sigma a^n f(a^n)$, are both convergent or both divergent.*

Group the terms of $\Sigma f(n)$ as follows :

$$\{f(1) + f(2) + \ldots + f(a)\},$$
$$+ \{f(a+1) + f(a+2) + \ldots + f(a^2)\},$$
$$+ \{f(a^2+1) + f(a^2+2) + \ldots + f(a^3)\} + \ldots,$$

where, if v_n denotes the sum of the terms of the nth group,

$$v_n = f(a^{n-1}+1) + f(a^{n-1}+2) + \ldots + f(a^n).$$

The number of these terms is $a^n - a^{n-1}$, and since $f(n)$ is a decreasing function,

$$(a^n - a^{n-1})\, f(a^n) \le v_n \le (a^n - a^{n-1})\, f(a^{n-1});$$

$$\therefore \qquad \frac{1}{a}(a-1)a^n f(a^n) \le v_n \le (a-1)\, a^{n-1} f(a^{n-1}).$$

Now, if $\Sigma a^n f(a^n)$ is convergent, so also is Σv_n; and therefore $\Sigma f(n)$ converges (Ch. 16, 5). If $\Sigma a^n f(a^n)$ is divergent, so also is Σv_n; and therefore $\Sigma f(n)$ diverges. But a series of positive terms must converge or diverge, therefore $\Sigma f(n)$ and $a^n \Sigma f(a^n)$ are both convergent or both divergent.

■ **2. An Important Test Series:** *The series $\Sigma \dfrac{1}{n(\log n)^p}$ is convergent if $p > 1$ and divergent if $p \le 1$.*

If a is any positive integer, this series is convergent or divergent according as the series Σv_n is convergent or divergent, where

$$v_n = \frac{a^n}{a^n (\log a^n)^p} = \frac{1}{(\log a)^p} \cdot \frac{1}{n^p}.$$

Since $(\log a)^p$ is constant, the given series is convergent or divergent according as $\Sigma 1/n^p$ is convergent or divergent, that is according as $p > 1$ or $p \le 1$.

---

* In any case, the terms of the decreasing function $f(n)$ are ultimately of the same sign (Exercise XXV, 1).

■ **3. Kummer's Test:** *Let Σu_n and $\Sigma 1/d_n$ be two series of positive terms and suppose that $\Sigma 1/d_n$ is divergent; also suppose that $v_n = d_n u_n / u_{n+1} - d_{n+1}$; then*

    (*i*) *if a fixed number k can be found so that after a certain stage, say for $n \geq m$, $v_n > k > 0$, the series Σu_n is convergent;*

    (*ii*) *if $v_n < 0$ for $n \geq m$, Σu_n is divergent.*

Proof. (*i*) Putting $m, m+1, \ldots, n-1$ in succession for n in the given condition, we have

$$d_m u_m - d_{m+1} u_{m+1} > k u_{m+1},$$
$$d_{m+1} u_{m+1} - d_{m+2} u_{m+2} > k u_{m+2},$$
$$\ldots\ldots\ldots\ldots\ldots\ldots\ldots\ldots\ldots\ldots$$
$$d_{n-1} u_{n-1} - d_n u_n > k u_n;$$

whence $u_{m+1} + u_{m+2} + \ldots + u_n < \dfrac{1}{k}\,(d_m u_m - d_n u_n) < \dfrac{1}{k}\,d_m u_m.$

Thus if s_r is the sum to r terms of Σu_n,

$$s_n < s_m + \frac{1}{k}\,d_m u_m,\ \text{a fixed number;}$$

therefore Σu_n is convergent (Ch.16, 7).

(*ii*) Since $d_n \dfrac{u_n}{u_{n+1}} - d_{n+1} < 0$ for $n \geq m$, we have

$$d_m u_m < d_{m+1} u_{m+1} < \ldots < d_n u_n,$$

therefore $u_n > d_m u_m \cdot \dfrac{1}{d_n}.$

Now $d_m u_m$ is a fixed number and $\Sigma 1/d_n$ is divergent, therefore Σu_n is divergent.

Note: If $v_n \to 0$ we cannot find k so that $v_n > k > 0$ for $n \geq m$, and the test fails. In part (*i*) of the test, $\Sigma 1/d_n$ need not diverge; in fact, d_n may be any positive function of n.

■ **4. Raabe's Test:** *Let Σu_n be a series of positive terms, then*

(*i*) *if after a certain stage, say for $n \geq m$,*

$$n\left(\frac{u_n}{u_{n+1}} - 1\right) > k > 1,$$

where k is a fixed number, then Σu_n is convergent.

(*ii*) If $n\left(\dfrac{u_n}{u_{n+1}} - 1\right) < 1$, *then Σu_n is divergent.*

This follows at once from Kummer's test by taking $d_n = n$, giving

$$v_n = n\left(\frac{u_n}{u_{n+1}} - 1\right).$$

In particular, if $\lim n \left(\dfrac{u_n}{u_{n+1}} - 1 \right) = l$, *then* Σu_n *converges when* $l > 1$ *and diverges when* $l < 1$: *but if* $l = 1$, *the test fails.*

This test may be tried when D'Alembert's test fails.

■ **EXAMPLE :** *Consider the convergence or divergence of* $x + \dfrac{1}{2} \dfrac{x^3}{3} + \dfrac{1 \cdot 3}{2 \cdot 4} \dfrac{x^5}{5} + \ldots$

Here $\qquad \dfrac{u_n}{u_{n+1}} = \dfrac{2n(2n+1)}{(2n-1)^2} \cdot \dfrac{1}{x^2} ; \quad \therefore \lim \dfrac{u_n}{u_{n+1}} = \dfrac{1}{x^2} .$

By D'Alembert's test, Σu_n converges if $|x| < 1$ and diverges if $|x| > 1$. If $|x| = 1$, the test fails; with Raabe's test, however, we have

$$\lim n \left(\frac{u_n}{u_{n+1}} - 1 \right) = \lim \frac{6n^2 - n}{(2n-1)^2} = \frac{3}{2} > 1 ;$$

therefore Σu_n is convergent.

Note: This series is the expansion of $\sin^{-1} x$, and was discovered by Newton. From it, he deduced the series for $\sin x$.

■ **5. Gauss's Test:** *Let* Σu_n *be a series of positive terms and suppose that* u_n/u_{n+1} *can be expressed in the form*

$$\frac{u_n}{u_{n+1}} = 1 + \frac{a}{n} + \frac{b_n}{n^p},$$

where $p > 1$ *and* $|b_n| < a$ *fixed number* k *or (in particular)* b_n *tends to a finite limit as* $n \to \infty$, *then* Σu_n *converges if* $a > 1$ *and diverges if* $a \le 1$.

Proof. First suppose that $a \ne 1$, then

$$n \left(\frac{u_n}{u_{n+1}} - 1 \right) = a + \frac{b_n}{n^{p-1}} \to a \text{ as } n \to \infty.$$

for $|b_n| < k$ and $p > 1$. Therefore, by Raabe's test, Σu_n converges if $a > 1$ and diverges if $a < 1$.

If $a = 1$, Raabe's test fails and we apply Kummer's. Taking $d_n = n \log n$ in Art. 3, we have

$$v_n = n \log n \cdot \frac{u_n}{u_{n+1}} - (n+1) \log (n+1)$$

$$= n \left(1 + \frac{1}{n} + \frac{b_n}{n^p} \right) \log n - (n+1) \log (n+1)$$

$$= (n+1) \log \frac{n}{n+1} + \frac{\log n}{n^{p-1}} \cdot b_n.$$

Now $(n + 1) \log \dfrac{n}{n+1} = (n + 1) \log \left(1 - \dfrac{1}{n+1}\right) \to -1$ as $n \to \infty$

and $\dfrac{\log n}{n^{p-1}} \cdot b_n \to 0$, for $p > 1$ [Ch. 19, 7, (5)], and $\mid b_n \mid < k$.

Therefore $v_n \to -1$ and, after a certain stage, $v_n < 0$. Hence Σu_n is divergent.

In many cases u_n/u_{n+1} can be expanded in a series of powers of $1/n$, and we have the following rule:

If
$$\dfrac{u_n}{u_{n+1}} = 1 + \dfrac{a}{n} + \dfrac{b}{n^n} + \dots,$$

the series being convergent for sufficiently large values of n, then Σu_n is convergent if $a > 1$ and divergent if $a \leq 1$.

For in this case b_n is the sum of a convergent series.

This is the most generally useful test for series of positive terms, and in cases where it applies, it is better to use it at once instead of beginning with D'Alembert's and Raabe's tests.

■ **EXAMPLE** : *Discuss the convergence or divergence of* $\dfrac{1^2}{2^2} + \dfrac{1^2 \cdot 3^2}{2^2 \cdot 4^2} + \dfrac{1^2 \cdot 3^2 \cdot 5^2}{2^2 \cdot 4^2 \cdot 6^2} + \dots.$

By division we can show that $\dfrac{u_n}{u_{n+1}} = \left(\dfrac{2n+2}{2n+1}\right)^2 = 1 + \dfrac{1}{n} + \dfrac{b_n}{n^2},$

where $b_n - = \dfrac{1}{4}\left(1 + \dfrac{1}{n}\right) \Big/ \left(1 + \dfrac{1}{2n}\right)^2 \to -\dfrac{1}{4}$ as $n \to \infty$.

Therefore by Gauss's test the series is divergent. Here the tests of D'Alembert and Raabe both fail.

CONTINUED FROM CH. 16, 25

■ **6. Binomial Series:** *When $x = -1$, the series*

$$1 + nx + \dfrac{n(n-1)}{\lfloor 2} x^2 + \dfrac{n(n-1)(n-2)}{\lfloor 3} x^3 + \dots$$

converges if $n > 0$ and diverges if $n < 0$.

When $x = -1$, after a certain stage all the terms have the same sign. Denoting the series by $u_0 + u_1 + u_2 + \dots$, we can show by division that

$$\dfrac{u_r}{u_{r+1}} = -\dfrac{r+1}{n-r} = 1 + \dfrac{n+1}{r} + \dfrac{b_r}{r^2},$$

where $b_r = n(n+1) \Big/ \left(1 - \dfrac{n}{r}\right) \to n(n+1)$ as $r \to \infty$.

Hence, by Gauss's test, the series converges or diverges according as $n + 1 \gtrless 1$, that is according as $n \gtrless 0$.

Apparently the test says that the series diverges when $n = 0$, but here the reasoning fails because u_r/u_{r+1} is of the form 0/0.

◼ **7. The Hyper-geometric Series:** The series

$$1 + \frac{\alpha \cdot \beta}{1 \cdot \gamma} x + \frac{\alpha (\alpha + 1) \beta (\beta + 1)}{1 \cdot 2 \cdot \gamma (\gamma + 1)} x^2 + \dots$$

is known as the *hyper-geometric series,* and is of great importance.

If none of the constants α, β, γ is a negative integer, the series is endless, and we shall prove that

   (*i*) *the series is absolutely convergent if $|x| < 1$, and is divergent if $|x| < 1$;*

   (*ii*) *when $x = 1$, the series converges if $\gamma > \alpha + \beta$, and not otherwise;*

   (*iii*) *when $x = -1$, the series converges if $\gamma + 1 > \alpha + \beta$, and not otherwise.*

Proof. Denoting the series by $u_0 + u_1 + \dots + u_n + \dots$ where

$$u_n = \frac{\alpha (\alpha + 1) \dots (\alpha + n - 1) \beta (\beta + 1) \dots (\beta + n - 1)}{1 \cdot 2 \dots n. \gamma (\gamma + 1) \dots (\gamma + n - 1)} x^n,$$

we have $\quad \dfrac{u_n}{u_{n+1}} = \dfrac{(1 + n)(\gamma + n)}{(\alpha + n)(\beta + n)} x = \dfrac{\left(1 + \dfrac{1}{n}\right)\left(1 + \dfrac{\gamma}{n}\right)}{\left(1 + \dfrac{\alpha}{n}\right)\left(1 + \dfrac{\beta}{n}\right)} x \to x.$

   (*i*) Hence the series converges absolutely if $|x| < 1$, and diverges if $|x| > 1$.

   (*ii*) *When* $x = 1$, $u_n/u_{n+1} \to 1$. Hence, after a certain stage, the terms have the same sign.

Applying Gauss's test, we can show by division that

$$\frac{u_n}{u_{n+1}} = 1 + (\gamma + 1 - \alpha - \beta)\frac{1}{n} + \frac{b_n}{n^2},$$

where $\qquad b_n = \left(A + \frac{B}{n}\right)\Big/\left(1 + \frac{\alpha}{n}\right)\left(1 + \frac{\beta}{n}\right)$

and A, B are fixed numbers. Thus $b_n \to A$ and Σu_n converges or diverges according as $\gamma + 1 - \alpha - \beta > 1$ or ≤ 1, that is according as

$$\gamma > \alpha + \beta \quad \text{or} \quad \gamma \leq \alpha + \beta$$

(*iii*) *When* $x = -1$, after a certain stage, the terms are alternately positive and negative.

Also $|u_{n+1}| < |u_n|$ if $(\alpha + n)(\beta + n) < (1 + n)(\gamma + n)$; that is, if

$$\alpha\beta + n(\alpha + \beta) < \gamma + n(1 + \gamma),$$

or if $\qquad\qquad 1 + \gamma - \alpha - \beta > (\alpha\beta - \gamma)\dfrac{1}{n}.$

This is true for sufficiently large values of n if $\gamma + 1 > \alpha + \beta$.

Again, as in the preceding, $-\dfrac{u_n}{u_{n+1}} = 1 + \dfrac{a_n}{n}$ where $a_n \to \gamma + 1 - \alpha - \beta$ as $n \to \infty$. Hence by Ch. 16, 24, $u_n \to 0$ if $\gamma + 1 > \alpha + \beta$, but not otherwise. Therefore by Ch. 16, 14, the series converges if $\gamma + 1 > \alpha + \beta$, but not otherwise.

▪ **8. De Morgan's and Bertrand's Test:** *Let Σu_n be a series of positive terms, and suppose that u_n/u_{n+1} can be expressed in the form*

$$\frac{u_n}{u_{n+1}} = 1 + \frac{1}{n} + \frac{a_n}{n \log n},$$

then (*i*) *if after a certain stage $a_n > 1 + k$ where k is a fixed positive number, Σu_n is convergent.*

(*ii*) *If $a_n \le 1$, Σu_n is divergent.*

Using Kummer's test, we take $d_n = n \log n$. Then, as in Art. 5,

$$v_n = n \log n . \frac{u_n}{u_{n+1}} - (n + 1) \log (n + 1)$$

$$= (n + 1) \log \left(1 - \frac{1}{n+1}\right) + a_n.$$

Therefore $v_n \to a_n - 1$ as $n \to \infty$. Hence Σu_n converges if $a_n > 1 + k$ and diverges if $a_n < 1$.

If $a_n = 1$, $v_n < 0$, for $\log \left(1 - \dfrac{1}{n+1}\right) < -\dfrac{1}{n+1}$, therefore Σu_n is divergent.

SERIES WITH TERMS POSITIVE OR NEGATIVE

▪ **9. Theorem:** *If Σu_n is a convergent series of positive terms and $a_1, a_2, a_3, \ldots$ is a sequence of real numbers, positive or negative, and all numerically less than some fixed number k, then $\Sigma a_n u_n$ is absolutely convergent.*

For
$$| a_{m+1}u_{m+1} | + | a_{m+2}u_{m+2} | + \ldots + | a_{m+p}u_{m+p} |$$
$$= | a_{m+1} | u_{m+1} + | a_{m+2} | u_{m+2} + \ldots + | a_{m+p} | u_{m+p}$$
$$< k (u_{m+1} + u_{m+2} + \ldots + u_{m+p}).$$

Since Σu_n is convergent, we can find m such that for all values of p,

$$u_{m+1} + u_{m+2} + \ldots + u_{m+p} < \varepsilon/k,$$

where ε is any positive number, as small as we choose. Therefore

$$| a_{m+1}u_{m+1} | + | a_{m+2}u_{m+2} | + \ldots + | a_{m+p}u_{m+p} | < \varepsilon.$$

Hence $\Sigma a_n u_n$ is absolutely convergent.

▌ EXAMPLE : *If Σu_n is a convergent series of positive terms, $\Sigma u_n \cos n\theta$ and $\Sigma u_n \sin n\theta$ are absolutely convergent for all values of θ.*

▪ **10. Abel's Inequality:** *If $u_1, u_2, u_3, \ldots$ is a sequence of real numbers such that*
$$l < u_1 + u_2 + u_3 + \ldots + u_r < h$$
for the values 1, 2, 3, … n of r, and if $a_1, a_2, a_3, \ldots$ is a decreasing sequence of positive terms, then $a_1 l < a_1 u_1 + a_2 u_2 + \ldots + a_n u_n < a_1 h$.

For let $\quad s_n = u_1 + u_2 + \dots + u_n,\quad$ and $\quad t_n = a_1 u_1 + a_2 u_2 + \dots + a_n u_n.$

Then $u_1 = s_1,\ u_2 = s_2 - s_1,\ u_3 = s_3 - s_2,$ etc., therefore

$$t_n = a_1 s_1 + a_2 (s_2 - s_1) + \dots + a_n (s_n - s_{n-1})$$
$$= (a_1 - a_2)s_1 + (a_2 - a_3)s_2 + \dots + (a_{n-1} - a_n)s_{n-1} + a_n s_n.$$

Now the factors $a_1 - a_2,\ a_2 - a_3,\ \dots a_{n-1} - a_n,\ a_n$ are positive or zero, and their sum is $a_1.$

Also $s_1,\ s_2,\ \dots s_n$ are all greater than l and less than $h.$

Therefore $\qquad\qquad\qquad\qquad\qquad a_1 l < t_n < a_1 h.$

■ **11. Dirichlet's Test:** *If Σu_n converges or oscillates between finite limits and $a_1,$ $a_2,\ a_3,\ \dots$ is a decreasing sequence of positive terms which tends to zero as a limit, then $\Sigma a_n u_n$ is convergent.*

Let $\qquad\qquad\qquad\qquad s_n = u_1 + u_2 + \dots + u_n.$

Since Σu_n converges or oscillates finitely, we can find numbers $h,\ l$ so that, for all values of m and $p,$

$$l < s_{m+p} - s_m < h,$$

that is to say,

$$l < u_{m+1} + u_{m+2} + \dots + u_{m+p} < h.$$

Therefore, by Abel's inequality

$$a_{m+1} l < a_{m+1} u_{m+1} + a_{m+2} u_{m+2} + \dots + a_{m+p} u_{m+p} < a_{m+1} h.$$

Now $a_{m+1} \to 0$ as $m \to \infty,$ therefore for any ε we can find m so that

$$|\, a_{m+1} u_{m+1} + a_{m+2} u_{m+2} + \dots + a_{m+p} u_{m+p} \,| < \varepsilon;$$

hence $\qquad\qquad\qquad\qquad \Sigma a_n u_n$ is convergent.

■ **EXAMPLE :** *If (a_n) is a decreasing sequence of positive terms which tends to zero as a limit, both the series $\Sigma a_n \cos n\theta$ and $\Sigma a_n \sin n\theta$ are convergent unless $\theta = 0$ or $2k\pi,$ in which case $\Sigma a_n \cos n\theta$ diverges.*

For if θ is not a multiple of $2\pi,$ the series $\Sigma \cos n\theta$ and $\Sigma \sin n\theta$ oscillate between finite limits (Exercise XXVII, 24, 25).

Hence the result follows by Dirichlet's test.

■ **12. Abel's Test:** *If Σu_n converges and $a_1,\ a_2,\ a_3,\ \dots$ is a decreasing sequence of positive terms, then $\Sigma a_n u_n$ is convergent.*

For as n tends to infinity, a_n tends to some limit $l,$ therefore

$$a_n - l \to 0.$$

Hence by Dirichlet's test, $\Sigma(a_n - l)\, u_n$ is convergent; that is to say, $\Sigma a_n u_n - l\, \Sigma u_n$ is convergent.

But Σu_n is convergent, therefore $\Sigma a_n u_n$ is convergent.

POWER SERIES

A series of the type $\Sigma a_n x^n$ is called a *Power Series*. In Arts. 13–15, x may be real or complex.

■ **13.** *If $\Sigma a_n x^n$ converges (absolutely or conditionally) for $x = x_1,$ it converges absolutely for all values of x such that $|\, x \,| < |\, x_1 \,|.$*

For since $\Sigma a_n x_1^{\ n}$ is convergent, $a_n x_1^{\ n} \to 0$. Therefore a positive number k exists such that $\qquad | a_n x_1^{\ n} | < k$ for every n. Hence if $| x | < | x_1 |$,

$$| a_n x^n | = | a_n x_1^{\ n} | . \left| \frac{x}{x_1} \right|^n < k \left| \frac{x}{x_1} \right|^n$$

Therefore every term of $\Sigma | a_n x^n |$ is less than the corresponding term of the convergent series

$$k + k \left| \frac{x}{x_1} \right| + k \left| \frac{x}{x_1} \right|^2 + \ \dots \ .$$

Therefore $\Sigma | a_n x^n |$ is convergent and $\Sigma a_n x^n$ is absolutely convergent.

■ **14.** *If $\Sigma a_n x^n$ is non-convergent for $x = x_1$, it is non-convergent for every x such that $| x | > | x_1 |$.*

For if the series converge for $x = x_2$ where $| x_2 | > | x_1 |$, by the last theorem it converges for $x' = x_1$, which is contrary to the hypothesis.

■ **15.** *With regard to the series $\Sigma a_n x^n$, either (i) it converges for $x = 0$ and for no other value of x, or (ii) it converges absolutely for all values of x, or (iii) a positive number R exists such that $\Sigma a_n x^n$ converges absolutely when $| x | < R$ and is non-convergent when $| x | > R$.*

Proof. Suppose that there is a value of x other than zero for which $\Sigma a_n x^n$ converges. Also, suppose that there is a value x' of x for which it is non-convergent. Choose a positive number a greater than $| x' |$, then, by Art. 14, the series is non-convergent for $x = a$.

Divide the real numbers in the interval $(0, a)$ into two classes.

The lower class is to contain every real number r such that $\Sigma a_n x^n$ converges if $| x | = r$. *The upper class* is to contain every real number r', such that $\Sigma a_n x^n$ does not converge if $| x | = r'$. Both classes exist, and it follows from Arts. 13 and 14 that every r is less than any r'.

Therefore the classification defines a real positive number R which separates the classes and may be assigned to either class. This number R is such that $\Sigma a_n x^n$ converges absolutely if $| x | < R$, and is non-convergent if $| x | > R$. Nothing is said as to what happens when $| x | = R$.

By writing $R = 0$, $R = \infty$ in the exceptional cases (i) and (ii) respectively, these may be included in (iii).

All that has been said applies to any power series, real or complex.

■ **16.** For a real series $\Sigma a_n x^n$, the interval $(- R, R)$ is called the *interval of convergence*. The series converges absolutely or is non-convergent according as x is within or outside the interval. At either end point $\Sigma a_n x^n$ may converge, diverge or oscillate.

For a complex series $\Sigma a_n z^n$, R is called the *radius of convergence*. The circle (O, R) with centre at the origin and radius R is called the *circle of convergence*. The series converges absolutely or is non-convergent according as the point z is inside or outside the circle. If z is on the circumference, the character of the series is not determined.

■ **17.** Let $\Sigma a_n z^n$ be any series, real or complex, and suppose that

$$\lim |a_{n+1}/a_n| = l.$$

Then
$$\lim \left| \frac{a_{n+1} z^{n+1}}{a_n z^n} \right| = l\,|z|,$$

and it follows from D'Alembert's test that $R = 1/l$.

Again, if $\lim |a_n|^{\frac{1}{n}} = l$, then $\lim |a_n z^n|^{\frac{1}{n}} = l\,|z|$, and by Cauchy's test $R = 1/l$.

■ **18.** In considering the behaviour of a complex series *at points on its circle of convergence,* the following theorem is often useful.

If (a_n) is a decreasing sequence of positive numbers such that $a_{n+1}/a_n \to 1$ and $a_n \to 0$, then, (i) if Σa_n converges, $\Sigma a_n z^n$ converges absolutely at every point on its circle of convergence, (ii) if Σa_n diverges, $\Sigma a_n z^n$ converges (though not absolutely) at every point on the circle, except at the point $z = 1$, where it diverges, (iii) the same statements hold for the series $\Sigma(-1)^n a_n z^n$, except that it diverges at the point $z = -1$, when $\Sigma(-1)^n a_n$ is divergent.

For $1/R = \lim a_{n+1}/a_n = 1$, and at a point on the circle of convergence

$$a_n z^n = a_n (\cos\theta + \iota \sin\theta)^n = a_n (\cos n\theta + \iota \sin n\theta).$$

Moreover, (*i*) if Σa_n converges, both $\Sigma a_n \cos n\theta$ and $\Sigma a_n \sin n\theta$ are absolutely convergent (Art. 9, Ex. 1); therefore $\Sigma a_n z^n$ is absolutely convergent when $|z| = 1$. (*ii*) If Σa_n diverges, both $\Sigma a_n \cos n\theta$ and $\Sigma a_n \sin n\theta$ converge unless $\theta = 0$ or $2k\pi$; in which case $\Sigma a_n \cos n\theta$ diverges (Art. 11, Ex. 1). Hence the statement (*ii*) follows.

(*iii*) This follows from (*i*) and (*ii*) on substituting $-z$ for z.

■ **19.** If $\Sigma a_n z^n$ converges when $z = z_1$, by Art. 13, it is absolutely convergent when $|z| < |z_1|$. Denoting its sum by $f(z)$, we shall prove that

$$f(z) = a_0 + a_1 z + a_2 z^2 + \ldots + (a_n + \eta)z^n, \quad \text{where } |\eta| \to 0 \text{ as } |z| \to 0.$$

We have
$$\eta z^n = a_{n+1} z^{n+1} + a_{n+2} z^{n+2} + \ldots ,$$

and
$$|\eta z^n| \le |a_{n+1} z^{n+1}| + |a_{n+2} z^{n+2}| + \ldots .$$

As in Art. 13, we can find a fixed number k, such that for every r,

$$|a_r z_1^r| < k; \text{ and therefore } |a_r z^r| < k \left| \frac{z}{z_1} \right|^r; \text{ and thus, if } |z| < |z_1|,$$

$$|\eta z^n| < k \left\{ \left| \frac{z}{z_1} \right|^{n+1} + \left| \frac{z}{z_1} \right|^{n+2} + \ldots \right\} = k.\left| \frac{z}{z_1} \right|^{n+1} \cdot \frac{|z_1|}{|z_1| - |z|}.$$

Hence if $z \ne 0, \eta < \dfrac{k}{|z_1|^n} \cdot \dfrac{|z_1|}{|z_1| - |z|}.$

Therefore $|\eta| \to 0$ as $|z| \to 0$.

▇ **20. Criterion for the Identity of Power Series:** *If $\Sigma a_n z^n = \Sigma b_n z^n$ for every value of z whose modulus is less than some number* μ, *then* $a_n = b_n$ *for every value of n.*

For by Art. 19, for every n,

$$a_0 + a_1 z + a_2 z^2 + \dots + (a_n + \eta)\, z^n = b_0 + b_1 z + b_2 z^2 + \dots + (b_n + \eta')\, z^n,$$

where η and η' tend to zero as $|z| \to 0$.

By making $|z| \to 0$, it will be seen that $|a_0 - b_0| <$ any positive number, however small, and since a_0, b_0 are constants, it follows that $a_0 = b_0$.

Since the above equation holds for values of z other than zero, we can divide by z, therefore

$$a_1 + a_2 z + \dots + (a_n + \eta)z^{n-1} = b_1 + b_2 z + \dots + (b_n + \eta)z^{n-1}.$$

Making $z \to 0$ as before, we have $a_1 = b_1$, and by continuing the process, $a_n = b_n$ for every n.

▇ **21. Binomial Series:** We consider the convergence of the series

$$1 + nz + \frac{n\,(n-1)}{\underline{|2}}\, z^2 + \dots + \frac{n\,(n-1)\dots(n-r+1)}{\underline{|r}}\, z^r + \dots \, ,$$

where n is real and z complex.

Denoting the series by $u_0 + u_1 z + u_2 z^2 + \dots$, we have

$$R = \lim_{r \to \infty} \left| \frac{u_r}{u_{r+1}} \right| = \lim_{r \to \infty} \left| \frac{r+1}{n-r} \right| = 1.$$

Therefore, it follows from Art. 17 that

(*i*) *The series converges absolutely if* $|z| < 1$, *and is non-convergent if* $z > 1$.

When $|z| = 1$, the point z is on the circle of convergence, and we shall prove that

(*ii*) *If* $n > 0$, *the series converges absolutely at every point on its circle of convergence.*

(*iii*) *If* $-1 < n < 0$, *the series converges* (*though not absolutely*) *at every point on the circle, except at the point* $z = -1$, *where it diverges.*

Let accents indicate absolute values, so that $u_r' = |u_r|$, then if $r > n$,

$$\frac{u'_{r+1}}{u_r'} = \frac{r-n}{r+1}.$$

Therefore $u'_{r+1} \gtreqless u_r'$, according as $r - n \gtreqless r + 1$, that is according as $n \lesseqgtr -1$.

If $n \le -1$, u_r' increases with r, or is constant and *the series cannot be convergent.*

If $n > -1$, we apply the theorem of Art. 18. After a certain stage, u_r' is a decreasing sequence, $u'_{r+1}/u_r' \to 1$ and, as shown in Ch. 16, 25, $u_r' \to 0$.

These are the conditions required in Art. 18, to apply the theorem of this article, and we have to consider the convergence of $\Sigma u_r'$.

After a certain stage, the terms of Σu_r are alternately positive and negative, so that $\Sigma u_r'$ is convergent or divergent, according as $\Sigma (-1)^r u_r$ is convergent or divergent, that is according as $n \gtreqless 0$. (See Art. 6.)

Hence the statements (*ii*) and (*iii*) follow immediately from Art. 18.

MULTIPLICATION OF SERIES

■ **22.** *If* $u_1 + u_2 + u_3 + \ldots$ *and* $v_1 + v_2 + v_3 + \ldots$ *are convergent series of positive terms, or if they are absolutely convergent real or complex series, their sums being s and t respectively, then the series*

$$u_1 v_1 + (u_1 v_2 + u_2 v_1) + (u_1 v_3 + u_2 v_2 + u_3 v_1) + \ldots$$

is absolutely convergent and its sum is st.

Multiply the terms of $u_1 + u_2 + u_3 + \ldots$ by $v_1, v_2, v_3, \ldots$, and arrange the products as below:

$$\left.\begin{array}{lll} u_1 v_1, & u_2 v_1, & u_3 v_1, \ldots \\[2mm] u_1 v_2, & u_2 v_2, & u_3 v_2, \ldots \\[2mm] u_1 v_3, & u_2 v_3, & u_3 v_3, \ldots \\[2mm] \multicolumn{3}{c}{\cdots\cdots\cdots\cdots\cdots\cdots\cdots} \end{array}\right\} \qquad \ldots\text{(A)}$$

This array extends to infinity on the right and below, and we shall consider two ways of arranging the terms so as to form simply infinite series.

Let s_n, t_n be the sums to n terms of Σu_n, Σv_n respectively; then if σ_n is the sum of the first n terms of the first n rows of (A), we have

$$\sigma_n = s_n \, t_n \to st.$$

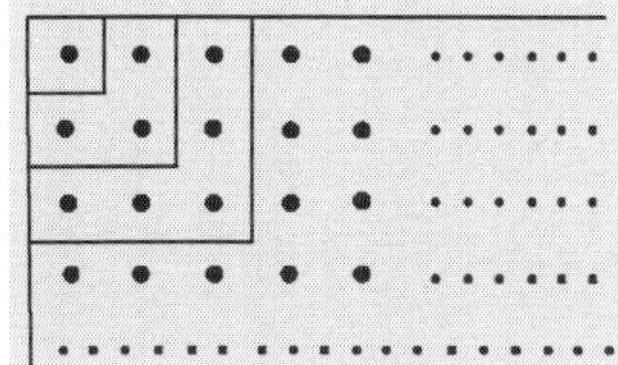

Fig. 62	**Fig. 63**

Draw a set of squares, as in Fig. 62, where the terms are represented by dots and the nth square just contains the terms in σ_n. Then the terms in (A) can be arranged to form the infinite series

$$u_1 v_1 + (u_1 v_2 + u_2 v_2 + u_2 v_1) + (u_1 v_3 + u_2 v_3 + u_3 v_3 + u_3 v_2 + u_3 v_1) + \ldots, \qquad \ldots\text{(B)}$$

where the nth term is the sum of the terms of (A) between the nth and the $(n-1)$th squares.

The sum to n terms of this series is σ_n, and its sum to infinity is st.

Again, the terms of (A) can be arranged to form the infinite series

$$u_1 v_1 + (u_1 v_2 + u_2 v_1) + (u_1 v_3 + u_2 v_2 + u_3 v_1) + \ldots, \qquad \ldots\text{(C)}$$

where the nth term is the sum of the terms between the nth and the $(n-1)$th diagonals, drawn as in Fig. 63.

If the brackets are removed in (B) and (C), we have two series which differ only in the arrangement of terms. The first series is absolutely convergent and its sum is st. this is therefore also true for the second series (Chap. 16, 5, 6 and 19).

▮ **EXAMPLE :** *If $E(x) = 1 + x + \dfrac{x^2}{\underline{2}} + \dfrac{x^3}{\underline{3}} + \ldots$, prove that, for all values of x and y,*

$$E(x) \cdot E(y) = E(x + y).$$

The series equivalent to $E(x)$ and $E(y)$ are absolutely convergent, therefore

$$E(x) \cdot E(y) = 1 + d_1 + d_2 + \ldots + d_n + \ldots ,$$

where
$$d_n = \frac{x^n}{\underline{n}} + \frac{x^{n-1}y}{\underline{n-1}\,\underline{1}} + \frac{x^{n-2}y^2}{\underline{n-2}\,\underline{2}} + \ldots + \frac{y^n}{\underline{n}} = \frac{(x+y)^n}{\underline{n}};$$

$\therefore$
$$E(x) \cdot E(y) = E(x + y).$$

In Arts. 23 and 24, the series (C) will be denoted by Σd_n, where

$$d_n = u_1 v_n + u_2 v_{n-1} + u_3 v_{n-2} + \ldots + u_n v_1;$$

it should be noted that, in every term of d_n, the sum of the suffixes is $n + 1$.

■ **23.** *If Σu_n and Σv_n are convergent with sums s and t respectively and one of these series, say Σu_n, is absolutely convergent, then Σd_n is convergent and its sum is st.* (Mertens.)

Let s_n, t_n, D_n be the sums to n terms of Σu_n, Σv_n, Σd_n respectively.

Considering the array (A) in Art. 22, it will be seen that

$$s_n t_n - D_n = u_2 w_1 + u_3 w_2 + u_4 w_3 + \ldots + u_n w_{n-1}, \qquad \ldots\text{(B)}$$

where
$$w_1 = v_n, \quad w_2 = v_{n-1} + v_n, \quad w_3 = v_{n-2} + v_{n-1} + v_n,$$

and generally
$$w_m = v_{n-m+1} + v_{n-m+2} + \ldots + v_n;$$

lastly
$$w_{n-1} = v_2 + v_3 + \ldots + v_n.$$

The series Σv_n is convergent; and hence (*i*) we can find a positive number μ such that $|v_r + v_{r+1} + v_{r+2} + \ldots + v_n| < \varepsilon$ for $r > \mu$, where ε is an arbitrarily small positive number; and therefore

$$|w_1|, |w_2|, \ldots |w_m|, \text{ are all} < \varepsilon, \text{ provided that } n - m + 1 > \mu; \qquad \ldots\text{(C)}$$

(*ii*) the sequence (t_n) is bounded; hence we can find a positive number k such that $|t_n| < k$ for every n, and consequently

$$|w_{m+1}|, |w_{m+2}|, \ldots |w_{n-1}| \text{ are all} < k.$$

We therefore consider separately the sum of the first m terms of the series in (B), and the sum of the remaining terms, where m is to be properly chosen.*

Let
$$P = u_2 w_1 + u_3 w_2 + \ldots + u_{m+1} w_m,$$
$$Q = u_{m+2} w_{m+1} + u_{m+3} w_{m+2} + \ldots + u_n w_{n-1};$$

then
$$|P| \leq |u_2||w_1| + |u_3||w_2| + \ldots + |u_{m+1}||w_m|$$
$$< \{|u_2| + |u_3| + \ldots + |u_{m+1}|\}\,\varepsilon,$$

with the condition (C). Hence, if s' is the sum of the convergent series $\Sigma |u_n|$, then

$$|P| < s'\,\varepsilon.$$

Also
$$|Q| \leq |u_{m+2}||w_{m+1}| + |u_{m+3}||w_{m+2}| + \ldots + |u_n||w_{n-1}|$$
$$< \{|u_{m+2}| + |u_{m+3}| + \ldots + |u_n|\}\,k,$$

---

* This artifice is often useful.

and since $\Sigma \mid u_n \mid$ is convergent, we can find a positive number μ' such that

$$\mid u_{m+2} \mid + \mid u_{m+3} \mid + \ldots + \mid u_n \mid < \varepsilon, \text{ provided that } m + 2 > \mu', \qquad \ldots\text{(D)}$$

and then $\qquad\qquad\qquad\qquad \mid Q \mid < k\varepsilon.$

The conditions (C) and (D) are satisfied if $n > 2m$ and $m >$ either of the two, $\mu - 1$ and $\mu' - 2$. Hence, as m, and consequently n, tend to infinity,

$$\mid P \mid \to 0, \quad \mid Q \mid \to 0 \quad \text{and} \quad s_n t_n - D_n \to 0,$$

which proves the theorem.

■ **24.** *If Σu_n and Σv_n are convergent with sums s and t respectively, and if Σd_n is convergent, then its sum is st.*   (Abel.)

Let s_n, t_n, D_n be the sums to n terms of Σu_n, Σv_n, Σd_n respectively, then

$$D_1 = u_1 t_1, \quad D_2 = u_2 t_1 + u_1 t_2, \quad D_3 = u_3 t_1 + u_2 t_2 + u_1 t_3, \text{ etc.}$$

Hence $\quad D_1 + D_2 + D_3 + \ldots + D_n = s_n t_1 + s_{n-1} t_2 + s_{n-2} t_3 + \ldots + s_1 t_n.$

Now $s_n \to s$ and $t_n \to t$, therefore by Ch. 15, 9, (4),

$$\frac{1}{n} (s_n t_1 + s_{n-1} t_2 + \ldots + s_1 t_n) \to st.$$

Also, if D is the sum of Σd_n, which is convergent, by Ch. 15, 9, (3),

$$\frac{1}{n} (D_1 + D_2 + \ldots + D_n) \to D.$$

Therefore $\qquad\qquad\qquad D = st.$

■ **EXAMPLE :** *If $- 1 < x \le 1$, prove that*

$$\{\log (1 + x)\}^2 = d_1 x^2 - d_2 x^3 + d_3 x^4 - \ldots,$$

where $\qquad\qquad d_n = \dfrac{2}{n+1}\left(1 + \dfrac{1}{2} + \dfrac{1}{3} + \ldots + \dfrac{1}{n}\right).$

Under suitable conditions, the rule for the multiplication of series gives

$$\left(x - \frac{x^2}{2} + \frac{x^3}{3} - \ldots\right)^2 = d_1 x^2 - d_2 x^3 + d_3 x^4 - \ldots,$$

where $\qquad\qquad d_n = \dfrac{1}{n \cdot 1} + \dfrac{1}{(n-1)\,2} + \dfrac{1}{(n-2)\,3} + \ldots + \dfrac{1}{1 \cdot n}$

$$= \frac{2}{n+1}\left(1 + \frac{1}{2} + \frac{1}{3} + \ldots + \frac{1}{n}\right).$$

If $- 1 < x < 1$, the series for $\log (1 + x)$ is absolutely convergent and the result in question holds by Art. 22.

If $x = 1$, the series $1 - \frac{1}{2} + \frac{1}{3} - \frac{1}{4} + \ldots$ is convergent; and so is $d_1 - d_2 + d_3 - \ldots$, for it is easily shown that $d_n < d_{n-1}$, and $d_n \to 0$ by Ch. 15, 9, (3). Hence Art. 24 applies, and the statement is true in this case also.

EXERCISE XXXIII

Prove that the series:

1. $1 + \dfrac{1}{2} + \dfrac{1 \cdot 3}{2 \cdot 4} + \dfrac{1 \cdot 3 \cdot 5}{2 \cdot 4 \cdot 6} + \ldots$ is divergent.

2. $1 + \dfrac{a}{b} + \dfrac{a\,(a+1)}{b\,(b+1)} + \dfrac{a\,(a+1)\,(a+2)}{b\,(b+1)\,(b+2)} + \ldots$ is convergent if $b - 1 > a > 0$, its sum being

$$\dfrac{b-1}{b-a-1}.$$

3. $1 + \dfrac{1}{2} \cdot \dfrac{a}{b} + \dfrac{1 \cdot 3}{2 \cdot 4} \cdot \dfrac{a\,(a+1)}{b\,(b+1)} + \dfrac{1 \cdot 3 \cdot 5}{2 \cdot 4 \cdot 6} \cdot \dfrac{a\,(a+1)\,(a+2)}{b\,(b+1)\,(b+2)} + \ldots$ is convergent if

$$2\,(b - a) > 1.$$

4. If $\dfrac{u_n}{u_{n+1}} = \dfrac{n^p + an^{p-1} + bn^{p-2} + \ldots + k}{n^p + a'n^{p-1} + b'n^{p-2} + \ldots + k'}$, show that Σu_n is convergent if $a - a' > 1$

and divergent if $a - a' \leq 1$.

5. The series $\Sigma x^n \cos n\theta$ and $\Sigma x^n \sin n\theta$ are absolutely convergent if $|x| < 1$.

6. The series $\Sigma \dfrac{1}{n^2} \cos n\theta$ and $\Sigma \dfrac{1}{n^2} \sin n\theta$ are absolutely convergent.

7. The series $\Sigma \dfrac{1}{n} \cos n\theta$ and $\Sigma \dfrac{1}{n} \sin n\theta$ are convergent excepting that the first series

diverges when θ is an even multiple of π.

8. The series $\Sigma \dfrac{1}{n} \cos n\theta$ and $\Sigma \dfrac{1}{n} \sin n\theta$ are not absolutely convergent excepting the last

series, when θ is a multiple of π.

[Use the following:

$|\cos n\theta| \geq \cos^2 n\theta \geq \frac{1}{2}\,(1 + \cos 2n\theta)$ and $|\sin n\theta| \geq \sin^2 n\theta \geq \frac{1}{2}\,(1 - \cos 2n\theta)$.]

9. If $u_1, u_2, u_3, \ldots$ is a decreasing sequence of positive terms tending to zero as a limit, the series

$$u_1 - \tfrac{1}{2}\,(u_1 + u_2) + \tfrac{1}{3}\,(u_1 + u_2 + u_3) - \ldots$$

is convergent.

[Denote the series by $v_1 - v_2 + v_3 - \ldots$ Using Ch. 15, 9, (3), show that $v_1, v_2, v_3, \ldots$ is a decreasing sequence tending to zero as a limit.]

10. Discuss the convergence of Σu_n where

$$u_n = \dfrac{1}{n}\left(1 + \dfrac{1}{2} + \dfrac{1}{3} + \ldots + \dfrac{1}{n-1}\right) \sin n\theta.$$

11. Show that, if $0 < p < \frac{1}{2}$, the rule for multiplication fails to obtain a convergent series for

$$\left(1 - \frac{1}{2^p} + \frac{1}{3^p} - \frac{1}{4^p} + \ldots\right)^2$$

[Show that the series is $1 - d_1 + d_2 - \ldots$ where

$$d_n = \sum_{r=1}^{r=n} \frac{1}{r^p (n - r + 1)^p}$$

and

$$r(n - r + 1) \le \tfrac{1}{4} (n + 1)^2.]$$

12. If $a_n = \left(1 + \dfrac{x}{n}\right) e^{-\frac{n}{x}}$, show that the series whose nth term is $a_n - 1$ is absolutely convergent for all values of x.

$$[a_n - 1 = -\frac{1}{2}\frac{x^2}{n^2} + \frac{1}{3}\frac{x^3}{n^3} - \ldots; \text{ therefore } n^2. \, | \, a_n - 1 \, | \rightarrow \tfrac{1}{2} \, | \, x \, |^2 \text{ as } n \rightarrow \infty. \text{ Also } \Sigma 1/$$

n^2 is convergent.]

13. If $F(\alpha, \beta, \gamma, x) = 1 + \dfrac{\alpha\beta}{1 \cdot \gamma} x + \dfrac{\alpha(\alpha + 1)\beta(\beta + 1)}{1 \cdot 2 \cdot \gamma(\gamma + 1)} x^2 + \ldots$ and $|x| < 1$, prove that

$\gamma F(\alpha, \beta, \gamma, x) - (\gamma - \alpha) F(\alpha, \beta + 1, \gamma + 1, x) = \alpha(1 - x) F(\alpha + 1, \beta + 1, \gamma + 1, x)$.
Under what circumstances is this true (*i*) when $x = 1$, and (*ii*) when $x = -1$?

■■■

CHAPTER 21

Binomial and Multinomial Theorems

■ **1. A General Statement:** The Binomial theorem asserts that the sum of the series

$$1 + nx + \frac{n(n-1)}{\lfloor 2} \, x^2 + \dots + \frac{n(n-1)\dots(n-r+1)}{\lfloor r} \, x^r + \dots,$$

when convergent, is one of the values of $(1+x)^n$. (Of course, in a complete statement, this value must be specified.)

The series terminates if, and only if, n is a positive integer; in which case the sum is $(1+x)^n$.

This leads to the following theorem, which enables us to deal with the case in which n is any rational number.

■ **2. Vandermonde's Theorem:** The following notation is often used: we write

$$x_r = x(x-1)(x-2)\dots(x-r+1),$$

where x is any number whatever. Vandermonde's theorem asserts that *if m and n are any numbers whatever, then*

$$(m+n)_r = m_r + C^r_1 \, m_{r-1} n_1 + C^r_2 \, m_{r-2} n_2 + \dots + n_r. \qquad \dots(A)$$

Proof. First suppose that m and n are positive integers; then

$$(1+x)^m = 1 + \frac{m_1}{\lfloor 1} x + \frac{m_2}{\lfloor 2} x^2 + \dots + \frac{m_r}{\lfloor r} \, x^r + \dots,$$

$$(1+x)^n = 1 + \frac{n_1}{\lfloor 1} x + \frac{n_2}{\lfloor 2} x^2 + \dots + \frac{n_r}{\lfloor r} \, x^r + \dots.$$

The coefficient of x^r in the product of these series is therefore equal to the coefficient of x^r in $(1+x)^{m+n}$, that is to say,

$$\frac{(m+n)_r}{\lfloor r} = \frac{m_r}{\lfloor r} + \frac{m_{r-1} n_1}{\lfloor r-1 \lfloor 1} + \frac{m_{r-2} n_2}{\lfloor r-2 \lfloor 2} + \dots + \frac{n_r}{\lfloor r}.$$

Multiplying by $\lfloor r$, we obtain equation (A).

This equation holds for all positive integral values of m and n; and since each side is a *polynomial* in m, n it is true for *all* values of m, n.

■ **3. Case of a Rational Index:** *If n is any rational number and $-1 < x < 1$, the sum of the series*

$$1 + nx + \frac{n(n-1)}{\lfloor 2} x^2 + \dots + \frac{n(n-1)\dots(n-r+1)}{\lfloor r} x^r + \dots$$

is the real positive value of $(1 + x)^n$.

We first give what is essentially *Euler's* proof, although in the last step we have to assume a property of infinite series which is not proved till later.

Euler's proof. The series is absolutely convergent if $|x| < 1$; denote its sum by $f(n)$, and let

$$n_r = n(n-1)(n-2)\dots(n-r+1),$$

so that

$$f(n) = 1 + n_1 x + \frac{n_2}{\lfloor 2} x^2 + \dots + \frac{n_r}{\lfloor r} x^r + \dots .$$

For all values of m and n, the series corresponding to $f(m)$, $f(n)$ are absolutely convergent, therefore by the rule for the multiplication of series,

$$f(m) \cdot f(n) = 1 + d_1 x + d_2 x^2 + \dots + d_r x^r + \dots ,$$

where

$$d_r = \frac{m_r}{\lfloor r} + \frac{m_{r-1}\, n_1}{\lfloor r-1 \lfloor 1} + \frac{m_{r-2}\, n_2}{\lfloor r-2 \lfloor 2} + \dots + \frac{n_r}{\lfloor r} .$$

Hence, by Vandermonde's theorem,

$$d_r = \frac{(m+n)_r}{\lfloor r} ,$$

and

$$f(m) \cdot f(n) = 1 + (m+n)_1 x + \frac{(m+n)_2}{\lfloor 2} x^2 + \dots,$$

so that

$$f(m) \cdot f(n) = f(m+n). \qquad \text{...(A)}$$

Hence $\quad f(m) \cdot f(n) \cdot f(p) = f(m+n) \cdot f(p) = f(m+n+p),$

a similar result holding for any number of factors.

Positive index. If $n = p/q$ where p, q are positive integers, by the preceding,

$$f\left(\frac{p}{q}\right) \cdot f\left(\frac{p}{q}\right) \dots \text{ to } q \text{ factors} = f\left(q \cdot \frac{p}{q}\right) = f(p).$$

Hence

$$f\left(\frac{p}{q}\right) \text{ is a } q\text{-th root of } (1+x)^p. \qquad \text{...(B)}$$

Thus, x being real, if the real positive qth root of $(1+x)^p$ is denoted by $(1+x)^{\frac{p}{q}}$,

we conclude that $f\left(\dfrac{p}{q}\right) = \pm (1+x)^{\frac{p}{q}}$, that is $f(n) = \pm (1+x)^n$.

With regard to sign, it will be shown in another volume that $f(n)$ *is a continuous function of x in the interval* $-1 < x < 1$.

Now $f(n)$ does not vanish for any value of x in the interval, and consequently its sign is the same throughout the interval.

But when $x = 0, f(n) = 1$, therefore the sign is positive and $f(n)$ is equal to the real positive value of $(1 + x)^n$.

Negative index. Let n be a positive rational. Putting $m = -n$ in equation (B),

$$f(n) \cdot f(-n) = f(0) = 1; \qquad \qquad \text{...(C)}$$

therefore
$$f(-n) = \frac{1}{(1+x)^n} = (1 + x)^{-n}.$$

Notes: (i) Equations (A) – (C) hold when x is complex, provided that $|x| < 1$. Hence *if x is complex, $|x| < 1$ and n rational, $f(x)$ is one of the values of $(1 + x)^n$.* In particular, if n is a positive integer, $f(-n) = (1 + x)^{-n}$.

(ii) *If n and x are real and $|x| < 1$, it can be shown that $f(n)$ is continuous for all values of n. Hence if $(1 + x)^n$ is defined as in Ch. 13, 7, when n is irrational, we have $(1 + x)^n = f(n)$.*

■ **4. Second Proof for the Case of a Real Index:** The following proof depends on very elementary considerations and, although it is unsuitable for reproduction *in toto*, the reader will find it a very useful exercise to work through the details. We shall prove that *if n and x are real and $|x| < 1$, then*

$$(1 + x)^n = 1 + nx + \frac{n(n-1)}{\lfloor 2} x^2 + \dots + \frac{n(n-1)\dots(n-r+1)}{\lfloor r} x^r + \dots,$$

where $(1 + x)^n$ has its real positive value.

Proof. Let $t_{n,r}$ be the rth term and $s_{n,r}$ the sum to r terms of the series. Also let
$$y_{n,r} = (1 + x)^n - s_{n,r}.$$
First suppose that $0 < x < 1$, then

$$y_{n,1} = (1 + x)^n - 1 \gtrless 0 \quad \text{according as} \quad n \gtrless 0. \qquad \text{...(A)}$$

Also
$$\frac{dy_{n,r}}{dx} = ny_{n-1,\,r-1} \quad \text{and } y_{n,\,r} = 0 \text{ when } x = 0. \qquad \text{...(B)}$$

If for any assigned values of n and r it can be shown that $y_{n-1,\,r-1}$ is of invariable sign for $0 < x < 1$, then by (B), $y_{n,r}$ must have the same sign as $ny_{n-1,\,r-1}$.

Repeating this reasoning when $n - 1, n - 2,\dots$ are substituted for n and observing that, by (A), $y_{n-r+1,\,1}$ has the same sign as $n - r + 1$, it follows that the three expressions.

$y_{n,\,r}, \quad n(n-1)\dots(n-r+2)\,y_{n-r+1,1}$, and $n(n-1)\dots(n-r+1)$ have the same sign. Therefore $y_{n,\,r}$ *has the same sign as* $t_{n,\,r+1}$.

After a certain stage, say for $r >$ some fixed number k, the signs of $t_{n,\,r+1}$, and consequently those of $y_{n,\,r}$ are alternately $+$ and $-$, as r increases. Therefore $(1 + x)^n$ lies between $s_{n,\,r}$ and $s_{n,\,r+1}$ for $r > k$.

But since the series is convergent, $s_{n,\,r+1} - s_{n,r} \to 0$ as $r \to \infty$.

Therefore $s_{n,\,r} \to (1 + x)^n$ as $r \to \infty$,

which proves this case of the theorem.

Next let $x < 0$. Changing the sign of x, it has to be shown that *if $0 < x < 1$, then*

$$(1 - x)^n = 1 - nx + \frac{n(n-1)}{\underline{2}} x^2 - \ldots + (-1)^r \frac{n(n-1)\ldots(n-r+1)}{\underline{r}} x^r + \ldots .$$

Here $\qquad\qquad y_{n,r} = (1 - x)^n - s_{n,r},$

so that $\qquad\qquad y_{n,1} = (1 - x)^n - 1 \gtrless 0$ according as $n \lessgtr 0;$ $\qquad$ ...(C)

also $\qquad\qquad \dfrac{dy_{n,r}}{dx} = -ny_{n-1,\,r-1}$ and $y_{n,\,r} = 0$ when $x = 0.$ $\qquad$ ...(D)

Reasoning as in the previous case and observing that by (C), $y_{n-r+1,\,1}$ and $n - r + 1$ have *opposite* signs, it follows that the three expressions

$$y_{n,\,r},\ (-1)^{r-1} n(n-1)\ldots(n-r+2)y_{n-r+1,\,1}$$

and $\qquad\qquad (-1)^r n(n-1)\ldots(n-r+1)$

have the same sign. Hence, as before,

$\qquad\qquad y_{n,\,r}$ *has the same sign as* $t_{n,\,r+1}.$ $\qquad$ ...(E)

After a certain stage, say for $r > k$, the sign of $t_{n,\,r+1}$ is invariable, and we proceed thus:

Let $\qquad\qquad z_{n,\,r} = y_{n,\,r} - t_{n,\,r+1}(1 - x)^{-1}.$

For every r, we have

$$t_{n,\,r+1} - xt_{n,\,r} = t_{n,\,r+1}\left(1 + \frac{r}{n-r+1}\right) = t_{n,\,r+1}\,\frac{n+1}{n-r+1} = t_{n+1,\,r+1}.$$

Therefore

$$(1 - x)\,s_{n,\,r} = s_{n+1,\,r} - xt_{n,r} \quad\text{and}\quad (1 - x)\,y_{n,\,r} = y_{n+1,\,r} + xt_{n,\,r}. \qquad \text{...(F)}$$

Hence $\qquad (1 - x)z_{n,\,r} = (1 - x)\,y_{n,\,r} - t_{n,\,r+1}$

$$= y_{n+1,\,r} + xt_{n,\,r} - t_{n,\,r+1}$$

$$= y_{n+1,\,r} - t_{n+1,\,r+1}$$

$$= y_{n+1,\,r+1}. \qquad\qquad \text{...(G)}$$

Hence by (E), $z_{n,\,r}$ has the same sign as $t_{n+1,\,r+2}.$ Now

$$t_{n+1,\,r+2} = -\frac{n+1}{r+1}\,r\cdot t_{n,\,r+1},$$

so that by (E)

$\qquad\qquad z_{n,r}$ *has the same sign as* $-(n+1)\,y_{n,r}.$ $\qquad$ ...(H)

(*i*) *If $n > -1$* it follows that $y_{n,\,r}$ and $z_{n,r}$ have opposite signs. Therefore $y_{n,r}$ lies between 0 and $t_{n,\,r+1}\cdot(1 - x)^{-1}$. But $t_{n,\,r+1} \to 0$ as $r \to \infty$, therefore $y_{n,r} \to 0$ and $s_{n,r} \to (1 - x)^n$.

(*ii*) *If $n < -1$*, then $-m - 1 < n \le -m$ where m is a positive integer. If $y_{n+1,\,r} \to 0$ as $r \to \infty$, since $t_{n,\,r} \to 0$ it follows from (F) that $y_{n,\,r} \to 0$.

Now $n + m > -1$, and therefore, by case (*i*), $y_{n+m,\,r} \to 0$. Hence in succession it follows that $y_{n+m-1,\,r},\,y_{n+m-2,\,r},\ \ldots\ y_{n,r}$ all tends to zero.

Hence $s_{n,r} \to (1 - x)^n$, and the theorem is proved in all cases.

■ **5.** We shall prove that *if n is a negative integer, z complex and $|z| < 1$, then*

$$(1 + z)^n = 1 + nz + \frac{n(n-1)}{\underline{|2}} z^2 + \dots .$$

Using the same notation as before, we see that the series is convergent (Chap. 20, 21) and $t_{n,\,r} \to 0$ as $r \to \infty$. Also $(1 + z)\,y_{n,r} = y_{n+1,\,r} - zt_{n,r}$ [compare equation (F)]. Hence as before, if $y_{n+1,\,r} \to 0$, then $y_{n,r} \to 0$.

Now when $n = -1$,

$$y_{n,r} = \frac{1}{1+z} - (1 - z + z^2 - \dots + (-1)^{r-1}z^{r-1}) = (-1)^r \frac{z^r}{1+z} \to 0.$$

Hence in succession $y_{n,\,r} \to 0$ for $n = -2, -3, -4, \dots$. So that

$$s_{n,\,r} \to (1 + z)^n$$

when n is a negative integer.

The theorem is obviously true when n is a positive integer. The case in which n is a fraction is considered later.

■ **6.** The reader should be familiar with the following particular instances; in each case it is assumed that $|x| < 1$:

$$(1 - x)^{-1} = 1 + (-1)(-x) + \frac{(-1)(-2)}{\underline{|2}} (-x)^2 + \dots = 1 + x + x^2 + x^3 + \dots,$$

and in a similar way

$$(1 - x)^{-2} = 1 + 2x + 3x^2 + 4x^3 + \dots ,$$

$$(1 - x)^{-3} = 1 + 3x + \frac{3 \cdot 4}{1 \cdot 2} x^2 + \frac{3 \cdot 4 \cdot 5}{1 \cdot 2 \cdot 3} x^3 + \frac{3 \cdot 4 \cdot 5 \cdot 6}{1 \cdot 2 \cdot 3 \cdot 4} x^4 + \dots$$

$$= \tfrac{1}{2}\,[1 \cdot 2 + 2 \cdot 3x + 3 \cdot 4x^2 + 4 \cdot 5x^3 + 5 \cdot 6x^4 + \dots],$$

$$(1 - x)^{-n} = 1 + nx + \frac{n(n+1)}{\underline{|2}} x^2 + \frac{n(n+1)(n+2)}{\underline{|3}} x^3 + \dots .$$

$$(1 - x)^{-\frac{1}{2}} = 1 + \frac{1}{2}x + \frac{1 \cdot 3}{2 \cdot 4} x^2 + \frac{1 \cdot 3 \cdot 5}{2 \cdot 4 \cdot 6} x^3 \dots .$$

▎**EXAMPLE 1.** *Expand $1/(3 + 2x)^2$ to three terms (i) in ascending powers of x; (ii) in descending powers of x; (iii) say for what values of x each expansion is valid.*

(i)
$$\frac{1}{(3 + 2x)^2} = \frac{1}{9}\left(1 + \frac{2}{3}x\right)^{-2} = \frac{1}{9}\left\{1 - 2\left(\frac{2}{3}x\right) + 3\left(\frac{2}{3}x\right)^2 - \dots\right\};$$

∴
$$1/(3 + 2x)^2 = \frac{1}{9} - \frac{4}{27}x + \frac{4}{27}x^2 \text{ to three terms.}$$

(ii)
$$\frac{1}{(2x + 3)^2} = \frac{1}{4x^2}\left(1 + \frac{3}{2x}\right)^{-2} = \frac{1}{4x^2}\left\{1 - 2\left(\frac{3}{2x}\right) + 3\left(\frac{3}{2x}\right)^2 - \dots\right\};$$

$$\therefore \qquad \frac{1}{(2x+3)^2} = \frac{1}{4x^2} - \frac{3}{4x^3} + \frac{27}{16x^4} \text{ to three terms.}$$

(*iii*) The first expansion is valid if $\left|\frac{2}{3}x\right| < 1$, that is if $|x| < \frac{3}{2}$.

The second is valid if $\left|\frac{3}{2x}\right| < 1$, that is if $|x| > \frac{3}{2}$.

◼ **EXAMPLE 2.** *Sum of series* $1 + \dfrac{5}{8} + \dfrac{5\cdot 8}{8\cdot 12} + \dfrac{5\cdot 8\cdot 11}{8\cdot 12\cdot 16} + \dots \text{ to } \infty.$

To see if this can be done by the Binomial theorem, take some particular term, say the fourth, and proceed thus:

$$\frac{5\cdot 8\cdot 11}{8\cdot 12\cdot 16} = \frac{\frac{5}{3}\cdot\frac{8}{3}\cdot\frac{11}{3}}{2\cdot 3\cdot 4}\left(\frac{3}{4}\right)^3 = 2\cdot\frac{\frac{2}{3}\cdot\frac{5}{3}\cdot\frac{8}{3}\cdot\frac{11}{3}}{1\cdot 2\cdot 3\cdot 4}\cdot\left(\frac{3}{4}\right)^4,$$

which is equal to $2n(n+1)(n+2)(n+3)x^4/\lfloor 4$, where $n = \frac{2}{3}$ and $x = \frac{3}{4}$. Let r be the sum of the series; treating every term as above, we have

$$s = 2\left\{\frac{\frac{2}{3}}{1}\cdot\frac{3}{4} + \frac{\frac{2}{3}\cdot\frac{5}{3}}{1.2}\left(\frac{3}{4}\right)^2 + \frac{\frac{2}{3}\cdot\frac{5}{3}\cdot\frac{8}{3}}{1.2.3}\left(\frac{3}{4}\right)^3 + \dots\right\}$$

$$= 2\left[\left(1-\tfrac{3}{4}\right)^{-\frac{2}{3}} - 1\right] = 4\sqrt[3]{2} - 2.$$

◼ **7. Numerically Greatest Term in the Expansion of $(1 + x)^n$ where $|x| < 1$:**

Let u_r be the rth term, and let accents indicate numerical values so that $x' = |x|$ and $u_r' = |u_r|$; *if n is positive,*

$$\frac{u_{r+1}}{u_r'} = \frac{|n-r+1|}{r}\cdot x',$$

Let k be the positive integer such that $k < n + 1 < k + 1$.

If $\qquad r \geq k + 1, \dfrac{u_{r+1}}{u_r'} = \dfrac{r-n-1}{r}\, x' < 1;$

therefore $u_{k+1}, u_{k+2}, \dots$ is a decreasing sequence, and the numerically greatest term is among the first $k + 1$ terms.

If $r \leq k$, then $u_{r+1} \gtreqless u_r'$ according as $(n + 1 - r)\, x' \gtreqless r$,

that is according as $r \lesseqgtr \dfrac{(n+1)\, x'}{1+x'}.$

(*i*) If $\dfrac{(n+1)\, x'}{1+x'} = i + f$, where i is a positive integer and f is a positive proper fraction, the $(i + 1)$th term is numerically the greatest term.

(ii) If $\dfrac{(n+1)\,x'}{1+x'} = i$, where i is a positive integer, then, numerically, the ith term

is equal to the $(i+1)$th, and each is greater than any other term.

Next, let n be negative and equal to $-m$, then

$$\frac{u_{r+1}}{u_r{}'} = \frac{m+r-1}{r}\cdot x';$$

therefore $u_{r+1} \gtreqless u_r{}'$ according as $(m+r-1)\,x' \gtreqless r,$

that is according as $\qquad r \lesseqgtr \dfrac{(m-1)\,x'}{1-x'}.$

It follows that: *(i)* If $m \le 1$, the first term is the greatest.

(ii) If $\dfrac{(m-1)\,x'}{1-x'} = i + f,$ where i is a positive integer or zero and f is a positive

proper fraction, the $(i+1)$th term is the greatest.

(iii) If $\dfrac{(m-1)\,x'}{1-x'} = i,$ where i is a positive integer, the ith term is equal to the

$(i+1)$th, and each of these is greater than any other.

We have thus proved the following important property of a binomial series: *If $|\,x\,| < 1$, the terms in the expansion of $(1+x)^n$ either, decrease numerically from the beginning of the series or else they increase numerically up to a greatest term (or to two equal terms) and then decrease numerically.*

■ **EXAMPLE :** *Which is the numerically greatest term (or terms) in the expansion of*

$(1+x)^{\frac{25}{2}}$ *when* $x = \frac{4}{5}$?

Let u_r denote the rth term;

$$\frac{u_{r+1}}{u_r} = \frac{\frac{25}{2}-r+1}{r}\cdot\frac{4}{5} = \frac{54-4r}{5r}. \qquad\qquad \text{...(A)}$$

First consider the terms for which u_{r+1}/u_r is negative. This will be the case if $4r > 54$ or if $r \ge 14$. For such values of r,

$$\left|\frac{u_{r+1}}{u_r}\right| = \frac{4r-54}{5r} < 1;\ \therefore\ |\,u_{r+1}\,| < |\,u_r\,|.$$

Hence the greatest term or terms must be included among the first 14 terms of the expansion.

If $r \le 13.\ u_{r+1}/u_r$ is positive, and from (A) it follows that

$$|\,u_{r+1}\,| \gtreqless |\,u_r\,|\ \text{according as}\ 54 - 4r \gtreqless 5r,$$

that is according as $r \lesseqgtr 6.$

Hence, the sixth term is equal to the seventh, and each of these is greater than any other term.

■ **8. Approximate Values:** The Binomial series can be used to obtain approximate values, as follows:

■ **EXAMPLE 1.** *Calculate $\sqrt[3]{2}$ to six places of decimals.*

The perfect cubes are 1, 8, 27, 64, 125,.... . Search for two of these such that one is roughly twice the other. We have

$$2 \times 64 = 125 + 3; \quad \therefore 2 = \frac{125}{64}\left(1 + \frac{3}{125}\right); \quad \therefore \sqrt[3]{2} = \frac{5}{4}\left(1 + \frac{3}{5^3}\right)^{\frac{1}{3}};$$

$$\therefore \quad \sqrt[3]{2} = \frac{5}{4}\left\{1 + \frac{1}{3} \cdot \frac{3}{5^3} + \frac{\frac{1}{3}\left(\frac{1}{3}-1\right)}{1 \cdot 2} \cdot \frac{3^2}{5^6} + \frac{\frac{1}{3}\left(\frac{1}{3}-1\right)\left(\frac{1}{3}-2\right)}{1 \cdot 2 \cdot 3} \cdot \frac{3^3}{5^9} + ...\right\}$$

$$= \frac{5}{4} + \frac{1}{4.5^2} - \frac{1}{4.5^5} + \frac{1}{12.5^7} \frac{1}{6.5^{10}} + ... \qquad \qquad ...(A)$$

Denoting this series by $u_1 + u_2 - u_3 + u_4 - u_5 + ...$, we have

$$u_1 = \frac{5}{4} = 1.25,$$

$$u_2 = \frac{1}{10^2} = 0.01,$$

$$u_3 = \frac{u_2}{5^3} = 0.000 \ 08$$

$$u_4 = \frac{u_3}{3 \cdot 5^2} = \frac{0.000 \ 001 \ 07}{1.260 \ 001 \ 07},$$

$$u_5 = \frac{2u_4}{5^3} = \frac{0.000 \ 000 \ 02}{0.000 \ 080 \ 02}.$$

Therefore, $\qquad \sqrt[3]{2} = 1.260 \ 001 \ \ 07 - 0.000 \ 080 \ 02$ nearly

$$= 1.259 \ 921 \ \ 05 \text{ nearly.} \qquad \qquad ...(B)$$

To estimate the possible error in this result, let s_n be the sum to n terms of the series (A) and R_n the remainder after n terms. The error arises from two causes:

(i) The error caused by taking s_5, as the value of $\sqrt[3]{2}$ is R_5. Now, in the series (A), the terms beginning with the second are alternately positive and negative, also each is numerically less than the preceding; therefore R_5 is positive and less than u_6.

Now $\qquad \qquad u_6 = -\frac{\frac{1}{3}-4}{5} \cdot \frac{3}{5^3} u_5 = \frac{11}{5^4} u_5 < 0.000 \ 000 \ 000 \ 4;$

$\therefore$ The error is less than 0.000 000 000 4.

(ii) The error in the calculated value of s_5 *arises from the errors in the calculated values of* u_4 *and* u_5. These cannot exceed 0.000 000 005, hence the error in the calculated value of s_5 cannot exceed 0.000 000 01.

Adding the results of (i) and (ii), we see that the error in the result (B) is numerically less than 0.000 000 010 4. Thus, to six places of decimals,

$$\sqrt[3]{2} = 1.259921.$$

■ **EXAMPLE 2.** *If x is small and n > 1, obtain an approximation to* $(1+x)^{\frac{1}{n}}$ *in the*

form $\dfrac{p+qx}{p'+q'x}$ *where p, q, p', q' are independent of x.*

We have

$$(1+x)^{\frac{1}{n}} = 1 + \frac{x}{n} - \frac{n-1}{\lfloor 2} \left(\frac{x}{n}\right)^2 + \frac{(n-1)(2n-1)}{\lfloor 3} \left(\frac{x}{n}\right)^3 - \ldots, \qquad \ldots(A)$$

and $\dfrac{1+ax}{1+bx} = 1 + \dfrac{(a-b)x}{1+bx} = 1 + (a-b)x - (a-b)bx^2 + (a-b)b^2x^3 - \ldots, \quad \ldots(B)$

Choose a and b so that

$$a - b = 1/n \text{ and } (a-b)b = (n-1)/2n^2;$$

$$\therefore \qquad b = (n-1)/2n \text{ and } a = (n+1)/2n.$$

Substituting these values for a and b in (B), we have

$$\frac{2n+(n+1)x}{2n+(n-1)x} = 1 + \frac{x}{n} - \frac{n-1}{2}\cdot\left(\frac{x}{n}\right)^2 + \left(\frac{n-1}{2}\right)^2\left(\frac{x}{n}\right)^3 - \ldots \qquad \ldots(C)$$

If x is small, the series (A) and (C) are both convergent, and since they are identical as far as the terms involving x^2, it follows that

$$(1+x)^{\frac{1}{n}} = \frac{2n+(n+1)x}{2n+(n-1)x} \quad nearly.$$

■ **9. Theorem:** *If n > 1 and 0 < x < 1, then* $\dfrac{2n+(n+1)x}{2n+(n-1)x}$ *is an approximate value*

of $(1+x)^{\frac{1}{n}}$, *with an error in defect less than* $\dfrac{n^2-1}{12n^3} x^3.$

By multiplying each side of equation (A) in the last example by

$$2n + (n-1)x,$$

it can be shown that

$$\left(2n+\overline{n-1}.x\right)(1+x)^{\frac{1}{n}} - \left(2n+\overline{n+1}.x\right)$$

$$= \frac{n^2-1}{6n^2} x^3 \left\{1 - \frac{2n-1}{4n}\cdot 2x + \frac{(2n-1)(3n-1)}{4n.5n}\cdot 3x^2 - \ldots\right\}.$$

Denoting the series in brackets by

$$u_1 - u_2 + u_3 - \ldots + (-1)^r u_r + \ldots,$$

we have

$$\frac{u_r}{u_{r-1}} = \frac{rn-1}{(r+2)n} \cdot \frac{r}{r-1} \cdot x.$$

Now $(rn-1)r < (r+2)(r-1)n$ provided that $r \geq 2$, for

$$(rn-1)r - (r+2)(r-1)n = -r-(r-2)n < 0;$$

and since $0 < x < 1$, it follows that $u_r < u_{r-1}$.

The sum of the series is therefore positive and less than unity;

$$\therefore \; 0 < (1+x)^{\frac{1}{n}} - \frac{2n+(n+1)x}{2n+(n-1)x} < \frac{n^2-1}{6n^2} \cdot \frac{x^3}{2n+(n-1)\,x};$$

and, since $n > 1$ and x is positive, the last expression is less than

$$(n^2-1)\,x^3/12n^3,$$

whence the result follows.

■ **EXAMPLE** : *Show that 635/504 is an approximation to $\sqrt[3]{2}$ with an error in defect less than 0.0000005.*

Here $\sqrt[3]{2} = \dfrac{5}{4}\left(1+\dfrac{3}{125}\right)^{\frac{1}{3}}$; hence, an approximation is

$$\frac{5}{4} \cdot \frac{750+12}{750+6}, \;\; \text{or} \;\; \frac{635}{504},$$

with an error in defect $< \dfrac{5}{4} \cdot \dfrac{8}{12.27} \cdot \dfrac{3^3}{125^3} < \dfrac{43}{10^8} < \dfrac{5}{10^7}$.

EXERCISE XXXIV

1. If x is positive, which is the first negative term in the expansion of $(1+x)^{\frac{27}{5}}$?

2. If x is positive, the terms in the expansion of $(1-x)^{\frac{19}{2}}$, beginning with a certain term, are all positive. Prove this, and find which is the first of this series of positive terms. Which is the numerically greatest term or terms in the expansions of

3. $(1+x)^{\frac{35}{2}}$ when $x = \frac{3}{4}$?

4. $(1-x)^{-\frac{3}{2}}$ when $x = \frac{9}{10}$?

5. $(1-x)^{-13}$ when $x = \frac{1}{3}$?

6. $(1+x)^{-\frac{5}{3}}$ when $x = \frac{15}{17}$?

7. Between what numbers must x lie in order that, in the expansion of $(1-x)^{-\frac{3}{2}}$, the third term may be the greatest?

8. If $x > 0$, which is the first negative term in the expansion of

$$(11 - 13x)/(1-x)^3?$$

9. If $0 < x < 1$, show that

(i) $\sqrt{1+x} = 1 + \frac{1}{2}x - \frac{1}{8}x^2$ *nearly*, the error being in defect and less than $\dfrac{1}{16}x^3$.

(ii) $\sqrt{1-x} = 1 - \frac{1}{2}x - \frac{1}{8}x^2$ *nearly*, the error being in excess and numerically less

than $\dfrac{1}{16}x^3/(1-x)$.

10. If x is large and positive, show that

$$\sqrt[3]{x^3+1} = x + \frac{1}{3}\cdot\frac{1}{x^2} - \frac{1}{9}\cdot\frac{1}{x^5}\ \ nearly,$$

the error being in defect and less than $5/81x^8$.

11. If $0 < x < 1$ and R_n is the remainder after n terms in the expansion of $1/(1-x)^2$, show that

$$R_n < (n+1)x^n/(1-x)^2.$$

12. If $0 < x < 0.01$, show that

(i) $1 + 2x + 3x^2$ is an approximate value of $1/(1-x)^2$ with an error in defect less than 0.000005.

(ii) $1 - 2x + 3x^2$ is an approximate value of $1/(1+x)^2$ with an error in excess less than 0.000004.

13. If $0 < x < 1$, and R_n is the remainder after n terms in the expansion of $1/(1-x)^3$, show that

$$R_n < \tfrac{1}{2}(n+1)(n+2)x^n/(1-x)^3.$$

14. If $0 < x < 0.01$, show that

(i) $1 + 3x + 6x^2$ is an approximate value of $1/(1-x)^3$ with an error in defect less than 0.000011.

(ii) $1 - 3x + 6x^2$ is an approximate value of $1/(1+x)^3$ with an error in excess less than 0.00001.

15. If ε is small, show that, neglecting cubes and higher powers of ε,

(i) $\dfrac{1}{2}\left(\dfrac{x}{\sqrt{x-\varepsilon}} + \sqrt{x} - \varepsilon\right) = \sqrt{x} + \dfrac{\varepsilon^2}{2\sqrt{x}}.$

(ii) $\dfrac{1}{3}\left\{\dfrac{x}{\left(\sqrt[3]{x-\varepsilon}\right)^2} + 2\left(\sqrt[3]{x}-\varepsilon\right)\right\} = \sqrt[3]{x} + \dfrac{\varepsilon^2}{\sqrt[3]{x}}.$

16. *If we have tables giving the square roots and cube roots of numbers to f significant figures, show that the roots can be found by division only, to 2f significant figures as follows.*

If $\sqrt{x} = a$ to f significant figures, then $\sqrt{x} = \dfrac{1}{2}\left(\dfrac{x}{a} + a\right)$ *to 2f significant figures.*

If $\sqrt[3]{x} = a$ to f significant figures, then $\sqrt[3]{x} = \dfrac{1}{3}\left(\dfrac{x}{a^2} + 2a\right)$ *to 2f significant figures.*

Show also that in each case the error is in excess.

[Put $a = \sqrt{x-\varepsilon}$, and use Ex. 15.]

17. Given $\sqrt{2} = 1.41421$ (approx.), show, by division, that

$$\sqrt{2} = 1.41421356237 \text{ (approx.).}$$

18. Given $\sqrt[3]{10} = 2.1544$ (approx.), show, by division, that

$$\sqrt[3]{10} = 2.154434690 \text{ (approx.).}$$

19. Show that

(*i*) if x is small, $\sqrt{(x^2 + 4)} - \sqrt{(x^2 + 1)} = 1 - \dfrac{1}{4}x^2 + \dfrac{7}{64}x^4$ *nearly*;

(*ii*) if x is large, $\sqrt{(x^2 + 4)} - \sqrt{(x^2 + 1)} = \dfrac{3}{2x}\left(1 - \dfrac{5}{4x^2} + \dfrac{21}{8x^4}\right)$ *nearly*,

20. If x is small, show that approximately

(*i*) $\sqrt{(4 - 3x)} \cdot \sqrt[3]{(8 + 3x)} \div \left(9 - \frac{2}{3}x\right)^{\frac{3}{2}} = \frac{4}{27}\left(1 - \frac{5}{36}x\right)$;

(*ii*) $\dfrac{(1 - 2x)^{-\frac{1}{2}} - (1 + 2x)^{\frac{1}{2}}}{(1 - x)^{-\frac{1}{2}} - (1 + x)^{\frac{1}{2}}} = 4 + 2x$;

(*iii*) $\left(1 + 4x + 10x^2\right)^{\frac{1}{2}} = 1 + 2x + 3x^2 - 6x^3$;

(*iv*) $\left(1 + 4x + 10x^2\right)^{-\frac{1}{2}} = 1 - 2x + x^2 + 10x^3$;

(*v*) $\left(1 + x + x^2 + x^3\right)^{\frac{1}{5}} = 1 + \frac{1}{5}x + \frac{3}{25}x^2 + \frac{11}{125}x^3$.

In each of the examples 21–26, (*i*) prove the given equality; (*ii*) by expanding the right-hand side to four terms, deduce an approximate value of the given surd; (*iii*) find an upper limit to the numerical value of the error.

21. $\sqrt[3]{1001} = 10(1 + 0.001)^{\frac{1}{3}}$.　　　　　　　**22.** $\sqrt{2} = \frac{7}{5}(1 - 0.02)^{-\frac{1}{2}}$.

23. $\sqrt[3]{4} = \frac{8}{5}(1 + 0.024)^{-\frac{1}{3}}$.　　　　　　　**24.** $\sqrt[3]{5} = \frac{5}{3}(1 + 0.08)^{\frac{1}{3}}$.

25. $\sqrt[5]{3} = \frac{5}{4}(1 - 0.01696)^{\frac{1}{5}}$.　　　　　　　**26.** $\sqrt[5]{4} = \frac{4}{3}\left(1 - \frac{13}{256}\right)^{\frac{1}{5}}$.

27. If a is nearly equal to b, show that

$$\sqrt[n]{\dfrac{a}{b}} = \dfrac{(n + 1)\,a + (n - 1)b}{(n - 1)\,a + (n + 1)b} \ \ \textit{nearly},$$

and that if $a > b$, the error is in defect and is less than

$$\dfrac{n^2 - 1}{12n^3}\left(\dfrac{a - b}{b}\right)^3.$$

[Assume the result of Art. 9, and put $x = (a - b)/b$.]

28. Show that $\sqrt[5]{\dfrac{543}{540}} = \dfrac{2709}{2706}$ *nearly*, the error being in defect and less than $3/10^9$.

29. If $a - b = h$, where h is small, show that

(i) $\quad\sqrt{\dfrac{a}{b}} = 1 + \dfrac{1}{2}\dfrac{h}{b} - \dfrac{1}{8}\dfrac{h^2}{b^2} + \dfrac{1}{16}\dfrac{h^3}{b^3} - \dfrac{5}{128}\dfrac{h^4}{b^4} + \ldots .$

(ii) $\quad\dfrac{a}{a+b} = \dfrac{1}{2} + \dfrac{1}{4}\dfrac{h}{b} - \dfrac{1}{8}\dfrac{h^2}{b^2} + \dfrac{1}{16}\dfrac{h^3}{b^3} - \dfrac{1}{32}\dfrac{h^4}{b^4} + \ldots .$

(iii) Hence show that if a is nearly equal to b, then

$$\sqrt{\dfrac{a}{b}} = \dfrac{a}{a+b} + \dfrac{1}{4}\dfrac{a+b}{b} \quad nearly,$$

the error being approximately equal to $(a - b)^4/128b^4$.

■ 10. Application of the Binomial Theorem to the Summation of Series:

(1) Many series may be summed by expanding some function of x as a convergent power series in two different ways and equating coefficients.

■ **EXAMPLE 1***. *If* $\qquad u_r = r\,(r + 1)\,(r + 2)\,\ldots\,(r + q - 1)$

and $\qquad v_r = (n - r)\,(n - r + 1)\,(n - r + 2)\,\ldots\,(n - r + p - 1),$

prove that

$$u_1 v_1 + u_2 v_2 + u_3 v_3 + \ldots + u_{n-1} v_{n-1} = \frac{\lfloor p \lfloor q}{\lfloor p + q + 1} \cdot \frac{\lfloor p + q + n - 1}{\lfloor n - 2} .$$

It will be easily seen that

$$u_1 = \lfloor q, \qquad\qquad\qquad v_{n-1} = \lfloor p,$$

$$u_2 = \lfloor q.(q + 1), \qquad\qquad v_{n-2} = \lfloor p\,(p + 1),$$

$$u_3 = \lfloor q \cdot \frac{(q + 1)(q + 2)}{\lfloor 2}, \qquad v_{n-3} = \lfloor p \cdot \frac{(p + 1)(p + 2)}{\lfloor 2} .$$

$$\ldots\ldots\ldots\ldots\ldots\ldots\ldots . \qquad\qquad \ldots\ldots\ldots\ldots\ldots\ldots\ldots .$$

$$u_{n-1} = \lfloor q \cdot \frac{(q + 1)(q + 2)\ldots(q + n - 2)}{\lfloor n - 2} . \qquad v_1 = \lfloor p \cdot \frac{(p + 1)(p + 2)\ldots(p + n - 2)}{\lfloor n - 2} .$$

therefore $\qquad u_1 + u_2 x + u_3 x^2 + \ldots + u_{n-1} x^{n-2} + \ldots = \lfloor q\,(1 - x)^{-(q+1)},$

and $\qquad v_{n-1} + v_{n-2} x + v_{n-3} x^2 + \ldots + v_1 x^{n-2} + \ldots = \lfloor p\,(1 - x)^{-(p+1)}.$

Hence if S is the sum of the series in question,

$$S/\lfloor p \lfloor q = \text{coefft. of } x^{n-2} \text{ in } (1 - x)^{-(p+1)} . (1 - x)^{-(q+1)}$$

$$= \text{coefft. of } x^{n-2} \text{ in } (1 - x)^{-(p+q+2)}$$

$$= \lfloor p + q + n - 1/\left(\lfloor p + q + 1 \lfloor n - 2\right) .$$

---

* This equality is required in the Theory of Probabilities.

(2) Theorem. *If $f(x) = a_0 + a_1x + a_2x^2 + a_3x^2 + \ldots + a_nx^n + \ldots$ for $|x| < k$, where k is any positive number, then $a_0 + a_1 + a_2 + \ldots + a_n$ is equal to the coefficient of x^n in the expansion of $f(x)/(1-x)$.*

For $\dfrac{f(x)}{1-x} = (a_0 + a_1x + a_2x^2 + \ldots)(1 + x + x^2 + \ldots)$,

and the series are absolutely convergent for sufficiently small values of x.

(3) In what follows, we suppose that $|x| < 1$ except when n is a positive integer, so that all the series which occur are convergent.

We shall write $(1 + x)^n = c_0 + c_1x + c_2x^2 + c_3x^3 + \ldots + c_rx^r + \ldots$;

then,
$$n(1 + x)^{n-1} = n + \frac{n(n-1)}{\lfloor 1} x + \frac{n(n-1)(n-2)}{\lfloor 2} x^2 + \ldots$$
$$= c_1 + 2c_2x + 3c_3x^2 + \ldots + rc_r x^{r-1} + \ldots,$$

and, by a succession of similar steps,*
$$n(n-1)(1 + x)^{n-2} = 1 \cdot 2c_2 + 2 \cdot 3c_3x + \ldots + r(r-1)c_r x^{r-2} + \ldots,$$
$$n(n-1)(n-2)(1 + x)^{n-3} = 1 \cdot 2 \cdot 3c_3 + \ldots + r(r-1)(r-2)c_r x^{r-3} + \ldots,$$

and so on, as far as required.

Hence *we can sum the series* $\displaystyle\sum_{r=0}^{\infty} u_r c_r x^r$ *where u_r is a polynomial in r.*

For, dividing u_r and successive quotients by $r, r-1, r-2, \ldots$ in this order we can find constants $A_0, A_1, A_2, A_3, \ldots$ such that
$$u_r = A_0 + A_1r + A_2r(r-1) + A_3r(r-1)(r-2) + \ldots\,;$$

then, $\displaystyle\sum_{r=0}^{\infty} u_r c_r x^r = A_0(1 + x)^n + A_1nx(1 + x)^{n-1} + A_2n(n-1)x^2(1 + x)^{n-2}$
$$+ A_3n(n-1)(n-2)x^3(1 + x)^{n-3} + \ldots\,.$$

■ **EXAMPLE 2.** *Find* $\displaystyle\sum_{r=0}^{n} (r+1)^3 c_r x^r$.

Denoting the sum by S and proceeding as above, we find that
$$(r+1)^3 = 1 + 7r + 6r(r-1) + r(r-1)(r-2);$$
$$\therefore \qquad S = (1+x)^n + 7nx(1+x)^{n-1} + 6n(n-1)x^2(1+x)^{n-2}$$
$$+ n(n-1)(n-2)x^3(1+x)^{n-3}$$
$$= (1+x)^{n-3}\{1 + (7n+3)x + (6n^2 + 8n + 3)x^2(n+1)^3x^3\}.$$

(4) Again, by expanding $(1 + x)^{n+1}$, we can show that

$$\frac{(1+x)^{n+1}}{n+1} = \frac{1}{n+1} + c_0x + \frac{c_1}{2}x^2 + \ldots + \frac{c_r}{r+1}x^{r+1} + \ldots,$$

and, by similar steps,

$$\frac{(1+x)^{n+2}}{(n+1)(n+2)} = \frac{1}{(n+1)(n+2)} + \frac{1}{n+1}x + \frac{c_0}{1\cdot 2}x^2 + \ldots + \frac{c_r}{(r+1)(r+2)}x^{r+2} + \ldots,$$

---

* These results can also be obtained by differentiating with regard to x, although so far the process has not been justified in case of an infinite series.

$$\frac{(1+x)^{n+3}}{(n+1)(n+2)(n+3)} = \frac{1}{(n+1)(n+2)(n+3)} + \frac{1}{(n+1)(n+2)}x + \frac{1}{2(n+1)}x^2$$

$$+ \frac{c_0}{1\cdot2\cdot3}x^3 + ... + \frac{c_r}{(r+1)(r+2)(r+3)}x^{r+3} + ... \ ;$$

and so on. These results can also be obtained by integration.

Hence, *if u_r is a polynomial in r, we can sum the series*

$$\sum_{r=0}^{\infty} \frac{u_r}{(r+1)(r+2)(r+3)...(r+k)}\, c_r x^r,$$

For, dividing u_r and successive quotients by $r+k$, $r+k-1$,...$r+2$, $r+1$ in this order, we can find an equation of the form

$$\frac{u_r}{(r+1)(r+2)...(r+k)} = v_r + \frac{A_1}{r+1} + \frac{A_2}{(r+1)(r+2)} + ... \frac{A_k}{(r+1)(r+2)...(r+k)},$$

where $A_k, A_{k-1}, ... A_1$ are the remainders in the successive divisions and v_r is the final quotient, and then the sum of the series is

$$\Sigma v_r c_r x^r + \frac{A_1}{x}\Sigma \frac{c_r}{r+1}x^{r+1} + \frac{A_2}{x^2}\Sigma \frac{c_r}{(r+1)(r+2)}x^{r+2} + ... \ .$$

The value of $\Sigma v_r c_r x^r$ is found by § (3), and the other sums are given by the above equations.

Further, *we can sum the series*

$$\sum_{r=0}^{\infty} \frac{u_r}{(r+a)(r+b)...(r+k)}\, c_r x^r,$$

where a, b,... k are unequal positive integers of which k is the greatest.

For this can be reduced to the preceding by multiplying numerator and denominator of the fraction by such factors as will convert the denominator into $(r+1)(r+2)$ $(r+3)... (r+k)$.

■ **EXAMPLE 3.** *Find $S = \displaystyle\sum_{r=0}^{\infty} \frac{1}{r+3} c_r x^r.$*

We have

$$\frac{1}{r+3} = \frac{r^2 + 3r + 2}{(r+1)(r+2)(r+3)} = \frac{1}{r+1} - \frac{2}{(r+1)(r+2)} + \frac{2}{(r+1)(r+2)(r+3)};$$

hence, $\quad S = \dfrac{1}{x}\Sigma \dfrac{c_r}{r+1}x^{r+1} - \dfrac{2}{x^2}\Sigma \dfrac{c_r}{(r+1)(r+2)}x^{r+2} + \dfrac{2}{x^3}\dfrac{c_r}{(r+1)(r+2)(r+3)}x^{r+3}$

$$= \frac{1}{x}\cdot\frac{(1+x)^{n+1}-1}{n+1} - \frac{2}{x^2}\left\{\frac{(1+x)^{n+2}-1}{(n+1)(n+2)} - \frac{x}{n+1}\right\}$$

$$+ \frac{2}{x^3}\left\{\frac{(1+x)^{n+3}-1}{(n+1)(n+2)(n+3)} - \frac{x}{(n+1)(n+2)} - \frac{x^2}{2(n+1)}\right\}.$$

$$= \frac{(1+x)^{n+1}}{(n+1)\,x} \left\{ 1 - \frac{2}{n+2} \cdot \frac{1+x}{x} + \frac{2}{(n+2)\,(n+3)} \cdot \left(\frac{1+x}{x} \right)^2 \right\}$$

$$- \frac{1}{x} \cdot \frac{1}{n+1} + \frac{2}{x^2} \left\{ \frac{1}{(n+1)\,(n+2)} + \frac{x}{n+1} \right\}$$

$$- \frac{2}{x^3} \left\{ \frac{1}{(n+1)\,(n+2)\,(n+3)} + \frac{x}{(n+1)\,(n+2)} + \frac{x^2}{2(n+1)} \right\},$$

$$\therefore \quad S = \frac{(1+x)^{n+1}}{(n+1)\,(n+2)\,(n+3)\,x} \left\{ n^2 + 3n + 2 - \frac{2(n+1)}{x} + \frac{2}{x^2} \right\}$$

$$- \frac{2}{(n+1)\,(n+2)\,(n+3)\,x^3}.$$

■ **EXAMPLE 4.** *Find* $\displaystyle\sum_{r=0}^{\infty} \frac{r^2+1}{r+3}\, c_r x^r$.

We have

$$\frac{r^2+1}{r+3} = r - 3 + \frac{10}{r+3}$$

and $\qquad \Sigma(r-3)\,c_r x^r = nx\,(1+x)^{n-1} - 3(1+x)^n;$

∴ The required sum $= (1+x)^{n-1}\{(n-3)x - 3\} + 10S$, where S is the same as in Ex. 3.

■ **11. Multinomial Theorem for any Real Index:**

If $b, c,...\, k$ are any m numbers, n is real, and
$$f\,(x) = (1 + bx + cx^2 +...+ kx^m)^n,$$
then (*i*) *f* (*x*) *can be expanded in the form*
$$1 + u_1 x + u_2 x^2 + ... + u_r x^r +... ,$$
where the series terminates if n is a positive integer, and in other cases is absolutely convergent for sufficiently small values of x.

(*ii*) *The coefficient* u_r *of* x^r *is the sum of all terms of the form*

$$\frac{n(n-1)\,(n-2)\,...\,(\alpha+1)}{\lfloor\beta\,\lfloor y\,...\,\lfloor \kappa}\, b^\beta c^\gamma\,...\, k^\kappa,$$

where $\beta, \gamma,\ ...\ \kappa$ *have any positive integral or zero values such that*
$$\beta + 2\gamma + ... + m\kappa = r,$$
and the corresponding values of α *are given by* $\alpha + \beta + \gamma +... + \kappa = n$.

For let $y = bx + cx^2 + ... + kx^m$, then by the Binomial theorem

$$f\,(x) = (1+y)^n = 1 + \sum_{s=1}^{s=\infty} \frac{n(n-1)\,...\,(n-s+1)}{\lfloor s}\, y^s, \quad ...(A)$$

provided that $|y| < 1$, and by Ch. 17, 3, (1), we can find a positive number ρ such that $|y| < 1$ when $|x| < \rho$.

By substituting $bx + cx^2 + \ldots$ for y in (A) and expanding the terms, we obtain an expansion of $f(x)$ in powers of x which is convergent if $|x| < \rho$.

In order that the terms of this series may be arranged in ascending powers of x, it is sufficient that the series is convergent when all the terms are made positive. This will be the case if $b'x' + c'x'^2 + \ldots + k'x'^m < 1$, where x', b', c', ... stand for $|x|$, $|b|$, $|c|$, ...; and we can find a positive number σ such that this is the case if $|x| < \sigma$.

This concludes the proof of the statement (*i*).

Again, s being any positive integer, by Ch. 3, 16, the coefficient of x^r in $(bx + cx^2 + \ldots + kx^m)^s$ is the sum of all terms of the form where β, γ, ... κ have any positive integral or zero values such that

$$\beta + 2\gamma + \ldots + m\kappa = r, \qquad\qquad \ldots(B)$$

and
$$\beta + \gamma + \ldots + \kappa = s. \qquad\qquad \ldots(C)$$

It follows from (A) that the coefficient of x^r in the expansion of $f(x)$ is the sum of all terms of the form

$$\frac{n(n-1)\ldots(n-s+1)}{\lfloor\beta\,\lfloor\gamma\ldots\,\lfloor\kappa}\, b^\beta c^\gamma \ldots k^\kappa,$$

where the values of β, γ, ... are given by (B) and the corresponding values of s by (C). The statement (*ii*) follows on writing $\alpha = n - s$.

■ **Example :** *Find the coefficient of x^4 in the expansion of $(1 + 2x - 4x^2 - 2x^3)^{-\frac{1}{2}}$.*

Here $n = -\frac{1}{2}$, $r = 4$, and the coefficient is the sum of all terms of the form

$$\frac{\left(-\frac{1}{2}\right)\left(-\frac{1}{2}-1\right)\left(-\frac{1}{2}-2\right)\ldots(\alpha+1)}{\lfloor\beta\,\lfloor\gamma\,\lfloor\delta}\,.2^\beta\,(-4)^\gamma\,(-2)^\delta,$$

where β, γ, δ have positive integral or zero values such that $\beta + 2\gamma + 3\delta = 4$, and the corresponding values of α are given by $\alpha + \beta + \gamma + \delta = -\frac{1}{2}$.

The possible values of α, β, γ, δ are shown in the margin.

The term corresponding to the first set of values is

$$\frac{\left(-\frac{1}{2}\right)\left(-\frac{3}{2}\right)}{\lfloor 1\,\lfloor 0\,\lfloor 1}\,.\,2\,.\,(-2)\quad = -3$$

α	β	γ	δ
$-2\frac{1}{2}$,	1,	0,	1
$-3\frac{1}{2}$,	2,	1,	0
$-4\frac{1}{2}$,	4,	0,	0
$-2\frac{1}{2}$,	0,	2,	0

Similarly it will be found that the other terms are 15, $\dfrac{35}{8}$, 6, so that the required coefficient is

$$-3 + 15 + 4\tfrac{3}{8} + 6 = 22\tfrac{3}{8}.$$

EXERCISE XXXV

1. By considering the expansion of $(1 - x)^{-n}$, show that

$$1 + n + \frac{n(n+1)}{\underline{2}} + \dots \frac{n(n+1)\dots(n+r-1)}{\underline{r}} = \frac{(n+1)(n+2)\dots(n+r)}{\underline{r}}.$$

2. Show that, if $z = \dfrac{x-y}{x+y}$, and $|z| < 1$, then

$$(x+y)^n = \frac{2^{n+1} x^n y^n}{x^n - y^n} \left\{ nz + \frac{n(n+1)(n+2)}{\underline{3}} z^2 + \dots \right\}.$$

3. If n, r are positive integers (including zero), show that the coefficient of x^{n+r-1} in the expansion of $(1+x)^n/(1-x)^2$ is $(n+2r)\, 2^{n-1}$.

[Expand $\{2 - (1-x)\}^n/(1-x)^2$.]

4. If n is a positive integer and

$$\frac{(1+x)^n}{(1-x)^3} = a_0 + a_1 x + a_2 x^2 + \dots a_n x^n + \dots ,$$

show that $a_0 + a_1 + a_2 + \dots + a_{n-1} = \frac{1}{3} n (n+2)(n+7) 2^{n-4}$.

5. If n is a positive integer, show that

$$1 + 3n + \frac{3\cdot 4}{1\cdot 2}\frac{n(n-1)}{\underline{2}} + \frac{4\cdot 5}{1\cdot 2}\frac{n(n-1)(n-2)}{\underline{3}} + \dots + \frac{(n+1)(n+2)}{2}$$

$$= \text{Coefficient of } x^n \text{ in } \{2 - (1-x)\}^n/(1-x)^3 = (n^2 + 7n + 8)2^{n-3}.$$

6. If n is a positive integer, show that

$$2^n - (n-1)\, 2^{n-2} + \frac{(n-2)(n-3)}{\underline{2}} 2^{n-4} - \frac{(n-3)(n-4)(n-5)}{\underline{3}} 2^{n-6} + \dots = n+1.$$

[From the expansion of $\{1 - x(2-x)\}^{-1} = (1-x)^{-2}$.]

7. Prove that if n is a positive integer,

$$1 - n\cdot\frac{1+x}{1+nx} + \frac{n(n-1)}{\underline{2}}\cdot\frac{1+2x}{(1+nx)^2} + \frac{n(n-1)(n-2)}{\underline{3}}\cdot\frac{1+3x}{(1+nx)^3} + \dots = 0.$$

8. Show that, for all values of m and n,

$$C_r^n - C_{r-1}^n C_1^m + C_{r-2}^n C_2^{m+1} - C_{r-3}^n C_3^{m+2} - \dots + (-1)^r C_r^{m+r-1}$$

$$= \text{The coefficient of } x^r \text{ in the expansion of } (1+x)^{n-m}.$$

9. If $p + q = 1$ and n, r are positive integers $(n > r)$, show that

$$p^n + C_1^n p^{n-1} q + C_2^n p^{n-2} q^2 + \dots + C_{n-r}^n p^r q^{n-r}$$

$$= p^r \left\{ 1 + rq + \frac{r(r+1)}{\underline{2}} q^2 + \dots + \frac{r(r+1)\dots(n-1)}{\underline{n-r}} q^{n-r} \right\}.$$

[Divide by p^r, put $1 - q$ for p, and show that the coefficient of q^k on the left hand side when expanded is

$$C_k^n - C_{k-1}^n \, C_2^{n-r-k+1} + C_{k-2}^n \, C_2^{n-r-k+2} - \ldots + (-1)^k \, C_k^{n-r}$$

$$= \text{Coefficient of } x^k \text{ in } (1 + x)^n \times (1 + x)^{-(n-r-k+1)}.]$$

In Exs. 10-13, if $(1 + x)^n = c_0 + c_1 x + c_2 x^2 + \ldots + c_r x^r + \ldots$ where n may have any value,

find the sum of the series, $\displaystyle\sum_{r=0}^{\infty} mc_r x^r$, when m is

10. *$(r + 1)^2$;

11. $\dfrac{r+1}{r+2}$;

12. $\dfrac{1}{(r+1)(r+3)}$;

13. * $\dfrac{r^2 - r + 1}{(r+1)(r+3)}$.

14. If $u_r = A_0 + A_1 r + A_2 r \, (r - 1) + \ldots + A_h r \, (r - 1) \ldots (r - h + 1)$ and
$\phi \, (x) = A_0 x^h + A_1 n x^{h-1} + A_2 n \, (n - 1) \, x^{h-2} + \ldots + A_h n \, (n - 1) \ldots (n - h + 1)$,
prove that

$$\Sigma u_r c_r x^r = (1 + x)^{n-h} \left[x^h \, \phi \, (1) + \frac{x^{h-1}}{\underline{1}} \, \phi' \, (1) + \frac{x^{h-2}}{\underline{2}} \, \phi'' \, (1) + \ldots \right].$$

15. Prove that $C_r^r \, C_{13-r}^{52-r} + C_r^{r+1} \, C_{13-r}^{51-r} + C_r^{r+2} \, C_{13-r}^{50-r} + \ldots$ to 40 terms $= C_{24}^{53}$.
   [Consider the product $(1 - x)^{-(r+1)} \times (1 - x)^{-(14-r)}.]$

16. Show that

$$\Sigma \frac{1}{\underline{2p+1} \, \underline{2q+1} \, \underline{2r+1}} = \frac{3^{2n+3} - 3}{4 \, \underline{2n+3}},$$

the summation being taken for all positive integral values, including zero, of p, q, r such that $p + q + r = n$. [Consider the expansion of

$$(a + b + c)^{2n+3} - (b + c - a)^{2n+3} - (c + a - b)^{2n+3} - (a + b - c)^{2n+3}.]$$

---

* Check the answer by putting $n = 1$.

■ ■ ■

CHAPTER 22

Rational Fractions, Recurring Series and Difference Equations

■ **1. Expansion of a Rational Fraction:** Any rational proper fraction in x, whose denominator does not contain x as a factor, may be written in the form

$$\frac{P}{Q} = \frac{a_0 + a_1 x + a_2 x^2 + \ldots + a_m x^m}{1 + p_1 x + p_2 x^2 + \ldots + p_r x^r} \quad \text{where } m < r.$$

Various ways of expanding such a fraction in a series of the form $\Sigma u_n x^n$ are given below.

If it can be shown that the series obtained by any process is convergent for all values of x in any interval including zero, then by Ch. 20, 20, *any other method of expansion yields identically the same series.*

■ **2. Theorem:** *The rational fraction P/Q can be expanded in a convergent series of the form*

$$u_0 + u_1 x + u_2 x^2 + \ldots + u_n x^n + \ldots ,$$

if and only if $|x| < |\lambda|$, *where* λ *is the root of Q = 0 with the least modulus.*

For, by the method of partial fractions, P/Q can be expressed as the sum of several terms of the form $A(x - \alpha)^{-r}$, where A is a constant, r a positive integer and α is any root, real or imaginary, of the equation $Q = 0$.

All of these terms can be expanded in a convergent series of ascending powers of x if, and only if, $|x| < |\lambda|$, where λ is the root of $Q = 0$ with the least modulus.

When this is the case, the expansion of P/Q can be obtained by adding the various series.

▌ **EXAMPLE.** *For what values of x can the fraction*
$$1/ (x^2 + 2x - 1)\ (x^2 - x + 5)$$
be expanded in a series of ascending powers of x?

Equating the denominator to zero, the roots are

$$-1 \pm \sqrt{2} \quad \text{and} \quad \tfrac{1}{2}(1 \pm \iota \sqrt{19})$$

and their moduli are $\sqrt{2} \pm 1$, $\sqrt{5}$, $\sqrt{5}$.

Hence the expansion is possible if and only if $|x| < \sqrt{2} - 1$.

■ **3. Methods of Expansion:**

(1) Suppose that

$$P/Q = u_0 + u_1 x + u_2 x^2 + \ldots$$

where the series is assumed to be convergent and P, Q are as described in Art. 1. Multiplying by Q,

$$a_0 + a_1 x + a_2 x^2 + \ldots + a_m x^m = (1 + p_1 x + p_2 x^2 + \ldots + p_r x^r)(u_0 + u_1 x + u_2 x^2 + \ldots).$$

Expanding the right-hand side and equating coefficients, the values of $u_0,\, u_1,\, u_2, \ldots$ are determined by the equations

$$u_0 = a_0, \quad u_1 + p_1 u_0 = a_1, \quad u_2 + p_1 u_1 + p_2 u_0 = a_2, \quad \ldots ,$$

$$u_{r-1} + p_1 u_{r-2} + p_2 u_{r-3} + \ldots + p_{r-1} u_0 = a_{r-1},$$

and for $n \geq r$, $\qquad u_n + p_1 u_{n-1} + p_2 u_{n-2} + \ldots + p_r u_{n-r} = 0.$

Hence, after a certain stage, any r successive coefficients are connected by a linear relation. A series having this property is called a *Recurring Series.*

(2) The process just described is equivalent to *synthetic division* carried out as in the next example. (See Ch. 3, 3)

■ **EXAMPLE 1.** *Show that, for sufficiently small values of x,*

$$\frac{1 + x + x^2 + x^3}{1 - 2x + 3x^2 - 4x^3} = 1 + 3x + 4x^2 + 4x^3 + 8x^4 + 20x^5 + \ldots .$$

The reckoning is as follows.

$$1 - (2 - 3 + 4) \ \left|\begin{array}{l} 1 + 1 + 1 + 1 \\[4pt] \qquad\qquad + 4 + 12 + 16 + 16 + 32 \quad (a) \\[4pt] \qquad\quad\; - 3 - 9 - 12 - 12 - 24 \qquad (b) \\[4pt] \underline{\quad + 2 + 6 + 8 + 8 + 16 \qquad\qquad (c)} \\[4pt] \quad 1 + 3 + 4 + 4 + 8 + 20 \qquad\quad (d) \end{array}\right.$$

The first term in the quotient (d) is 1; 1 $(2 - 3 + 4) = 2 - 3 + 4$; put $2, -3, +4$ diagonally in (c), (b), (a); $2 + 1 = 3$, the next term in (d); $3(2 - 3 + 4) = 6 - 9 + 12$; put $6, -9, +12$ diagonally in (c), (b), (a); $6 - 3 + 1 = 4$, the next term in (d).

If the reader will do this example by the method in Art. 3, (1), he will see that the reckoning is essentially the same as that just described.

(3) *The method of Partial Fractions.*

■ **EXAMPLE 2.** *If $1/(1 - 2x + 5x^2) = u_0 + u_1 x + \ldots + u_n x^n + \ldots$ find the value of u_n.*

We have
$$\frac{1}{1 - 2x + 5x^2} = \frac{1}{(1 - \alpha x)(1 - \beta x)} = \frac{1}{\alpha - \beta}\left\{\frac{\alpha}{1 - \alpha x} - \frac{\beta}{1 - \beta x}\right\}$$

where α, β are the roots of $x^2 - 2x + 5 = 0$, so that we may take $\alpha = 1 + 2\iota$, $\beta = 1 - 2\iota$. Hence
$$\frac{1}{1 - 2x + 5x^2} = \frac{1}{4\iota}\{\alpha(1 + \alpha x + \alpha^2 x^2 + \ldots) - \beta(1 + \beta x + \beta^2 x^2 + \ldots)\};$$

$$\therefore \qquad u_n = \frac{1}{4\iota}(\alpha^{n+1} - \beta^{n+1}).$$

Now $\alpha = \sqrt{5}\,(\cos\theta + \iota\sin\theta)$, $\beta = \sqrt{5}\,(\cos\theta - \iota\sin\theta)$, where $\theta = \tan^{-1}2$,

$$\therefore \qquad \alpha^{n+1} - \beta^{n+1} = 5^{\frac{1}{2}(n+1)}\,.\,2\iota\sin(n+1)\,\theta;$$

hence $\qquad u_n = \dfrac{1}{2}\,.\,5^{\frac{1}{2}(n+1)}\,\sin(n+1)\,\theta$, where $\theta = \tan^{-1}2$.

(4) A rational fraction can also be expanded by the *Multinomial Theorem*. Using this method and that of Partial Fractions and equating coefficients, we can obtain many important results, as in the following exercise.

EXERCISE XXXVI

1. Use synthetic division to expand

   (i) $\dfrac{1}{(1-x)^2\,(1-2x)}$; $\qquad\qquad$ (ii) $\dfrac{1-4x+x^2}{1-6x+11x^2-6x^3}$,

   as far as the terms containing x^5. Find also the coefficient of x^n in each expansion.

2. Show that, if p and q are real numbers, the condition that $1/(1 + px + qx^2)$ can be expanded in a convergent series of ascending powers of x is as follows:

   (i) if $p^2 > 4q$, then $|x| < \left| \left(p' - \sqrt{p^2 - 4q}\right)/2q \right|$, where $p' = |p|$;

   (ii) if $p^2 < 4q$, then $|x| < 1/\sqrt{q}$.

3. If $|x|$ lies between $|\alpha|$ and $|\beta|$, show that the fraction $1/(x - \alpha)(x - \beta)$ can be expanded in the form $\ldots + b_2x^{-2} + b_1x^{-1} + a_0 + a_1x + a_2x^2 + \ldots$, where the series extends to ∞ on the right and left.

   [Let $|\alpha| < |\beta|$. The fraction $= (\alpha - \beta)\,\{1/(x - \alpha) - 1/(x - \beta)\}$.

   If $|\alpha| < |x| < |\beta|$, prove that

   $$1/(x - \alpha) = x^{-1}\,(1 + \alpha x^{-1} + \alpha^2 x^{-2} + \ldots),$$

   and $\qquad 1/(x - \beta) = -\beta^{-1}\,(1 + \beta^{-1}x + \beta^{-2}x^2 + \ldots).]$

4. If H_n is the sum of the homogeneous products of n dimensions which can be formed with $a,\,b,\,c$, show that

   $$a^{n+2}\,(b - c) + b^{n+2}\,(c - a) + c^{n+2}\,(a - b) = -H_n\,(b - c)\,(c - a)\,(a - b).$$

   [H_n is the coefficient of x^n in the product

   $$(1 + ax + a^2x^2 + \ldots)\,(1 + bx + b^2x^2 + \ldots)\,(1 + cx + c^2x^2 + \ldots),$$

   i.e. in the expansion of $1/(1 - ax)\,(1 - bx)\,(1 - cx)$. The result is obtained by the method of partial fractions.]

5. Prove that, for sufficiently small values of x,

   $$\frac{1}{1 - px + qx^2} = 1 + a_1x + a_2x^2 + \ldots + a_nx^n + \ldots,$$

   where $\qquad a_n = p^n - (n - 1)\,p^{n-2}q + \dfrac{(n - 2)\,(n - 3)}{\lfloor 2}\,p^{n-4}q^2 - \ldots,$

   the $(r + 1)$th term being $(-1)^r\,C_r^{n-r}\,p^{n-2r}q^r$.

6. If $\alpha + \beta = p$, $\alpha\beta = q$, use the identity $\dfrac{1}{1-\alpha x} + \dfrac{1}{1-\beta x} = \dfrac{2-px}{1-px+qx^2}$

to prove that
$$\alpha^n + \beta^n = p^n - \frac{n}{\underline{|1}}\, p^{n-2}q + \frac{n\,(n-3)}{\underline{|2}}\, p^{n-4}q^2 - \;...\;,$$

the $(r+1)$th term being $\quad (-1)^r\, \dfrac{n\,(n-r-1)\,(n-r-2)\,...\,(n-2r+1)}{\underline{|r}}\, p^{n-2r}q^r.$

[Expand both sides, equate coefficients and use Ex. 5.]

7. If $\alpha + \beta = p$, $\alpha\beta = q$, use the identity $\dfrac{1}{1-\alpha x} - \dfrac{1}{1-\beta x} = \dfrac{(\alpha-\beta)x}{1-px+qx^2}$

to prove that
$$\frac{\alpha^n - \beta^n}{\alpha - \beta} = p^{n-1} - \frac{n-2}{\underline{|1}}\, p^{n-3}q + \frac{(n-3)(n-4)}{\underline{|2}}\, p^{n-5}q^2 - \;...\;,$$

the $(r+1)$th team being $(-1)^r\, \dfrac{(n-r-1)\,(n-r-2)\,...\,(n-2r)}{\underline{|r}}\, p^{n-2r-1}q^r.$

[Expand both sides and use Ex. 5.]

8. If n is a positive integer, prove that

(i) $2\cos n\theta = (2\cos\theta)^n - \dfrac{n}{\underline{|1}}\,(2\cos\theta)^{n-2} + \dfrac{n\,(n-3)}{\underline{|2}}\,(2\cos\theta)^{n-4} - \;...\;,$

the $(r+1)$th term being $(-1)^r\, \dfrac{n(n-r-1)\,...\,(n-2r+1)}{\underline{|r}}\,(2\cos\theta)^{n-2r}.$

(ii) $\dfrac{\sin n\theta}{\sin\theta} = (2\cos\theta)^{n-1} - \dfrac{n-2}{\underline{|1}}\,(2\cos\theta)^{n-3} + \dfrac{(n-3)(n-4)}{\underline{|2}}\,(2\cos\theta)^{n-5} - \;...\;,$

the $(r+1)$th term being $(-1)^r\, \dfrac{(n-r-1)\,(n-r-2)\,...\,(n-2r)}{\underline{|r}}\,(2\cos\theta)^{n-2r-1}.$

[In Exs. 6, 7 put $\alpha = z$, $\beta = z^{-1}$ where $z = \cos\theta + \iota\sin\theta.$]

9. Show that (i) the coefficient of x^{3n} in the expansion of $\dfrac{1+x}{(1+x+x^2)^3}$ is $\frac{1}{2}(n+1).$

$(3n+2)$.

(ii) The coefficient of x^{4n} in the expansion of $\dfrac{1}{(1-x)(1-x^2)(1-x^4)}$ is $(n+1)^2.$

▨ **4. Recurring Series:** (1) Let $u_0 + u_1 x + u_2 x^2 + \;...$ be a series in which any $r+1$ successive coefficients are connected by the equation

$$u_n + p_1 u_{n-1} + p_2 u_{n-2} + \;...\; + p_r u_{n-r} = 0, \qquad\qquad ...(A)$$

where r is a fixed number and $p_1, p_2, ...\, p_r$ are constants. Such a series is called a *Recurring Series of the r-th order.*

Some authors call equation (A) the *Scale of Relation* of the series; others use this term to denote the polynomial $1 + p_1 x + p_2 x^2 + \;...\; + p_r x^r$. We shall take it to mean *either* of these things.

If the coefficients u_0, u_1, ... u_{r-1} are known, the subsequent coefficients can be found in succession by equation (A). Thus a recurring series of the rth order depends on $2r$ constants.

Hence if the first $2r$ coefficients are given, in general the series can be continued as a recurring series of the rth order, and in one way only. Also it can be continued as a recurring series of the $(r + 1)$th order in a doubly infinite number of ways. For to do this we can give u_r and u_{r+1} any values.

It may happen that the first $2r$ terms belong to a recurring series of order $r - 1$. Two conditions are necessary that this may be so, and then the series can be continued as a recurring series of the $(r - 1)$th order.

▌ **EXAMPLE 1.** *Discuss the question of continuing $\Sigma u_n x^n$ as a recurring series when the first six terms are known.*

Let the scale of relation be $u_n + pu_{n-1} + qu_{n-2} + ru_{n-3} = 0$.

Then p, q, r are given by

$$u_3 + pu_2 + qu_1 + ru_0 = 0, \quad u_4 + pu_3 + qu_2 + ru_1 = 0, \quad u_5 + pu_4 + qu_3 + ru_2 = 0.$$

Now $\Sigma u_n x^n$ can be continued as a recurring series of the third order if p, q, r have definite values and $r \neq 0$. This will be the case if

$$\begin{vmatrix} u_0 & u_1 & u_2 \\ u_1 & u_2 & u_3 \\ u_2 & u_3 & u_4 \end{vmatrix} \neq 0 \quad \text{and} \quad \begin{vmatrix} u_1 & u_2 & u_3 \\ u_2 & u_3 & u_4 \\ u_3 & u_4 & u_5 \end{vmatrix} \neq 0. \qquad \text{...(A)}$$

In case either determinant is zero, suppose that the scale is $u_n + pu_{n-1} + qu_{n-2} = 0$; then $u_2 + pu_1 + qu_0 = 0, \quad u_3 + pu_2 + qu_1 = 0,$
$u_4 + pu_3 + qu_2 = 0, \quad u_5 + pu_4 + qu_3 = 0.$

That these equations may be consistent, each of the determinants in (A) must vanish, so that *two* conditions are necessary that $\Sigma u_n x^n$ may be a recurring series of the second order.

If it is one of the first order, the scale being $u_n + pu_{n-1} = 0$, then

$$u_1 + pu_0 = 0, \quad u_2 + pu_1 = 0,.... \qquad u_5 + pu_4 = 0,$$

so that the *four* conditions $u_1^2 - u_0 u_2 = 0, \quad u_2^2 - u_1 u_3 = 0, \quad u_3^2 - u_2 u_4 = 0,$
$u_4^2 - u_3 u_5 = 0$ must be satisfied.

(2) *Any recurring series $\Sigma u_n x^n$ is convergent for sufficiently small values of x.*

For let the scale be

$$u_n + p_1 u_{n-1} + p_2 u_{n-2} + ... + p_r u_{n-r} = 0 \qquad \text{...(A)}$$

and let $|u_0|, |u_1|, ... |p_1|, |p_2|, ...$ be denoted by u_0', u_1', ... p_1', p_2', ...

Then by (A),

$$|u_n| = |p_1 u_{n-1} + p_2 u_{n-2} + ... + p_r u_{n-r}|$$

$$\leq p_1' u'_{n-1} + p_2' u'_{n-2} + ... + p_r' u'_{n-r}$$
$$\leq gu'_{n-s},$$

where g is the greater of the numbers $p_1' + p_2' + ... + p_r'$ and 1, and u'_{n-s} is the greatest of the set $u'_{n-1}, u'_{n-2}, ... u'_0$. In the same way

$$|u_{n-s}| \leq gu'_{n-t},$$

where u'_{n-t} is the greatest of $u'_{n-s-1}, u'_{n-s-2}, ... u_0'$.

Continuing thus, by multiplication we can show that $|u_n| \le g^n A$, where A is the greatest of $u'_{n-1}, u'_{n-2}, \ldots u'_0$.

Thus
$$|u_n x^n| \le |gx|^n \cdot A,$$

and $\Sigma u_n x^n$ is convergent if $|x| < 1/g$, for then $|u_n x^n|$ is less than the $(n+1)$th term of a convergent series of positive terms.

(3) *The sum to infinity of any recurring series* $\Sigma u_n x^n$, *when convergent, is a rational proper fraction, which is called the* generating function *of the series.*

Let the scale be
$$u_n + p_1 u_{n-1} + p_2 u_{n-2} + \ldots + p_r u_{n-r} = 0, \qquad \ldots\text{(A)}$$
and let
$$s = u_0 + u_1 x + u_2 x^2 + \ldots \text{ to } \infty. \qquad \ldots\text{(B)}$$

Multiplying (B) by $p_1 x, p_2 x^2, \ldots p_r x^r$ and adding, we find that
$$s\,(1 + p_1 x + p_2 x^2 + \ldots + p_r x^r) = a_0 + a_1 x + a_2 x^2 + \ldots + a_{r-1} x^{r-1}, \qquad \ldots\text{(C)}$$

where $a_0 = u_0,\ a_1 = u_1 + p_1 u_0,\ a_2 = u_2 + p_1 u_1 + p_2 u_0, \ldots ,$
$$a_{r-1} = u_{r-1} + p_1 u_{r-2} + \ldots p_{r-1} u_0.$$

the coefficients of powers of x higher than x^{r-1} vanishing on account of equation (A).

Thus s is equal to a rational fraction, of which the denominator is
$$Q = 1 + p_1 x + p_2 x^2 + \ldots + p_r x^r.$$

It follows from Art. 2 that *the recurring series is convergent if* $|x| < |\lambda|$, *where* λ *is the root of* $Q = 0$ *which has the least modulus.*

(4) *If* u_n *is a positive integral function of n of degree r, then* $\Sigma u_n x^n$ *is a recurring series of which the scale is* $(1-x)^{r+1}$.

To prove this, we can proceed as in Ch. 8, 7.

Let
$$s = u_0 + u_1 x + u_2 x^2 + \ldots + u_n x^n \ldots ,$$
then
$$s\,(1-x) = u_0 + v_1 x + v_2 x^2 + \ldots + v_n x^n + \ldots ,$$

where $v_n = u_n - u_{n-1}$, so that v_n is a polynomial in n of degree $r - 1$. By r steps of this kind, we find that
$$s(1-x)^r = u_0 + k(x + x^2 + x^3 + \ldots),$$

where k is independent of x, and therefore
$$s(1-x)^{r+1} = u_0 + (k - u_0)x,$$

which proves the theorem,

■ **EXAMPLE 2.** *Find the sum to infinity of*
$$1^2 + 2^2 x + 3^2 x^2 + 4^2 x^3 + \ldots, \text{ when } |x| < 1.$$

This is a convergent recurring series, and the scale is $(1-x)^3$. Denoting the sum by S,

$$S = 1^2 + 2^2 x + \qquad 3^2 x^2 + \qquad 4^2 x^3 + \ldots ,$$
$$-3xS = \quad -3\cdot 1^2 x - 3\cdot \quad 2^2 x^2 - 3\cdot \quad 3^2 x^3 - \ldots ,$$
$$3x^2 S = \qquad\qquad 3\cdot \quad 1^2 x^2 + 3\cdot \quad 2^2 x^3 + \ldots ,$$
$$-x^2 S = \qquad\qquad\qquad\qquad 1^2 x^3 - \ldots ,$$

$$\therefore S\,(1-x)^3 = 1 + x \text{ and } S = \frac{1+x}{(1-x)^3}.$$

(5) *The sum to n terms* of a recurring series can be found by a process similar to that in § (3), or in the last example.

■ **EXAMPLE 3.** *Find the sum to n terms of $u_0 + u_1 x + u_2 x^2 + \ldots$, where $u_n + pu_{n-1} + qu_{n-2} = 0$ for $n > 1$.*

Let $\qquad s_n = u_0 + u_1 x + u_2 x^2 + \ldots + u_{n-1} x^{n-1}$

then $\qquad pxs_n = pu_0 x + pu_1 x^2 + \ldots + pu_{n-2} x^{n-1} + pu_{n-1} x^n,$

and $\qquad qx^2 s_n = qu_0 x^2 + \ldots + qu_{n-3} x^{n-1} + qu_{n-2} x^n + qu_{n-1} x^{n+1},$

Since $\qquad u_r + pu_{r-1} + qu_{r-2} = 0$ for $r > 1$, by addition,

$$s_n (1 + px + qx^2) = u_0 + (u_1 + pu_0) x - u_n x^n + qu_{n-1} x^{n+1},$$

which determines the value of s_n, unless x is a root α of the equation $1 + px + qx^2 = 0$. In this case the sum may be deduced from the general result by supposing that $x \to \alpha$.

(6) Given the scale $u_n + p_1 u_{n-1} + \ldots + p_r u_{n-r} = 0$, and the values of u_0, u_1, … u_{r-1}, we can find the value of u_n by (*i*) finding the generating function; (*ii*) expressing this as the sum of partial fractions of the form $A(1 - \alpha x)^{-m}$, where α is a root of $x^r + p_1 x^{r-1} + p_2 x^{r-2} + \ldots + p_r = 0$; and (*iii*) expanding the partial fractions and collecting the coefficients of x^n.

Examples are given in the next article.

■ **5. Examples of Finite Difference Equations:** Suppose that any $r + 1$ consecutive terms of the sequence (u_n) are connected by the equation

$$u_n + p_1 u_{n-1} + p_2 u_{n-2} + \ldots + p_r u_{n-r} = 0, \qquad \ldots\text{(A)}$$

where p_1, p_2, … p_r are constants. From this point of view, equation (A) is called a *Linear Finite Difference Equation with constant coefficients.*

So far as equation (A) is concerned, the first r terms of the sequence may have any values whatever, that is to say they are *arbitrary constants.*

Thus the process described in the last section enables us to find the general value of u_n which satisfies equation (A), and shows that this value involves r arbitrary constants. The general value of u_r is called the *general solution,* and the process of finding it is called *solving the equation.*

The process leads to the following results:

(*i*) *The general solution of* $u_n + pu_{n-1} + qu_{n-2} = 0$ *is* $u_n = A\alpha^n + B\beta^n$ *or* $(A + nB)\alpha^n$, *where* α, β *are the roots of* $x^2 + px + q = 0$ *and A, B are arbitrary constants, the first or second form being taken according as* $\alpha \neq \beta$ *or* $\alpha = \beta$.

(*ii*) *The general solution of* $u_n + pu_{n-1} + qu_{n-2} + ru_{n-3} = 0$ *is*

$$u_n = A\alpha^n + B\beta^n + C\gamma^n,$$

where α, β, γ *are the roots of* $x^3 + px^2 + qx + r = 0$ *and A, B, C are arbitrary constants, provided that no two of the roots are equal:*

If $\alpha = \beta \neq \gamma$, *the solution is* $u_n = (A + nB)\alpha^n + C\gamma^n$.

If $\alpha = \beta = \gamma$, *then* $u_n = (A + nB + n^2 C)\alpha^n$.

(*iii*) *For any other value of r the general solution of equation* (A) *has a similar form.*

We give the proof of (*ii*).

If u_0, u_1, u_2 have any values, then

$$u_0 + u_1 x + u_2 x^2 + \ldots + u_n x^n + \ldots = \frac{a + bx + cx^2}{1 + px + qx^2 + rx^3} = \frac{P}{Q} \text{ (say)},$$

where a, b, c are functions of u_0, u_1, u_2. Also
$$1 + px + qx^2 + rx^3 = (1 - \alpha x)(1 - \beta x)(1 - \gamma x),$$
where α, β, γ are the roots of
$$x^3 + px^2 + qx + r = 0.$$

(*i*) If no two of α, β, γ are equal,

$$\Sigma u_n x^n = \frac{P}{Q} = \frac{A}{1 - \alpha x} + \frac{B}{1 - \beta x} + \frac{C}{1 - \gamma x},$$

where $A \cdot B$, C are arbitrary constants.

Expanding the fractions on the right and equating coefficients,
$$u_n = A\alpha^n + B\beta^n + C\gamma^n.$$

(*ii*) If $\alpha = \beta \neq \gamma$, then

$$\frac{P}{Q} = \frac{A}{1 - \alpha x} + \frac{B}{(1 - ax)^2} + \frac{C}{1 - \gamma x}$$

and
$$u_n = (A + nB)\alpha^n + C\gamma^n.$$

(*iii*) If $\alpha = \beta = \gamma$, then

$$\frac{P}{Q} = \frac{A}{1 - \alpha x} + \frac{B}{(1 - \alpha x)^2} + \frac{C}{(1 - \alpha x)^3}$$

and
$$u_n = A\alpha^n + Bn\alpha^n + C \cdot \frac{1}{2}(n + 1)(n + 2)\alpha^n$$

$$= (A + nB' + n^2 C')\alpha^n,$$

where B', C' are arbitrary constants.

▌ **EXAMPLE 1.** *The first term of a sequence is 1, the second is 2, and every other term is the sum of the two preceding terms. Find the n-th term.*

Denoting the sequence by u_1, u_2, ... we have
$$u_n - u_{n-1} - u_{n-2} = 0 \text{ and } u_1 = 1, \ u_2 = 2.$$

Hence $u_n = A\alpha^n + B\beta^n$, $2 = A\alpha^2 + B\beta^2$, $1 = A\alpha + B\beta$, where α, β are the roots of $x^2 - x - 1 = 0$.

Eliminating A, B from these equations, we find that
$$(\alpha - \beta)u_n = (2 - \beta)\alpha^{n-1} - (2 - \alpha)\beta^{n-1}.$$

Now $\alpha + \beta = 1$, $\alpha\beta = -1$ and if $\alpha > \beta$, $\alpha - \beta = \sqrt{5}$, hence $2 - \beta = 1 + \alpha = \alpha^2$ (for $\alpha^2 - \alpha - 1 = 0$), and similarly $2 - \alpha = \beta^2$. Therefore

$$u_n = \frac{1}{\sqrt{5}}(\alpha^{n+1} - \beta^{n+1}) = \frac{1}{2^{n+1}\sqrt{5}}\{(1 + \sqrt{5})^{n+1} - (1 - \sqrt{5})^{n+1}\}.$$

EXAMPLE 2. *Find the nth term of a recurring series of which the first four terms are*
$$1 + 2x + 7x^2 + 20x^3.$$

Find also the sum of the first n terms when x = – 1.

Denote the series by $u_0 + u_1 x + u_2 x^2 + \dots$, and let the scale be
$$u_n + p u_{n-1} + q u_{n-2} = 0.$$

Then $7 + 2p + q = 0$ and $20 + 7p + 2q = 0$, giving $p = -2$, $q = -3$, and the scale is $u_n - 2u_{n-1} - 3u_{n-2} = 0$.

The roots of $x^2 - 2x - 3 = 0$ are $3, -1$, and therefore
$$u_n = A \cdot 3^n + B (-1)^n,$$

where $A + B = 1$, $3A - B = 2$. Hence $A = \frac{3}{4}$, $B = \frac{1}{4}$.

Thus the *n*th term $= u_{n-1} x^{n-1} = \frac{1}{4}[3^n + (-1)^{n-1}]x^{n-1}$;

$$\therefore \text{ Sum to } n \text{ terms} = \frac{3}{4} \cdot \frac{1-3^n x^n}{1-3x} + \frac{1}{4} \cdot \frac{1-(-x)^n}{1+x}.$$

Now
$$\lim_{x \to -1} \frac{1-(-x)^n}{1+x} = \lim_{y \to 1} \frac{1-y^n}{1-y} = n;$$

$\therefore$ The sum to n terms of the series $1 - 2 + 7 - 20 + \dots$

$$= \frac{3}{4} \cdot \frac{1-(-3)^n}{4} + \frac{n}{4} = \frac{1}{16}\{4n + 3 + (-3)^{n+1}\}.$$

EXAMPLE 3. *Find the general solution of* $u_n - u_{n-1} + u_{n-2} = 0$ *in a real form.*

The solution is $u_n = C\alpha^n + D\beta^n$ where α, β are the roots of $x^2 - x + 1 = 0$ and C, D are arbitrary constants. Changing the constants, this may be written
$$u_n = A(\alpha^n + \beta^n) - \iota B(\alpha^n - \beta^n).$$

The values of α, β are
$$\frac{1}{2} \pm \iota \frac{\sqrt{3}}{2} = \cos \frac{\pi}{3} \pm \iota \sin \frac{\pi}{3}.$$

Hence the values of α^n, β^n are $\cos \dfrac{n\pi}{3} \pm \iota \sin \dfrac{n\pi}{3}$, and so

$$\alpha^n + \beta^n = 2 \cos \frac{n\pi}{3}, \quad \alpha^n - \beta^n = 2\iota \sin \frac{n\pi}{3},$$

and the solution is $\quad u_n = A \cos \dfrac{n\pi}{3} + B \sin \dfrac{n\pi}{3}.$

EXERCISE XXXVII

1. Find the nth term and the sum to n terms of the recurring series:

   (i) $1 + 2 + 5 + 14 + ...$;   (ii) $1 + 2 + 5 + 12 + ...$.

2. Show that the recurring series

$$2 + \frac{7}{5} + 1 + \frac{91}{125} + ...$$

   is convergent, and that the sum to infinity is $7\frac{1}{2}$.

3. Show that, if $u_n - 2u_{n-1} + 4u_{n-2} - 3u_{n-3} = 0$, the series $\Sigma u_n x^n$ is convergent if $|x| <$
   $\frac{1}{\sqrt{3}}$; and if the first three terms are $1 + 2x + 3x^2$, the sum to infinity is

$$(1 + 2x^2) / (1 - 2x + 4x^2 - 3x^3).$$

   [The moduli of the roots of $x^3 - 2x^2 + 4x - 3 = 0$ are 1, $1/\sqrt{3}$, $1/\sqrt{3}$.]

4. If $3u_n - 7u_{n-1} + 5u_{n-2} - u_{n-3} = 0$ and $u_1 = 1$, $u_2 = 8$, $u_3 = 17$, find the values of u_n and $u_1 + u_2 + u_3 + ... + u_n$.

5. Find the nth term of the recurring series $1 + 2 + 3 + 5 + 7 + 9 + ...$.

6. If u_n is the nth term of the recurring series $1 + 2 + 3 + 8 + 13 + 30 + ...$,

   show that $u_n = \frac{1}{9}\{(3n - 4)(-1)^n + 2^{n+2}\}$, and find the sum to infinity of the series

$$u_1 + u_2 / \underline{|1} + u_3 / \underline{|2} + u_4 / \underline{|3} + ...$$

7. If $u_n - pu_{n-1} + qu_{n-2} = 0$, where $p^2 < 4q$, and $-\frac{1}{2}\pi < \theta < \frac{1}{2}\pi$, show that $u_n \sin$

$$\theta = u_2 q^{\frac{n-2}{2}} \sin(n-1)\theta - u_1 q^{\frac{n-1}{2}} \sin(n-2)\theta, \text{ where } \tan\theta = \frac{\sqrt{(4q - p^2)}}{p}.$$

8. If $\Sigma u_n x^n$, $\Sigma v_n x^n$ are recurring series of which the scales are $1 + px + qx^2$ and $1 + p'x + q'x^2$, then $\Sigma(u_n + v_n)x^n$ is a recurring series whose scale is
$$(1 + px + qx^2)(1 + p'x + q'x^2).$$

9. If Σu_n, Σv_n are recurring series of which the scales are
$$u_n + pu_{n-1} + qu_{n-2} = 0 \text{ and } v_n + p'u_{n-1} + q'u_{n-2} = 0,$$
   where $p^2 \neq 4q$ and $p'^2 \neq 4q'$, then $\Sigma u_n v_n$ is a recurring series whose scale is
$$u_n - pp'u_{n-1} + (p^2 q' + p'^2 q - 2qq')u_{n-2} - pp'qq'u_{n-3} + q^2 q'^2 u_{n-4} = 0.$$
   [If α, β, and α', β', are the roots of $x^2 + px + q = 0$ and $x^2 + p'x + q' = 0$ respectively, then $u_n v_n = A\alpha^n \alpha'^n + B\alpha^n \beta'^n + C\alpha'^n \beta^n + D\alpha'^n \beta'^n$,
   where A, B, C, D are constants. Hence the scale of $\Sigma u_n v_n x^n$ is
$$(1 - \alpha\alpha'x)(1 - \alpha\beta'x)(1 - \alpha'\beta x)(1 - \alpha'\beta'x).]$$

10. If $\Sigma u_n x^n$, $\Sigma v_n x^n$ are recurring series whose scales are $1 + px + qx^2$ and $1 - 2kx + k^2 x^2$, then $\Sigma u_n v_n x^n$ is a recurring series whose scale is
$$(1 + pkx + qk^2 x^2)^2.$$
    Hence show that $\Sigma n u_n x^n$ is a recurring series whose scale is $(1 + px + qx^2)^2$.

11. If $\Sigma u_n x^n$ is a recurring series whose scale is $1 + px + qx^2$, then $\Sigma u_n^2 x^n$ is a recurring series whose scale is $[1 - (p^2 - 2q)x + q^2 x^2]\,(1 - qx)$.

[If α, β are the roots of $x^2 + px + q = 0$, then $u_n^2 = A(\alpha^2)^n + B(\beta^2)^n + Cq^n$, where A, B, C are constants. Hence the scale of $\Sigma u_n^2 x^n$ is

$$(1 - \alpha^2 x)\,(1 - \beta^2 x)\,(1 - qx).]$$

12. If $\dfrac{1}{1 + px + qx^2} = \displaystyle\sum_{n=0}^{\infty} u_n x^n$, use Ex. 11 to show that

$$\Sigma u_n^2 x^n = \frac{1 + qx}{[1 - (p^2 - 2q)\,x + q^2 x^2]\,(1 - qx)},$$

13. Prove that each of the series

$$\cos\alpha + x\cos(\theta + \alpha) + x^2\cos(2\theta + \alpha) + \dots ,$$
$$\sin\alpha + x\sin(\theta + \alpha) + x^2\sin(2\theta + \alpha) + \dots ,$$

is a recurring series, the scale being $1 - 2x\cos\theta + x^2$ in each case. Show also that the sum to n terms of the first series is

$$\frac{\cos\alpha - x\cos(\alpha - \theta) - x^n\cos(\alpha + n\theta) + x^{n+1}\cos(\alpha + \overline{n-1}\theta)}{1 - 2x\cos\theta + x^2},$$

and that the sum to n terms of the second series may be obtained from this expression by changing cos to sin in the numerator.

■ **6. Finite Difference Equations:** Let (u_n) be a sequence in which some or all of the terms u_n, u_{n-1}, u_{n-2}, ..., u_{n-r} are connected by an equation which holds for all values of $n \ge r$, where r is a fixed number. Such an equation is called a *Finite Difference Equation.*

An important class of these equations has been considered in Art. 5. Here we give methods which apply to various cases which occur in ordinary algebra. The general theory belongs to the Calculus of Finite Differences.

To *solve* a difference equation is to express u_n as a function of n in the most general form.

The result is called the *general solution*, and from what has been said in Art. 5 it is clear that this solution must involve one or more arbitrary constants.

A *particular solution* is obtained by giving these constants special values.

If u_n is a given function of n containing one or more arbitrary constants, a difference equation may be obtained as in the following examples:

▌ **EXAMPLE 1.** *If $u_n = Cn - 2$ where C is an arbitrary constant, obtain the corresponding difference equation.*

The required equation is found by eliminating C from

$$u_n = Cn - 2 \quad\text{and}\quad u_{n-1} = C(n - 1) - 2,$$

giving
$$(n - 1)\,u_n - nu_{n-1} = 2.$$

▌ **EXAMPLE 2.** *If $u_n = A\alpha^n + B\beta^n$ where A, B are arbitrary constants and α, β are given unequal numbers neither of which is zero, prove that*

$$u_n - (\alpha + \beta)u_{n-1} + \alpha\beta u_{n-2} = 0.$$

The result is found by eliminating A, B from

$$u_n = A\alpha^n + B\beta^n, \quad u_{n-1} = A\alpha^{n-1} + B\beta^{n-1}, \quad u_{n-2} = A\alpha^{n-2} + B\beta^{n-2}.$$

■ 7. Solution of Some Elementary Types:

(1) *If $u_n = au_{n-1}$, then $u_n = Ca^n$ where C is an arbitrary constant.*

For $u_n/u_{n-1} = u_{n-1}/u_{n-2} = ... = u_2/u_1 = a$, and u_1 may have any value.

(2) *If $u_n = a_n u_{n-1}$ where a_n is a given function of n, then $u_n = Ca_1 a_2 ... a_n$ where C is an arbitrary constant.*

For
$$\frac{u_n}{u_{n-1}} \cdot \frac{u_{n-1}}{u_{n-2}} ... \frac{u_2}{u_1} = a_n a_{n-1} ... a_2,$$

therefore
$$u_n = \frac{u_1}{a_1} \cdot a_1 a_2 ... a_n,$$

and $u_1 = a_1$ may have any value we choose.

(3) *If $u_n - a_n u_{n-1} = b_n$ where a_n and b_n are given functions of n, the general solution is*

$$u_n = a_1 a_2 ... a_n \left\{ \frac{b_1}{a_1} + \frac{b_2}{a_1 a_2} + ... + \frac{b_n}{a_1 a_2 ... a_n} + C \right\},$$

where C is an arbitrary constant.

Let $v_n = a_1 a_2 ... a_n$, and divide each side of the given equation by v_n.
The result is
$$u_n/v_n - u_{n-1}/v_{n-1} = b_n/v_n.$$
In this, writing $n-1$, $n-2$, ... 2 in succession for n, we have
$$u_{n-1}/v_{n-1} - u_{n-2}/v_{n-2} = b_{n-1}/v_{n-1}.$$
$$...$$
$$u_2/v_2 - u_1/v_1 = b_2/v_2.$$

Whence by addition,
$$u_n/v_n - u_1/v_1 = b_1/v_1 + b_2/v_2 + ... + b_n/v_n - b_1/v_1.$$

This gives the result in question, for the value of $u_1/v_1 - b_1/v_1$ is arbitrary.

In examples of this type, it is better to apply the method just described rather than to use the actual result.

■ **EXAMPLE 1.** *Solve the equation $(n-1)u_n - nu_{n-1} = 2$.*

Dividing by $n(n-1)$, we have
$$u_n/n - u_{n-1}/(n-1) = 2/n(n-1).$$
Writing $n-1$, $n-2$, ... 2 in succession for n and adding,
$$\frac{u_n}{n} - \frac{u_1}{1} = 2 \left\{ \frac{1}{1.2} + \frac{1}{2.3} + ... \frac{1}{(n-1)n} \right\} = 2 \left(1 - \frac{1}{n} \right).$$

Hence $u_n = C_n - 2$ where $C = 2 + u_1$, and since u_1 is an arbitrary constant, so also is C. (See Art. 6, Ex. 1.)

■ **EXAMPLE 2.** *Find the general solution of $u_n - nu_{n-1} = \lfloor n$.*

Dividing by $\lfloor n$, the equation becomes $u_n/\lfloor n - u_{n-1}/\lfloor n-1 = 1$, and proceeding as before, we find that $u_n = \lfloor n \ (n + C)$ where C is an arbitrary constant.

(4) *The general solution of $u_n - \alpha u_{n-1} = c\beta^n$ is*

$$A\alpha^n + c\beta^{n+1}/(\beta - \alpha) \quad or \quad (A + cn)\alpha^n,$$

according as α and β are unequal or equal.

For dividing by α^n, we have

$$\frac{u_n}{\alpha^n} - \frac{u_{n-1}}{\alpha^{n-1}} = c\left(\frac{\beta}{\alpha}\right)^n.$$

Writing $n - 1, n - 2, \ldots 2$ in succession for n and adding,

$$\frac{u_n}{\alpha^n} - \frac{u_1}{\alpha} = c\left\{\frac{\beta}{\alpha} + \frac{\beta^2}{\alpha^2} + \ldots + \frac{\beta^n}{\alpha^n} - \frac{\beta}{\alpha}\right\}.$$

Therefore

$$u_n = c\alpha^n \left\{\frac{\beta}{\alpha} + \frac{\beta^2}{\alpha^2} + \ldots + \frac{\beta^n}{\alpha^n}\right\} + B\alpha^n.$$

where $B(= u_1/\alpha - c\beta/\alpha)$ is an arbitrary constant.

Hence, if $\alpha \neq \beta$, $\qquad u_n = c\beta . \dfrac{\beta^n - \alpha^n}{\beta - \alpha} + B\alpha^n,$

which may be written $\quad u_n = A\alpha^n + C\beta^{n+1}/(\beta - \alpha),$

where $A(= B - c\beta/(\beta - \alpha))$ is an arbitrary constant.

If $\alpha = \beta$, then $u_n = cn\alpha^n + B\alpha^n$.

(5) *The general solution of $u_n - (\alpha + \beta) u_{n-1} + \alpha\beta u_{n-2} = 0$ is*

$$u_n = A\alpha^n + B\beta^n \quad or \quad u_n = (A + nB)\alpha^n,$$

according as α and β are unequal or equal, where A, B are arbitrary constants.

For the equation may be written

$$u_n - \alpha u_{n-1} = \beta (u_{n-1} - \alpha u_{n-2}).$$

Hence by §(1), $\qquad u_n - \alpha u_{n-1} = C\beta^n,$

where C is an arbitrary constant, and therefore, by § (4),

$$u_n = A\alpha^n + C\beta^{n+1}/(\beta - \alpha) \quad or \quad (A + Cn)\alpha^n,$$

according as $\alpha \neq \beta$ or $\alpha = \beta$. This establishes the result in question, for $C\beta/(\beta - \alpha)$ is an arbitrary constant if $\alpha \neq \beta$. (See also Art. 5)

■ **8. Linear Difference Equations:**

(1) An equation of the form

$$au_n + bu_{n-1} + cu_{n-2} + \ldots + ku_{n-r} = l \qquad \text{...(A)}$$

is called a *linear difference equation*. Here r is a fixed number, and $a, b,\ldots k, l$ are given functions of n or constants. All the equations in Art. 7 are of this type.

With regard to equation (A), it should be noticed that

(*i*) *the general solution involves r arbitrary constants*, for $u_{r+1}, u_{r+2}, \ldots$ can be found in succession in terms of $u_1, u_2, \ldots u_r$, which may have any values.

(*ii*) *if v_n is the general solution of*

$$au_n + bu_{n-1} + cu_{n-2} + \ldots + ku_{n-r} = 0,$$

and w_n is any particular solution of equation (A), then $v_n + w_n$ is the general solution of (A).

For it is obviously a solution, also it is the general solution, because v_n involves r arbitrary constants.

(*iii*) the method of dealing with such equations as

$$au_n + bu_{n-1} + \ldots + ku_{n-r} = l_n,$$

where $a, b, \ldots k, r$ are constants and l_n is a given function of n, is illustrated in the next example.

■ **EXAMPLE** : *Solve the equation $u_n - 5u_{n-1} + 6u_{n-2} = n^2 + 5^n + 2^n$.*

The general solution of $u_n - 5u_{n-1} + 6u_{n-2} = 0$ is

$$u_n = A \cdot 2^n + B \cdot 3^n \qquad \ldots(A)$$

Next search for a particular solution of

$$u_n - 5u_{n-1} + 6u_{n-2} = n^2.$$

Assume as a trial solution $u_n = a + bn + cn^2$, then for all values of n,

$$a + bn + cn^2 - 5 \{a + b(n-1) + c(n-1)^2\} + 6 \{a + b(n-2) + c(n-2)^2\} = n^2.$$

Equating coefficients of powers of n, we find that

$$2c = 1, \quad 2b - 14c = 0, \quad 2a - 7b + 19c = 0.$$

giving $c = \frac{1}{2}$, $b = \frac{7}{2}$, $a = \frac{15}{2}$, and so a solution is

$$u_n = \tfrac{1}{2} (n^2 + 7n + 15). \qquad \ldots(B)$$

Again, take the equation

$$u_n - 5u_{n-1} + 6u_{n-2} = 5^n,$$

and assume as a trial solution $u_n = a \cdot 5^n$, then

$$a(5^n - 5 \cdot 5^{n-1} + 6 \cdot 5^{n-2}) = 5^n, \text{ giving } a = \frac{25}{6}.$$

Thus a solution is

$$u_n = \tfrac{1}{6} \cdot 5^{n+2}. \qquad \ldots(C)$$

Finally, take the equation

$$u_n - 5u_{n-1} + 6u_{n-2} = 2^n.$$

Since 2 is a root of $x^2 - 5x + 6 = 0$, it is clear that $a \cdot 2^n$ is not a solution. Trying $u_n = an \cdot 2^n$, we have

$$a\{n \cdot 2^n - 5(n-1)2^{n-1} + 6(n-2)2^{n-2}\} = 2^n,$$

giving $a = -2$. Thus a solution is

$$u_n = -n \cdot 2^{n+1}. \qquad \ldots(D)$$

The general solution of the given equation is obtained by adding the right-hand sides of equations (A)–(D), and is

$$u_n = \tfrac{1}{2}(n^2 + 7n + 15) + \tfrac{1}{6} \cdot 5^{n+2} + (A - 2n) \cdot 2^n + B' \cdot 3^n.$$

EXERCISE XXXVIII

1. The first term of a sequence is 1, and every other term is the sum of all the preceding terms. Find the nth term.

2. If $2u_n - u_{n-1} = na$, show that

$$u_n = \frac{u_1}{2^{n-1}} + a\left\{n - 1 + \frac{1}{2^{n-2}}\right\}.$$

3. The first term of a sequence is a, the second is b, and every other term is the arithmetic mean of the two preceding terms. Show that the nth term is

$$\tfrac{1}{3}(a + 2b) - \tfrac{1}{3}(a - b)(-\tfrac{1}{2})^{n-2},$$

thus proving that the nth term tends to $\tfrac{1}{3}(a + 2b)$ as a limit as $n \to \infty$.

4. The first term of a sequence is 1, the second is a, and every other term is the geometric mean of the two preceding terms.

Show that the nth term is a^x where $x = \tfrac{1}{3}\{2 + (-\tfrac{1}{2})^{n-2}\}$.

5. If $u_n = pu_{n-1} + qu_{n-2}$ and u_n tends to a limit other than zero as $n \to \infty$, show that $|q| < 1$, and the roots of $x^2 - px - q = 0$ are 1 and $-q$; also show

that
$$\lim u_n = \frac{u_2 + qu_1}{1 + q}.$$

6. Show that the equation

$$u_n - (\alpha + \beta + \gamma)u_{n-1} + (\beta\gamma + \gamma\alpha + \alpha\beta)u_{n-2} - \alpha\beta\gamma u_{n-3} = 0$$

can be written in the form

$$u_n - \alpha u_{n-1} - (\beta + \gamma)(u_{n-1} - \alpha u_{n-2}) + \beta\gamma(u_{n-2} - \alpha u_{n-3}) = 0.$$

Hence prove that

$$u_n - \alpha u_{n-1} = B\beta^n + C\gamma^n \text{ or } (B + Cn)\beta^n,$$

according as $\beta \neq \gamma$ or $\beta = \gamma$, where B, C are arbitrary constants. Deduce the general solution of the equation as given in Art. 5, (*ii*).

7. If $u_n = nu_{n-1} + (-1)^n$ and $u_0 = 1$, show that $u_n/\lfloor n \to e^{-1}$ as $n \to \infty$.

$$[u_n/\lfloor n = u_{n-1}/\lfloor n-1 + (-1)^n/\lfloor n .]$$

8. If $u_n = n(u_{n-1} + u_{n-2})$, find u_n in terms of u_0 and u_1, and show that

$$u_n/\lfloor n + 1 \to u_0 + (u_1 - 2u_0)e^{-1} \text{ as } n \to \infty.$$

[Write the equation $u_n - (n + 1)u_{n-1} = -(u_{n-1} - nu_{n-2}).$]

9. If $u_n = (n - 1)(u_{n-1} + u_{n-2})$ and $u_1 = 0$, $u_2 = 1$, prove that

$$u_n = \left\{ \frac{1}{\lfloor 2} - \frac{1}{\lfloor 3} + \frac{1}{\lfloor 4} - \ldots + \frac{(-1)^n}{\lfloor n} \right\}.$$

[Write the equation in the form

$$u_n - nu_{n-1} = -\{u_{n-1} - (n - 1)u_{n-2}\}.]$$

10. Obtain a particular solution of

$$u_n - 5u_{n-1} + 6u_{n-2} = n$$

by assuming $u_n = an + b$.

Hence find the general solution of the equation.

11. Find the general solution of $u_n - 5u_{n-1} + 6u_{n-2} = 7^n$.

[Assume $u_n = a.\, 7^n$ as a trial solution.]

12. Find the general solution of

$$u_n - u_{n-1} - u_{n-2} = n^2.$$

[Take as a trial solution $u_n = an^2 + bn + c$.]

13. Find a particular solution of

$$u_n + u_{n-1} = \sin n\alpha,$$

by assuming $u_n = a \cos n\alpha + b \sin n\alpha$.

Hence show that the general solution is

$$u_n = C(-1)^n + \frac{\sin(n+1)\alpha + \sin n\alpha}{2(1 + \cos \alpha)}.$$

14. If $u_n u_{n-1} = \dfrac{a}{n}$, find u_n in terms of u_1.

[$nu_n = (n - 1)u_{n-2}$; note that there are two forms.]

15. Show that the general solution of the equation

$$u_n u_{n-1} + au_n + bu_{n-1} + c = 0$$

can be written in the form

$$u_n = \frac{A\alpha^n + \beta^n}{A\alpha^{n-1} + \beta^{n-1}} - a,$$

where A is an arbitrary constant and α, β are the roots of

$$x^2 - (a - b)x + c - ab = 0;$$

unless $\alpha = \beta$, in which case

$$u_n = \frac{a - b}{2(A + n)} - \frac{1}{2}(a + b).$$

[Write the equation as $u_n(u_{n-1} + a) + bu_{n-1} + c = 0$. Put $u_{n-1} + a = \dfrac{v_{n-1}}{v_{n-2}}$,

so that $u_n + a = \dfrac{v_n}{v_{n-1}}$. The equation then reduces to

$$v_n + (b - a)v_{n-1} + (c - ab)v_{n-2} = 0,$$

unless $v_{n-1} = 0$.]

16. If $u_n u_{n-1} + 3u_n - 4u_{n-1} - 2 = 0$, prove that

$$u_n = \frac{A \cdot 2^n + 5^n}{A \cdot 2^{n-1} + 5^{n-1}} - 3.$$

17. If $u_n u_{n-1} + 5u_n + u_{n-1} + 9 = 0$, then

$$u_n = \frac{2}{A + n} - 3.$$

18. If $u_{n+1} = \dfrac{n(n+1)}{2n - u_n}$ and $u_1 = a$, prove that

$$u_{n+1} = (n + 1) \cdot \frac{a + n(1 - a)}{a + (n+1)(1 - a)}$$

and
$$\lim_{n \to \infty} \frac{u_1 u_2 \dots u_{n+1}}{\lfloor n} = \frac{a}{1 - a}.$$

[Put $u_n = nv_n$.]

■ ■ ■

CHAPTER 23

The Operators Δ, E, D: Interpolation

1. The Operator Δ: Let u_n be a function of the positive integral variable n, and consider the sequence $u_1, u_2, u_3, \ldots, u_n, \ldots$.

(*i*) The meaning of the symbol Δ is defined by

$$\Delta u_n = u_{n+1} - u_n,$$

and Δ is regarded as denoting the *operation* which when applied to u_n produces the difference $u_{n+1} - u_n$. Here n may have any value, so that

$$\Delta u_1 = u_2 - u_1, \ \Delta u_2 = u_3 - u_2, \text{ etc.}$$

(*ii*) *If* $\Delta u_n = v_n$, *then* $v_1 + v_2 + v_3 + \ldots + v_n = u_{n+1} - u_1$.

For $v_n = u_{n+1} - u_n$, $v_{n-1} = u_n - u_{n-1}$, $\ldots$ $v_1 = u_2 - u_1$,

whence the result follows by addition.

(*iii*) *If* u_n, v_n *are functions of* n, *then* $\Delta (u_n + v_n) = \Delta u_n + \Delta v_n$.

For
$$\Delta (u_n + v_n) = (u_{n+1} + v_{n+1}) - (u_n + v_n)$$
$$= (u_{n+1} - u_n) + (v_{n+1} - v_n) = \Delta u_n + \Delta v_n.$$

In this connection it should be noted that if

$$s_n = u_1 + u_2 + u_3 + \ldots + u_n,$$

then Δs_n is *not equal to* $\Delta u_1 + \Delta u_2 + \ldots + \Delta u_n$ unless $u_1 = 0$, for

$$\Delta s_n = (u_1 + u_2 + \ldots + u_n + u_{n+1}) - (u_1 + u_2 + \ldots + u_n) = u_{n+1},$$

and
$$\Delta u_1 + \Delta u_2 + \ldots + \Delta u_n = (u_2 - u_1) + (u_3 - u_2) + \ldots + (u_{n+1} - u_n)$$
$$= u_{n+1} - u_1.$$

(*iv*) Following the index notation of Algebra, the symbol Δ^2 denotes that the operation Δ is to be repeated, and we write

$$\Delta^2 u_n = \Delta(\Delta u_n), \ \Delta^3 u_n = \Delta(\Delta^2 u_n), \text{ and so on; thus}$$
$$\Delta^2 u_n = \Delta u_{n+1} - \Delta u_n$$
$$= (u_{n+2} - u_{n+1}) - (u_{n+1} - u_n) = u_{n+2} - 2u_{n+1} + u_n,$$
$$\Delta^3 u_n = \Delta^2 u_{n+1} - \Delta^2 u_n$$
$$= (u_{n+3} - 2u_{n+2} + u_{n+1}) - (u_{n+2} - 2u_{n+1} + u_n)$$
$$= u_{n+3} - 3u_{n+2} + 3u_{n+1} - u_n.$$

This process may be continued to any extent and, from the way in which the successive differences are formed, it is obvious that the numerical coefficients in the expressions for $\Delta^2 u_n$, $\Delta^3 u_n$, $\ldots$ are the same as those in the expansions of $(1 - x)^2$, $(1 - x)^3$, $\ldots$. Hence we conclude that

$$\Delta^r u_n = u_{n+r} - C_1^r u_{n+r-1} + C_2^r u_{n+r-2} - \ldots + (-1)^n u_n, \quad \ldots(A)$$

a result which is proved more concisely in Art. 4.

For the sequence $u_0, u_1, u_2, \ldots$ a *table of differences* is conveniently written thus:

$$u_0 \qquad u_1 \qquad u_2 \qquad u_3 \ldots$$
$$\Delta u_0 \qquad \Delta u_1 \qquad \Delta u_2 \ldots$$
$$\Delta^2 u_0 \qquad \Delta^2 u_1 \ldots$$
$$\ldots\ldots\ldots\ldots\ldots$$

where $\Delta u_0 = u_1 - u_0$, $\Delta u_1 = u_2 - u_1$, $\Delta^2 u_0 = \Delta u_1 - \Delta u_0$, etc.

Thus, for the sequence $0^2, 1^2, 2^2, 3^2, \ldots$, where $u_n = n^2$, we write

u	0	1	4	9	16...
Δu		1	3	5	7
$\Delta^2 u$			2	2	2
$\Delta^3 u$				0	0

and $\Delta 0^2 = 1$, $\Delta^2 0^2 = 2$, $\Delta^3 0^2 = 0$.

(*v*) If u_x is a function of *any* variable x, the meaning of the operation Δ applied to u_x is *defined* by $\Delta u_x = u_{x+1} - u_x$.

All that has been said about the symbol Δ still holds, for in finding the successive differences of u_x we are merely concerned with the sequence $u_x, u_{x+1}, u_{x+2} \cdots ,$

Illustrations. (*i*) If $u_n = n(n-1)(n-2)$, then

$$\Delta u_n = (n+1)\, n\, (n-1) - n\, (n-1)\, (n-2) = 3n\, (n-1),$$
$$\Delta^2 u_n = 3(n+1)n - 3n\, (n-1) = 6n, \quad \Delta^3 u_n = 6(n+1) - 6n = 6$$

and
$$\Delta^r u_n = 0 \text{ for } r > 3.$$

(*ii*) If $u_n = n^3$, since $n^3 = n(n-1)(n-2) + 3n(n-1) + n$, by the preceding, $\Delta u_n = 3n(n-1) + 6n + 1 = 3n^2 + 3n + 1$, $\Delta^2 u_n = 6n + 6$, $\Delta^3 u_n = 6$, $\Delta^4 u_n = 0$.

■ **2. Special Cases:** (*i*) The following notation is often used; we write

$$x_r = x(x-1)(x-2) \ldots (x-r+1) \text{ and } x_{-r} = \frac{1}{x(x+1)(x+2)\ldots(x+r-1)}\,.$$

If x is variable and Δ applies to x, then
$$\Delta x_r = (x+1)\, x(x-1) \ldots (x-r+2) - x(x-1)(x-2) \ldots (x-r+1)$$
$$= x\, (x-1) \ldots (x-r+2)\, \{(x+1) - (x-r+1)\} = rx\, (x-1) \ldots (x-r+2),$$

that is $\Delta x_r = r x_{r-1}.$ \hfill ...(A)

Similarly we can show that $\Delta x_{-r} = -r x_{-(r+1)},$ \hfill ...(B)

it being assumed that *no denominator vanishes*.

These formulae may be compared with

$$\frac{d}{dx} x^r = r x^{r-1} \text{ and } \frac{d}{dx} x^{-r} = -r x^{-(r+1)}.$$

■ **EXAMPLE 1.** Show *that*

$$n(n-1) + (n-1)(n-2) + \ldots + 2 \cdot 1 = \tfrac{1}{3}\, (n+1)n\, (n-1),$$

$$n(n-1)(n-2) + (n-1)(n-2)(n-3) + \ldots + 3 \cdot 2 \cdot 1 = \tfrac{1}{4}(n+1)n(n-1)(n-2),$$

and so on.

If S is the sum of the first series,

$$S = n_2 + (n-1)_2 + (n-2)_2 + \ldots:$$

and since $\Delta n_3 = 3n_2$, by Art. 1, (ii),

$$S = \frac{1}{3}(n+1)_3 + C \text{ where } C \text{ is a constant.}$$

Putting $n = 2$, we find that $C = 0$. Hence the result. Similarly for the second series. Of course, the rule in Ch. 8, 3, might have been applied.

(ii) **A Polynomial.** If u_x is a polynomial in x of degree r, dividing by x, $x - 1$, $x - 2$, ... in succession, we can find $a_0, a_1, \ldots a_r$, independent of x, such that

$$u_x = a_0 + a_1 x_1 + a_2 x_2 + \ldots + a_r x_r,$$

and then

$$\Delta u_x = a_1 + 2a_2 x_1 + 3a_3 x_2 + \ldots + ra_r x_{r-1},$$

$$\Delta^2 u_x = 2 \cdot 1 a_2 + 3 \cdot 2 a_3 x_1 + \ldots + r(r-1) a_r x_{r-2},$$

and so on. Finally

$$\Delta^r u_x = a_r \cdot \lfloor r \text{ and } \Delta^{r+1} u_x = 0.$$

■ **EXAMPLE 2.** *If* $m < r$, *then* $\Delta^m u_0 = \lfloor m \cdot a_m$.

For $\Delta^m u_x = \lfloor m \cdot a_m +$ Terms containing x as a factor.

■ **3. The Operator E:** The effect of the operation denoted by E, when applied to u_n, is to increase n by unity. Thus for all values of n,

$$E u_n = u_{n+1}.$$

If a is a constant, the following equations define the meaning of $E + a$, $a + E$, Ea and aE, regarded as operators.

$$(E + a) u_n = u_{n+1} + au_n = (a + E) u_n,$$

$$(Ea) u_n = E(au_n) = au_{n+1} = (aE) u_n.$$

Further, we write

$$E^2 u_n = E(E u_n) = E u_{n+1} = u_{n+2},$$

$$E^3 u_n = E(E^2 u_n) = E u_{n+2} = u_{n+3},$$

and so on; thus, if r is a positive integer,

$$E^r u_n = u_{n+r}.$$

Hence, if p, q are positive integers,

$$E^p E^q u_n = E^p(E^q u_n) = E^p u_{n+q} = u_{n+p+q},$$

so that

$$E^p E^q u_n = E^{p+q} u_n = E^q E^p u_n.$$

Again, if a, b are constants,

$$(E + a)(E + b) u_n = (E + a)(u_{n+1} + bu_n)$$

$$= u_{n+2} + bu_{n+1} + au_{n+1} + abu_n$$

$$= \{E^2 + (a+b)E + ab\} u_n.$$

Thus, *so far as addition (subtraction) and multiplication are concerned, the operator E combines with itself and with constants according to the laws of Algebra.*

The same is true for Δ, *for*

$$\Delta u_n = u_{n+1} - u_n = (E - 1)\, u_n.$$

More generally, if u_x is a function of *any* variable x, the operation E is defined by

$$Eu_x = u_{x+1}.$$

■ **4. Fundamental Theorems:** *If* u_x *is a function of the variable* x, *and* r *is a positive integer, then*

$$(i)\ \ \Delta^r u_x = u_{x+r} - C_1^r\, u_{x+r-1} + C_2^r\, u_{x+r-2} - \ldots + (-1)^r u_x, \qquad \ldots (A)$$

$$(ii)\ \ u_{x+r} = u_x + C_1^r\, \Delta u_x + C_2^r\, \Delta^2 u_x + \ldots + \Delta^r u_x. \qquad \ldots (B)$$

Proofs. The first theorem follows from the reasoning in Art 1, (iv), on writing x for n.

More easily thus:

(i)
$$\Delta^r u_x = (E - 1)^r u_x$$

$$= \{E^r - C_1^r\, E^{r-1} + C_2^r\, E^{r-2} - \ldots + (-1)^r\} u_x$$

$$= E^r u_x - C_1^r\, E^{r-1} u_x + C_2^r\, E^{r-2} u_x - \ldots + (-1)^r u_x$$

$$= u_{x+r} - C_1^r\, u_{x+r-1} + C_2^r\, u_{x+r-2} - \ldots + (-1)^r u_x.$$

Again, (ii) $u_{x+r} = E^r u_x = (1 + \Delta)^r u_x$, and therefore

$$u_{x+r} = (1 + C_1^r\, \Delta + C_2^r\, \Delta^2 + \ldots + \Delta^r)\, u_x$$

$$= u_x + C_1^r\, \Delta u_x + C_2^r\, \Delta^2 u_x + \ldots + \Delta^r u_x.$$

(iii) *If* u_x *is a polynomial in* x *of degree* $r - 1$, *then*

$$u_{x+r} - C_1^r\, u_{x+r-1} + C_2^r\, u_{x+r-2} - \ldots + (-1)^r u_x = 0. \qquad \ldots (C)$$

For the left-hand side $= \Delta^r u_x = 0$ [Art. 2, (ii)].

(iv) *If* s_n *is the sum to n terms of the series* $u_1 + u_2 + u_3 + \ldots$, *then*

$$s_n = nu_1 + \frac{n(n-1)}{\lfloor 2} \Delta u_1 + \frac{n(n-1)(n-2)}{\lfloor 3} \Delta^2 u_1 + \ldots + \Delta^{n-1} u_1. \qquad \ldots (D)$$

For $u_n = E^{n-1} u_1 = (1 + \Delta)^{n-1} u_1$, therefore

$$u_n = u_1 + (n-1)\, \Delta u_1 + \frac{(n-1)(n-2)}{\lfloor 2} \Delta^2 u_1 + \ldots + \Delta^{n-1} u_1.$$

Similarly,

$$u_{n-1} = u_1 + (n-2)\Delta u_1 + \frac{(n-2)(n-3)}{\lfloor 2} \Delta^2 u_1 + \ldots + \Delta^{n-2} u_1,$$

$$\ldots\ldots\ldots\ldots\ldots\ldots\ldots\ldots\ldots\ldots\ldots$$

$$u_2 = u_1 + \Delta u_1,$$

$$u_1 = u_1.$$

Adding and using the results of Art. 2, Ex. 1, the result follows at once.

EXAMPLE 1. *Find a cubic function of n which has the values 1, −3, −1, 13 when n = 1, 2, 3, 4 respectively.*

Denoting the function by u_n, we have

$$
\begin{array}{c|cccc}
n & 1 & 2 & 3 & 4 \\
\hline
u_n & 1 & -3 & -1 & 13 \\
\Delta u_n & & -4 & 2 & 14 \\
\Delta^2 u_n & & & 6 & 12 \\
\Delta^3 u_n & & & & 6
\end{array}
$$

and $\Delta^r u_n = 0$ for $r > 3$, since u_n is a cubic function.

Also $$u_n = E^{n-1} u_1 = (1 + \Delta)^{n-1} u_1;$$

$$\therefore \; u_n = u_1 + (n - 1)\, \Delta u_1 + \frac{(n-1)(n-2)}{\lfloor 2} \Delta^2 u_1 + \frac{(n-1)(n-2)(n-3)}{\lfloor 3} \Delta^3 u_1$$

$$= 1 + (-4)(n-1) + 6\,\frac{n^2 - 3n + 2}{\lfloor 2} + 6\,\frac{n^3 - 6n^2 + 11n - 6}{\lfloor 3}$$

$$= 5 - 2n - 3n^2 + n^3.$$

EXAMPLE 2. *Sum the series*
$$2 \cdot 3 + 3 \cdot 6 + 4 \cdot 11 + \ldots + (n + 1)(n^2 + 2).$$

Here $u_n = (n + 1)(n^2 + 2)$. This is a cubic function of n, so that $\Delta^r u_n = 0$ for $n > 3$. Writing down the first four terms, we find, as on the right, that $u_1 = 6$, $\Delta u_1 = 12$, $\Delta^2 u_1 = 14$, $\Delta^2 u_1 = 6$.

$$
\begin{array}{c|cccc}
u_n & 6 & 18 & 44 & 90 \\
\Delta u_n & & 12 & 26 & 46 \\
\Delta^2 u_n & & & 14 & 20 \\
\Delta^3 u_n & & & & 6
\end{array}
$$

Hence by (*iv*),

$$s_n = 6n + 12 \cdot \frac{n(n-1)}{\lfloor 2} + 14\,\frac{n(n-1)(n-2)}{\lfloor 3} + 6\,\frac{n(n-1)(n-2)(n-3)}{\lfloor 4}$$

$$= \frac{1}{12} n (3n^3 + 10n^2 + 21n + 38). \quad (\text{Cf. Ch. 8, 4, Ex. 1.})$$

EXAMPLE 3. *If $S = x^n - C_1^r (x + 1)^n + C_2^r (x + 2)^n - \ldots + (-1)^r (x + r)^n$ where r, n are positive integers, then S = 0 if r > n and S = $(-1)^n \lfloor n$ if r = n.*

For if Δ applies to x, $S = (1 - E)^r x^n = (-1)^r \Delta^r x^n$, and the result follows by Art. 2, (*ii*).

EXAMPLE 4.* (*Bernoulli's Numbers.*) *If B_r is defined as in Ch. 8, 8, (5), then*

$$(-1)^r B_r = \left\{ 1 - \frac{\Delta}{2} + \frac{\Delta^2}{3} - \ldots + (-1)^r \frac{\Delta r}{r+1} \right\} 0^r . \qquad \ldots(A)$$

For $n^r = E^n 0^r = (1 + \Delta)^n 0^r$ and $\Delta^m 0^r = 0$ if $m > r$;

$$\therefore \qquad n^r = \left\{ n\Delta + \frac{n(n-1)}{\lfloor 2} \Delta^2 + \ldots + \frac{n(n-1)\ldots(n-r+1)}{\lfloor r} \Delta^r \right\} 0^r. \qquad \ldots(B)$$

Hence *if $S_r = 1^r + 2^r + \ldots + n^r$*, we have

$$Sr = \left\{ \frac{(n+1)^n}{\lfloor 2} \Delta + \frac{(n+1)n(n-1)}{\lfloor 3} \Delta^2 + \ldots + \frac{(n+1)n(n-1)\ldots(n-r+1)}{\lfloor r+1} \Delta^r \right\} 0^r; \qquad \ldots(C)$$

Now, by definition, B_r = Coefficient of n in S_r, hence

$$B_r = \left\{ \frac{\Delta}{\lfloor 2} - \frac{\lfloor 1 \Delta^2}{\lfloor 3} + \frac{\lfloor 2 \Delta^3}{\lfloor 4} - \ldots + (-1)^{r-1} \frac{\lfloor r-1}{\lfloor r+1} \Delta^r \right\} 0^r ;$$

$$\therefore \qquad B_r = \left\{ \Delta \left(1 - \frac{1}{2} \right) - \Delta^2 \left(\frac{1}{2} - \frac{1}{3} \right) + \ldots + (-1)^{r-1} \left(\frac{1}{r} - \frac{1}{r+1} \right) \Delta^r \right\} 0^r. \qquad \ldots(D)$$

Again, equating the coefficients of n in (B),

$$\left\{ \Delta - \frac{\Delta^2}{2} + \frac{\Delta^3}{3} - \ldots + (-1)^{r-1} \frac{\Delta r}{r} \right\} 0^r = 0 \ \text{if } r > 1. \qquad \ldots(E)$$

Hence if $r > 1$,

$$B_r = \left\{ 1 - \frac{\Delta}{2} + \frac{\Delta^2}{3} - \ldots + (-1)^r \frac{\Delta^r}{r+1} \right\} 0^r. \qquad \ldots(F)$$

Moreover, $\Delta 0^r = 1$, $B_1 = \dfrac{1}{2}$ and $(-1)^r B_r = B_r$ for $r > 1$ [Ch. 8, 8, (5)].

Hence equation (A) holds for $r > 0$.

EXAMPLE 5. *Apply equation (A) of Ex. 4 to show that $B_2 = \dfrac{1}{6}$.*

$$B_2 = \left(1 - \frac{\Delta}{2} + \frac{\Delta^2}{3} \right) 0^2 = 0 - \frac{1}{2} + \frac{2}{3} = \frac{1}{6}.$$

■ **5. Operators in a Fractional Form:** Many writers give the following 'proof' of the theorem in Art. 4, (*iv*):

$$s_n = u_1 + u_2 + \ldots + u_n$$
$$= (1 + E + E^2 + \ldots + E^{n-1}) u_1$$

---

*   This result was obtained by Cayley by a much more difficult process (Vol. IX, pp. 259-262 of his collected works).

$$= \frac{E^n - 1}{E - 1} \, u_1 = \frac{(1 + \Delta)^n - 1}{\Delta} \, u_1$$

$$= \left(n + \frac{n(n-1)}{\lfloor 2} \Delta + \frac{n(n-1)(n-2)}{\lfloor 3} \Delta^2 + ... + \Delta^{n-1} \right) u_1$$

$$= nu_1 + \frac{n(n-1)}{\lfloor 2} \Delta u_1 + \frac{n(n-1)(n-2)}{\lfloor 3} \Delta^2 u_1 + ... + \Delta^{n-1} u_1.$$

Here the operator is written in a fractional form in order to transform a polynomial in E into a polynomial in Δ.

The result, when obtained, can be verified by multiplication and addition, this fact justifies the process.

But at every stage *the operator must be regarded as a whole.* Thus the equation

$$s_n = \frac{E^n - 1}{E - 1} \, u_1$$

does not imply that $(E{-}1)s_n = (E^n - 1)u_1.$

In fact the last statement is not true, for

$$(E - 1) \, s_n = s_{n+1} - s_n = u_{n+1} \text{ and } (E^n - 1)u_1 = u_{n+1} - u_1.$$

We conclude that, in such transformations, *the operator can be put in the form $f(\Delta) / \phi(\Delta)$, where $f(\Delta)$ and $\phi(\Delta)$ are polynomials in Δ, provided that $f(\Delta)$ is divisible by $\phi(\Delta)$, and $f(\Delta)$, $\phi(\Delta)$ are not regarded as separate operators.*

■ **6. Interpolation:** Suppose that y is a function of x whose value has been determined by experiment, or otherwise, for $x = a, b, c,...$.

The problem of interpolation is to find approximate values of y for intermediate values of x.

The graphical method is to plot the points (x, y) for $x = a, b, c, ...,$ and draw a 'smooth' curve through the points. We assume that this curve represents as nearly as possible the graph of the function y.

Let us suppose that the equation to the curve so drawn is $y = f(x)$. It is natural to take for $f(x)$ the simplest function of x which satisfies the conditions. If y is known for n values of x, we assume that

$$y = p_0 + p_1 x + p_2 x^2 + ... + p_{n-1} x^{n-1},$$

where the n constants $p_0, p_1, ...$ are to be determined by making the curve go through the n points.

If only two values of y are known, the curve is the straight line joining the corresponding points, and we have the rule of *proportional parts* as used in working with tables of logarithms and trigonometric functions.

■ **7. Lagrange's Formula:** If $y = y_1, y_2, y_3$ for $x = a, b, c$ respectively, we assume that

$$y = y_1 \frac{(x-b)(x-c)}{(a-b)(a-c)} + y_2 \frac{(x-a)(x-c)}{(b-a)(b-c)} + y_3 \frac{(x-a)(x-b)}{(c-a)(c-b)}$$

For y is a quadratic function of x of which the values are y_1, y_2, y_3 when $x = a, b, c.$

Similarly, if $y = y_1, y_2, y_3, y_4$ when $x = a, b, c, d$ we take

$$y = y_1 \frac{(x-b)(x-c)(x-d)}{(a-b)(a-c)(a-d)} + \text{Three similar terms;}$$

with similar formulae when more than four values of y are given.

■ **8.** If the values of y are known for values of x *at equal intervals*, instead of applying Lagrange's formulae, we use the operators E and Δ.

For example, suppose that $y = y_0, y_1, y_2, y_3$ for $x = x_0, x_1, x_2, x_3$. Assume that y is a cubic function of x; then $\Delta^r y_0 = 0$ for $r \geq 4$, and if n is a positive integer,

$$y_n = E^n y_0 = (1 + \Delta)^n y_0,$$

therefore

$$y_n = y_0 + n\Delta y_0 + \frac{n(n-1)}{\underline{|2}}\Delta^2 y_0 + \frac{n(n-1)(n-2)}{\underline{|3}}\Delta^3 y_0.$$

Hence the equation

$$y = y_0 + x\Delta y_0 + \frac{x(x-1)}{\underline{|2}}\Delta^2 y_0 + \frac{x(x-1)(x-2)}{\underline{|3}}\Delta^3 y_0$$

represents a cubic curve which passes through the four points (x_0, y_0) (x_1, y_1), ... and gives an approximate value of y corresponding to any value of x within the specified range.

■ **EXAMPLE 1.** *Given that*
$$sin\ 45° = 0.7071,\ sin\ 50° = 0.7660,\ sin\ 55° = 0.8192,\ sin\ 60° = 0.8660,$$
find sin 52°.

Let $u_x = 10^4 \sin(45 + 5x)°$. Construct a table of differences as below.

	u	Δu	$\Delta^2 u$	$\Delta^3 u$
$u_0 = 7071$				
		589		
$u_1 = 7660$			-57	
		532		-7
$u_2 = 8192$			-64	
		468		
$u_3 = 8660$				

$$\therefore\ \Delta u_0 = 589,\quad \Delta^2 u_0 = -57,\quad \Delta^3 u_0 = -7.$$

Assuming that $\Delta^r u_0 = 0$ for $r > 3$, approximate values of u_x are given by

$$u_x = u_0 + x\Delta u_0 + \frac{x(x-1)}{\underline{|2}}\Delta^2 u_0 + \frac{x(x-1)(x-2)}{\underline{|3}}\Delta^3 u_0.$$

Let $52 = 45 + 5x$, so that $x = 1.4$ and

$$u_0 = 7071$$
$$x\Delta u_0 = 1.4 \times 589 = 824.6$$

$$\frac{x\,(x-1)}{\underline{|2}}\,\Delta^2 u_0 = -0.7 \times 0.4 \times 57 = -15.96$$

$$\frac{x\,(x-1)\,(x-2)}{\underline{|3}}\,\Delta^3 u_0 = 1.4 \times 0.4 \times 0.7 = \frac{0.392}{7895.992}$$

$$u_x = \frac{15.96}{7880.032}$$

∴ $\quad\quad\quad \sin 52° = 0.7880$ to four places.

The correct seven-place value in 0.7880108.

If we only use the values of sin 50 and sin 55, the rule of proportional parts would give

$$\sin 52° = 0.7660 + \tfrac{2}{5} \times 0.0532 = 0.7873.$$

If u_0 and several other terms of the sequence u_0, u_1, u_2, ... are known and approximate values of the missing terms are required, we may proceed as follows.

■ **EXAMPLE 2.** *If $u_0 = 4.3315$, $u_1 = 4.7046$, $u_3 = 5.6713$, $u_5 = 7.1154$, find approximate values of u_2 and u_4.*

Four points on the graph of u_x are known. We therefore assume that it can be represented by a cubic function of x. This is the same as assuming that $\Delta^4 u_x = 0$. Hence we have approximately

$$\Delta^4 u_0 = u_4 - 4u_3 + 6u_2 - 4u_1 + u_0 = 0,$$
$$\Delta^4 u_1 = u_5 - 4u_4 + 6u_3 - 4u_2 + u_1 = 0;$$

∴ $\quad u_4 + 6u_2 = 4\,(u_1 + u_3) - u_0 = 37.1721,$

$$u_4 + u_2 = \frac{1}{4}\,(u_5 + 6u_3 + u_1) = 11.4619.$$

∴ $\quad\quad u_2 = 5.1420,\ u_4 = 6.3199$ (approx.).

Note: In this example $u_0 = \tan 77°$, $u_1 = \tan 78°$, $u_3 = \tan 80°$, $u_5 = \tan 82°$, so we have found that $\tan 79° = 5.1420$, $\tan 81° = 6.3199$ (approx.).

The correct values are 5.1446 and 6.3138.

It should be noticed that the approximations just found are much closer than those given by

$$u_2 = \tfrac{1}{2}\,(u_1 + u_3) = 5.1879,\ u_4 = \tfrac{1}{2}\,(u_3 + u_5) = 6.3933.$$

■ **9. Bessel's Formula:** In using Mathematical Tables, the effect of second differences may be easily and accurately allowed for by using a formula due to Bessel.

Let u_{-1}, u_0, u_1, u_2 be four consecutive values of a function u_x, as given in the tables, and suppose that values of u_x are required for values of x between 0 and 1.

Denoting first and second differences by a, b, c and d, e, as shown on the right, Bessel's formula is

$$u_{-1}$$
$$a$$
$$u_0 \qquad d$$
$$b \qquad f$$
$$u_1 \qquad e$$
$$c$$
$$u_2$$

$$u_x = u_0 + xb + \frac{x(x-1)}{2} \cdot \frac{d+e}{2}, \qquad \ldots(A)$$

which is the same as

$$u_x = u_0 + xb + \frac{x(x-1)}{4}(c-a). \qquad \ldots(B)$$

The values of u_0, a, b, c can be taken from the tables, so that very little labour is required in using the formula.

To show that this gives a good approximation, and to estimate the error, if we neglect differences of the fourth and higher orders, we have

$$u_x = (1+\Delta)^{x+1} u_{-1} = u_{-1} + (x+1)a + \frac{(x+1)x}{2}d + \frac{(x+1)x(x-1)}{6}f.$$

Now $u_0 - u_{-1} = a$, $b - a = d$, $e - d = f$; and therefore

$$u_x = u_0 + xb + \frac{x(x-1)}{2}d + \frac{(x+1)x(x-1)}{6}f$$

$$= u_0 + xb + \frac{x(x-1)}{2}\left(\frac{d+e}{2} - \frac{e-d}{2}\right) + \frac{(x+1)x(x-1)}{6}f,$$

so that if v_x is the approximation given by (A),

$$u_x = v_x - \frac{x(x-1)}{2} \cdot \frac{f}{2} + \frac{(x+1)x(x-1)}{6}f$$

$$= v_x + \frac{x(x-1)(2x-1)}{12}f.$$

Hence the error is approximately $\frac{1}{12}x(x-1)(2x-1)f$.

■ **EXAMPLE :** *Use tables of square roots to find $\sqrt{12344}$.*

Using Barlow's tables, we take

$$u_{-1} = \sqrt{12330} = \ldots\ldots\ldots$$

$$.045020 = a$$

$$u_0 = \sqrt{12340} = 111.085553$$

$$.045001 = b$$

$$u_1 = \sqrt{12350} = \ldots\ldots\ldots$$

$$.044983 = c.$$

$$u_2 = \sqrt{12360} = \ldots\ldots\ldots .$$

(The differences a, b, c are given in the tables.) For $\sqrt{12344}$, we have

$$x = .4, \quad \frac{1}{4}x(x-1) = -.06, \quad c - a = -.000037.$$

$$u_0 = 111.085553$$
$$xb = .0180004$$
$$\frac{1}{4}x(x-1)(c-a) = \underline{\quad.00000224\quad}$$
$$\sqrt{12344} = 111.103556$$

The result is correct to the last figure.

■ **10. The Operator D:** When we write

$$\frac{du}{dx} = \frac{d}{dx}u, \frac{d^n u}{dx^n} = \left(\frac{d}{dx}\right)^n u,$$

the symbol $\dfrac{d}{dx}$ is regarded as an *operator*, which is sometimes denoted by D. Thus

$$D^m D^n u = D^{m+n} u = D^n D^m u.$$

If a is a constant, the following equations define the meaning of $D + a, a + D, Da$ and aD, regarded as operators.

$$(D + a)u = Du + au = (a + D)u, \quad (Da)u = D(au) = aDu = (aD)u.$$

Hence if a, b are constants, $\quad (D + a)(D + b)u = (D + a)(Du + bu)$,

and $(D + a)(D + b)u = D^2 u + bDu + aDu + abu = \{D^2 + (a + b)D + ab\}u.$

Thus, so far as addition, subtraction and multiplication are concerned, the operator D combines with itself and with *constants* according to the laws of Algebra.

■ **11. The n-th Derivative of a Product:** If u, v are functions of x,

$$\frac{d}{dx}(uv) = \frac{du}{dx}v + u\frac{dv}{dx}.$$

Let the symbols D_1, D_2 denote differentiation with regard to x, and suppose that D_1 only operates on u and its differential coefficients, and that D_2 operates only on v and its differential coefficients, so that

$$D_1(uv) = \frac{du}{dx}v, \quad D_2(uv) = u\frac{dv}{dx}; \text{ and hence } \frac{d}{dx}(uv) = (D_1 + D_2)uv.$$

As D_1 and D_2 are independent of each other, they obey the laws of addition and multiplication, therefore

$$\left(\frac{d}{dx}\right)^n (uv) = (D_1 + D_2)^n (uv) = (D_1{}^n + C_1{}^n D_1{}^{n-1} D_2 + C_2{}^n D_1{}^{n-2} D_2{}^2 + \dots + D_2{}^n) (uv);$$

that is, $\dfrac{d^n}{dx^n}(uv) = \dfrac{d^n u}{dx^n} v + C_1^n \dfrac{d^{n-1}u}{dx^{n-1}} v + C_2^n \dfrac{d^{n-2}u}{dx^{n-2}} \dfrac{d^2 v}{dx^2} + \dots + u \dfrac{d^n v}{d^n x}.$

This result is known as *Leibniz' Theorem*, and is often written in the form

$$(uv)_n = u_n v + C_1^n u_{n-1}v_1 + C_2^n u_{n-2}v_2 + \dots + uv_n,$$

where the suffixes indicate differentiation with regard to x.

■ **EXAMPLE :** *If $y = u + v$ where $u^m = x + (x^2 - k)^{\frac{1}{2}}$, $v^m = x - (x^2 - k)^{\frac{1}{2}}$, then*

$$(x^2 - k) y_{n+2} + (2n + 1) xy_{n+1} + \left(n^2 - \frac{1}{m^2}\right) y_n = 0, \text{ for } n = 0, 1, 2, \dots .$$

Differentiating with regard to x, we have

$$mu^{m-1}u_1 = 1 + x (x^2 - k)^{-\frac{1}{2}} = u^m (x^2 - k)^{-\frac{1}{2}}; \text{ hence } (x^2 - k)^{\frac{1}{2}} u_1 = u/m.$$

Differentiating again,

$$(x^2 - k)^{\frac{1}{2}} u_2 + x (x^2 - k)^{-\frac{1}{2}} u_1 = u_1/m = (x^2 - k)^{-\frac{1}{2}} \cdot u/m^2;$$

$$\therefore \qquad (x^2 - k) u_2 + xu_1 - u/m^2 = 0.$$

Similarly, $\qquad\qquad (x^2 - k) v_2 + xv_1 - v/m^2 = 0,$

whence by addition, $\qquad (x^2 - k) y_2 + xy_1 - y/m^2 = 0.$

Differentiating n times by Leibniz' theorem,

$$(x^2 - k) y_{n+2} + n \cdot 2x \cdot y_{n+1} + \frac{n(n-1)}{2} \cdot 2y_n$$

$$+ \qquad xy_{n+1} + \qquad\qquad ny_n$$

$$-y_n/m^2 = 0;$$

$$\therefore \quad (x^2 - k) y_{n+2} + (2n + 1) xy_{n+1} + \left(n^2 - \frac{1}{m^2}\right) y_n = 0.$$

This example leads to a remarkable theorem on cubic equations, given in Ch. 29.

EXERCISE XXXIX

1. Find the polynomial of the third degree in n which has the values 2, 11, 32, 71, when $n = 1, 2, 3, 4$ respectively. (Use Art. 4)

2. Use the method of Art. 4, Ex. 2 to find the sum to n terms of the series
    (*i*) $1^3 + 3^3 + 5^3 + \dots$;
    (*ii*) $1^2 \cdot 2^2 + 2^2 \cdot 3^2 + 3^2 \cdot 4^2 + \dots$.

3. If $u_n = x^n$ and Δ applies to n, show that
    (*i*) $\Delta^r u_n = x^n (x - 1)^r$;

$$(ii)\ \ 1 + x + x^2 + \dots + x^{n-1} = n + \frac{n(n-1)}{\underline{|2}}(x-1) + \frac{n(n-1)(n-2)}{\underline{|3}}(x-1)^2 + \dots +$$

$$(x-1)^{n-1}.$$

4. Show that

$$(i)\ \ x^n - C_1^n(x-y)^n + C_2^n(x-2y)^n - \dots + (-1)^n(x-ny)^n = y^n\,\underline{|n}\,.$$

$$(ii)\ \ xy - C_1^n(x-1)(y-1) + C_2^n(x-2)(y-2) - \dots + (-1)^n(x-n)(y-n)$$

$$= x + y - 1,\ 2\ \text{or}\ 0,\ \text{according as}\ n = 1,\ n = 2\ \text{or}\ n > 2.$$

$$(iii)\ \ (n-1)^2\,C_1^n + (n-3)^2\,C_3^n + (n-5)^2\,C_5^n + \dots = 2^{n-3}\,n\,(n+1),\ \text{if}\ n > 2.$$

[In each case let. S be the sum of the series.

(i) Let $u_r = (x - ry)^n$, then $S = (1 - E)^n u_0 = (-1)^n \Delta^n u_0$.

Also $\Delta^n u_r = (-y)^n\,\underline{|n}\,$, etc.

(ii) Let $u_r = (x - r)(y - r)$, then $S = (-1)^n\,\Delta^n u_0$. Also

$\Delta u_0 = (x-1)(y-1) - xy = -x - y + 1,\ \Delta^2 u_0 = 2,\ \Delta^n u_r = 0$ if $n > 2$.

(iii) Let $u_r = r^2$, then $S = \frac{1}{2}\{(E+1)^n - (E-1)^n\}u_0$;

$\therefore\ S = \frac{1}{2}\{(2+\Delta)^n - \Delta^n\}\,u_0$. Also $u_0 = 0,\ \Delta\,u_0 = 1,\ \Delta^2 u_0 = 2$, and $\Delta^s u_r = 0$ if $s > 2$.]

5. If $x_r = x(x-1)(x-2)\dots(x-r+1)$, show that

$$(x+n)_r = x_r + C_1^n\,r x_{r-1} + C_2^n\,r\,(r-1)\,x_{r-2} + \dots,$$

the last term being $r\,(r-1)\dots(r-n+1)x_{r-n}$ if $r > n$ and $C_r^n\,\underline{|r}$ if $r \le n$.

Verify that, if $n = 3$ and $r = 2$, the identity becomes

$$(x+3)(x+2) = x(x-1) + 6x + 6.$$

6. If $x^r = x_r + ax_{r-1} + bx_{r-2} + \dots$, prove that

$$a = \tfrac{1}{2}r\,(r-1),\ \ b = \tfrac{1}{24}r\,(r-1)\,(r-2)\,(3r-5).$$

[Equate to 0 the coefficients of x^{r-1} and x^{r-2} on the right.]

7. Show that

$$(i)\ \ \Delta^r x^{r+1} = (x + \tfrac{1}{2}r)\,\underline{|r+1}\,;$$

$$(ii)\ \ \Delta^r x^{r+2} = \tfrac{1}{2}[x^2 + rx + \tfrac{1}{12}r\,(3r+1)]\,.\,\underline{|r+2}\,.$$

[Use Ex. 6.]

8. Prove that

$$x^n - C_1^r(x+1)^n + C_2^r(x+2)^n - \dots + (-1)^r(x+r)^n = (-1)^r(x + \tfrac{1}{2}r)\,\underline{|r+1}$$

when

$$n = r + 1,$$

and

$$= (-1)^r\tfrac{1}{2}\left\{x^2 + rx + \tfrac{1}{12}r\,(3r+1)\right\}\underline{|r+2}$$

when $n = r + 2$. [Use Ex. 7.]

9. If none of the denominators vanish, prove that

$$(i) \quad \Delta^n \frac{1}{x} = \frac{(-1)^n \lfloor n}{x(x+1)(x+2)\dots(x+n)}.$$

Deduce that

$$(ii) \quad \frac{\lfloor n}{x(x+1)(x+2)\dots(x+n)} = \frac{1}{x} - C_1^n \frac{1}{x+1} + C_2^n \frac{1}{x+2} - \dots + (-1)^n \frac{1}{x+n};$$

$$(iii) \quad \frac{1}{x+1} + \frac{1}{x+2} + \frac{1}{x+3} + \dots + \frac{1}{x+n} = n \frac{1}{x+1} - \frac{n(n-1)}{2} \frac{1}{(x+1)(x+2)}$$

$$+ \frac{n(n-1)(n-2)}{3} \frac{1}{(x+1)(x+2)(x+3)} - \dots.$$

10. Prove that

$$\frac{1}{m} + n \frac{2}{m+1} + \frac{n(n-1)}{1\cdot2} \frac{2^2}{m+2} + \dots + \frac{2^n}{m+n}$$

$$= \frac{3^n}{m} - \frac{n\cdot2\cdot3^{n-1}}{m(m+1)} + \frac{n(n-1)\cdot2^2\cdot3^{n-2}}{m(m+1)(m+2)} - \dots + (-1)^n \frac{n(n-1)\dots1\cdot2^n}{m(m+1)\dots(m+n)}.$$

11. If $u_n = 1/(an + b)$, show that

$$(i) \quad \Delta^r u_n = (-1)^r \lfloor r \frac{a^r}{(an+b)(a\cdot\overline{n+1}+b)\dots(a\cdot\overline{n+r}+b)};$$

$$(ii) \quad \frac{1}{b} + \frac{1}{a+b} + \frac{1}{2a+b} + \dots + \frac{1}{(n-1)a+b}$$

$$= \frac{n}{b} - \frac{n(n-1)}{b(a+b)}\cdot\frac{a}{2} + \frac{n(n-1)(n-2)}{b(a+b)(2a+b)}\cdot\frac{a^2}{3} - \dots.$$

12. If $x_{-r} = 1/x(x+1)(x+2)\dots(x+r-1)$, show that

$$(x+n)_{-r} = x_{-r} - C_1^n\, r x_{-(r+1)} + C_2^n\, r(r+1)\, x_{-(r+2)} - \dots \text{ to } n+1 \text{ terms.}$$

13. If n is a positive integer and y is any number such that no denominator vanishes, prove that

$$(i) \quad 1 + \frac{n}{y} + \frac{n(n-1)}{y(y-1)} + \frac{n(n-1)(n-2)}{y(y-1)(y-2)} + \dots = \frac{y+1}{y-n+1};$$

$$(ii) \quad 1 + 2\frac{n}{y} + 3\frac{n(n-1)}{y(y-1)} + 4\frac{n(n-1)(n-2)}{y(y-1)(y-2)} + \dots = \frac{(y+1)(y+2)}{(y-n+1)(y-n+2)};$$

$$(iii) \quad 1 + 3\frac{n}{y} + 6\frac{n(n-1)}{y(y-1)} + 10\frac{n(n-1)(n-2)}{y(y-1)(y-2)} + \dots$$

$$= \frac{(y+1)\,(y+2)\,(y+3)}{(y-n+1)\,(y-n+2)\,(y-n+3)};$$

each series being continued to $n+1$ terms.

[These results follow from Ex. 12 by putting successively $r = 1$, $x = -(y+1)$; $r = 2$, $x = -(y+2)$; $r = 3$, $x = -(y+3)$.]

14. Show that

$$(i) \quad \frac{1}{x(x+1)} - C_1^n \frac{1}{(x+1)(x+2)} + C_2^n \frac{1}{(x+2)(x+3)} - \dots$$

$$+ (-1)^n \frac{1}{(x+n)(x+n+1)} = \frac{\lfloor n+1}{x\,(x+1)\,(x+2)\dots(x+n+1)}$$

$$(ii) \quad \frac{1}{x\,(x+1)\,(x+2)} - C_1^n \frac{1}{(x+1)(x+2)(x+3)} + C_2^n \frac{1}{(x+2)(x+3)(x+4)}$$

$$- \dots \text{ to } n+1 \text{ terms} = \frac{\lfloor n+2}{x(x+1)(x+2)\dots(x+n+2)}.$$

15. If $u_n = \dfrac{a(a-1)\,(a-2)\dots(a-n+1)}{b\,(b-1)\,(b-2)\dots(b-n+1)}$, prove that

$$\Delta u_n = \frac{a-b}{b} \cdot \frac{a\,(a-1)\dots(a-n+1)}{(b-1)\,(b-2)\dots(b-n)},$$

$$\Delta^2 u_n = \frac{a-b}{b} \cdot \frac{a-b+1}{b-1} \cdot \frac{a\,(a-1)\dots(a-n+1)}{(b-2)\,(b-3)\dots(b-n-1)},$$

$$\Delta^r u_n = \frac{a-b}{b} \cdot \frac{a-b+1}{b-1} \dots \frac{a-b+r-1}{b-r+1} \cdot \frac{a\,(a-1)\dots(a-n+1)}{(b-r)\,(b-r-1)\dots(b-r-n+1)}.$$

16. If $u_0 = 1$, $u_1 = \dfrac{a}{b}$, $u_2 = \dfrac{a(a-1)}{b(b-1)}$, etc., as in Ex. 15, prove that

$$\Delta^r u_0 = \frac{a-b}{b} \cdot \frac{a-b+1}{b-1} \cdot \frac{a-b+2}{b-2} \dots \frac{a-b+r-1}{b-r+1}.$$

[Putting $n = 1$ in Ex. 15, find Δu_0, $\Delta^2 u_0$, etc., in succession from the equations
$$\Delta u_0 = u_1 - u_0, \quad \Delta^2 u_0 = \Delta u_1 - \Delta u_0, \quad \Delta^3 u_0 = \Delta^2 u_1 - \Delta^2 u_0, \text{ etc.]}$$

17. Prove that

$$\left(1 - \frac{a}{b}\right)\left(1 - \frac{a}{b-1}\right)\left(1 - \frac{a}{b-2}\right)\dots\left(1 - \frac{a}{b-n+1}\right)$$

$$= 1 - n\frac{a}{b} + \frac{n\,(n-1)}{\lfloor 2} \frac{a\,(a-1)}{b\,(b-1)} - \frac{n(n-1)\,(n-2)}{\lfloor 3} \frac{a(a-1)\,(a-2)}{b\,(b-1)\,(b-2)} + \dots$$

to $n+1$ terms.

[Using Ex. 16, the left-hand side $= (-1)^n \Delta^n u_0 = (1 - E)^n u_0$.]

18. Show that the sum of the products two together of

$$\frac{1}{x}, \frac{1}{x+1}, \frac{1}{x+2}, ..., \frac{1}{x+n-1},$$

is equal to

$$\frac{n\,(n-1)}{2x\,(x+1)}\left\{1-\frac{n-2}{\underline{|1}}\,\frac{1}{x+2}+\frac{(n-2)\,(n-3)}{\underline{|2}}\cdot\frac{1}{(x+2)\,(x+3)}-...\,to\;n-1\;terms\right\}.$$

[Equate the coefficients of a^2 in Ex. 17 and put $x = -b$.]

19. In Art. 4, Ex. 4, it has been shown that

$$(-1)^r\,B_r = \left(1-\frac{\Delta}{2}+\frac{\Delta^2}{3}-...+(-1)^r\,\frac{\Delta^r}{r+1}\right)0^r.$$

Prove also that

$$B_r = \left(1-\frac{\Delta}{2}+\frac{\Delta^2}{3}-...+(-1)^r\,\frac{\Delta^r}{r+1}\right)1^r.$$

and use each of these formulae to show that $B_4 = -\dfrac{1}{30}$.

[Equation (D) of Ex. 4 can be written

$$B_r = \left(1-\frac{\Delta}{2}+\frac{\Delta^2}{3}-...\right)(1+\Delta)\,0^r \quad\text{and}\quad (1+\Delta)0^r = E0^r = 1^r.]$$

20. Given that

$$\sqrt{12500} = 111.803399,$$
$$\sqrt{12510} = 111.848111,$$
$$\sqrt{12520} = 111.892806,$$
$$\sqrt{12530} = 111.937483,$$

show that

$$\sqrt{12516} = 111.874929.$$

21. Given that

$$\sec 89° \; 4' = 61.3911,$$
$$\sec 89° \; 5' = 62.5072,$$
$$\sec 89° \; 6' = 63.6646,$$
$$\sec 89° \; 7' = 64.8657,$$

show that Bessel's formula gives

$$\sec 89° \; 5' \; 40'' = 63.2741.$$

22. If $S_n = 1^n + \dfrac{2^n}{\underline{|1}} + \dfrac{3^n}{\underline{|2}} + ... + \dfrac{(m+1)^n}{\underline{|m}} + ...$ to ∞, prove that

(i) $\Delta^r S_n = \dfrac{2^n \cdot 1^r}{\underline{|1}} + \dfrac{3^n \cdot 2^r}{\underline{|2}} + \dfrac{4^n \cdot 3^r}{\underline{|3}} + ...$.

(ii) $\Delta^r S_0 = 1^{r-1} + \dfrac{2^{r-1}}{\lfloor 1} + \dfrac{3^{r-1}}{\lfloor 2} + \dots = S_{r-1}.$

(iii) $S_n = S_{n-1} + C_1^n S_{n-2} + C_2^n S_{n-3} + \dots + C_{n-1}^n S_0 + S_0.$

(iv) $S_n = S_{n-1} + C_1^n S_{n-1} - C_2^n S_{n-2} - \dots + (-1)^{n-1} S_0.$

(v) Use (iii) or (iv) to find S_1, S_2, $\dots$ S_7, showing that for

$n = 1, 2, 3, 4, 5, 6, 7,$

$S_n/e = 2, 5, 15, 52, 203, 877, 4140.$

$[(iii)$ and (iv), we have $S_n = (\Delta + 1)^n S_0$ and $S_{n-1} = (E - 1)^n S_0.]$

23. If $T_n = 1^n - \dfrac{2^n}{\lfloor 1} + \dfrac{3^n}{\lfloor 2} - \dots + (-1)^m \dfrac{(m+1)^n}{\lfloor m} + \dots$ to ∞, prove that

(i) $\Delta^r T_0 = -T_{r-1}.$

(ii) $-T_n = T_{n-1} + C_1^n T_{n-2} + C_2^n T_{n-3} + \dots + C_{n-1}^n T_0 - T_0.$

(iii) Using (ii), show that for

$n = 1, \quad 2, 3, 4, 5,$

$T_n \cdot e = 0, -1, -1, 2, 9.$

■■■

CHAPTER 24

Continued Fractions

■ **1. Definitions:** Any expression of the form

$$a_1 \pm \cfrac{b_2}{a_2 \pm \cfrac{b_3}{a_3 \pm \ldots}} \qquad \ldots\text{(A)}$$

is called a *continued fraction*, and is written $a_1 \pm \cfrac{b_2}{a_2 \pm} \, \cfrac{b_3}{a_3 \pm} \, \ldots\ldots$ ...(B)

The signs $\pm$ may be supposed to be attached to the b's, and so any continued fraction may be expressed in the form

$$F = a_1 + \cfrac{b_2}{a_2 +} \, \cfrac{b_3}{a_3 +} \, \ldots \, \cfrac{b_n}{a_n +} \, \ldots, \qquad \ldots\text{(C)}$$

where the a's and b's may be positive or negative numbers. The quantities a_1, $b_2/a_2, \ldots b_n/a_n \ldots$ are called the *elements* of F, and the fraction obtained by stopping at any particular stage is called a *convergent*. Thus the first, second, third, ..., convergents are

$$\frac{a_1}{1}, \; a_1 + \frac{b_2}{a_2}, \; a_1 + \cfrac{b_2}{a_2 +} \, \cfrac{b_3}{a_3}, \; \ldots \; .$$

It is obvious that the fraction is unaltered if b_n, a_n, b_{n+1} ($n = 2, 3,...$) are all multiplied by any number k or if the signs of b_n, a_n and b_{n+1} are all changed. Thus any continued fraction may be written in the form (C) where $a_2, a_3, \ldots$ are all positive.

▌ **EXAMPLE :** *Prove that, if each fraction contains n elements,*

$$\cfrac{2}{4 +} \, \cfrac{2}{4 +} \, \cfrac{2}{4 +} \, \ldots = \cfrac{1}{2 +} \, \cfrac{1}{4 +} \, \cfrac{1}{2 +} \, \cfrac{1}{4 +} \, \ldots \; .$$

The fraction on the left is equal to $\cfrac{1}{2 +} \, \cfrac{1}{4 +} \, \cfrac{2}{4 +} \, \cfrac{2}{4 +} \, \ldots = \cfrac{1}{2 +} \, \cfrac{1}{4 +} \, \cfrac{1}{2 +} \, \cfrac{1}{4 +} \, \ldots .$

■ **2. Formation of Convergents:**

(1) Let u_n denote the nth convergent of the continued fraction,

$$F = a_1 + \cfrac{b_2}{a_2 +} \, \cfrac{b_3}{a_3 +} \, \ldots \, \cfrac{b_n}{a_n +} \, \ldots, \quad \text{so that } u_1 = \frac{a_1}{1}, \; u_2 = a_1 + \frac{b_2}{a_2} = \frac{a_2 a_1 + b_2}{a_2} .$$

We shall write $u_1 = p_1/q_1$, $u_2 = p_2/q_2$ where
$$p_1 = a_1, \; p_2 = a_2 a_1 + b_2, \; q_1 = 1, \; q_2 = a_2.$$
Observing that u_3 may be obtained from u_2 by changing a_2 into $a_2 + b_3/a_3$, we have

$$u_3 = \frac{(a_2 + b_3/a_3)a_1 + b_2}{a_2 + b_3/a_3} = \frac{a_3(a_2 a_1 + b_2) + b_3 a_1}{a_3 a_2 + b_3} = \frac{a_3 p_2 + b_3 p_1}{a_3 q_2 + b_3 q_1},$$

and so $u_3 = p_3/q_3$, where $p_3 = a_3 p_2 + b_3 p_1$ and $q_3 = a_3 q_2 + b_3 q_1$.

Proceeding thus we can show that, if p_n and q_n are defined for successive values of n by the equations

$$p_n = a_n p_{n-1} + b_n p_{n-2}, \quad q_n = a_n q_{n-1} + b_n q_{n-2}, \qquad \text{...(A)}$$

then p_n/q_n is the nth convergent of F.

Equations (A) are called the *recurrence formulae*. It will be found convenient to write

$$p_0 = 1 \quad \text{and} \quad q_0 = 0, \qquad \text{...(B)}$$

and it will be seen that $p_2 = a_2 p_1 + b_2 p_0$ and $q_2 = a_2 q_1 + b_2 q_0$, so that equation (A) hold for $n \geq 2$.

(2) It is important to state all this rather more precisely:

Definition. *For the continued fraction*

$$F = a_1 + \cfrac{b_2}{a_2 +} \; \cfrac{b_3}{a_3 +} \cdots \cfrac{b_n}{a_n +} \cdots$$

the quantities p_n and q_n are defined by the equations

$$p_n = a_n p_{n-1} + b_n p_{n-2}, \; q_n = a_n q_{n-1} + b_n q_{n-2}, \qquad \text{...(A)}$$

with the initial values $p_0 = 1$, $q_0 = 0$, $p_1 = a_1$, $q_1 = 1$.

Theorem. *If p_n and q_n are defined in this way, then p_n/q_n is the nth convergent of F.*

Proof. Assume that this holds for the values 2, 3, ... r of n, and denote the nth convergent by u_n, then

$$u_r = \frac{p_r}{q_r} = \frac{a_r p_{r-1} + b_r p_{r-2}}{a_r q_{r-1} + b_r q_{r-2}}.$$

Now $p_{r-1}, q_{r-1}, p_{r-2}, q_{r-2}$ are independent of a_r and b_r; also u_r may be transformed into u_{r+1} by writing $a_r + b_{r+1}/a_{r+1}$ for a_r, therefore

$$u_{r+1} = \frac{\left(a_r + \dfrac{b_{r+1}}{a_{r+1}}\right) p_{r-1} + b_r p_{r-2}}{\left(a_r + \dfrac{b_{r+1}}{a_{r+1}}\right) q_{r-1} + b_r q_{r-2}} = \frac{a_{r+1}(a_r p_{r-1} + b_r p_{r-2}) + b_{r+1} p_{r-1}}{a_{r+1}(a_r q_{r-1} + b_r q_{r-2}) + b_{r+1} q_{r-1}}.$$

Hence
$$u_{r+1} = \frac{a_{r+1} p_r + b_{r+1} p_{r-1}}{a_{r+1} q_r + b_{r+1} q_{r-1}} = \frac{p_{r+1}}{q_{r+1}}.$$

Thus the theorem holds for $n = r + 1$; but it holds for $n = 2$, and therefore for $n = 3, 4, 5,...$ in succession.

(3) For the fraction

$$\frac{b_1}{a_1 +} \frac{b_2}{a_2 +} \cdots \frac{b_n}{a_n +} \cdots,$$

p_n and q_n are defined by equation (A) with the initial values

$$p_0 = 0, \; q_0 = 1, \; p_1 = b_1, \; q_1 = a_1.$$

Again, the fraction

$$a_1 - \frac{b_2}{a_2 -} \frac{b_3}{a_3 -} \cdots \frac{b_n}{a_n -} \cdots$$

is obtained from the fraction in § (2) by changing the signs of the b's, so that the recurrence formulae are

$$p_n = a_n p_{n-1} - b_n p_{n-2}, \quad q_n = a_n q_{n-1} - b_n q_{n-2}.$$

In the following articles we shall call the reader's attention to certain types of continued fractions which will be considered in detail later.

■ **EXAMPLE :** *Prove that* $(1 - x)^{-2} = 1 + \dfrac{2x}{1-} \dfrac{3x}{2+} \dfrac{x}{3-} \dfrac{2x}{1}.$

Calculating the convergents of the continued fraction, we have

$$\frac{p_1}{q_1} = \frac{1}{1}, \; \frac{p_2}{q_2} = \frac{1+2x}{1}, \; \frac{p_3}{q_3} = \frac{2(1+2x) - 3x \cdot 1}{2 \cdot 1 - 3x \cdot 1} = \frac{2+x}{2-3x};$$

$$\frac{p_4}{q_4} = \frac{3(2+x) + x(1+2x)}{3(2-3x) + x \cdot 1} = \frac{6 + 4x + 2x^2}{6 - 8x};$$

$$\frac{p_5}{q_5} = \frac{1 \cdot (6 + 4x + 2x^2) - 2x(2+x)}{1.(6-8x) - 2x(2-3x)} = \frac{1}{(1-x)^2},$$

which proves the statement in question. *Notice that no fractions are reduced to lower terms, until the final stage.*

■ **3. Infinite Continued Fractions:** With regard to the fraction

$$F = a_1 + \frac{b_2}{a_2 +} \frac{b_3}{a_3 +} \cdots \frac{b_n}{a_n +} \cdots,$$

if the number of elements is finite, F is called a *terminating* continued fraction.

We may, however, suppose the b's and a's to be determined by a rule of such a kind that there is no last element. In this case F stands for an *infinite* continued fraction, *which is to be regarded as defining the sequence of convergents*

$$p_1/q_1, \quad p_2/q_2, \quad \ldots, \quad p_n/q_n, \quad \ldots,$$

and we say that *the continued fraction is convergent, divergent or oscillatory according as this sequence converges, diverges or oscillates.*

If the sequence converges, the *value* of the fraction F is defined as $\lim p_n/q_n$, and we write

$$F = a_1 + \frac{b_2}{a_2 +} \frac{b_3}{a_3 +} \ldots = \lim p_n/q_n.$$

Illustrations. Let x stand for $\dfrac{3}{2-} \dfrac{3}{2-} \dfrac{3}{2-} \ldots$ to ∞ , where every element after the first is $-3/2$. *If we assume that x may be treated as a number*, we have $x = 3/(2 - x)$ leading to $x = 1 \pm \sqrt{2}\,\iota$, a result which is clearly meaningless. Again, if x stands for

$$\frac{3}{2+} \frac{3}{2+} \frac{3}{2+} \ldots \text{ to } \infty,$$

and we assume that x is a definite number, we have

$$x = 3/(2 + x); \quad \therefore \ x^2 + 2x - 3 = 0; \quad \therefore \ x = 1 \quad \text{or} \quad -3.$$

We have thus proved that if *the fraction has a value*, its value must be unity, but *we have not shown that a value exists*.

In neither of these examples can the first step be justified without knowing that the continued fraction is convergent.

▨ **4. Simple Continued Fractions:** An expression of the form

$$a_1 + \frac{1}{a_2 +} \frac{1}{a_3 +} \ldots \frac{1}{a_n +} \ldots,$$

where every a is a positive integer, except that a_1 may be zero, is called a *simple continued fraction.*

Theorem. *Any rational number can be expressed as a simple terminating continued fraction which can be arranged so as to have either an odd or an even number of quotients.*

Let A/B be the number, A and B being positive integers. In the process of finding the G.C.M. of A and B, let $a_1, a_2, \ldots$ be the quotients and $r_1, r_2, \ldots$ the remainders, so that

$$A = a_1 B + r_1, \quad B = r_1 a_2 + r_2, \quad r_1 = r_2 a_3 + r_3, \quad \text{etc.,}$$

and therefore

$$\frac{A}{B} = a_1 + \frac{1}{B/r_1}, \quad \frac{B}{r_1} = a_2 + \frac{1}{r_1/r_2}, \quad \frac{r_1}{r_2} = a_3 + \frac{1}{r_2/r_3}, \quad \text{etc.}$$

This process terminates, and if a_n is the last quotient,

$$\frac{A}{B} = a_1 + \frac{1}{a_2 +} \frac{1}{a_3 +} \ldots \frac{1}{a_n}.$$

Again, if $a_n > 1$, we can write

$$\frac{1}{a_n} = \frac{1}{a_n - 1 +} \frac{1}{1},$$

and if $a_n = 1$,

$$\frac{1}{a_{n-1} +} \frac{1}{a_n} = \frac{1}{a_{n-1} + 1}.$$

Thus matters can always be arranged so that the continued fraction has either an odd, or an even, number of quotients, as required.

■ **EXAMPLE 1.** *Express* $\dfrac{157}{68}$ *at a simple continued fraction with (i) an even and (ii) an odd number of quotients.*

$$
\begin{array}{c|cc|c}
a_2 = 3 & 68 & 157 & 2 = a_1 \\
 & 63 & 136 & \\
a_4 = 5 & 5 & 21 & 4 = a_3 \\
 & 5 & 20 & \\
 & 0 & 1 &
\end{array}
$$

i.e. $\quad \dfrac{157}{68} = 2 + \dfrac{1}{3+}\ \dfrac{1}{4+}\ \dfrac{1}{5}$

or $\quad 2 + \dfrac{1}{3+}\ \dfrac{1}{4+}\ \dfrac{1}{4+}\ \dfrac{1}{1}.$

Note: The fact that the number of quotients can be made odd or even as required is of importance. For instance, the penultimate convergents to 157/68 in the two forms above are 30/13 and 127/55 ; and it will be found that (127, 55) and (30,13) are the *least solutions* in positive integers of

$$68x - 157y = + 1 \quad \text{and} \quad 68x - 157y = - 1$$

respectively. (See Art. 9 and Ch. 25, 3.)

■ **EXAMPLE 2.** *Use the H.C.F. process to express $(1 - x)^{-2}$ as a continued fraction.* The H.C.F. process is

$$
\begin{array}{c|c|c|c}
1/2x & 1 - 2x + x^2 & 1 & 1 \\
 & 1 - x/2 & 1 - 2x + x^2 & \\ \hline
-9/2x & -x\cdot 3/2 + x^2 & 2x - x^2 & -4/3 \\
 & -x\cdot 3/2 & 2x - x^2 \cdot 4/3 & \\ \hline
 & x^2 & x^2 \cdot 1/3 & 1/3
\end{array}
$$

giving

$$
\frac{1}{(1-x)^2} = 1 + \frac{1}{1/2x+}\ \frac{1}{-4/3+}\ \frac{1}{-9/2x+}\ \frac{1}{1/3}
$$

$$
= 1 + \frac{2x}{1+}\ \frac{2x}{-4/3+}\ \frac{1}{-9/2x+}\ \frac{1}{1/3}
$$

$$
= 1 + \frac{2x}{1-}\ \frac{3x}{2+}\ \frac{-3/2}{-9/2x+}\ \frac{1}{1/3}
$$

$$
= 1 + \frac{2x}{1-}\ \frac{3x}{2+}\ \frac{x}{3-}\ \frac{2x}{1}.
$$

In the last three steps, the statement at the end of Art. 1 has been used.

■ **5. Recurring Continued Fractions:** If, after a certain stage, the elements recur in the same order, we have a '*recurring*' continued fraction.

The recurring elements form the '*recurring period*' or the '*cycle*,' and the non-recurring elements, if such exist, form the '*acyclic part*' of the fraction.

The cycle is usually denoted by putting asterisks under the first and last of the recurring elements, thus

$$\frac{1}{1+}\ \frac{1}{2+}\ \frac{1}{3+}\ \frac{1}{4+}\ \frac{1}{3+}\ \frac{1}{4+}\ \dots \text{ is denoted by } \frac{1}{1+}\ \frac{1}{2+}\ \frac{1}{3+}\ \frac{1}{4}.$$

Here $\dfrac{1}{3+}\ \dfrac{1}{4+}$ is the cycle and $\dfrac{1}{1+}\ \dfrac{1}{2}$ the acyclic part.

■ **EXAMPLE 1.** (*i*) *If* α, β *are the roots of* $x^2 - ax + b = 0$, *prove that the n-th convergent of*

$$a - \frac{b}{a-}\ \frac{b}{a-}\ \dots\ \frac{b}{a-}\ \dots$$

is equal to $(\alpha^{n+1} - \beta^{n+1})/(\alpha^n - \beta^n)$ *or* $(1 + 1/n)\ \alpha$ *according as* $\alpha \neq \beta$ *or* $\alpha = \beta$.

(*ii*) *Hence show that the continued fraction is convergent if* $a^2 \geq 4b$, *and that its value is the numerically greater root of* $x^2 - ax + b = 0$.

(*i*) Here p_n and q_n are given by

$$p_n = ap_{n-1} - bp_{n-2}, \quad q_n = aq_{n-1} - bq_{n-2},$$

together with $p_0 = 1$, $p_1 = a$, $q_0 = 0$, $q_1 = 1$.

First suppose that $\alpha \neq \beta$, then by Ch. 22, 5, (*i*),

$$p_n = A\alpha^n + B\beta^n, \quad q_n = A'\alpha^n + B'\beta^n,$$

where A, B, A', B' are constants. Therefore

$$p_1 = A\alpha + B\beta = a = \alpha + \beta, \quad q_1 = A'\alpha + B'\beta = 1,$$
$$p_0 = A + B = 1, \quad q_0 = A' + B' = 0,$$

giving
$$A/\alpha = -B/\beta = A' = -B' = 1/(\alpha - \beta);$$

thus
$$p_n = (\alpha^{n+1} - \beta^{n+1})/(\alpha - \beta) \text{ and } q_n = (\alpha^n - \beta^n)/(\alpha - \beta),$$

leading to the first result.

Next let $\alpha = \beta$, so that $a = 2\alpha$ and $b = \beta^2$, then by Ch. 22, 5, (*i*),

$$p_n = (A + nB)\alpha^n, \quad q_n = (A' + nB')\ \alpha^n.$$

Proceeding as before, we find that $A = B = 1$, $A' = 0$, $B' = 1/\alpha$, hence

$$p_n = (n + 1)\ \alpha^n, \quad q_n = n\alpha^{n-1}, \quad p_n/q_n = (1 + 1/n)\alpha.$$

(*ii*) If $a^2 > 4b$, then α, β are real. Let $|\alpha| > |\beta|$, then $\beta^n/\alpha^n \to 0$ as $n \to \infty$ and

$$\frac{p_n}{q_n} = \frac{\alpha^{n+1}}{\alpha^n} \cdot \frac{1 - (\beta/\alpha)^{n+1}}{1 - (\beta/\alpha)^n} \to \alpha.$$

If $a^2 = 4b$, then $\alpha = \beta$ and $p_n/q_n = (1 + 1/n)\ \alpha \to \alpha$.

Hence, in both cases, the continued fraction is convergent and its value is α.

■ **EXAMPLE 2.** *If p_n/q_n is the n-th convergent of the recurring fraction*

$$\frac{1}{a+}\ \frac{1}{b+}\ \frac{1}{a+}\ \frac{1}{b+}\ \dots\ ,$$

prove that $\qquad\qquad p_n - (ab+2)\,p_{n-2} + p_{n-4} = 0$

and $\qquad\qquad q_n - (ab+2)\,q_{n-2} + q_{n-4} = 0.$

If n is *even*, the recurrence formulae give

$$p_n = bp_{n-1} + p_{n-2},$$
$$p_{n-1} = ap_{n-2} + p_{n-3},$$
$$p_{n-2} = bp_{n-3} + p_{n-4},$$

whence eliminating p_{n-1} and p_{n-3}, we have

$$p_n - (ab+2)\,p_{n-2} + p_{n-4} = 0.$$

If n is odd, we have the same equations as before, except that a and b are inter changed, leading to the same result. And so for the q's.

■ **EXAMPLE 3.** *If p_n/q_n is the n-th convergent of*

$$\frac{1}{1+}\ \frac{1^2}{2+}\ \frac{3^2}{2+}\ \frac{5^2}{2+}\ \dots\ \frac{(2n-3)^2}{2+}\ \dots$$

prove that $\qquad p_n/q_n = 1 - \dfrac{1}{3} + \dfrac{1}{5} - \dots + (-1)^{n-1}\dfrac{1}{2n-1}.$

Hence show that the fraction is convergent and that its value is $\pi/4$.

If v_n denotes either p_n or q_n, we have

$$v_n = 2v_{n-1} + (2n-3)^2\,v_{n-2};$$
$$\therefore\quad v_n - (2n-1)\,v_{n-1} = -(2n-3)\,\{v_{n-1} - (2n-3)v_{n-2}\},$$

and this holds if we write $n-1,\ n-2,\ \dots\ 2$ in succession for n, therefore

$$v_n - (2n-1)v_{n-1} = (-1)^{n-1}(2n-3)\,(2n-5)\,\dots\,3\cdot 1\,(v_1 - v_0).$$

Replacing the v's by p's and observing that $p_0 = 0$ and $p_1 = 1$, we have

$$p_n - (2n-1)p_{n-1} = (-1)^{n-1}\,1\cdot 3\cdot 5\,\dots\,(2n-3);$$

$$\therefore\quad \frac{p_n}{1\cdot 3\dots(2n-1)} - \frac{p_{n-1}}{1\cdot 3\dots(2n-3)} = (-1)^{n-1}\frac{1}{2n-1};$$

$$\therefore\quad \frac{p_n}{1\cdot 3\dots(2n-1)} = A + \frac{1}{1} - \frac{1}{3} + \frac{1}{5} - \dots + (-1)^{n-1}\frac{1}{2n-1},$$

where A is independent of n. Putting $n=1$, we find that $A=0$.

Next, replacing the v's by q's and observing that $q_0 = 1$ and $q_1 = 1$, we have

$$q_n - (2n-1)q_{n-1} = 0,$$

whence we find that $q_n = 1\cdot 3\cdot 5\,\dots\,(2n-1)$, leading to the first result.

Again, the series $1 - 1/3 + 1/5 - \dots$ is convergent and its sum is $\pi/4$; therefore the fraction converges and its value is $\pi/4$.

EXERCISE XL

1. Find the first four convergents (unreduced) of

   (i) $\dfrac{1}{2+}\dfrac{2}{3+}\dfrac{3}{4+}\dfrac{4}{5}$;

   (ii) $\dfrac{1}{1-}\dfrac{2}{2-}\dfrac{3}{3-}\dfrac{4}{4}$.

2. Without calculating the convergents, prove that

$$\frac{2}{2+}\frac{3}{3+}\frac{4}{4+}\frac{5}{5}=\frac{1}{1+}\frac{1}{2+}\frac{2}{3+}\frac{3}{4}.$$

3. Express $\frac{233}{151}$ as a simple continued fraction having (i) an even and (ii) an odd number of quotients.

4. The reckoning on the right shows that

$$\frac{85}{52}=2-\frac{1}{3-}\frac{1}{4-}\frac{1}{5}.$$

   $$
   \begin{array}{r|rll}
   3 & 52 & 85 & 2 \\
     & 57 & 104 & \\
   5 & 5 & 19 & 4 \\
     & 5 & 20 & \\
   \hline
     & 0 & 1 &
   \end{array}
   $$

   Explain this.

5. Express $\frac{203}{85}$ in the form $a_1-\dfrac{1}{a_2-}\dfrac{1}{a_3-}\ldots$, where the a's are integers greater than unity.

6. Express $(2a^2-a-1)/(2a^2-3a)$ in the form $1+\dfrac{1}{f_1+}\dfrac{1}{f_2+}\ldots$, and verify by calculating the convergents.

7. Show that the nth convergent of

$$2-\frac{1}{2-}\frac{1}{2-}\frac{1}{2-}\ldots$$

   is $(n+1)/n$, and that the value of the infinite fraction is unity.

8. Show that the nth convergent of

$$\frac{1}{1-}\frac{1}{2-}\frac{1}{2-}\frac{1}{2-}\ldots$$

   is n, and that the infinite fraction is divergent.

9. There being $(n+1)$ elements in all, show that

$$2-\frac{1}{2-}\frac{1}{2-}\ldots\frac{1}{2-}\frac{1}{x}=\frac{(n+1)\,x-n}{nx-n+1}.$$

10. If $a^2>4b$, the numerically smaller root of $x^2-ax+b=0$ is equal to

$$\frac{b}{a-}\frac{b}{a-}\frac{b}{a-}\ldots\frac{b}{a-}\ldots.$$

   [This follows from Art. 5, Ex. 1.]

11. For the fraction

$$\frac{b}{a+}\ \frac{b}{a+}\ \frac{b}{a+}\ \dots +\ \frac{b}{a+}\ \dots\ ,$$

prove that $p_n = -\alpha\beta(\alpha^n - \beta^n)/(\alpha - \beta)$, $q_n = (\alpha^{n+1} - \beta^{n+1})/(\alpha - \beta)$ where α, β are the roots of $x^2 - ax - b = 0$. Hence show that if a, b are positive, the fraction is convergent and equal to the positive root of $x = b/(a + x)$.

12. If p_n/q_n is the (unreduced) nth convergent of

$$a_1 + \frac{b_2}{a_2+}\ \frac{b_3}{a_3+}\ \dots -\ \frac{b_n}{a_n+}\ \dots\ ,$$

and the *number of terms* in p_n, q_n respectively are denoted by u_n, v_n respectively, then u_n/v_n is the nth convergent of

$$1 + \frac{1}{1+}\ \frac{1}{1+}\ \dots\ .$$

Hence show that,

$$u_{n-1} = v_n = \frac{1}{2^{n-1}}\left(C_1^n + 5C_3^n + 5^2 C_5^n + \dots\right).$$

13. Prove that

(i) $\dfrac{1}{1-}\ \dfrac{3x}{1+}\ \dfrac{x}{1-}\ \dfrac{2x}{3+}\ \dfrac{x}{2-}\ \dfrac{x}{1} = \dfrac{1}{(1-x)^3}\ .$

(ii) $\dfrac{1}{1+}\ \dfrac{3x}{1-}\ \dfrac{2x}{1+}\ \dfrac{x}{3-}\ \dfrac{5x}{2+}\ \dfrac{x}{5-}\ \dfrac{3x}{1} = (1-x)^3.$

14. Prove that, each fraction containing n elements,

$$\frac{1}{1-}\ \frac{1}{4-}\ \frac{1}{1-}\ \frac{1}{4-}\ \dots = 2\left(\frac{1}{2-}\ \frac{1}{2-}\ \frac{1}{2-}\ \dots\right).$$

Hence show the nth convergent of the first fraction reduces to $2n/(n + 1)$.

15. Prove that

$$\frac{1}{1-}\ \frac{1^2}{3-}\ \frac{2^2}{5-}\ \frac{3^2}{7-}\ \dots\ \frac{(n-1)^2}{2n-1} = 1 + \frac{1}{2} + \frac{1}{3} + \dots + \frac{1}{n}\ ,$$

showing that the infinite continued fraction is divergent.

$[p_n = (2n - 1)\, p_{n-1} - (n - 1)^2 p_{n-2}; \quad \therefore\ p_n - np_{n-1} = (n - 1)\, \{p_{n-1} - (n - 1)\, p_{n-2}\}.]$

16. Prove that

$$\frac{a^2}{2a+1-}\ \frac{(a+1)^2}{2a+3-}\ \frac{(a+2)^2}{2a+5-}\ \dots\ \frac{(a+n-1)^2}{2a+2n-1-}\ \dots = a\ ,$$

showing incidentally that, if $k_n = a\,(a + 1)\,(a + 2)\ \dots\ (a + n)$, then

$$\frac{q_n}{k_n} = \frac{1}{a} + \frac{1}{a+1} + \frac{1}{a+2} + \dots + \frac{1}{a+n}\quad \text{and}\quad p_n/q_n = a - k_n/q_n.$$

Use this to show that the fraction in Ex. 15 is divergent.

$[\text{Here } q_n - (a + n)q_{n-1} = (a + n - 1)\{q_{n-1} - (a + n - 1)\, q_{n-2}\}.]$

17. Putting $a = -m$, in Ex. 16, where m is a positive integer, show that

$$\frac{m}{2m-1-}\,\frac{(m-1)^2}{2m-3-}\,\frac{(m-2)^2}{2m-5-}\,\cdots\,\frac{1^2}{1} = 1 .$$

Also prove this result by induction.

18. For the fraction $\dfrac{1}{1+}\,\dfrac{2}{2+}\,\dfrac{3}{3+}\cdots\dfrac{n}{n+}\cdots$, prove that

$$\frac{p_n}{\lfloor n+1}=1-\frac{q_n}{\lfloor n+1}=\frac{1}{\lfloor 2}-\frac{1}{\lfloor 3}+\frac{1}{\lfloor 4}-\,\cdots\text{ to }n\text{ terms.}$$

Hence show that the fraction is convergent and equal to $1/(e-1)$.

19. If n is a positive integer, prove by induction that

$$\frac{n}{n+}\,\frac{n-1}{n-1+}\,\frac{n-2}{n-2+}\cdots\frac{2}{2+}\,\frac{1}{1+}\,\frac{1}{2}=\frac{n+1}{n+2} .$$

20. For the fraction

$$\frac{1}{1+}\,\frac{1^2}{1+}\,\frac{2^2}{1+}\,\frac{3^2}{1+}\cdots\frac{(n-1)^2}{1+}\cdots,$$

prove that $\qquad \dfrac{p_n}{q_n}=1-\dfrac{1}{2}+\dfrac{1}{3}-\cdots+(-1)^{n-1}\dfrac{1}{n} .$

Hence show that the fraction is convergent and equal to $\log 2$.

■ **6. Simple Continued Fractions** are of the form

$$x_1 = a_1 + \frac{1}{a_2+}\,\frac{1}{a_3+}\cdots\frac{1}{a_n+}\cdots,$$

where the a's are positive integers, except that a_1 may be zero.

Here $a_1, a_2,\ldots$ are called the *(partial) quotients* and if

$$x_2 = a_2 + \frac{1}{a_3+}\,\frac{1}{a_4+}\cdots,\ x_3 = a_3 + \frac{1}{a_4+}\,\frac{1}{a_5+}\cdots,\text{ etc.}$$

then $x_2, x_3, \ldots$ are called the *complete quotients*.

Thus we have

$$x_1 = a_1 + \frac{1}{a_2+}\cdots\frac{1}{a_{n-1}+}\,\frac{1}{x_n};$$

that is to say, *the continued fraction x_1 is obtained from its n-th convergent by substituting the complete quotient x_n for the quotient a_n.*

■ **7.** As in Art. 2; p_n and q_n are defined by the equations

$$p_n = a_n p_{n-1} + p_{n-2}\ \text{ and }\ q_n = a_n q_{n-1} + q_{n-2}, \qquad \ldots\text{(A)}$$

with the initial values $p_0 = 1,\ p_1 = a_1,\ q_0 = 0,\ q_1 = 1$.

The convergents are $\left(\dfrac{1}{0}\right), \dfrac{a_1}{1}, \dfrac{a_1 a_2 +1}{a_2}, \dfrac{a_1 a_2 a_3 + a_3 + a_1}{a_2 a_3 +1}$, etc., and it will be seen that q_n *is the same function of* a_2, a_3, ... a_n *as* p_{n-1} *is of* a_1, a_2, ... a_{n-1}.

■ **8.** We have $p_2 > p_1$, and if $n \geq 3$, $p_n - p_{n-1} = (a_n - 1) p_{n-1} + p_{n-2} \geq p_{n-2} \geq 1$; and so for the q's, since $q_2 \geq q_1$, thus p_n *and* q_n *increase with n, and in the case of an infinite continued fraction, they tend to infinity with n.*

■ **9.** From the recurrence formulae (A), we have

$$p_n q_{n-1} - p_{n-1} q_n = (a_n p_{n-1} + p_{n-2}) q_{n-1} - p_{n-1} (a_n q_{n-1} + q_{n-2})$$
$$= - (p_{n-1} q_{n-2} - p_{n-2} q_{n-1}).$$

This holds if we write $n - 1$, $n - 2$, ...2 in succession for n, also

$$p_1 q_0 - p_0 q_1 = - 1,$$

therefore $\qquad\qquad p_n q_{n-1} - p_{n-1} q_n = (-1)^n.$ $\qquad\qquad$ ...(B)

Again, we have

$$p_n q_{n-2} - p_{n-2} q_n = (a_n p_{n-1} + p_{n-2}) q_{n-2} - p_{n-2} (a_n q_{n-1} + q_{n-2})$$
$$= a_n (p_{n-1} q_{n-2} - p_{n-2} q_{n-1});$$

therefore $\qquad\qquad p_n q_{n-2} - p_{n-2} q_n = (-1)^{n-1} a_n.$ $\qquad\qquad$ ...(C)

From (B) and (C) we obtain the formulae

$$\frac{p_n}{q_n} - \frac{p_{n-1}}{q_{n-1}} = (-1)^n \frac{1}{q_n q_{n-1}} \quad \text{and} \quad \frac{p_n}{q_n} - \frac{p_{n-2}}{q_{n-2}} = (-1)^{n-1} \frac{a_n}{q_n q_{n-2}} . \qquad ...(D)$$

■ **10.** *Every convergent* p_n / q_n *is in its lowest terms, also* p_n *is prime to* p_{n-1} *and* q_n *to* q_{n-1}.

For if the numbers belonging to any of the pairs (p_n, q_n), (p_n, p_{n-1}), (q_n, q_{n-1}) have a common factor g, then on account of equation (B), g would divide ± 1.

■ **11.** From the recurrence formulae (A), we have

$$\frac{p_n}{p_{n-1}} = a_n + \frac{1}{p_{n-1} / p_{n-2}} \quad \text{and} \quad \frac{q_n}{q_{n-1}} = a_n + \frac{1}{q_{n-1} / q_{n-2}} .$$

The first of these holds if we write $n - 1$, $n - 2$,... 2 for n.

The second of these holds if we write $n - 1$, $n - 2$,... 3 for n.

Also $p_1/p_0 = a_1$ and $q_2/q_1 = a_2$; therefore

$$\frac{p_n}{p_{n-1}} = a_n + \frac{1}{a_{n-1} +} \frac{1}{a_{n-2} +} ... \frac{1}{a_2 +} \frac{1}{a_1}, \qquad\qquad ...(E)$$

$$\frac{q_n}{q_{n-1}} = a_n + \frac{1}{a_{n-1} +} \frac{1}{a_{n-2} +} ... \frac{1}{a_2}. \qquad\qquad ...(F)$$

It follows that *if* p_r/q_r, p_r'/q_r' *are respectively the r-th convergents of*

$$a_1 + \frac{1}{a_2 +} ... \frac{1}{a_n} \quad \text{and} \quad a_n + \frac{1}{a_{n-1} +} ... \frac{1}{a_1} ,$$

where in the second the quotients occur in the reverse order, then

$$p_n' = p_n, \ q_n' = p_{n-1}, \ p'_{n-1} = q_n, \ q'_{n-1} = q_{n-1}. \qquad\qquad ...(G)$$

For by the preceding, $p_n'/q_n' = p_n/p_{n-1}$ and $p'_{n-1}/q'_{n-1} = q_n/q_{n-1}$, and by Art. 10 these fractions are in their lowest terms.

■ **12.** It has been shown that if p'/q', p/q are consecutive convergents of a simple continued fraction, then $pq' - p'q = \pm 1$.

Conversely, let $pq' - p'q = \pm 1$ where $q' \leq q$, and suppose that p/q is expressed as a simple continued fraction with an even or an odd number of quotients according as the arbitrary sign is $+$ or $-$. We can prove that p'/q' *is the convergent immediately preceding* p/q.

For if p''/q'' is the last convergent but one, by hypothesis and equation (B) of Art. 9,

$pq' - p'q = pq'' - p''q$, and therefore $p(q' - q'') = q(p' - p'')$.

Now p is prime to q, for $pq' - p'q = \pm 1$, therefore q divides $q' - q''$ or else $q' = q''$.

Also $q' \leq q$ and $q'' \leq q$, therefore q cannot divide $q' - q''$, and consequently $q' = q''$ and $p' = p''$, which proves the theorem.

■ **13. Sequences of Odd and Even Convergents.** Let $u_n = p_n/q_n$, then, from equation (D) of Art. 9, it follows that

$$u_n - u_{n-1} \quad \text{and} \quad u_n - u_{n-2}$$

are of opposite sign. Therefore u_n lies between u_{n-1} and u_{n-2}, that is to say, *every convergent lies between the two preceding convergents.*

Now $u_1 < u_2$ and consequently $u_1 < u_3 < u_2$; $u_3 < u_4 < u_2$; $u_3 < u_5 < u_4$, and so on, therefore

$$u_1 < u_3 < u_5 \ldots < u_6 < u_4 < u_2.$$

Thus the odd convergents form an increasing sequence, the even convergents form a decreasing sequence, and every odd convergent is less than any even convergent.

Consequently, as n tends to infinity, both u_{2n-1} and u_{2n} tend to finite limits, and these limits are the same, for by equation (D) of Art. 9,

$$u_{2n} - u_{2n-1} = 1/q_{2n}q_{2n-1} \to 0.$$

Hence u_n tends to a limit x such that for every n,

$$u_{2n-1} < x < u_{2n}.$$

In other words, (*i*) *every infinite simple continued fraction is convergent, and* (*ii*) *its value is greater than any odd convergent and* (*iii*) *less than any even convergent.*

■ **14.** *Each convergent is a closer approximation to the value of the continued fraction than the preceding.*

For let $z = a_1 + \dfrac{1}{a_2 +} \cdots \dfrac{1}{a_n +} \dfrac{1}{a_{n+1} +} \cdots$ and $z' = a_{n+1} + \dfrac{1}{a_{n+2} +} \cdots,$

so that z' is the $(n + 1)$th complete quotient, then

$$p_{n+1}/q_{n+1} = (a_{n+1}p_n + p_{n-1})/(a_{n+1}q_n + q_{n-1})$$

and therefore $\qquad z = (z'p_n + p_{n-1})/(z'q_n + q_{n-1}).$

Consequently,

$$z'(zp_n - p_n) = -(zq_{n-1} - p_{n-1})$$

and $\qquad\qquad z - \dfrac{p_n}{q_n} = -\dfrac{q_{n-1}}{z'q_n}\left(z - \dfrac{p_{n-1}}{q_{n-1}}\right).$

Now $q_n > q_{n-1}$ and $z' \geq 1$, therefore

$$\left| z - \frac{p_n}{q_n} \right| < \left| z - \frac{p_{n-1}}{q_{n-1}} \right|,$$

that is to say, z is nearer to p_n/q_n than to p_{n-1}/q_{n-1}.

■ **15.** *Any convergent p/q of a simple continued fraction is a closer approximation to the value of the fraction than any rational fraction r/s with a denominator less than q.*

For let $z = a_1 + \dfrac{1}{a_2 +} \cdots$, and let p'/q' be the convergent immediately preceding

p/q. By Art. 14, if r/s is nearer to z than p/q, it is also nearer to z than p'/q'. Moreover, z lies between p/q and p'/q'; therefore r/s lies between p/q and p'/q', and

$$\left| \frac{r}{s} - \frac{p'}{q'} \right| < \left| \frac{p}{q} - \frac{p'}{q'} \right|, \quad i.e. \quad \left| \frac{r}{s} - \frac{p'}{q'} \right| < \frac{1}{qq'}.$$

Therefore $q \cdot | rq' - sp' | < s$; and, since $rq' - sq'$ is an integer, it follows that

$$s > q.$$

■ **16.** *If ε is the numerical value of the error, in representing the value z of the continued fraction, by p_n/q_n, then*

$$a_{n+2}/q_n q_{n+2} < \varepsilon < 1/q_n q_{n+1}, \qquad \qquad ...(H)$$

and within wider limits

$$1/q_n (q_n + q_{n+1}) < \varepsilon < 1/q_n^2. \qquad \qquad ...(I)$$

For if $u_n = p_n/q_n$, then by Art. 13,

$$u_n, \ u_{n+2}, \ z, \ u_{n+1},$$

is an increasing or a decreasing sequence, therefore

$$| u_n - u_{n+2} | < | u_n - z | < | u_n - u_{n+1} |.$$

Also by equation (D),

$$|u_n - u_{n+2}| = a_{n+2}/q_n q_{n+2} \quad \text{and} \quad |u_n - u_{n+1}| = 1/q_n q_{n+1},$$

whence we have the inequalities (H).

Again, $q_{n+2} = a_{n+2} q_{n+1} + q_n$, therefore

$$q_{n+2}/a_{n+2} = q_{n+1} + q_n/a_{n+2} \leq q_{n+1} + q_n$$

and

$$1/q_n (q_n + q_{n+1}) \leq a_{n+2}/q_n q_{n+2}.$$

Also $q_{n+1} > q_n$, and therefore $1/q_n q_{n+1} < 1/q_n^2$.

Hence the inequalities (I) follow from (H).

■ **17.** *Any positive irrational z can be expressed as a simple continued fraction, and that in only one way.*

Let $z = a_1 + \dfrac{1}{z_2}$ where a_1 is the integral part of z (which may be zero), then $1/z_2$

< 1 and $z_2 > 1$. Continuing the process, let

$$z_2 = a_2 + 1/z_3, \ z_3 = a_3 + 1/z_4, \ \ldots \ z_n = a_n + 1/z_{n+1},$$

where $a_2, a_3, \ldots$ are integers such that, for every r,

$$a_r < z_r < a_r + 1,$$

since $z_2 > 1$, it follows in succession that $a_2, a_3, \ldots$ are *positive* integers and we have

$$z = a_1 + \cfrac{1}{a_2 +} \ \ldots \ \cfrac{1}{a_n +} \ \cfrac{1}{z_{n+1}}.$$

The process can be continued indefinitely, and leads to the infinite continued fraction

$$F = a_1 + \cfrac{1}{a_2 +} \ \ldots \ \cfrac{1}{a_n +} \ \ldots \ .$$

Again, we have $a_2 < z_2$, $a_3 < z_3$, etc., therefore

$$a_1 + \cfrac{1}{a_2} > z, \ a_2 + \cfrac{1}{a_3} > z_2, \ \text{etc.}$$

and
$$a_1 + \cfrac{1}{a_2 +} \ \cfrac{1}{a_3} < z, \quad a_2 + \cfrac{1}{a_3 +} \ \cfrac{1}{a_4} < z_2, \ \text{etc.}$$

Continuing thus, it can be shown that if u_r is the rth convergent of F, then

$$u_{2n-1} < z < u_{2n}.$$

Now as n tends to infinity, u_{2n-1} and u_{2n} tend to the same limit, which must therefore be equal to z. That is to say, the fraction F converges to z as a limit, and we write

$$z = a_1 + \cfrac{1}{a_2 +} \ \ldots \ \cfrac{1}{a_n +} \ \ldots .$$

It remains to show that this mode of expression is unique. Suppose that

$$z = a_1 + \cfrac{1}{a_2 +} \ \cfrac{1}{a_3 +} \ \ldots = b_1 + \cfrac{1}{b_2 +} \ \cfrac{1}{b_3 +} \ \ldots \ .$$

The integral parts of these continued fractions are equal, therefore

$$a_1 = b_1 \quad \text{and} \quad a_2 + \cfrac{1}{a_3 +} \ \ldots = b_2 + \cfrac{1}{b_3 +} \ \ldots \ .$$

Hence, by similar reasoning,

$$a_2 = b_2 \quad \text{and} \quad a_3 + \cfrac{1}{a_4 +} \ \ldots = b_3 + \cfrac{1}{b_4 +} \ \ldots,$$

and so on indefinitely. Thus the continued fractions are identical.

When z is expressed in this way, any convergent of the continued fraction will be called a *convergent* to z.

■ **18.** *The value of any infinite simple continued fraction is an irrational number.*

For any rational number can be expressed as a finite simple continued fraction.

■ **Example :** *Express $\left(\sqrt{37}+8\right)/9$ as a simple continued fraction.*

The integral part of this surd is 1, and

$$\frac{\sqrt{37}+8}{9}=1+\frac{\sqrt{37}-1}{9}=1+\frac{4}{\sqrt{37}+1}, \quad \frac{\sqrt{37}+1}{4}=1+\frac{\sqrt{37}-3}{4}=1+\frac{7}{\sqrt{37}+3},$$

$$\frac{\sqrt{37}+3}{7}=1+\frac{\sqrt{37}-4}{7}=1+\frac{3}{\sqrt{37}+4}, \quad \frac{\sqrt{37}+4}{3}=3+\frac{\sqrt{37}-5}{3}=3+\frac{4}{\sqrt{37}+5},$$

$$\frac{\sqrt{37}+5}{4}=2+\frac{\sqrt{37}-3}{4}=2+\frac{7}{\sqrt{37}+3}, \quad \frac{\sqrt{37}+3}{7}=1+\text{ etc.}$$

The complete quotient $\left(\sqrt{37}+3\right)/7$ has occurred before, and the subsequent work consists of repetitions of the last three steps, thus, with the notation of Art. 5:

$$\frac{\sqrt{37}+8}{9}=1+\frac{1}{1+}\frac{1}{1+}\frac{1}{3+}\frac{1}{2},$$

the continued fraction being a recurring fraction.

■ **19. Approximations:**

(1) If a given number x is expressed as a simple continued fraction of which p_r/q_r is the rth convergent, then p_n/q_n is an approximate value of x in defect or excess, according as n is odd or even, and the error is numerically less than $1/q_n q_{n+1}$, and a *fortiori* less than $1/q_n^2$. Now $q_{n+1}=a_{n+1}q_n+q_{n-1}$, hence *if a_{n+1} is large, p_n/q_n is a specially good approximation.*

Or again, if we require *an approximation with an error numerically less than a given fraction $1/a$,* we calculate the convergents until we find one p_n/q_n such that $q_n q_{n+1}\geq a$. This is the required approximation. Of course, the conditions are satisfied if $q_n \geq \sqrt{a}$.

■ **Example :** *It is required to find good approximations to the value of π in the form of vulgar fractions, given that*

$$\pi = 3 + \frac{1}{7+}\frac{1}{15+}\frac{1}{1+}\frac{1}{292+}\frac{1}{1+}\frac{1}{1+}\;\cdots.$$

The convergents are

$$\frac{3}{1},\frac{22}{7},\frac{333}{106},\frac{355}{113},\frac{103993}{33102},\;\cdots.$$

The denominators 106, 33102 are large compared with the preceding ones, therefore 22/7 and 355/113 are specially good approximations, and they are both in excess. *For 355/113, the error is less than*

$$\frac{1}{113\times 33102}<\frac{1}{3740000}<0.00000027.$$

■ **20.** The problem may be of such a kind that no suitable approximation can be found among the convergents (as in Art. 22, Ex. 2).

We may then proceed as follows: Choose two convergents p/q, p'/q', one an odd and the other an even convergent. If m, m' are positive integers, the fraction

$$\frac{P}{Q} = \frac{mp + m'p'}{mq + m'q'}$$

lies between p/q and p'/q'. Also x lies between these convergents, therefore P/Q is a closer approximation than p/q, and perhaps also than p'/q'; and it may be possible to choose m, m' to suit the conditions of the problem in question.

■ **21.** The most important fractions of this kind are the so-called *intermediate convergents*. Let p_n/q_n be the nth convergent of

$$x = a_1 + \frac{1}{a_2 +} \frac{1}{a_3 +} \cdots .$$

Consider the sequence

$$\frac{p_n}{q_n}, \ \frac{p_n + p_{n+1}}{q_n + q_{n+1}}, \ \frac{p_n + 2p_{n+1}}{q_n + 2q_{n+1}}, \cdots \frac{p_n + a_{n+2}\, p_{n+1}}{q_n + a_{n+2}\, q_{n+1}} \left(= \frac{p_{n+2}}{q_{n+2}} \right),$$

the $(r + 1)$th term being P_r/Q_r where $P_r = p_n + rp_{n+1}$, $Q_r = q_n + rq_{n+1}$.

The terms of this sequence (excluding the first and last) are called the *intermediate convergents* between p_n/q_n and p_{n+2}/q_{n+2}. The sequence possesses the following properties:

(i) It is an increasing or a decreasing sequence according as n is odd or even, and the denominators form an increasing sequence.

(ii) Every fraction P_r/Q_r is in its lowest terms.

(iii) Any fraction which lies between P_r/Q_r and P_{r+1}/Q_{r+1} has a denominator greater than either Q_r or Q_{r+1}.

(iv) Limits to the numerical value of the error in taking P_r/Q_r as the value of x are given by

$$\frac{a_{n+2} - r}{q_{n+2}\, Q_r} < \left| x - \frac{P_r}{Q_r} \right| < \frac{1}{q_{n+1}\, Q_r}$$

For we have $P_{r+1}Q_r - P_r Q_{r+1} = p_{n+1}q_n - p_n q_{n+1} = (-1)^{n+1}$,

whence the statements (i)–(iii) follow: and because

$$P_r/Q_r, \ p_{n+2}/q_{n+2}, \ x, \ p_{n+1}/q_{n+1},$$

is either an increasing or a decreasing sequence, therefore

$$\left| \frac{p_{n+2}}{q_{n+2}} - \frac{P_r}{Q_r} \right| < \left| x - \frac{P_r}{Q_r} \right| < \left| \frac{p_{n+1}}{q_{n+1}} - \frac{P_r}{Q_r} \right|$$

Since $r < a_{n+2}$, this proves the statement.

Now consider the sequence formed by writing down the odd convergents

$$p_1/q_1, \ \ldots, \ p_3/q_3, \ \ldots, \ p_5/q_5 \ \ldots , \qquad \qquad \ldots(A)$$

and inserting all the intermediate convergents, then

  (*i*) This is an *increasing* sequence of fractions in their lowest terms, and each fraction is a closer approximation in defect to x than the preceding.

  (*ii*) The denominators form an increasing sequence.

  (*iii*) If P/Q, P'/Q' are successive terms, any fraction which lies between them has a denominator greater than Q or Q'.

Again, the sequence

$$p_2/q_2, \ldots, p_4/q_4, \ldots, p_6/q_6, \ldots, \qquad \ldots\text{(B)}$$

formed by the even convergents and their intermediates, has the same properties as the preceding, except that it is a *decreasing* sequence in which each term is a closer approximation *in excess* to x than the preceding.

■ **22. Problem:** *It is required to find the fraction which, of all those with denominators less than a given number a, is the closest approximation (in defect or in excess) to a given number x.*

To do this we express x as a simple continued fraction, and calculate the convergents p_1/q_1, p_2/q_2, ... until we find p_n/q_n such that

$$q_n \leq a < q_{n+1}.$$

If $q_n = a$, then p_n/q_n is the required fraction.

If $q_n < a$, we first consider approximations in defect. Taking the sequence (A) formed by the odd convergents and their intermediates, we find successive terms P/Q, P'/Q' such that $Q \leq a < Q'$; then P/Q is the closest approximation in defect of all fractions with denominators less than a.

For if h/k is nearer to x than P/Q, then h/k must coincide with a term to the right of P/Q in the sequence, or else it must lie between two such terms, and in either case $k \geq Q' > a$.

In the same way, by considering the sequence (B) formed by the even convergents and their intermediates, we can find the fraction with denominator less than a which is the closest approximation in excess. Of these two fractions, the one which is nearer to x is the number required.

■ **EXAMPLE 1.** *Find the fraction with denominator less than 500 which is nearest to* $\sqrt{14}$.

As in Art. 18, Ex. 1, we can show that $\sqrt{14} = 3 + \cfrac{1}{1+} \cfrac{1}{2+} \cfrac{1}{1+} \cfrac{1}{6}$,

and we find that

$$\frac{p_7}{q_7} = \frac{333}{89}, \frac{p_8}{q_8} = \frac{449}{120}, \frac{p_9}{q_9} = \frac{3027}{809} .$$

For the approximation in defect, we insert intermediate convergents between p_7/q_7 and p_9/q_9. These are of the form $P_r/Q_r = (333 + 449r)/(89 + 120r)$.

The conditions $Q_r \leq 500 < Q_{r+1}$ give $r = 3$, and $P_3/Q_3 = 1680/449$.

For the approximation in excess, we consider the convergents p_8/q_8, p_{10}/q_{10}. Since $a_{10} = 1$, there are no intermediate convergents and the approximation in excess is $p_8/q_8 = 449/120$.

Finally, we can show that $14 - (1680/449)^2 < (449/120)^2 - 14$, so that the required approximation is 1680/449.

EXAMPLE 2. *The mean tropical year is 365.2422642 days. It is required to find a practically convenient method of correcting the calendar, the year being taken as 365 days.*

We have

$$0.2422642 \;=\; 0 + \cfrac{1}{4+}\,\cfrac{1}{7+}\,\cfrac{1}{1+}\,\cfrac{1}{4+}\,\cfrac{1}{1+}\,\cfrac{1}{6+}\,\cfrac{1}{11+}\, \cdots,$$

and the convergents are

$$\frac{0}{1},\,\frac{1}{4},\,\frac{7}{29},\,\frac{8}{33},\,\frac{39}{161},\,\frac{47}{194},\,\frac{321}{1325}\, \cdots .$$

Take 1/4 as a first approximation. This amounts to adding 1 day every 4 years, and is an over-correction.

The fraction 7/29 is inconvenient; but 8/33 suggests an addition of 24 days in every 99 years, *i.e.* make every fourth year a leap-year, except at the end of a century. This is also an over-correction.

None of the other convergents give convenient approximations; and the same is true for the intermediate convergents in the first few intervals. We therefore proceed as in Art. 20. We choose the convergents 1/4 and 39/161, the first an even and the second an odd convergent; then, if m, m' are positive integers,

$$\frac{1}{4} < \frac{m \cdot 1 + m' \cdot 39}{m \cdot 4 + m' \cdot 161} < \frac{39}{161}.$$

The values $m = 19$, $m' = 2$ give the approximation.

$$\frac{19 \cdot 1 + 2 \cdot 39}{19 \cdot 4 + 2 \cdot 161} = \frac{97}{400}.$$

This gives one more day to be added in 400 years than the fraction 8/33 ; *i.e.* it makes every fourth year a leap-year, except those at the end of a century, with the further correction that every fourth century is to be counted as a leap-year.

The fact that this approximation, 97/400, can be obtained from the convergents 1/4 and 47/194, which are both in excess, shows that this also is an over-correction.

The error is less than a day in 4000 years.

MISCELLANEOUS THEOREMS

23. Let p/q be a fraction in its lowest terms, which is an approximate value of some quantity z. It may be important to know *under what circumstances p/q is a convergent to z*. The answer to this is contained in the following;

Let p/q be expressed as a simple continued fraction with an even or an odd number of quotients according as $p/q \gtrless z$, and let p'/q' be the last convergent but one.

$$\text{If} \qquad \left| z - \frac{p}{q} \right| < \frac{1}{q\,(q+q')}, \; \text{or a fortiori if} \; \left| z - \frac{p}{q} \right| < \frac{1}{2q^2},$$

then p/q is a convergent to z.

For suppose that

$$\frac{p}{q} = a_1 + \frac{1}{a_2 +} \cdots \frac{1}{a_n},$$

where n is even or odd according as $p/q \gtrless z$, and let z' be determined by

$$z = a_1 + \frac{1}{a_2 +} \cdots \frac{1}{a_n +} \frac{1}{z'}.$$

It is sufficient to prove that $z' > 1$, for then z' can be expressed as a simple continued fraction in which the first quotient is not zero. We have

$$z = \frac{pz' + p'}{qz' + q'} \quad \text{and therefore} \quad \frac{p}{q} - z = \frac{pq' - p'q}{q(qz' + q')},$$

Now $pq' - p'q = 1$ or -1, according as $p/q \gtrless z$; therefore in both cases $q(qz' + q')$ is *positive* and

$$\frac{1}{q(qz' + q')} = \left| \frac{p}{q} - z \right| < \frac{1}{q(q + q')}.$$

Hence $qz' + q' > q + q'$ and $z' > 1$, which proves the theorem.

■ **24.** *If p_r/q_r is the r-th convergent of the simple continued fraction of which the value is z, then $z^2 \gtrless p_n p_{n-1}/q_n q_{n-1}$, according as n is even or odd.*

For if k is the complete $(n + 1)$th quotient,

$$z = (kp_n + p_{n-1}) / (kq_n + q_{n-1}).$$

Therefore $z^2 \gtrless p_n p_{n-1}/q_n q_{n-1}$, according as

$$q_n q_{n-1} (kp_n + p_{n-1})^2 - p_n p_{n-1} (kq_n + q_{n-1})^2 \gtrless 0.$$

The expression on the left is equal to

$$(k^2 p_n q_n - p_{n-1} q_{n-1}) (p_n q_{n-1} - p_{n-1} q_n) = (-1)^n (k^2 p_n q_n - p_{n-1} q_{n-1}),$$

and this is positive or negative according as n is even or odd, for $k \geq 1$, $p_n > p_{n-1}$, and $q_n > q_{n-1}$, which proves the theorem.

■ **25.** *If p/q is an even convergent, and X/Y an odd convergent, of a simple continued fraction whose value is z, then*

$$pX \gtrless z^2 qY,$$

according as p/q precedes or follows X/Y in the sequence of convergents.

(i) If p/q precedes X/Y, let X'/Y' be the convergent which *immediately* precedes X/Y. Then X'/Y' is an even convergent, and it occurs later than p/q or else coincides with it, therefore

$$\frac{X}{Y} < z < \frac{X'}{Y'} \leq \frac{p}{q};$$

and since X/Y is an odd convergent immediately preceded by X'/Y',

$$z^2 < \frac{XX'}{YY'} \quad \text{and} \quad z^2 < \frac{pX}{qY}.$$

(*ii*) *If p/q follows X/Y, let p'/q' be the convergent which immediately precedes p/q. Then p'/q' is an odd convergent, and it occurs later than X/Y or else coincides with it, therefore*

$$\frac{X}{Y} \le \frac{p'}{q'} < z < \frac{p}{q};$$

and since p/q is an even convergent immediately preceded by p'/q',

$$z^2 > \frac{pp'}{qq'} \quad \text{and} \quad z^2 > \frac{pX}{qY}.$$

■ **26. Symmetric Continued Fractions:** A simple finite continued fraction, in which the quotients equidistant from the beginning and end are equal, is said to be *symmetric.* For instance, the fractions,

$$\frac{247}{77} = 3 + \frac{1}{4+} \frac{1}{1+} \frac{1}{4+} \frac{1}{3}, \frac{425}{132} = 3 + \frac{1}{4+} \frac{1}{1+} \frac{1}{1+} \frac{1}{4+} \frac{1}{3},$$

are symmetric, the former having an odd number of quotients, and the latter an even number.

The following is an important property of a fraction of this kind with an even number of quotients.

(1) *Let* $\dfrac{p}{q} = a_1 + \dfrac{1}{a_2+} \dots \dfrac{1}{a_r}$ *and* $\dfrac{P}{Q} = a_1 + \dfrac{1}{a_2+} \dots \dfrac{1}{a_r+} \dfrac{1}{a_r+} \dfrac{1}{a_{r-1}+} \dots \dfrac{1}{a_1},$

the fractions p/q, P/Q being in their lowest terms, and let p'/q', P'/Q' be the convergents immediately preceding p/q and P/Q, then

$$P = p^2 + p'^2, \quad Q' = q^2 + q'^2 \quad and \quad Q^2 + 1 = PQ'.$$

For by Art. 11, $P/P' = P/Q$, therefore $P' = Q$, also

$$\frac{p}{p'} = a_r + \frac{1}{a_{r-1}+} \dots \frac{1}{a_1} \quad \text{and} \quad \frac{q}{q'} = a_r + \frac{1}{a_{r-1}+} \dots \frac{1}{a_2};$$

therefore

$$\frac{P}{Q} = a_1 + \frac{1}{a_2+} \dots \frac{1}{a_r+} \frac{p'}{p} = \frac{p^2 + p'^2}{pq + p'q'},$$

$$\frac{P'}{Q'} = a_1 + \frac{1}{a_2+} \dots \frac{1}{a_r+} \frac{q'}{q} = \frac{pq + p'q'}{q^2 + q'^2}.$$

If $X = p^2 + p'^2$, and $Y = pq + p'q'$, we have

$$Yp - Xq = (-1)^r p' \quad \text{and} \quad Xq' - Yp' = (-1)^r p;$$

therefore any common factor of X and Y divides p and p', which are prime to one another, therefore X/Y is in its lowest terms. So also is P/Q, therefore $P = X$ and $Q = Y$, and consequently

$$P = p^2 + p'^2 \text{ and } Q' = q^2 + q'^2.$$

Moreover, $PQ' - P'Q = (-1)^{2r}$ and $P' = Q$, therefore $Q^2 + 1 = PQ'$.

■ **EXAMPLE 1.** *If p_r/q_r is the r-th convergent of the continued fraction*

$$\frac{P}{Q} = a_1 + \frac{1}{a_2 +} \cdots \frac{1}{a_r +} \frac{1}{a_r +} \cdots \frac{1}{a_1},$$

prove that the remainders in the process of finding the G.C.M. of Q, P are

$$p_{2r-2},\ p_{2r-3},\ \ldots,\ p_1,\ 1.$$

For $P = p_{2r}$, $Q = p_{2r-1}$ and $p_{2r} = a_1 p_{2r-1} + p_{2r-2}$ with similar equations, the last being $p_2 = a_1 p_1 + 1$. Hence the G.C.M. process is that shown on the right.

$$
a_2 \ \left|\ \begin{array}{c} p_{2r-1} \\ \dfrac{a_2 p_{2r-2}}{p_{2r-3}} \\ \cdots\cdots \\ p_1 \end{array} \right|\ \left.\begin{array}{c} p_{2r} \\ \dfrac{a_1 p_{2r-1}}{p_{2r-2}} \\ \cdots\cdots \\ \dfrac{p_2}{a_1 p_1} \\ 1 \end{array}\right|\ \begin{array}{c} a_1 \\ \\ a_3 \\ \\ a_1 \end{array}
$$

■ **EXAMPLE 2.** *Given that $218^2 + 1 = 25 \cdot 1901$, express 1901 as the sum of two squares.*

The reckoning shows that

$$\frac{1901}{218} = 8 + \frac{1}{1+} \frac{1}{2+} \frac{1}{1+} \frac{1}{1+} \frac{1}{2+} \frac{1}{1+} \frac{1}{8}.$$

Hence, if p_r/q_r is the rth convergent,

$$1901 = p_3^2 + p_4^2 = 26^2 + 35^2. \text{ (See Ex. 1 and Art. 26.)}$$

$$
\begin{array}{c|c} \begin{array}{c}1\\1\\2\\8\end{array} & \begin{array}{c}218\\61\\\dfrac{26}{8}\end{array} \end{array}
\quad
\begin{array}{c|c} \begin{array}{c}1901\\157\\\dfrac{35}{9}\\1\end{array} & \begin{array}{c}8\\2\\1\\1\end{array} \end{array}
$$

■ **27. Application to the Theory of Numbers:** *If Q is any positive integer, any factor of $Q^2 + 1$ can be expressed as the sum of two squares.*

For let $Q^2 + 1 = PR$ where $P > R$, then obviously $P > Q$ and P is prime to Q. Let

$$\frac{P}{Q} = a_1 + \frac{1}{a_2 +} \cdots \frac{1}{a_{2r}};\qquad\qquad \ldots\text{(A)}$$

then if P'/Q' is the convergent immediately preceding P/Q,

$$PQ' - P'Q = 1 = PR - Q^2.$$

Therefore $P(R - Q') = Q(Q - P')$ and, since P is prime to Q, it is a divisor of $Q - P'$ or else $Q = P'$. The first alternative is impossible, for $P > Q$ and $P > P'$, therefore $Q = P'$ and $R = Q'$.

Consequently $\dfrac{P}{Q} = \dfrac{P}{P'} = a_{2r} + \dfrac{1}{a_{2r-1} +} \cdots \dfrac{1}{a_2 +} \dfrac{1}{a_1}$, and therefore $a_{2r} = a_1$, $a_{2r-1} = a_2$, etc., and the continued fraction is symmetric.

Denoting the mth convergent of (A) by p_m/q_m, we have, by Art. 26,

$$P = p_r^2 + p_{r-1}^2, \quad R = Q' = q_r^2 + q_{r-1}^2,$$

which proves the theorem.

■ **28.** *Conversely, if $P = x^2 + y^2$ where x is prime to y and $x > y$, then an integer Q can be found so that $Q^2 + 1$ is divisible by P.*

Let $\dfrac{x}{y} = a_r + \dfrac{1}{a_{r-1} +} \cdots \dfrac{1}{a_1}$, and $\dfrac{X}{Q} = a_1 + \dfrac{1}{a_2 +} \dfrac{1}{a_r +} \dfrac{1}{a_r +} \dfrac{1}{a_{r-1} +} ,\cdots \dfrac{1}{a_1}$;

then, if p_m/q_m is the mth convergent to the last continued fraction,

$$X = p_r^2 + p_{r-1}^2 = x^2 + y^2 = P \text{ and } Q = p_{2r-1}.$$

Also $X q_{2r-1} - Q p_{2r-1} = 1$, therefore $Q^2 + 1 = P q_{2r-1}$.

■ **Example 1.** *Given that $3637 = 46^2 + 39^2$, find a solution in positive integers of the equation*

$$x^2 + 1 = 3637y.$$

Using Art. 28, we have $\dfrac{46}{39} = 1 + \dfrac{1}{5+} \dfrac{1}{1+} \dfrac{1}{1+} \dfrac{1}{3}$, and writing down the continued

fraction

$$3 + \dfrac{1}{1+} \dfrac{1}{1+} \dfrac{1}{5+} \dfrac{1}{1+} \dfrac{1}{1+} \dfrac{1}{5+} \dfrac{1}{1+} \dfrac{1}{1+} \dfrac{1}{3},$$

of which the last two convergents are $\dfrac{1027}{290}, \dfrac{3637}{1027}$, we have $3637 \times 290 - 1027^2 = 1$, and the required solution therefore is $(1027, 290)$.

Or thus: the last two convergents of $3 + \dfrac{1}{1+} \dfrac{1}{1+} \dfrac{1}{5+} \dfrac{1}{1}$ are $\dfrac{39}{11}$ and $\dfrac{46}{13}$. Therefore

$$y = 11^2 + 13^2 = 290 \quad \text{and} \quad x = \sqrt{(3637 \cdot 290 - 1)} = 1027.$$

■ **29.** *If x is prime to y, any factor of $x^2 + y^2$ is the sum of two squares.*

For we can find an integer Q such that $x^2 + y^2$ is a factor of $Q^2 + 1$ and *every* factor of $Q^2 + 1$ is the sum of two squares.

■ **30. Simple Recurring Continued Fractions:** These may be grouped as follows:

(*i*) Fractions which have *no* acyclic part.

(*ii*) Those with an acyclic part consisting of *a single* quotient.

(*iii*) Those with an acyclic part consisting of *more than one* quotient.

The reasons for this classification will appear later. The conclusions are of fundamental importance, and we proceed to consider these types in order.

■ **31. No Acyclic Part:**

If $\qquad \alpha = b_1 + \dfrac{1}{b_2 +} \dfrac{1}{b_3 +} \cdots \dfrac{1}{b_n +} \dfrac{1}{b_1 +} \dfrac{1}{b_2 +} \cdots,$ $\qquad$ ...(A) the fraction having *no acyclic part, then* α *is a root of a quadratic equation with integral coefficients. Moreover, if* β *is the second root of the equation, then*

$$-\frac{1}{\beta} = b_n + \frac{1}{b_{n-1}+} \frac{1}{b_{n-2}+} \cdots \frac{1}{b_1+} \frac{1}{b_n+} \frac{1}{b_{n-1}+} \cdots, \qquad \text{...(B)}$$

where the order of the quotients is reversed, whence it follows that

$$-1 < \beta < 0.$$

Denoting the rth convergents of the fractions (A), (B) by p_r/q_r and p_r'/q_r' respectively, we have

$$\alpha = b_1 + \frac{1}{b_2+} \cdots \frac{1}{b_n+} \frac{1}{\alpha} = \frac{\alpha p_n + p_{n-1}}{\alpha q_n + q_{n-1}},$$

therefore α is the positive root of

$$q_n x^2 - (p_n - q_{n-1})\, x - p_{n-1} = 0; \qquad \text{...(C)}$$

and in the same way the fraction (B) is the positive root of

$$q_n' x^2 - (p_n' - q_{n-1}')\, x - p_{n-1}' = 0,$$

which by Art. 11 is the same equation as

$$p_{n-1} x^2 - (p_n - q_{n-1})\, x - q_n = 0. \qquad \text{...(D)}$$

Now equation (D) can be transformed into equation (C) by writing $-1/x$ for x. Therefore $-1/\beta$ is the positive root of (D), and is consequently equal to the fraction (B).

Also, from (B), β is negative, and $-\dfrac{1}{\beta} > 1$; from which it follows that $\beta > -1$.

■ **32. A Single Acyclic Quotient:**

If $\qquad \alpha = a_1 + \dfrac{1}{b_1+} \dfrac{1}{b_2+} \cdots \dfrac{1}{b_n+} \dfrac{1}{b_1+} \dfrac{1}{b_2+} \cdots, \qquad \text{...(A)}$

where the single quotient a_1 *does not recur, then* α *is a root of a quadratic equation whose second root cannot lie between 0 and* -1.

Let $x = a_1 + \dfrac{1}{b_1+} \dfrac{1}{b_2+} \cdots \dfrac{1}{b_n+} \dfrac{1}{b_1+} \cdots$ and $y = b_1 + \dfrac{1}{b_2+} \cdots \dfrac{1}{b_n+} \dfrac{1}{b_1+} \cdots,$

then $x - a_1 = 1/y$, and if p_r/q_r is the rth convergent of the second fraction, as in Art. 31,

$$q_n y^2 - (p_n - q_{n-1})\, y - p_{n-1} = 0, \qquad \text{...(B)}$$

and therefore $\qquad p_{n-1}(x - a_1)^2 + (p_n - q_{n-1})\,(x - a_1) - q_n = 0. \qquad \text{...(C)}$

Denoting the roots of (B) by α, β where α is positive, and the corresponding roots of (C) by α', β', we have

$$\alpha' = a_1 + 1/\alpha, \quad \beta' = a_1 + 1/\beta.$$

The root α' is the value of the fraction (A), and by Art. 31, we have

$$\beta' = a_1 - \left(b_n + \frac{1}{b_{n-1}+} \frac{1}{b_{n-2}+} \cdots \right) = a_1 - b_n - f,$$

where f is a positive proper fraction.

Now a_1 and b_n are unequal, for if $a_1 = b_n$ then a_1 would belong to the cycle, which is contrary to the hypothesis. Consequently $\beta' = I - f$ or $- I - f$ where I is a positive integer, and therefore β' cannot lie between 0 and -1.

■ **33. More than one Quotient Acyclic:**

$$If\ \alpha = a_1 + \cfrac{1}{a_2 +} \cdots \cfrac{1}{a_m +} \cfrac{1}{b_1 +} \cfrac{1}{b_2 +} \cdots, \cfrac{1}{b_n +} \cfrac{1}{b_1 +} \cfrac{1}{b_2 +} \cdots, \qquad \text{...(A)}$$

in which at least two quotients do not recur, then α is a root of a quadratic equation with rational coefficients whose second root is positive.

Let $\qquad x = a_1 + \cfrac{1}{a_2 +} \cdots \cfrac{1}{a_m +} \cfrac{1}{b_1 +} \cfrac{1}{b_2 +} \cdots$ and $y = b_1 + \cfrac{1}{b_2 +} \cdots \cfrac{1}{b_n +} \cdots,$

and let P_r/Q_r and p_r/q_r be the rth convergents of the first and second fractions respectively, then

$$x = a_1 + \cfrac{1}{a_2 +} \cdots \cfrac{1}{a_m +} \cfrac{1}{y} = \frac{yP_m + P_{m-1}}{yQ_m + Q_{m-1}}, \qquad \text{...(B)}$$

and $\qquad q_n y^2 - (p_n - q_{n-1})y - p_{n-1} = 0.$ $\qquad$ ...(C)

If α, β are the roots of (C), α being positive, then by Art. 31,

$$\alpha = b_1 + \cfrac{1}{b_2 +} \cdots \cfrac{1}{b_n +} \cfrac{1}{b_1 +} \cdots. \quad \text{and} \quad -\frac{1}{\beta} = b_n + \cfrac{1}{b_{n-1} +} \cdots \cfrac{1}{b_1 +} \cfrac{1}{b_n +} \cdots. \qquad \text{...(D)}$$

Also if we eliminate y from equations (B), (C) we obtain a quadratic in x with rational coefficients, whose roots α', β' are found by substituting α, β respectively for y in (B). Thus α' is equal to the fraction (A) and

$$\beta' = a_1 + \cfrac{1}{a_2 +} \cdots \cfrac{1}{a_{m-1} +} \cfrac{1}{a_m + 1/\beta}.$$

From equation (D) it follows that $a_m + 1/\beta = a_m - b_n - f$, where f is a positive proper fraction. Now a_m is not equal to b_n, for if $a_m = b_n$ then a_m would belong to the cycle, which is contrary to the hypothesis. It follows that

$$a_m + 1/\beta = I - f \text{ or } - I - f,$$

where I is a positive integer. Thus

$$a_{m-1} + \cfrac{1}{a_m + 1/\beta} = a_{m-1} + \cfrac{1}{I - f} \quad \text{or} \quad a_{m-1} - \cfrac{1}{I + f},$$

and since a_{m-1} is a positive integer, the expression on the left, and therefore also β', is positive in all cases.

> **Note:** The argument depends on the existence of a_{m-1}, and fails if there is not more than one quotient in the acyclic part.

■ **34. Summary:** It has been proved that *any simple recurring continued fraction is a root of a quadratic equation with rational coefficients.*

Also the second root β of the equation is restricted as follows:

   (*i*) *If there is no acyclic part, then* $-1 < \beta < 0$.

  (*ii*) *If the acyclic part consists of a single quotient, then* $\beta < -1$ *or* $\beta > 0$.

 (*iii*) *If the acyclic part contains at least two quotients, then* $\beta > 0$.

◼ **Example :** *Find the value of* $1 + \dfrac{1}{1+}\dfrac{1}{1+}\dfrac{1}{3+}\dfrac{1}{2}$.

Denoting the value of the fraction by x, we have

$$x = 1 + \frac{1}{1+}\frac{1}{y} \quad \text{where } y = 1 + \frac{1}{3+}\frac{1}{2+}\frac{1}{y};$$

$\therefore$
$$x = (2y + 1)/(y + 1) \text{ and } y = (9y + 4)/(7y + 3),$$

giving
$$y = \left(\sqrt{37}+3\right)/7 \text{ and } x = \left(\sqrt{37}+8\right)/9.$$

EXERCISE XLI

1. For the fraction $\dfrac{1}{a+}\dfrac{1}{a+}\dfrac{1}{a+}\ldots$, prove that $p_n = q_{n-1}$ and

$$1/(1 - ax - x^2) = 1 + q_1 x + q_2 x^2 + \ldots + q_n x^n + \ldots,$$

provided that $|x| < \frac{1}{2}\left(-a + \sqrt{a^2 + 4}\right)$. (See Ch. 22, 2)

2. For the fraction $\dfrac{1}{a_1+}\dfrac{1}{a_2+}\ldots\dfrac{1}{a_n+}\ldots$, prove that

  (*i*) $p_n q_{n-3} - q_n p_{n-3} = (-1)^{n-1}(a_n a_{n-1} + 1)$,

 (*ii*) $p_n q_{n-4} - q_n p_{n-4} = (-1)^{n-1}(a_n a_{n-1} a_{n-2} + a_n + a_{n-2})$.

3. Prove that

$$a + \frac{1}{b+}\frac{1}{2a+}\frac{1}{b+}\frac{1}{2a+}\ldots = \sqrt{\frac{a}{b}(ab+2)}.$$

Deduce the following:

$$\sqrt{a(a+1)} = a + \frac{1}{2+}\frac{1}{2a+}\frac{1}{2+}\frac{1}{2a+}\ldots, \quad \sqrt{a(a+2)} = a + \frac{1}{1+}\frac{1}{2a+}\frac{1}{1+}\frac{1}{2a+}\ldots,$$

$$\sqrt{a^2-1} = a - 1 + \frac{1}{1+}\frac{1}{2(a-1)}\ldots, \qquad\qquad \sqrt{a^2+2} = a + \frac{1}{a+}\frac{1}{2a+}\ldots.$$

4. Prove that

$$\sqrt{a^2+1} = a + \frac{1}{2a+}\frac{1}{2a+}\frac{1}{2a+}\ldots,$$

$$\sqrt{a^2-2} = a - 1 + \frac{1}{1+}\frac{1}{a-2+}\frac{1}{1+}\frac{1}{2(a-1)}.$$

5. Prove that

$(i)\ \ a+\cfrac{1}{b+}\cfrac{1}{c} = \dfrac{abc+c+a-b+\sqrt{\{(abc+a+b+c)^2+4\}}}{2(bc+1)}$

$(ii)\ \ a+\cfrac{1}{b+}\cfrac{1}{c+}\cfrac{1}{b+}\cfrac{1}{2a} = \sqrt{\left(a+\dfrac{1}{b}\right)\left(a+\dfrac{c}{bc+2}\right)}.$

6. Show that $449/120$ differs from $\sqrt{14}$ by less than $1/90,000$. (See Art. 22, Ex. 1)

7. A straight ruler is graduated in inches and centimetres, the ratio of an inch to a centimetre being N. If N is expressed as a simple continued fraction and p/q is any convergent, show that the distance between the graduations 'p centimetres' and 'q inches' is less than the distance between any two preceding graduations corresponding to a convergent of the fraction.

8. Find the fractions which, of all those with denominators less than 500, are the closest approximations (i) in defect and (ii) in excess to 0.2422642. (See Art. 22, Ex. 2)

9. Given that 1 metre = 3.2809 ft., obtain the two approximations, 8 kilometres = 5 miles, 103 kilometres = 64 miles; and find an upper limit of error in each case.

10. Given that 1 kilogramme = 2.2046 pounds, show that 44 kilogrammes is slightly greater than 97 pounds, the error being less than half a grain.

Obtain also the usual approximation, 7 grammes = 108 grains.

11. If $p/q, p'/q', p''/q''$ are consecutive convergents to a simple continued fraction F, prove

that F lies between $\sqrt{pp'/qq'}$ and $\sqrt{p'p''/q'q''}$.

12. For the fraction $F = \cfrac{1}{a+}\cfrac{1}{b+}\cfrac{1}{a+}\cfrac{1}{b+}\ldots$, prove that

$(i)\ \ q_n = p_{n-1};$

$(ii)\ \ F = \dfrac{1}{2a}\left\{-ab+\sqrt{ab(ab+4)}\right\}.$

Also if α, β are the roots of $x^2 - (ab+2)x + 1 = 0$ and $d_n = \alpha^n - \beta^n$, then

$(iii)\ \ ap_{2n} = bq_{2n-1} = abd_n/d_1;$

$(iv)\ \ p_{2n-1} = q_{2n} = (d_{n-1} - d_n)/d_1;$

$(v)\ \ $ if $c = ab + 2$, and $x^2 < \tfrac{1}{2}\left(-c+\sqrt{c^2+4}\right)$, then

$$\dfrac{1+bx-x^2}{1-cx^2+x^4} = p_1 + p_2 x + \ldots + p_{n+1}x^n + \ldots,$$

$$\dfrac{1+ax-x^2}{1-cx^2+x^4} = 1 + q_1 x + q_2 x^2 + \ldots + q_n x^n + \ldots,$$

[If u_n stands for p_{2n} or for p_{2n-1}, then, by Art. 5, Ex. 2,

$$u_n - cu_{n-1} + u_{n-2} = 0, \therefore u_n = A\alpha^n + B\beta^n.$$

If $u_n = p_{2n}$, we have $A + B = p_0 = 0$, $A\alpha + B\beta = p_2 = b$,

giving $\qquad A = -B = b/d_1$, $\therefore p_{2n} = bd_n/d_1$.

If $u = p_{2n-1}$, since $c = \alpha + \beta$, then $A\alpha + B\beta = p_1 = 1$, $\quad A\alpha^2 + B\beta^2 = p_3 = c - 1$,

giving $A = (1 - \alpha^{-1})/d_1$, $B = -(1 - \beta^{-1})/d_1$, $\quad \therefore p_{2n-1} = (d_n - d_{n-1})/d_1$.

Finally, it is easy to show that

$$1/(1 - cx^2 + x^4) = (d_1 + d_2 x^2 + \ldots\, d_{n-1}x^{2n} + \ldots\,)/d_1.$$

and then the results in (v) can be obtained by equating coefficients.]

13. If p_n/q_n is the nth convergent to $\cfrac{1}{a_1 +}\cfrac{1}{a_2 +}\ \ldots\ \cfrac{1}{a_n}$, then

$$p_{n-1}\,q_{s-1} - q_{n-1}\,p_{s-1}$$

is the numerator of the $(n - s)$th convergent to $\cfrac{1}{a_n +}\cfrac{1}{a_{n-1} +}\cfrac{1}{a_{n-2} +}\ \ldots\ .$

14. Prove that the fraction $a - \cfrac{1}{a -}\cfrac{1}{a -}\ \ldots\ \cfrac{1}{a -}\cfrac{1}{a - x}$, in which a is equal to -1 and is repeated

any number of times, must have one of three values, and that if x satisfies the equation
$2x^3 + 3x^2 - 3x - 2 = 0$, the fraction satisfies this equation.

15. If $a_1 + \cfrac{1}{a_2 +}\ \ldots\ \cfrac{1}{a_{r-1} +}\cfrac{1}{a_r +}\cfrac{1}{a_{r-1} +}\ \ldots\ \cfrac{1}{a_2 +}\cfrac{1}{a_1} = \dfrac{P}{Q}$ and P/Q is in its lowest terms,

prove that $Q^2 - 1$ is divisible by P.

16. If P, Q, R are positive integers such that $Q^2 - 1 = PR$ and $P > R$, then P/Q. can be
expressed in the form

$$\frac{P}{Q} = a_1 + \cfrac{1}{a_2 +}\ \ldots\ \cfrac{1}{a_{r-1} +}\cfrac{1}{a_r +}\cfrac{1}{a_{r-1} +}\ \ldots\ \cfrac{1}{a_2 +}\cfrac{1}{a_1},$$

and if p_m/q_m is the mth convergent, then

$$P = p_r\,(p_r + p_{r-2}), \quad Q = p_r q_{r-1} + p_{r-1} q_{r-2}, \quad R = q_{2r} = q_{r-1}\,(q_r + q_{r-2}).$$

Verify by taking $P = 33$, $Q = 23$.

■ ■ ■

CHAPTER 25

Indeterminate Equations of the First Degree

In this chapter a, b, c,... stand for positive integers.

■ **1. Under this heading** we consider integral solutions, and more particularly positive integral solutions, of a single linear equation in two or more variables or of a system of m linear equations in n variables ($m < n$).

■ **2. A Single Equation of the Form $ax \pm by = \pm c$:** It is obvious that we need only consider the forms

$$ax - by = c \quad \text{and} \quad ax + by = c.$$

If either of these equations is satisfied by integral values of x, y, then any common divisor of a, b must divide c. Hence *if a, b have a common divisor which does not divide c, no solution in integers exists.*

If a, b have a common divisor which divides c, this can be removed by division. Hence *there is no loss of generality in supposing that a is prime to b, and we shall assume this to be the case.*

■ **3. The Equation $ax - by = 1$ (a prime to b):**

(*i*) *To find a solution in positive integers,* express a/b as a simple continued fraction with an *even* number of quotients (Ch. 24, 4). The last convergent is a/b. Let p/q be the convergent immediately preceding a/b. Then since a/b is an even convergent, $aq - bp = 1$ (Ch. 24, 9), and therefore (q, p) is a solution.

(*ii*) *To find the general solution in integers,* suppose that (x, y) is *any* solution in integers, then

$$ax - by = 1 = aq - bp.$$

Therefore $a(x - q) = b(y - p)$. Hence $a(x - q)$ is divisible by b, and since a is prime to b, $x - q$ must be divisible by b. Consequently

$$(x - q)/b = (y - p)/a = t,$$

where t is an integer or zero. Hence

$$x = q + bt, \quad y = p + at \quad \text{where} \quad t = 0, \pm 1, \pm 2, \quad \text{etc.}$$

This is called the *general solution in integers.*

(*iii*) Since p/q is a convergent which precedes a/b, it follows that $p < a$ and $q < b$, and from the form of the general solution we conclude that (q, p) *is the only solution in positive integers such that $p < a$ and $q < b$.* We shall call this *the 'least' solution in positive integers.*

(*iv*) *Since* $a(b - q) - b(a - p) = -1$, the 'least' solution in positive integers of $ax - by = -1$ is $(b - q, a - p)$.

■ **4. The Equation $ax - by = c$ (a prime to b):**

(*i*) *To find a solution in positive integers*, express a/b as a simple continued fraction with an even number of quotients, and let p/q be the convergent immediately preceding a/b. Then

$$aq - bp = 1 \text{ and } aqc - bpc = c.$$

Hence (qc, pc) is a solution.

(*ii*) *To find the general solution in integers.* The equation may be written $a(x - qc) = b(y - pc)$, and since a is prime to b, $x - qc$ must be divisible by b. Therefore $(x - qc)/b = (y - pc)/a = t$, where t is an integer or zero.

Hence the general solution is

$$x = qc + bt, \quad y = pc + at \ (t = 0, \pm 1, \pm 2, \ldots).$$

(*iii*) Let m be the integral part of the smaller of pc/a, qc/b, then the *least solution in positive integers* is (α, β) where $\alpha = qc - bm$, $\beta = pc - am$, and *the general solution in positive integers is*

$$x = \alpha + bt, y = \beta + at \ (t = 0, 1, 2, \ldots).$$

Thus *the equation has infinitely many positive integral solutions.*

■ **EXAMPLE :** *Find the general solution in positive integers of*
(*i*) $13x - 17y = 5$; (*ii*) $13x - 17y = -5$; (*iii*) $18x - 17y = 996$.
We find that

$$\frac{13}{17} = 0 + \frac{1}{1+} \frac{1}{3+} \frac{1}{4} = 0 + \frac{1}{1+} \frac{1}{3+} \frac{1}{3+} \frac{1}{1}.$$

The convergents of the continued fractions are

$$\frac{0}{1}, \frac{1}{1}, \frac{3}{4}, \frac{13}{17}; \quad \frac{0}{1}, \frac{1}{1}, \frac{3}{4}, \frac{10}{13}, \frac{13}{17};$$

and
$$13 \cdot 4 - 17 \cdot 3 = 1, \ 13 \cdot 13 - 17 \cdot 10 = -1.$$

Hence *for the first equation* a solution is $(20, 15)$. The general solution in integers is

$$x = 20 + 17t, \quad y = 15 + 13t \quad (t = 0, \pm 1, \pm 2, \ldots).$$

The least solution in positive integers (given by $t = -1$) is $(3, 2)$, and the general solution in positive integers is $x = 3 + 17t, y = 2 + 13t \ (t = 0, 1, 2, \ldots)$.

Similarly *for the second equation* a solution is given by $x = 13 \cdot 5 = 65, y = 10$. $5 = 50$, and the general solution in positive integers is

$$x = 14 + 17t, \quad y = 11 + 13t \quad (t = 0, 1, 2, \ldots).$$

In case of *the third equation,* the reckoning may be shortened thus : Dividing 996 by 13 (the smaller of the coefficients), we find that $996 = 13 \cdot 76 + 8 = 13 \cdot 77 - 5$.

Hence the equation may be written $13 (x - 77) - 17y = -5$, and by the preceding the least positive integral solution is given by $x - 77 = 14, y = 11$.

Therefore the general solution in positive integers is

$$x = 91 + 17t, \quad y = 11 + 13t \ (t = 0, 1, 2, \ldots).$$

Note: In this way the solution of $ax - by = \pm c$ can be made to depend on that of an equation of the same form in which c is less than or equal to the smaller of $\frac{1}{2}a$, $\frac{1}{2}b$.

■ **5. The Equation $ax + by = c$ (a prime to b):**

(*i*) *General solution in integers.* Let a/b be expressed as a simple continued fraction with an even number of quotients, and let p/q be the last convergent but one. Then $aq - bp = 1$, and the equation may be written

$$ax + by = c\,(aq - bp) \quad \text{or} \quad a(cq - x) = b(y + cp).$$

Since a is prime to b, $cq - x$ must be divisible by b, and therefore

$$(qc - x)/b = (y + pc)/a = t,$$

where t is an integer or zero.

Hence the general solution in integers is

$$x = qc - bt, \quad y = at - pc \quad (t = 0, \pm 1, \pm 2, \text{etc.}).$$

(*ii*) *Positive integral solutions.* The values of t (if such exist) for which x and y are positive are given by the conditions, $pc/a < t < qc/b$.

Let m, n have positive integral or zero values such that

$$pc/a = m + f, \quad qc/b = n + g, \quad \text{where} \quad 0 \le f < 1 \quad \text{and} \quad 0 < g \le 1;$$

then x, y will be positive integers if $t = m + 1, m + 2, \dots n$.

Hence if N is the number of solutions in positive integers,

$$N = n - m = \frac{qc}{b} - \frac{pc}{a} + f - g = \frac{c}{ab} + f - g.$$

Now $-1 \le f - g < 1$, and therefore

$$\frac{c}{ab} - 1 \le N < \frac{c}{ab} + 1.$$

Hence, *if c is not divisible by ab*, the number (N) of solutions in positive integers is one of the integers next to c/ab, and is greater than or less than c/ab according as $f \gtrless g$.

If c is divisible by ab, then pc/a and qc/b are integers; therefore

$$f = 0, \ g = 1 \ \text{and} \ N = \frac{c}{ab} - 1.$$

■ **EXAMPLE :** *Find the positive integral solutions of $13x + 17y = 3000$. Use the rule just given, to show that there are 13 such solutions.*

As in Art. 4, Ex. 1, we have $13 \cdot 4 - 17 \cdot 3 = 1$, and the equation may be written

$$13x + 17y = 3000(13 \cdot 4 - 17 \cdot 3) \quad \text{or} \quad 13(12000 - x) = 17(y + 9000):$$

and the integral solutions are given by $x = 12000 - 17t$, $y = 13t - 9000$.

Moreover, $\dfrac{12000}{17} = 705\dfrac{15}{17}, \quad \dfrac{9000}{13} = 692\dfrac{4}{13};$

therefore the positive integral solutions are given by $t = 693, 694, \dots 705$. These solutions are (15, 165), (32, 152), ... (219, 9).

Now, $13 < \dfrac{3000}{13 \cdot 17} < 14$; also $f = \dfrac{4}{13}$, $g = \dfrac{15}{17}$; so that $f < g$; hence the number of solutions is 13. Since $705 - 693 + 1 = 13$, this result agrees with the preceding.

■ 6. Two Equations with Three Unknowns.

(1) The general solution in integers, and all the positive integral solutions, if any exist, may be found as in the next example.

| EXAMPLE : *Find all the positive integral solutions of the pair of equations,*
$$2x - 2y + 3z = 7,\ -4x + 7y + 3z = 19.$$

Eliminating one of the unknowns (z), we have $2x - 3y = -4$; and the general solution in integers of this equation is $x = 3m - 2$, $y = 2m$, $(m = 0, \pm 1, \pm 2 \ldots)$.

Substituting these values of x, y in one of the given equations, say in the first, we find that $2m + 3z = 11$; and the general solution in integers of this equation is

$m = 4 - 3t$, $z = 1 + 2t$, and therefore $x = 3m - 2 = 10 - 9t$, $y = 2m = 8 - 6t$.

Hence the general solution in integers is $x = 10 - 9t$, $y = 8 - 6t$, $z = 1 + 2t$.

The only values of t for which x, y, z are positive are 0 and 1, therefore the only positive integral solutions are $(10, 8, 1)$ and $(1, 2, 3)$.

(2) *If (α, β, γ) is a solution in integers of the two equations*
$$ax + by + cz = d,\ a'x + b'y + c'z = d',$$
it is required to find the general solution in integers.

(i) If none of the three (bc'), (ca'), (ab') is zero, by Ch. 10, 6, (4), the equations are equivalent to

$$\frac{x - \alpha}{(bc')} = \frac{y - \beta}{(ca')} = \frac{z - \gamma}{(ab')}. \qquad \ldots(A)$$

Denote each member of (A) by t/f where t may be any integer or zero and f is an integer to be chosen later, then

$$x - \alpha = (bc')t/f,\ y - \beta = (ca')t/f,\quad z - \gamma = (ab')t/f.$$

Thus x, y, z will be integers for *all* integral values of t if and only if f is a common factor of (bc'), (ca'), (ab').

Now every common factor of these numbers is a factor of their G.C.M., hence *all integral solutions are given by*

$$\frac{x - \alpha}{(bc')} = \frac{y - \beta}{(ca')} = \frac{z - \gamma}{(ab')} = \frac{t}{g},$$

where g is the G.C.M. of (bc'), (ca'), (ab') and $t = 0, \pm 1, \pm 2, \ldots$.

(ii) If $(bc') = 0$, the equations are equivalent to

$$x = \alpha,\ \frac{y - \beta}{(ca')} = \frac{z - \gamma}{(ab')},$$

and it is easily seen that *the general solution in integers is*

$$x = \alpha,\ \frac{y - \beta}{(ca')} = \frac{z - \gamma}{(ab')} = \frac{t}{g_1},$$

where g_1 is the G.C.M. of (ca'), (ab').

■ **7. A Single Equation of the Form** $ax + by + cz + \ldots = d.$

(1) If a, b, c, ... are all positive, then all the positive integral solutions (if any exist) may be found as follows:

▌ **EXAMPLE 1.** *Find the positive integral solutions of* $7x + 11y + 26z = 123.$

The greatest coefficient is 26, and $7 + 11 + 26z \leq 123$, hence $z \leq 4$. Therefore the only possibilities are

$z = 1,$	$7x + 11y = 97,$	of which there is one solution $(6, 5),$
$z = 2,$	$7x + 11y = 71,$	of which there is one solution $(7, 2),$
$z = 3,$	$7x + 11y = 45,$	of which there is no solution $(7, 2),$
$z = 4,$	$7x + 11y = 19,$	of which there is no solution $(7, 2).$

Thus, there are two solutions in positive integers, namely $(6, 5, 1)$ and $(7, 2, 2)$.

▌ **EXAMPLE 2.** *Find all the positive integral solutions of*
$$3x_1 + 4x_2 + 7x_3 + 11x_4 = 67.$$

Here $11x_4 \leq 67 - (3 + 4 + 7)$, therefore $x_4 < 5$, and we may have

$x_4 = 4,$	$x_3 = 1,$	$3x_1 + 4x_2 = 16,$	$x_1 = 4, x_2 = 1,$	
	$x_3 = 2,$	$3x_1 + 4x_2 = 9,$	no solution,	
$x_4 = 3,$	$x_3 = 1,$	$3x_1 + 4x_2 = 27,$	$x_1 = 5, 1,$	$x_2 = 3, 6,$
	$x_3 = 2,$	$3x_1 + 4x_2 = 20,$	$x_1 = 4,$	$x_2 = 2,$
	$x_3 = 3,$	$3x_1 + 4x_2 = 13,$	$x_1 = 3,$	$x_2 = 1,$
	$x_3 = 4,$	$3x_1 + 4x_2 = 6,$	no solution.	

The other possibilities, namely $x_4 = 2, 1$, may be treated in the same way. There are 28 solutions in all.

(2) If (α, β, γ) is an integral solution of $ax + by + cz = d$ (where a, b, c are positive or negative integers), *other solutions are given by*
$$x = \alpha + bw - cv, \quad y = \beta + cu - aw, \quad z = \gamma + av - bu,$$
where u, v, w have any integral or zero values.

This is obviously true. Further, we can show that *if any two of the three a, b, c are prime to each other, then every integral solution can be so expressed.*

For if $x - \alpha$, $y - \beta$, $z - \gamma$ are denoted by X, Y, Z respectively, the equation becomes
$$aX + bY + cZ = 0. \qquad\qquad \ldots\text{(A)}$$

Let a be prime to b, then integers u, v can be found so that, whatever integral value Z may have,
$$av - bu = Z.$$

If, then, X, Y, Z are integers which satisfy (A),
$$a(X + cv) + b(Y - cu) = aX + bY + cZ = 0;$$
and since a is prime to b, we must have
$$X + cv = bw, \quad Y - cu = -aw,$$
where w is some integer or zero. This proves the statement in question.

EXERCISE XLII

1. Find the least solution in positive integers of

   (*i*) $68x - 157y = 1$;　　　　　　　　　(*ii*) $68x - 157y = -1$.

   In Exs. 2–7, find the general solution in integers and the least positive integral values of x, y.

2. $8x - 27y = 125$.　　　　　　　　　3. $27x - 8y = 125$.

4. $29x - 13y = 1$.　　　　　　　　　5. $17x - 41y = 1$.

6. $12x - 7y = 211$.　　　　　　　　　7. $199x - 225y = 20$.

   Find all the positive integral solutions of equations 8–10.

8. $7x + 8y = 50$.　　　　　　　　　9. $11x + 7y = 151$.

10. $12x + 29y = 700$.

11. Express 68/77 as the sum of two proper fractions whose denominators are 7 and 11.

12. Show that $1/(13 \times 17)$ can be expressed in two ways as the difference between two proper fractions whose denominators are 13 and 17; and express the given fraction in these ways.

13. AB and CD are two rods, each one foot in length. AB is scaled off into 13 equal parts, and CD into 17 equal parts. If the two rods are placed side by side, with the scales in contact, and the ends of the rods in alignment, show that the distance between one scale division of AB and one of CD can never be less than $1/(13 \times 17)$ of a foot, and that there are two cases in which the distance has this value. Which are the pairs of divisions in these cases?

14. The sum of two positive integers is 100. If one is divided by 7 the remainder is 1, and if the other is divided by 9 the remainder is 7. Find the numbers.

15. If c is increased by kab, then the number of positive integral solutions of $ax + by = c$ is increased by k. [For c/ab is increased by k and f, g are unaltered.]

16. The least value of c for which the equation $ax + by = c$ has N solutions is given by

$$c = a + b + (N - 1)ab.$$

   [This follows from Ex. 15, coupled with the fact that $a + b$ is the least value of c for which the equation has one solution.]

17. Find the least positive integer, a, for which the equation $3x + 4y = a$ has six positive integral solutions.

18. Find the least value of c in order that the equation $13x + 17y = c$ may have 13 solutions in positive integers, showing that the required value is 2682. Verify by finding the number of solutions for $c = 2682, 2681$.

19. The number of solutions in positive integers of the equation $x + by = c$ is the integral part of c/b.

   Find all the positive integral solutions of equations 20–23.

20. $\left.\begin{array}{l} x + 3y - 4z = 8, \\ 2x + y + 3z = 39. \end{array}\right\}$　　　　　　21. $\left.\begin{array}{l} 7x + 11y + 13z = 201, \\ 11x + 13y + 7z = 209. \end{array}\right\}$

22. $9x + 16y + 25z = 109$.

23. $17x + 25y + 23z = 270$.

24. Find the least number which when divided by 7, 11, 20, leaves the remainders 1, 5, 9, respectively. Find also the general form of all such numbers.

25. A certain set of stamps is made up of four different sorts: a collector has one of each sort and some duplicates. The values of the stamps are 2d., 3d., 9d; Is., respectively; and the lot are worth 3s. 9d. Find all the different ways in which the batch could have been made up.

26. A man has in his pocket three half-crowns, seven other silver coins which are either shillings or sixpences, and eightpence in coppers; he paid a bill of 9s. 7d. Find the possible number of different ways of doing so.

27. Find the sum of money which is expressed in farthings by the same digits written in the same order as those which are required to express it in pounds, shillings and pence.

How many solutions are there to this problem?

28. If (p_1, p_2, p_3, p_4) is an integral solution of $ax_1 + bx_2 + cx_3 + dx_4 = e$, show that other solutions are given by

$$x_1 = p_1 + bw - cv - du',$$
$$x_2 = p_2 + cu - dv' - aw,$$
$$x_3 = p_3 + dw' + av - bu,$$
$$x_4 = p_4 + au' + bv' - cw',$$

where u, v, w, u', v', w' have any integral or zero values.

29. If $ax + by + c = 0$, $a'x + b'y + c' = 0$ and x is an integer, prove that y is an integer, unless both b and b' are divisible by $ab' - a'b$.

[For $a(bc') + b(ca') + c(ab') = 0$ and $a'(bc') + b'(ca') + c'(ab') = 0$.]

■■■

CHAPTER 26

Theory of Numbers

■ 1. Congruence:

(1) Let n be any positive whole number, which we shall call the *modulus*. If a, b are any numbers, positive or negative, such that $a - b$ is divisible by n, then a and b are said to be *congruent* with respect to the modulus n, and each is said to be a *residue* of the other.

This is expressed by writing $a \equiv b$ (mod n) or $a - b \equiv 0$ (mod n), where the symbol $\equiv$ is an abbreviation for '*is congruent with.*' When no doubt exists as to the modulus in question, we simply write $a \equiv b$. Any statement of this kind is called a *congruence*.

Every residue of a to the modulus n is of the form $a + qn$, where q may be positive or negative. If r is the remainder when a is divided by n, we have

$$a \equiv r \text{ (mod } n) \text{ and } 0 \leq r < n;$$

thus r is the *least positive residue* of a.

Again, $a \equiv r \equiv -(n - r)$ (mod n), and either r or $n - r \leq \frac{1}{2} n$. Hence we can always find r' so that $a \equiv r'$ (mod n) and $| r' | \leq \frac{1}{2} n$, where r' may be positive or negative. This number r' is called the *absolute least residue* of a.

For example, if 5 is the modulus, $34 \equiv 4 \equiv -1$ and $-34 \equiv -4 \equiv 1$. Thus the least positive residues of 34 and −34 are 4 and 1, and the absolute least residues are −1 and 1 respectively.

(2) Fundamental Theorems: *If $a \equiv b(mod\ n)$ and $a' \equiv b'$ (mod n), then (i) $a + a' \equiv b + b'$. (ii) $a - b \equiv a' - b'$. (iii) $aa' \equiv bb'$, and in particular $ac \equiv bc$. (iv) If a and b are divisible by d, a number prime to n, then $a/d \equiv b/d$.*

For by hypothesis $a - b$ and $a' - b'$ are divisible by n, hence

$$(a + a') - (b + b') \text{ and } (a - a') - (b - b') \text{ are divisible by } n.$$

Again, $\qquad aa' - bb' = (a - b)a' + (a' - b')\, b,$

therefore $aa' - bb'$ is divisible by n. This proves the theorems (i)–(iii).

Lastly, $a - b = qn$, where q is a whole number, then since a and b are divisible by d, so also is qn. Now d is prime to n, therefore q/d is a whole number, and since $a/d - b/d = q/d\,.\,n$, it follows that $a/d \equiv b/d$ (mod n).

In particular, *if $a \equiv b(mod\ p)$ where p is a prime and if d is any common divisor of a and b, then $a/d \equiv b/d$ (mod p), except when $d \equiv 0$ (mod p).*

Thus a close analogy exists between 'congruences to a prime modulus' and 'equalities.'

(3) The following are immediate consequences of the theorems in § (2):

(i) If to any modulus $a \equiv a'$, $b \equiv b'$, ... $k \equiv k'$, then to the same modulus $ab ... k \equiv a'b' ... k'$, and in particular $a^m \equiv a'^m$.

(ii) If $f(x, y, z,...)$ is a polynomial in $x, y, z,...$ with integral coefficients and $x \equiv x'$, $y \equiv y'$, $z \equiv z'$, ... to any modulus, then to the same modulus $f(x, y, z, ...) \equiv f(x', y', z', ...)$.

▮ **EXAMPLE :** *Give a test as to the divisibility of a number by 7, 11 or 13.*

Any number N may be written in the form

$$N = a_0 + a_1 t + a_2 t^2 + ... + a_n t^n \text{ where } t = 1000 \text{ and } 0 \le a_r < t.$$

Let $s = a_0 - a_1 + a_2 - ... + (-1)^n a_n$, then we have $t + 1 = 1001 = 7. 11. 13$; hence $t \equiv -1$ and $t^r \equiv (-1)^r$, with regard to each of the moduli 7, 11 and 13. Consequently $N \equiv s$ (mod 7, 11 or 13).

Thus if $N = 12345671$, then $s = 671 - 345 + 12 = 338$. Now 338 is divisible by 13 but not by 7 or 11; therefore 12345671 is divisible by 13 but not by 7 or 11.

▮ **2. The Numbers Less than a Given Number and Prime to it:** If n is any number greater than 1, the number of positive integers less than n and prime to it is denoted by $\varphi(n)$: moreover, we *define* the value of $\varphi(1)$ as 1. Here, it should be noted that 1 is to be regarded as prime to every other number. Thus the numbers less than 12 and prime to it are 1, 5, 7, 11, and $\varphi(12) = 4$.

(1) *If a is prime to n, the number of terms of the arithmetical progression*

$$x, \ x + a, x + 2a, \ ... \ x + (n-1)a$$

which are prime to n is $\varphi(n)$.

For if these numbers are divided by n, the remainders are $0, 1, 2, ... n - 1$, taken in a certain order (Ch. 1, 10), and if a number is prime to n, so also is its remainder. Consequently, as many terms in the progression are prime to n as there are numbers less than n and prime to it.

(2) *If m is prime to n, then $\varphi(mn) = \varphi(m) . \varphi(n)$.*

To enumerate the numbers less than mn and prime to it, we arrange the numbers $1, 2, 3, ... mn$ in n rows and m columns as below:

1	...	2	...	k	...	m
$m + 1$	...	$m + 2$	...	$m + k$	...	$2m$
$2m + 1$	...	$2m + 2$	...	$2m + k$	...	$3m$
....						
$(n-1)m + 1$	$(n-1)m + 2...$		$(n-1)m + k ...$		$nm.$

Since m is prime to n, the numbers in question are those which are prime to both m and n. Now every member of the kth column is or is not prime to m, according as k is or is not prime to m.

Thus the numbers in question, being prime to m, are members of the $\varphi(m)$ columns headed by numbers prime to m.

Moreover, every column contains $\varphi(n)$ numbers which are also prime to n, for the numbers in each column form an arithmetical progression of n terms with a common difference m which is prime to n.

It follows that there are $\varphi(m) \cdot \varphi(n)$ numbers less than mn and prime to both m and n, that is to mn, and so $\varphi(mn) = \varphi(m) \cdot \varphi(n)$.

(3) *If* m_1, m_2, m_3 ... m_r *are prime to one another, then*

$$\varphi(m_1\, m_2\, m_3\, ...\, m_r) = \varphi(m_1) \cdot \varphi(m_2) \cdot \varphi(m_3) ... \varphi(m_r).$$

For by (2), $\varphi(m_1 m_2) = \varphi(m_1), \varphi(m_2)$. Also m_3 is prime to both m_1 and m_2, therefore it is prime to $m_1 m_2$. Hence

$$\varphi(m_1 m_2 m_3) = \varphi(m_1 m_2) \cdot \varphi(m_3) = \varphi(m_1) \cdot \varphi(m_2) \cdot \varphi(m_3),$$

and so on, for any number of steps.

(4) *If p is a prime, then* $\varphi(p^r) = p^r(1 - 1/p)$.

The proposition is true when $r = 1$, for $\varphi(p) = p - 1$.

If $r > 1$, since p is a prime, of the numbers $1, 2, 3,... p^r$, those which are not prime to p^r are $p, 2p, 3p,... p^r$, and their number is p^{r-1}.

All the rest are prime to p^r, and their number is $p^r - p^{r-1}$, which proves the theorem.

(5) *If* $n = p^a \cdot q^b \cdot r^c$ ... *where p, q, r,... are the prime factors of n, then*

$$\varphi(n) = n \left(1 - \frac{1}{p}\right)\left(1 - \frac{1}{q}\right)\left(1 - \frac{1}{r}\right) ... .$$

For p^a, q^b, r^c ... are prime to one another, and the result follows from (3) and (4).

■ **EXAMPLE 1.** *Find the number of integers less than n and prime to it, when*
$$n = 1024, 1025,\ 1026.$$

We have $1024 = 2^{10}$, hence $\varphi(1024) = 1024(1 - \tfrac{1}{2}) = 512$.

$$1025 = 5^2 \cdot 41 \text{ and } \varphi(1025) = 1025(1 - \tfrac{1}{5})(1 - \tfrac{1}{41}) = 800.$$

$$1026 = 2 \cdot 3^3 \cdot 19 \text{ and } \varphi(1026) = 1026(1 - \tfrac{1}{2})(1 - \tfrac{1}{3})(1 - \tfrac{1}{19}) = 324.$$

This illustrates the irregularity in the variation of the function $\varphi(n)$.

■ **EXAMPLE 2.** *If $n \geq 2$, the sum of the integers less than n and prime to it is* $\tfrac{1}{2}\, n\varphi(n)$.

This is obviously true when $n = 2$. If $n > 2$ and x is any number less than n and prime to it, so also is $n - x$.

Hence the numbers less than n and prime to it can be arranged in pairs such that each pair has a sum n.

Thus $\varphi(n)$ must be even, the number of pairs is $\tfrac{1}{2}\varphi(n)$ and the sum of the numbers in question is $\tfrac{1}{2} n\,\varphi(n)$.

(6) Theorem: *If $d_1(= 1)$, d_2, d_3, ... $d_r(= n)$ are the divisors of any number n, then*
$$\varphi(d_1) + \varphi(d_2) + \varphi(d_3) + ... + \varphi(d_r) = n.$$

Let $n = p^a q^b r^c$ ... where p, q, r, ... are primes and let
$$S = \varphi(d_1) + \varphi(d_2) + ... \varphi(d_r),$$
then every divisor of n is of the form $p^x q^y r^z$ ... where $0 \leq x \leq a, 0 \leq y \leq b, 0 \leq z \leq c,$
etc., so that
$$S = \Sigma\varphi(p^x q^y r^z ...) = \Sigma\{\varphi(p^x). \varphi(q^y). \varphi(r^z) ... \},$$
where $x, y, z,...$ have all values subject to the above conditions. Henoe S is equal to
the product
$$[1 + \varphi(p) + \varphi(p^2) + ... + \varphi(p^a)][1 + \varphi(q) + ... + \varphi(q^b)]$$
$$[1 + \varphi(r) + ... + \varphi(r^c)]... .$$

Now $1 + \varphi(p) + \varphi(p^2) + ... + \varphi(p^a)$
$$= 1 + (p - 1) + (p^2 - p) +... + (p^a - p^{a-1}) = p^a,$$
and similarly for the other series in the brackets. Therefore
$$S = p^a. \, q^b. \, r^c ... = n.$$

For instance, the divisors of 24 are 1, 2, 3, 4, 6, 8, 12, 24; and $\varphi(1) + \varphi(2) + \varphi(3)$
$+ \varphi(4) + \varphi(6) + \varphi(8) + \varphi(12) + \varphi(24) = 1 + 1 + 2 + 2 + 2 + 4 + 4 + 8 = 24.$

■ **3. Fermat's Theorem:** *If p is a prime and a is any number prime to p, then*
$a^{p-1} -1$ *is divisible by p.*

For since a is prime to p, if $a, 2a, 3a, ... (p - 1)a$ are divided by p, the remainders
are 1, 2, 3,... $p - 1$ taken in a certain order.

Also the product of the numbers $a, 2a, 3a, ...$ is congruent (mod p) with the product
of the remainders, therefore $a^{p-1} \lfloor p-1 \equiv \lfloor p-1$; and since $\lfloor p-1$ is prime to p,
therefore $a^{p-1} \equiv 1$ (mod p), which proves the theorem.

Corollary 1. *If p is a prime and a is any number whatever, then $a^p - a$ is divisible
by p.*

For $a^p - a = a (a^{p-1} - 1)$, and if a is prime to p, $a^{p-1} -1$ is divisible by p, which
is a prime number. Hence in all cases $a^p - a$ is divisible by p.

Corollary 2. *If p is an odd prime and a is prime to p, then*
$$a^{\frac{1}{2}(p-1)} \equiv \pm 1 (\text{mod } p).$$

For $a^{p-1} - 1 = (a^{\frac{1}{2}(p-1)} -1) (a^{\frac{1}{2}(p-1)} + 1)$, and $a^{p-1} - 1$ is divisible by p, therefore
p is a divisor of $a^{\frac{1}{2}(p-1)} - 1$ or of $a^{\frac{1}{2}(p-1)} + 1,$ since p is a prime number.

▌ **EXAMPLE 1.** *The fifth power of any number N has the same right-hand digit as N.
For $N^5 - N$ is divisible by 5 and by 2.*

▌ **EXAMPLE 2.** *The ninth power of any number N is of one of the forms $19m$,
$19m \pm 1$.*

If N is prime to 19, since $\frac{1}{2}(19 - 1) = 9$, by Cor. 2 we have $N^9 \equiv \pm 1$ (mod 19).
Otherwise we must have $N = 19m$.

■ **4. Euler's Extension of Fermat's Theorem:** *If n is any number and a is prime
to n, then $a^{\varphi(n)} \equiv 1 \ (mod \ n)$.*

Let $a_1(= 1), a_2, a_3, ... a_{\varphi(n)}$ be the numbers, in ascending order, less than n and
prime to it, and consider the products

$$aa_1,\ aa_2,\ aa_3,\ \ldots\ aa_{\varphi(n)}. \hspace{2cm} \ldots(A)$$

If these are divided by n, the remainders are all different. For if we suppose that two products as $aa_r,\ aa_s\ (r > s)$ give the same remainder, then $a(a_r - a_s)$ would be divisible by n. This is impossible, for a is prime to n and $a_r - a_s < n$.

Also the remainders are all prime to n, for the factors of any product are both prime to n.

Hence the remainders are the numbers $a_1, a_2,\ldots a_{\varphi(n)}$ taken in some order or other. Now the product of the numbers in the set (A) is congruent (mod n) with the product of the remainders, therefore

$$a^{\varphi(n)}.\ a_1 a_2\ \ldots\ a_{\varphi(n)} \equiv a_1\ a_2\ \ldots\ a_{\varphi(n)}\ (\text{mod } n),$$

and dividing by $a_1 a_2\ \ldots\ a_{\varphi(n)}$, which is prime to n, we have the result in question.

■ **5. Wilson's Theorem:** *If p is a prime number, then* $\lfloor\underline{p-1} + 1$ *is divisible by p.*

For if a is any one of the numbers $1, 2, 3,\ldots p - 1$, and if the products $1.\ a, 2a, 3a,\ \ldots (p-1)a$ are divided by p, the remainders are the numbers $1, 2, 3,\ldots p - 1$, taken in some order or other. Hence, for every a, there is one number a' and one only, such that $aa' \equiv 1\ (\text{mod } p)$.

If $a' = a$, then $a^2 - 1$ must be divisible by p, and since p is a prime and $a < p$, it follows that either $a + 1 = p$ or $a - 1 = 0$.

Thus 1 and $p - 1$ are the only values of a for which a' can be equal to a.

If then a is any one of the $p - 3$ numbers $2, 3, \ldots p - 2$, then a' is not equal to a.

Consequently, these numbers can be arranged in $\frac{1}{2}(p - 3)$ pairs, such that the product of each pair is congruent with 1.

Therefore $2 \cdot 3 \ldots (p - 2) \equiv 1\ (\text{mod } p)$ and $\lfloor\underline{p-1} \equiv p - 1\ (\text{mod } p)$.

Hence $\lfloor\underline{p-1} + 1 \equiv 0(\text{mod } p)$, which is the result in question.

Conversely, if $\lfloor\underline{p-1} + 1$ *is divisible by p, then p is a prime number.*

For otherwise p would have a factor q which would divide $\lfloor\underline{p-1}$ and $\lfloor\underline{p-1} + 1$.

■ **EXAMPLE :** *Prove that* $\lfloor\underline{28} + 233$ *is divisible by 899.*

We have $899 = 29 \cdot 31$, also $233 \equiv 1\ (\text{mod } 29)$ and $233 \equiv 16\ (\text{mod } 31)$.

By Wilson's theorem, $\lfloor\underline{28} + 1 \equiv 0\ (\text{mod } 29)$, therefore $\lfloor\underline{28} + 233 \equiv 0\ (\text{mod } 29)$.

Again, to the modulus 31, we have $\lfloor\underline{30} + 1 \equiv 0$, therefore $30 \cdot 29.\ \lfloor\underline{28} + 1 \equiv 0$, and so $(-1)(-2)\ \lfloor\underline{28} + 32 \equiv 0$. Since 2 is prime to 31, it follows that $\lfloor\underline{28} + 16 = 0$, and therefore $\lfloor\underline{28} + 233 \equiv 0$. Hence $\lfloor\underline{28} + 233$ is divisible by 29 and 31, and therefore by 899.

■ **6. Lagrange's Theorem:** *If p is a prime and r is any number less than $p - 1$, the sum of the products of the numbers $1, 2, 3,\ldots, p - 1$ taken r together is divisible by p.*

Let $f(x) = (x + 1)(x + 2) \dots (x + p - 1)$; then if a_r denotes the sum of the products of $1, 2, 3, \dots, p - 1$ taken r together,

$$f(x) = x^{p-1} + a_1 x^{p-2} + a_2 x^{p-3} + \dots + a_{p-1}. \quad \dots(A)$$

Also we have the identity $(x + p) f(x) = (x + 1) f(x + 1)$, that is,

$$(x + p)(x^{p-1} + a_1 x^{p-2} + a_2 x^{p-3} + \dots + a_{p-1})$$
$$= (x + 1)^p + a_1 (x + 1)^{p-1} + a_2(x + 1)^{p-2} + \dots + a_{p-1}(x + 1). \quad (B)$$

Equating the coefficients of $x^{p-2}, x^{p-3}, \dots x$, we find that

$$pa_1 = C_2^p + C_1^{p-1} \cdot a_1,$$

$$pa_2 = C_3^p + C_2^{p-1} a_1 + C_1^{p-2} a_2,$$

$$\dots\dots\dots\dots\dots\dots\dots\dots\dots\dots\dots\dots\dots$$

$$pa_{p-2} = C_{p-1}^p + C_{p-2}^{p-1} a_1 + \dots C_1^2 a_{p-2},$$

Now, since p is a prime, C_r^p is divisible by p, for $r < p$; and C_r^{p-1}, C_r^{p-2}, etc., are all prime to p. It follows in succession that $a_1, a_2, \dots a_{p-2}$ are all divisible by ${}^s p$, which is the theorem in question. For another proof, see Art. 11, (6).

Note: We can immediately deduce the theorems of Wilson and Fermat, thus:
(*i*) Equating the terms independent of x in the identity (B),
$$pa_{p-1} = 1 + a_1 + a_2 + \dots + a_{p-1},$$
and since $a_1, a_2, \dots, a_{p-2}$ are divisible by p, so also is $a_{p-1} + 1$, that is $\lfloor p - 1$
+ 1 is divisible by p, which is Wilson's theorem.
(*ii*) If x is any number prime to p, one of the numbers $x + 1, x + 2, \dots, x + p - 1$
must be divisible by p, hence $f(x) \equiv 0 \pmod{p}$.
Also $f(x) \equiv x^{p-1} + a_{p-1} \equiv x^{p-1} - 1 \pmod{p}$, therefore $x^{p-1} - 1 \equiv 0 \pmod{p}$, which is Fermat's theorem.

▌ **EXAMPLE 1.** *If p is a prime, greater than 3, prove that*

$$\left(\frac{1}{1} + \frac{1}{2} + \dots + \frac{1}{p-1}\right) \cdot \lfloor p - 1 \equiv 0 \pmod{p^2}.$$

If, in the identity (A), we put x equal to p and $-2p$ in succession, we have
$$(p + 1)(p + 2) \dots (2p - 1) = a_{p-1} + a_{p-2} \cdot p + a_{p-3} \cdot p^2 + \dots + p^{p-1},$$
$$(2p - 1)(2p - 2) \dots (p + 1) = a_{p-1} - a_{p-2} \cdot 2p + a_{p-3} \cdot 4p^2 + \dots + (2p)^{p-1};$$
and since the left-hand sides are identical, we have by subtraction,
$$3p \cdot a_{p-2} - 3p^2 \cdot a_{p-3} + 9p^3 \cdot a_{p-4} - \dots - (2^{p-1} - 1)p^{p-1} = 0.$$

Now, since p is a prime greater than 3, a_{p-3} exists; and, by Lagrange's theorem, it is divisible by p.

Hence, $3p \cdot a_{p-2}$ is divisible by p^3; and, since 3 is prime to p, it follows that a_{p-2} is divisible by p^2.

▌ **EXAMPLE 2.** *If p is a prime, greater than 3, prove that, for all integral values of m,*

$$\lfloor mp - \lfloor m \cdot (\lfloor p)^m$$

is divisible by p^{m+3}.

Substituting rp for x in the identity (A), we have, as in Ex. 1,

$$(rp + 1)(rp + 2)...(rp + p - 1) \equiv a_{p-1} \pmod{p^3},$$

where $r = 1, 2, 3, ... , m - 1$, and p is prime and greater than 3.

Hence, $\qquad \lfloor (r+1)p / \left(\lfloor rp \cdot \lfloor p \right) \equiv r + 1 \pmod{p^3};$

and, by giving r the values $1, 2,..., m - 1$, it follows that

$$\lfloor mp / \left(\lfloor p \right)^m \equiv \lfloor m \pmod{p^3};$$

hence, $\qquad \lfloor mp - \lfloor m \left(\lfloor p \right)^m \equiv 0 \pmod{p^{m+3}}.$

EXERCISE XLIII

1. If $\quad N = a_0 + a_1 t + a_2 t^2 + ...+ a_n t^n$, and $S = a_0 - a_1 + a_2 - ... + (-1)^n t^n$, where $t = 10000$ and $0 \le a_r < t$, prove that $N \equiv S \pmod{73 \text{ or } 137}$. Hence state a rule for the divisibility of a number by 73 or 137. Apply it to 30414 and 81103.

2. If p is a prime and $x_1, x_2, x_3, ... x_a$ are any numbers, show that
    $$(x_1 + x_2 + ... + x_a)^p \equiv x_1^p + x_2^p + ... + x_a^p \pmod{p},$$
    and in particular $(ax)^p \equiv ax^p \pmod{p}$, where x is any number.

    Consequently, $a^p \equiv a \pmod{p}$, and if a is not divisible by p, $a^{p-1} \equiv 1 \pmod{p}$, which is Fermat's theorem.

3. Prove that, if n is a prime greater than 7, then $n^6 - 1$ is divisible by 504.

4. Prove that, if n is a prime greater than 13, then $n^{12} - 1$ is divisible by 65520.

5. If p and q are different primes, then $p^{q-1} + q^{p-1} - 1$ is divisible by pq.

6. If p is a prime of the form $4m + 1$, show that $\lfloor \frac{1}{2}(p - 1)$ is a solution of the congruence,

    $x^2 + 1 \equiv 0 \pmod{p}$.

    [Show that $1 \cdot 2 \cdot 3. ... .2m \equiv (4m) . (4m - 1) . (4m - 2)... (2m + 1) \pmod{p}$.]

7. If p is a prime of the form $4m - 1$, show that $\lfloor \frac{1}{2}(p - 1)$ is a solution of the congruence,

    $x^2 - 1 \equiv 0 \pmod{p}$.

8. If the congruence, $x^2 \equiv -1 \pmod{p}$, has no solution, show that

    (i) $\lfloor p - 1 \equiv (-1)^{\frac{1}{2}(p-1)}$;　and　(ii) that p is of the form $4m - 1$.

9. Show that $\lfloor 18 + 1$ is divisible by 437.

10. If p is a prime, $2 \lfloor p - 3 + 1$ is divisible by p,

11. If n is any odd number, show that

    $$\lfloor n - 1 \left[1 + \frac{1}{2} + \frac{1}{3} + ... + \frac{1}{n-1} \right] \equiv 0 \pmod{n}.$$

12. Arrange the numbers $2, 3, 4, ..., 15$ in pairs, a and a', such that for each pair, $aa' \equiv 1 \pmod{17}$.

13. Let $n = a^p b^q c^r$ ..., where a, b, c, ... are different primes, and consider the groups

$$\left.\begin{array}{l} a, 2a, 3a, ... n/a \cdot a \\ b, 2b, 3b, ... n/b \cdot b \\ .... \end{array}\right\}, \qquad ...(A)$$

$$\left.\begin{array}{l} ab, 2ab, 3ab, ... n/ab \cdot ab \\ bc, 2bc, 3bc, ... n/bc \cdot bc \\ .... \end{array}\right\}, \qquad ...(B)$$

together with groups (C), (D), etc., formed in a similar way from the combinations of a, b, c, ... taken 3, 4, ... together. If s_1, s_2, s_3, ... denote the sums of the rth powers of the numbers in the groups (A), (B), (C), ... respectively, show that the sum of the rth powers of the numbers not greater than n and not prime to it is $s_1 - s_2 + s_3 - s_4 + ...$.

14. If S_2 is the sum of the squares, and S_3 the sum of the cubes, of the numbers less than n and prime to it, show that

$$(i)\ \ S_2 = \frac{1}{3}\, n^3 \left(1 - \frac{1}{a}\right)\left(1 - \frac{1}{b}\right) ... + \frac{1}{6}\, n\, (1 - a)\, (1 - b)\, (1 - c)\, ...,$$

$$(ii)\ \ S_3 = \frac{1}{4}\, n^4 \left(1 - \frac{1}{a}\right)\left(1 - \frac{1}{b}\right) ... + \frac{1}{4}\, n^2\, (1 - a)\, (1 - b)\, (1 - c)\, ...,$$

where a, b, c, ... are the different prime factors of n.

[Use the last example to show that the sum of the squares of the numbers not greater than n and not prime to it is $\dfrac{1}{3}\, n^3 \left(\Sigma \dfrac{1}{a} - \Sigma \dfrac{1}{ab} + ...\right) + \dfrac{1}{2}\, n^2 + \dfrac{1}{6}\, n\, (\Sigma a - \Sigma ab + ...).$]

■ **7. Roots of a Congruence:** If $f(x)$ is a polynomial of the rth degree in x, with integral coefficients, any value of x which satisfies the condition $f(x) \equiv 0 \pmod{n}$ is called a *solution* or *root* of the congruence $f(x) \equiv 0$, which is said to be of the rth degree.

If x_0 is a root of the congruence $f(x) \equiv 0 \pmod{n}$, so also is any number congruent with x_0 to the modulus n.

Two roots will be regarded as *distinct* only when they are incongruent with respect to the modulus, and when we say that a congruence has r roots we mean that it has r distinct roots.

❚ **EXAMPLE 1.** *It follows from Fermat's theorem that, if p is a prime, the roots of the congruence $x^{p-1} - 1 \equiv 0 \pmod{p}$ are 1, 2, 3, ... $p - 1$.*

❚ **EXAMPLE 2.** *Every square number is of the form $5n$ or $5n \pm 1$, thus no square can be congruent with 2 to the modulus 5. In other words, the congruence $x^2 \equiv 2 \pmod{5}$ has no solution.*

> **Note:** It is obvious that, in dealing with the congruence $f(x) \equiv 0 \pmod{n}$, we may replace any coefficient by any number congruent with it to the modulus n. Thus the congruence $13x \equiv 17 \pmod{5}$ is identical with $3x \equiv 2$ or with $3x \equiv -3$ or with $x \equiv -1 \pmod{5}$.

■ **8. The Linear Congruence:**

(1) *If a is prime to n, the congruence $ax \equiv b \pmod{n}$ has just one distinct root.*

For if the terms of the arithmetical progression $0, a, 2a, \dots (n-1)\,a$ are divided by n, since a is prime to n, the remainders are the numbers $0, 1, 2, \dots n-1$, in some order or other. Hence there is just one value of x such that $ax \equiv b \pmod{n}$ and $0 \le x < n$.

To solve $ax \equiv b(mod\ n)$ *when a is prime to n,* let a/n be expressed as a simple continued fraction with an even number of quotients, and let y_0/x_0 be the last convergent but one.

Then $ax_0 - ny_0 = 1$, so that $ax_0 \equiv 1 \pmod{n}$. Therefore $ax \equiv b \equiv abx_0$, and since a is prime to n, $x \equiv bx_0$, which is the required solution.

If a is small, or the product of small primes, it is generally easier to proceed as in Exs. 1–3 below.

(2) *If a is not prime to n, and g is the greatest common divisor of these numbers, there are g incongruent solutions of* $ax \equiv b(mod\ n)$, *provided that b is divisible by g. If b is not divisible by g, there is no solution.*

For let $a = ga'$, $n = gn'$, so that a' is prime to n'. We require values of x such that $ga'x - b$ is divisible by gn'. No such value exists unless b is divisible by g. If this condition is satisfied and $b = gb'$, then $a'x - b'$ is divisible by n' and the given congruence is equivalent to $a'x \equiv b' \pmod{n'}$.

Since a' is prime to n', the last congruence has one solution $\alpha < n'$, and g solutions $< n$, namely, $\alpha,\ \alpha + n',\ \alpha + 2n',\dots \alpha + (g-1)n'$, which are the g distinct roots of $ax \equiv b \pmod{n}$.

(3) The expression $b/a(mod\ n)$ is used to denote *any* solution of

$$ax \equiv b(mod\ n).$$

If a and n have a common divisor which does not divide b, then $b/a \pmod{n}$ has no meaning : in every other case it has infinitely many values. If a is prime to n, these values are all congruent $\pmod{n}$: if a is not prime to n, they are all congruent $\pmod{n/g}$ where g is the greatest common divisor of a and n.

Such expressions possess many of the properties of ordinary fractions, and it is easy to prove the following:

    (*i*) If $a \equiv a'$ and $b \equiv b' \pmod{n}$, the expressions $a/b \pmod{n}$ and $a'/b' \pmod{n}$ are equivalent.

    (*ii*) $am/bm \pmod{mn}$ is equivalent to $a/b \pmod{n}$.

    (*iii*) $ak/bk \pmod{n}$ is equivalent to $a/b \pmod{n}$, if k is prime to n.

▮ **EXAMPLE 1.** *Solve (i)* $5x \equiv 2\ (mod\ 7)$; *(ii)* $15x \equiv 6\ (mod\ 21)$.

    (*i*) Here $5x \equiv 2 - 7 \equiv -5 \pmod{7}$, therefore $x \equiv -1 \equiv 6 \pmod{7}$.

    (*ii*) Since 3 is a divisor of 15, 6, 21, the congruence is equivalent to $5x \equiv 2 \pmod{7}$, so that $x \equiv 6 \pmod{7}$ and $x \equiv 6, 13, 20 \pmod{21}$.

▮ **EXAMPLE 2.** *Solve* $36x \equiv 7(mod\ 157)$.

    We have

$$12x \equiv \frac{7}{3} \equiv \frac{-150}{3} \equiv -50, \text{ therefore } x \equiv \frac{-50}{12} \equiv \frac{-25}{6} \equiv \frac{132}{6} \equiv 22 \pmod{157}.$$

■ **EXAMPLE 3.** *Solve 17x ≡ 11 (mod 43).*

We have $\dfrac{43}{17} = 2 + \dfrac{1}{1+}\dfrac{1}{1+}\dfrac{1}{8}$ and the third convergent is 5/2, thus

$$1 = 43 \cdot 2 - 17. \quad 5 \equiv -17 \cdot 5 (\text{mod } 43) \text{ and } \tfrac{1}{17} \text{ (mod } 43) \equiv -5.$$

Therefore $\quad x \equiv \tfrac{11}{17} (\text{mod } 43) \equiv -55 \equiv 31 \text{ (mod } 43).$

(4) Simultaneous Congruences: (*i*) *It is required to find x so that*
$$x \equiv a \ (\text{mod } \alpha) \quad and \quad x \equiv b \ (mod \ \beta).$$
We have $x = a + \alpha y$ where y is given by
$$a + \alpha y \equiv b(\text{mod } \beta) \text{ or } \alpha y \equiv b - a \ (\text{mod } \beta).$$
Let g be the greatest common divisor of α and β. If $b - a$ is not divisible by g there is no solution. Otherwise, there is just one value y_1 of y less than β/g, which satisfies the last congruence, and the general value of y is given by $y = y_1 + \beta/g \cdot t$, so that the general solution is
$$x = x_1 + \alpha\beta/g \cdot t \quad \text{where} \quad x_1 = a + \alpha y_1.$$
Thus a solution x_1 of the given congruences exists if, and only if, $b - a$ is divisible by g, the greatest common divisor of α, β, and then the congruences are equivalent to the single congruence $x \equiv x_1 \ (\text{mod } l)$, where l is the L.C.M. of α and β.

It should be observed that solutions always exist when α is prime to β.

(*ii*) *To find x so as to satisfy a number of relations of the form*
$$x \equiv a(mod \ \alpha), \quad x \equiv b(mod \ \beta), \ x \equiv c(mod \ \gamma), \ ...$$
We replace the first two by $x \equiv x_1(\text{mod } l)$ as in the last section. Taking this with the third, the first three congruences are equivalent to
$$x \equiv x_2(\text{mod } m),$$
where x_2 is any solution (found as before) and m is the L.C.M. of l and γ, and therefore of α, β, γ. Continuing thus, it will be seen that *the given congruences are together equivalent to $x \equiv \xi(mod \ L)$ where ξ is any solution and L is the L.C.M. of α, β, γ, ... Moreover, a solution always exists if α, β, γ, ... are prime to one another.*

■ **EXAMPLE 4.** *Given that 35x ≡ 4 (mod 9) and 55x ≡ 2 (mod 12), find the general value of x.*

We have $x = \dfrac{4}{35} (\text{mod } 9) \equiv \dfrac{4}{-1} \equiv 5(\text{mod } 9)$, and $x = \dfrac{2}{55} (\text{mod } 12) \equiv \dfrac{14}{7} \equiv 2 \text{ (mod } 12).$

Hence $x = 5 + 9y \equiv 2(\text{mod } 12)$, giving $y = \dfrac{-3}{9} (\text{mod } 12) \equiv \dfrac{-1}{3} (\text{mod } 4) \equiv 1(\text{mod } 4).$

Thus $y = 1 + 4t$ and $x = 5 + 9 (1 + 4t) = 14 + 36t$, which is the general solution.

(5) We can apply these methods to solve a linear congruence when the modulus is a composite number, as in the next example.

■ **EXAMPLE 5.** *Solve 19x ≡ 1 (mod 140).*

We have $140 = 4 \cdot 5 \cdot 7$, and therefore $19x \equiv 1$ for the moduli 4, 5 and 7 and since these numbers are prime to one another, any value of x which satisfies these conditions is a solution. Thus we have

$$x \equiv \frac{1}{19} \ (\text{mod } 4) \equiv \frac{1}{-1} \equiv -1 \ (\text{mod } 4), \ x \equiv \frac{1}{19} \ (\text{mod } 5) \equiv -1 \ (\text{mod } 5),$$

and $\quad x \equiv \dfrac{1}{19} \ (\text{mod } 7) \equiv \dfrac{8}{-2} \equiv -4 \ (\text{mod } 7).$

Hence $x = -1 + 4y \equiv -1 \ (\text{mod } 5)$, so that $y = 5z$.

Therefore $\quad x = -1 + 20z \equiv -4 \ (\text{mod } 7)$ and $z \equiv \dfrac{-3}{20} \ (\text{mod } 7) \equiv 3 \ (\text{mod } 7)$. Thus

$z = 3 + 7t$, $y = 15 + 35t$ and $x = 59 + 140t$, which is the general solution.

In this case the solution can be found more easily by the method of § (3), Ex. 3,

thus $\dfrac{140}{19} = 7 + \dfrac{1}{2+} \dfrac{1}{1+} \dfrac{1}{2+} \dfrac{1}{2}$. The fourth convergent is $\dfrac{59}{8}$, and $59 \cdot 19 - 140 \cdot 8 = 1$.

Therefore $\quad x \equiv \dfrac{1}{19} \ (\text{mod } 140) \equiv 59 \ (\text{mod } 140)$.

■ **9. A Theorem on Fractions:** *If n is the product of factors a, b, c, d, ... l, which are prime to one another, and m is prime to n, the fraction m/n can be expressed uniquely in the form*

$$\frac{m}{n} = \frac{\alpha}{a} + \frac{\beta}{b} + \frac{\gamma}{c} + \dots + \frac{\lambda}{l} \pm k, \qquad \dots \text{(A)}$$

where α, β, ... λ, k are positive integers and $\alpha < a$, $\beta < b$, etc.

Let $a' = bcd... \, l$, then a is prime to a', and therefore integers x, x' can be found so that $a'x + ax' = m$. Moreover, x' is prime to a', for any common factor of these two would divide m, which is prime to a'. Thus we have

$$\frac{m}{aa'} = \frac{x}{a} + \frac{x'}{a'} \quad \text{and} \quad \frac{m}{n} = \frac{x}{a} + \frac{x'}{bcd \dots l} \, .$$

Proceeding in this way, we can obtain a relation of the form

$$\frac{m}{n} = \frac{x}{a} + \frac{y}{b} + \frac{z}{c} + \dots + \frac{w}{l},$$

where x, y, z, w are integers, and then m/n can be put in the required form by writing $x = \alpha + k_1 a$, $y = \beta + k_2 b$, etc, where α, β, ... are the least positive residues of $x, y,...$ with regard to the moduli $a, b,...$ respectively.

Again, if we multiply each side of the equality (A) by n, it will be seen that $\alpha \cdot bc \dots l \equiv m(\text{mod } a)$, $\beta \cdot ac \dots l \equiv m(\text{mod } b)$, etc. Since $bc... \, l$ is prime to a, α has just one positive value less than a. Similarly β has just one value less than b, and so on. Hence the expression found for m/n is unique.

▌ **EXAMPLE :** *Express* $\dfrac{101}{1540}$ *in the form* $\dfrac{x}{4} + \dfrac{y}{5} + \dfrac{z}{7} + \dfrac{u}{11} + k$, *the fractions being positive proper fractions and k an integer.*

Here x, y, z, u are determined by

$5 \cdot 7 \cdot 11x \equiv 101 \ (\text{mod } 4), \qquad 1 \cdot (-1)(-1) \, x \equiv 1, \qquad x \equiv 1 \ (\text{mod } 4);$

$4 \cdot 7 \cdot 11y \equiv 101 \ (\text{mod } 5), \qquad (-1) \cdot 2 \cdot 1 \cdot y \equiv 1 \equiv -4, \qquad y \equiv 2 (\text{mod } 5);$

$$4 \cdot 5 \cdot 11z \equiv 101 \ (\text{mod } 7), \qquad (-3)(-2) \cdot 4z \equiv 3, \qquad z \equiv 1(\text{mod } 7);$$
$$4 \cdot 5 \cdot 7u \equiv 101 \ (\text{mod } 11), \qquad 4u \equiv 1 \equiv 12, \qquad u \equiv 3(\text{mod } 11).$$

With $x = 1$, $y = 2$, $z = 1$, $u = 3$, we find that $\frac{1}{4} + \frac{2}{5} + \frac{1}{7} + \frac{3}{11} = 1\frac{101}{1540}$; so that $k = -1$.

■ **10. The General Congruence:** In this article $f(x)$, $\varphi(x)$, ... will denote polynomials with integral coefficients.

(1) If every coefficient of a polynomial $f(x)$ is divisible by n, then $f(x)$ is divisible by n for all values of x, and we say that $f(x)$ is *identically congruent* with zero to the modulus n. This is expressed by writing

$$f(x) \equiv 0(\text{mod } n).$$

Again, $f(x)$ is said to be identically congruent with $\varphi(x)$ (mod n) if $f(x) - \varphi(x) \equiv 0$ (mod n), which is also written as $f(x) \equiv \varphi(x)$ (mod n).

(2) Division (mod n). If $f(x)$ is divided by $\varphi(x)$, a polynomial of lower degree than $f(x)$, we obtain an identical relation of the form

$$f(x) = Q'(x) \cdot \varphi(x) + R'(x),$$

where the remainder $R'(x)$ is of lower degree than $\varphi(x)$.

Suppose that any coefficient which occurs in the division is replaced by any number congruent with it (mod n). The process modified in this way is called *division (mod n)*, and leads to an identical congruence of the form

$$f(x) \equiv Q(x) \cdot \varphi(x) + R(x) \ (\text{mod } n),$$

where $R(x)$ is of lower degree than $\varphi(x)$.

If $R(x) \equiv 0$ (mod n), then $f(x)$ is said to be *divisible (mod n)* by $\varphi(x)$.

If $f(x)$ is divisible (mod n) by $(x - \alpha)^r$, the congruence $f(x) \equiv 0$ (mod n) is said to have r roots equal to α.

■ **EXAMPLE :** *Prove that 2, 4, 6 are roots of $5x^3 + 3x^2 - 4x - 2 \equiv 0$ (mod 7).*

Division (mod 7) by $x - 2$ and $x - 4$ in succession is shown on the right. In the third line 13 is replaced by -1 and -6 by $+1$. In the last line 19 is replaced by 5 and 21 by 0. The reckoning shows that

$$5x^3 + 3x^2 - 4x - 2 \equiv 5(x - 2)(x - 4)(x + 1) \ (\text{mod } 7),$$

$$
\begin{array}{r|rrrr}
1-2 & 5 + 3 & - 4 & - 2 \\
    &   10 & - 2 & + 2 \\
\hline
1-4 & 5 - & 1 & + 1 & + 0 \\
    &   20 & + 20 \\
\hline
    & 5 + & 5 & + 0
\end{array}
$$

so that the congruence holds when $x = 2$, 4 or 6.

■ **11. Congruences to a Prime Modulus:** The analogy between congruences and equalities is carried a step further by the following theorems, in which the modulus p is a prime.

(1) *If $MN \equiv 0$ (mod p) where p is a prime, then either $M \equiv 0$ (mod p), or $N \equiv 0$ (mod p).* For since MN is divisible by the prime p, either M or N is so divisible.

(2) *If α is a root of $f(x) \equiv 0$ (mod p), then $f(x)$ is divisible (mod p) by $x - \alpha$.* For by division (mod p) we have $f(x) \equiv Q(x) \cdot (x - \alpha) + R$, where R is independent of x. Putting $x = \alpha$ we have $R \equiv f(\alpha) \equiv 0$ (mod p); therefore $f(x)$ is divisible (mod p) by $x - \alpha$.

(3) *If $f(x)$ is divisible (mod p) by $(x - \alpha)^r$ and by $(x - \beta)^s$ where $\alpha \neq \beta$, then it is divisible (mod p) by $(x - \alpha)^r (x - \beta)^s$.* For any common divisor of $x - \alpha$ and $x - \beta$ is a divisor of $\alpha - \beta$ which is independent of x.

(4) *If $f(x)$ is a polynomial of degree n, the congruence $f(x) \equiv 0$ (mod p) cannot have more than n roots, any root which occurs r times being counted as r distinct roots.* This follows at once from the preceding theorems.

(5) *If $f(x) \equiv 0$ (mod p) is a congruence of the n-th degree with n roots and $\varphi(x)$ is a factor of $f(x)$ of degree r, then the congruence $\varphi(x) \equiv 0$ (mod p) has r roots.* This is an immediate consequence of (4).

(6) *If p is a prime, it follows from Fermat's theorem that the roots of the congruence $x^{p-1} - 1 \equiv 0 (mod\ p)$ are 1, 2, 3, ... , p − 1, and therefore*

$$x^{p-1} - 1 \equiv (x - 1)(x - 2) \ldots (x - p + 1)\ (mod\ p).$$

Hence if a_r denotes the sum of the products r together of $1, 2, 3, \ldots p - 1$, it follows

that $\qquad a_1 x^{p-2} - a_2 x^{p-3} + \ldots + a_{p-2} x - \left\{ \lfloor p-1 + 1 \rfloor \right\} \equiv 0$ (mod p).

Hence $a_r \equiv 0$ (mod p) for $r < p - 1$, which is Lagrange's theorem (Art. 6); also

$\lfloor p - 1 \rfloor + 1 \equiv 0$ (mod p), which is Wilson's theorem.

▮ **EXAMPLE :** *Solve $x^2 + x + 1 \equiv 0$ (mod 7).*

Here $4x^2 + 4x + 4 \equiv 0$, or $(2x + 1)^2 \equiv -3 \equiv 4$; and $2x + 1 \equiv \pm 2$. Hence $x \equiv 2$ or 4 (mod 7).

In treatises on the Theory of Numbers the notion of congruence is extended to complex numbers, thus completing the analogy referred to at the head of this article.

▮ **12. Reduction of a Congruence:** Let $f(x) \equiv 0$ (mod p) be a congruence of the nth degree to a prime modulus p. Let the roots of the congruence be $\alpha_1, \alpha_2, \ldots \alpha_r$, where some of these may be multiple roots.

By a modified H.C.F. process we can find a congruence $R(x) \equiv 0$ (mod p) of degree r of which the roots are $\alpha_1, \alpha_2, \ldots \alpha_r$.

For the roots of $x^{p-1} - 1 \equiv 0$ (mod p) are $1, 2, 3, \ldots p - 1$, hence every solution of $f(x) \equiv 0 (\text{mod } p)$ is a solution of $x^{p-1} - 1 \equiv 0 (\text{mod } p)$.

Let the H.C.F. process be carried out for the functions $f(x)$ and $x^{p-1} - 1$ with the modification that any coefficient may be replaced by any number congruent with it (mod p); then the last divisor $R(x)$ is the highest common divisor (mod p) of $f(x)$ and $x^{p-1} - 1$.

The truth of this statement depends on the following. Let u, v be any two consecutive remainders in the process. If w is the next remainder, $w = au + bv$ where neither a nor b is divisible by p. Hence the common divisors (mod p) of u and v are the same as those of v and w.

It follows that $R(x) \equiv 0 \pmod{p}$ is the congruence referred to above.

■ **EXAMPLE :** *Show that the congruence $3x^4 - 2x^3 - 4x^2 - x - 1 \equiv 0$ (mod 7) has only one distinct root and solve it.*

Carrying out the modified H.C.F. process for the left-hand side and $x^6 - 1$, it is found that the last divisor is $x + 2$, showing that -2 is the *only* distinct root.

By division (mod 7) we find that $3x^4 - 2x^3 - 4x^2 - x - 1 \equiv 3(x+2)^2 (x^2 - 3)$ (mod 7). Now $x^2 \equiv 3$ (mod 7) has no solution, hence -2 is the only distinct root.

EXERCISE XLIV

1. Find the form of x, if
      (*i*) $29x \equiv 1$ (mod 13);   (*ii*) $2x \equiv 3$ (mod 7) and $3x \equiv 5$ (mod 11);
      (*iii*) $8x \equiv 1$ (mod 7), $3x \equiv 4$ (mod 11), and $7x \equiv 3$ (mod 20).

2. Solve
      (*i*) $78x \equiv 1$ (mod 179);   (*ii*) $78x \equiv 13$ (mod 179).

3. Solve
      (*i*) $16x \equiv 31$ (mod 1217);   (*ii*) $30x \equiv 31$ (mod 1861);
      (*iii*) $15x \equiv 28$ (mod 1009);   (*iv*) $26x \equiv 35$ (mod 1901).

4. If $x \equiv a$ (mod 22) and $x \equiv b$ (mod 40), show that $a - b$ is even and $x \equiv 100a - 99b$ (mod 440).

5. If $x \equiv a$ (mod 16) $\equiv b$ (mod 5) $\equiv c$ (mod 11), prove that
   $$x \equiv 385a + 176b - 560c \pmod{880}.$$

6. If $x \equiv a$ (mod 7) $\equiv b$ (mod 11) $\equiv c$ (mod 13), prove that
   $$x \equiv -286a + 364b - 77c \pmod{1001}.$$

7. If $x \equiv a$ (mod 3) $\equiv b$ (mod 5) $\equiv c$ (mod 7), prove that
   $$x \equiv -35a + 21b + 15c \pmod{105}.$$

8. Express $\dfrac{1}{1760}$ in the form $\dfrac{x}{32} + \dfrac{y}{5} + \dfrac{z}{11} \pm k$, where x, y, z, k are positive integers and $x < 32, y < 5, z < 11$.

9. Express $\dfrac{1}{1001}$ in the form $\dfrac{x}{7} + \dfrac{y}{11} + \dfrac{z}{13} \pm k$, where x, y, z, k are positive integers and $x < 7, y < 11, z < 13$.

10. Find the form of numbers, of which the first, second and third powers are of the forms $3n + 1$, $4n + 1$ and $5n + 1$ respectively.

11. If $n^2 - 2$ is divisible by 7, and $n^2 - 3$ is divisible by 11, find the form of n.

12. If $x^2 + 4x + 2$ is divisible by 23, find the form of x.

13. If n^2 and $(n + 1)^2$ are of the forms $11m + 4$ and $12m + 4$, find the form of n.

14. Find four consecutive numbers divisible by 5, 7, 9, 11 respectively.

15. Prove that the congruence $3x^4 - 4x^3 - x^2 + 3x - 4 \equiv 0$ (mod 7) has only two incongruent solutions, and find them.

16. Prove that the congruence $4x^4 \equiv 5x^3 + x^2 + 2x + 5$ (mod 11) has four incongruent solutions, and find them.

17. Prove that the congruence $11x^9 + 1 \equiv 0$ (mod 29) has only one distinct solution, and find it.

18. *Bicycle gear as a revolution counter.* A bicycle with no free wheel has 19 teeth in the back chain-wheel, 62 teeth in the front chain-wheel, 121 links in the chain, and the circumference of the back wheel of the bicycle is known to be almost exactly 7 feet.

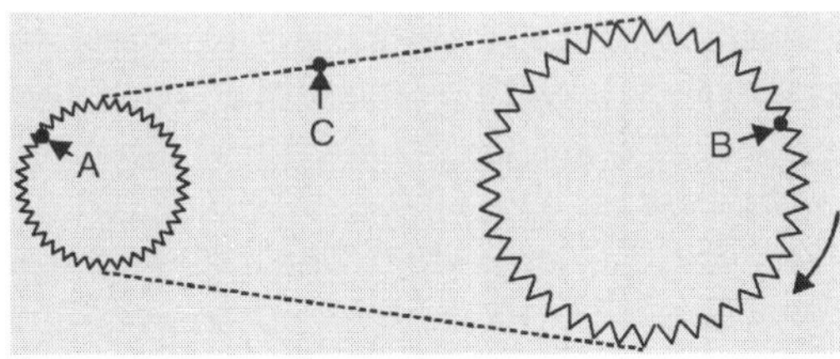

Fig. 64

Prove the following rule to find the number of revolutions (R) of the back wheel (and so the distance travelled) in any journey of not more than 10 miles.

Mark a tooth A in the back chain-wheel, a tooth B in the front chain-wheel, and a link C in the chain. At the start count the teeth or links from A to B and from A to C in the direction of turning. At the finish count these again.

Take the increase from A to B or 62 minus the decrease, and denote it by a.

Take the increase from A to C or 121 minus the decrease, and denote it by b.

Then the number R is the remainder in the division $(3025a + 2108b) \div 7502$.

[If x is the number of revolutions of the back wheel,

$$19x \equiv a(\text{mod } 62) \quad \text{and} \quad 19x \equiv b(\text{mod } 121).$$

Hence show that $x \equiv 49a \ (\text{mod } 62)$ and $x \equiv 51b(\text{mod } 121)$,

and therefore $\qquad\qquad x \equiv 3025a + 2108b \ (\text{mod } 7502).]$

■ ■ ■

Residues of Powers of a Number, Recurring Decimals

Residues of Powers of a Number

■ **1.** *If a and g are prime to any modulus n, the least positive residues of successive terms of the geometrical progression a, ag, ag^2, ... recur, and if t is the number of terms in the recurring period, $t \leq \varphi(n)$ and is independent of a.*

Denote the residues by r_0, r_1 r_2, etc. Each of these is one of the $\varphi(n)$ numbers $< n$ and prime to it, for since a and g are prime to n, so is every residue. Hence, of the numbers $r_0, r_1, r_2, ... r_{\varphi(n)}$, at least two, say r_s and r_{s+t}, are equal. Therefore $ag^s \equiv ag^{s+t}$ (mod n); and, since a is prime to n, hence $g^t \equiv 1$ (mod n). It follows that $ag^t \equiv a$, $ag^{t+1} \equiv ag$, $ag^{t+2} \equiv ag^2$, etc., and so $r_t = r_0$, $r_{t+1} = r_1$, $r_{t+2} = r_2$, etc., where $t \leq \varphi(n)$.

Moreover, *the number of terms in the period is the least index for which $g^t \equiv 1$* (mod n).

It should be noticed that: (*i*) No two residues in the period are equal. (*ii*) If $r_\mu = 1$, then $\mu \equiv 0$ (mod t), and conversely. (*iii*) If $r_\mu = r_v$, then $\mu \equiv v$ (mod t), and conversely. (*iv*) $r_{\mu+1} \equiv gr_\mu$ (mod n). (*v*) If $n - 1$ occurs as a residue, the subsequent residues are $n-r_1$, $n - r_2$, $n - r_3$, etc.; for they are congruent (mod n) with $-g$, $-g^2$, $-g^3$, etc.

Thus *for the modulus* 19, the residues of 2^0, 2^1, 2^2, ... are

Index: 0, 1, 2, 3, 4, 5, 6, 7, 8, 9, 10, 11, 12, 13, 14, 15, 16, 17, 18 ... .

Residue: 1, 2, 4, 8, 16, 13, 7, 14, 9, 18, 17, 15, 11, 3, 6, 12, 5, 10, 1 ... .

Having found $r_4 = 16$, we have $r_5 \equiv 16 \times 2 \equiv 13$ (mod 19). Again, $r^9 = 18 = 19 - 1$, and the subsequent residues are $19 - 2$, $19 - 4$, etc.

Here $t = 18 = \varphi(19)$, and *every* number not divisible by 19 is congruent (mod 19) with some power of 2.

For the modulus 13, the residues of 5^0, 5^1, 5^2, ... are

Index: 0, 1, 2, 3, 4, 5, 6, 7, 8, 9, 10, 11, 12.

Residue: 1, 5, 12, 8,1, 5, 12, 8, 1, 5, 12, 8, 1.

Here $t = 4$, which is a divisor of $\varphi(13) = 12$.

■ **2.** *If g is prime to n and t is the least index other than zero for which $g^t \equiv 1$ (mod n), then $t = \varphi(n)$ or is a divisor of $\varphi(n)$.*

For $g^{\varphi(n)} \equiv 1 \pmod{n}$, so that t may be equal to $\varphi(n)$, and by the preceding $t \le \varphi(n)$.   If $t < \varphi(n)$, let $\varphi(n) = at + b$, where $0 \le b < t$, then

$$g^{\varphi(n)} = g^{at+b} = (g^t)^a \cdot g^b \equiv g^b \pmod{n}.$$

Therefore $g^b \equiv 1 \pmod{n}$, and consequently $b = 0$, for $b < t$. Hence, if $t < \varphi(n)$, it is a divisor of $\varphi(n)$.

Second proof. Let the least positive residues of $g^0, g^1, g^2, \ldots g^{t-1}$ be

$$a_0, a_1, a_2, \ldots a_{t-1}. \qquad \ldots\text{(A)}$$

Each of these is one of the $\varphi(n)$ numbers less than n and prime to it. Also no two of them are equal; for if $a_r = a_s$ where $s < r < t$, then

$$g^r \equiv g^s \pmod{n} \text{ and } g^{r-s} \equiv 1,$$

which is not the case, for $r - s < t$.

Let S denote the set of numbers less than n and prime to it. If the set (A) includes all the numbers in S, then $t = \varphi(n)$.

If some number b of the set S does not occur in the set (A), consider the residues of $bg^0, bg^1, bg^2, \ldots bg^{t-1}$, denoting them by

$$b_0, b_1, b_2, \ldots b_{t-1}. \qquad \ldots\text{(B)}$$

These are all included in the set S, for b is prime *to n. Also they are different from one another and from those of the set* (A).

For if $b_r = b_s$ where $s < r < t$, then $bg^r \equiv bg^s$, and since b is prime to n, $g^{r-s} \equiv 1$, which contradicts the hypothesis that t is the least number such that $g^t \equiv 1$.

Again, if $b_r = a_s$ ($r < t$, $s < t$), then $bg^r \equiv g^s$, and therefore $b \equiv g^{s-r}$ or $b \equiv g^{t-(r-s)}$ according as $s \gtrless r$. In both cases b would belong to the set (A), which is contrary to supposition.

If the sets (A) and (B) include all the numbers in S, then $2t = \varphi(n)$.

Otherwise $2t < \varphi(n)$, and then some number c of the set S does not occur in (A) or (B). We then consider the residues of $cg^0, cg^1, cg^2, \ldots cg^{t-1}$, denoting them by

$$c_0, c_1, c_2, \ldots c_{t-1}. \qquad \ldots\text{(C)}$$

As before, we can show that these numbers are all included in S, and that they are different from one another and from those in the sets (A) and (B).

Thus $3t \le \varphi(n)$; and, if $3t < \varphi(n)$, the process can be continued until the numbers in S have been arranged in groups of t numbers. Hence t is a divisor of $\varphi(n)$.

Incidentally, this proof establishes the truth of Euler's generalisation of Fermat's theorem.

■ **3. An Odd Prime Modulus: Primitive Roots:** Let p be an odd prime, g any number not divisible by p, and t the least index for which $g^t \equiv 1 \pmod{p}$, then g is said to *belong to the index t.*

Since $\varphi(p) = p - 1$, t is equal to or is a divisor of $p - 1$. If $t = p - 1$, so that the period of residues includes every number less than p, the period is said to be *complete* and g is called a *primitive root of the modulus p.* If t is a divisor of $p - 1$, the period is *incomplete* and g may be called a *subordinate root.*

Hence *if g is a primitive root of p, every number prime to p is congruent with some power of g.*

It can be proved that, for any odd prime modulus, primitive roots exist; for the present we assume this to be the case. When one has been found, the others can be found as in the following illustration.

Illustration. In the margin, for the modulus 13 the least positive residues of 2^0, 2^1, 2^2, ... 2^{11} are written down round the circumference of a circle.

Every number less than 13 occurs, and so 2 is a primitive root of 13.

Hence, without further reckoning, we can write down the residues (mod 13) of powers of *any* number.

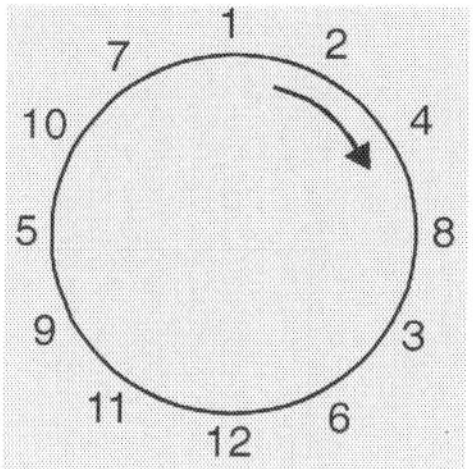

Fig. 65

Thus for powers of 6, since $6 \equiv 2^5$, $6^2 \equiv 2^{10}$, etc., the residues of 6^0, 6^1, 6^2, ... are found by starting with 1 and taking every *fifth* number. The residues are

$$1, 6, 10, 8, 9, 2, 12, 7, 3, 5, 4, 11.$$

Every number less than 13 appears, the reason being that 5 is prime to 12.

Hence 6 is a primitive root; and, in the same way, if μ is any number less than 12 and prime to it, 2^μ is congruent (mod 13) with a primitive root.

Since $\varphi(12) = 4$, there are *four* primitive roots of 13; they are congruent with 2^1, 2^5, 2^7, 2^{11}, and are equal to 2, 6, 11, 7, the first, fifth, seventh and eleventh numbers in order round the circle, *counting from* 1.

Again, we have $5 \equiv 2^9$, and the residues of 5^0, 5^1, 5^2, ... are obtained by counting every ninth number, or every third number in the reverse order.

Since 3 is the greatest common divisor of 9 and 12 and $12 = 3 \cdot 4$, after four counts we reach the number, 5, from which we started.

Thus 5 is a subordinate root; it belongs to the index 4, and the period is 1, 5, 12, 8.

■ **4.** If p is an odd prime, g any number > 1 and not divisible by p, t the least index for which $g^t \equiv 1 \pmod{p}$ and r_0, r_1, r_2, ... the least positive residues of g^0, g^1, g^2, ..., then

$$r_0^t + r_1 + r_2 + ... + r_{t-1} \equiv 1 + g^1 + g^2 + ... + g^{t-1}$$
$$\equiv (g^t - 1)/(g - 1) \bmod p$$
$$\equiv 0 \pmod{p},$$

that is to say, *the sum of the residues which form the period is divisible by p.*

Thus if $p = 13$ and $g = 5$, then $t = 4$ and

$$r_0 + r_1 + r_2 + r_3 = 1 + 5 + 12 + 8 = 13 \times 2.$$

Again, suppose that e is a factor of t (not t itself) and $t = ef$. Starting with any residue r_m and taking every eth residue, we have the sequence

$$r_m, \ r_{m+e}, \ r_{m+2e}, \ \dots \ .$$

These residues recur, and they form a sub-period containing f residues;
for $g^x \equiv 1 (\mathrm{mod}\ p)$ if $x = ef$, and for no smaller value of x. Also

$$r_m + r_{m+e} + r_{m+2e} + \dots + r_{m+(f-1)e} \equiv g^m (1 + g^e + g^{2e} + \dots + g^{(f-1)e}) \ (\mathrm{mod}\ p)$$
$$\equiv g^m (g^{ef} - 1)/ (g^e - 1) \ (\mathrm{mod}\ p)$$
$$\equiv 0 (\mathrm{mod}\ p).$$

Hence *the sum of the residues forming the sub-period is divisible by p.*

Thus if $p = 19$ and $g = 2$ [See Art. 1, (v)], then $t = 18$ and

$$r_0 + r_2 + r_4 + \dots + r_{16} = 1 + 4 + 16 + 7 + 9 + 17 + 11 + 6 + 5 = 4 \times 19,$$
$$r_1 + r_4 + r_7 + r_{10} + r_{13} + r_{16} = 2 + 16 + 14 + 17 + 3 + 5 = 3 \times 19.$$

◼ **5. Recurring Decimals:** Let $m < n$ and prime to n, then

(1) *If n is prime to 10, m/n is equal to a pure recurring decimal with a period of t figures, where t is the least index other than zero for which $10^t \equiv 1 \ (\mathrm{mod}\ n)$; so that $t = \varphi(n)$ or is a divisor of $\varphi(n)$.*

For the remainders $r_1, r_2, r_3, \dots$, in the process of expressing m/n as a decimal, are the least positive residues $(\mathrm{mod}\ n)$ of $m, 10m, 10^2 m, \dots$. Hence, by Arts. 1 and 2, the remainders recur, with a period of t figures.

Also if $m/n = 0 \cdot a_1 a_2 \dots a$, then $r_s/n = 0 \cdot a_{s+1} a_{s+2} \dots a_t a_1 \dots a_s$ so that all the fractions $m/n, r_1/n, r_2/n, \dots$ have the same period. It follows that

(i) if $t = \varphi(n)$ and m is any number $< n$ and prime to it, all the fractions m/n have the same period.

(ii) if $\varphi(n) = et$ where $e > 1$, the fractions m/n can be arranged in e groups of t fractions, such that the periods are the same for fractions in the same group and different for those in different groups.

This follows from the reasoning in the last part of Art. 2.

(iii) if $n = p$, *any prime except 2 or 5, then $t = p - 1$, or t is a divisor of $p - 1$, according as 10 is or is not a primitive root of p.**

If $n = p^r$, then $\varphi(n) = p^{r-1}(p - 1)$ and t is equal to or is a divisor of $p^{r-1}(p - 1)$. The case in which $n = p^2$ deserves further consideration.

(iv) *if p is a prime, not 2 or 5, and $1/p$ has a period of t figures, then $1/p^2$ has a period of t or pt figures.*

For let T be the number of figures in the period of $1/p^2$, and let

$$F(t) = 1 + 10^t + 10^{2t} + \dots + 10^{(p-1)t}.$$

Then $F(t) \equiv 0 (\mathrm{mod}\ p)$, for $10^t \equiv 1 \ (\mathrm{mod}\ p)$, and therefore

$$10^{pt} - 1 = (10^t - 1) \, F(t) \equiv 0 \ (\mathrm{mod}\ p^2).$$

Hence T is pt, or a divisor of pt, and, since p is a prime, $T = t$ or $p\tau$, where τ is t, or a divisor of t,

---

* The primes less than 100 of which 10 is a primitive root are 7, 17, 19, 23, 29, 47, 59, 61, 97.

If $\tau < t$, we have

$$1 + 10^{\tau} + 10^{2\tau} + \ldots + 10^{(p-2)\tau} = (10^{(p-1)\tau} - 1) / (10^{\tau} - 1) \equiv 0 \pmod{p},$$

for 10^{τ} is not $\equiv 1$. Hence $F(\tau) \equiv 10^{(p-1)\tau} \equiv 1 \pmod{p}$.

Now
$$10^{p\tau} - 1 = (10^{\tau} - 1) \, F(\tau),$$

and therefore $10^{p\tau} - 1$ is not $\equiv 0 \pmod{p^2}$.

Consequently τ is not $< t$ and $T = t$ or pt.

Note: The only known cases in which $1/p^2$ has a period of t figures are when $p = 3$ or 487.

(2) *If $F = \dfrac{m}{2^{\mu}5^{\nu}} < 1$, where m is prime to 10, and k is the greater of the two μ, ν, then F is a terminating decimal with k figures.*

For $10^k F$ is an integer with not more than k digits, the last of which is not zero.

(3) *If $F = \dfrac{m}{2^{\mu}5^{\nu}n} < 1$, where m, n and 10 are prime to one another, then F is a mixed recurring decimal, with a period of t figures, where t is the number of figures in the period of $1/n$.*

For let k be the greater of the two μ, ν, then $10^k F = m'/n$, where m' is prime to n and less than 10^k. Therefore $10^k F = I + f$, where I is an integer of not more than k digits and f is a pure recurring decimal with a period of t figures.

(4) *If $F = \dfrac{m}{p^a q^b r^c \ldots}$ where p, q, r,... are different primes other than 2 and 5 and m is prime to p, q, r, ..., then if t, t', t'' ... are the numbers of figures in the periods of the decimal equivalents of F, $1/p^a$, $1/q^b$, etc., respectively, t is the L.C.M. of t', t'', t''', etc.*

For let $n = p^a q^b \ldots$, then $10^t \equiv 1 \pmod{n}$, therefore $10^t \equiv 1 \pmod{p^a}$ and, since t' is the least index such that $10^{t'} \equiv 1 \pmod{p^a}$, t must be a multiple of t'; similarly, t is a multiple of t'', t''', etc.

Therefore t is a common multiple of t', t'', etc.

Again, if τ is *any* common multiple of t', t'', ... we have $10^{\tau} \equiv 1$ for each of the moduli p^a, q^b, ... , and since p^a, q^b, ... are prime to one another, $10^{\tau} \equiv 1 \pmod{n}$.

It follows that t is the *least* common multiple of t' t'', etc.

Given a table showing the periods of decimals equivalent to fractions $1/n$, where n is a prime or a power of a prime, we can easily express *any* fraction as a decimal by using the theorem of Ch. 26, 9.

■ **5.** In practice, the following theorem is useful.

Let $m/n = 0 \,.\, a_1 a_2 a_3 \ldots$ where n is prime to 10 and m prime to n. If, after r steps of division, the remainder $n - m$ occurs, the subsequent figures of the decimal are $9 - a_1, 9 - a_2, 9 - a_3, \ldots$, and the period contains 2r figures.

For let $m, m_1, m_2, \ldots$ be the least positive residues of $m, 10m, 10^2m, \ldots$,

then
$$m_r = n - m \equiv -m \pmod{n}.$$

Hence the subsequent residues are congruent with $-m_1, -m_2$, etc., and are equal to $n - m_1, n - m_2$, etc.

Moreover, a_{r+1} is the greatest integer in $10(n-m)/n$;

hence $\quad a_{r+1} = 1 \left(10 - \overline{a_1 + f}\right) = 9 - a_1,$

where f is a positive proper fraction. Similarly $a_{r+2} = 9 - a_2$, etc.

■ **EXAMPLE :** *To express 1/73 as a decimal, the division need not be carried beyond the stage shown in the reckoning, for since the remainder 72(=73 −1) occurs, the subsequent remainders are 73 − 10, 73 − 27, etc., the corresponding figures in the decimal being 9 − 0, 9 − 1, etc. Thus*

$$73 \,)\, \underline{100} \,(\, .0136$$
$$\underline{270}$$
$$\underline{510}$$
$$720$$

$$1/73 = 0.01369863.$$

Again, the number of figures in the period is 8 and $\varphi(73) = 72$. Hence for values of m less than 73, the fractions $m/73$ can be arranged in $72/8 = 9$ groups such that the period is the same for those in the same group. By the preceding, the complete set of remainders for 1/73 is

$$1, 10, 27, 51, 72, 63, 46, 22, \qquad \qquad \text{...(A)}$$

and one group is obtained by giving m these values. To get a second group, choose any number less than 73 not included in the set (A), say 2. The remainders for 2/73 are congruent (mod 73) with 2×1, 2×10, 2×27, etc., and are therefore

$$2, 20, 54, 29, 71, 53, 19, 44. \qquad \qquad \text{...(B)}$$

A second group is obtained by giving m these values. A third group can be found in the same way by choosing a number less than 73 and not included in the sets (A), (B) – and so on.

EXERCISE XLV

Note: The prime factors of $f(t) = 10^t - 1$ for $t = 1, 2, 3, \ldots 9$ are given in the following table:

$$f(1) = 3^2, \qquad f(2) = 3^2 \cdot 11, \qquad f(3) = 3^3 \cdot 37,$$
$$f(4) = 3^2 \cdot 11 \cdot 101, \qquad f(5) = 3^2 \cdot 41 \cdot 271, \qquad f(6) = 3^3 \cdot 7 \cdot 11 \cdot 13 \cdot 37,$$
$$f(7) = 3^2 \cdot 239 \cdot 4649, \qquad f(8) = 3^2 \cdot 11 \cdot 73 \cdot 101 \cdot 137, \quad f(9) = 3^4 \cdot 37 \cdot 333667.$$

1. For the modulus 19, given the least positive residues of 2^0, 2^1, 2^2, etc. (Art. 1), write down those of 10^0, 10^1, 10^2, etc. and 7^0, 7^1, 7^2, etc. What are the primitive roots of 19?

2. For the modulus 17, find the least positive residues of 3^0, 3^1, 3^2, etc. What are the primitive roots of 17?

3. If p is an odd prime, g belongs to the index t, and r_0, r_1, r_2, etc., are the least positive residues of g^0, g^1, g^2, etc., prove that

    (*i*) $p - 1$ occurs as a residue if, and only if, t is even.

(ii) $r_0 r_1 r_2 \dots r_{t-1} \equiv -1$ or $+1 (\bmod\ p)$ according as t is even or odd. Verify when $p = 13$, $g = 2, 3, 5$.

(iii) if g is a primitive root of p, show that theorem (ii) becomes Wilson's theorem.

   [(i) If t is even, $g^{t/2} \equiv -1 \ (\bmod\ p)$.

   (ii) $r_0 r_1 r_2 \dots r_{t-1} \equiv g^s (\bmod\ p)$. where $s = 0 + 1 + 2 + \dots + (t - 1)$.]

RECURRING DECIMALS

Here n is prime to 10, p is a prime not 2 or 5, t is the number of figures in the period of $1/n$ or $1/p$; r_0, r_1, r_2 etc., are the least positive residues of $10^0, 10^1, 10^2$, etc. (mod

n or p), and we write $\dfrac{1}{n}$ or $\dfrac{1}{p} = 0 \cdot a_1 a_2 a_3 \dots$.

4. Prove that $0 \cdot a_1 a_2 \dots a_t = a_1 a_2 \dots a_t / (10^t - 1)$ where $a_1 a_2 \dots a_t$ is the number of which the digits are $a_1, a_2, \dots$.

$$\left[\text{The decimal} = a_1 a_2 \dots a_t \left(\frac{1}{10^t} + \frac{1}{10^{2t}} + \frac{1}{10^{3t}} + \dots \right). \right]$$

5. If $\dfrac{1}{n} = 0 \cdot a_1 a_2 \dots a_r \dots$ and $\dfrac{n-1}{n} = 0 \cdot b_1 b_2 \dots b_r \dots$, prove that

$$a_1 + b_1 = a_2 + b_2 = \dots a_r + b_r = \dots .$$

6. If $1/n = 0. \, a_1 a_2 \dots a_t$ and r is the remainder immediately preceding the remainder 1 in the division, then r and a_t are the least positive integers x, y such that

$$10x - ny = 1$$

and $$a_1 a_2 \dots a_i \times r = a_t \, a_1 a_2 \dots a_{t-1},$$

where the a's are digits of the numbers indicated. Having found r and a_t, all the figures in the period may be found as in Ex. 8.

[$1/n = 0 \cdot a_1 \, a_2 \dots a_t a_1 \dots$, $r/n = 0 \cdot a_t a_1 \dots a_{t-1} \dots$; explain why no figure has to be carried when a_1 is multiplied by r.]

7. If p is a prime and $1/p = 0 \cdot a_1 a_2 \dots a_r a_{r+1} \dots a_{2r}$, prove that

$$a_{r+1} + a_1 = a_{r+2} + a_2 = \dots = a_{2r} + a_r = 9,$$

and give an example to show that such relations do not hold if p is a composite number.

8. *Express $\frac{1}{13}$ as a decimal by using the equality* $10 \times 4 - 13 \times 3 = 1$.

(i) Put down 3 (the last figure of the period) and multiply by 4 as follows:

076923

4

$4 \times 3 = 12,$	put 2 to the left of 3 and carry 1.
$4 \times 2 + 1 = 9,$	put 9 to the left of 2.
$4 \times 9 = 36,$	put 6 to the left of 9 and carry 3.
$4 \times 6 + 3 = 27,$	put 7 to the left of 6 and carry 2.
$4 \times 7 + 2 = 30,$	put 0 to the left of 7.

If the process is continued, the figures recur and $\frac{1}{13} = 0.076923$.

(*ii*) Or thus–in Ex. 7, $a_t a_1 a_2 \ldots a_{t-1} / r = a_1 a_2 \ldots a_t$.

Hence the following : Put down 3 and divide by 4 as below:

$$4\overline{)3}$$
$$\overline{076923.}$$

4 into 3 is 0, which is the first figure of the period.

4 into 30 is 7 and 2 over.	Put 7 to the right of 0.
4 into 27 is 6 and 3 over.	Put 6 to the right of 7.
4 into 36 is 9.	Put 9 to the right of 6.
4 into 9 is 2 and 1 over.	Put 2 to the right of 9.
4 into 12 is 3.	Put 3 to the right of 2.

4 into 3 is 0, and the figures recur.

9. Cases in which $1/p$ can be quickly expressed as a decimal by the methods of Ex. 8 are $p = 17, 19, 23, 29, 43, 59, 79, 89$.

10. Justify the following:

$$\text{If} \qquad A = 0588$$
$$B = 2352$$
$$C = 9408$$
$$\cdots\cdots\cdots\cdots\cdots\cdots\cdots$$

$$\text{then} \qquad \frac{1}{17} = \overline{.0588235294} \ldots\ldots$$

$$17\overline{)\,100\,} \,(.0588$$
$$\underline{150}$$
$$\underline{140}$$
$$4$$

where $B = 4A$, $C = 4B$, etc.

[The division shows that $10^4 = 588 \times 17 + 4$;

$$\therefore \qquad \frac{1}{17} = \frac{588}{10^4}\left(1 - \frac{4}{10^4}\right)^{-1} = 0.0588\left(1 + \frac{4}{10^4} + \frac{4^2}{10^8} + \ldots\right) \cdot \,]$$

11. By the method of Ex. 10 show that

$$\frac{1}{31} = 0.032258064516129.$$

[After 6 steps of division, the remainder 2 occurs.]

12. Show that for values of $m < 41$ the fractions $m/41$ can be arranged in 8 groups such that all the fractions in any group have the same period. What are the values of m for the group to which $2/41$ belongs?

13. Prove that $\dfrac{1}{81} = 0.012345679$, and show that the values of m for which $m/81$ has this period are given by $m \equiv 1 \pmod 9$.

14. If n is any number prime to 10, prove that a number n' can be found such that every digit in the product nn' is 1.

Find n' when $n = 41$ and when $n = 123$.

$$\left[\text{Express } \frac{1}{n} \text{ or } \frac{1}{9n} \text{ as a decimal, according as } n \text{ is or is not prime to } 3. \right]$$

15. If n is a prime or a power of a prime, prove that the values of n for which the decimal equivalent of $1/n$ has a period of t figures are as follows:

t	1	2	3	4	5	6	7	8	9
n	3, 9	11	27, 37	101	41, 271	7, 13	239, 4649	73, 137	333667

[The values of n for which $1/n$ has a period of t figures are included among the divisors of $10^t - 1$. See note at the head of this Exercise.]

16. Show that the only prime p for which the decimal equivalent of $1/p$ has (*i*) a period of 10 figures is 9091, and (*ii*) a period of 12 figures is 9901.

Also express $1/9091$ and $1/9901$ as decimals, explaining why only 5 and 6 steps of division respectively are necessary.

■ **6. Primitive Roots (continued):** The process of finding a primitive root is one of trial. The orthodox methods of shortening the reckoning are given in another volume; but in the example below we give a method, founded on the assumption of Art. 3, which is usually successful. For convenience in solving the congruences in Ex. XLVI, we here give *the least primitive root, g, for any prime modulus, p, less than* 100.

$g = 2,$ $p = 3, 5, 11, 13, 19, 29, 37, 53, 59, 61, 67, 83;$

$g = 3,$ $p = 7, 17, 31, 43, 79, 89;$

$g = 5,$ $p = 23, 47, 73, 79;$

$g = 6,$ $p = 41; \; g = 7, \; p = 71$

■ **EXAMPLE :** *Find the primitive roots of 67.*

If $1/67$ is converted into a decimal fraction, the reckoning shows that 10 is a subordinate root of modulus 67, with a period of 33 figures. Art. 3 suggests that this period consists of every alternate figure of the period of *some* primitive root, g; and, consequently, that $g^2 \equiv 10 (\bmod\ 67)$ will lead to a primitive root.

We find that a value of g is ± 12; and, converting $12/67$ into a decimal fraction, we obtain, as remainders in the process, the residues of the odd powers of 12. When these are interpolated between the remainders in the reckoning for $1/67$, we find that all the numbers less than 67 are included; thus 12 is a primitive root, and its period is

$$1, \textit{12}, 10, \textit{53}, 33, \textit{61}, 62, 7, 17, \textit{3}, 36, \textit{30}, 25, \textit{32}, 49, \textit{15}, 21, \text{ etc.}$$

From this, as in Art. 3, all the other primitive roots can be immediately written down, together with their periods; thus we find that other primitive roots are 61, 7, 32, 51, 41, 13, 2, etc., these being the residues of 12^k, where k is prime to 66.

■ **7. The Congruence $x^n \equiv a (\bmod\ p)$:** It is supposed that p is an odd prime.

(1) *If n is prime to $p - 1$, there is a single solution.*

For, if $x^n \equiv a$, then $x^{n\lambda} \equiv a^\lambda$; and, since n is prime to $p - 1$, there is, by Chap. 25, 3, a single value of λ, less than $p - 1$, such that

$$\lambda n \equiv 1 \ (\bmod\ p - 1).$$

If λ has this value, $x^{n\lambda} = x. (x^{p-1})^k \equiv x \pmod{p}$; and thus, $x \equiv a^{\lambda} \pmod{p}$, which is the only solution.

■ **EXAMPLE 1.** *Solve the congruence,* $x^{13} \equiv 21 \pmod{23}$.

Here $x^{13\lambda} \equiv 21^{\lambda} \pmod{23}$; and $13 \cdot 17 \equiv 1 \pmod{22}$; hence we have

$$x \equiv 21^{17} \equiv (-2)^{17} \equiv (-2) \cdot (256)^2 \equiv (-2). (3)^3 \equiv 5 \pmod{23}.$$

(2) *If n is not prime to $p-1$, there are several solutions, or there is no solution.*

Let d be the highest common divisor of n and $p-1$; and let g be a primitive root of p. Then, since the period of g contains all the numbers less than p, we can find r, such that $g^r \equiv a \pmod{p}$.

Let $x \equiv g^s \pmod{p}$; then $g^{ns} \equiv a \equiv g^r \pmod{p}$, and therefore

$$ns \equiv r \pmod{p-1}. \qquad \qquad ...(A)$$

The congruence (A), and therefore also the congruence $x^n \equiv a$, has d solutions, or no solution, according as r is, or is not, divisible by d; for d is a factor of n and $p-1$ (Chap. 26, 8). Also, the solutions of $x^n \equiv a \pmod{p}$ are given by $x \equiv g^s \pmod{p}$, where s has any value which satisfies (A).

■ **EXAMPLE 2.** *Solve the congruence,* $x^{15} \equiv 31 \pmod{37}$.

A primitive root of 37 is 2; and we find that $2^9 \equiv 31$. Hence, if $x \equiv 2^s$, then $2^{15s} \equiv 2^9$, and $15s \equiv 9 \pmod{36}$; hence $s \equiv 3, 15, 27 \pmod{36}$, and

$$x = 2^3, 2^{15}, 2^{27}, i.e.\ 8, 23, 6.$$

EXERCISE XLVI

Solve the congruences in Exs. 1–6:

1. $x^3 \equiv 1 \pmod{19}$.

2. $x^5 \equiv 2 \pmod{17}$.

3. $x^7 \equiv 8 \pmod{13}$.

4. $x^{11} \equiv 5 \pmod{31}$.

5. $x^6 \equiv 31 \pmod{41}$.

6. $x^{21} \equiv 2 \pmod{31}$.

7. Find the general solution in integers of $x^2 = 19y + 5$.

8. Find three solutions in integers of $x^3 = 73y + 3$.

9. Show, without calculation, that a primitive root of 237 will satisfy one at least of the congruences, $g \equiv 10$, $g^2 \equiv 10$, $g^4 \equiv 10$.

■ ■ ■

CHAPTER 28

Numerical Solution of Equations

■ **1. The Problem** under consideration is *to find the actual or approximate values of the real roots of any equation with numerical coefficients.*

This question is quite distinct from that of finding an algebraical solution. In fact, no algebraical solution of the general equation of the fifth or higher degree has been discovered.

If any rational or multiple roots exist they can be found and removed from the equation (Ch. 18, 1, and 6, 14). It may be convenient not to remove the multiple roots.

Thus we are only concerned with irrational roots. The usual procedure is as follows:

(*i*) We find an interval which contains *all* the roots. Ways of doing this have been given in Ch. 6, 12. A method due to Newton which involves more calculation, but which yields closer limits, is given in the next article.

(*ii*) We *separate* the roots, that is to say we find intervals each of which contains a single root or a multiple root.

(*iii*) Taking any interval which contains a single root, by a process of approximation, we find smaller and smaller intervals which contain the root.

■ **2. Newton's Method of finding an Upper Limit to the Roots:** This depends on the following theorem:

If h is a number such that $f(x)$ and all its derivatives are positive when $x = h$, then h is an upper limit to the roots of $f(x) = 0$.

For if n is the degree of $f(x)$, then

$$f(x + h) = f(h) + xf'(h) + \ldots + \frac{1}{\lfloor n} x^n f^{(n)}(h).$$

Hence if $f(h), f'(h), \ldots f^{(n)}(h)$ are all positive, $f(x + h) > 0$, for $x \geq 0$, that is to say $f(x) > 0$ for $x \geq h$, which proves the theorem.

Note: In applying this theorem, we suppose that the coefficient of the highest power of x in $f(x)$ is positive, so that $f^{(n)}(x) > 0$.

We then find the least integer h_1 such that $f^{(n-1)}(x) > 0$ when $x = h_1$. If this value of x does not make $f^{(n-2)}(x) > 0$, we find by trial a greater integer h_2 such that $f^{(n-2)}(h_2) > 0$.

Continuing thus, we can find the least integer h which makes $f(x)$ and all its derivatives positive.

It should be observed that if we have found a number h_r such that
$$f^{(n)}(x), f^{(n-1)}(x), \dots f^{(r)}(x)$$
are all positive, then these functions are all positive for $x > h_r$. This follows from the above theorem, for the derivatives of $f^{(r)}(x)$ are $f^{(r+1)}(x)$, $\dots f^{(n-1)}(x), f^{(n)}(x)$.

■ **EXAMPLE** : *Find the integral part of the greatest root of*
$$f(x) = x^3 - 10x^2 - 11x - 100 = 0. \text{ (Cf. Ex. 1, p. 87.)}$$
Here $f'(x) = 3x^2 - 20x - 11$, $f''(x) = 2(3x - 10)$, $f'''(x) = 6$.

The least integral values of x for which $f''(x)$, $f'(x)$ and $f(x)$ are positive are respectively 4, 8 and 12. Thus 12 is the least integer which exceeds the greatest root, the integral part of which is therefore 11.

EXERCISE XLVII

In Exs. 1–5 find an upper limit to the roots by Newton's method.
1. $x^3 - 20x^2 - 31x + 1609 = 0.$
2. $x^3 - 2x^2 - 51x - 110 = 0.$
3. $x^4 - 4x^3 - 3x + 23 = 0.$
4. $x^5 + 4x^4 - 2x^3 + 10x^2 - 2x = 962.$
5. $x^5 + 6x^4 - 10x^3 - 112x^2 - 207x = 110.$

■ **3. Separation of the Roots:** Rules for this purpose were given by Fourier and by Sturm. Fourier's rule, though often convenient, is incomplete, while that of Sturm definitely separates the roots, but its application may be very laborious. We require the following theorems.

Subsidiary Theorems.

(1) If α is a root of $f(x) = 0$, as the variable x increases through the value α, the functions $f(x)$ and $f'(x)$ have opposite signs just before $x = \alpha$ and the same signs just after $x = \alpha$.

For since $f(\alpha) = 0$, we have

$$f(\alpha + h) = hf'(\alpha) + \frac{h^2}{\lfloor 2} f''(\alpha) + \dots + \frac{h^r}{\lfloor r} f^{(r)}(\alpha) + \dots , \qquad \dots(A)$$

$$f'(\alpha + h) = f'(\alpha) + hf''(\alpha) + \dots + \frac{h^{r-1}}{\lfloor r-1} f^{(r)}(\alpha) + \dots \qquad \dots(B)$$

Suppose that $f^{(r)}(\alpha)$ is the first term of the sequence $f'(\alpha), f''(\alpha), \dots$ which is not zero. For sufficiently small values of h, the signs of $f(\alpha + h)$ and $f'(\alpha + h)$ are respectively the signs of the first terms in (A) and (B) which do not vanish; they are therefore the same as the signs of $h^r f^{(r)}(\alpha)$ and $h^{r-1} f^{(r)}(\alpha)$. Hence $f(\alpha + h)$ and $f'(\alpha + h)$ have the same or opposite signs according as $h \gtrless 0$, which proves the theorem.

(2) *If α is an r-multiple root of $f(x) = 0$, then, as the variable x increases through the value α, the signs of the functions*

$$f(x),\, f'(x),\, f''(x),\, \ldots f^{(r-1)}(x),\, f^{(r)}(x) \qquad \ldots(C)$$

are alternately + and −, or − and +, just before x = α, and all the functions have the same sign just after x = α.

For, when $x = \alpha$, all the terms of the sequence (C) except the last vanish, and the result follows by applying theorem (1) to every two consecutive terms. It follows that *as x increases through the value α, r changes of sign are lost in the sequence (C).*

It is important to observe that since $f^{(r)}(\alpha) \neq 0$ and $f^{(r)}(x)$ is continuous at $x = \alpha$, this function has the same sign just before and just after $x = \alpha$.

■ **4. Sturm's Functions:** Let $f(x)$ be a polynomial and $f_1(x)$ its first derivative. *Let the operation of finding the H.C.F. of $f(x)$ and $f_1(x)$ be performed with this alteration:*

The sign of each remainder is to be changed before it is used as a divisor: the sign of the last remainder is also to be changed.

Denote the *modified* remainders by $f_2(x),\, f_3(x),\, \ldots f_r(x)$; then $f(x),\, f_1(x),\, f_2(x),\, \ldots f_r(x)$ are called *Sturm's functions* and $f_1(x),\, f_2(x),\, \ldots f_r(x)$ are known as the *auxiliary functions.*

In the ordinary H.C.F. process we can multiply (or divide) any remainder by any constant. *In the modified process it is essential that such multipliers should be positive.*

Sturm's functions are connected by the equations

$$\left.\begin{aligned}
f(x) &= q_1 f_1(x) - f_2(x), \\
f_1(x) &= q_2 f_2(x) - f_3(x), \\
&\;\;\cdots\cdots\cdots\cdots\cdots\cdots\cdots\cdots \\
f_{r-2}(x) &= q_{r-1} f_{r-1}(x) - f_r(x),
\end{aligned}\right\} \qquad \ldots(A)$$

where $q_1, q_2, \ldots$ are the quotients in the process just described, these quotients being functions of x.

It is important to notice that the relation between $f(x)$ and $f_1(x)$ is essentially different from that connecting $f_1(x)$ and $f_2(x)$; $f_2(x), f_3(x)$; etc.

■ **5. Sturm's Theorem:** *If $f(x)$ is a polynomial and a, b are any real numbers (a < b), the number of distinct roots of $f(x) = 0$ which lie between a and b (any multiple root which may exist being counted once only) is equal to the excess of the number of changes of sign in the sequence of Sturm's functions*

$$f(x),\, f_1(x),\, f_2(x),\, \ldots f_r(x), \qquad \ldots(S)$$

when x = a over the number of changes of sign in the sequence when x = b.

Proof for the case in which $f(x) = 0$ has no repeated root.

Equations (A) lead to the following conclusions.

(*i*) Since $f(x) = 0$ has no repeated root, $f(x)$ and $f_1(x)$ have no common factor, and consequently $f_r(x)$ is independent of x.

(*ii*) No two consecutive terms of the sequence (S) can vanish for the same value of x, for, if this happened, all the subsequent terms including $f_r(x)$ would vanish.

(*iii*) If any term of (S) except the first is zero, the terms which precede and follow it have opposite signs. Thus if $f_s(x) = 0$, then

$$f_{s-1}(x) = -f_{s+1}(x).$$

A similar remark applies if one of the q's is zero. Thus if $q_s = 0$, then $f_{s-1}(x) = -f_{s+1}(x)$.

Suppose now that x increases continuously from $x = a$ to $x = b$. As x varies, no one of Sturm's functions can change its sign unless x passes through a value which makes that function vanish, for these functions are polynomials.

Suppose that x passes through a value α which makes *just one* of Sturm's functions vanish.

(*i*) If α is a root of $f(x) = 0$, one change of sign is lost in the sequence (S). For $f(x)$ and $f_1(x)$ have opposite signs just before $x = \alpha$, and they have the same signs just after $x = \alpha$.

(*ii*) If α is a root of $f_m(x)$ where $m = 1, 2, \ldots$ or $r - 1$, no change of sign is gained or lost. For $f_m(\alpha) = 0$, and therefore $f_{m-1}(\alpha) = -f_{m+1}(\alpha)$.

If $m = 1$, $f_0(\alpha)$ stands for $f(\alpha)$. Moreover, $f_{m-1}(x)$ and $f_{m+1}(x)$ are continuous at $x = \alpha$, so that each of these is of invariable sign near $x = \alpha$.

Hence, just before and also just after $x = \alpha$, the signs of $f_{m-1}(x), f_m(x), f_{m+1}(x)$ are either $++-$ or $+--$ or $-++$ or $--+$, showing that no change of sign is lost or gained.

If x passes through a value α which makes more than one of Sturm's functions vanish, no two of them can be consecutive functions.

If $f(x)$ is one of them, then by the preceding, one change of sign is lost between $f(x)$ and $f_1(x)$, and no change is gained or lost in the sequence $f_1(x), f_2(x), \ldots f_r(x)$. Thus one change is lost in the sequence (S).

If $f(x)$ is not one of them, no change of sign is gained or lost.

Thus, as x increases, a change of sign in the sequence (S) is lost whenever x passes through a root of $f(x) = 0$, and under no other circumstance is a change gained or lost. This proves the theorem.

Remarks. In applying Sturm's theorem, labour may be saved by the following considerations.

(*i*) If there is no repeated root, the last function $f_r(x)$ is independent of x and its *sign only* is required.

Let x have a value such that $f_{r-1}(x) = 0$, then $f_r(x) = -f_{r-2}(x)$. Thus, if the last three functions are $a'x^2 + b'x + c'$, $a''x + b''$, a''',

$$\text{then} \qquad a''' = -\left(a' \cdot \frac{b''^2}{a''^2} - b' \cdot \frac{b''}{a''} + c' \right);$$

and from this the sign of a''' may often be found without much calculation.

(*ii*) If, at any stage, we arrive at a function $f_s(x)$ such that all the roots of $f_s(x) = 0$ are imaginary, the H.C.F. process need not be continued, and we can use $f(x)$, $f_1(x), \ldots f_s(x)$, instead of the complete set of functions.

For the essential property of the last function is that it should remain of invariable sign for all real values of x, and $f_s(x)$ has this property.

(*iii*) Suppose that one of Sturm's functions as $f_m(x)$ vanishes when $x = a$; then, in counting the number of changes of sign in the sequence $f(a), f_1(a), \ldots f_r(a)$, we may regard the sign of $f_m(a)$ as either $+$ or $-$. For if $f_m(a) = 0$, then $f_{m-1}(a)$ and $f_{m+1}(a)$ have opposite signs.

EXAMPLE 1. *Find the number and position, relative to numbers on the scale of integers, of the real roots of*

$$f(x) = x^4 - 3x^3 - 2x^2 + 7x + 3 = 0.$$

Here $f_1(x) = 4x^3 - 9x^2 - 4x + 7$, and the modified H.C.F. process is as follows:

$$
\begin{array}{l|llll|l}
f_1(x) = & 4 - \quad 9 - \quad 4 + \quad 7 & 4 - 12 - \ 8 + \quad 28 + \quad 12 = f(x) & 1 \\
 & & 4 - 9 - \ 4 + \quad 7 \\
 & & \quad -3 - \ 4 + \quad 21 + \quad 12 \\
 & & \qquad\qquad\qquad\qquad\quad 4 \\
 & & \overline{\quad -12 - 16 + \quad 84 + \quad 48} & -3 \\
 & \qquad\qquad\qquad 43 & -12 + 27 + \quad 12 - \quad 21 \\
4 & \overline{172 - 387 - \ 172 + \ 301} & \overline{\qquad 43 - \quad 72 - \ 69 = f_2(x)} \\
 & 172 - 288 - \ 276 & \qquad\qquad\qquad\qquad\qquad 83 \\
 & \overline{\quad - \ 99 + \ 104 + \ 301} & \overline{3569 - 5976 - 5727} & 43 \\
 & \qquad\qquad\qquad 43 & 3569 - 8213 \\
-99 & \overline{-4257 + 4472 + 12943} & \overline{\qquad 2237 - 5727} \\
 & -4257 + 7128 + \ 6831 & \qquad\qquad 83 \\
 & \overline{\quad 32\,)\,2656 - \ 6112} & \overline{185671 - 47\,5341} & 2237 \\
f_3(x) = & \qquad\quad 83 - \quad 191 & 185671 + 427267 \\
 & & \overline{f_4(x) = \qquad + \ 48074}
\end{array}
$$

Thus $f_2(x) = 43x^2 - 72x - 69$, $f_3(x) = 83x - 191$, $f_4(x) = + 48074$.

The sign of $f_4(x)$ is found more easily by Remark (*i*). If $f_3(x) = 0$, then $f_4(x) = - f_2(x)$.

When $x = \dfrac{191}{83}$, $f_2(x)$ is negative. Therefore $f_4(x)$ is positive.

The number of changes of sign in the sequence of Sturm's functions for various values of x are shown below:

x	$-\infty$	-2	-1	0	2	3	∞
$f(x)$	+	+	−	+	+	+	+
$f_1(x)$	−	−	−	+	−	+	+
$f_2(x)$	+	+	+	−	−	+	+
$f_3(x)$	−	−	−	−	−	+	+
$f_4(x)$	+	+	+	+	+	+	+
Number of changes of sign	4	4	3	2	2	0	0

Hence the equation $f(x) = 0$ has one root between -2 and -1, one between -1 and 0 and two between 2 and 3.

■ **EXAMPLE 2.** *Find the number and position of the real roots of*
(i) $x^5 - x - 1 = 0$;   (ii) $x^6 - 2x^2 + 3x - 4 = 0$.

(i) Taking $f(x) = x^5 - x - 1$, we hare $f_1(x) = 5x^4 - 1$, and the modified H.C.F. process is

$$\begin{array}{r|l|l} 5x^4-1 & 5x^5 - 5x - 5 & x. \\ & \underline{5x^5 - x} & \\ & f_2(x) = 4x + 5 & \end{array}$$

Also, when $f_2(x) = 0$, $f_3(x) = -f_1(x)$, therefore

$$f_3(x) = -\{5(-\tfrac{5}{4})^4 - 1\} < 0.$$

For various values of x, the signs of Sturm's functions are:

x	$-\infty$	0	1	2	∞
$f(x)$	$-$	$-$	$-$	$+$	$+$
$f_1(x)$	$+$	$-$	$+$	$+$	$+$
$f_2(x)$	$-$	$+$	$+$	$+$	$+$
$f_3(x)$	$-$	$-$	$-$	$-$	$-$
Number of changes of sign	2	2	2	1	1

showing that there is one real root between 1 and 2, and that the other roots are imaginary.

(ii) Here $f(x) = x^6 - 2x^2 + 3x - 4$, $f_1(x) = 6x^5 - 4x + 3$.

$$\begin{array}{r|l|l} 6x^5 - 4x + 3 & 6x^6 - 12x^2 + 18x - 24 & x. \\ & \underline{6x^6 - 4x^2 + 3x} & \\ & f_2(x) = 8x^2 - 15x + 24 & \end{array}$$

The roots of $f_2(x) = 0$ are imaginary, for $15^2 < 4 \cdot 8 \cdot 24$, and we need not proceed further.

x	$-\infty$	-2	-1	1	2	∞
$f(x)$	$+$	$+$	$-$	$-$	$+$	$+$
$f_1(x)$	$-$	$-$	$+$	$+$	$+$	$+$
$f_2(x)$	$+$	$+$	$+$	$+$	$+$	$+$

Thus there is one real root between -2 and -1, another between 1 and 2, the other roots being imaginary.

■ **6. Sturm's Theorem (Multiple Roots):** It remains to consider the case in which the equation $f(x) = 0$ has repeated roots. Suppose that

$$f(x) = (x - \alpha)^p (x - \beta)^q (x - \gamma)^r \ldots$$

where $p, q, r, \ldots$ are positive integers and $\alpha, \beta, \gamma, \ldots$ are all different (real or imaginary), then

$$f_1(x) = u \cdot \{p(x - \beta)(x - \gamma) \ldots + q(x - \alpha)(x - \gamma) \ldots + r(x - \alpha)(x - \beta) \ldots + \ldots\},$$

where $\qquad u = (x - \alpha)^{p-1}(x - \beta)^{q-1}(x - \gamma)^{r-1} \ldots$.

Hence u is the H.C.F. of $f(x)$ and $f_1(x)$, and all of Sturm's functions,

$$f(x), f_1(x), f_2(x), \ldots f_r(x), \qquad\qquad \text{...(S)}$$

are divisible by u. Denote the quotients by

$$\phi(x), \psi_1(x), \psi_2(x), \ldots \psi_r(x), \qquad\qquad \text{...(S')}$$

so that $\qquad\qquad \phi(x) = (x - \alpha)(x - \beta)(x - \gamma) \ldots ,$

and $\psi_r(x)$ is independent of x. Also for any value of x, the number of changes of sign in the sequence (S) is the same as that in the sequence (S').

Dividing equations (A) of Art. 4 by u, we have the identities

$$\phi(x) = q_1\psi_1(x) - \psi_2(x), \ \psi_1(x) = q_2\psi_2(x) - \psi_3(x), \text{ etc.}$$

Whence, as in Art. 5, it follows that no change of sign is gained or lost as x passes through a value which makes one of the set $\psi_1(x), \psi_2(x), \ldots \psi_r(x)$ vanish.

Again, $\qquad \psi_1(\alpha) = p(\alpha - \beta)(\alpha - \gamma) \ldots = p\phi_1(\alpha),$

therefore $\psi_1(x)$ and $\phi_1(x)$ have the same sign when $x = \alpha$, and since these functions are continuous, they have the same signs for values of x near $x = \alpha$. Hence, by Art. 5, as x increases through any root α of $\phi(x)$ a change of sign is lost in the sequence (S').

It follows that the number of real roots of $\phi(x) = 0$ in the interval (a, b), where $a < b$ is equal to the excess of the number of changes of sign in the sequence (S') when $x = a$ over the number of changes of sign when $x = b$. That is, the number of *distinct* roots of $f(x)$ in the interval (a, b) is equal to the excess of the number of changes of sign in the sequence (S) when $x = a$ over the number of changes of sign when $x = b$.

Conditions for the Reality of all the Roots: Let $f(x)$ be a polynomial of degree n with its leading coefficient positive. In general, the number of Sturm's functions is $n + 1$.

The necessary and sufficient conditions that the roots of $f(x) = 0$ may be all real and different are: (i) *the number of Sturm's functions must be $n + 1$. and* (ii) *the leading coefficients of all these functions must be positive.*

For, as x passes from $-\infty$ to $+\infty$, n changes of sign will be lost in the sequence of Sturm's functions if and only if both of these conditions hold.

▪ **7. Application of Sturm's Theorem:** For the biquadratic,

$$f(z) = z^4 + 6Hz^2 + 4Gz + K,$$

where $K = a^2I - 3H^2$ [Ch. 12, 9; 10 (6)], Sturm's functions are $f(z)$ and

$$f_1(z) = z^3 + 3Hz + G,$$
$$f_2(z) = -3Hz^2 - 3Gz - K,$$

$$f_3(z) = -(2HI - 3aJ)\, z - GI,$$
$$f_4(z) = I^3 - 27\, J^2 = \Delta.$$

In proving this we require the identity [Ch. 12, 10, (6)]

$$G^2 + 4H^3 = a^2(HI - aJ). \qquad\qquad \text{...(A)}$$

Sturm's process is

$$f_3(z) = -Lz - GI \text{ where } L = 2HI - 3aJ.$$

Again, if $f_3(z) = 0$, then $f_4(z) = -f_2(z)$, therefore

$$L^2 f_4(z) = 3H\,(GI)^2 - 3G(GI)L + KL^2$$
$$= (a^2 I - 3H^2)\,(3aJ - 2HI)^2 + 3G^2 I\,(3aJ - HI)$$
$$= 9a^4 I J^2 - 12a^3 H I^2 J + a^2 H^2 (4I^3 - 27J^2)$$
$$\qquad + 3I(G^2 + 4H^3)\,(3aJ - HI)^*$$
$$= a^2 H^2\,(4I^3 - 27J^2) - 3a^2 H^2 I^3$$
$$= a^2 H^2 (I^3 - 27\, J^2).$$

Disregarding square factors, we may therefore take

$$f_4(z) = I^3 - 27 J^2.$$

Hence if Δ is positive and both H and $2HI - 3a\,J$ are negative, all the roots are real. If Δ is positive and at least one of the two, H and $2HI - 3aJ$, is positive, all the roots are imaginary.

For, as x passes from $-\infty$ to $+\infty$, in the first case four changes of sign are lost in the sequence of Sturm's functions, and in the second case no change of sign is lost.

If Δ is negative, two roots are real and two imaginary.

All this has been proved more easily in Ch. 12.

■ **8. Fourier's Theorem:** *Let $f(x)$ be a polynomial of the n-th degree and let $f_1(x)$, $f_2(x)$, ... $f_n(x)$ be its successive derivatives. Let R be the number of real roots of $f(x) = 0$ which lie between a and b, where a, b are any real numbers, of which a is less than b, and an r-multiple root is counted r times. Let N, N' respectively denote the number of changes of sign in the sequence*

---

* The identity (A) is used in these steps.

$$f(x),\ f_1(x),\ f_2(x),\ \dots\ f_n(x) \qquad \dots(F)$$

when x = a and when x = b, then

$$N \ge N'\ \text{ and }\ R \le N - N';$$

also $(N - N') - R$ is an even number or zero.

In particular if $N - N' = 1$, there is just one real root in the interval, and if $N - N' = 0$ no real root lies between a and b.

In this connection $f(x)$ and its derivatives are called Fourier's functions.

Proof. Let x increase continuously from $x = a$ to $x = b$. As x varies, no change of sign can be gained or lost in the sequence (F) unless x passes through a value which makes one of its terms vanish.

(1) Suppose that x passes through a root α of $f(x) = 0$.

(*i*) If α is an unrepeated root, a change of sign is lost between $f(x)$ and $f_1(x)$. This has been proved in Art. 3.

(*ii*) If α is an r-multiple root, r changes of sign are lost in the sequence $f(x)$, $f_1(x),\ \dots\ f_r(x)$. (See Art. 3.)

(2) Suppose that x passes through a value α' which makes one or more of the functions $f_1(x),\ f_2(x),\ \dots\ f_n(x)$ vanish.

(*i*) If $f_m(\alpha') = 0$, but neither $f_{m-1}(\alpha')$ nor $f_{m+1}(\alpha')$ is zero, there is no gain or loss of changes of sign when $f_{m-1}(\alpha')$ and $f_{m+1}(\alpha')$ have opposite signs: but if these have the same sign, two changes are lost.

For $f_m(x)$ and $f_{m+1}(x)$ have opposite signs just before $x = \alpha'$, and they have the same sign just after $x = \alpha'$.

(*ii*) Suppose that when $x = \alpha'$ the r functions

$$f_m(x),\ f_{m+1}(x),\ \dots\ f_{m+r-1}(x)$$

all vanish, but neither $f_{m-1}(x)$ nor $f_{m+r}(x)$ is zero.

Then if r is odd there is a loss of $r - 1$ or $r + 1$ changes of sign according as $f_{m-1}(\alpha')$ and $f_{m+r}(\alpha')$ have opposite signs or the same sign, while if r is even there is a loss of r changes of sign in either case.

Thus, under this heading, no change of sign is ever gained, and if any are lost, their number is even.

▮ **Example :** *Apply Fourier's method to separate the roots of*

$$f(x)\ =\ 2x^5 + 7x^4 - 40x^3 - 23x^2 + 38x - 4 = 0.$$

Here
$$f_1(x)\ =\ 2(5x^4 + 14x^3 - 60x^2 - 23x + 19),$$
$$f_2(x)\ =\ 2(20x^3 + 42x^2 - 120x - 23),$$
$$f_3(x)\ =\ 8(15x^2 + 21x - 30),$$
$$f_4(x)\ =\ 8(30x + 21),$$
$$f_5(x)\ =\ 240.$$

By Newton's method we find that all the roots lie in the interval $(-7, 4)$. For various values of x the signs of Fourier's functions are as below:

x	−7	−6	−2	−1	0	1	3	4
$f(x)$	−	+	+	−	−	−	−	+
$f_1(x)$	+	+	−	−	+	−	+	+
$f_2(x)$	−	−	+	+	−	−	+	+
$f_3(x)$	+	+	−	−	−	+	+	+
$f_4(x)$	−	−	−	−	+	+	+	+
$f_5(x)$	+	+	+	+	+	+	+	+
Number of changes of sign	5	4	4	3	3	1	1	0

These results show that there is one root in each of the intervals $(-7, -6)$, $(-2, -1)$, $(3, 4)$, and either two roots or none at all in the interval $(0, 1)$.

Subdividing the last interval, we find that when $x = 0.5$, $f(x)$ is positive. Hence there is a root in each of the intervals $(0, 0.5)$ and $(0.5, 1)$.

It should be noticed that we need not find the signs of $f_1(x)$, $f_2(x)$, etc., when $x = 0.5$, for it has been shown that either two roots, or none at all, lie between 0 and 1.

Remarks. (1) In applying Fourier's theorem, as stated above, the question arises as to *what sign is to be attached to one of Fourier's functions which may happen to vanish when $x = a$ or b.*

This difficulty is met by observing that, for sufficiently small values of h, the number of roots between a and b $(a < b)$ is the same as the number of roots between $a + h$ and $b - h$. We may therefore put $a + h$ instead of a and $b - h$ instead of b, where $h \rightarrow 0$.

(2) **Disadvantage of Fourier's Method.** Suppose that (as in the last example) all the roots of an equation are accounted for except two, which, if real, must lie in a certain interval (a, b). If these roots are nearly equal, it may require a large number of trials to separate them. If we fail to do this the possibilities are:

(*i*) The roots may be very nearly equal, in which case they can be separated by further trials.

(*ii*) The roots may be equal, and if we begin by testing for equal roots it would be better to use Sturm's process.

(*iii*) The roots may be imaginary, and no matter how many trials are made, we may not be able to discover whether this is or is not the case.

(3) If the roots of $f(x)$ are diminished by a, the transformed equation is

$$f(a) + xf_1(a) + \frac{x^2}{\underline{|2}} f_2(a) + \ldots + \frac{x^n}{\underline{|n}} f_n(a) = 0,$$

so that *N is the number of changes of sign in this equation. Similarly N′ is the number of changes of sign in the equation obtained by diminishing the roots of $f(x) = 0$ by b. If N and N′ are defined in this way, we obtain Budan's statement of Fourier's* theorem.

Imaginary Roots. As x increases from $-\infty$ to $+\infty$, n changes of sign are lost in the sequence of Fourier's functions. If it is known that a certain interval contains no root of $f(x) = 0$, and that, as x increases through the interval, s changes are lost, then the equation cannot have more than $n - s$ real roots, and must therefore have at least s imaginary roots.

EXERCISE XLVIII

1. From Fourier's theorem deduce (*i*) Descartes' rule of signs; (*ii*) Newton's method of finding an upper limit to the roots.

   In Exs. 2-11, find intervals (a, b) where a, b are consecutive integers which contain the real roots of the following equations.

   In Exs. 2-4 use Fourier's method; and in Exs. 5-11 use Sturm's method.

2. $x^3 - 3x^2 - 4x + 11 = 0.$
3. $x^3 - 20x^2 - 31x + 1609 = 0.$

4. $2x^3 + 11x^2 - 10x + 1 = 0.$
5. $x^3 + 2x^2 - 51x + 110 = 0.$

6. $2x^3 - 15x^2 - 8x + 166 = 0.$
7. $x^4 - 4x^3 - 3x + 23 = 0.$

8. $3x^4 - 10x^2 - 6x + 16 = 0.$
9. $x^4 - x^3 - 4x^2 + 4x + 1 = 0.$

10. $x^5 - 3x^4 - 24x^3 + 95x^2 - 46x - 101 = 0.$
11. $x^5 + 3x^4 + 2x^3 - 3x^2 - 2x - 2 = 0.$

12. Use Sturm's theorem to show that the equation $x^3 + 3Hx + G = 0$ has three real and distinct roots if and only if $G^2 + 4H^3 < 0$.

13. If $e < 0.55$, prove that the equation $x^4 + 4x^3 + 6x^2 - 4x + e = 0$ has two real and two imaginary roots. $[I = e + 7, J = -4]$

■ **9. Newton's Method of Approximating to a Root:** Suppose that the equation $f(x) = 0$ has a single unrepeated root in a small interval (α, β), so that $f'(x) \neq 0$ within the interval. Let $\alpha + h$ be the actual value of the root, then by the second mean value theorem,

$$f(\alpha + h) = f(\alpha) + hf'(\alpha) + \tfrac{1}{2} h^2 f''(\alpha + \theta h) = 0,$$

where $0 < \theta < 1$. Hence

$$h = -\frac{f(\alpha)}{f'(\alpha)} + \varepsilon \text{ where } \varepsilon = -\tfrac{1}{2} h^2 \frac{f''(\alpha + \theta h)}{f'(\alpha)}. \qquad \text{...(A)}$$

It follows, that if k is the greatest value of $f''(x)$ in the interval, and
$$\alpha_1 = \alpha - f(\alpha)/f'(\alpha),$$
the value of the root is $\alpha_1 + \varepsilon$ where

$$|\varepsilon| < \tfrac{1}{2} h^2 \frac{k}{|f'(\alpha)|} < \tfrac{1}{2} (\beta - \alpha)^2 \cdot \frac{k}{|f'(\alpha)|}. \qquad \text{...(B)}$$

Thus, if $\beta - \alpha$ is a small number of the first order, the error in taking α_1 as the value of the root is of the second order of smallness, unless $k/f'(\alpha)$ is large. By repeating the process, the approximation can be carried to any required degree of accuracy.

Note: In the preceding, $f(x)$ may be *any* function of x such that $f(x), f'(x)$ and $f''(x)$ are continuous in the interval (α, β).

■ **EXAMPLE 1.** *Find an approximate value of a, such that* $a^{\sqrt{a}} = 3$.

Put $a = x^2$; then $f(x) = x^{2x} - 3 = 0$; and $f'(x) = x^{2x}(2 + 2\log_e x)$.

Also $f(x) = -2$ when $x = 1$, and $f(x) = +0.375$ when $x = 1.5$.

Starting with $\alpha = 1.5$, the work proceeds as follows.

$$f(\alpha) = 0.375, f'(\alpha) = 9.487, f(\alpha)/f'(\alpha) = 0.04, \text{ and } \alpha_1 = 1.46;$$

$$f(\alpha_1) = 0.0193, f'(\alpha_1) = 8.324, f(\alpha_1)/f'(\alpha_1) = 0.0023, \text{ and } \alpha_2 = 1.4577;$$

$$f(\alpha_2) = 0.000255, f'(\alpha_2) = 8.262, f(\alpha_2)/f'(\alpha_2) = 0.000031, \text{ and } \alpha_3 = 1.457669;$$

hence, the value of a required is $(1.457669)^2 = 2.12480$.

This method is often employed in cases where the interval (α, β) is not very small, and then it is important to know under what circumstances α_1 is a closer approximation to the root than α.

Fourier's Rule. We shall assume that neither $f'(x)$ nor $f''(x)$ is zero for any value of x in the interval (α, β), so that each of these functions retains the same sign for all values of x with which we are concerned. Now

$$f(\alpha + h) = f(\alpha) + hf'(\alpha + \theta_1 h) = 0,$$

where $0 < \theta_1 < 1$, so that

$$h = -f(\alpha)/f'(\alpha + \theta_1 h).$$

Hence if $t = -f(\alpha)/f'(\alpha)$, it follows that t has the same sign as h.

Now α_1 is certainly a better approximation than α if α_1 lies between α and $\alpha + h$, and since $\alpha_1 = \alpha + t$, this is the case if $|t| < |h|$, that is if

$$f'(\alpha) > f'(\alpha + \theta_1 h).$$

This condition is satisfied if $f(\alpha)$ and $f''(\alpha)$ have the same sign. For $f'(x)$ increases with x or decreases as x increases according as $f''(x) \gtrless 0$, and $h \gtrless 0$ according as $f(\alpha)$ and $f'(\alpha)$ have different signs or the same sign.

Thus we have Fourier's rule, which is as follows: *If neither $f'(x)$ nor $f''(x)$ is zero near the point α, and $f(\alpha)$ and $f''(\alpha)$ have the same sign, then α_1 is a better approximation than α.*

The various cases which can arise are illustrated in Fig. 66, $(i) - (iv)$. It is to be noted that the tangent to the graph of $f(x)$ at the point where $x = \alpha$ cuts the x-axis at the point α_1.

In the figures on the left, $f(\alpha)$ and $f''(\alpha)$ have the sane sign; in those on the right they have opposite signs and Fourier's rule does not apply.

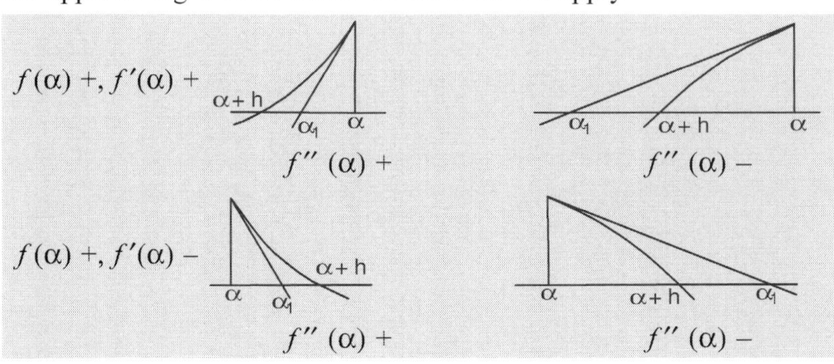

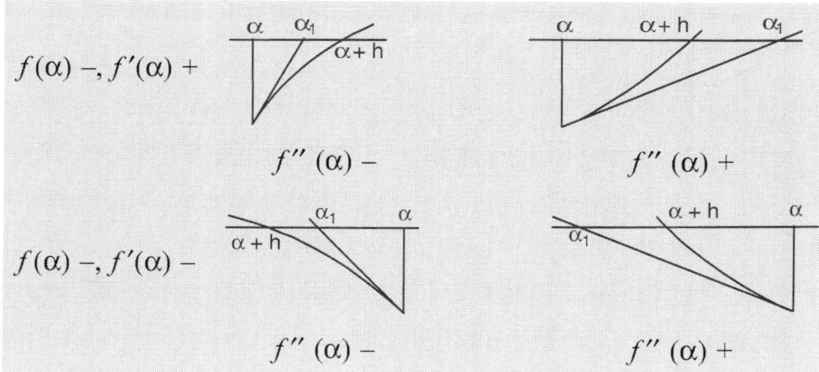

Fig. 66

We shall now show that α_1 is a better approximation than α, *provided that the interval is sufficiently small.*

The true value of the root $= \alpha + h = \alpha_1 + \varepsilon$, and therefore α_1 is a better approximation than α if $|\varepsilon| < |h|$, that is if

$$\frac{1}{2}h^2 \left| \frac{f''(\alpha + \theta h)}{f'(\alpha)} \right| < |h| \quad \text{or if} \quad |h| < 2 \left| \frac{f'(\alpha)}{f''(\alpha + \theta h)} \right|.$$

If $f''(x) \neq 0$ in the interval (α, β), we can find a positive number l such that $|l| < |f''(\alpha + \theta h)|$, and we draw the following conclusion:

If neither $f'(x)$ nor $f''(x)$ is zero in the interval (α, β) and

$$|h| < \frac{2|f'(\alpha)|}{l},$$

then α_1 is a better approximation than α.

This is certainly the case if $|\beta - \alpha| < \dfrac{2|f'(\alpha)|}{l}$.

Moreover, in cases where Fourier's rule does not apply, $|h| < |t|$, and α_1 is a closer approximation than α if

$$|t| < \frac{2|f'(\alpha)|}{l}, \quad \text{that is, if} \quad l\,|f(\alpha)| < 2\,\{f'(\alpha)\}^2 \qquad \text{...(C)}$$

In this case, by (A),

$$|\varepsilon| < \frac{1}{2}t^2 \cdot \frac{k}{|f'(\alpha)|}, \quad \text{that is} \quad |\varepsilon| < \frac{1}{2}\frac{\{f(\alpha)\}^2}{|f'(\alpha)|^3} \cdot k. \qquad \text{...(D)}$$

▪ **EXAMPLE 2.** *Find the real root of $x^3 = 2x + 5$ to nine significant figures.*

Let $f(x) = x^3 - 2x - 5$, then $f'(x) = 3x^2 - 2$, $f''(x) = 6x$.

Since $f(2) < 0 < f(3)$, the root lies between 2 and 3. Also $f(2) = -1$, $f'(2) = 10$, $f''(2) = 12$.

Taking $\alpha = 2$ and applying Newton's method, $\alpha_1 = 2 - \dfrac{-1}{10} = 2.1$.

Here $f(2)$ and $f''(2)$ have opposite signs, and Fourier's rule does not apply. But $h < 0.1$, and the least value of $|f''(x)|$ in the interval $(2,3)$ is 12. Hence

$$|h| < \frac{2|f''(\alpha)|}{l},$$

and α_1 is a better approximation than α.

Moreover, $f''(x)$ increases with x, and therefore

$$f''(\alpha + \theta h) < f''(\alpha + h) = f''(2.1) = 12.6,$$

so that $\quad |\varepsilon| = \frac{1}{2}h^2 \dfrac{f''(\alpha + \theta h)}{f'(\alpha)} < \frac{1}{2}(0.1)^2 \cdot \dfrac{12.6}{10} < 0.007.$

Next, taking $\alpha = 2.1$,* we have $f(2.1) = 0.061$, $f'(2.1) = 11.23$, $f''(2.1) = 12.6$.

Hence $\qquad \alpha_1 = 2.1 - \dfrac{0.061}{11.23} = 2.094569 \ldots$

Also the value of h is equal to that of ε in the first approximation, so that

$$|h| < 0.007,$$

and consequently $\qquad |\varepsilon| < \dfrac{1}{2}(0.007)^2 \dfrac{f''(2.1)}{f'(2.1)} < 0.00003.$

Moreover, ε is negative, and so the root lies between 2.09457 and 2.09453. Hence the error in taking 2.09455 as the root is numerically less than 0.00002.

Again, taking $\alpha = 2.09455$, we find that

$$f(\alpha) = -0.000016557 \ldots \quad \text{and } f'(\alpha) = 11.11607,$$

giving $\qquad \alpha_1 = \alpha - f(\alpha)/f'(\alpha) = 2.094551483 \ldots .$

Also $\qquad |\varepsilon| < \dfrac{1}{2}(0.00002)^2 \cdot \dfrac{12.6}{11} < 10^{-11}.24.$

Moreover, ε is negative, and so the value of the root to nine significant figures is 2.09455148.

◼ **EXAMPLE 3.** *A donkey is tethered to a point A in the circumference of a circular field of radius a by a rope of length l. Find the value of l in order that he may be able to graze over exactly half of the field.*

In Fig. 67 AB is the stretched rope, $\angle AOB = x$, $\angle OAB = x'$ measured in radians.

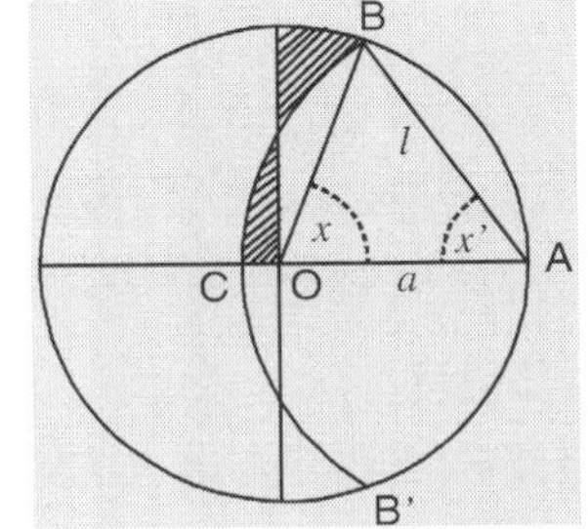

Fig. 67

Thus $2x' + x = \pi, \quad l = 2a \sin \dfrac{x}{2}$.

Area donkey can graze over

$= \text{Segment } ABB' + \text{Segment } CBB'$

$= a^2 (x - \frac{1}{2} \sin 2x) + l^2(x' - \frac{1}{2} \sin 2x').$

Equating this to $\frac{1}{2}a^2\pi$, it will be found that x is given by

$$f(x) = x - \tan x + \pi \left(\tfrac{1}{2} \sec x - 1 \right) = 0.$$

Hence, $f'(x) = \tan x \left(\frac{\pi}{2} \sec x - \tan x \right)$, and $f''(x) \equiv \frac{\pi}{2} \sec x (1 + 2 \tan^2 x) - 2 \tan x \sec x.$

---

* This approximation might have been obtained graphically.

In a question of this kind it is important to start with a really good approximation, to be found graphically or otherwise. We draw the figure so that the shaded areas are nearly equal, and then by measurement

$$\angle AOB = 71° = 1.239 \text{ radians (nearly)}.$$

Taking $\alpha = 1.239$, we find that, approximately, $f(\alpha) = 0.018$, $f'(\alpha) = 5.578$, and the next approximation is

$$\alpha_1 = \alpha - f(\alpha)\,/f'(\alpha) = 1.239 - 0.003 = 1.236.$$

If x increases from 0 to $\pi/2$, $f''(x)$ increases from $\pi/2$ to ∞ ; therefore $f''(\alpha)$ is positive. Hence $f(\alpha)$ and $f''(\alpha)$ have the same sign and, by Fourier's rule, α_1 is a closer approximation than α. Further, we can show that $f''(\alpha) < 34$; therefore $f''(x) < 34$ if $x < \alpha$. Hence, if $\alpha_1 + \varepsilon$ is the true value of x, by Art. 9 (B),

$$|\,\varepsilon\,| < \tfrac{1}{2}\,(0.003)^{2.}\,\tfrac{34}{5} < 0.00004.$$

Thus the following results are correct to the last figure:

$$\angle AOB = 1.236 \text{ radians} = 70° \ 50', \quad \text{and} \quad l/a = 2 \sin 35° \ 25' = 1.16.$$

Hence $\qquad l = 7a/6$, very nearly.

▨ **10. Nearly equal Roots:** Suppose that the equation $f(x) = 0$ has two nearly equal roots within the small interval (α, β), and let $\alpha + h$ be the true value of one of them, then

$$f(\alpha + h) = a + bh + ch^2 + dh^3 + \ldots = 0,$$

where $\quad a = f(\alpha), \quad b = f'(\alpha), \quad c = \tfrac{1}{2} f''(\alpha), \quad$ etc.

Neglecting squares of h, we have $a + bh = 0$, giving Newton's rule. But the roots are separated by a root of $f'(x) = 0$, so that $f'(x)$ is zero for some value of x in the interval. Consequently $f'(\alpha)$ is small, and the term ch^2 cannot be neglected in comparison with bh, so that Newton's rule does not apply. Neglecting cubes of h, we have $a + bh + ch^2 = 0$.

This equation has nearly equal roots, and is therefore approximately the same as $(a + \tfrac{1}{2} bh)^2 = 0$. Hence we have

$$h = -\frac{2a}{b} = -\frac{b}{2c} \quad \text{(approx.)}.$$

This result is used in Horner's method (Art. 11, Ex. 3) to *suggest* a trial value of h. It is only applicable when $2a/b$ and $b/2c$ are nearly equal.

In the same way, if the equation has three nearly equal roots in the interval (α, β), the value of one of them being $\alpha + h$, neglecting h^4, we have $a + bh + ch^2 + dh^3 = 0$; which is nearly the same as $(a + \tfrac{1}{3} bh)^3 = 0$. Hence,

$$h = -\frac{3a}{b} = -\frac{b}{c} = -\frac{c}{3d} \quad \text{(nearly)}.$$

■ **11. Horner's Method:** The process described in this article, due to Horner, is the most convenient way of finding approximate values of the irrational roots of equations of the type $f(x) = 0$ where $f(x)$ is a polynomial. Any rational roots which may exist can also be found in the same way. One great merit of the method is the very concise and orderly way in which the reckoning can be arranged. The root is evolved as a decimal, the figures of which are obtained in succession.

In what follows, we are only concerned with positive roots. To find the negative roots of $f(x) = 0$, we find the positive roots of $f(-x) = 0$.

Consider the equation $f(x) = a_0 x^n + a_1 x^{n-1} + \dots + a_{n-1} x + a_n = 0$,   ...(A)

and suppose that this has a single root α in the interval $(p, p+1)$, where p is a positive integer. (The case of two nearly equal roots will be considered later.) Let $\alpha = p.qrs$ ... where q, r, s ... are the figures in the decimal part of the root: then p can be found as in Ch. 28, 2, or preferably figure by figure as in Ex. 2 below.

Decrease the roots of (A) by p, and let the transformed equation be

$$a_0 x^n + b_1 x^{n-1} + \dots + b_{n-1} x + b_n = 0. \qquad \text{...(B)}$$

This has a single root between 0 and 1, namely $0.qrs$ ... . In finding the value of q, we avoid the use of decimals by multiplying the roots of (B) by 10, obtaining the equation

$$\phi(x) = a_0 x^n + 10 b_1 x^{n-1} + \dots + 10^{n-1} b_1 x + 10^n b_n = 0. \qquad \text{...(C)}$$

This has a single root between 0 and 10, namely $q.rs$ ... . *Moreover, q is the greatest number such that if the roots of (C) are decreased by q, the sign of the last term of the transformed equation is the same as that of the last term of (C).* This follows from the fact that $\phi(x) = 0$ has a root between q and $q+1$, and there is no root between 0 and q, so that $\phi(0)$, $\phi(q)$ and $-\phi(q+1)$ have the same signs.

In practice *the value of q is suggested and, after a certain stage, is definitely given by Newton's method.*

For, regarding p as an approximate value of α, we have

$$0.qrs \ \dots \ = -f(p)/f'(p) = -b_n/b_{n-1} \text{ (approximately)}.$$

Having found the correct value of q, we decrease the roots of (C) by q, multiply the roots of the resulting equation by 10, so obtaining an equation of which $r.s$ ... is a root.

Continuing as above, we can find as many figures as we choose in the decimal representation of α.

When a certain number of figures have been found, about as many more can be found by a contracted process explained in the next example.

■ **EXAMPLE 1.** *Find to ten significant figures the real root of*

$$f(x) = x^3 - 111 = 0 \qquad \text{...(A)}$$

The work is arranged below in a form suitable for explanation. In practice it would be arranged as in Ex. 2. The explanation is given on the opposite page.

$$\dfrac{17}{48} = 0.9 \ldots$$

$$\dfrac{4}{69} = 0.05 \ldots$$

$$\dfrac{62}{69} = 0.8\ldots$$

$$\dfrac{66}{69} = 0.9\ldots$$

$$
\begin{array}{llll}
1 & + & 0 + 0 - 111\ (4 & \ldots(a) \\
& & 4 + 16 - 47 & \\
& & 8 + 48 & \\
& & 12 & \\
1 & + & 120 + 4800 - 47000\ (8 & \ldots(b) \\
& & 128 + 5824 - 408 & \\
& & 136 + 6912 & \\
& & 144 & \\
1 & + & 1440\ \ + 691200\ \ \ \ - 408000\ (0 & \ldots(c) \\
1 & + & 14400 + 69120000 - 408000000\ (5 & \ldots(d) \\
& & 14405 + 69192025 - 62039875 & \\
& & 14410 + 69264075 & \\
& & 14415 & \\
& & 144{,}15 + 6926407{,}5 - 62039875\ (8 & \ldots(e) \\
& & 6927560\ \ \ - 6619395 & \\
& & 6928713 & \\
& & 1{,}44\ \ + 692871{,}3 - 6619395\ (9 & \ldots(f) \\
& & 692884\ \ \ \ - 383439 & \\
& & 692897 & \\
& & 69289{,}7\ \ \ \ \ \ \ - 383439\ \ (5 & \ldots(g) \\
& & 6928{,}9\ \ \ \ - 36990\ \ (5 & \\
& & 692{,}8\ \ \ \ - 2345\ \ (3 & \\
& & 69{,}2\ \ \ \ - 267\ \ (3 & \\
& & 6{,}9\ \ \ \ - 59\ \ (8 & \\
& & - 4 & \\
\end{array}
$$

Hence the required root is 4.8058955338, with possibly an error in the last figure.

Note: Denoting the root by α, the successive steps show that

$$4 < \alpha < 5, \quad 4.8 < \alpha < 4.9, \quad 4.80 < \alpha < 4.81, \quad \text{etc.}$$

Also that for the values 4, 4.8, 4.805 … 4.8058955338 of x,

$$f(x) = -47, \qquad -0.408, \qquad -0.062 \ldots, \qquad \ldots -0.000000004\ldots\ .$$

EXPLANATION OF THE SUCCESSIVE STEPS

(*a*) Decreasing the roots by 4, the resulting equation is

$$x^3 + 12x^2 + 48x - 47 = 0. \qquad \qquad ...(B)$$

If $\alpha = 4 + h$, by Newton's method

$$h = -\frac{f(4)}{f'(4)} = \frac{47}{48} = 0.9 \text{ ... approx.}$$

(*b*) Multiplying the roots of (B) by 10, the resulting equation is

$$\phi(x) = x^3 + 120x^2 + 4800x - 47000 = 0. \qquad \qquad ...(C)$$

Taking 9 as a trial figure, we find that this is too large, for decreasing the roots of (C) by 9 *changes the sign of the last term*; this shows that $\phi(0)$ and $\phi(9)$ have opposite signs, and thus a root of $\phi(x)$ lies between 0 and 9.

Trying 8, we find that this is the correct figure, and decreasing the roots by 8, the resulting equation is

$$x^3 + 144x^2 + 6912x - 408 = 0. \qquad \qquad ...(D)$$

Denoting the root of (C) by $8 + k$, we have

$$k = -\frac{\phi(8)}{\phi'(8)} = \frac{408}{6912} = 0.05 \text{ approx.}$$

Hence we may expect that the next two figures of the root are 0 and 5, and this proves to be correct.

(*c*) Here the roots of (D) are multiplied by 10, and those of the resulting equation decreased by 0: if they were decreased by 1, the *sign of the last term would be changed,* so 0 is the next figure of the root.

(*d*) The roots of the equation obtained in (*c*) are multiplied by 10, and we find that 5 is the next figure.

(*e*) The equation found in (*d*) is

$$x^3 + 14415x^2 + 69264075x - 62039875 = 0, \qquad \qquad ...(E)$$

and the next figure of the root is 8. We have to multiply the roots of (E) by 10 and decrease those of the resulting equation by 8.

In doing this the important figures will retain the same relative position if, instead of introducing 0's, we write down the last term of (E) *as it stands, cut off one figure from the coefficient of x, two figures from that of x^3, and neglect the coefficient of x^3.*

In the case of an equation of higher degree, we should cut off three figures from the coefficient of x^3, and so on.

Note: *In the multiplication allowance should be made for the figures cut off.*

(*f*) This step is similar to (*e*).

(*g*) Only the last two terms remain, and the process is simply contracted division.

■ **EXAMPLE 2.** *Find to nine significant figures the real root of $x^3 - x^2 - x - 2000 = 0$.*

It is easily seen that the root lies between 10 and 20, and the first step is to decrease the roots by 10. In practice, the work is arranged thus:

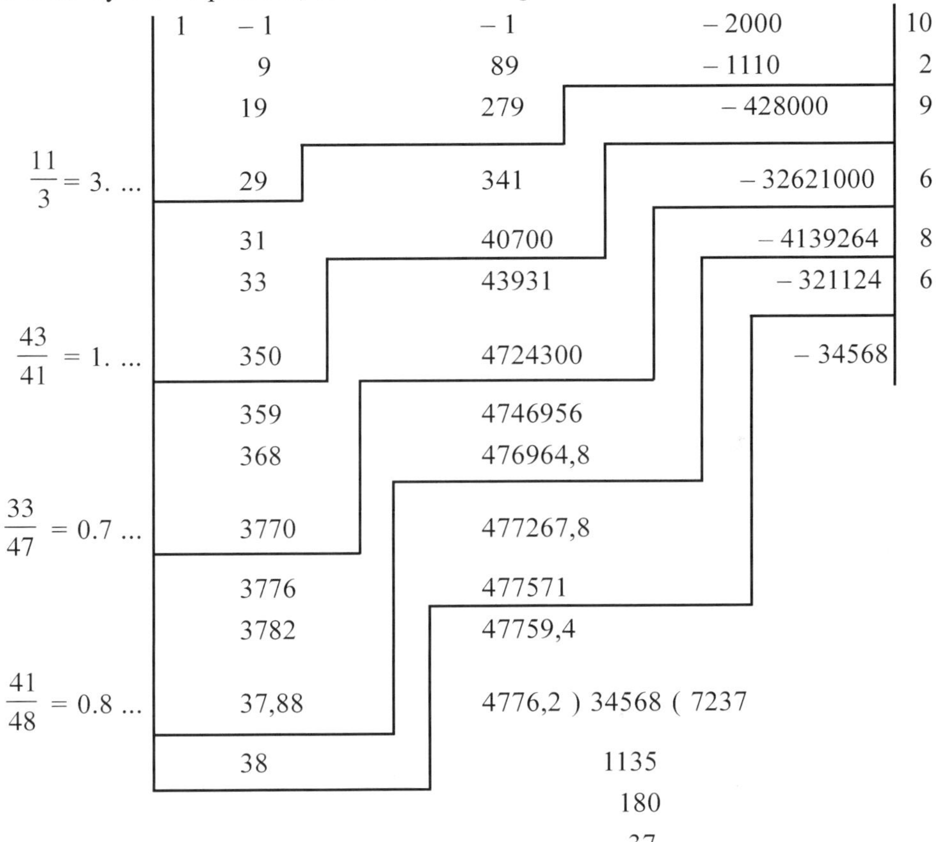

The required root is 12.9686724, to nine significant figures.

Note: It will be seen that Newton's approximation gives 1 more than the correct figure until we come to the 8.

▮ **EXAMPLE 3.** *Find approximate values of the two roots of*
$$3x^4 - 61x^3 + 127x^2 + 220x - 520 = 0$$
which lie between 2 and 3.

The reckoning is as follows, and the explanation is on the opposite page.

```
3 −     61 +    127 +    220 − 520 (2                    ...(a)
   −     55 +     17 +    254 −  12
   −     49 −     81 +     92
   −     43 −    167
   −     37

3 −    370 −  16700 + 92000 − 120000 (2                 ...(b)
   −    364 −  17428 + 57144 −   5712
   −    358 −  18144 + 20856
   −    352 −  18848
   −    346
```

```
3 − 3460 − 1884800 + 20856000 − 57120000 (5        ...(c)
  − 3445 − 1902025 + 11345875 −   390625
  − 3430 − 1919175 + 1750000
  − 3415 − 1936250
  − 3400
  −   3,4 − 19362,5 +   175000 − 390625 (4        ...(d)
        − 19376  +    97496 −   641
        − 19390  +    19936
        − 19404
```

For the smaller root: For the greater root:

```
− 194,04 + 1993,6 − 641  (0 ...(e) |  − 194,0 + 1993,6 −  641   (9 ...(e′)
−   1,94 + 199,3  − 641  (3 ...(f) |          + 247     + 1582
          + 193   − 62             |          − 1499
          + 187                    |  − 1,9   − 149,9  + 1582   (9 ...(f′)
          18, 7   − 62  (3 ...(g)  |          − 167    +   79
          1,8     −  6  (3 ...(h)  |          − 184
                                   |  −         18,4       79   (4 ...(h′)
                                   |                        5
                                   |  −          1,8        5   (2
```

Thus the roots are 2.2540333 ... and 2.2549942 ... , where the last figure in each case is not to be relied upon.

EXPLANATION OF THE SUCCESSIVE STEPS

(a) Decreasing the roots by 2, the resulting equation is
$$3x^4 - 37x^3 - 167x^2 + 92x - 12 = 0.$$
By the question, this has two roots between 0 and 1.

(b) Suspecting two nearly equal roots, we apply the rule of Art. 10, and the figure 2 is suggested by
$$-\frac{2\,(-12)}{92} = 0.2... \quad \text{and} \quad -\frac{92}{2\,(-167)} = 0.2... \ .$$

Decreasing the roots by 2, the sign of the last term of the resulting equation is − .

If the roots are decreased by 3, it will be found that again the sign of the last term is − . So far, then, we have no assurance that 2 is the proper figure.

But when the roots are decreased by 2, the signs of the terms are + − − + −, and when they are decreased by 3, the signs are + − − − −, so that two changes of sign are lost.   Hence, by Art. 8, the two roots lie between 2.2 and 2.3.

(c) and (d). Similar remarks apply to these steps. The figure 5 in (c) is suggested by
$$-\frac{2\,(-5712)}{20856} = 0.5 \quad \text{and} \quad \frac{20856}{2\,(-18848)} = 0.5.... $$

Thus the roots lie between 2.254 and 2.255.

(e) and (e'). Here the rule of Art. 10 fails, and the separation of the roots is a matter of trial. If at (e) we decrease the roots by 1 (instead of by 0), the signs of the terms are $+ - - + +$, one change of sign being lost. The same is true if, as at (e'), we decrease the roots by 9. Thus one root lies between 2.2540 and 2.2541, and the other between 2.2549 and 2.255; and the roots are separated.

Immediately after the roots have been separated, *i.e.* beginning at the stages (f), (f'), the proper figure is suggested by Nowton's rule.

EXERCISE XLIX

Find by Newton's method to five significant figures:

1. The two positive roots of $x^3 - 7x + 7 = 0$.
2. The positive root of $2x^3 - 3x - 6 = 0$.

Find the roots indicated below correct to seven places of decimals by Horner's method.

3. The real root of $x^3 + 29x - 97 = 0$.
4. The cube root of 4129.
5. The three real roots of $x^3 - 3x + 1 = 0$.
6. The two roots of $2x^3 + 11x^2 - 10x + 1 = 0$ in the interval $(0, 1)$.
7. The two roots of $x^3 - 20x^2 - 31x + 1609 = 0$ in the interval $(13, 15)$.
8. The two roots of $2x^3 - 15x^2 - 8x + 166 = 0$ in the interval $(5, 6)$.
9. The greatest and least roots of
$$3x^4 - 61x^3 + 127x^2 + 220x - 520 = 0.$$
10. The two roots of $3x^4 - 10x^2 - 6x + 16 = 0$ in the interval $(1, 2)$.
11. The two roots of $3x^4 - 2x^2 - 34x + 40 = 0$ in the interval $(1, 2)$.
12. The two root of $7x^4 - 11x^2 - 32x + 40 = 0$ in the interval $(1, 2)$.
13. The two roots of $x^4 - 2x^3 - 13x^2 - 11x + 133 = 0$ in the interval $(3, 4)$.
14. The two roots of $3x^4 - 4x^3 - 12x^2 + 12x + 9 = 0$ in the interval $(1, 2)$.
15. The two roots of $2x^4 - 16x^3 - 17x^2 + 392x - 782 = 0$ in the interval $(4, 5)$.
16. The two roots of $x^4 + 58x^2 - 388x + 571 = 0$ in the interval $(2, 3)$.
17. The two roots of $2x^4 - 27x^3 + 25x^2 + 179x - 275 = 0$ in the interval $(2, 3)$.
18. The positive root of $2x^4 + 5x^3 + 4x^2 + 3x = 8002$.
19. The greatest root of the following:
    (*i*) $x^5 + 4x^4 - 2x^3 + 10x^2 - 2x = 962$.
    (*ii*) $x^5 + 6x^4 - 10x^3 - 112x^2 - 207x = 110$.
    (*iii*) $x^5 + 12x^4 + 59x^3 + 150x^2 + 201x = 207$.
20. If $f(x) = e^x - 1 - 2x$, show that the equation $f(x) = 0$ has just one solution other than zero, the solution being $x = 1.260$ ....
    Show also that $f(x) < 0$ if $0 < x < 1.260$.
21. If $f(x) = e^x - e^{-x} - 4x$, show that the equation $f(x) = 0$ has just one positive solution given by $x = 2.177$ ... , and that $f(x) < 0$ if $0 < x < 2.177$.
22. Find an approximate solution of $x^x = 10$, showing that $x = 2.50616$ (nearly), where the last figure is not to be relied on. Verify that for this value of x, $x^x = 9.9995$ (approx.). [Here x is given by $f(x) = x^{-1} - \log_{10} x = 0$. By drawing the curves $y = x^{-1}$ and $y = \log_{10} x$, it will be found that $x = 2.5$ is a good approximation. Five applications of Newton's process give $x = 2.50616$.]

CHAPTER 29

Implicit Functions, Curve Tracing

■ **1. Implicit Functions:** If two variables x and y are connected by an equation represented by $f(x, y) = 0$, we say that y is an *implicit function* of x.

In particular, if $f(x, y)$ is a polynomial in x, y and $f(x, y) = 0$, then y is called an *algebraic function* of x.

Taking as a simple case the equation

$$y^2 - 2xy + 2x^2 = 1,$$

this gives $y = x \pm \sqrt{(1 - x^2)}$, and so defines y as a two-valued function of x.

But if, as generally happens, we are unable to express y explicitly in terms of x, we are not justified in assuming without further enquiry that y is really a function of x in the sense hitherto understood.

The general question involves difficult analysis, and cannot be considered here. Some exercise in tracing curves from their equations will, however, convince the student that, at any rate in a large number of cases, the equation $f(x, y) = 0$ *does* define y as a single or multiple-valued function of x.

The general form of a curve represented by an algebraic equation can generally be found by pure algebra, and affords excellent practice in finding approximate values.

■ **2. A Rule for Approximations:** The following process is of great practical use.

Suppose that $y = a + xf(y) + x^2 \phi(y) + x^3 \psi(y) + \ldots,$ ...(A)

where x is small and $f(y)$, $\phi(y)$, etc., do not involve x. *It is required to approximate to the value of y, in ascending powers of x.*

Let
$$\left.\begin{aligned} f(y + h) &= f(y) + hf_1(y) + h^2 f_2(y) + \ldots \\ \phi(y + h) &= \phi(y) + h\phi_1(y) + h^2 \phi_2(y) + \ldots \\ &\text{\dots\dots\dots\dots\dots\dots\dots\dots\dots\dots} \end{aligned}\right\}$$ ...(B)

The first approximation is obtained by putting $x = 0$, giving

$$y = a$$ ...(C)

For the second approximation, let $y = a + \alpha$ where α is $O(x)$,* then

$$y = a + xf(a + \alpha) + x^2\phi(a + \alpha) + \ldots$$
$$= a + x\{f(a) + \alpha f_1(a) + \ldots\} + x^2\{\phi(a) + \ldots\} + \ldots,$$

---

* This means 'of the same order as x.' (Ch, 2, 7.)

478

and neglecting small quantities of the second order, *i.e.* terms containing $x\alpha$, x^2, etc., *the second approximation is*

$$y = a + xf(a), \qquad \text{...(D)}$$

so that $\qquad \alpha = xf(a).$

Next let $\qquad y = a + \alpha + \beta$ where β is $O(x^2)$, then

$$y = a + xf(a + \alpha + \beta) + x^2\phi(a + \alpha + \beta) + ...$$
$$= a + x\{f(a + \alpha) + \beta f_1(a + \alpha) + \beta^2 f_2(a + \alpha) + ...\}.$$
$$+ x^2\{\phi(a) + (\alpha + \beta)\,\phi_1(a) + ...\} + ...,$$

and neglecting small quantities of the third order,

$$y = a + xf(a + \alpha) + x^2\phi(\alpha), \qquad \text{...(E)}$$

which gives *the third approximation*, namely

$$y = a + x\{f(a) + xf(a).\,f_1(a)\} + x^2\phi(a).$$

Thus $\qquad \beta = x^2\{f(a).\,f_1(a) + \phi(a)\}.$

Similarly, if γ is $O(x^3)$, neglecting small quantities of the fourth order, we shall find that

$$y = a + xf(a + \alpha + \beta) + x^2\phi(a + \alpha) + x^3\psi(a), \qquad \text{...(F)}$$

which gives *the fourth approximation.*

Hence the following *Rule*: Having found r successive approximations to the value of y, to find the equation giving the next approximation, in equation (A) substitute the result of

the rth approximation in $xf(y)$,
the $(r - 1)$th approximation in $x^2\phi(y)$,
the $(r - 2)$th approximation in $x^3\psi(y)$,

.... ,

the first approximation in the term containing x^r,

and neglect terms containing powers of x higher than x^r.

▪ **EXAMPLE 1.** *Newton used the expansion of* $(1 - x^2)^{-\frac{1}{2}}$ *to prove that*

$$\sin^{-1} x = x + \frac{1}{2}.\frac{x^3}{3} + \frac{1\cdot 3}{2\cdot 4}.\frac{x^5}{5} + ... \qquad \text{...(A)}$$

Putting $x = \sin y$, he deduced that

$$\sin y = y - \frac{1}{\lfloor 3}y^3 + \frac{1}{\lfloor 5}y^5 - ... . \qquad \text{...(B)}$$

To do this, write equation (A) in the form

$$\sin y = y - \tfrac{1}{6}\sin^3 y - \tfrac{3}{40}\sin^5 y - \tfrac{5}{112}\sin^7 y - ... . \qquad \text{...(C)}$$

If y is small, so is $\sin y$, and the first approximation is

$$\sin y = y. \qquad \text{...(D)}$$

Substituting y for $\sin y$ in the second term of (C) and neglecting higher terms, the second approximation is

$$\sin y = y - \tfrac{1}{6} y^3. \qquad \qquad \text{...(E)}$$

Following the rule, the third approximation is given by

$$\sin y = y - \tfrac{1}{6} (y - \tfrac{1}{6} y^3)^3 - \tfrac{3}{40} y^5,$$

and neglecting the powers of y higher than y^5,

$$\sin y = y - \tfrac{1}{6} y^3 + \tfrac{1}{120} y^5. \qquad \qquad \text{...(F)}$$

The fourth approximation is given by

$$\sin y = y - \tfrac{1}{6} (y - \tfrac{1}{6} y^3 + \tfrac{1}{120} y^5)^3 - \tfrac{3}{40} (y - \tfrac{1}{6} y^3)^5 - \tfrac{5}{112} y^7.$$

Expanding and neglecting powers of y higher than y^7, this gives

$$\sin y = y - \tfrac{1}{6} y^3 + \tfrac{1}{120} y^5 - \tfrac{1}{5040} y^7,$$

and so on.

■ **Example 2.** *If μ is small, find approximate values of y which satisfy the equation*
$$y^3 - y + \mu = 0.$$

If μ is small, so is $y(y - 1)(y + 1)$, therefore y is nearly equal to 0 or ± 1. Write the equation

$$y = \mu + y^3 \qquad \qquad \text{...(A)}$$

(*i*) If y is small, the first approximation is $y = \mu$.

For the second, substitute μ^3 for y^3 in (A), giving $y = \mu + \mu^3$.

For the third, $y = \mu + (\mu + \mu^3)^3$. giving $y = \mu + \mu^3 + 3\mu^5$.

For the fourth, $y = \mu + (\mu + \mu^3 + 3\mu^5)^3$. giving $y = \mu + \mu^3 + 3\mu^5 + 12\mu^7$.

Further approximations can be found in the same way.

(*ii*) If $y - 1$ is small, put $Y = y - 1$, then equation (A) may be written

$$Y = - \tfrac{1}{2} \mu - \tfrac{3}{2} Y^2 - \tfrac{1}{2} Y^3.$$

The first approximation is $y = - \tfrac{1}{2} \mu$.

The second is $\qquad Y = - \tfrac{1}{2} \mu - \tfrac{3}{2} (-\tfrac{1}{2} \mu)^2 = -\tfrac{1}{2} \mu - \tfrac{3}{8} \mu^2.$

For the third, $Y = - \tfrac{1}{2} \mu - \tfrac{3}{2} (-\tfrac{1}{2} \mu - \tfrac{3}{8} \mu^2)^2 - \tfrac{1}{2} (-\tfrac{1}{2} \mu)^3,$

giving $\qquad Y = - \tfrac{1}{2} \mu - \tfrac{3}{8} \mu^2 - \tfrac{1}{2} \mu^3,$

that is, $\qquad y = 1 - \tfrac{1}{2} \mu - \tfrac{3}{8} \mu^2 - \tfrac{1}{2} \mu^3.$

(*iii*) Similarly we can show that if $y + 1$ is small, then approximately

$$y = - 1 - \tfrac{1}{2} \mu + \tfrac{3}{8} \mu^2 - \tfrac{1}{2} \mu^3.$$

This result also follows from the preceding, for the sum of the roots of equation (A) is zero.

■ **3. Cubic Equation with Three Real Roots:** If all the roots of $ax^3 + 3bx^2 + 3cx + d = 0$ are real, the equation can be reduced to

$$y^3 - y + \mu = 0 \quad \text{where} \quad \mu^2 < 4/27 \qquad \text{...(A)}$$

by a transformation of the form $x = p + qy$, where p, q are real numbers (Ch. 12, 5, Ex. 1). The last example suggests the possibility of expressing the roots of equation (A) as power series in μ.

Theorem. *If* $\mu^2 < 4/27$, *the roots of* $y^3 - y + \mu = 0$ *are*

$$F(\mu) - F(-\mu), \ 1 - F(\mu), \ -1 + F(-\mu),$$

where

$$F(\mu) = \frac{\mu}{2} + \frac{3}{\lfloor 2} \left(\frac{\mu}{2}\right)^2 + \frac{4 \cdot 6}{\lfloor 3} \left(\frac{\mu}{2}\right)^3 + \frac{5 \cdot 7 \cdot 9}{\lfloor 4} \left(\frac{\mu}{2}\right)^4 + \frac{6 \cdot 8 \cdot 10 \cdot 12}{\lfloor 5} \left(\frac{\mu}{2}\right)^5 + \dots$$

$$+ \frac{(n+1)(n+3)(n+5)\dots(3n-3)}{\lfloor n} \left(\frac{\mu}{2}\right)^n + \dots$$

Write the equation in the form $y^3 - y = 2x$,   ...(A)

where $x = -\mu/2$ and $x^2 < 1/27$.

Assuming that the values of y can be expressed as power series in x, by Taylor's theorem, these values are given by

$$y = (y)_0 + x(y_1)_0 + \frac{x^2}{\lfloor 2} (y_2)_0 + \dots + \frac{x^n}{\lfloor n} (y_n)_0 + \dots , \qquad \text{...(B)}$$

where $y_n = \dfrac{d^n y}{dx_n}$ and $(y)_0, (y_1)_0, (y_2)_0, \dots$ are the values of $y, y_1, y_2, \dots$

when $x = 0$. (Of course, x is to be made zero *after* differentiation.)

The assumption will be justified if the series is convergent.

Let $y = u + v$, then $y^3 - 3uvy = u^3 + v^3$. This equation will be identical with (A) if $u^3 + v^3 = 2x$ and $3uv = 1$.

Thus we may take $u^3 = x + (x^2 - k)^{\frac{1}{2}}$, $v^3 = x - (x^2 - k)^{\frac{1}{2}}$, where $k = 1/27$.

Writing $u_n = \dfrac{d^n u}{dx^n}$, $v_n = \dfrac{d^n v}{dx^n}$, and differentiating with regard to x,

we have $3u^2 u_1 = 1 + x(x^2 - k)^{-\frac{1}{2}} = u^3(x^2 - k)^{-\frac{1}{2}}$;

therefore $(x^2 - k)^{\frac{1}{2}} u_1 = u/3$; and similarly $(x^2 - k)^{\frac{1}{2}} v_1 = -v/3$.

Differentiating again, we have $(x^2 - k)^{\frac{1}{2}} u_2 + xu_1 (x^2 - k)^{-\frac{1}{2}} = u_1/3$, giving $\qquad (x^2 - k)u_2 + xu_1 = (x^2 - k)^{\frac{1}{2}} \cdot u_1/3 = u/9.$

Similarly, $\quad (x^2 - k)v_2 + xv_1 = v/9.$

Since $y = u + v$, it follows that $(x^2 - k) y_2 + xy_1 = y/9.$ $\qquad$ ...(C)*

Differentiating n times by Leibniz' theorem (Ch. 23, 11),

$$(x^2 - k) y_{n+2} + (2n + 1) xy_{n+1} + (n^2 - \tfrac{1}{9}) y_n = 0. \qquad ...(D)$$

Putting $x = 0$ and $k = 1/27$, we have for $n = 0, 1, 2, ...$

$$(y_{n+2})_0 = 3(3n + 1)(3n - 1)(y_n)_0. \qquad ...(E)$$

The possible values of $(y)_0$ are $0, \pm 1$.

(*i*) If $(y)_0 = -1$, since $y_1 = 2/(3y^2 - 1)$ we have $(y_1)_0 = 1$.

Hence from (E) in succession,

$$(y_2)_0 = 3 \cdot 1 \cdot (-1) \ (y)_0 = 3, \quad (y_3)_0 = \frac{6 \cdot 4 \cdot 2}{2} \ (y_1)_0 = 6 \cdot 4,$$

$$(y_4)_0 = \frac{9 \cdot 7 \cdot 5}{3} \ (y_2)_0 = 9 \cdot 7 \cdot 5, \quad (y_5)_0 = \frac{12 \cdot 10 \cdot 8}{2} \ (y_3)_0 = 12 \cdot 10 \cdot 8 \cdot 6$$

and if $n \geq 2$, $(y_n)_0 = (3n - 3)(3n - 5) \ ... \ (n + 3)(n + 1)$.

Thus $y = -1 + F(x)$ where

$$F(x) = x + \frac{3}{\lfloor 2} x^2 + \frac{4 \cdot 6}{\lfloor 3} x^3 + ... + \frac{(n+1)(n+3) \ ... \ (3n-3)}{\lfloor n} x^n + ...,$$

for such values of x as make the series convergent.

Denoting the series by $u_1 + u_2 + u_3 + ...$, we have

$$\frac{u_{n+2}}{u_n} = \frac{3(3n+1)(3n-1)}{(n+2)(n+1)} x^2 \rightarrow 27x^2 \text{ as } n \rightarrow \infty.$$

Hence the series Σu_{2n}, Σu_{2n+1}, and consequently also Σu_n, are convergent if $x^2 < 1/27$.

(*ii*) If $(y)_0 = 1$, then $(y_1)_0 = 1$. Hence the numerical value of $(y_n)_0$ is the same as before, the sign being $+$ or $-$ according as n is odd or even.

Therefore

$$y = 1 - F(-x).$$

(*iii*) Since the sum of the roots of (A) is zero, the root corresponding to $(y)_0 = 0$ is given by

$$y = F(-x) - F(x).$$

Putting $x = -\mu/2$, we have the results stated above.

A more general theorem, which includes this, is given in Ex. L, 7.

---

* An alternative proof of this result is indicated in Ex. 6 of the next Exercise.

EXERCISE L

1. Having given that

$$x = \tan x - \tfrac{1}{3} \tan^3 x + \tfrac{1}{5} \tan^5 x - \tfrac{1}{7} \tan^7 x + \ldots ,$$

prove that $\tan x = x + \tfrac{1}{3} x^3 + \tfrac{2}{15} x^5 + \tfrac{17}{315} x^7 + \ldots .$

2. If $y = x + ay^n$ where $x,\ y$ are both small, prove that approximately

$$y = x + ax^n + na^2 x^{2n-1} + \tfrac{1}{2} n\,(3n-1)\,a^3 x^{3n-2}.$$

3. If $y = x + ay^2 + by^3$ and $x,\ y$ are both small, then approximately
$$y = x + ax^2 + (b + 2a^2)x^3 + 5a(b + a^2)x^4.$$

4. If $y = 1 + xa^y$ and x is small, then approximately

$$y = 1 + xa + x^2 a^2 \log a + \tfrac{3}{2} x^3 a^3 \,(\log a)^2.$$

5. If $s,\ t$ are positive integers $(s > t)$ and λ is small, the equation
$$y^s - y^t + (s - t)\lambda = 0$$
has a root nearly equal to

$$1 - \lambda - (-1 + s + t)\,\frac{\lambda^2}{\underline{2}} - (-1 + 2s + t)\,(-1 + s + 2t)\,\frac{\lambda^3}{\underline{3}} .$$

6. If $y^3 - y = 2x$, show that
(i) $4(27x^2 - 1) = (3y^2 - 1)^3 - 3(3y^2 - 1)^2$;

(ii) $y_2 = - \dfrac{24\,y}{(3y^2 - 1)^3} = - \dfrac{12(y - 3xy_1)}{(3y^2 - 1)^2},$

where $y_1 = \dfrac{dy}{dx},\ y_2 = \dfrac{d^2 y}{dx^2}$. Deduce that $\left(x^2 - \tfrac{1}{27}\right) y_2 + xy_1 = \tfrac{1}{9} y.$

7. If $y = u + v$ where $u^m = x + \sqrt{(x^2 - k)},\ v^m = x - \sqrt{(x^2 - k)}$, m being a positive integer, then
(i) x and y are connected by the equation

$$2x = y^m - \frac{m}{\underline{1}}\, y^{m-2} k^{\frac{1}{m}} + \frac{m(m - 3)}{\underline{2}}\, y^{m-4} k^{\frac{2}{m}}$$

$$- \frac{m(m - 4)\,(m - 5)}{\underline{3}}\, y^{m-6} k^{\frac{3}{m}} + \ldots \qquad \ldots(A)$$

(ii) If x is sufficiently small,

$$y = (y)_0 + x\,(y_1)_0 + \frac{x^2}{\underline{2}}\,(y_2)_0 + \ldots , \qquad \ldots(B)$$

where $(y)_0$ is a value of y obtained by putting $x = 0$ in (A), and $(y_1)_0$, $(y_2)_0$, ... are the

corresponding values of $\dfrac{dy}{dx}, \dfrac{d^2 y}{dx^2}, \dots$.

(*iii*) Using Ch. 23, 11, Ex. 1, show that

$$k\,(y_{n+2})_0 = \left(n^2 - \frac{1}{m^2} \right) (y_n)_0.$$

Hence prove that equation (B) holds if $x^2 < k$.

(*iv*) If m is odd, one value of $(y)_0$ is 0, and then

$$y = (-1)^{\frac{m-1}{2}} \cdot \frac{2}{mk^{\frac{m-1}{2m}}} \left\{ x + \frac{x^3}{\lfloor 3} \cdot \frac{1}{k} \left(1^2 - \frac{1}{m^2} \right) + \frac{x^5}{\lfloor 5} \cdot \frac{1}{k^2} \left(1^2 - \frac{1}{m^2} \right) \left(2^2 - \frac{1}{m^2} \right) + \dots \right\}.$$

(*v*) If $m = 5$ and $k^{\frac{1}{5}} = \frac{1}{5}$ equation (A) becomes

$$y^5 - y^3 + \tfrac{1}{5} y = 2x,$$

and the values of $(y)_0$ are $0, \pm \sqrt{\left(\dfrac{1}{2} \pm \dfrac{1}{2\sqrt{5}} \right)}$.

Thus all the roots are real, and they can be expressed as power series in x if $x^2 < \dfrac{1}{5^5}$.

(*vi*) If $m = 7$ and $k^{\frac{1}{7}} = \frac{1}{7}$, the equation is $\;\; y^7 - y^5 + \tfrac{2}{7} y^3 - \tfrac{1}{49} y = 2x.$

■ **4. Tangents and Asymptotes:** If P is a point on the curve and Q a point on it near P, the tangent at P is defined as the limiting position of the secant PQ as Q moves up to P. Thus as Q moves to P, the distance of Q from the tangent at P becomes indefinitely small, so that near P the curve is indefinitely close to the tangent.

If P moves to an infinite distance along a branch of the curve, and the tangent at P tends to a limiting position, this limiting position is called an *asymptote* to the curve. It may happen that an infinite branch has no asymptote. We then find the simplest equation which nearly represents the branch at ∞, and say that this is the equation to a *curvilinear asymptote.*

In the examples which follow, tangents and asymptotes are found by a process of approximation.

■ **5. Intersections of a Straight Line and Curve:** If $u = 0$ is the equation to a curve of the nth degree and $y = \mu x + v$ that of any straight line, the abscissae of the points of intersection are given by an equation of the form

$$a_0 x^n + a_1 x^{n-1} + a_2 x^{n-2} + \dots + a_n = 0, \qquad \dots \text{(A)}$$

obtained by substituting $\mu x + v$ for y in $u = 0$.

(*i*) If this equation has two equal roots, the line meets the curve in 'two coincident points' and, except in special cases [see (*iii*)], touches the curve. If three roots are equal, the line meets the curve in three coincident points. Except in special cases [see (*iii*)], the point of contact is a *point of inflexion* (Ch. 17, 20). At such a point, the curve *crosses the tangent.*

(*ii*) If $a_0 = 0$, one root of equation (A) is infinite, and the line meets the curve in *one point at infinity.*

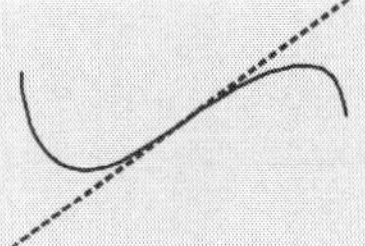

Fig. 68

If $a_0 = 0$ and $a_1 = 0$, the line meets the curve in *two points at infinity* and, except in special cases [see (*iv*)], is an asymptote.

(*iii*) Any point (*D*) where a curve crosses itself is called a *double point* or *node.* At such a point there are two tangents, each meeting the curve in *three* consecutive points. Any other straight line through *D* meets the curve in *two* consecutive points. (Fig. 69).

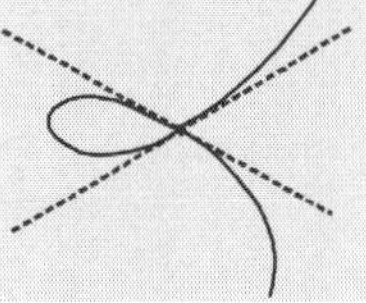

Fig. 69

If the double point is at infinity, the tangents are parallel straight lines. So in this case the curve has two parallel asymptotes, each meeting the curve in three points at infinity. *Any* straight line parallel to the asymptotes meets the curve in two points at infinity.

If the tangents at a double point coincide, the point is called a *cusp.* There are two kinds of cusp, as shown in Fig. 70.

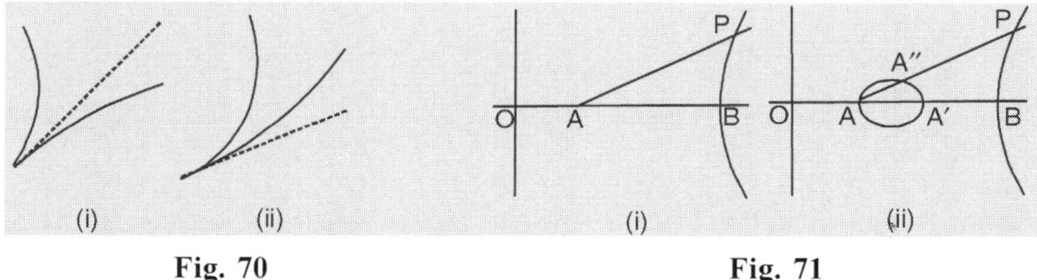

Fig. 70   Fig. 71

In tracing a curve, it is important to be able to account for *all the n points* in which any straight line meets the curve, remembering that imaginary points occur in pairs.

(*iv*) Fig. 71, (*i*), is a sketch of the curve whose equation is

$$y^2 = (x - a)^2 (x - b).$$

The axis of *x* meets the curve in *A* (*a*, 0) and *B*(*b*, 0). The equation to any straight line *AP* through *A* is $y = m(x - a)$; and, on eliminating *y*, we have $(x - a)^2 (x - b - m^2) = 0$; hence, *AP* meets the curve in *two coincident points at A*, and at a point *P*, given by $x = b + m^2$. Therefore, *A* is the *only point* on the curve to the left of the line $y = b$; and thus *A* may be regarded as *a double point at which the tangents are imaginary.*

A point on a curve, such as *A*, is called a *conjugate point*, if no other point in its immediate neighbourhood is on the curve.

The conjugate point at *A*, in Fig. 71, (*i*), may be considered as the limiting form, when $a = a'$, of the oval in Fig. 71, (*ii*), which is a sketch of the curve whose equation is

$$y^2 = (x - a) (x - a') (x - b),$$

■ **6. Curve Tracing:** To find the general shape of a curve represented by a given equation, proceed as follows: (*i*) Consider any symmetry which the curve may possess, *e.g.* symmetry with regard to either axis, or in opposite quadrants. (*ii*) Mark any points on the curve which can be found easily, *e.g.* the points (if any) where it cuts the axes. (*iii*) Find the approximate form of the curve near some of these points.

Near (0, 0) (if on the curve) *x, y* are small, *but not necessarily of the same order.* To find the form near any other point, move the origin to that point.

The quickest way of finding the tangent at any point is to find the value of $\dfrac{dy}{dx}$.

But, without further differentiation, we cannot say on which side of the tangent the curve lies, or discover any peculiarities which the point may possess. So, for a *few* points, it is better to use a process of approximation, as in Ex. 1.

(*iv*) Find any regions in which there is no part of the curve. (Exs. 1, 2)

(*v*) Find the form of the curve near infinity, *i.e.* where one or both of *x* and *y* are large, but not necessarily of the same order.

In doing this, we find the (linear) asymptotes (if there are any), and the approximate form of any 'parabolic' branches.

The reader should be able to draw the 'parabolas' represented by $y^m = x^n$, where *m* and *n* are positive integers. For example,

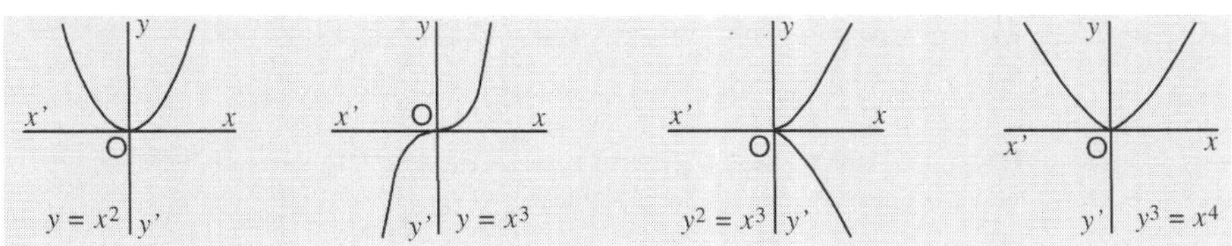

Fig. 72

(*vi*) By considering the value of $\dfrac{dy}{dx}$, find the tangents (if any exist) which are parallel to the axes. Notice that, at a point of inflexion, $\dfrac{d^2y}{dx^2} = 0$.

(*vii*) It may be possible to find *y* explicitly in terms of *x* or *x* in terms of *y*; or again, we may be able to express *x* and *y* as functions of some variable *t*. We can then find any number of points on the curve and the gradients at these points.*

---

* The examples given in Art. 7 and in Exercise LI have been chosen so that one of these methods can be used. Thus in each case the curve can be 'plotted' to any degree of accuracy required: so that the student can verify the conclusions arrived at by the methods just described, and gain confidence in using them. However, when only the *characteristic form* of the curve is required, the advantages of the above method over the method of plotting, even when the latter can be done, will be obvious. [See Ex. 22 of Exercise LI]

■ 7. Examples on Curve Tracing:

EXAMPLE 1. *Trace the curve* $(y^2 - 1)y = (x^2 - 4) x$. ...(A)

The equation is unaltered by writing $-x, -y$ for x, y. Hence there is symmetry in opposite quadrants. The points $(0, 0)$, $(0, \pm 1)$, $(\pm 2, 0)$, $(\pm 2, \pm 1)$ are on the curve.

Write the equation $y^3 - x^3 - y + 4x = 0$ ...(A)

Near $(0, 0)$, x and y are small and of the same order.

1st approx., $y = 4x$.

This is the equation to the tangent at $(0, 0)$. 2nd approx., $y = 4x + 63x^3$, obtained by putting $(4x)^3$ for y^3 in (A), showing that the curve is above the tangent in the first quadrant and below it in the third quadrant. Thus $(0,0)$ is a *point of* inflexion.

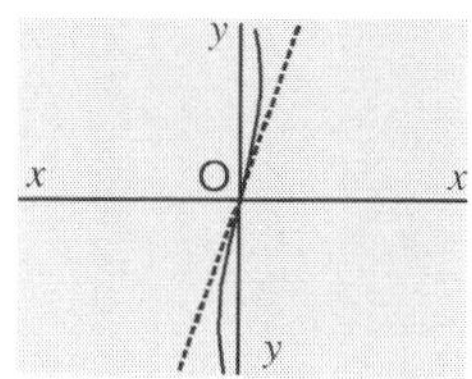

Fig. 73

Near $(0,1)$, putting $Y = y - 1$, equation (A) becomes $2Y + 3Y^2 + Y^3 = (x^2 - 4)x$.

1st approx., $2Y = -4x$ or $y = 1 - 2x$,

which is the equation to the tangent at $(0,1)$.

2nd approx, $2Y + 3(-2x)^2 = -4x$, giving $y = 1 - 2x - 6x^2$, which shows that the curve is below the tangent.

Near $(2, 0)$, putting $X = x - 2$, we find that $y = -8x - 6x^2$ (approx.), showing that the equation to the tangent is $y = -8(x - 2)$ and that the curve is below it.

Again, $dy/dx = (3x^2 - 4)/(3y^2 - 1)$. Therefore, at each of the points $(\pm 2, \pm 1)$, the gradient of the tangent is 4. The tangent is parallel to Ox where $x = \pm 2 / \sqrt{3} = \pm 1.16$; and the tangent is parallel to Oy, where $y = \pm 1/\sqrt{3} = \pm 0.58$.

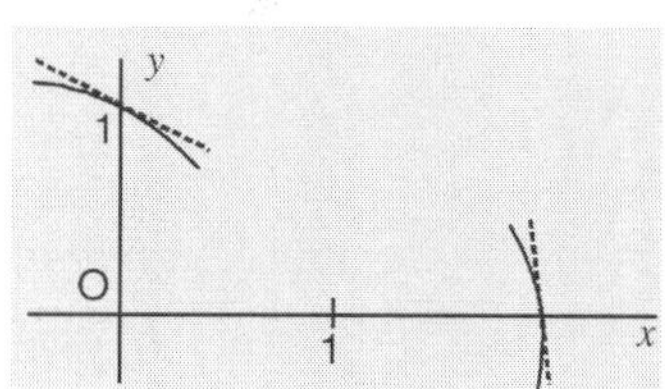

Fig. 74

Near (∞, ∞), x and y are both large and of the same order; and approximately

$$y^3 - x^3 = 0, \text{ giving } y - x = 0.$$

2nd approx. From (A), $\quad y - x = \dfrac{y - 4x}{x^2 + xy + y^2} = \dfrac{-3x}{3x^2} = -\dfrac{1}{x}.$

This shows that, as $x \to \infty$, the curve becomes indefinitely close to the line $y - x = 0$, which is therefore an asymptote.

Again, on the curve $y = x - \dfrac{1}{x}$ (approx.), and on the asymptote $y = x$; showing that the curve is *below* the asymptote on the right and *above* it on the left. Thus the curve is as given in Fig. 75.

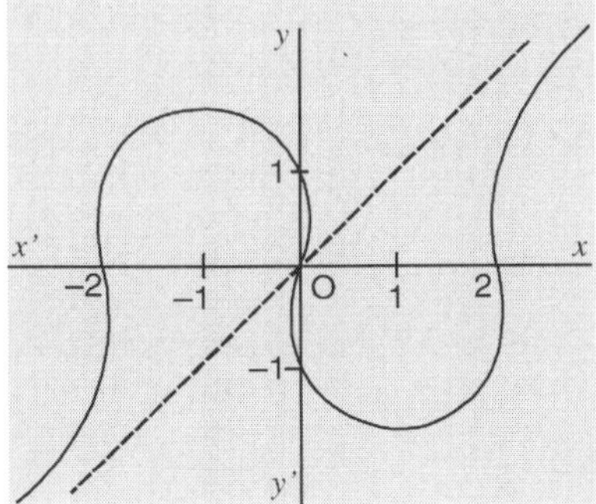

Fig. 75

Putting $y = tx$ in equation (A), it will be seen that any number of points on the curve can be found by giving any values to t in the equations $y = tx$, $x^2 = (4 - t)/(1 - t^3)$.

If $1 < t < 4$, the corresponding point is *imaginary*; thus no part of the curve lies in the space swept out by a line turning round O from the position $y = x$ to the position $y = 4x$.

■ **EXAMPLE 2.** *Trace the curve $(x - 1)(x - 2)y^2 = 2x^2$.* ...(A)

The equation is unaltered by writing $-y$ for y; hence the curve is symmetrical with regard to Ox. Also y is imaginary if $1 < x < 2$; hence no part of the curve lies between the lines $x = 1$, $x = 2$. Write the equation in the form

$$x^2 y^2 - 3xy^2 + 2(y^2 - x^2) = 0. \qquad \text{...(A)}$$

Near $(0, 0)$, the curve is approximately given by

$$y^2 - x^2 = 0 \text{ or } y = \pm x.$$

Taking $y = x$ as a first approximation, from (A)

$$2(y - x) = (3xy^2 - x^2 y^2)/(y + x);$$

giving as a second approximation

$$y - x = \frac{1}{2} \cdot \frac{3x^3}{2x} = \frac{3}{4} x^2,$$

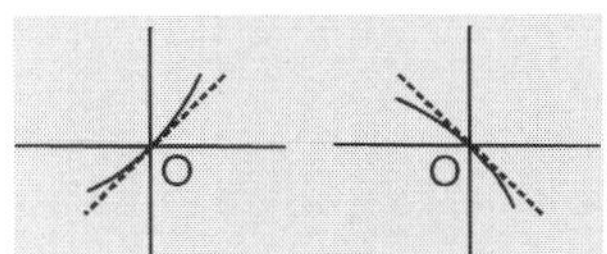

Fig. 76

showing that $y - x = 0$ is a tangent at $(0, 0)$ and that the curve is above the tangent. Similarly $y + x = 0$ is a tangent at $(0, 0)$ which is a *double* point.

$$\text{Near } (1, \infty), \ x - 1 = \frac{2x^2}{(x - 2)y^2} = \frac{2 \cdot 1^2}{(1 - 2)y^2} = -\frac{2}{y^2}.$$

$$\text{Near } (2, \infty), \ x - 2 = \frac{2x^2}{(x - 1)y^2} = \frac{2 \cdot 2^2}{(2 - 1)y^2} = \frac{8}{y^2},$$

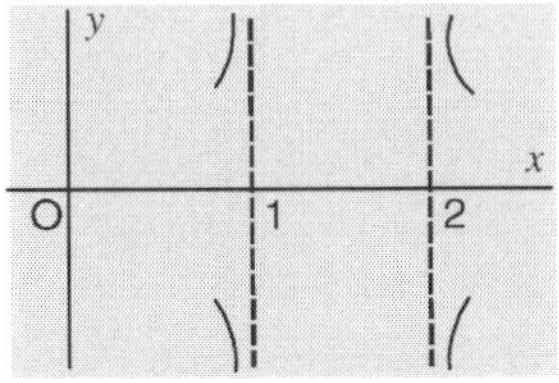

Fig. 77

showing that $x - 1 = 0$ and $x - 2 = 0$ are asymptotes, the curve approaching the first from the left and the second from the right.

Again, writing equation (A) in the form $x^2(y^2 - 2) = 3xy^2 - 2y^2$, it will be seen that

near $(\infty, \sqrt{2})$, $y - \sqrt{2} = \dfrac{3}{\sqrt{2} \cdot x}$; and near $(\infty, -\sqrt{2})$, $y + \sqrt{2} = -\dfrac{3}{\sqrt{2} \cdot x}$

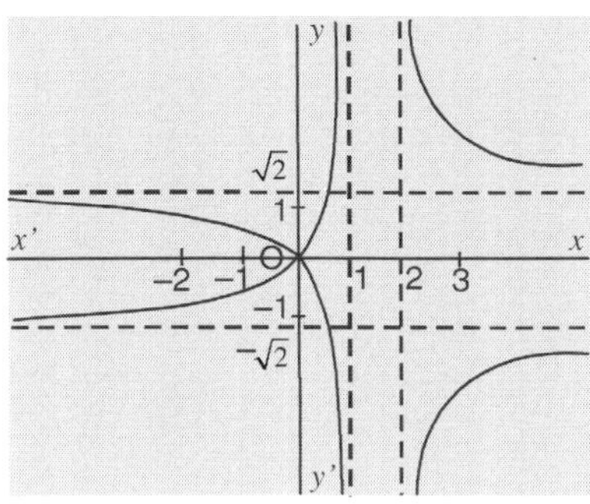

Fig. 78

We can now sketch the curve, as in Fig. 78.

Note: The asymptotes parallel to Ox can be found *by equating to zero the coefficient of the highest power of y in* (A): similarly for those parallel to Oy. Also, for all values of k except ± 1, the line $y = k$ meets the curve in two finite points and in two points at ∞.
The same is true for the line $x = l$, where $l < 1$ or > 2.

■ **8.** In the next example we consider the possibility of finding the approximate form of a curve near the origin, or near infinity, *by retaining only certain terms in its equation.*

▌ **EXAMPLE :** *Trace the curve $x^3 + y^3 - 3axy = 0$ ($a > 0$).* ...(A)

Omitting the first term, we have $y^3 - 3axy = 0$.

Disregarding the factor y, we consider the equation $y^3 - 3ax = 0$. ...(B)

If y is small, x is small and $O(y^2)$; so that x^3 is $O(y^6)$, and can therefore be neglected in comparison with the other terms in (A). Hence (B) approximately represents the curve or part of the curve near $(0, 0)$.

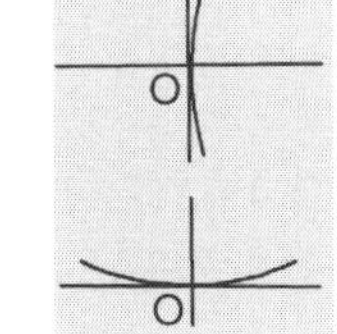

Similarly, by omitting the second term, we find that part of the curve near $(0, 0)$ is represented by $x^2 - 3ay = 0$.

Fig. 79

Omitting the third term, we get $x^3 + y^3 = 0$.

This represents the curve near (∞, ∞); for if y is $O(x)$, then $3axy$ is $O(x^2)$, and can be neglected in comparison with the other terms of (A).

Thus near (∞, ∞), as a first approximation $y + x = 0$.

For a second approximation, $y + x = 3a \cdot \dfrac{xy}{x^2 - xy + y^2} = 3a \cdot \dfrac{-x^3}{3x^2} = -a.$

A third approximation is obtained by substituting $-x-a$ for y, thus

$$y + x = 3a\,\frac{xy}{x^2 - xy + y^2} = 3a\,.\,\frac{-x\,(x+a)}{x^2 + x\,(x+a) + (x+a)^2} = -a\left(1 - \frac{a^3}{3x^2} + ...\right),$$

(the last step being obtained by long division). Hence the third approximation is

$$y + x = -a + \frac{a^3}{3x^2}\,,$$ showing that $y + x = -a$ is an asymptote, and that the curve

approaches it from above at both ends.

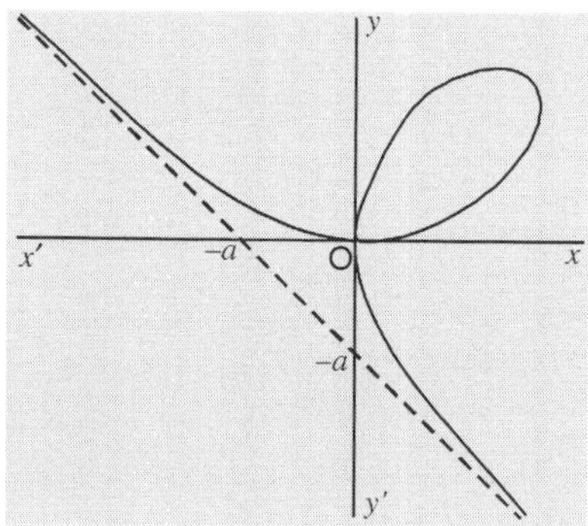

Fig. 80

The curve is symmetrical with regard to the line $y = x$, for its equation is unaltered if x and y are interchanged. Further, if $y = tx$, then $x = 3at/(1 + t^3)$, so that any number of points can be found by giving different values to t; e.g. $t = 1$, $x = y = 3a/2$.

■ **9. Newton's Parallelogram:** Consider the curve represented by $u = 0$ where u is a polynomial in x, y.

Let v be the expression obtained by retaining only certain terms of u. A rule which goes by the name of '*Newton's Parallelogram*' enables us to choose these terms so that $v = 0$ may represent the curve, or part of it, near the origin or near infinity, when such parts exist.

Let $ax^m y^n$, $bx^p y^q$, $cx^r y^s$, ... be any terms of u. Represent these by points $A, B, C,...$ whose coordinates referred to any axes are (m, n), (p, q), (r, s), ...

The figure so formed is '*Newton's Diagram.*'

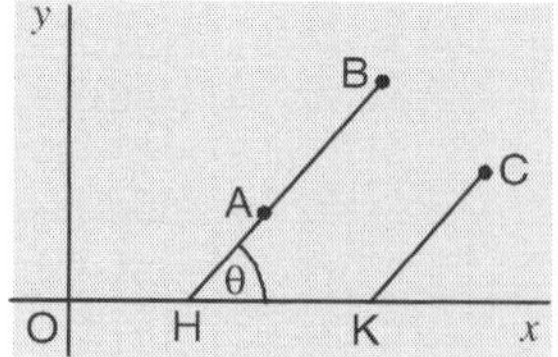

Fig. 81

Suppose that $ax^m y^n$, $bx^p y^q$ are small or large numbers of the same order; then y^{n-q} is $O(x^{p-m})$, and therefore y is $O(x^{(p-m)/(n-q)})$. Join AB, cutting Ox in H, and let the angle $xHA = \theta$; then y is $O(x^{-\cot\theta})$ and $ax^m y^n$ is $O(x^{m-n\cot\theta})$. Also, $m - n\cot\theta = OH$; therefore the terms represented by A, B are $O(x^{OH})$.

Draw CK parallel to AB to meet Ox in K ; then, for similar reasons, the term $cx^r y^s$ represented by C is $O(x^{OK})$. Thus the term C is of an order lower or higher than the terms A, B according as C is on the same side of AB as O, or on the opposite side. Hence the following rule.

Rule. Mark on Newton's diagram points A, B, C,... representing all the terms of the polynomial u. Choose any two of these, say A, B.

Let v be the expression whose terms are those represented by A, B and any points which may happen to lie on the line AB.

If all the other points lie on the side of AB remote from the origin O, then $v = 0$ approximately represents the curve $u = 0$ (or part of it) near O (if on the curve).

If all the other points lie on the same side of AB as O, then $v = 0$ represents the curve $u = 0$ (or part of it) near infinity.

> **Note:** If the straight line AB passes through 0, the rule fails; for the corresponding terms cannot be of the same order.

In the following examples the points representing the 1st, 2nd, 3rd, ... terms of the equation are marked 1, 2, 3, ... in the diagram.

■ **EXAMPLE 1.** $x^3 + y^3 - 3axy = 0$.

The terms 1, 3 and 2, 3 give the form of the curve near the origin.

The terms 1, 2 give the form near infinity. (See Ex. 1, Art. 8)

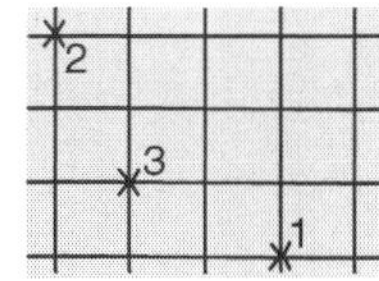

Fig. 82

■ **EXAMPLE 2.** $(ax - by)^2 = ax^2 y + y^4$.

Write the equation in the form,
$$a^2 x^2 - 2abxy + b^2 y^2 - ax^2 y - y^4 = 0.$$

The form near the origin is given by terms 1, 2, 3 and near infinity by the terms 1, 4 and 4, 5.

Thus *near* $(0,0)$, $(ax - by)^2 = 0$.

Near infinity, $a^2 x^2 - ax^2 y = 0$ and $ax^2 y + y^4 = 0$, giving $y - a = 0$ and $y^3 + ax^2 = 0$.

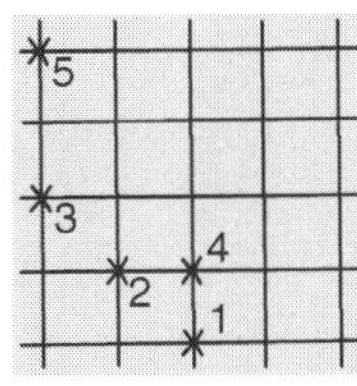

Fig. 83

EXERCISE LI

Trace the curves represented by the following equations and work out the accompanying details. (For hints for *plotting*, see Ex. 22)

1. $y^3 - y + x = 0$.   [Fig. 84, (*i*)]

Near $(0, 0)$, $y = x + x^3$.   Near $(0, 1)$, $y = 1 - \frac{1}{2} x - \frac{3}{8} x^2$.

Near (∞, ∞), $y = - x^{\frac{1}{3}} - \frac{1}{3} x^{-\frac{1}{3}}$.

The tangent is perpendicular to Ox at the points $\pm\dfrac{2}{3\sqrt{3}},\ \pm\dfrac{1}{\sqrt{3}}$.

2. $y^2 = x^2(x+1)$ [Fig. 84, (*ii*)]. Near $(0, 0)$, $y = \pm(x + \tfrac{1}{2}x^2)$; near ∞, $y^2 = x^3$.

3. $x^3 + y^3 = a^2x$ [Fig. 84, (*iii*)]. Near $(0, 0)$, $y^3 = a^2x$; near (∞, ∞), $y = -x + \dfrac{a^2}{3x}$.

4. $x^3 + y^3 = 3ax^2$ [Fig. 84, (*iv*)].

Near $(0, 0)$, $y^3 = 3ax^2$; near (∞, ∞), $y = -x + a + \dfrac{a^2}{x}$.

5. $(x - a)y^2 = a^2x$ [Fig. 84, (*v*)]. Near $(0, 0)$, $y^2 + ax = 0$;

near (a, ∞), $x = a + \dfrac{a^3}{y^2}$; near $(\infty, \pm a)$, $y = \pm\left(a + \dfrac{a^2}{2x}\right)$.

6. $y^2(x - a) = x^2(x + a)$ [Fig. 84, (*vi*)]. *O* is on the curve, but the curve passes through no real point near *O*: consequently *O* is a *'conjugate point.'*

Near $(\infty. \infty)$, $y = \pm\left(x + a + \dfrac{a^2}{2x}\right)$. If $\dfrac{dy}{dx} = 0$, $x = \tfrac{1}{2}a\left(\sqrt{5} + 1\right) = 1.6$ approx.

7. $x(y - x)^2 = a^2y$ [Fig. 84, (*vii*)]. Near $(0, 0)$, $x^3 = a^2y$. Near $(0, \infty)$, $x = \dfrac{a^2}{y}$.

Near (∞, ∞), $y = x \pm a$, closer approximations being $y = x \pm a + \dfrac{a^2}{2x}$.

Tangents are parallel to Ox where $y = 3x$.

8. $x(y - x)^2 = ay^2$ [Fig. 84, (*viii*)]. Near $(0, 0)$, $x^3 - ay^2 = 0$.

Near (a, ∞), $x = a + \dfrac{2a^2}{y}$. Near (∞, ∞), 1st approx., $y = x \pm \sqrt{ax}$.

2nd approx., $y = x \pm \sqrt{ax} + a$. (A). 3rd approx., $y = x \pm \sqrt{ax} + a \pm \sqrt{\dfrac{a^3}{x}}$.

This shows that the curve becomes indefinitely close to that represented by (A), *i.e.* to the parabola $(y - x - a)^2 = ax$. This is therefore a parabolic asymptote.

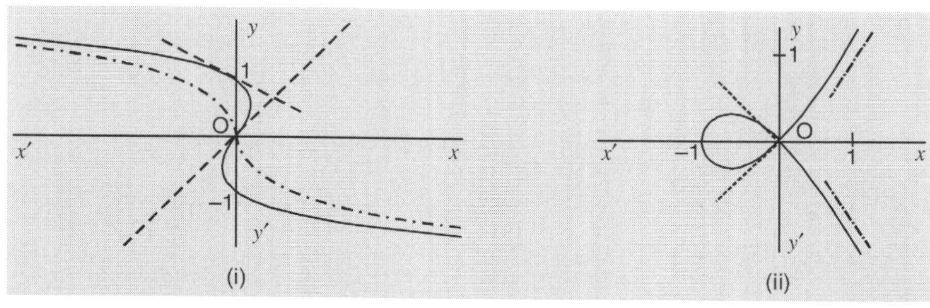

(i)   (ii)

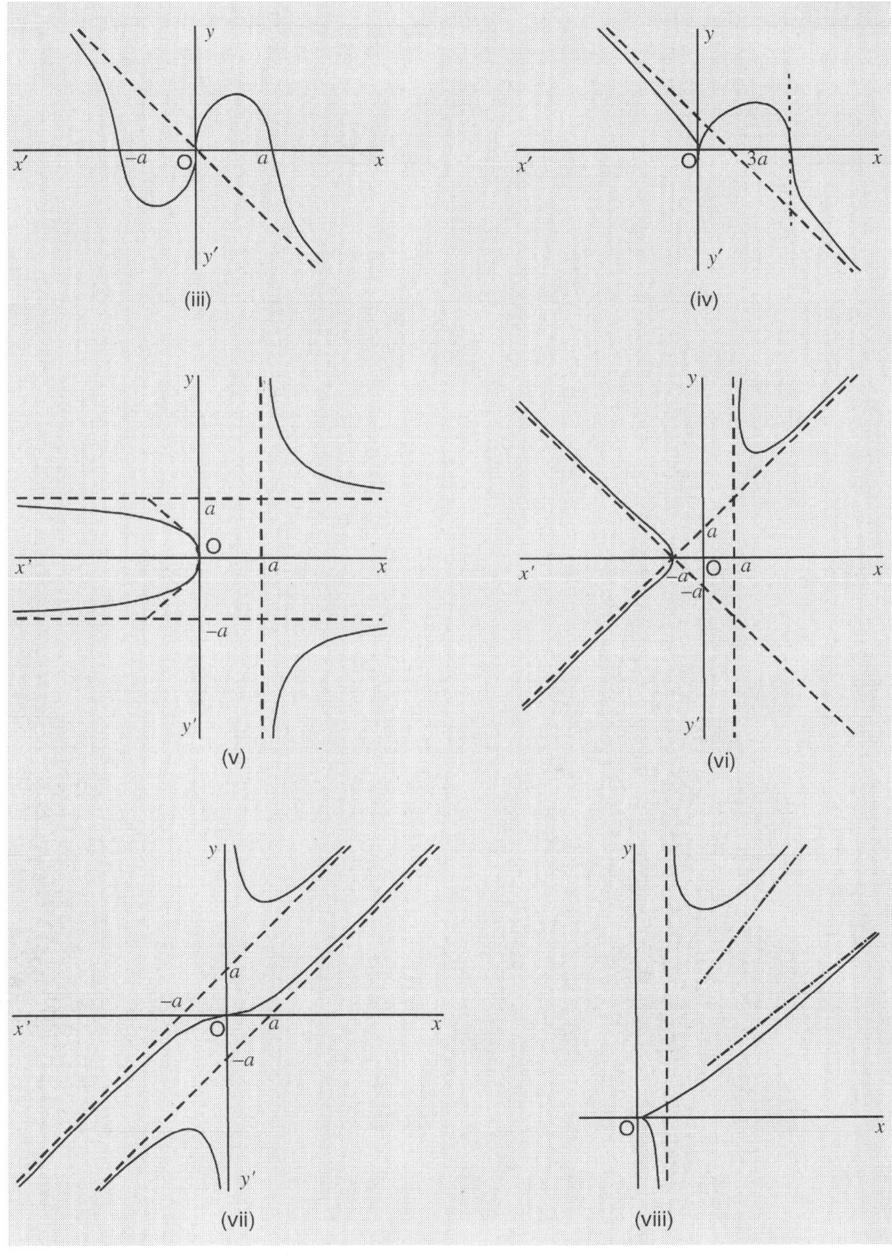

Fig. 84

9. $x^4 - 3xy^2 + 2y^3 = 0$ [Fig. 85, (i)]. Near $(0, 0)$, $x^3 - 3y^2 = 0$ and $y = \dfrac{3}{2}x - \dfrac{2}{9}x^2$.

Near (∞, ∞),   1st approx., $x^4 + 2y^3 = 0$, *i.e.* $y = -\dfrac{x^{\frac{4}{3}}}{2^{\frac{1}{3}}}$.   ...(A)

2nd approx., $y = -\dfrac{x^{\frac{4}{3}}}{2^{\frac{1}{3}}} + \dfrac{x}{2}$.   ...(B)

Equation (A) is represented by –. –. –. in the diagram; it gives the general trend of the curve, but (B) shows that it is not a parabolic asymptote.

If $y = tx$, then $x = t^2(3 - 2t)$. Mark points on the curve corresponding to $t = 1, 2, -\frac{1}{2}, -1$.

10. $ax(y - x)^2 = y^4$ [Fig. 85, (ii)]. Near $(0, 0)$, $y - x = 0$ and $ax = y^2$.

With $y = x$ as 1st approx., the 2nd is $y = x \pm \dfrac{x^{\frac{3}{2}}}{a^{\frac{1}{2}}}$. Near (∞, ∞), $ax^3 - y^4 = 0$.

11. $x^4 - a^2xy - b^2y^2 = 0$ [Fig 85, (iii)]. Near $(0, 0)$, $y = -\dfrac{a^2}{b^2} x - \dfrac{x^3}{a^2}$ and $x^3 - a^2y = 0$.

Near (∞, ∞), by solving for y and expanding in descending powers of x, show that approximately

$$y = \pm \frac{x^2}{b} - \frac{a^2x}{2b^2} \mp \frac{a^4}{8b^3}.$$

The equations $y = \pm \dfrac{x^2}{b}$ are represented by –. –. – in the diagram.

12. $(3x - 2y)^2 - 3x^2y - y^4 = 0$ [Fig. 85, (iv)]. ...(A)

The points $(0, 0)$ $(0, \pm 2)$ are on the curve. Near $(0, 0)$, $3x - 2y = 0$.

2nd approx., $y = \dfrac{3x}{2} \pm \dfrac{3}{2\sqrt{2}} x^{\frac{3}{2}}$, showing that the point O is a cusp, the tangent being

$3x - 2y = 0$, which meets the curve at *one* other point, *viz.* $(-\frac{8}{9}, -\frac{4}{3})$.

Near $(\infty, 3)$, $y = 3 - \dfrac{12}{x}$. The asymptote $y = 3$ cuts the curve again at $(-\frac{5}{4}, 3)$.

Near (∞, ∞), $3x^2 + y^3 = 0$ (represented by –. –. –). ...(B)

2nd approx., $y = -3^{\frac{1}{3}} x^{\frac{2}{3}} - 1$, showing that the curve (A) is below (B) and that $(y + 1)^3 + 3x^2 = 0$ is a parabolic asymptote.

If $3x^2 + y^3 = 0$, then $(3x - 2y)^2 = 0$, showing that the curves (A), (B) *touch* at the point $\left(-\frac{8}{9}, -\frac{4}{3}\right)$.

By solving for x, show that $y < -1$ or else $0 < y < 4$. Also show that the lines $y = -1$ and $y = 4$ touch the curve.

Figure 85, (v), (vi), (vii), (viii), are curves whose equations are given in Exs. 13–16. These are of the form $xy^2 - 2ky = px^3 + qx^2 + rx + s.$* (See Exs. 17, 18.)

13. $xy^2 - 12y = x^3 - 15x^2 + 63x - 85$.

14. $xy^2 - 16y = x^3 - 14x^3 + 60x - 104$.

15. $xy^2 - 24y = x^3 - 10x^2 - 5x + 150$.

16. $xy^2 - 8y = x^3 - 8x^2 - 3x - 8$.

---

* Curves having an equation of this form belong to the first of four classes into which Newton divided cubic curves (*Enumeratio Linearum tertii ordinis,* 1706).

17. Show that every curve whose equation is $xy^2 - 2ky = px^3 + qx^2 + rx + s$ has three asymptotes, or only one, according as p is positive or negative.

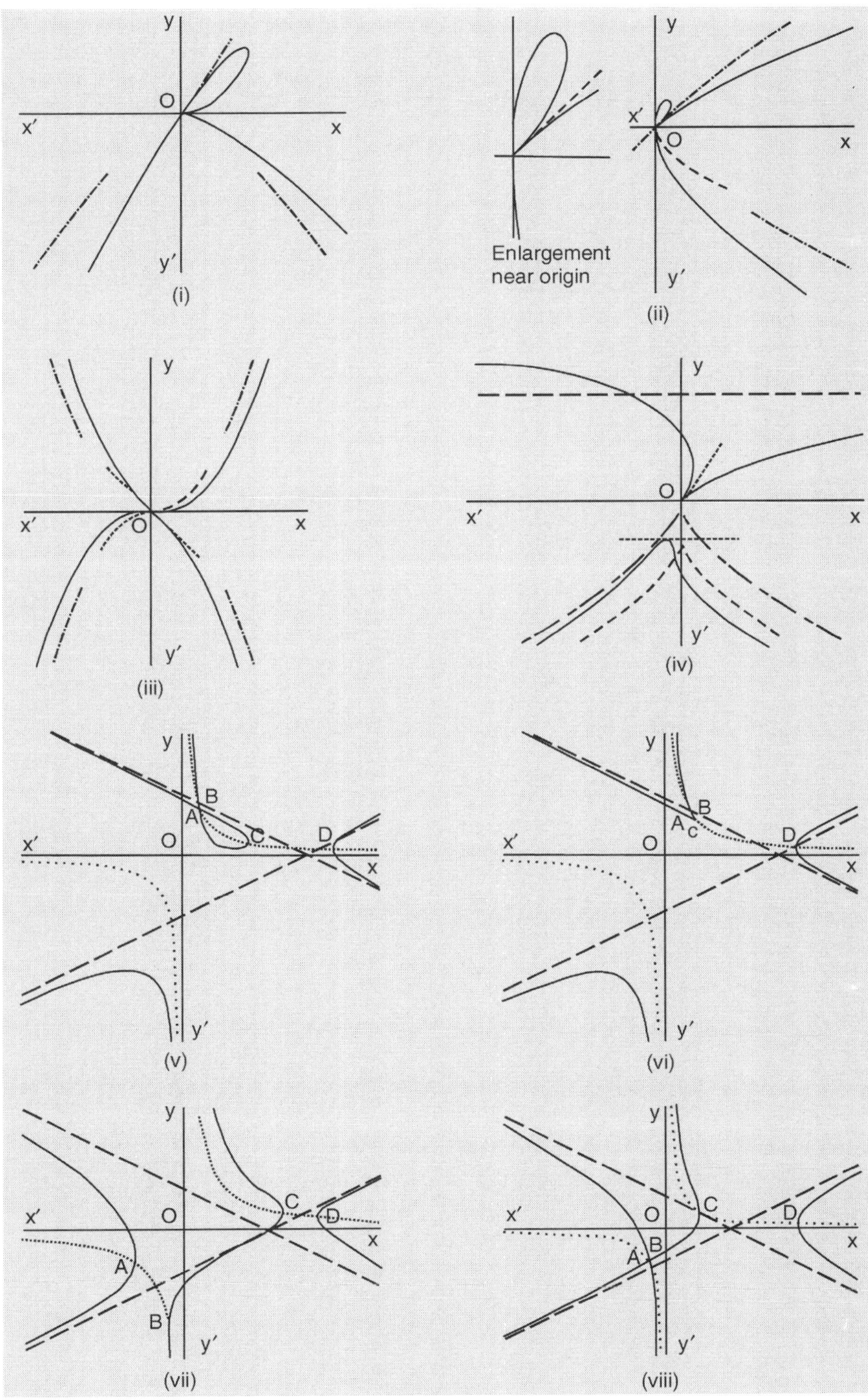

Fig. 85

18. (*i*) Show that, if $xy^2 - 2ky = px^3 + qx^2 + rx + s$, where $p \neq 0$, then, for values of x other than zero, y is given by an equation of the form

$$(xy - k)^2 = p\,(x - a)\,(x - b)\,(x - c)\,(x - d).$$

(*ii*) In Exs. 13–16 employ the methods of the previous examples to find first and second approximations to the curve at infinity; and, by the help of the points, *A, B, C, D,* which correspond to $x = a, b, c, d$, and of the points where the curves cut the asymptotes, trace the curves.

> **Note:** Diagrams (*v*) and (*vi*) show that a cusp is of a higher order of singularity than a node; while (*vii*) and (*viii*) show the 'change of partners' that generally occurs when two of the points, *A, B, C, D,* approach one another, coincide, and then separate again, owing to a change in the coefficients, *p, q, r, s.*

Points of Inflexion

19. Show that the points of inflexion of the cubic curve

$$y(a'x^2 + 2b'\,x + c') = ax^3 + 3bx^2 + 3cx + d \qquad \text{...(A)}$$

lie on the straight line whose equation is

$$\tfrac{4}{3}\,y\,(a'c' - b'^2) = (ac' - 2bb' + ca')\,x + (bc' - 2cb' + da').$$

[From (*i*), by differentiation,

$$\tfrac{1}{3}\frac{dy}{dx}\,(a'x^2 + 2b'x + c') + \tfrac{2}{3}y\,(a'x + b') = ax^2 + 2bx + c. \qquad \text{...(B)}$$

Differentiating again, and putting $\dfrac{d^2y}{dx^2} = 0$, we have

$$\tfrac{2}{3}\frac{dy}{dx}\,(a'x + b') + \tfrac{1}{3}ya' = ax + b. \qquad \text{...(C)}$$

The coordinates of a point of inflexion satisfy equations (A), (B) and (C). Multiplying equations (A), (B), (C) respectively by

$$a', \; -2(a'x + b'), \; a'x^2 + 2b'x + c',$$

the result is obtained by addition.

20. Show that the curve (p. 487), whose equation is $(y^2 - 1)\,y = (x^2 - 4)x$, has three points of inflexion which very nearly lie on the line $x = 64y$.

[If $p = dy/dx$, show that, at a point of inflexion, $x^2 = (4p^6 - p^4)/(p^6 - 1)$, and $y^2 = (4p^2 - 1)/(p^6 - 1)$; so that $p^4 + 8p^3 - 2p - 4 = 0$.]

21. Prove that the points of inflexion on the curve $y^2 = x^2(x^2 + 2ax + b)$ are determined by the equation $2x^3 + 6ax^2 + 3\,(a^2 + b)x + 2ab = 0$.

[At a point of inflexion $y\dfrac{dy}{dx} = 2x^3 + 3ax^2 + bx$, and $\left(\dfrac{dy}{dx}\right)^2 = 6x^2 + 6ax + b$.]

22. Plot some of the curves in Examples 1–12, as follows:

For Ex. 1, plot $x = -y^3$, and $x = y$, and add corresponding abscissae.

For Exs. 2, 5, 6, 7, 8, 11, express y in terms of x.

For Exs. 3, 4, 9, 10, use $y = tx$, or $x = ty$.

For Ex. 12, express x in terms of y.

■ ■ ■

CHAPTER 30

Infinite Products

■ **1.** If $a_1, a_2,...a_n...$ is a sequence of numbers, real or complex, the product $a_1a_2...$ a_n will be denoted by P_n or by $\prod_{r=1}^{r=n} a_r$. Thus we write

$$P_n = \prod_{r=1}^{r=n} a_r = a_1a_2...a_n.$$

Definition. If P_n tends to a finite limit P, *different from zero*, as $n\to\infty$, except when a factor of P_n is zero, we say that the infinite product $a_1a_2a_3...$ is *convergent and that its value is P*, or that it *converges to P*, and we write

$$P = a_1a_2a_3 \text{ ... to } \infty = \prod_{n=1}^{n=\infty} a_n, \text{ or simply} = \Pi a_n.$$

If $P_n \to \infty$ or $-\infty$, or if $P_n \to 0$ when no factor of P_n is zero, the infinite product is said to *diverge* or to be *divergent*.

If P_n oscillates, the infinite product is said to *oscillate*.

The reader may ask why we say that the product *diverges* (to zero) when $P_n\to 0$. The reason is as follows. According to the definition, if $a_1, a_2, ...$ are real and positive, *the infinite product Πa_n and the infinite series $\Sigma \log a_n$* both converge, both diverge or both oscillate.

For $\log P_n = \log a_1 + \log a_2 +...+ \log a_n$, and P_n tends to a finite limit, to ∞, to zero or oscillates according as $\log P_n$ tends to a finite limit, to ∞, to $-\infty$ or oscillates.

If $P_n \to$ a finite limit, then $P_{n-1} \to$ the same limit, and if this limit is not zero $a_n = P_n/P_{n-1} \to 1$. Thus *if Πa_n is convergent, then $a_n \to 1$*.

It is generally convenient to write an infinite product in the form
$$P = (1 + u_1)(1 + u_2)...(1 + u_n)... .$$

A necessary (but not sufficient) condition for convergence is that $u_n\to 0$, and so, after a certain stage, $|u_n|<1$.

The character of P, as regards convergence, will not be affected by removing any number of factors from the beginning of the product. Consequently there will be no loss of generality in supposing that $|u_n| < 1$ for every n.

The simplest and at the same time the most important case is when all the u's have the same sign.

■ **2. Theorem:** *If $u_1, u_2, u_3, ...$ are positive and less than unity, each of the infinite products*

$$(1 + u_1)(1 + u_2)(1 + u_3)... , \quad (1 - u_1)(1 - u_2)(1 - u_3)... ,$$

is convergent or divergent according as the series Σu_n is convergent or divergent.

Proof. The series Σu_n is convergent or divergent.

Suppose that Σu_n is convergent and that its sum is s; then m can be found so that $u_m + u_{m+1} + u_{m+2} + \ldots$ to $\infty < 1$.

Also the omission of the first $m-1$ factors of a product does not affect its convergency.

Hence there is no loss of generality in assuming that $s < 1$ and consequently $s_n < s < 1$ for every n.

By Ch. 15, 1, we have

$$P_n = (1 + u_1)(1 + u_2) \ldots (1 + u_n) < \frac{1}{1 - s_n} < \frac{1}{1 - s},$$

$$Q_n = (1 - u_1)(1 - u_2) \ldots (1 - u_n) > 1 - s_n > 1 - s.$$

Also P_n increases and Q_n decreases as n increases. Therefore P_n and Q_n tend to positive limits as $n \to \infty$, and both of the products $\Pi\,(1 + u_n)$ and $\Pi\,(1 - u_n)$ are convergent.

Next suppose that Σu_n is divergent. By Ch. 15, 1,

$$P_n = (1 + u_1)(1 + u_2) \ldots (1 + u_n) > 1 + s_n \to \infty,$$

$$Q_n = (1 - u_1)(1 - u_2) \ldots (1 - u_n) < \frac{1}{1 + s_n} \to 0.$$

Moreover, $Q_n > 0$. Therefore $P_n \to \infty$ and $Q_n \to 0$, so that $\Pi(1 + u_n)$ diverges to ∞ and $\Pi(1 - u_n)$ diverges to zero.

■ **3. General Condition for Convergence:** Let (a_n) be a sequence of numbers, real or complex, none of which is zero. The necessary and sufficient condition for the convergence of the infinite product

$$P = a_1 a_2 a_3 \ldots$$

is as follows: *Corresponding to any positive number* ε, *however small, an integer* m *must exist such that for all values of* p

$$\mid a_{m+1}\, a_{m+2} \ldots a_{m+p} - 1 \mid < \varepsilon,$$

that is to say, $\mid \dfrac{P_{m+p}}{P_m} - 1 \mid < \varepsilon$ *for* $p = 1, 2, 3, \ldots .,$ $\qquad\qquad$ ...(A)

Proof. First suppose that P is convergent. Then P_n tends to a limit which is not zero, and consequently a positive number k exists such that

$$\mid P_n \mid > k \text{ for every } n.$$

Also, by Ch. 15, 6, we can find m such that

$$\mid P_{m+p} - P_m \mid < k\varepsilon \text{ for } p = 1, 2, 3, \ldots .$$

Therefore

$$\mid \frac{P_{m+p}}{P_m} - 1 \mid < \frac{k\varepsilon}{\mid P_m \mid} < \frac{k\varepsilon}{k} = \varepsilon,$$

so that condition (A) holds.

Conversely, if condition (A) holds, then

$$- \varepsilon < \frac{P_{m+p}}{P_m} - 1 < \varepsilon \quad \text{and} \quad 1 - \varepsilon < \frac{P_{m+p}}{P_m} < 1 + \varepsilon. \qquad \text{...(B)}$$

Taking $\varepsilon < 1$, we have $| P_{m+p} | > (1 - \varepsilon) \cdot | P_m |$, so that $| P_n | > $ some fixed positive number when $n > m$. Consequently P_n cannot tend to zero. Also by (B), P_{m+p} and P_m have the same sign, and therefore

$$| P_{m+p} - P_m | = | P_{m+p} | - | P_m | < \varepsilon \cdot | P_m |.$$

Hence $| P_{m+p} - P_m | < \varepsilon'$, where ε' is positive and as small as we choose. Hence $P_n \to$ a limit different from zero and the product P is convergent. (See Ch. 15, 6)

■ **4. Absolute Convergence:** If $u_1, u_2, u_3, \ldots$ are real or complex numbers whose absolute values, or moduli, are denoted by $u_1', u_2', u_3', \ldots$, the infinite product

$$P = (1 + u_1)(1 + u_2)(1 + u_3) \ldots$$

is said to be *absolutely convergent* when the product

$$P' = (1 + u_1')(1 + u_2')(1 + u_3') \ldots$$

is convergent.

Hence the product $\Pi(1 + u_n)$ is absolutely convergent if, and only if, the series Σu_n is absolutely convergent.

Theorem. *An absolutely convergent infinite product* (P) *is convergent.*

Let P and P' be defined as above and suppose that P' is convergent. Then, by Art. 3, we can find m so that

$$| (1 + u'_{m+1}) (1 + u'_{m+2}) \ldots (1 + u'_{m+p}) - 1 | < \varepsilon \text{ for } p = 1, 2, 3, \ldots .$$

Now $| (1 + u_{m+1}) (1 + u_{m+2}) \ldots (1 + u_{m+p}) - 1 |$

$$\leq | (1 + u'_{m+1}) (1 + u'_{m+2}) \ldots (1 + u'_{m+p}) - 1 |.$$

This follows on expanding each side and remembering that for any numbers $a, b, c, \ldots$

$$| a + b + c + \ldots | \leq | a | + | b | + | c | + \ldots .$$

Therefore

$$| (1 + u_{m+1}) (1 + u_{m+2}) \ldots (1 + u_{m+p}) - 1 | < \varepsilon \text{ for } p = 1, 2, 3, \ldots .$$

Hence, by Art. 3, P is convergent.

■ **5. Derangement of Factors:** We shall prove that *if the factors of an absolutely convergent infinite product* (P) *are rearranged in any way, the product remains absolutely convergent and its value is unaltered.*

Consider the infinite products

$$P = (1 + u_1) (1 + u_2) \ldots, \quad Q = (1 + v_1) (1 + v_2) \ldots ,$$

where P is absolutely convergent and Q is formed by rearranging the factors of P, so that every v is a u and every u is a v.

Since P is absolutely convergent, so is the series Σu_n. This series remains absolutely convergent when its terms are rearranged. Therefore Σv_n is absolutely convergent, and consequently the product Q is absolutely convergent.

It remains to be shown that $P = Q$. Let $a_r = 1 + u_r$, $b_r = 1 + v_r$.

For any suffix m, it is possible to find n so that all the factors of
$P_m = a_1 a_2 \ldots a_m$ are included among the factors of $Q_n = b_1 b_2 \ldots b_n$,
and therefore

$$Q_n / P_m = a_p a_q \ldots a_s,$$

where p, q,... s are all $> m$. Also, as m tends to ∞, so does n, and since P is convergent, it follows from Art. 3 that

$$| a_p a_q \ldots a_s - 1 | \to 0, \quad \text{and therefore } Q_n / P_m \to 1.$$

Thus P_m and Q_n tend to the same limit, that is to say, $P = Q$.

■ **6. Expansion of an Infinite Product as a Series:** Consider the infinite product:

$$P = (1 + xu_1) (1 + xu_2) (1 + xu_3) \ldots ,$$

where u_1, u_2, u_3, ... and x are any numbers, real or complex.

Let p_r be the sum of the products of u_1, u_2, ... u_n taken r together, so that

$$P_n = (1 + xu_1) (1 + xu_2) \ldots (1 + xu_n) = 1 + p_1 x + p_2 x^2 + \ldots + p_n x^n.$$

We shall prove that *if the series Σu_n is absolutely convergent, then*

(i) *As $n \to \infty$, $p_r \to$ a finite limit l_r for $r =$ 1, 2, ... n.*

(ii) *For all values of x,*

$$P = (1 + xu_1)(1 + xu_2)(1 + xu_3) \ldots \text{ to } \infty = 1 + l_1 x + l_2 x^2 + \ldots \text{ to } \infty.$$

Let $u_r' = | u_r |$, $x' = | x |$, and

$$P_n' = (1 + x'u_1') (1 + x'u_2') \ldots (1 + x'u_n') = 1 + p_1' x' + p_2' x'^2 + \ldots + p_n' x'^n,$$

so that p_r' is the sum of the products of u_1', u_2', ... u_n', taken r together.

Each of these products multiplied by $\lfloor r$ occurs in the expansion of

$$(u_1' + u_2' + \ldots + u_n')^r,$$

therefore $p_r' \cdot \lfloor r$ occurs in the expansion, and it follows that

$$p_r' < (u_1' + u_2' + \ldots + u_n')^r / \lfloor r .$$

Since Σu_n is absolutely convergent, $\Sigma u_n'$ converges to a sum s', and therefore

$$p_r' < s'^r / \lfloor r.$$

Also p_r' increases with n, therefore, as $n \to \infty$, $p_r' \to$ a limit l_r' such that

$$0 < l_r' < s'^r / \lfloor r.$$

If $r \to \infty$, then $s'^r / \lfloor r \to 0$, and consequently $l_r' \to 0$, so that l_r' is finite, however, great r may be.

Thus as $n \to \infty$, p_r' becomes an infinite series, which is convergent; and p_r becomes an absolutely convergent series. Hence $p_r \to$ a finite limit l_r, and we may write $l_r = (1 + \varepsilon_r) p_r$ where $\varepsilon_r \to 0$.

Let $Q_n = 1 + l_1 x + l_2 x^2 + \ldots + l_n x^n$, then

$$| Q_n - P_n | = | \varepsilon_1 p_1 x + \varepsilon_2 p_2 x^2 + \ldots + \varepsilon_n p_n x^n |$$
$$\leq | \varepsilon_m | \cdot (P_1' x' + p_2' x'^2 + \ldots + p_n' x'^n)$$
$$\leq | \varepsilon_m | \cdot (P_n' - 1),$$

where ε_m is the greatest of $| \varepsilon_1 |$, $| \varepsilon_2 |$, etc.

Therefore $Q_n - P_n \to 0$, and since $P_n \to P$, it follows that, for all values of x,
$$P = 1 + l_1 x + l_2 x^2 + \dots \text{ to } \infty.$$

▌ **EXAMPLE 1.** *Find the coefficient of z^r in the expansion of*
$$(1 + zx)\,(1 + zx^2)\,(1 + zx^3) \dots (1 + zx^n).$$

Assume that
$$(1 + zx)\,(1 + zx^2)\dots(1 + zx^n) = 1 + p_1 z + p_2 z^2 + \dots + p_n z^n.$$

In this identity put zx for z, therefore
$$(1 + zx^2)\,(1 + zx^3) \dots (1 + zx^{n+1}) = 1 + p_1 zx + p_2 z^2 x^2 + \dots + p_n z^n x^n\ ;$$

therefore $\quad (1 + zx^{n+1})\,(1 + p_1 z + p_2 z^2 + \dots + p_n z^n)$
$$= (1 + zx)\,(1 + p_1 zx + p_2 z^2 x^2 + \dots + p_n z^n x^n).$$

Equating the coefficients of z^r,
$$p_r + p_{r-1} x^{n+1} = p_r x^r + p_{r-1} x^r\ ;$$

$$\therefore \qquad p_r = \frac{1 - x^{n-r+1}}{1 - x^r} \cdot x^r p_{r-1}.$$

Putting $r - 1,\ r - 2,\ \dots 2,\ 1$ in succession for r,

$$p_{r-1} = \frac{1 - x^{n-r+2}}{1 - x^{r-1}} \cdot x^{r-1} p_{r-2},$$

$$\dots\dots\dots\dots\dots\dots\dots\dots\dots\dots$$

$$p_1 = \frac{1 - x^n}{1 - x} \cdot x.$$

Whence by multiplication

$$p_r = \frac{(1 - x^n)(1 - x^{n-1})(1 - x^{n-2})\dots(1 - x^{n-r+1})}{(1 - x)(1 - x^2)(1 - x^3)\dots(1 - x^r)} \cdot x^{\frac{1}{2} r(r+1)}.$$

▌ **EXAMPLE 2.** *If $|x| < 1$, show that for all values of z,*
$$(1 + zx)\,(1 + zx^2)\,(1 + zx^3) \dots \text{ to } \infty = 1 + l_1 z + l_2 z^2 + \dots + l_r z^r + \dots\ ,$$

where $\qquad l_r = \dfrac{1}{(1 - x)(1 - x^2)(1 - x^3)\dots(1 - x^r)} \cdot x^{\frac{1}{2} r(r+1)}.$

The series $x + x^2 + x^3 \dots$ is absolutely convergent, therefore the infinite product is convergent, and denoting its value by P, by the theorem just proved,
$$P = 1 + l_1 z + l_2 z^2 + \dots + l_r z^r + \dots\ ,$$

where $l_r = \lim\limits_{n \to \infty} p_r$ and p_r is the same as in Ex. 1. Also $x^n, x^{n-1}, \dots x^{n-r+1}$ all tend to zero as $n \to \infty$, therefore l_r has the value stated above.

▊ **7. Another Method:** When the factors are real and positive, the theory of infinite products can be made to depend on that of infinite series by the use of logarithms. Let $P_n = (1 + u_1)\,(1 + u_2) \dots (1 + u_n)$, where $u_1,\ u_2,\ \dots$ are real and $-1 < u_r < 1$ for every r. As explained in Art. 1, the last assumption involves no loss of generality.

Then
$$\log P_n = \log (1 + u_1) + \log (1 + u_2) + \ldots + \log (1 + u_n).$$

Hence the infinite product P and the infinite series $\Sigma \log (1 + u_n)$ both converge, both diverge or both oscillate. By Ch. 19, 7, (4),

if $\qquad x > -1$ and $\neq 0$, then $\dfrac{1}{2} \dfrac{x^2}{h} < x - \log (1 + x) < \dfrac{1}{2} \dfrac{x^2}{l}$, $\qquad$ ...(A)

where l is the smaller and h the greater of the numbers 1 and $1 + x$.

Let
$$u_{m+1} + u_{m+2} + \ldots + u_n = r_{m,n},$$
$$u^2_{m+1} + u^2_{m+2} + \ldots + u_n^2 = s_{m,n},$$
$$\log (1 + u_{m+1}) + \log (1 + u_{m+2}) + \ldots + \log (1 + u_n) = t_{m,n}.$$

Putting $u_{m+1}, u_{m+2}, \ldots u_n$ for x in (A), by addition we find that

$$\frac{1}{2} \frac{s_{m,n}}{H} < r_{m,n} - t_{m,n} < \frac{1}{2} \frac{s_{m,n}}{L},$$

where L is the least and H the greatest of the numbers $1, 1 + u_1, 1 + u_2, \ldots, 1 + u_n$.

(i) *Suppose that Σu_n^2 is convergent*, then by the general principle of convergence we can find m so that for all values of n, $s_{m,n} < 2 L\varepsilon$, and therefore

$$0 < r_{m,n} - t_{m,n} < \varepsilon,$$

where ε is positive and as small as we please.

Hence by the general principle of convergence, *P converges, diverges to ∞, diverges to 0, or oscillates, as Σu_n converges, diverges to ∞, diverges to $-\infty$, or oscillates.*

(ii) *Suppose that Σu_n^2 diverges*, then since $r_{m,n} - t_{m,n} > s_{m,n}/2H$, it follows that, for sufficiently large values of n,

$$r_{m,n} - t_{m,n} > N$$

for any positive value of N, however great. Hence

(a) *If Σu_n converges or oscillates between finite limits, then P diverges to 0.*

(b) *If Σu_n diverges to $+ \infty$ or oscillates so that its upper limit is $+ \infty$, then P may converge.*

This includes all possible cases, and when the logarithm of a complex number has been defined and its properties investigated, the same method can be applied when the factors of P are complex numbers.

<h3 align="center">EXERCISE LII</h3>

Obtain the results in Exs. 1–5 by considering the value of P_n in each case.

1. $\left(1 + \dfrac{1}{2}\right) \left(1 + \dfrac{1}{3}\right) \left(1 + \dfrac{1}{4}\right) \ldots \to \infty$. $\left[\, P_n = \dfrac{1}{2}(n + 1). \,\right]$

2. $\left(1 - \dfrac{1}{2}\right) \left(1 - \dfrac{1}{3}\right) \left(1 - \dfrac{1}{4}\right) \ldots \to 0$. $\left[\, P_n = \dfrac{1}{n} . \,\right]$

3. $\left(1 - \dfrac{1}{2^2}\right)\left(1 - \dfrac{1}{3^2}\right)\left(1 - \dfrac{1}{4^2}\right) \cdots \to \dfrac{1}{2}. \ \left[\ P_n = \dfrac{1}{2}.\dfrac{n+1}{n}.\ \right]$

4. $\left(1 - \dfrac{1}{2}\right)\left(1 + \dfrac{1}{3}\right)\left(1 - \dfrac{1}{4}\right)\left(1 + \dfrac{1}{5}\right) \cdots \to \dfrac{1}{2}. \ \left[\ P_{2n+1} = \dfrac{1}{2}\ \text{and}\ P_n/P_{n-1} \to 1.\ \right]$

5. $\left(1 + \dfrac{1}{2}\right)\left(1 - \dfrac{1}{3}\right)\left(1 + \dfrac{1}{4}\right)\left(1 - \dfrac{1}{5}\right) \cdots \to 1. \ \left[P_{2n+1}=1. \right]$

6. Show that

$$(1 + x)(1 + x^2)(1 + x^4)(1 + x^8) \ldots (1 + x^{2^{n-1}}) = \frac{1 - x^{2^n}}{1 - x}.$$

Hence if $|x| < 1$,

$$(1 + x)(1 + x^2)(1 + x^4)\ldots \text{ to } \infty = \frac{1}{1 - x}.$$

7. Show that

$$\sin x = 2^n \sin\frac{x}{2^n}\ \cos\frac{x}{2}\ \cos\frac{x}{2^2}\ \cdots\ \cos\frac{x}{2^n}\ ;$$

Remembering that $\displaystyle\lim_{\theta \to 0} \frac{\sin\theta}{\theta} = 1$, deduce that

$$\cos\frac{x}{2}\ \cos\frac{x}{2^2}\ \cos\frac{x}{2^3}\ \ldots \text{ to } \infty = \frac{\sin x}{x}.$$

8. Show that $(1 + x)\left(1 + \dfrac{x}{2}\right)\left(1 + \dfrac{x}{3}\right) \ldots$ diverges to ∞ or to 0 according as $x \gtrless 0$.

9. Show that the infinite products

$$\left(1 + \frac{x}{1^2}\right)\left(1 + \frac{x}{2^2}\right)\left(1 + \frac{x}{3^2}\right)\left(1 + \frac{x}{4^4}\right) \ldots \text{ and}$$

$$\left(1 + \frac{x}{1^2}\right)\left(1 - \frac{x}{2^2}\right)\left(1 + \frac{x}{3^2}\right)\left(1 - \frac{x}{4^2}\right)\ldots$$

are convergent for all values of x.

10. If $|x| < 1$ and $|zx| < 1$, show that

$$\frac{1}{(1 - zx)(1 - zx^2)\ldots(1 - zx^n)} = 1 + q_1 z + q_2 z^2 + \ldots + q_r z^r + \ldots ,$$

where $q_r = \dfrac{(1 - x^n)(1 - x^{n+1})\ldots(1 - x^{n+r-1})}{(1 - x)(1 - x^2)\ldots(1 - x^r)} . \ x^r.$

[Proceed as in Art. 6, Ex. 1.]

11. If n, r are positive integers, show that
$$(1 - x^n)\,(1 - x^{n+1})\, ... \,(1 - x^{n+r-1})$$
is exactly divisible by $(1 - x)\,(1 - x^2)\, ... \,(1 - x^r)$.

12. If $|x| < 1$ and $|zx| < 1$, show that
$$\frac{1}{(1 - zx)(1 - zx^2)(1 - zx^3)...\text{to } \infty} = 1 + t_1 z + t_2 z^2 + ... + t_r z^r + ... \text{ to } \infty,$$

where $t_r = x^r/(1 - x)\,(1 - x^2)\, ... \,(1 - x^r)$.

[Having explained why such an expansion is possible, put zx for z and show that
$$(1 - zx)\,(1 + t_1 z + t_2 z^2 + ...) = 1 + t_1 zx + t_2 z^2 x^2 + ... \,.$$
Hence by equating the coefficients of z^r,
$$(1 - x^r)\,t_r = x \cdot t_{r-1}.]$$

13. If $(1 + zx)\,(1 + zx^3)\, ... \,(1 + zx^{2n-1}) = 1 + p_1 z + p_2 z^2 + ... + p_n z^n$, show that
$$p_r = \frac{(1 - x^{2n})(1 - x^{2(n-1)})...(1 - x^{2(n-r+1)})}{(1 - x^2)(1 - x^4)...(1 - x^{2r})} \cdot x^{r^2}.$$

14. If $|x| < 1$, show that for all values of z,
$$(1 + zx)\,(1 + zx^3)\,(1 + zx^5)\, ... = 1 + l_1 z + l_2 z^2 + ... + l_r z^r + ... \,,$$

where $l_r = \dfrac{1}{(1 - x^2)(1 - x^4)...(1 - x^{2r})} \cdot x^{r^2}.$

15. If $|x| < 1$ and $|zx| < 1$, show that

(i) $\dfrac{1}{(1 - zx)(1 - zx^3)...(1 - zx^{2n-1})} = 1 + q_1 z + q_2 z^2 + ... \,,$

where $q_r = \dfrac{(1 - x^{2n})\,(1 - x^{2(n+1)})\, ... \,(1 - x^{2(n+r-1)})}{(1 - x^2)\,(1 - x^4)\, ... \,(1 - x^{2r})} \cdot x^r.$

(ii) $\dfrac{1}{(1 - zx)\,(1 - zx^3)\,(1 - zx^5)\, ... \text{ to } \infty} = 1 + t_1 z + t_2 z^2 + ... \text{ to } \infty,$

where $t_r = x^r/(1 - x^2)\,(1 - x^4)\, ... \,(1 - x^{2r}).$

Permutations, Combinations and Distributions

■ **1. Combinations with Repetitions:** The number of combinations of n things r together, when each may be taken as often as we please, is denoted by H_r^n, and is the same as the number of homogeneous products of degree r which can be formed with n letters $a, b, c, \dots k$.

The *sum* of these products is the coefficient of x^r in the expansion of

$$(1 + ax + a^2x^2 + \dots)(1 + bx + b^2x^2 + \dots) \dots (1 + kx + k^2x^2 + \dots).$$

When $a, b, \dots k$ are all equal to unity, we obtain the *number* of the products. Thus

$$H_r^n = \text{Coefft. of } x^r \text{ in } (1 + x + x^2 + \dots)^n$$

$$= \text{Coefft. of } x^r \text{ in } (1 - x)^{-n};$$

therefore

$$H_r^n = \frac{n(n+1)\dots(n+r-1)}{\lfloor r} = C_r^{n+r-1}.$$

■ **2. Combinations and Permutations of n Things, not all different:** Let C be the number of combinations, and P the number of permutations, of n letters $aa \dots bb \dots cc \dots$ taken k together, then

(1) *C is equal to the coefficient of x^k in the product*

$$(1 - x)^{-f}(1 - x^{p+1})(1 - x^{q+1})(1 - x^{r+1}) \dots ,$$

where f is the number of different letters, p is the number of a's, q the number of b's, r the number of c's, and so on.

For the *sum* of the combinations in question is the coefficient of x^k in the product

$$(1 + ax + a^2x^2 + \dots + a^px^p)(1 + bx + b^2x^2 + \dots + b^qx^q) \dots$$

$$\dots (1 + cx + c^2x^2 + \dots + c^rx^r) \dots \text{ to } f \text{ factors;}$$

and their *number* is the coefficient of x^k in the expression

$$(1 + x + x^2 + \dots + x^p)(1 + x + x^2 + \dots + x^q)(1 + x + x^2 + \dots x^r) \dots .$$

which is obtained from the preceding by putting $a, b, c, \dots$ equal to unity. Whence the result follows.

(2) *P is $\lfloor k$ times the coefficient of x^k in the product*

$$\left(1+\frac{x}{\lfloor 1}+\frac{x^2}{\lfloor 2}+...+\frac{x^p}{\lfloor p}\right)\left(1+\frac{x}{\lfloor 1}+\frac{x^2}{\lfloor 2}+...+\frac{x^q}{\lfloor q}\right)\cdot\left(1+\frac{x}{\lfloor 1}+\frac{x^2}{\lfloor 2}+...+\frac{x^r}{\lfloor r}\right)\ ...\ \textit{to f factors.}$$

For the number of permutations involving λa's, μb's, $v c$'s, ... is

$$\lfloor \lambda+\mu+v+.../\lfloor\lambda\lfloor\mu\lfloor v....$$

Hence

$$\frac{P}{\lfloor k}\ =\ \Sigma\frac{1}{\lfloor\lambda\lfloor\mu\lfloor v....},$$

where the summation is to include every set of values of λ, μ, v, ... chosen (with repetitions) from 0, 1, 2, 3, ... such that

$$\lambda + \mu + v + ... = k.$$

Therefore $P\,/\,\lfloor k$ is the coefficient of x^k in the above product.

■ **EXAMPLE :** *Find the number of combinations (C) and the number of permutations (P) of the letters of the word parallel taken four together.*

We have by the preceding,

$$C\ =\ \text{Coefficient of } x^4 \text{ in } (1-x)^{-5}(1-x^3)(1-x^4)(1-x^2)^3,$$

$$P/\lfloor 4\ =\ \text{Coefficient of } x^4 \text{ in } \left(1+\frac{x}{\lfloor 1}+\frac{x^2}{\lfloor 2}\right)\left(1+\frac{x}{\lfloor 1}+\frac{x^2}{\lfloor 2}+\frac{x^3}{\lfloor 3}\right)\left(1+\frac{x}{\lfloor 1}\right)^3,$$

whence we find that $C = 22$, $P = 286$.

■ **3. Distributions:** If n things are divided into r lots or classes, the result is called a *distribution*. A number of different cases arise, thus:

(*i*) Some of the things may be alike, or they may be all different.

(*ii*) The *order* of the things in each class may, or may not; have to be considered. In the first case the classes are called *groups*; in the second, they are called *parcels*.

(*iii*) When a distribution has been made, it may, or may not, be necessary to consider the order in which the classes stand.

In the first case the *classes are said to be different* (meaning that they have to be distinguished from one another); in the second case they are said to be *indifferent*.

(*iv*) Blank lots may, or may not, be admissible, a 'blank lot' being a 'lot' which contains none of the things.

For example, consider what is implied in the following questions.

■ **EXAMPLE 1.** *How many different signals can be made by flying five flags a, b, c, d, e on three masts?*

Denoting the masts by 1, 2, 3, the following arrangements give different signals:

	1	2	3			1	2	3			1	2	3
(A)	*a*	*d*	*e*		(B)	*b*	*d*	*e*		(C)	*d*	*a*	*e*
	b					*a*					*b*		
	c					*c*					*c*		

Because such arrangements as (A) and (B) are different, it is a case of distribution in *groups*.

Because (A), (C) are different, the *groups are said to be different.*

Further *blank lots are admissible,* for all the flags may be flown on only one or only two masts. (See Art. 4, Ex. 1.)

■ **EXAMPLE 2.** *In how many ways can five books a, b, c, d, e be divided among three people denoted by 1, 2, 3 ?*

Such distributions as (A), (B) in Ex. 1 are identical, so it is a case of distribution in *parcels*. Also the distributions (A), (C) are different, and so the *parcels are said to be different.*

Further, it is implied that each person gets at least one book, and so blank lots are not allowed. (See Art. 5, Ex. 1)

■ **EXAMPLE 3.** *In how many ways can five books be tied up in three bundles?*

Here the order of the books in a bundle does not matter, so the distribution is in *parcels*. Also the bundles are not to be distinguished from one another, and so *the parcels are indifferent.* (See Art. 5, Ex. 2)

■ **4. Arrangement in Groups:** *The number of ways in which n different things can be arranged in r different groups is*

$$r(r+1)(r+2) \ldots (r+n-1) \quad \text{or} \quad \lfloor n \, C_{r-1}^{n-1},$$

according as blank groups are or are not admissible.

Proof. (*i*) Let n letters $a, b, \ldots k$ be written in a row in any order. All the arrangements of the letters in r groups, blank groups being admissible, can be obtained thus: place among the letters $r - 1$ marks of partition, and arrange the $n + r - 1$ things (consisting of letters and marks) in all possible orders.* Since $r - 1$ of the things are alike, the number of different arrangements is

$$\lfloor n+r-1 / \lfloor r-1 \ = \ r(r+1)(r+2) \ldots (r+n-1).$$

(*ii*) All the arrangements of the letters in r groups, none of the groups being blank, can be obtained as follows: (*a*) Arrange the letters in all possible orders. This can be done in $\lfloor n$ ways. (*b*) In every such arrangement, place $(r-1)$ marks of partition in $(r-1)$ out of the $(n-1)$ spaces between the letters. This can be done in C_{r-1}^{n-1} ways.

Hence the required number is $\lfloor n \, C_{r-1}^{n-1}$.

■ **EXAMPLE :** *How many different signals can be made by flying five different flags on three masts?*

The required number $= 3 \cdot 4 \cdot 5 \cdot 6 \cdot 7 = 2520$.

■ **5. Distribution in Parcels:**

(1) *The number of ways in which n different things can be distributed into r different parcels, blank lots being admissible, is r^n.*

For each of the n things can be assigned to any one of the r parcels.

---

* Thus one arrangement of 7 letters in 5 groups, the second blanks, is indicated by $a \parallel bcd \mid ef \mid g$.

(2) *The number of ways in which n different things can be distributed into r different parcels, there being no blank lots, is*

$$r^n - C_1^r(r-1)^n + C_2^r(r-2)^n - \ldots + (-1)^{r-1} C_{r-1}^r,$$

which is $\underline{n}$ *times the coefficient of* x^n *in the expansion of* $(e^x - 1)^r$.

Proof. In any distribution, denote the parcels by $\alpha_1, \alpha_2, \ldots \alpha_r$, and consider the distributions in which blanks are allowed.

The total number of these is $\qquad$ r^n.

The number in which α_1 is blank is $\qquad$ $(r-1)^n$.

Therefore the number in which α_1 is not blank is $\qquad$ $r^n - (r-1)^n$.

Of these last, the number in which α_2 is blank is $\qquad$ $(r-1)^n - (r-2)^n$.

Therefore the number in which α_1, α_2 are not blank is $\qquad$ $r^n - 2(r-1)^n + (r-2)^n$.

Of these last, the number in which α_3 is blank is $\qquad$ $(r-1)^n - 2(r-2)^n + (r-3)^n$.

Therefore the number in which α_1, α_2, α_3 are not blank is $\qquad$ $r^n - 3(r-1)^n + 3(r-2)^n - (r-3)^n$.

This process can be continued as far as we like, and it is obvious that the coefficients are formed as in a binomial expansion.

Hence *the number of distributions in which no one of x assigned parcels is blank is*

$$r^n - C_1^x(r-1)^n + C_2^x(r-2)^n - \ldots + (-1)^x (r-x)^n.$$

When $x = r$, this gives the result in question.

A *complete* proof is as follows:

Let u_x denote the number of distributions in which no one of the parcels α_1, $\alpha_2, \ldots \alpha_x$ is blank. Of these, the number in which α_{x+1} is blank is obtained by changing r into $r-1$ in u_x (which is a function of r). This number is represented by $E^{-1}u_x$, where E^{-1} denotes the operation of changing r into $r-1$.

Therefore the number of distributions in which no one of α_1, $\alpha_2, \ldots \alpha_{x+1}$ is blank is $u_x - E^{-1} u_x$, that is to say, $u_{x+1} = (1 - E^{-1})u_x$.

Also we have $\qquad u_1 = r^n - (r-1)^n = (1 - E^{-1})r^n$.

Hence $\qquad u_x = (1 - E^{-1})^x r^n$

$$= (1 - C_1^x E^{-1} + C_2^x E^{-2} - \ldots + (-1)^x E^{-x})r^n$$

$$= r^n - C_1^x (r-1)^n + C_2^x (r-2)^n - \ldots + (-1)^x (r-x)^n,$$

and when $x = r$, this gives the result in question.

■ **EXAMPLE 1.** *In how many ways can five different books be distributed among three persons, if each person is to have at least one book?*

The required number $= 3^5 - 3 \cdot 2^5 + 3 \cdot 1^5 = 150$.

EXAMPLE 2. *In how many ways can five different books be tied up in three bundles?*

$$\text{The required number } = \frac{1}{\underline{|3}} (3^5 - 3 \cdot 2^5 + 3 \cdot 1^5) = \frac{150}{6} = 25.$$

(3) *The number of ways in which n things of the same sort can be distributed into r different parcels is C_{r-1}^{n+r-1} or C_{r-1}^{n-1}, according as blank lots are or are not admissible.*

Proof. *If there are no blank lots,* any such distribution can be effected as follows: place the n things in a row and put marks of partition in a selection of $r - 1$ out of the $n - 1$ spaces between them. This can be done in C_{r+1}^{n-1} ways, which is therefore the number of such distributions.

If blank lots are allowed, the number of distributions is the same as that of $n + r$ things of the same sort into r parcels with no blank lots. For such a distribution can be effected thus: put one of the $n + r$ things into each of the r parcels, and distribute the remaining n things into r parcels, blank lots being allowed. Hence the number in question is C_{r-1}^{n+r-1}.

These theorems can also be proved as in Ex. 2 of the next section.

(4) *The number of ways in which n things of the same sort can be distributed into r different parcels, where no parcel is to contain more than k things and there are no blank lots, is the coefficient of x^n in the expansion of*

$$(x + x^2 + x^3 + \dots x^k)^r \quad or \quad of \quad (1 - x^{k+1})^r (1 - x)^{-r}.$$

For the product of r factors, each of which is $x + x^2 + x^3 + \dots + x^r$, may be denoted by $\Sigma x^\lambda x^\mu x^\nu \dots$, where $x^\lambda x^\mu x^\nu \dots$ are any terms selected from $x, x^2, \dots x^k$, the summation to include every such selection.

The coefficient of x^n in the product is therefore the number of ways in which r numbers $\lambda, \mu, \dots$ can be chosen from $1, 2, 3, \dots k$, repetitions being allowed, such that their sum may be n, and this is the required number.

Distributions of things of the same sort into *indifferent parcels* are considered in Art. 8.

EXAMPLE 1. *In how many ways can three persons, each throwing a single die once, make a total score of 11?*

The required number is equal to the co. of x^{11} in $(x + x^2 + x^3 + \dots + x^6)^3$.

$$= \text{Co. of } x^{11} \text{ in } x^3 (1 - x^6)^3 (1 - x)^{-3}$$
$$= \text{Co. of } x^8 \text{ in } (1 - x)^{-3} (1 - 3x^6)$$
$$= \text{Co. of } x^8 \text{ in } \left(\frac{3 \cdot 4}{2} x^2 + \frac{9 \cdot 10}{2} x^8 \right) (1 - 3x^6)$$
$$= 45 - 18 = 27.$$

EXAMPLE 2. *Apply this method to prove the theorems of the last section.*

The number of distributions of n like things into r parcels is the coefficient of x^n in $(1 + x + x^2 + \dots \text{ to } \infty)^r$ or in $(x + x^2 + x^3 + \dots \text{ to } \infty)^r$, according as blank lots are or are not allowed.

These expressions are respectively equal to $(1 - x)^{-r}$ and $x^r (1 - x)^{-r}$, etc.

■ **6. Derangements:** Any change in the order of the things in a group is called a *derangement.*

If n things are arranged in a row, the number of ways in which they can be deranged so that no one of them occupies its original place is

$$\lfloor\underline{n}\left\{1 - \frac{1}{\lfloor\underline{1}} + \frac{1}{\lfloor\underline{2}} - \frac{1}{\lfloor\underline{3}} + \ldots + (-1)^n \cdot \frac{1}{n}\right\}. \qquad \ldots\ldots\ldots\ldots (A)$$

Proof. Denote the groups of things by $a_1, a_2, \ldots a_n$. The total number of arrangements is $\lfloor\underline{n}$; and of these.

The number in which a_1 stands first, *i.e.* is *in situ,* is	$\lfloor\underline{n-1}.$
Therefore the number in which a_1 is not *in situ* is	$\lfloor\underline{n} - \lfloor\underline{n-1}.$
Of these last, the number in which a_2 is *in situ* is	$\lfloor\underline{n-1} - \lfloor\underline{n-2}.$
Therefore the number in which neither a_1 nor a_2 is *in situ* is	$\lfloor\underline{n} - 2\lfloor\underline{n-1} + \lfloor\underline{n-2}.$
Of these last, the number in which a_3 is *in situ* is	$\lfloor\underline{n-1} - 2\lfloor\underline{n-2} + \lfloor\underline{n-3}.$
Therefore the number in which no one of a_1, a_2, a_3 is *in situ* is	$\lfloor\underline{n} - 3\lfloor\underline{n-1} + 3\lfloor\underline{n-2} - \lfloor\underline{n-3}.$

This process can be continued as far as we like, and it is obvious that the coefficients are formed as in a binomial expansion.

Hence, *the number of arrangements in which no one of x assigned letters is* in situ *is*

$$\lfloor\underline{n} - C_1^x\lfloor\underline{n-1} + C_2^x\lfloor\underline{n-2} - \ldots + (-1)^x\lfloor\underline{n-x}.$$

If $x = n$, the last term is $(-1)^n$, and we obtain the result in question.

A *complete* proof is as follow:

Let v_x denote the number of arrangements in which no one of $a_1, a_2, \ldots a_x$ is *in situ.*

Of these, the number in which a_{x+1} is *in situ* is obtained by changing n into $n - 1$ in v_x (which is a function of n).

This number is represented by $E^{-1} v_x$, where E^{-1} denotes the operation of changing n into $n - 1$.

Therefore the number of arrangements in which no one of $a_1, a_2, \ldots a_{x+1}$ is *in situ* is $v_x - E^{-1} v_x$, that is to say,

$$v_{x+1} = (1 - E^{-1})v_x.$$

Also, as in the preceding, we have

$$v_1 = \lfloor\underline{n} - \lfloor\underline{n-1} = (1 - E^{-1})\lfloor\underline{n}.$$

Hence it follows that

$$v_x = (1 - E^{-1})^x\lfloor\underline{n}$$

$$= (1 - C_1^x E^{-1} + C_2^x E^{-2} - \dots + (-1)^x E^{-x})\lfloor n$$

$$= \lfloor n - C_1^x \lfloor n-1 + C_2^x \lfloor n-3 - \dots + (-1)^x \lfloor n-x.$$

If $x = n$, the last term is $(-1)^n$, and we obtain the result in question.

Notes: (i) The expression (A) is sometimes called *subfactorial n,* and is denoted by $\lVert n.$

It is easily shown that $\lVert n$ is the integer nearest to $\lfloor n / e.$

(ii) The values of $\lVert n$ for $n = 1, 2, 3, \dots$ can be calculated thus: Begin with 1 and multiply successively by 1, 2, 3, ... , increasing any even product by 1 and decreasing every odd product by 1 before the next multiplication. Thus

$$\lVert 1 = 1 \cdot 1 - 1 = 0 \qquad\qquad \lVert 2 = 0 \cdot 2 + 1 = 1,$$

$$\lVert 3 = 3 \cdot 1 - 1 = 2, \qquad\qquad \lVert 4 = 4 \cdot 2 + 1 = 9.$$

■ **7. A General Theorem:** In Arts. 5 and 6 we have instances of a theorem, stated by Whitworth* as follows:

If there are N sets of letters and, if out of r assigned letters a, b, c, ..., each letter occurs in N_1 sets, each combination of two letters occurs in N_2 sets, each combination of three letters occurs in N_3 sets, and so on; and finally all the r letters occur in N_r sets, then the number of sets free from all the letters is

$$N - C_1^r N_1 + C_2^r N_2 - C_3^r N_3 + \dots + (-1)^r N_r$$

Or more generally: *If there are N events and r possible conditions such that every single condition is satisfied in N_1 of the events, every two of the conditions are simultaneously satisfied in N_2 of the events, ... and finally all the r conditions are simultaneously satisfied in N_r of the events, then the number of events free from all the conditions is as stated above.*

The proof is exactly similar to that in the last article.

■ **8. Partition of Numbers:**

(1) In this section, we consider a number as formed by the addition of other numbers.

Any selection of the numbers 1, 2, 3, 4, ... (with repetitions), of which the sum is n, is called a *partition of n* ; if the selection contains p numbers, it is called a *partition of n into p parts* or, shortly, a *p-partition of n.*

Thus $7 = 3 + 2 + 1 + 1$, and we say that 3211 is a 4-partition of 7.

▌ **EXAMPLE 1.** *Supposing that all the partitions of 1, 2, 3, 4, 5 have been written down, explain how to find all the partitions of 6.*

Take first 6; then 5 followed by 1; then 4, followed by all the partitions of 2 (that is, 2, 11); then 3, followed by all the partitions of 3 (that is, 3, 21, 111); then 2, followed by the partitions of 4 *which contain no part greater than 2* (that is, 22, 211, 1111); lastly, 111111. The complete set is

6, 51, 42, 411, 33, 321, 3111, 222, 2211, 21111, 111111.

---

* Whitworth, *Choice and Chance,* p. 73.

(2) If Π_p^n is the number of p-partitions of n, then*

$$\Pi_p^n = \Pi_{p-1}^{n-1} + \Pi_p^{n-p} \qquad \ldots (A)$$

To make the p-partitions of n is to distribute n elements into p indifferent parcels, blanks being inadmissible.

This can be done by placing one of the elements in each of the p parcels and then distributing the remaining $n - p$ elements into 1 or 2 or 3 ... or p parcels. Therefore

$$\Pi_p^n = \Pi_1^{n-p} + \Pi_2^{n-p} + \Pi_3^{n-p} + \ldots + \Pi_p^{n-p}. \qquad \ldots (B)$$

Changing n into $n - 1$ and p into $p - 1$, we have

$$\Pi_{p-1}^{n-1} = \Pi_1^{n-p} + \Pi_2^{n-p} + \Pi_3^{n-p} + \ldots + \Pi_{p-1}^{n-p},$$

whence the result (A) follows.

If $p > \dfrac{1}{2}n$, then $\quad \Pi_p^n = \Pi_{p-1}^{n-1}.$

For in this case $p > n - p$, and therefore $\Pi_p^{n-p} = 0$.

(3) Construction of a table of values of Π_p^n. We have

$$\Pi_1^n = 1, \quad \Pi_2^n = \frac{1}{2}(n - 1) \text{ or } \frac{1}{2}n,$$

according as n is odd or even. Also $\quad \Pi_n^n = 1, \ \Pi_{n-1}^n = 1.$

In the following table, denote the nth space of the mth row by (n, m): call the set of spaces $(n, 1), (n + 1, 2), (n + 2, 3), \ldots$ *the n-th diagonal.*

Table of *p*-Partitions of *n* [*i.e.* Values of *P* (*n, p, **)]

$n =$	1	2	3	4	5	6	7	8	9	10	11	12	13	14	15	16	17	18	19	20
$p = 1$	1	1	1	1	1	1	1	1	1	1	1	1	1	1	1	1	1	1	1	1
2		1	1	2	2	3	3	4	4	5	5	6	6	7	7	8	8	9	9	10
3			1	1	2	3	4	5	7	8	10	12	14	16	19	21	24	27	30	33
4				1	1	2	3	5	6	9	11	15	18	23	27	34	39	47	54	64
5					1	1	2	3	5	7	10	13	18	23	30	37	47	57	70	84
6						1	1	2	3	5	7	11	14	20	26	35	44	58	71	90
7							1	1	2	3	5	7	11	15	21	28	38	49	65	82
8								1	1	2	3	5	7	11	15	22	29	40	52	70
9									1	1	2	3	5	7	11	15	22	30	41	54
10										1	1	2	3	5	7	11	15	22	30	42

---

* Π_p^n is subsequently denoted by P (*n, p, **).

We can fill up the first and second rows and the first and second diagonals, and using § (2), we have

$$\Pi_1^{n+1} = \Pi_1^n, \quad \Pi_2^{n+2} = \Pi_1^n + \Pi_2^n, \quad \Pi_3^{n+3} = \Pi_1^n + \Pi_2^n + \Pi_3^n, \text{ etc.}$$

Hence the numbers in the $(n + 1)$th diagonal are the sums of one, two, three, ... terms of the nth column, starting from the top.

The second column is 1, 1, 0, 0, ... , therefore the third diagonal is 1, 2, 2, 2, ... The third column is 1, 1, 1, 0, 0 ... , and the fourth diagonal 1, 2, 3, 3, 3, ... ; and so on to any extent.

(4) Various classes of partitions of n may be considered: the number of those in any particular class is denoted by a symbol of the form $P(n, ,)$ where the number and the nature of the parts are respectively indicated in the first and second spaces following n.

In the first space, p means that there are p parts; and $\leq p$ means that the number of parts does not exceeds p.

In the second space, q means that the greatest of the parts is q; and $\leq q$ means that no part exceeds q.

An *asterisk* means that no restriction is placed on the number or the nature of the parts, as the case may be.

If the parts are to be unequal, than Q is used instead of P. Thus:

$P(n, p, q)$ is the number of partitions of n into p parts, the greatest of which is q.

$P(n, p, *)$ is the number of p-partitions of n.

$P(n, \leq p, *)$ is the number of partitions of n into p or any smaller number of parts.

$Q(n, *, \leq q)$ is the number of partitions of n into unequal parts, none of which exceeds q.

(5) The values of $P(n, \leq p, *)$ can be found from the table on p. 494. For instance, the number of partitions of 10 into not more than 4 parts is the sum of the first four numbers in the 10th column, and is therefore the 4th number in the 11th diagonal, that is 23.

That is to say that

$$P(10, \leq 4, *) = P(14, 4, *),$$

and, in general, we have

$$P(n, \leq p, *) = P(n + p, p, *). \qquad \qquad \text{... (C)}$$

(6) We have the following relations.

$$P(n, p, q) = P(n - q, p - 1, \leq q)$$

and

$$Q(n, p, q) = Q(n - q, p - 1, \leq q). \qquad \qquad \text{... (D)}$$

For consider any partition of n into p parts, of which the greatest is q. If we remove the part q, we have a partition of $n - q$ into $p - 1$ parts, none of which exceeds q. Also from every partition of the latter kind we can derive one of the former.

(7) Conjugate Partitions. Consider a partition of 8 into 3 parts of which the greatest is 4, say 4 3 1.

This may be written as in the margin, and if we sum the columns	1111
instead of the rows, we have 3 2 2 1, a partition of 8	111
into 4 parts of which 3 is the greatest.	1

Clearly 4 3 1 can be derived from 3 2 2 1 by the same process, and these are called *conjugate partitions.*

Thus *to every partition of n into p parts of which the greatest is q, there corresponds a partition of n into q parts of which the greatest is p.*

Consequently, we have

$$P(n, p, q) \;=\; P(n, q, p), \qquad\qquad \ldots\text{(E)}$$
$$P(n, p, *) \;=\; P(n, *, p),$$
$$P(n, \;\leq p, q) \;=\; P(n, q, \;\leq p), \qquad\qquad \ldots\text{(F)}$$
$$P(n, \;\leq p, *) \;=\; P(n, *, \;\leq p). \qquad\qquad \ldots\text{(G)}$$

Thus by § 5,

$$P(n, *, \;\leq p) \;=\; P(n + p, p, *). \qquad\qquad \ldots\text{(H)}$$

(8) Euler's use of Series in the Enumeration of Partitions.

(*i*) $Q(n, *, \;\leq q)$ *is the coefficient of x^n in the expansion of*

$$(1 + x)\,(1 + x^2)\,(1 + x^3)\,\ldots\,(1 + x^q).$$

For this coefficient is the number of ways in which n can be obtained by adding a selection of the numbers, 1, 2, 3, ... q.

(*ii*) $P(n, *, \;\leq q)$ *is the coefficient of x^n in the expansion of*

$$(1 - x)^{-1}\,(1 - x^2)^{-1}\,(1 - x^3)^{-1}\,\ldots\,(1 - x^q)^{-1}.$$

For consider the product

$$(1 + x + x^2 + \ldots)\,(1 + x^2 + x^4 + \cdots)\,(1 + x^3 + x^6 + \ldots)\,\ldots\,(1 + x^q + x^{2q} + \ldots).$$

Any terms out of the first, second, ... qth factors may be represented by x^α, $x^{2\beta}$, $x^{3\gamma}$ ... $x^{q\theta}$ where α, β, γ ... θ are any of the numbers 0, 1, 2, ... q. If the product of these terms is x^n, we have

$$\alpha + 2\beta + 3\gamma + \ldots + q\theta = n.$$

Hence the coefficient of x^n in the product is the number of ways in which n can be made up by adding any selection of the numbers 1, 2, 3, ... q, *repetitions being allowed.* Hence the result follows.

(*iii*) $Q(n, p, \;\leq q)$ *is the coefficient of $z^p x^n$ in the expansion of*

$$(1 + zx)\,(1 + zx^2)\,(1 + zx^3)\,\ldots\,(1 + zx^q),$$

and is therefore the coefficient of $x^{n - 1/2 p(p+1)}$ in the expansion of

$$\frac{(1 - x^q)(1 - x^{q-1})\ldots(1 - x^{q-p+1})}{(1 - x)(1 - x^2)\ldots(1 - x^p)}.$$

For the coefficient of $z^p x^n$ in the first product is the number of ways in which n can be obtained by adding a selection of p of the numbers 1, 2, 3, ... q.

The rest follows from Ch. 30, 6, Ex. 1.

(*iv*) It follows that $Q(n, p, *)$ is the coefficient of $x^{n - \frac{1}{2} p(p+1)}$ in the expansion of $(1 - x)^{-1}\,(1 - x^2)^{-1}\,(1 - x^3)^{-1}\,\ldots\,(1 - x^p)^{-1}$.

Hence by (*ii*), $Q(n, p, *) \;=\; P(n - \tfrac{1}{2} p(p + 1), *, \;\leq p)$,

whence by § (5), $Q(n, p, *) \;=\; P(n - \tfrac{1}{2} p(p - 1), p, *)$;

thus the number of partitions of n into p unequal parts can be found from the table of values of $P(n, p, *)$ in (3).

▌**Example :** *In how many ways can 17 be made up by adding four of the numbers 1, 2, 3, ... , 9 ?*

We have $\qquad Q(17, 4, *) = P(17 - \tfrac{1}{2}.4.3, 4, *) = P(11, 4, *) = 11.$

Thus the total number of ways of forming 17 by adding four of the numbers 1, 2, 3, ..., is 11.

From this number we must subtract the number of partitions with a part greater than 9.

These partitions are 10, 4, 2, 1; 11, 3, 2, 1; and their number is 2.

Or thus, $\qquad Q(17, 4, 10) = Q(7, 3, \leq 10) = Q(7, 3, *) = P(4, 3, *) = 1,$

and similarly $\qquad Q(17, 4, 11) = 1.$

Hence the required number is $11 - 2 = 9.$

(v) The reader should have no difficulty in varifying the following:

$$(1 - zx)^{-1}(1 - zx^2)^{-1} \ldots (1 - zx^9)^{-1} = 1 + \Sigma P(n, p \leq q)z^p x^n,$$

$$(1 - zx)^{-1}(1 - zx^2)^{-1}(1 - zx^3)^{-1} \ldots \text{ to } \infty = 1 + \Sigma P(n, p, *)\, z^p x^n,$$

$$(1 - z)^{-1}(1 - zx)^{-1} \ldots (1 - zx^9)^{-1} = 1 + \Sigma P(n, \leq p, \leq q)\, z^p x^n,$$

$$(1 - z)^{-1}(1 - zx)^{-1}(1 - zx^2)^{-1} \ldots \text{ to } \infty = 1 + \Sigma P(n, \leq p, *)\, z^p x^n.$$

EXERCISE LIII

1. Find the number of ways of displaying 5 flags on 4 masts if all the flags are to be used, and (i) one or more of the masts need not be used; (ii) all the masts are to be used.

2. Find the number of ways in which n things, of which r are alike, can be arranged in circular order.

3. Show that a selection of 10 balls can be made from an unlimited number red, white, blue and green balls in 286 different ways, and that 84 of these contain balls of all four colours.

4. Show that the number of different selections of 5 letters which can be made from five a's, four b's, there c's, two d's and one e is 71.

5. Show that a selection of $(n - 1)$ things can be made from $3n$ things, of which n are alike of one sort, n are alike of another sort and n are alike of a third at, in $n \cdot 2^{n-1}$ different ways.

6. In the expansion of $(a_1 + a_2 + \ldots + a_p)^n$ where n is any positive integer not rather than p, prove that the coefficient of any term in which none of the numbers $a_1, a_2, \ldots\ldots a_p$ occurs more than once is $\lfloor n$.

7. In how many ways can a white balls, b black balls and c red balls be put in m in different bags, if one or more of the bags may remain empty?

8. In an examination, the maximum mark for each of three papers is n; the fourth paper it is $2n$. Prove that the number of ways in which a candidate n get $3n$ marks is

$$\frac{1}{6}(n+1)(5n^2 + 10n + 6).$$

Number = Coefficient of x^{3n} in $(1 - x^{n+1})^3 (1 - x^{2n+1})(1 - x)^{-4}.]$

9. In how many ways can 5 rings be worn on the 4 fingers of one hand?

10. In how many ways can an examiner assign 30 marks to 8 questions, ring not less than 2 marks to any question?

11. The number of ways in which $2n$ things of one sort, $2n$ of another sort and of a third sort can be divided between two persons so that each may have things is $3n^2 + 3n + 1$.

    [Number of selections of $3n$ things = Coefficient of x^{3n} in $(1 - x^{2n+1})^3 (1 - x)^{-3}$.]

12. The total number of permutations of n things taken 1, 2, 3 ... or n together the integer nearest to $e \lfloor n - 1$.

    $$\left[\text{Total number} = e \lfloor n - 1 - \left(\frac{1}{n+1} + \frac{1}{n+1 \cdot n+2} + ...\right)\right]$$

13. Three are 5 letters and 5 directed envelopes, (*i*) In how many ways can all the letters be put into the wrong envelopes? (*ii*) In how many ways can 2 letters be rightly placed and 3 letters wrongly placed?

14. The number of arrangements of the n terms $a_1, a_2, ... a_n$ in which no one of r sequences $a_1 a_2, a_2 a_3, ..., a_r a_{r+1}$ occurs is

    $$\lfloor n - C_1^r \lfloor n - 1 + C_2^r \lfloor n - 2 - ... + (-1)^r C_r^r \lfloor n - r.$$

    [For the total number of arrangements is $\lfloor n.$ Any one of the sequences occurs in $\lfloor n - 1$ of these; any two in $\lfloor n - 2$; and so on.]

15. Show that the number of arrangements of the letters *abcd* which involve none of the sequences *ab, bc, cd* is 11. Verify by writing down the arrangements.

16. The number of arrangements of the letters *abcd* in which neither *a, b* nor *c, d* come together is

    $$\lfloor 4 - C_1^2 \cdot 2 \lfloor 3 + C_2^2 \cdot 2^2 \lfloor 2 = 8.$$

    Explain this, and write down the arrangements in question.

17. A party of 10 consists of 2 Englishmen, 2 Scotsmen, 2 Welshmen and 4 men of other nationalities (all different). In how many ways can they stand in a row so that no two men of the same nationality are next one another?

    In how many ways can they sit at a round table?

18. The number of combinations n together of $3n$ letters of which n are a and n are b and the rest unlike is $(n + 2)2^{n-1}$.

    [Number = Coefficient of x^n in $(1 - x^{n+1})^2(1 + x)^n(1 - x)^{-2}$, or in $(2 - \overline{1 - x})^n (1 - x)^{-2}$

    = Coefficient of x^n in $2^n(1 - x)^{-2} - n \cdot 2^{n-1}(1 - x)^{-1}$, etc.]

19. Show that if $a_1, a_2 .... a_m$ are distinct prime numbers other than unity, the number of solutions in integers (excluding unity) of the equation.

    $$x_1 x_2 x_3 ... x_n = a_1 a_2 a_3 ... a_n$$

    is n^m. Show also that the number of solutions in which at least one x is unity is

    $$C_1^n (n - 1)^m - C_2^n (n - 2)^m + ... + (-1)^{n-1} C_{n-1}^n \cdot 1^m.$$

    [Solution follows from the general theorem of Art. 7.]

20. Given n pairs of gloves, in how many ways can each of n persons take a right-handed and a left-handed glove without taking a pair?

21. (*i*) If n things are arranged in circular order, the number of ways of selecting three of the things no two of which are next each other is

    $$\frac{1}{6}n (n - 4)(n - 5).$$

(*ii*) If the n things are arranged in a row, the number of such sets of three is

$$\frac{1}{6}(n-2)(n-3)(n-4).$$

22. Find the number of positive integral solutions of $x + y + z + w = 20$ under the following conditions:

    (*i*) Zero values of x, y, z, w are included.

    (*ii*) Zero values are excluded.

    (*iii*) No variable may exceed 10; zero values excluded.

    (*iv*) Each variable is an odd number.

    (*v*) x, y, z, w have different values (zero excluded).

23. Prove that the number of positive integral solutions (zero values excluded) of the equation

$$x + 2y + 3z = n$$

is

$$\frac{1}{12}(n-1)(n-5) + \frac{1}{8}(-1)^n + c,$$

where $c = \dfrac{11}{24}$ or $\dfrac{1}{8}$ according as n is or is not a multiple of 3. Verify when $n = 9$, 10.

[The required number is co. of x^n in

$$(x + x^2 + x^3 + ...)(x^2 + x^4 + x^6 + ...)(x^3 + x^6 + x^9 + ...).]$$

24. The number of independent solutions in unequal positive integers, zero included, of

$$x + y + z + u = 2p$$

is the integer nearest to $\dfrac{1}{36}p^2(2p-3)$.

25. Show that $\Pi_3^{6n} = 3n^2$.

$[\Pi_3^{6n} - \Pi_3^{6n-3} = 3n - 1,\ \Pi_3^{6n-3} - \Pi_3^{6n-6} = 3n - 2,\ ...\ \Pi_3^6 - \Pi_3^3 = 2,\ \Pi_3^3 = 1.]$

26. A necklace is made up of 3 beads of one sort and $6n$ of another, those of each sort being similar. Show that the total number of possible arrangements of the beads is $3n^2 + 3n + 1$.

[Having put on the three beads, the number of arrangements of the $6n$ beads in the spaces between the three (which are indifferent) is

$$\Pi_3^{6n} + \Pi_2^{6n} + \Pi_1^{6n} = 3n^2 + 3n + 1.]$$

27. Find the values of (*i*) $P(18, 5\ *)$, (*ii*) $P(12,\ \le 5,\ *)$, (*iii*) $P(12,\ *,\ 4)$.

28. In how many ways can 20 be expressed as the sum of four unequal positive integers?

29. Show that the total number of partitions of n is $P(2n, n, *)$. Find the total number of partitions of 10.

30. Show that $P(2n + r, n + r, *) = P(2n, n, *)$. Hence use the table of Art. 8, (3), to find $P(25, 18, *)$.

31. Show that $P(n, p, \le q)$ is equal to the coefficient of x^{n-p} in the expansion of

$$\frac{(1-x^q)(1-x^{q+1})...(1-x^{q+p-1})}{(1-x)(1-x^2)...(1-x^p)}.$$

Hence show that

$$Q(n, p, \le q) = P\{n - \tfrac{1}{2}p(p-1),\ p,\ \le (q - p + 1)\}.$$

32. Show that

$$\Pi_p^n = \text{Co. of } x^{n-p} \text{ in } \frac{1}{(1-x)(1-x^2)...(1-x^p)}.$$

EXERCISE LIV*

1. There are n points in a plane, no three being collinear except p of them, which are collinear. How many triangles can be drawn with their vertices at three of the points?

2. There are n points in a plane, no three being collinear. Show that the number of n-sided polygons with their vertices at these points is $\frac{1}{2}\lfloor n-1$. Verify if $n = 4$.

3. If m points in a straight line are joined in all possible ways to n points in another straight line, then, excluding the $(m + n)$ points, the number of points of intersection is

$$\frac{1}{2}mn \, (m - 1) \, (n - 1).$$

4. If n straight lines of indefinite length are drawn in a plane, no two of them being parallel and no three meeting in a point, then

    (i) the number of points of intersection is $\frac{1}{2}n(n-1)$;

    (ii) the plane is divided into $\frac{1}{2}(n^2+n+2)$ parts.

[(ii) If u_n is the number of parts, show that $u_n = u_{n-1} + n$.]

5. Show that, in general, n planes divide space into $\frac{1}{6}(n^3+5n+6)$ regions. What are the exceptional cases?

[If no three planes have a common line of intersection and no four meet in a point, then $v_n = v_{n-1} + u_{n-1}$ where v_n is the required number and u_n is the same as in Ex. 4.]

6. The number of squares formed on a piece of squared paper by m horizontal lines and n vertical lines ($m < n$) is

$$\frac{1}{6}m(m-1)(3n-m-1).$$

7. (i) If m, n are positive integers such that $m \le n$ and $m + n = a$, where a is constant, prove that the greatest value of $m(m - 1)(3n - m - 1)$ is

$$\frac{1}{4}(a+1)(a-1)(a-3) \quad \text{or} \quad \frac{1}{4}a\,(a-1)(a-2),$$

according as a is odd or even.

(ii) If a straight rods of indefinite length are placed so as to form the greatest possible number of squares, prove that this number is

$$\frac{1}{24}(a+1)(a-1)(a-3) \quad \text{or} \quad \frac{1}{24}a(a-1)(a-2),$$

according as a is odd or even.

---

  * The results in Exs, 9–12 were given by Dudeney.

8. Show that the number of ways in which three numbers in arithmetical progression can be selected from $1, 2, 3, \ldots n$ is $\frac{1}{4}(n-1)^2$ or $\frac{1}{4}n(n-2)$, according as n is odd or even.

9. The sides of a triangle are a, b, c inches where a, b, c are integers and $a \le b \le c$. If c is given, show that the number of different triangles is $\frac{1}{4}(c+1)^2$ or $\frac{1}{4}c(c+2)$, according as c is odd or even.

10. Of the triangles in the last question, the number of those which are isosceles or equilateral is $\frac{1}{2}(3c-1)$ if c is odd and $\frac{1}{2}(3c-2)$ if c is even.

11. Each side of a triangle is an integral number of inches, no side exceeding c inches. Prove that the number of different (*i.e.* non-congruent) triangles which can be so formed is

$$\frac{1}{24}(c+1)(c+3)(2c+1) \quad \text{or} \quad \frac{1}{24}c(c+2)(2c+5),$$

according as c is odd or even.

12. Of the triangles in the last question, the number of those which are isosceles or equilateral is $\frac{1}{4}(3c^2+1)$ or $\frac{3}{4}c^2$, according as c is odd or even.

■ ■ ■

CHAPTER 32

Probability

1. First Principles: In speaking of the probability of an event, doubt is implied as to whether it will or will not happen. The degree of doubt depends on our knowledge of the controlling conditions. A complete knowledge of these enables us to say that the event is certain to happen or certain to fail.

Measurement of Probabilities. We are constantly forming rough estimates of probabilities, saying that some event is unlikely, likely, very likely or certain to happen. This implies that (at any rate in certain cases) probability can be measured and can be compared with *certainty,* which must be regarded as a degree of probability. Thus there is nothing to prevent us from choosing *certainty* as the *unit of probability,* and this is always done.

In what follows the words *probability* and *chance* are synonymous, both of them being used to denote *measure of probability.*

Suppose that n balls $A, B, \ldots K,$ all of the same sort, are placed in a bag, and that one of them is drawn. There is nothing to favour the drawing of one ball more than another, in other words it is *equally likely* that $A, B, \ldots$ or K will be drawn.

Thus if x is the chance that A is drawn, the chance that any other ball, say B, is drawn is also x.

Our notions with regard to probabilities therefore lead us to assert that the chance that *either A or B is drawn* is $2x$, and the chance that one of the n balls is drawn is nx. But it is certain that one ball will be drawn, and so $nx = 1$ and $x = 1/n$. Thus the probability of drawing A is $1/n$.

The following definition is a statement of *Laplace's First Principle.*

Definition. If an event can happen in a ways and fail in b ways, and there is nothing to lead us to believe that one of these ways should occur rather than another, that is to say, all the ways are equally likely, then

the chance that the event happens is	$a/(a + b),$
the chance that the event fails is	$b/(a + b),$
the odds in favour of the event are as	$a : b,$
the odds against the event are as	$b : a.$

If $b = 0$, the event is certain to happen, and its probability is 1; *if* $a = 0$, it is certain to fail, and the probability of its happening is 0.

If $a = b$, the event is as likely to happen as to fail, and we say that there is an *even chance* of its happening.

520

If p, q are the respective chances of the happening and failing of an event, then

$$p + q = 1.$$

Events of a type to which this definition does not directly apply will be considered later.

■ **EXAMPLES :** (*i*) *If a coin is tossed, the chance of 'heads' is* $\dfrac{1}{2}$.

(*ii*) *If a six-faced die is thrown, the chance that 'ace' turns up is* $\dfrac{1}{6}$.

(*iii*) *If a ball is drawn from a bag containing 3 white and 2 black balls, the chance of drawing white is* $\dfrac{3}{5}$; *the odds against black are as 3 : 2.*

(*iv*) *If a card is drawn from a well-shuffled pack,*

the chance that the card is ace of spades is $\dfrac{1}{52}$;

the chance that the card is a spade is $\dfrac{13}{52}$;

the chance that the card is an ace is $\dfrac{4}{52}$.

It is easy to make a mistake as to the meaning of '*equally likely.*'

■ **EXAMPLE :** *A man of much practical experience in the laying of odds said: 'If you toss three pennies simultaneously, it is an even chance that they will fall all alike, i.e., all heads or all tails.*

'For they must fall all heads, or all tails, or 2 heads and a tail, or 2 tails and a head. So they fall all alike in 2 out of 4 possible cases.'

As a matter of fact, *the odds are 3 to 1 against their falling alike.* For, if the pennies are denoted by a, b, c then a can fall head or tail, and so for b and c. Hence there are 8 *equally likely* cases of which 2 are favourable.

■ **2. Exclusive Events:** Events are said to be *mutually exclusive* when the happening of any one of them precludes the happening of any other. In the case of mutually exclusive events, *the chance that one or other of them occurs is the sum of the chances of the separate events.*

This statement is known as *Laplace's second principle.* It can also be stated as follows: *If an event can happen in several different ways, one only of which can occur on the same occasion, the chance of its happening is the sum of the chances of its happening in the several different ways.*

For consider n events of which the chances are a_1/g, a_2/g, ... , a_n/g respectively, where the letters denote integers. Out of g equally likely ways, the events can happen in a_1, a_2, ... a_n ways respectively, and since the events cannot concur, one or other of them can happen in $a_1 + a_2 + ... + a_n$ ways. The chance of this is therefore $(a_1 + a_2 + ... + a_n)/g$, which is the sum of the chances of the events.

■ **EXAMPLE 1.** *In a single cast with two dice, what is the chance of throwing (i) two aces, (ii) doublets, (iii) five-six (i.e. one die turns up five and the other six)?*

Any face of either die may turn up, so there are 6×6 equally likely cases.

These are made up of the six throws (1, 1), (2, 2), ... (6, 6), together with the thirty throws (1, 2), (2, 1), ... (5, 6), (6, 5).

We count (5, 6), (6, 5) as the same throw, and call it *five-six*.

Thus

$$\text{the chance of throwing two aces} = \frac{1}{36};$$

$$\text{the chance of throwing doublets} = \frac{6}{36};$$

$$\text{the chance of throwing fix-six} = \frac{2}{36}.$$

■ **EXAMPLE 2.** *In a single cast with two dice, what are the odds against throwing 7, i.e. against two numbers of which the sum is 7?*

Out of 36 possible cases, those in favour of the event are (6, 1), (1, 6), (5, 2), (2, 5), (4, 3), (3, 4) and the number of these is 6, therefore

$$\text{the chance of throwing 7 is } \frac{6}{36} = \frac{1}{6},$$

and the odds against throwing 7 are as 5 : 1.

■ **EXAMPLE 3.** *In a single cast with three dice, what is the chance of throwing (i) four-five-six, (ii) eleven, (iii) less than eleven and (iv) more than ten?*

The number of possible cases of 6^3.

(*i*) Among these *four-five-six* occurs $\lfloor 3$ times;

$$\therefore \text{ Chance of throwing four-five-six} = \frac{\lfloor 3}{6^3} = \frac{1}{36}.$$

(*ii*) Let s be the number of cases in which 11 is thrown, then

$$s = \text{Coefft. of } x^{11} \text{ in } (x + x^2 + x^3 + ... + x^6)^3 = 27.$$

$$\therefore \text{ Chance of throwing 11 is } \frac{27}{6^3} = \frac{1}{8}.$$

(*iii*) Let t be the number of cases in which less than 11 is thrown. These cases are those in which exactly 3, 4, 5, ... 10 are thrown. Therefore t is the sum of the coefficients of $x^3, x^4, x^5, ... x^{10}$ in the expansion of

$$(x + x^2 + x^3 + ... + x^6)^3;$$

hence,

$$t = \text{Coefficient of } x^{10} \text{ in } x^3 (1 - x^6)^3 (1 - x)^{-4}$$

$$= \text{Coefficient of } x^7 \text{ in } (1 - x)^{-4} (1 - 3x^6)$$

$$= \text{Coefficient of } x^7 \text{ in } \left(4x + \frac{8 \cdot 9 \cdot 10}{1 \cdot 2 \cdot 3} x^7 \right) (1 - 3x^6)$$

$$= 120 - 12 = 108;$$

$$\therefore \text{ Chance of throwing less than 11 is } \frac{108}{6^3} = \frac{1}{2}.$$

(*iv*) The total number of possible cases is made up of those in which more than 10 are thrown and those in which less than 11 are thrown;

$$\therefore \quad \text{Chance of throwing more than 10} = 1 - \frac{1}{2} = \frac{1}{2}.$$

(In the old game of *passe-dix,* a player stakes on throwing more than 10.)

EXAMPLE 4. *A party of ten take their seats at a round table. What are the odds against two specified persons (A, B) sitting together?*

A having taken his place, *B* has a choice of 9 places, 2 of which are next to *A*, hence the odds against *B* sitting next to *A* are as 7 : 2.

EXAMPLE 5. *Four cards are drawn from a pack of 52 cards. What is the chance that one of each suit is drawn?*

Four cards can be selected from 52 in C_4^{52} ways; and this is the number of possible cases. One of each suit can be selected in 13^4 ways; and this is the number of favourable cases;

$$\therefore \text{ Required chance} = \frac{13^4}{C_4^{52}} = \frac{28561}{270725}.$$

EXAMPLE 6. *Five balls are drawn from a bag containing 6 white and 4 black balls. What is the chance that 3 white and 2 black balls are drawn?*

Five balls can be selected out of ten in C_5^{10} ways. This is the number of possible cases. Also three white balls and two black balls can be selected in $C_3^6 \cdot C_2^4$ ways, which is the number of favourable cases, therefore,

$$\text{The required chance} = \frac{C_3^6 \cdot C_2^4}{C_5^{10}} = \frac{10}{21}.$$

▨ **3. Independent Events:** Events are said to be *independent* when the happening of any one of them does not affect the happening of any of the others.

Laplace's third principle. The probability of the concurrence of several independent events is the product of their separate probabilities.

It is sufficient to consider two independent events A_1 and A_2, of which the probabilities are p_1 and p_2 respectively. Suppose that A_1 can happen in a_1 ways and fail in b_1 ways, all of which are equally likely. Also suppose that A_2 can happen in a_2 ways and fail in b_2 ways, are equally likely.

Then $p_1 = a_1/(a_1 + b_1)$, $p_2 = a_2/(a_2 + b_2)$, and there are the following possibilities:

A_1 and A_2 may both happen, and this may occur in $a_1 a_2$ ways.

A_1 may happen and A_2 fail, and this may occur in $a_1 b_2$ ways.

A_2 may happen and A_1 fail, and this may occur in $a_2 b_1$ ways.

A_1 and A_2 may both fail, and this may occur in $b_1 b_2$ ways.

The total number of ways in which both A_1 and A_2 are concerned is $(a_1 + b_1)(a_2 + b_2)$, and these are all equally likely. Hence the chance that both A_1 and A_2 occur is $a_1 a_2/(a_1 + b_1)(a_2 + b_2)$, which is equal to $p_1 p_2$.

It should also be noticed that

the chance that A_1 happens and A_2 fails is $p_1 (1 - p_2)$;

the chance that A_2 happens and A_1 fails is $p_2 (1 - p_1)$;

the chance that both A_1 and A_2 fails is $(1 - p_1)(1 - p_2)$.

Hence also the chance that *one and only one* of the events happens is

$$p_1(1 - p_2) + p_2(1 - p_1) = p_1 + p_2 - 2p_1 p_2.$$

Also the chance that *at least one* of the events happens (*i.e.* both do not fail) is

$$1 - (1 - p_1)(1 - p_2) = p_1 + p_2 - p_1 p_2.$$

■ **EXAMPLE 1.** *What is the chance of throwing ace with a single die in two trials?*

The chance of not throwing ace in one trial is $\dfrac{5}{6}$,

the chance of not throwing ace in two trials is $\left(\dfrac{5}{6}\right)^2$;

∴ The chance of throwing at least one ace $= 1 - \left(\dfrac{5}{6}\right)^2 = \dfrac{11}{36}$.

Or thus,

Chance of success in both trials $= \dfrac{1}{36}$.

Chance of success in the first trial and failure in the second $= \dfrac{1}{6} \cdot \dfrac{5}{6}$.

Chance of failure in the first and success in the second $= \dfrac{5}{6} \cdot \dfrac{1}{6}$.

Chance of at least one success $= \dfrac{1}{36} + 2 \cdot \dfrac{1}{6} \cdot \dfrac{5}{6} = \dfrac{11}{36}$.

■ **EXAMPLE 2.** *Referring to the last example, explain why the following reasoning is incorrect: The chance of throwing ace in the first trial is* $\dfrac{1}{6}$ *and the chance of ace in the second trial is* $\dfrac{1}{6}$*, therefore the chance of ace in two trials is* $\dfrac{2}{6}$*.*

The throwing of ace in the first trial does not exclude the possibility of throwing ace in the second trial. Thus the events are not mutually exclusive and Art. 2 does not apply. It should be noticed that, if the reasoning were correct, it would follow that the chance of throwing ace in 6 trials would be $\dfrac{6}{6}$, *i.e.* that ace would *certainly* occur in 6 trials; which is obviously false.

■ **EXAMPLE 3.** *In how many throws with a single die will it be an even chance that ace turns up at least once?*

The chance against ace turning up in n throws is $\left(\dfrac{5}{6}\right)^n$, and the chance that it turns up at least once is $1 - \left(\dfrac{5}{6}\right)^n$. If then n is the required number, we have

$$1 - \left(\dfrac{5}{6}\right)^n = \dfrac{1}{2}; \quad \therefore \left(\dfrac{5}{6}\right)^n = \dfrac{1}{2}; \quad \therefore n = 3.8 \ nearly.$$

Hence it is less than an even chance that ace turns up once in 3 trials, and more than an even chance that it turns up in 4 trials.

■ **4. Interdependent Events:** If two events A_1, A_2 are such that p_1 is the probability of A_1, and p_2 is the probability of A_2 *on the supposition that A_1 has happened,* then the probability that both A_1 and A_2 happen is $p_1 p_2$.

Also if p_3 is the probability of a third event A_3 *after A_1 and A_2 have happened*, then the probability that A_1, A_2 and A_3 will happen is $p_1 p_2 p_3$, and so on for any number of dependent events.

For the reasoning of the last article holds good under these conditions.

EXAMPLE 1. *One bag contains 3 white balls and 2 black balls; another contains 5 white and 3 black balls. If a bag is chosen at random and a ball is drawn from it, what is the chance that it is white?*

The chance that the first bag is chosen is $\dfrac{1}{2}$, and the chance of drawing a white ball from it is $\dfrac{3}{5}$;

∴ The chance of choosing the first bag and drawing a white is $\dfrac{1}{2} \cdot \dfrac{3}{5}$.

Similarly the chance of choosing the second bag and drawing a white is $\dfrac{1}{2} \cdot \dfrac{5}{8}$;

∴ The chance of drawing a white from a bag chosen at random is

$$\frac{1}{2} \cdot \frac{3}{5} + \frac{1}{2} \cdot \frac{5}{8} = \frac{49}{80}.$$

Note: If all the balls were put into one bag and a ball were drawn, the chance of its being white is $\dfrac{8}{13}$, which is not the same as before.

EXAMPLE 2. *A person draws a card from a pack, replace it, and shuffles the pack. He continues doing this until he draws a spade. What is the chance that he will have to make (i) at least three trials, (ii) exactly three trials?*

The chance of success at any particular trial is $\dfrac{1}{4}$. *In the first case,* he will have to fail at the first and second trials;

∴ Required chance $= \left(\dfrac{3}{4}\right)^2 = \dfrac{9}{16}$.

In the second case, he must fail at the first and second trials and succeed at the third;

∴ Required chance $= \left(\dfrac{3}{4}\right)^2 \cdot \dfrac{1}{4} = \dfrac{9}{64}$.

EXAMPLE 3. *Five balls are drawn, one by one, from a bag containing 6 white and 4 black balls. What is the chance that 3 white and 2 black balls are drawn?*

First method. It makes no difference to the result if we suppose the 5 balls drawn *simultaneously.* Hence, as in Art. 2,

the required chance $= C_3^6 \cdot C_2^4 / C_5^{10} = \dfrac{10}{21}$.

Second method. Suppose the balls drawn in the order indicated by *wwwbb.* The chance that this occurs is easily seen to be

$$\frac{6}{10}\cdot\frac{5}{9}\cdot\frac{4}{8}\cdot\frac{4}{7}\cdot\frac{3}{6}.$$

Also the chance that 3 white and 2 black balls are drawn in any other particular order is obtained from the preceding by merely altering the order of the numerators.

The number of orders in which this can occur is the number of permutations of the letters *wwwbb,* and is therefore $\lfloor 5 / \lfloor 3 \lfloor 2 = 10$

$$\therefore \text{ The required chance } = 10.\frac{6}{10}\cdot\frac{5}{9}\cdot\frac{4}{8}\cdot\frac{4}{7}\cdot\frac{3}{6},$$

which is equal to the result previously obtained.

■ **5. Another Way of Estimating Probabilities:** It is obvious that the definition of Art. 1 only applies to events of a restricted type; in particular, to games of chance. There are other cases in which probability is estimated by considering the frequency with which the event occurs, or is believed to occur, 'in the long run.' For example:

(*i*) It is an observed fact that about 51 per cent of the children born in Europe are boys. Hence we say that the probability that a child, about to be born, will be a boy is about 0.51.

(*ii*) According to 'tables of mortality,' such as are used by Insurance Companies, out of 81, 188 men 50 years old, 70,552 live to be age of 60. We therefore say that the chance that a man of age 50 will live another 10 years is $70552/81188 = 0.869$.

In general, if an event has happened in pN trials out of a total of N trials, where N is a large number, we say that p is the probability that the event will happen on a fresh trial.

This notion of probability is not at variance with that contained in Laplace's first principle (Art. 1).

If a six-faced die is thrown a large number of times, we expect one face to turn up about as often as another, and this has been verified experimentally. Thus in a large number of throws we may expect ace to turn up in about $\dfrac{1}{6}$ of the total number of trials, and from this point of view we say that the chance of throwing ace is $\dfrac{1}{6}$.

■ **6. Expectation:** Suppose that a person (A) has a ticket in a lottery which gives him a chance p of a prize of £a.

If the lottery were held N times, where N is a large number, we may expect him to get the prize about pN times, receiving £pNa. Thus we may say that *on an average* he receives £pa for a single lottery, and this is called his *expectation.*

Next suppose that A's ticket gives him a chance p_1 of receiving £a_1, a chance p_2 of receiving £a_2, and so on.

If the lottery were held N times, where N is large, we may expect him to receive £a_1 on about p_1N occasions, £a_2 on about p_2N occasions, and so on. Altogether he may be expected to get about £$N(p_1a_1 + p_2a_2 + ...)$.

Thus we may say that *on an average* he gets $£(p_1a_1 + p_2a_2 + ...)$ for a single lottery. This sum is the sum of the expectations arising from his chances of securing the separate sums, and is called his *expectation.*

Definitions. The *average value* of a quantity P, subject to risk, is the average value which P assumes in the long run. This value is also called the *expected value* of P or the *expectation* with regard to P.

If a quantity P can assume the values $P_1, P_2, P_3, ...$ and the chances that it has these values are respectively $p_1, p_2, p_3, ...$, the *average* or the *expected* value of P is $p_1P_1 + p_2P_2 + p_3P_3 + ...$.

This is merely a generalisation of what has been said in the case of a lottery.

It should be observed that every p denotes the chance that P has the corresponding value, so that *events need not be independent.*

■ **EXAMPLE 1.** *A man's expectation of life is usually taken to mean the average number of years which men of his age survive.*

If p_r is the chance that he will survive r years, dying before the end of the next year, his expectation of life is roughly $p_1 . 1 + p_2 . 2 + p_3 . 3 + ...$. A closer estimate will be given in another volume.

■ **EXAMPLE 2.** *A person draws 2 balls from a bag containing 3 white and 5 black balls. If he is to receive 10s. for every white ball which he draws and 1s. for every black ball, what is his expectation?*

The number of ways in which 2 balls can be drawn is $C_2^8 = 28$.

Of these the number of ways in which

2 white balls can be drawn is $\qquad\qquad\qquad\qquad\qquad\qquad C_2^3 = 3,$

1 white and 1 black ball can be drawn is $3 \cdot 5 \qquad\qquad\quad = 15,$

2 black balls can be drawn is $\qquad\qquad\qquad\qquad\qquad\qquad C_2^5 = 10.$

Hence his chances of receiving 20s. 11s. and 2s. are respectively 3/28, 15/28, 10/28;

$$\therefore \text{Expectation} = \left(\frac{3}{28} \cdot 20 + \frac{15}{28} \cdot 11 + \frac{10}{28} \cdot 2 \right) = 8\text{s. } 9\text{d.}$$

■ **7. Successive Events:** Let p, q be the respective chances of the happening and of the failing of an event at a single trial, then:

(1) *The chance of its happening exactly r times in n trial is $C_r^n p^r q^{n-r}$.*

For the chance of its happening r times and failing $n - r$ times *in a specified order* is $p^r q^{n-r}$.

Now the number of different orders in which these things can occur is the number of different arrangements of r things of one sort and $n - r$ things to another sort. This number is $\lfloor n / \lfloor r \lfloor n - r$, which is equal to C_r^n.

Thus the event in question can happen in C_r^n ways, which are mutually exclusive, and the chance that it happens in any one of them is $p^r q^{n-r}$; therefore the required chance is $C_r^n p^r q^{n-r}$.

(2) We have $C_r^n = C_{n-r}^n$, therefore *the chances that the event happens exactly n times, exactly (n – 1) times, exactly (n – 2) times, ... are respectively the first, second, third, ... terms in the expansion.*

$$(p + q)^n = p^n + C_1^n p^{n-1} q + C_2^n p^{n-2} q^2 + \; ... \; .$$

(3) The greatest term in the expansion is that in which r is the integral part of $\dfrac{(n+1)q/p}{1+q/p}$, that is of $(n+1)q$. Hence *the most probable case is that of $n - r$ successes and r failures* where r is the integral part of $(n + 1)q$.

If nq is an integer, then $r = nq$ and $n - r = np$, so that *in the most probable case the ratio of the number of successes to the number of failures is $p : q$.*

(4) It may be useful to state an important theorem due to Bernoulli: If the chance of an event happening on a single trial is p and if a number of independent trials are made, the probability that the ratio of the number of successes to the number of trials differs from p by less than any assigned number, however small, can be made as near to certainty as we choose, by making the number of trials sufficiently great.

The proof of this is beyond our present scope, and requires a use of approximate values of large factorials. For instance, we may use Stirling's theorem, which states that

$$\lfloor n = \sqrt{2\pi n} \cdot n^n e^{-n}\left(1 + \frac{1}{12n} + \frac{1}{288n^2} - ...\right).$$

(5) *If $u_{n,r}$ is the chance that the event happens* at least r *times in n trials, we shall prove that*

$$(i) \quad u_{n,r} = p^n + C_1^n p^{n-1} q + C_2^n P^{n-2} q^2 + \; ... \; + C_{n-r}^n p^r q^{n-r},$$

$$(ii) \quad u_{n,r} = p^r \left\{ 1 + rq + \frac{r(r+1)}{\lfloor 2} q^2 + ... + \frac{r(r+1)...(n-1)}{\lfloor n-r} q^{n-r} \right\}.$$

For if the event happens *at least r times,* then *(i)* it must happen *exactly n* times, or *exactly (n – 1)* times, or *exactly (n – 2)* times, ... or *exactly r* times, and the chances that these things occur are the successive terms of the first expression.

Or, *(ii)* it must happen *exactly r* times in *just r* trials, or in *just (r + 1)* trials, or in *just (r + 2)* trials, ... or in *just n* trials. Now that it may happen *exactly r* times in *just (r + s)* trials, it must happen in the last of these and also in $(r - 1)$ out of the preceding

$(r + s - 1)$ trials. The chance of this is $pC_{r-1}^{r+s-1} p^{r-1} q^s = C_{r-1}^{r+s-1} p^r q^s$; and therefore

$$u_{n,r} = p^r + C_{r-1}^r p^r q + C_{r-1}^{r+1} p^r q^2 + ... + C_{r-1}^{n-1} p^r q^{n-r}$$

$$= p^r \{ 1 + C_1^r q + C_2^{r+1} q^2 + ... + C_{n-r}^{n-1} q^{n-r} \}$$

which is the second of the expressions in question. That these two expressions are equal may be proved as in Exercise XXXV, Ex. 9.

▌ EXAMPLE 1. *If a die is thrown 5 times, what is the chance that ace turns up (i) exactly 3 times, (ii) at least 3 times?*

Let x and y be the required chances, then

$$x = C_3^5 \left(\frac{1}{6}\right)^3 \left(\frac{5}{6}\right)^2 = \frac{250}{6^5},$$

and

$$y = \left(\frac{1}{6}\right)^5 + C_1^5 \left(\frac{1}{6}\right)^4 \cdot \frac{5}{6} + C_1^5 \left(\frac{1}{6}\right)^3 \cdot \left(\frac{5}{6}\right)^2 = \frac{46}{6^4}$$

or by the second formula,

$$y = \left(\frac{1}{6}\right)^3 \left\{1 + 3 \cdot \frac{5}{6} + \frac{3 \cdot 4}{1 \cdot 2}\left(\frac{5}{6}\right)^3\right\} = \frac{46}{6^4}.$$

▮ **EXAMPLE 2.** *A and B throw alternately with a single die, A having the first throw. The person who first throws ace is to receive £1. What are their expectations?*

The chances that the stake is won at the first, second, third, ... throws are

$$\frac{1}{6}, \; \frac{5}{6} \cdot \frac{1}{6}, \; \left(\frac{5}{6}\right)^2 \cdot \frac{1}{6}, \; \left(\frac{5}{6}\right)^3 \cdot \frac{1}{6}, \ldots;$$

$$\therefore A\text{'s chance} = \frac{1}{6} + \left(\frac{5}{6}\right)^2 \cdot \frac{1}{6} + \left(\frac{5}{6}\right)^4 \cdot \frac{1}{6} + \ldots = \frac{1}{6} \cdot \frac{1}{1 - \left(\frac{5}{6}\right)^2} = \frac{6}{11};$$

$$B\text{'s chance} = 1 - \frac{6}{11} = \frac{5}{11}.$$

Hence, the expectations of A and B are $£\dfrac{6}{11}$ and $£\dfrac{5}{11}$ respectively.

▮ **EXAMPLE 3. 'The Problem of Points.'** *A and B play a series of games which cannot be drawn, and p, q are their respective chances of winning a single game. What is the chance that A wins m games before B wins n games?*

Here A must win at least m games out of $m + n - 1$, and the required chance is obtained by substituting these numbers for r and n in either of the formulae in (5).

(6) Several Alternatives. So far, at each trial there have been two alternatives—the event may succeed or it may fail.

Consider the case of an experiment which at any trial must produce one of three results denoted by A, B, C. Let the chances that A, B, C occur at any trial be p, q, r respectively. One of the three results has to occur, therefore, $p + q + r = 1$. Also the chance that in n trials A occurs α times, B occurs β times and C occurs γ times (where $\alpha + \beta + \gamma = n$) is

$$\frac{\lfloor n}{\lfloor \alpha \lfloor \beta \lfloor \gamma} p^\alpha q^\beta r^\gamma,$$

that is to say, the chance is the term containing $p^\alpha q^\beta r^\gamma$ in the expansion of

$$(p + q + r)^n.$$

For the chance that A may happen α times, B β times and C γ times *in a specified order* is $p^{\alpha}q^{\beta}r^{\gamma}$.

Now the number of orders in which these things can occur is the number of arrangements of α things of one sort (A), β things of another sort (B), and γ things of a third sort (C), placed in a row.

This number is $\lfloor n / \lfloor \alpha \rfloor \beta \rfloor \gamma$. Hence the result follows.

■ **EXAMPLE 4.** *A card is drawn from a pack, the card is replaced and the pack shuffled. If this is done six times, what is the chance that the cards drawn are 2 hearts, 2 diamonds and 2 black cards?*

At any trial, the chance that a heart or a diamond is drawn is $\dfrac{13}{52} = \dfrac{1}{4}$; also the

chance that a black card is drawn is $\dfrac{1}{2}$. Hence by the preceding,

$$\text{required chance} = \frac{\lfloor 6}{\lfloor 2 \rfloor 2 \rfloor 2} \cdot \left(\frac{1}{4}\right)^2 \cdot \left(\frac{1}{4}\right)^2 \cdot \left(\frac{1}{2}\right)^2 = \frac{45}{512}.$$

■ **EXAMPLE 5.** *A 'hand' of six cards is dealt from a pack in the ordinary way. Find the chance that it consists of 2 hearts, 2 diamonds and 2 black cards.*

The chance that the

first	card dealt is a heart	is	$\dfrac{13}{52}$,
second	card dealt is a heart	is	$\dfrac{12}{51}$,
third	card dealt is a diamond	is	$\dfrac{13}{50}$,
fourth	card dealt is a diamond	is	$\dfrac{12}{49}$,
fifth	card dealt is a black	is	$\dfrac{26}{48}$,
sixth	card dealt is a black	is	$\dfrac{25}{47}$,

Hence the chance that 2 hearts, 2 diamonds and 2 black cards are dealt *in this or in any specified order* is

$$\frac{13}{52} \cdot \frac{12}{51} \cdot \frac{13}{50} \cdot \frac{12}{49} \cdot \frac{26}{48} \cdot \frac{25}{47}.$$

The required chance is obtained by multiplying this number by $\lfloor 6 / \lfloor 2 \rfloor 2 \rfloor 2$; and is

therefore equal to $\dfrac{7605}{78302}.$

▌ **EXAMPLE 6.** *Find the chance that in a game of whist the dealer (A) has exactly two honours.*

The card turned up is, or is not, an honour; and the chances of these possibilities are respectively $\dfrac{4}{13}$ and $\dfrac{9}{13}$.

In the first case A certainly has one honour. Of the remaining 51 cards, 12 belong to *A* and 39 to the other players.

Let p to the chance that, of the 3 remaining honours, 1 is among the 12 cards and 2 are among the 39, then

$$p = C_1^3 \cdot \frac{12}{51} \cdot \frac{39}{50} \cdot \frac{38}{49}.$$

In the second case, where the card turned up is not an honour, if p' is the chance that 2 honours belong to *A* and 2 to the other players, we have

$$p' = C_2^4 \cdot \frac{12}{51} \cdot \frac{11}{50} \cdot \frac{39}{49} \cdot \frac{38}{48}.$$

∴ The required chance $= \dfrac{4}{13}p + \dfrac{9}{13}p' = \dfrac{2223}{8330}.$

▪ **8. Probability of Causes:**

(1) Suppose that an event has happened which must have arisen from one of a certain number of causes $C_1, C_2, \ldots$.

What is the probability that a specified cause, C_1, actually led to the event?

This question is said to be one of *inverse probability.*

▌ **EXAMPLE 1.** *Each of three A, B, C contains white and black balls, the numbers of which are as follows:*

		A	B	C
white	—	—	a_1	a_2	a_3
black	—	—	b_1	b_2	b_3

A boy is chosen at random, a ball is drawn from it and is found to be white.

It is required to find the probabilities Q_1, Q_2, Q_3 that the ball came from A, B, C respectively.

If the numbers of the balls are altered as below, the probabilities in question remain unchanged:

		A	B	C
white	—	—	$a_1 x$	$a_2 y$	$a_3 z$
black	—	—	$b_1 x$	$b_2 y$	$b_3 z$

when x, y, z are any numbers whatever. Choose these so that

$$x(a_1 + b_1) = y(a_2 + b_2) = z(a_3 + b_3).$$

The three bags now contain the same number of balls, therefore any one of the $(a_1 x + a_2 y + a_3 z)$ white balls is as likely to be drawn as another.

If the ball which was drawn came from *A,* then it belonged to the group $a_1 x$ of white balls, and so for the other possibilities.

Therefore $\qquad Q_1 : Q_2 : Q_3 = a_1 x : a_2 y : a_3 z = p_1 : p_2 : p_3,$

where $\qquad p_1 = a_1/(a_1 + b_1), p_2 = a_2/(a_2 + b_2), p_3 = a_3/(a_3 + b_3).$

Now a white ball is drawn, therefore

$$Q_1 + Q_2 + Q_3 = 1 \quad \text{and} \quad Q_1 = p_1/(p_1 + p_2 + p_3),$$

with similar values for $Q_2, Q_3.$

It is to be noticed that p_1 *is the probability that the event will occur on the supposition that the ball comes from A, and so for* $p_2, p_3.$

■ **EXAMPLE 2.** *The same as the preceding, except that there are* m_1 *bags such as A containing* a_1 *white and* b_1 *black balls,* m_2 *such as B, and* m_3 *such as C, as in Ex. 1. Also* Q_1, Q_2, Q_3 *are the chances that the ball came from an A, B, C bag respectively.*

Alter the numbers of the balls as in the preceding. Then any one of the

$$m_1a_1x + m_2a_2y + m_3a_3z$$

white balls is as likely to be drawn as another, and Q_1 is the chance that it comes from one of the m_1 groups of a_1x balls, so also for Q_2, Q_3 therefore

$$Q_1 : Q_2 : Q_3 = m_1a_1x : m_2a_2y : m_3a_3z$$
$$= m_1p_1 : m_2p_2 : m_3p_3,$$

and as before, $\qquad Q_1 + Q_2 + Q_3 = 1,$

therefore $\qquad Q_1 = m_1p_1/(m_1p_1 + m_2p_2 + m_3p_3),$

with similar values for $Q_2, Q_3.$

> **Note:** If P_1 is the chance estimated *before* the event that an A bag will be chosen and P_2, P_3 have similar meanings, we have
> $$P_1 : P_2 : P_3 = m_1 : m_2 : m_3 ;$$
> therefore $\qquad Q_1 : Q_2 : Q_3 = P_1p_1 : P_2p_2 : P_3p_3$
> and $\qquad Q_1 = P_1p_1/(P_1p_1 + P_2p_2 + P_3p_3),$ etc.
> Observe that P_1p_1 is the chance that a white ball is drawn, and that from an A bag.

(2) A General Statement. Suppose that an event has occurred which must have been due to one of the causes, $C_1, C_2, \dots C_n.$

Let p_r be the probability of the existence of the cause C_r, estimated *before* the event took place.

Let p_r be the probability of the event on the assumption that the cause C_r exists.

Then the probability Q_r *of the existence of the cause* C_r, *estimated* after *the event has occurred, is given by*

$$Q_r = P_rp_r/(P_1p_1 + P_2p_2 + \dots P_np_n).$$

For an argument similar to that in the last two examples shows that

$$Q_1 : Q_2 : Q_3 : \dots = P_1p_1 : P_2p_2 : P_3p_3 : \dots ,$$

and since the event *has* happened,

$$Q_1 + Q_2 + \dots + Q_n = 1,$$

whence the result follows.

> **Note:** It is usual to call $P_1, P_2, \dots P_n$ the *a priori* probabilities of the existence of the causes and $Q_1, Q_2, \dots Q_n$ the *a posteriori* probabilities.
> The product P_rp_r is the *antecedent probability* that the event will occur, and that from the rth cause.

The argument depends on the assumption (1) that $Q_1, Q_2, \ldots$ *are proportional to* $P_1 p_1, P_2 p_2, \ldots$, which is justified in (1).

In particular, if an event is due to one of *two causes,* the odds in favour of its having occurred from the first cause are as $P_1 p_1 : P_2 p_2$.

The way in which these principles are applied to determine the *Probability of Future Events* is illustrated in Ex. 4.

EXAMPLE : *A bag contains 5 balls, and of these it is equally likely that 0, 1, 2, 3, 4, 5 are white. A ball is drawn and is found to be white. What is the chance that this is the only white ball?*

There are 6 possible hypotheses; the number of white balls may be 0, 1, 2, 3, 4, 5.

Denoting these possibilities by $C_0, C_1, \ldots C_5$, and using a notation similar to the above, we have

$$P_0 = P_1 = \ldots = P_5 = \frac{1}{6} \; ; \text{ and } P_0 = 0, \quad p_1 = \frac{1}{5}, \quad p_2 = \frac{2}{5}, \ldots, p_5 = \frac{5}{5},$$

$$\therefore \qquad P_0 p_0 + P_1 p_1 + \ldots + P_5 p_5 = \frac{1}{6} \cdot \frac{1+2+3+4+5}{5} = \frac{1}{2};$$

$$\therefore \text{ The required chance} = Q_1 = P_1 p_1 \div \frac{1}{2} = \frac{1}{15}.$$

EXAMPLE 2. *If in the last example the ball which was drawn is replaced, what is the chance that a second drawing will give a white ball?*

$$Q_0 p_0 + Q_1 p_1 + \ldots + Q_5 p_5 = \frac{2}{6}(p_0^2 + p_1^2 + \ldots + p_5^2)$$

$$= \frac{2}{6 \cdot 5^2}(1^2 + 2^2 + 3^2 + 4^2 + 5^2).$$

Hence the required chance $\qquad = \dfrac{11}{15}.$

EXAMPLE 3. *A pack of cards is counted, face downwards, and it is found that one card is missing. Two cards are drawn and are found to be spades. What are the odds against the missing card being a spade?*

The 'event' is that two spades are drawn. There are two possible hypotheses:

(C_1) The missing card is a spade.

(C_2) The missing card is not a spade.

The *a priori* probabilities P_1, P_2 of C_1, C_2 are $P_1 = \dfrac{1}{4}$, $P_2 = \dfrac{3}{4}.$

The chances p_1, p_2 of the event under the hypotheses C_1, C_2 are

$$p_1 = \frac{12}{51} \cdot \frac{11}{50}, \quad p_2 = \frac{13}{51} \cdot \frac{12}{50}.$$

The odds against the missing card being a spade are as

$$P_2 p_2 : P_1 p_1 = \frac{3}{4} \cdot \frac{13}{51} \cdot \frac{12}{50} : \frac{1}{4} \cdot \frac{12}{51} \cdot \frac{11}{50} = 39 : 11.$$

■ **EXAMPLE 4.** *A bag contains m balls which are either white or black, all possible numbers being equally likely. If p white and q black balls have been drawn in p + q successive trials without replacement, the chance that another drawing will give a white ball is*

$$(p + 1)/(p + q + 2).$$

The possible hypotheses, all equally likely, are that the number of white balls are

$$m - q, \quad m - q - 1, \; ... \quad m - q - r + 1, \; ... \; p.$$

Denote these by $C_1, C_2, \; ... \; C_r, \; .... \; C_{m-q-p+1}.$

If $p_1, p_2, \; ...$ are the antecedent probabilities of the event under these hypotheses,

we have
$$p_r = C_p^{m-q-r+1} \cdot C_q^{q+r-1}/C_{p+q}^m = A u_r v_r,$$

where
$$u_r = r(r + 1) (r + 2) \; ... \; (r + q - 1),$$
$$v_r = (n - r) (n - r + 1) (n - r + 2) \; ... \; (n - r + p - 1),$$
$$n = m - q - p + 2,$$

and A is independent of r.

If $Q_1, Q_2, \; ...$ are the *a posteriori* probabilities of $C_1, C_2, \; ... \;$, then $Q_r = u_r v_r / S$,

where
$$S = \sum_{r=1}^{r=n-1} u_r v_r = \frac{\lfloor p \lfloor q}{\lfloor p+q+1} \cdot \frac{\lfloor m+1}{\lfloor n-2} \qquad \text{(by Ch. 21, 10, Ex. 1.)}$$

If p_r' is the chance that, with hypothesis C_r, another drawing gives a white ball,

$$p_r' = \frac{m - q - p - r + 1}{m - q - p} = \frac{n - 1 - r}{m - q - p};$$

thus $p_{n-1} = 0$, and the chance that another drawing gives a white ball is

$$\sum_{r=1}^{r=n-2} Q_r p_r' = S' / (m - q - p) S, \text{ where } S' = \sum_{r=1}^{r=n-2} u_r (n - 1 - r) (n - r) ...$$
$$(n - r + p).$$

Thus S' can be obtained from S by writing $n - 1$ for n and p for $p - 1$; and hence

$$S' = \frac{\lfloor p+q+2}{\lfloor p+1 \lfloor q} \cdot \frac{\lfloor m+1}{\lfloor n-3} \text{ and } \frac{S'}{S} = \frac{(p+1)(n-2)}{p+q+2}; \text{ the chance} = (p + 1)/(p + q + 2).$$

■ **9. Value of Testimony:** The theory of probability has been used to estimate the value of the testimony of witnesses. Such an application is open to adverse criticism. It rests on two assumptions which can hardly be justified, namely: (*i*) that to each witness there pertains a constant p (his *credibility*), which measures the average frequency with which he speaks the truth; (*ii*) that the statements of witnesses are independent of one another in the sense required in the theory of probability.

If we are prepared to make these assumptions, the procedure is as follows.

■ **EXAMPLE 1.** *If p is the probability that a statement made by A is true and p' has a similar meaning for B, what are the odds in favour of the truth of a statement which A and B concur in making?*

The 'event' is the agreement of A and B in making a certain statement. The possibilities are: (*i*) the statement is true; (*ii*) it is false.

If it is true, the chance that they both say it is true is pp'.

If it is false, the chance that they both say it is true is $(1 - p)(1 - p')$.

Thus the *antecedent probabilities* of the event on the hypotheses (*i*), (*ii*) are
$$pp' \quad \text{and} \quad (1 - p)(1 - p'),$$
and the odds in favour of the truth of the statement are as $pp' : (1 - p)(1 - p')$.

EXAMPLE 2. *A bag contains n balls, one of which is white. The probabilities that A and B speak the truth are p, p′ respectively. A ball is drawn from the bag, and A and B both assert that it is white. What are the odds in favour of its being white?*

The 'event' is the agreement of A and B in making a statement. The possibilities are (*i*) it is true, (*ii*) it is false.

The *a priori* probabilities P_1, P_2, of (*i*), (*ii*) are $P_1 = 1/n$, $P_2 = (n - 1)/n$.

If p_1, p_2 are the chances of the event under (*i*) and (*ii*), we have $p_1 = pp'$.

In the case of (*ii*), $(n - 1)$ balls remain in the bag, and one of these is white. The chance that A should choose this ball and wrongly assert that it was drawn from the bag is $(1 - p)(n - 1)$. The chance that B should do the same thing is $(1 - p')(n - 1)$;

hence
$$p_2 = (1 - p)(1 - p')(n - 1)^2,$$
and the odds in favour of a white ball having been drawn are as

$$P_1 p_1 : P_2 p_2 = \frac{1}{n}pp' : \frac{n-1}{n} \cdot (1 - p)(1 - p') \cdot \frac{1}{(n-1)^2}$$
$$= (n - 1)pp' : (1 - p)(1 - p').$$

■ **10. Geometrical Applications:** The following statements are axiomatic:

(*i*) If a point is taken at random on a given straight line AB, the chance that it falls on a particular segment PQ of the line is PQ/AB.

Or we may say that the total number of cases is represented by AB, and the number of favourable cases by PQ.

(*ii*) If a point is taken at random on an area S which includes an area σ, the chance that the point falls on σ is σ/S.

EXAMPLE 1. *Given that x + y = 2a where a is constant and that all values of x between 0 and 2a are equally likely, show that it is an even chance that* $xy > \dfrac{3}{4}a^2$.

Let AB be a diameter of a circle with centre O and radius a. Take a point P at random in AB.

Let $AP = x$, $PB = y$, then $x + y = 2a$, and all values of x between 0 and $2a$ are equally likely.

Draw the ordinate PQ, then $PQ^2 = AP \cdot PB = xy$.

If A', B' are the mid-points of OA, OB, the ordinates

at these points are equal to $a \cdot \sqrt{\dfrac{3}{4}}$.

Hence $PQ > a\sqrt{\dfrac{3}{4}}$ if, and only if, P lies in $A'B'$.

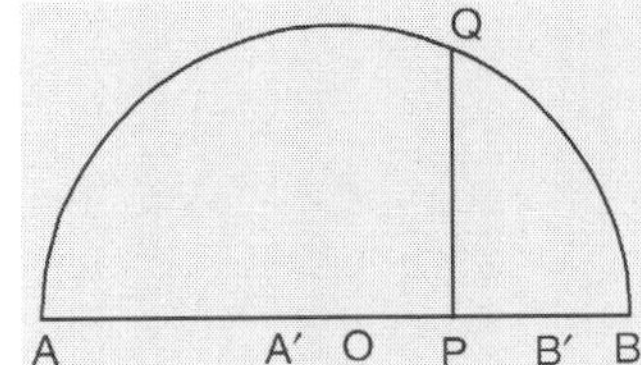

Fig. 86

Hence the chance that $xy > \dfrac{3}{4}a^2$ is $A'B'/AB$, that is $\dfrac{1}{2}$.

■ **EXAMPLE 2.** *Two points P, Q are taken at random on a straight line OA of length a. Show that the chance that PQ>b, where b<a, is $(a - b)^2/a^2$.*

The points are as likely to fall in the order O, P, Q, A as in the order O, Q, P, A.

We may therefore suppose that Q is to the right of P.

Draw OA' at right angles to OA and equal to it.

Complete the figure as in the diagram, where

$$OL = PR = b.$$

If δx is small, the number of cases in which the distance of P from O lies between x and $x + \delta x$ and Q is in PA, is represented by δx. PA, i.e. by the area of the shaded rectangle.

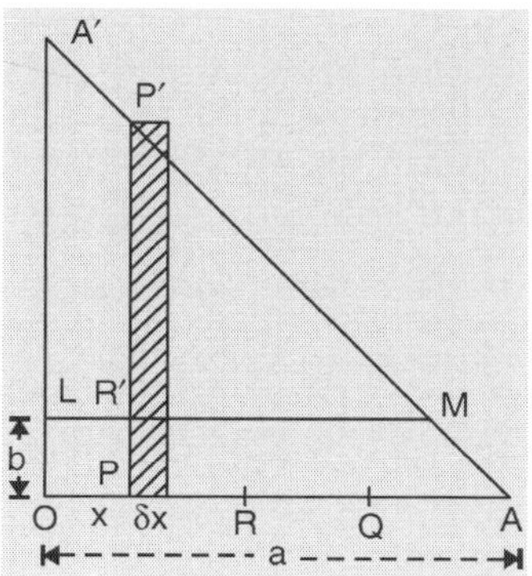

Fig. 87

Of these, the favourable cases are those in which Q lies in RA, and their number is represented by the upper part of the shaded rectangle cut off by LM.

Hence the total number of cases is represented by area of the triangle OAA', and the total number of favourable cases by the area of the triangle LMA';

$$\therefore \quad \text{The required chance} \ = \ \frac{\Delta LMA'}{\Delta OAA'} \ = \ \left(\frac{a-b}{a}\right)^2 .$$

EXERCISE LV

1. If two balls are drawn from a bag containing 2 white, 4 red and 5 black balls, what is the chance that

    (*i*) they are both red?

    (*ii*) one is red and the other black?

2. A bag contains m white and n black balls. If $p + q$ balls are drawn without replacement, the chance that these consist of p white and q black balls is $C_p^m \cdot C_q^n / C_{p+q}^{m+n}$.

3. If four balls are drawn from a bag containing 3 white, 4 red and 5 black balls, what is the chance that *exactly* two of them are black? (*i.e.* two are black and two are white or red.)

4. In the last question what is the chance that *at least* three of the four balls drawn from the bag are black?

5. Out of 20 consecutive numbers, two are chosen at random; prove that the chance that their sum is odd is $\dfrac{10}{19}$.

6. In a throw with three dice what is the chance that the dice fall (*i*) all alike, (*ii*) two alike and the third different, (*iii*) all different?

    Explain why the sum of these probabilities is unity, and verify the truth of this statement.

    [(*ii*) Any particular throw of this sort, as five-five-six, occurs in 3 cases, and there are $6 \cdot 5$ throws of this kind.]

7. What is the chance of throwing 10 in a single cast with three dice?

8. In a single cast with four dice what is the chance of throwing (*i*) two doublets, (*ii*) exactly 12, (*iii*) less than 12?

9. How many times must a pair of dice be thrown that it may be at least an even chance that double six may occur at least once?

10. If four cards are drawn from a pack, what is the chance that they are ace, king, queen, knave (of the same or of different suits)?

11. A letter is chosen at random out of 'assinine' and one is chosen at random out of 'assassin'. Show that the chance that the same letter is chosen on both occasions is $\frac{1}{4}$.

[The number of possible cases is $7 \cdot 8$. Of these 14 are favourable.]

12. If the letters of 'attempt' are written down at random, find the chance that (*i*) all the *t*'s are together, (*ii*) no two *t*'s are together.

13. A party of n men of whom A, B are two, form single rank. What is the chance that (*i*) A, B are next one another, (*ii*) that exactly m men are between them, (*iii*) that not more than m men are between them?

14. From a bag containing a balls, m balls are drawn simultaneously and replaced; then n balls are drawn. Show that the chance that exactly r balls are common to the two drawings is $C_r^m C_{n-r}^{a-m} / C_n^a$. Prove also that this expression is unaltered by interchanging m and n.

[Consider the second drawing. There are C_n^a possible cases. The favourable cases consist of selections of r balls out of the m already drawn, with selections of $n - r$ out of $a - m$ not already drawn.]

15. A squad of $4n$ men form fours. Find the chance that two specified men are next one another in the same four.

16. If two squares are chosen at random on a chess board, show that

(*i*) the chance that they have a side common is $\dfrac{1}{18}$;

(*ii*) the chance that they have contact at a corner is $\dfrac{7}{144}$.

[(*i*) Out of 64×63 possible cases, the number of favourable cases is $4 \cdot 2 + 24 \cdot 3 + 36.4$.]

17. One bag contains a white balls and b black balls; another contains a' white and b' black balls.

Let p be the chance of drawing a white ball from a bag chosen at random, and let p' be the chance of drawing a white ball from a bag containing $(a + a')$ white and $(b + b')$ black balls.

Find p, p', and show that $p \lessgtr p'$ according as $a + b - a' - b'$ and $ab' - a'b$ have the same or opposite signs.

18. If A's chance of winning a single game against B is $\frac{2}{3}$, find A's chance of winning (*i*) two games (at least) out of the first three; (*ii*) four games (at least) out of the first six.

19. A draws a card from a pack of n cards marked $1, 2, 3, \dots n$; the card is replaced in the pack and then B draws a card. Find the chance that A draws (*i*) the same card as B;

(*ii*) a higher card than B; (*iii*) a lower card than B. Verify that the sum of these chances is unity.

$$\left[\text{ For } (ii),\ \text{chance} = \frac{1}{n}\left\{\frac{1}{n}+\frac{2}{n}+\ldots+\frac{n-1}{n}\right\}.\right]$$

20. A bag contains n cards marked 1, 2, 3, ... n. If A is to draw all the cards in succession and is to receive 1 shilling for every card which comes out in its proper order, prove that his expectation is 1 shilling.

[The expectation arising from each ball is $\dfrac{1}{n}$ of a shilling, etc. The fact that the probabilities, on which the several expectations depend, are not independent does not affect the argument.]

21. A bag contains n tickets numbered 1, 2, 3, ... n. A person draws two tickets at once, and is to receive a number of shillings equal to the product of the numbers drawn. What is his expectation?

22. What is the chance that a hand of five cards contains (*i*) exactly two aces; (*ii*) at least two aces?

[For (*ii*), number of favourable cases $= C_2^4 \cdot C_3^{48} + C_3^4 \cdot C_2^{48} + 48.$]

23. If p_1, p_2, p_3, p_4 are the probabilities of four independent events, find the chance that (*i*) two and not more than two of the events happen; (*ii*) at least two of the events happen.

24. Two persons each make a single throw with a pair of dice. What is the chance that the throws are equal?

$$\left[\text{ If } (x + x^2 + \ldots + x^6)^2 = a_2 x^2 + \ldots + a_{12} x^{12},\ \text{then the chance} = \frac{1}{36^2}(a_2^2 + \ldots + a_{12}^2):\right.$$
$$\left.\text{and this is equal to } \left\{\text{coefficient of } x^{10} \text{ in } \left(\frac{1-x^6}{1-x}\right)^4\right\}/36^2.\right]$$

25. The decimal part of the logarithms of two numbers taken at random are found to seven places. What is the chance that the second can be subtracted from the first without 'borrowing'?

26. In a game of whist, find the chance that each of the four players should have an honour:

    (*i*) if the card turned up is an honour;

    (*ii*) if this card is not an honour;

    (*iii*) before the last card is turned up.

27. In a game of whist, find the chance that each party has two honours:

    (*i*) if the card turned up is an honour;

    (*ii*) if this card is not an honour;

    (*iii*) before the last card dealt is turned up.

28. Out of n persons sitting at a round table, three, A, B, C are chosen at random. Prove that the chance that no two of these three are sitting next one another is

$$(n-4)(n-5)/(n-1)(n-2).$$

[Relative to A, B can sit in $n-3$ places. For 2 of them, C can sit in $n-5$ places; for $n-5$ of them, C can sit in $n-6$ places. Hence, the number of favourable cases $= 2(n-5) + (n-5)(n-6).$]

29. Three cards are drawn from a bag containing n cards marked 1, 2, 3 ... n. Find the chance that (i) the three cards form a sequence; (ii) they contain a sequence of two; (iii) they involve no sequence.

30. In a set of tennis between A and B, the chance that the server wins a game is always p, the chance that he loses is q. The score is five all, they proceed to play deuce and vantage games and A has the service. Denoting A's chance of winning the set by x, show that $x = pq + (p^2 + q^2)x$, whence it follows that A and B have equal chances of winning the set.

31. A and B play with two dice on the condition that A wins if he throws 6 before B throws 7 and B wins if he throws 7 before A throws 6. Show that A's chance is to B's chance as 30 : 31. (Huyghens.)

32. Three players A, B, C of equal skill engage in a match consisting of a series of games in which A plays B, B plays C, C plays A, A plays B, and so on, in rotation. The player who first wins two consecutive games wins the match. Prove that the chances which A, B, C respectively have of winning the match are as 12 : 20 : 17.

[First show that if A loses the first game, his chance of winning the match is $\frac{1}{7}$: if he wins the first game, his chance is $\frac{17}{49}$.]

33. A bag contains 3 balls, and it is equally likely that 1, 2 or all of them are white. A ball is drawn and is found to be white.

    (i) What is the chance that this is the only white ball?

    (ii) What is the chance that another drawing will give a white ball?

34. The same question as the last, except that each ball is as likely as not to be white. [Proceeding as in Ex. 1, p. 513, we see that P_0, P_1, P_2, P_3 are the terms in the expansion of $\left(\frac{1}{2} + \frac{1}{2}\right)^3$. The rest proceeds as in Ex. 2, p. 514.]

35. A bag contains 5 balls, and it is not known how many of these are white. Two balls are drawn, and these are white. What is the chance that all are white?

36. The chances that A, B, C speak the truth are respectively p, p', p''. What are the odds in favour of an event actually having happened which

    (i) all three assert to have happened?

    (ii) A, B assert to have happened and C denies?

GEOMETRICAL

37. The sides of a rectangle are chosen at random, each less than a given length a, all such lengths being equally likely. Show that the chance that the diagonal is less than a is $\pi/4$.

[Draw a square, $OACB$, whose side is a; and the arc AB. If $ONPM$ is one of the rectangles, the favourable cases are when P lies in the quadrant OAB.]

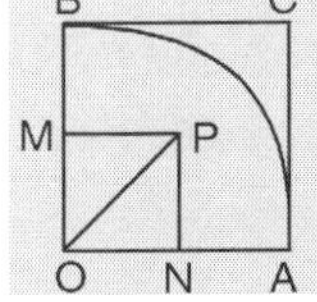

Fig. 88

38. A floor is paved with rectangular bricks, each of length a and breadth b. A circular disc of diameter c is thrown on the floor. Show that the chance that it falls entirely on one brick is $(a - c)(b - c)/ab$.

39. A point P is taken at random in a straight line AB. Show that the chance that the greater of the parts AP, PB is at least k times the smaller is $2/(k + 1)$.

40. If two points are taken at random on the circumference of a circle, the chance that their distance apart is greater than the radius of the circle is $\dfrac{2}{3}$.

[The unfavourable cases are represented by the thick arc in Fig. 89.]

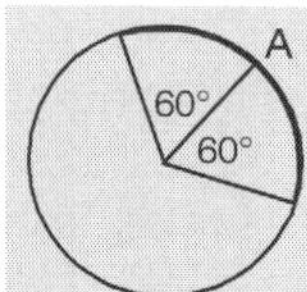

Fig. 89

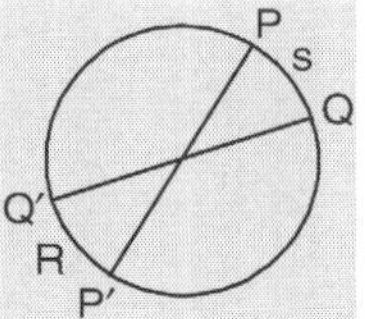

Fig. 90

41. If three points P, Q, R, are taken at random on the circumference of a circle, the chance that they do not lie on the same semicircle is $\dfrac{1}{4}$.

[Choose P. Choose Q on one semicircle on PP'; R must lie on $P'Q'$. (Fig. 90.)

If $PQ = s$, number of favourable cases $= \displaystyle\int_0^{\frac{1}{2}c} s\,ds = \dfrac{1}{8}c^2$; total number $= \dfrac{1}{2}c$, c, etc.]

42. On a straight line AB of length $a + b$, two segments PQ, $P'Q'$, of lengths a, b respectively, are measured at random. (*i*) If $a > b$, the chance that $P'Q'$ lies entirely within PQ is $(a - b)/a$. (*ii*) If c is less than a or b, the chance that the common part of PQ, $P'Q'$ is less than c is c^2/ab.

[(*ii*) Let $AP = x$. If $LQ = c$, $MQ = x$ and $P'Q'$ overlaps PQ towards B, P' must lie in LM. Hence, the number of favourable cases $= \displaystyle\int_0^c (c - x)\,dx = \dfrac{1}{2}c^2$;

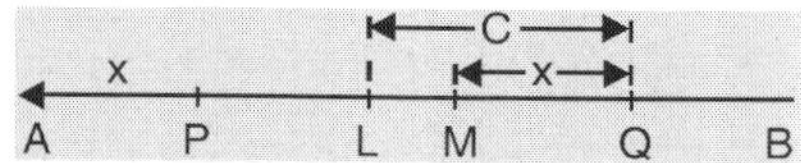

Fig. 91

... *total* number of favourable cases $= c^2$. Total number of cases $= ab$, etc.]

■ ■ ■

CHAPTER 33

Continued Fractions

Expression of a Quadratic Surd as a Simple Continued Fraction

■ **1. Surds of the Form $\pm (\sqrt{N} \pm b_1)/r_1$:** From Ch. 24, 33, it follows that *any simple recurring continued fraction is equal to a quadratic surd.* In other words, its value is of the form $\pm (\sqrt{N} \pm b_1)/r_1$, where N, b_1, r_1 are positive integers, except that b_1 may be zero, and N is not a perfect square.

We shall prove that conversely *any positive number of the form*

$$\pm (\sqrt{N} \pm b_1)/r_1$$

can be expressed as a simple recurring continued fraction.

In considering this theorem, it is to be observed that:

(*i*) *There is no loss of generality in assuming that $N - b_1^2$ is divisible by r_1.*

For $(\sqrt{N} \pm b_1)/r_1 = \left(\sqrt{N r_1^2} \pm b_1 r_1\right)\Big/ r_1^2$ and $N r_1^2 - (b_1 r_1)^2$ is divisible by r_1^2, hence $(\sqrt{N} \pm b_1)/r_1$ can always be replaced by an expression of the same form for which the above condition holds.

(*ii*) We need only consider surds which are greater than unity.

For if $N - b_1^2 = \pm r_1 r_2$, where r_2 is a positive integer, then

$$\frac{\sqrt{N} \pm b_1}{r_1} = 1 \div \frac{\sqrt{N} \mp b_1}{r_2}$$

It follows that every positive quadratic surd or its reciprocal is of one of four types considered in the next article.

■ **2. Types of Quadratic Surds:** In every case it is assumed that

(*i*) *The surd in question is greater than unity.*

(*ii*) $N - b_1^2$ is divisible by r_1.

(**A**) *The type $(\sqrt{N} + b_1)/r_1$ where $b_1^2 < N$.* This will be called the *normal* type, and is of special importance in the theory: it includes the forms $\sqrt{A/B}$ and $\sqrt{N}$.

To express $(\sqrt{N} + b_1)/r_1$ as a simple continued fraction, we form the equations

$$\frac{\sqrt{N} + b_1}{r_1} = a_1 + \frac{\sqrt{N} - b_2}{r_1} = a_1 + \frac{r_2}{\sqrt{N} + b_2}, \qquad \text{... (A)}$$

where a_1 is the integral part of $(\sqrt{N} + b_1)/r_1$ and

$$b_2 = a_1 r_1 - b_1, \quad r_1 r_2 = N - b_2^2. \qquad \text{... (B)}$$

We shall first show that b_2, r_2 *are positive integers.*

Since a_1 is the integral part of $(\sqrt{N} + b_1)/r_1$,

$$a_1 r_1 < \sqrt{N} + b_1 < a_1 r_1 + r_1, \qquad \ldots \text{(C)}$$

and therefore $\qquad\qquad b_2 < \sqrt{N} < b_2 + r_1. \qquad\qquad\qquad \ldots \text{(D)}$

If we suppose that $b_2 \leq 0$, it follows that $\sqrt{N} < r_1$, but $b_1 < \sqrt{N}$, therefore $b_1 < r_1$ and consequently $b_1 < a_1 r_1$. This is contrary to the supposition that $b_2 \leq 0$. Hence b_2 is positive, and it is obviously an integer.

Again, b_2 is a positive number less than $\sqrt{N}$, therefore $N - b_2^2 > 0$, and consequently r_2 is positive. Further, we have

$$N - b_2^2 = N - (a_1 r_1 - b_1)^2 = N - b_1^2 - r_1 (a_1^2 r_1 - 2a_1 b_1),$$

and since $N - b_1^2$ is divisible by r_1, so also is $N - b_2^2$. Hence r_2 is a positive integer. Continuing the process, we form the equations

$$\frac{\sqrt{N} + b_n}{r_n} = a_n + \frac{\sqrt{N} - b_{n+1}}{r_n} = a_n + \frac{r_{n+1}}{\sqrt{N} + b_{n+1}}, \qquad \ldots \text{(E)}$$

for $n = 2, 3, \ldots$ where a_n is an integer such that

$$a_n < (\sqrt{N} + b_n)/r_n < a_n + 1, \qquad \ldots \text{(F)}$$

and $\qquad\qquad b_{n+1} = a_n r_n - b_n, \quad r_n r_{n+1} = N - b_{n+1}^2, \qquad \ldots \text{(G)}$

leading to

$$\frac{\sqrt{N} + b_1}{r_1} = a_1 + \frac{1}{a_2 +} \cdots \frac{1}{a_n +} \cdots . \qquad \ldots \text{(H)}$$

By steps similar to the above, we can show that *every a, b, r is a positive integer.* For taking $n = 2$, by the preceding $(\sqrt{N} + b_2)/r_2 > 1$, therefore a_2 is a positive integer. Also $N - b_2^2$ is divisible by r_2. Reasoning as before, it will be seen that b_3 and r_3 are positive integers, and so for $n = 3, 4, 5, \ldots$.

Finally, *the fraction is periodic,* for the nth complete quotient is

$$(\sqrt{N} + b_n)/r_n,$$

and for every n, $b_n < \sqrt{N}$, therefore $r_n = (b_n + b_{n+1})/a_n < 2\sqrt{N}$.

Hence the fraction $(\sqrt{N} + b_n)/r_n$ cannot have more than $2N$ distinct values, and one of the complete quotients must occur again. From this stage, all the succeeding a's recur and in the same order : the same is true for the b's and r's. Thus *the continued fraction is periodic.*

It will now be shown that in the process of expressing a positive quadratic surd of any other type as a simple continued fraction, *a stage must occur at which the complete quotient is a surd of normal type. Whence it will follow that in every case the continued fraction is periodic.*

(B) *The type $(\sqrt{N} - b_1) / r_1$ where $b_1^2 < N$.* Here the second complete quotient is a surd of normal type.

For instance, $\qquad \dfrac{\sqrt{57} - 7}{2} = 1 \Big/ \dfrac{2}{\sqrt{57} - 7} = 1 \Big/ \dfrac{\sqrt{57} + 7}{4} = \dfrac{1}{3+}\dfrac{1}{1+}\dfrac{1}{1+}\dfrac{1}{1+}\dfrac{1}{1+}\dfrac{1}{7}$

and
$$\frac{\sqrt{57}-3}{2} = 2 + \frac{\sqrt{57}-7}{2} = 2 + 1 \bigg/ \frac{\sqrt{57}+7}{4}.$$

(C) *The form* $(b_1 + \sqrt{N})/r_1$ *where* $b_1{}^2 > N$. We form the equations

$$\frac{b_1 + \sqrt{N}}{r_1} = a_1 + \frac{b_2 + \sqrt{N}}{r_1} = a_1 + \frac{r_2}{b_2 - \sqrt{N}},$$

$$\frac{b_2 - \sqrt{N}}{r_2} = a_2 + \frac{b_3 - \sqrt{N}}{r_2} = a_2 + \frac{r_3}{b_3 + \sqrt{N}}, \text{ etc.,}$$

where a_1, a_2, ... are the integral parts of the fractions on the left and, for every n,

$$b_{n+1} = b_n - a_n r_n, \qquad r_n r_{n+1} = b_{n+1}^2 - N.$$

As in (A), every r is an integer and every b is an integer or zero, and by hypothesis $b_1 > \sqrt{N}$.

Suppose that b_1, b_2, ... b_n are all greater than $\sqrt{N}$; then r_1, r_2, r_n are all positive, and b_1, b_2, ... b_{n+1} is a decreasing sequence of integers. Hence a stage must occur at which

$$b_n > \sqrt{N} > b_{n+1}.$$

This being so, *n cannot be even* ; for in that case we should have

$$b_n - \sqrt{N} > a_n r_n,$$

and therefore $b_{n+1} > \sqrt{N}$.

Hence *n is odd*, and $b_n + \sqrt{N} > a_n r_n$. Consequently $\sqrt{N} + b_{n+1} > 0$; and since also $\sqrt{N} - b_{n+1} > 0$, it follows that $N - b_{n+1}^2 > 0$, and therefore r_{n+1} is negative.

Now $(n + 1)$ being even, the $(n + 1)$th complete quotient may be written

$$(\sqrt{N} - b_{n+1})/(-r_{n+1});$$

and since $b_{n+1}^2 < N$, this or the next quotient is a surd of normal type according as $b_{n+1} \lessgtr 0$.

(D) *The type* $(b_1 - \sqrt{N})/r_1$ *where* $b_1{}^2 > N$. Here the second complete quotient is a surd of type (C).

■ **3. Theorem:** *Let the surd* $(\sqrt{N} + b)/r$ *be expressed as a simple recurring fraction, then (i) if* $b < \sqrt{N} < b + r$, *the fraction has no acyclic part; (ii) if* $\sqrt{N} > b + r$, *it has an acyclic part consisting of a single quotient; (iii) if* $b > \sqrt{N}$, *it has an acyclic part of one or more quotients.*

For if $x = (\sqrt{N} + b)/r$, then $(rx - b)^2 = N$ and the second root of this quadratic is $(-\sqrt{N} + b)/r$; hence, from Chap. 24, 33, it follows that (i) *if* $-1 < (-\sqrt{N} + b)/r < 0$, the fraction has no acyclic part and these conditions are equivalent to $b < \sqrt{N} < b + r$; (ii) *if* $(-\sqrt{N} + b)/r < -1$, that is, if $\sqrt{N} > b + r$, the acyclic part consists of a single quotient; and (iii) *if* $(-\sqrt{N} + b)/r > 0$, that is, if $b > \sqrt{N}$, the acyclic part contains one or more quotients.

In particular, *if* $A > B$, *the simple continued fraction equivalent to* $\sqrt{A/B}$ *has a single non-recurring quotient.*

■ **4. Method of Reckoning:** In practice we replace the written work involving surds by an easy mental process, as in the next example.

■ **EXAMPLE :** *Express $(\sqrt{37} + 8)/9$ as a simple continued fraction.*

Here a_1 = integral part of the surd = 1, also $b_1 = 8$, $r_1 = 9$. The various quantities are now found in succession from the equations

$$b_n = a_{n-1} r_{n-1} - b_{n-1}, \quad r_n = (N - b_n^2)/r_{n-1}, \quad a_n = \text{Integral part of } (\sqrt{N} + b_n)/r_n,$$

giving the table :

n	1	2	:	3	4	5	:	6
b	8	1	:	3	4	5	:	3
r	9	4	:	7	3	4	:	7
a	1	1	:	1	3	2	:	1

The reckoning, if continued, is a repetition of the part between the dotted lines, and

$$\frac{\sqrt{37}+8}{9} = 1 + \cfrac{1}{1+}\cfrac{1}{1+}\cfrac{1}{3+}\cfrac{1}{2+}\cdots .$$

Note: The third complete quotient, $(\sqrt{37} + 3)/7$, marks the beginning of the recurring period, for $3 < \sqrt{37} < 3 + 7$. It will be shown in the last article of this chapter that, *from this stage, the above reckoning can be replaced by a G.C.M. process.*

■ **5. $\sqrt{A/B}$ as a Continued Fraction:**

(1) It is supposed that A, B are positive integers and that $A > B$. We have

$$\sqrt{\frac{A}{B}} = \frac{\sqrt{AB}}{B} = \frac{\sqrt{N}+b_1}{r_1}, \text{ where } N = AB, b_1 = 0, r_1 = B. \text{ The surd is therefore of the}$$

type discussed in Art. 3, and it is expressed as a simple continued fraction by constructing equations of the type.

$$\frac{\sqrt{N}+b_n}{r_n} = a_n + \frac{\sqrt{N}-b_{n+1}}{r_n} = a_n + \frac{r_{n+1}}{\sqrt{N}+b_{n+1}}, \text{ for values of } n = 1, 2, 3 \ldots ;$$

where a_n is an integer such that $a_n < (\sqrt{N} + b_n)/r_n < a_n + 1$,

and
$$b_{n+1} = a_n r_n - b_n, \quad r_n r_{n+1} = N - b_{n+1}^2,$$

leading to
$$\sqrt{\frac{A}{B}} = \frac{\sqrt{N}}{r_1} = a_1 + \cfrac{1}{a_2+}\cdots\cfrac{1}{a_n+}\cdots .$$

As in Art. 2, every a, b and r is a positive integer and the continued fraction is periodic. Also, it has been shown in Art. 3 that a_1 is the only non-recurring quotient.

(2) The following inequalities are required. Every complete quotient is greater than 1, therefore

$$\sqrt{N} + b_n > r_n \qquad \ldots \text{(A)}$$

Also $r_{n-1} = (N - b_n^2)/r_n = (\sqrt{N} - b_n)(\sqrt{N} + b_n)/r_n > \sqrt{N} - b_n$, therefore

$$\sqrt{N} < b_n + r_{n-1}. \qquad \ldots \text{(B)}$$

Again, if $n > 1$, the continued fraction equivalent to $(\sqrt{N} + b_n)/r_n$ has no acyclic part, therefore, as in Art. 3, $b_n < \sqrt{N} < b_n + r_n$ $(n \geq 2)$.   ... (C)

For any suffix m, $b_m < \sqrt{N}$, hence from (B) and (C),

$$b_m - b_n < r_{n-1} \quad \text{and} \quad b_m - b_n < r_n \ (n \geq 2). \qquad \text{... (D)}$$

(3) The Cycle of Quotients: Let c be the number of elements in the cycle, then

$$\frac{\sqrt{N}}{r_1} - a_1 = \frac{1}{a_2 +} \frac{1}{a_3 +} \ldots \frac{1}{a_{c+1}}.$$

Now $\sqrt{N}/r_1 - a_1$ is a root of $r_1{}^2 (x + a_1)^2 = N$, the other root of this equation being $-\sqrt{N}/r_1 - a_1$. Therefore by Ch. 24, 31,

$$\frac{\sqrt{N}}{r_1} + a_1 = a_{c+1} + \frac{1}{a_c +} \frac{1}{a_{c-1} +} \ldots \frac{1}{a_2}.$$

Hence $\displaystyle a_{c+1} + \frac{1}{a_c +} \ldots \frac{1}{a_2 +} \frac{1}{a_{c+1} +} \ldots = 2a_1 + \frac{1}{a_2 +} \ldots \frac{1}{a_{c+1} +} \frac{1}{a_2 +} \ldots,$

and therefore $\qquad a_{c+1} = 2a_1, \quad a_c = a_2, \quad a_{c-1} = a_3,$ etc.

Summary. It has been shown that $\sqrt{A/B}$ can be expressed as simple recurring fraction in which (*i*) *there is a single non-recurring element, a_1; (ii) the last partial quotient of the cycle is $2a_1$; (iii) for the rest of the cycle, the partial quotients equidistant from the beginning and end are equal.*

All this may be expressed by writing

$$\sqrt{\frac{A}{B}} = \frac{\sqrt{N}}{r_1} = a_1 + \frac{1}{a_2 +} \frac{1}{a_3 +} \ldots \frac{1}{a_3 +} \frac{1}{a_2 +} \frac{1}{2a_1 +} \ldots;$$

and we shall call the sequence, $a_2, a_3, \ldots a_3, a_2$, the *reciprocal part* of the cycle of quotients.

(4) The b and r Cycles. Let c be the number of elements in the cycle, so that

$$\frac{\sqrt{N}}{r_1} = a_1 + \frac{1}{a_2 +} \ldots \frac{1}{a_{c+1}}.$$

The recurrence is due to the fact that the second complete quotient appears again as the $(c + 2)$th. From this stage all the complete quotients recur in the same order; hence, $(\sqrt{N} + b_{c+m})/r_{c+m} = (\sqrt{N} + b_m)/r_m$, and therefore $b_{c+m} = b_m$ and $r_{c+m} = r_m$ for $m \geq 2$.

Remembering that $a_{c+1} = 2a_1$, $a_{c-m} = a_{2+m}$ for $m = 0, 1, 2, \ldots$, we have $r_{c+1} = (N - b_{c+2}^2)/r_{c+2} = (N - b_2{}^2)/r_2 = r_1$, $r_c = (N - b_{c+1}^2)/r_{c+1} = (N - b_2{}^2)/r_1 = r_2$, $b_{c+1} = a_{c+1}r_{c+1} - b_{c+2} = 2a_1r_1 - b_2 = b_2$, $b_c = a_c r_c - b_{c+1} = a_2 r_2 - b_2 = b_3$, and so on.

Hence, (*i*) *the cycle of b's is $b_2, b_3, \ldots b_{c+1}$ and is reciprocal, i.e. $b_{c+1} = b_2$, $b_c = b_3$, etc.; (ii) the cycle of r's is $r_1, r_2, \ldots r_c$, and is reciprocal after the first term, i.e. $r_c = r_2$, $r_{c-1} = r_3$, etc.*

The character of the recurrence and the reciprocity of the three sequences is exhibited below, where the recurring periods are enclosed in brackets.

$$\begin{matrix}
1 & 2 & 3 & c & c+1 & c+2, \\
b_1 & [b_2 & b_3 \,\ldots & b_3 & b_2] & b_2, \\
[r_1 & r_2 & r_3 \,\ldots & r_2] & r_1 & r_2, \\
a_1 & [a_2 & a_3 \,_{\ldots} & a_2 & 2a_1] & a_2.
\end{matrix}$$

It should be noticed that the a and b cycles correspond, and the *reciprocal* parts of the a and r cycles correspond.

(5) Calculation of the Quotients. In finding the values of b_n, r_n, a_n, *if we can tell when the middle of the b cycle has been reached,* no further calculation is necessary. This matter is settled by the following theorem.

If b_m, r_m, a_m are respectively equal to b_{n+1}, r_n, a_n, then b_m and b_{n+1}, are equidistant from the ends of the b cycle.

For
$$r_{n+1} = (N - b_{n+1}^2)/r_n = (N - b_m^2)/r_m = r_{m-1}.$$

Also
$$a_{m-1}a_{m-1} = b_{m-1} + b_m \text{ and } a_{n+1}r_{n+1} = b_{n+2} + b_{n+1}.$$

Subtracting and using the hypothesis, we find that

$$a_{m-1} - a_{n+1} = (b_{m-1} - b_{n+2})/r_{m-1} \qquad \ldots (A)$$

Now by the inequalities in (2), (D), both $(b_{m-1} - b_{n+2})/r_{n+1}$ and $(b_{n+2} - b_{m-1})/r_{m-1}$ are less than unity. Therefore the right-hand side of (A) is numerically less than untiy, and consequently $a_{m-1} = a_{n+1}$ and $b_{m-1} = b_{n+2}$.

We can show by a similar argument that b_{m-2}, r_{m-2}, a_{m-2} are respectively equal to b_{n+3}, r_{n+2}, a_{n+2} and so on.

Putting $m = n$ and $m = n + 1$ in succession, we have the following:

(i) If two consecutive b's namely b_m and b_{m+1}, are equal, then r_m and a_m are the mid-terms of the reciprocal parts of the r and a cycles. In this case c is even.

(ii) If $r_n = r_{n+1}$ and $a_n = a_{n+1}$ (so that two consecutive r's are equal and likewise the corresponding a's), then r_n, r_{n+1} and a_n, a_{n+1} are the mid-terms of the reciprocal parts of the r and a cycles, and b_{n+1} is the mid-term of the b cycle. In this case c is odd.

■ **EXAMPLE :** *Express $\sqrt{\dfrac{17}{11}}$ and $\sqrt{61}$ as simple continued fractions.*

(i) $\sqrt{\dfrac{17}{11}} = \sqrt{\dfrac{187}{11}}$. Hence $N = 187$, $r_1 = 11$, $a_1 = 1$, and proceeding as in Art. 4, we construct the following table by a mental process:

$$\begin{matrix}
b & 0 & [11 & 13 & 11 & 11 \\
r & & [11 & 6 & 3 & 22 \\
a & & 1 & [4 & 8 & 1 & 8 & 4 & 2].
\end{matrix}$$

Because $b_4 = b_5 = 11$, therefore a_4 is the mid-term of the reciprocal part of the a cycle. This cycle can therefore be completed, for its last term is $2a_1 = 2$, thus

$$\sqrt{\frac{17}{11}} = 1 + \cfrac{1}{4 +} \cfrac{1}{8 +} \cfrac{1}{1 +} \cfrac{1}{8 +} \cfrac{1}{4 +} \cfrac{1}{2}.$$

(*ii*) For $\sqrt{61}$ we have $r_1 = 1$, $a_1 = 7$ and the b, r, a, table is

b	0	[7	5	7	5	4	6				
r	[1	12	3	4	9	5	5				
a	7	[1	4	3	1	2	2	1	3	4	1	14].

Seeing that $a_6 = a_7$ and $r_6 = r_7$, we conclude that a_6, a_7 are the mid-terms of the reciprocal part of the a cycle. Also the last quotient of the cycle is 14.

Therefore
$$\sqrt{61} = 7 + \cfrac{1}{1+}\,\cfrac{1}{4+}\,\cfrac{1}{3+}\,\cfrac{1}{1+}\,\cfrac{1}{2+}\,\cfrac{1}{2+}\,\cfrac{1}{1+}\,\cfrac{1}{3+}\,\cfrac{1}{4+}\,\cfrac{1}{1+}\,\cfrac{1}{14}.$$

(6) Relations connecting the p's and q's. Suppose that
$$\sqrt{\frac{A}{B}} = \frac{\sqrt{N}}{r_1} = a_1 + \cfrac{1}{a_2 +}\,\cdots\,\cfrac{1}{a_n +}\cdots,$$
where $N = AB$, $r_1 = B$; and let p_r/q_r be the rth convergent, then
$$\frac{\sqrt{N}}{r_1} = a_1 + \cfrac{1}{a_2 +}\,\cdots\,\cfrac{1}{a_n +}\,\cfrac{1}{z}, \quad \text{where} \quad z = \frac{\sqrt{N} + b_{n+1}}{r_{n+1}};$$

therefore
$$\frac{\sqrt{N}}{r_1} = \frac{z p_n + p_{n-1}}{z q_n + q_{n-1}} = \frac{\left(\sqrt{N} + b_{n+1}\right) p_n + r_{n+1} p_{n-1}}{\left(\sqrt{N} + b_{n+1}\right) q_n + r_{n+1} q_{n-1}}$$

and
$$\sqrt{N}\left(\sqrt{N} q_n + b_{n+1} q_n + r_{n+1} q_{n-1}\right) = r_1\left(\sqrt{N} p_n + b_{n+1} p_n + r_{n+1} p_{n-1}\right).$$

Equating rational and irrational parts,
$$N q_n = r_1(b_{n+1} p_n + r_{n+1} p_{n-1}), \quad r_1 p_n = b_{n+1} q_n + r_{n+1} q_{n-1}.$$

Solving for b_{n+1} and r_{n+1} and putting $N = AB$ and $r_1 = B$, we find that
$$A q_n q_{n-1} - B p_n p_{n-1} = (-1)^n b_{n+1}, \qquad \text{...(A)}$$
$$B p_n^2 - A q_n^2 = (-1)^n r_{n+1}. \qquad \text{...(B)}$$

(7) Integral Solutions of $Bx^2 - Ay^2 = M$. If for some value of n it happens that $(-1)^n r_{n+1} = M$, then, by equation (B), (p_n, q_n) is a solution of $Bx^2 - Ay^2 = M$.

■ **EXAMPLE :** *Find two positive integral solutions of $11x^2 - 17y^2 = 3$.*

Expressing $\sqrt{\dfrac{17}{11}}$ as a continued fraction [§ (5), Ex. 1], we find that $r_3 = r_5 = 3$. Also $p_2/q_2 = 5/4$, $p_4/q_4 = 46/37$, and therefore $(5, 4)$, $(46, 37)$ are solutions.

■ **6. $\sqrt{N}$ as a Continued Fraction:**

(1) Here $r_1 = 1$ and all the conclusions of the last article hold. Additional theorems are the following:

(*i*) *In the cycle of r's, one member and only one is equal to 1, namely r_1.*

For suppose that $r_m = 1$, then, remembering that $b_2 = a_1$ and using the equations
$$a_m r_m = b_m + b_{m+1}, \quad r_m r_{m+1} = N - b_{m+1}^2,$$
we have $a_m = $ Integral part of $(\sqrt{N} + b_m)/r_m = a_1 + b_m = b_2 + b_m$.

Also $a_m = b_m + b_{m+1}$, therefore $b_{m+1} = b_2$ and
$$r_{m+1} = N - b_{m+1}^2 = N - b_2^2 = r_1 r_2 = r_2.$$

Therefore $(\sqrt{N} + b_{m+1})/r_{m+1} = (\sqrt{N} + b_2)/r_2$, and $r_m = 1$ marks the beginning of a new cycle of r's.

(ii) *If* $r_n \geq 2$, *then* $a_n \leq b_n$. For in this case $n > 1$ and $\sqrt{N} < b_n + r_n$ so that
$$a_n < (\sqrt{N} + b_n)/r_n < (2b_n + r_n)/r_n < b_n + 1 \leq b_n.$$

(iii) *If* $r_n = 2$, *the number (c) of elements in the cycle belonging to* $\sqrt{N}$ *is even. Also* a_n *is the mid-term of the reciprocal part of the 'a' cycle, and* $a_n = a_1$ *or* $a_1 - 1$.

For a_1 is the integral part of $\sqrt{N}$ and $\sqrt{N} > b_n$, therefore $a_1 \geq b_n$. Also a_n is the integral part of $(a_1 + b_n)/r_n$, and so if $r_n = 2$, we have
$$a_n > (a_1 + b_n)/2 - 1 > b_n - 1 \geq b_n.$$

But by the preceding $a_n \leq b_n$, therefore $a_n = b_n$. Also $2a_n = b_n + b_{n+1}$, therefore $b_n = b_{n+1}$.

Hence by § (5), (i), c is even and a_n is the mid-term of the reciprocal part of the a cycle. Finally a_n is the integral part of $(a_1 + a_n)/2$, so that $a_n = (a_1 + a_n)/2$ or $(a_1 + a_n - 1)/2$, and therefore $a_n = a_1$ or $a_1 - 1$.

(2) Equations (A) and (B) at the top of the page become
$$Nq_n q_{n-1} - p_n p_{n-1} = (-1)^n b_{n+1}, \qquad \text{...(A)}$$
$$p_n^2 - Nq_n^2 = (-1)^n r_{n+1}. \qquad \text{...(B)}$$

(3) Integral Solutions of $x^2 - Ny^2 = M$. If, for some value of n, it is found that $(-1)^n r_{n+1} = M$, then (p_n, q_n) is a solution of $x^2 - Ny^2 = M$.

In particular, if c is the number of elements in the cycle belonging to $\sqrt{N}$, the necessary and sufficient condition that r_{n+1} may be equal to 1 is that $n = tc$ where $t = 1, 2, 3, ...$; hence, it follows that

(i) *for the equation* $x^2 - Ny^2 = 1$, solutions are (p_{tc}, q_{tc}) when c is even and (p_{2tc}, q_{2tc}) when c is odd;

(ii) *for the equation* $x^2 - Ny^2 = -1$, solutions are $[p_{(2t-1)c}, q_{(2t-1)c}]$ when c is odd; the method giving no solution when c is even.

■ **EXAMPLE :** *Find a positive integral solution of $x^2 - 13y^2 = -1$.*

We find that
$$\sqrt{13} = 3 + \frac{1}{1+}\frac{1}{1+}\frac{1}{1+}\frac{1}{1+}\frac{1}{6},$$

so that $c = 5$. Also $p_5 = 18$, $q_5 = 5$ and (18, 5) is a solution.

(4) Calculation of Convergents. The following formulae are useful in this connection. If we multiply equations (A), (B) firstly by q_n and q_{n-1}, and secondly by p_n and p_{n-1}, we obtain by addition
$$p_n = r_{n+1} q_{n-1} + b_{n+1} q_n \quad \text{and} \quad Nq_n = r_{n+1} p_{n-1} + b_{n+1} p_n. \qquad \text{... (C)}$$

Again, changing n into $n - 1$ in equation (B) and using equation (A), we find that
$$p_{n-1} = r_n q_n - b_{n+1} q_{n-1} \quad \text{and} \quad Nq_{n-1} = r_n p_n - b_{n+1} p_{n-1}. \qquad \text{... (D)}$$

In general, the p's are considerably larger than the q's, and to find p_n/q_n we need only calculate the q's and then use the first of equations (C).

(5) The Convergent p_c/q_c: (i) *If* $m + n = c$, *where c is the number of elements in the cycle belonging to* $\sqrt{N}$, *then*
$$p_c = p_m q_{n+1} + p_{m-1} q_n \quad \text{and} \quad q_c = q_m q_{n+1} + q_{m-1} q_n. \qquad \text{... (E)}$$

For
$$\frac{p_c}{q_c} = a_1 + \frac{1}{a_2 +}\cdots\frac{1}{a_m +}\frac{1}{f},$$

where $\;f = a_{m+1} + \dfrac{1}{a_{m+2} +}\cdots\dfrac{1}{a_c} = a_{n+1} + \dfrac{1}{a_n +}\dfrac{1}{a_{n-1} +}\cdots\dfrac{1}{a_2} = \dfrac{q_{n+1}}{q_n}$;

therefore
$$\frac{p_c}{q_c} = \frac{fp_m + p_{m-1}}{fq_m + q_{m-1}} = \frac{p_m q_{n+1} + p_{m-1}q_n}{q_m q_{n+1} + q_{m-1}q_n}.$$

If X is the numerator and Y the denominator of the last fraction,
$$Xq_m - Yp_m = (-1)^{m-1}q_n \quad \text{and} \quad Xq_{m-1} - Yp_{m-1} = (-1)^m q_{n+1}.$$

Hence any common factor of X and Y is a factor of q_n and of q_{n+1}, and since these numbers are prime to one another, so also are X and Y. Thus the fractions p_c/q_c and X/Y are in their lowest terms, and therefore $p_c = X$ and $q_c = Y$, which are the results in question.

(*ii*) In particular, *if c is odd and* $n = \dfrac{1}{2}(c-1)$,

$$p_c = p_n q_n + p_{n+1}\, q_{n+1} \quad \text{and} \quad q_c = q_n^2 + q_{n+1}^2. \qquad \ldots \text{(F)}$$

If c is even and $n = \dfrac{1}{2}c$.

$$p_c = p_n q_{n+1} + p_{n-1}\, q_n \quad \text{and} \quad q_c = q_n(q_{n+1} + q_{n-1}). \qquad \ldots \text{(G)}$$

(*iii*) *If* $m + n = c$, *then*

$$p_n + \sqrt{N}q_n = (-1)^m (p_m - \sqrt{N}q_m)(p_c + \sqrt{N}q_c), \qquad \ldots \text{(H)}$$

which is equivalent to the two equations
$$p_m p_c - Nq_m q_c = (-1)^m p_n, \quad p_m q_c - q_m p_c = (-1)^m q_n \qquad \ldots \text{(I)}$$

For using equations (E), $p_m q_c - q_m p_c = (p_m q_{m-1} - q_m p_{m-1})q_n = (-1)^m q_n$.

Also we have the identity
$$(p_m p_c - Nq_m q_c)^2 - N(p_m q_c - q_m p_c)^2 = (p_m^2 - Nq_m^2)(p_c^2 - Nq_c^2).$$

Now
$$p_m^2 - Nq_m^2 = (-1)^m r_{m+1}, \quad p_n^2 - Nq_n^2 = (-1)^n r_{n+1};$$

and, since $m + n = c$, we have $r_{m+1} = r_{n+1}$; and therefore
$$p_n^2 - Nq_n^2 = (-1)^c (p_m^2 - Nq_m^2) = (p_m^2 - Nq_m^2)(p_c^2 - Nq_c^2).$$

Hence
$$(p_m p_c - Nq_m q_c)^2 = p_n^2.$$

It remains to consider the sign of $p_m p_c - Nq_m q_c$.

The convergent p_m/q_m precedes p_c/q_c. First suppose that c is even.

If m is even, $\sqrt{N} < p_c/q_c < p_m/q_m$ and $p_m p_c > Nq_m q_c$. If m is odd, it follows from Ch. 24, 25, that $p_m p_c < Nq_m q_c$. The same results follow in a similar way when c is odd, and thus $p_m p_c - Nq_m q_c = (-1)^m p_n$.

It should be noticed that *these results hold if c is replaced by tc.*

(*iv*) *If N is a prime and c is even and* $n = \dfrac{1}{2}c$, *then*

$$p_c + \sqrt{N}q_c = \frac{1}{2}(p_n + \sqrt{N}q_n)^2, \qquad \ldots \text{(J)}$$

which is equivalent to $p_c = \dfrac{1}{2}(p_n^2 + Nq_n^2)$ *and* $q_c = p_n q_n$.

Also $p_n^2 - Nq_n^2 = (-1)^n . 2$, *and the mid-term of the reciprocal part of the a cycle is* a_1 *or* $a_1 - 1$.

For by equations (I), $p_n p_c - Nq_n q_c = (-1)^n p_n$ and $p_n q_c - q_n p_c = (-1)^n q_n$,

therefore $\qquad\qquad\qquad p_c/q_c = (p_n^2 + Nq_n^2)/2p_n q_n$ $\qquad\qquad$ ... (K)

and $\qquad \left(\dfrac{p_n^2 - Nq_n^2}{2p_n q_n}\right)^2 = \left(\dfrac{p_c}{q_c}\right)^2 - N = \dfrac{1}{q_c^2}$;

hence $\qquad \dfrac{\left(p_n^2 - Nq_n^2\right)}{p_n q_n} = (-1)^n \dfrac{2}{q_c}$,

the sign being determined by the fact that $p_n \gtrless \sqrt{N}q_n$, according as n is even or odd. Since N is a prime and p_n is prime to q_n, the last fraction is in its lowest terms. Hence $p_n^2 - Nq_n^2 = (-1)^n . 2$ or $(-1)^n$, and the latter alternative is impossible.

Therefore $\quad p_n^2 - Nq_n^2 = (-1)^n . 2$ and $q_c = p_n q_n$

Hence, $r_{n+1} = 2$; and the rest follows by § (I), (*iii*), p. 527.

Note: *When N is not a prime, p_c and q_c may be found by using equation (K) and noting that p_c/q_c is in its lowest terms.*

Thus we find that $\qquad\qquad \sqrt{21} = 4 + \dfrac{1}{1+}\dfrac{1}{1+}\dfrac{1}{2+}\dfrac{1}{1+}\dfrac{1}{1+}\dfrac{1}{8}$,

so that for $\sqrt{21}$, $c = 6$, $p_3 = 9$, $q_3 = 2$, and

$$\frac{p_6}{q_6} = \frac{81 + 21 \cdot 4}{2 \cdot 9 \cdot 2} = \frac{55}{12}, \text{ and therefore } p_6 = 55, q_6 = 12.$$

(6) The Convergent p_{n+tc}/q_{n+tc}. (*i*) *If p_n/q_n is the n-th convergent of*

$$\sqrt{N} = a_1 + \frac{1}{a_2 +} \cdots \frac{1}{2a_1 +} \frac{1}{a_2 +} \cdots$$

and c is the number of quotients in the cycle, then

$$p_{n+tc} = p_n p_{tc} + Nq_n q_{tc} \text{ and } q_{n+tc} = p_n q_{tc} + q_n p_{tc}. \qquad \text{... (L)}$$

For the $(tc + 1)$th quotient is $2a_1$ and the corresponding complete quotient is $a_1 + \sqrt{N}$, therefore

$$\sqrt{N} = \frac{\left(a_1 + \sqrt{N}\right)p_{tc} + p_{tc-1}}{\left(a_1 + \sqrt{N}\right)q_{tc} + q_{tc-1}}.$$

Removing fractions and equating rational and irrational parts, we have

$$a_1 p_{tc} + p_{tc-1} = Nq_{tc} \text{ and } a_1 q_{tc} + q_{tc-1} = p_{tc}. \qquad \text{... (M)}$$

Again,

$$\frac{p_{n+tc}}{q_{n+tc}} = a_1 + \frac{1}{a_2 +} \cdots \frac{1}{2a_1 +} \frac{1}{a_2 +} \cdots \frac{1}{a_n} = a_1 + \frac{1}{a_2 +} \cdots \frac{1}{a_1 + p_n/q_n},$$

the last quotient being the $(tc + 1)$th. From this, by means of equation (I),

$$\frac{p_{n+tc}}{q_{n+tc}} = \frac{(a_1 + p_n/q_n)p_{tc} + p_{tc-1}}{(a_1 + p_n/q_n)q_{tc} + q_{tc-1}} = \frac{p_n p_{tc} + Nq_n q_{tc}}{p_n q_{tc} + q_n p_{tc}}.$$

If X is the numerator and Y the denominator of the last fraction,

$$p_{tc}X - Nq_{tc}Y = p_n(p_{tc}^2 - Nq_{tc}^2) \quad \text{and} \quad p_{tc}Y - q_{tc}X = q_n(p_{tc}^2 - Nq_{tc}^2).$$

Now $p_{tc}^2 - Nq_{tc}^2 = \pm 1$, by (5), (*iii*), therefore any common factor of X and Y divides p_n and q_n, which are prime to one another. Hence X/Y is in its lowest terms: so also is p_{n+tc}/q_{n+tc}. Therefore $p_{n+tc} = X$, $q_{n+tc} = Y$, which are the equations in question.

As a special case we have

$$P_{(s+t)c} = p_{sc}p_{tc} + Nq_{sc}q_{tc} \text{ and } q_{(s+t)c} = p_{sc}q_{tc} + q_{sc}p_{tc}; \qquad \text{... (N)}$$

in particular, $\qquad p_{2tc} = p_{tc}^2 + Nq_{tc}^2 \text{ and } q_{2tc} = 2p_{tc}q_{tc}.$ $\qquad$ ... (O)

(*ii*) Since $\sqrt{N}$ is irrational, equations (L) are *equivalent* to the single relation.

$$p_{n+tc} + \sqrt{N}q_{n+tc} = (p_n + \sqrt{N}q_n)(p_{tc} + \sqrt{N}q_{tc}).$$

Putting $t = 1$ and $n = c, 2c, 3c, \ldots$ in succession, we find that

$$p_{2c} + \sqrt{N}q_{2c} = (p_c + \sqrt{N}q_c)^2, \; p_{3c} + \sqrt{N}q_{3c} = (p_c + \sqrt{N}q_c)^3 \ldots ,$$

and generally, for every positive t, $\quad p_{tc} + \sqrt{N}q_{tc} = (p_c + \sqrt{N}q_c)^t;$ $\qquad$ ... (P)

and therefore $\qquad p_{n+tc} + \sqrt{N}q_{n+tc} = (p_n + \sqrt{N}q_n)(p_c + \sqrt{N}q_c)^t.$ $\qquad$ ... (Q)

▎ **EXAMPLE :*** *Find a solution in positive integers of $x^2 - 61y^2 = 1$.*

In Ex. 1, p. 526, $\sqrt{61}$ is expressed as a continued fraction, the number of quotients in the cycle being 11. Hence the smallest solution of the equation in question is (p_{22}, q_{22}). The mid-elements of the cycle are the 5th and 6th, and $p_5 = 164$, $q_5 = 21$, $p_6 = 453$, $q_6 = 58$. Hence by equation (F).

$$p_{11} = p_5 q_5 + p_6 p_6 = 164.21 + 453.58 = 29718, \; q_{11} = q_5^2 + q_6^2 = 21^2 + 58^2 = 3805.$$

Using equations (O) and observing that $p_{11}^2 - 61q_{11}^2 = -1$,

$$p_{22} = p_{11}^2 + 61q_{11}^2 = 2p_{11}^2 + 1 = 1766319049, \; q_{22} = 2p_{11}q_{11} = 226153980.$$

Thus the least solution is $(1766319049, 226153980)$.

▨ **7. The Cycle belonging to $z = (\sqrt{N} + b)/r$:** Suppose that

$$z > 1, \; b < \sqrt{N} < b + r, \; N - b^2 = rr',$$

where r' is a positive integer, then the simple continued fraction corresponding to z has no acyclic part (Art. 3), and the cycle can be calculated as follows.

Rule. *Let (x, y) be a solution in positive integers of any one of the equations.*

$$x^2 - Ny^2 = \pm 1 \text{ or } \pm 4, \qquad \qquad \text{...(A)}$$

the latter pair being used only when r, r' are both even and N, b are odd. Substitute x/y for $\sqrt{N}$ in the given surd. Express the result as a simple continued fraction with an even or an odd number of quotients according as the sign on the right of the chosen equation is $+$ or $-$. This fraction consists of one or more of the cycles belonging to $(\sqrt{N}+b)/r$.

---

* This problem was set by Fermat as a challenge to the English mathematicians of his time.

Proof. Since $N = b^2 + rr'$, the surd $(\sqrt{N} + b)/r$ is the positive root of

$$rz^2 - 2bz - r' = 0. \qquad \qquad ...(B)$$

(1) We shall prove that this equation can be written in the form

$$z = (pz + p')/(qz + q'), \text{ where } p, q, p', q' \text{ are positive integers} \qquad ...(C)$$

such that $\qquad \qquad pq' - p'q = \pm 1 \text{ and } q' \leq q. \qquad \qquad ...(D)$

For equation (C) is the same as $qz^2 - (p - q')z - p' = 0$, which is identical with (B)

if $\qquad \qquad q = ry, \quad p - q' = 2by, \quad p' = r'y, \qquad \qquad ...(E)$

where y may be any rational.

From (D), (E) we have $\dfrac{1}{4}(p + q')^2 = Ny^2 \pm 1 = x^2,$ $\qquad \qquad ...(F)$

where x is rational. All possible cases are included in the following:

(A) x, y are positive integers such that $x^2 - Ny^2 = \pm 1$.

(B) $x = \xi/2, y = \eta/2$ where ξ, η are odd integers such that $\xi^2 - N\eta^2 = \pm 4$.

In case (B), r, r' are both even, for $q = ry, p' = r'y$. Hence $N - b^2$ is divisible by 4, and if N is odd so also is b.

In both cases, from (E) and (F), we find that

$$p = x + by, \quad q' = x - by, \qquad \qquad ...(G)$$

and $\qquad \qquad p/q = (x/y + b)/r. \qquad \qquad ...(H)$

It remains to be showns that $q' \leq q$. This will be so if $x/y \leq b + r$.

Three cases must be considered:

(i) If $x^2 - Ny^2 = -1$ or $\xi^2 - N\eta^2 = -4$, then $x/y < \sqrt{N} < b + r$ and $q' < q$.

(ii) If $x^2 - Ny^2 = 1$, the least value of y is 1, and

$$x^2/y^2 = N + 1/y^2 \leq N + 1.$$

Now $N < (b + r)^2$, therefore $N \leq (b + r)^2 - 1$. Hence

$$x/y \leq \sqrt{N+1} \leq b + r \text{ and } q' \leq q$$

(iii) If $x = \xi/2, \quad y = \eta/2$ where $\xi^2 - N\eta^2 = 4$, the least value of η is 1 and

$$x/y \leq \sqrt{N+4}.$$

Let $N = (b + r)^2 - k$ so that k is a positive integer, then $rr' = r(2b + r) - k$ where r, r' are even. Therefore k is divisible by 4, and 4 is its least value. Hence

$$x/y \leq \sqrt{N+4} \leq b + r \text{ and } q' \leq q.$$

(2) It follows that if $\qquad \dfrac{p}{q} = a_1 + \dfrac{1}{a_2 +} \cdots \dfrac{1}{a_n},$ $\qquad \qquad ...(I)$

where n is even or odd according as $pq' - p'q = 1$ or -1, then p'/q' is the convergent immediately preceding p/q. (Ch.24, 12.) Now let z' be determined by

$$z = a_1 + \dfrac{1}{a_2 +} \cdots \dfrac{1}{a_n + z'},$$

then $\qquad \qquad z = \dfrac{pz' + p'}{qz' + q'} \text{ and } \dfrac{pz' + p'}{qz' + q'} = \dfrac{pz + p'}{qz + q'}.$

Hence $z' = z$ and *the fraction* (I) *consists of one or more of the cycles belonging to z.*

The rule can be used (as in Ex. 2) to express any quadratic surd as a continued fraction, beginning at the stage where recurrence commences.

■ **EXAMPLE 1.** *Express $(\sqrt{621} + 21)/18$ as a simple continued fraction.*

Since $21 < \sqrt{621} < 21 + 18$, there is no acyclic part. Also $18r' = 621 - 21^2 = 180$, giving $r' = 10$.

Thus, $r,\ r'$ are even and b is odd, also
$$25^2 - 621.1^2 = 4.$$

Using the rule, we substitute $25/1$ for $\sqrt{621}$ in the given surd, and express the result as a simple continued fraction with an *even* number of quotients, thus

$$\frac{25 + 21}{18} = \frac{23}{9} = 2 + \frac{1}{1+}\frac{1}{1+}\frac{1}{4},$$

and
$$\frac{\sqrt{621} + 21}{18} = 2 + \frac{1}{1+}\frac{1}{1+}\frac{1}{4}.$$

■ **EXAMPLE 2.** *Express $(\sqrt{13} + 7)/4$ as a simple continued fraction.*

$$
\begin{array}{c|cc}
b & 7, & 1 \\
r & 4, & 3 \\
a & 2, & 1
\end{array}
$$

Since $7 > \sqrt{13}$, there is an acyclic part. The reckoning on the right shows that the second complete quotient is $(\sqrt{13} + 1)/3$, and since $1 < \sqrt{13} < 1 + 3$, we can apply the rule. We have
$$3^2 - 13 \cdot 1^2 = -4 \quad \text{and} \quad 18^2 - 13 \cdot 5^2 = -1.$$

Since $r = 3$, an odd number, the first equation is useless. Taking the second, we substitute $18/5$ for $\sqrt{13}$ in $(\sqrt{13} + 1)/3$, and express the result as a simple continued fraction with an *odd* number of quotients, thus

$$\left(\frac{18}{5} + 1\right)\frac{1}{3} = \frac{23}{15} = 1 + \frac{1}{1+}\frac{1}{1+}\frac{1}{6+}\frac{1}{1}.$$

This is the cycle belonging to the given surd, and

$$\frac{\sqrt{13} + 7}{4} = 2 + \frac{1}{1+}\frac{1}{1+}\frac{1}{1+}\frac{1}{6+}\frac{1}{1}.$$

EXERCISE LVI

1. Verify some of the following results, where the values of $b,\ r,\ a$ are given as far as the middle of the first cycle:

(i) $\sqrt{13} \begin{cases} b.0.3.1.2 \\ r.1.4.3.3 \\ a.3.1.1.1 \end{cases}$

(ii) $\sqrt{19} \begin{cases} b.0.4.2.3.3 \\ r.1.3.5.2.5 \\ a.4.2.1.3.1 \end{cases}$

(iii) $\sqrt{31}\begin{cases} b.0.5.1.4.5.5 \\ r.1.6.5.3.2.3 \\ a.5.1.1.3.5.3 \end{cases}$ (iv) $\sqrt{43}\begin{cases} b.0.6.1.5.4.5.5 \\ r.1.7.6.3.9.2.9 \\ a.6.1.1.3.1.5.1 \end{cases}$

(v) $\sqrt{91}\begin{cases} b.0.9.1.8.7.7 \\ r.1.10.9.3.14.3 \\ a.9.1.1.5.1.5 \end{cases}$ (vi) $\sqrt{109}\begin{cases} b.0.10.8.7.5.9.7.8.10 \\ r.1.9.5.12.7.4.15.3.3 \\ a.10.2.3.1.2.4.1.6.6 \end{cases}$

(vii) $\sqrt{a^2+1}\begin{cases} b.0.a.a \\ r.1.1.1 \\ a.a.2a.2a \end{cases}$ (viii) $\sqrt{a^2-1}\begin{cases} b.0.a-1.a-1 \\ r.1.2(a-1).1 \\ a.a-1.1.2(a-1) \end{cases}$

(ix) $\sqrt{a^2+2}\begin{cases} b.0.a.a \\ r.1.2.1 \\ a.a.a.2a \end{cases}$ (x) $\sqrt{a^2-2}\begin{cases} b.0.a-1.a-2.a-2 \\ r.1.2a-3.2.2a-3 \\ a.a-1.1.a-2.1 \end{cases}$

$(a>2)$

(xi) $\sqrt{a(a+1)}\begin{cases} b.0.a.a \\ r.1.a.1 \\ a.a.2.2a \end{cases}$ (xii) $\sqrt{a^2+4}\begin{cases} b.0.a.a-2.2 \\ r.1.4.a.a \\ a.a.\frac{1}{2}(a-1).1.1 \end{cases}$

(xiii) $\sqrt{a(a+4)}\begin{cases} b.0.a+1.a-2.a.a \\ r.1.2a-1.4.a.4 \\ a.a+1.1.\frac{1}{2}(a-1).2.\frac{1}{2}(a-1) \end{cases}$ $(a>1 \text{ and odd})$

2. In certain cases, a solution of one of the equations $x^2 - Ny^2 = \pm 1$ or ± 4 can be written down at once; thus

$$\text{if } N = a^2 \pm 1, \quad \text{then} \quad a^2 - N\,1^2 = \mp 1;$$
$$\text{if } N = a^2 \pm 2, \quad \text{then} \quad (a^2 \pm 1)^2 - Na^2 = 1;$$
$$\text{if } N = a(a + 1), \quad \text{then} \quad (2a + 1)^2 - N.2^2 = 1;$$
$$\text{if } N = a^2 \pm 4, \quad \text{then} \quad a^2 - N.1^2 = \mp 4.$$

Apply this to write down the integral solutions of $x^2 - Ny^2 = 1$ when
$$N = 51, 47, 56.$$

3. Find a solution of $x^2 - 109y^2 = -4$, and apply the H.C.F. process to show that

$$\frac{\sqrt{109}+7}{10} = 1 + \cfrac{1}{1+}\cfrac{1}{2+}\cfrac{1}{1+}\cfrac{1}{9+}\cfrac{1}{1+}\cfrac{1}{2}.$$

4. Obtain the following:

(i) $\dfrac{\sqrt{17}+15}{8} = 2 + \cfrac{1}{2+}\cfrac{1}{1+}\cfrac{1}{1+}\cfrac{1}{3};$ (ii) $\dfrac{\sqrt{61}+7}{6} = 2 + \cfrac{1}{2+}\cfrac{1}{7};$

(iii) $\sqrt{\dfrac{13}{5}} = 1 + \cfrac{1}{1+}\cfrac{1}{1+}\cfrac{1}{1+}\cfrac{1}{1+}\cfrac{1}{2};$ (iv) $\dfrac{\sqrt{19}+1}{6} = 1 + \cfrac{1}{8+}\cfrac{1}{2+}\cfrac{1}{1+}\cfrac{1}{3+}\cfrac{1}{2};$

(v) $\sqrt{\dfrac{17}{3}} = 2 + \cfrac{1}{2+}\cfrac{1}{1+}\cfrac{1}{1+}\cfrac{1}{1+}\cfrac{1}{2+}\cfrac{1}{4}.$

5. Find two positive integral solutions of

(i) $5x^2 - 13y^2 = 7$; (ii) $5x^2 - 13y^2 = -7$; (iii) $3x^2 - 17y^2 = 7$.

6. Find positive integral solutions of $x^2 - 621y^2 = 4$ and of $x^2 - 621y^2 = 1$. Referring to Art. 7, verify that if $623/25$ and $7775/312$ are substituted for $\sqrt{621}$ in the fraction $(\sqrt{621} + 21)/18$, the first gives two cycles and the second three cycles of the fraction.

7. If $N = (2m + 3)^2 - 4$, prove that

$$\frac{\sqrt{N} + 2m - 1}{4} = m + \frac{1}{2+} \frac{1}{m+1+} \frac{1}{4(m+1)+} \frac{1}{1}.$$

8. If a is odd, find a positive integral solution of

(i) $x^2 - (a^2 + 4)y^2 = a$; (ii) $x^2 - (a^2 + 4)y^2 = -a$;

9. Prove that $r_n \leq 2a_1$, and that if $r_n = 2a_1$, then $a_n = 1$ and a_n is the mid-term of the reciprocal part of the a cycle.

Show also that if N is an odd prime and $r_n = 2a_1$, then N must be equal to 3. [If $r_n = 2a_1$, $a_n < (\sqrt{N} + b_n)/r_n \leq 1$, therefore $b_n + b_{n+1} = b_n r_n = 2a_1$, so that $b_n = b_{n+1}$. Hence a_n is the mid-term of the reciprocal part of the a cycle. If N is an odd prime, we must have $a_n = a_1$ or $a_1 - 1$.

Hene $a_1 = 1$ or 2, and $N = 3, 5, 7$. On trial it is found that $N = 3$.]

10. If $r_n = r_{n+1} = a_1$, then $a_n = a_{n+1} = 1$, so that a_n, a_{n+1} are the mid-terms of the reciprocal part of the a cycle.

[$a_n = 1$ $(a_1 + b_n)/a_1 = 1$ or 2, according as $b_n \neq a_1$ or $b_n = a_1$. If $b_n = a_1$, $b_{n+1} = 2a_1 - b_n = a_1$, so that $b_n = b_{n+1}$. This is impossible. Hence $a_n = 1$ and $a_{n+1} = I$ $(a_1 + b_{n+1})/a_1 = 1$.]

11. If a_1 is an odd prime, or a power of such a number, and $r_n = a_1$, then $a_n = 1$.

[For $r_n r_{n+1} = N - b_{n+1}^2$, $r_{n-1} r_n = N - b_n^2$, therefore b_n and b_{n+1} are solutions of $x^2 \equiv N \pmod{a_1}$, and since each $\leq a_1$, and a_1 is an odd prime, $b_n + b_{n+1} = a_1$, therefore $a_n = (b_n + b_{n+1})/r_n = a_1/a_1 = 1$.]

12. If, after a certain stage, the simple continued fractions which are equivalent to two irrationals z and z' are identical, prove that

$$z' = (pz + p')/(qz + q') \text{ where } pq' - p'q = \pm 1$$

where p, p', q, q' are integers. Show also that the converse is true.

13. If the number (c) of elements in the cycle belonging to $\sqrt{N}$ is even,

(i) prove that $p_{c/2}/q_{c/2} = (p_c \pm 1)/q_c$, according as c is or is not divisible by 4.

(ii) By taking $N = 21$, illustrate the fact that equations (J) of Art. 6, (5), (iv), are not necessarily true unless N is a prime, and verify the equations for $N = 19$.

14. If the number (c) of elements in the cycle belonging to $\sqrt{N}$ is odd:

(i) Prove that N is the sum of two squares which are prime to one another. Also, by talking $N = 205$, show that the converse of this is not generally true.

(ii) If $m = \frac{1}{2}(c - 1)$ and $n = \frac{1}{2}(c + 1)$, prove that

$$p_m^2 + p_n^2 = N(q_m^2 + q_n^2) = Nq_c,$$
$$p_m^2 + Nq_m^2 = (2p_m q_m p_c + 1)/q_c,$$
$$p_m^2 - Nq_m^2 = (p_m q_n + p_n q_m)/q_c,$$

and verify when $N = 29$.

[(i) N is a factor of $p_c^2 + 1$.]

MISCELLANEOUS EXERCISES (A)

1. Show that the product of any two numbers, as 19 and 35, can be found as on the right.* In the left-hand column, each number is the quotient when the preceding number is divided by 2. Any remainders are disregarded. In the right-hand column, each number is twice the preceding.

$$
\begin{array}{r|l}
19 & 35 \\
9 & 70 \\
4 & \cancel{140} \\
2 & \cancel{280} \\
1 & 560
\end{array}
$$

$$19 \times 35 = 665$$

The numbers 140 and 280 which are opposite *even* numbers on the left are crossed out, and the sum of the other numbers on the right is 19 × 35.

[The reckoning shows that

$$19 \times 35 = (1 + 2 + 2^2 \cdot 0 + 2^3 \cdot 0 + 2^4)\, 35 = 35 + 70 + 560.]$$

2. Show that n^3 is the sum of n consecutive odd numbers, and find the middle or the two middle numbers.

3. Take any sum of money less than £12, say £9 8s. 7d. Reverse the order of pence and pounds, obtaining £7 8s. 9d. The difference between these sums is £1 19s. 10d. Reversing the order of pence and pounds, we obtain £10 19s. 1d. Adding the last two sums, we have £12 18s. 11d. Prove that the result will be the same whatever sum less than £12 is chosen.

4. If a/b is a proper fraction and $q_1, q_2, \dots q_n$ are the quotients and $r_1, r_2 \dots r_n$ the remainders when b is divided by a, $r_1, r_2 \dots r_{n-1}$, respectively, show that

   (i) $$\frac{a}{b} = \frac{1}{q_1} - \frac{1}{q_1 q_2} + \frac{1}{q_1 q_2 q_3} - \dots + (-1)^{n-1} \frac{1}{q_1 q_2 \dots q_n} + R_n,$$

   where $R_n = (-1)^n \dfrac{1}{q_1 q_2 \dots q_n} \cdot \dfrac{r_n}{b}.$

   (ii) For some value of n, $R_n = 0$, so that a/b can always be expressed in the form

   $$\frac{1}{p} - \frac{1}{q} + \frac{1}{r} - \frac{1}{s} + \dots,$$

   where $p, q \dots$ are positive integers, their number being finite.

   (iii) Express in this way $\dfrac{32}{51}$ and $\dfrac{25}{53}$.

5. Prove that $a + b + c + d$ and $a + b - c - d$ are factors of

   $$2(a^4 + b^4 + c^4 + d^4) - (a^2 + b^2 + c^2 + d^2)^2 + 8abcd,$$

   and find the remaining factors.

6. Solve $$\left(\frac{x-1}{x+2}\right)^3 = \frac{x-4}{x+5}.$$

---

* Known as the Russian peasants' method of multiplication.

7. If x, y, z are real, prove that
$$(x^2 + y^2 + z^2)/(yz + zx + xy)$$
cannot lie between 1 and -2.

8.   (i) Find the simplest equation with integral coefficients which has
$$-\sqrt{\frac{2}{3}} + \sqrt{\frac{3}{2}} \quad \text{and} \quad -\sqrt{\frac{2}{3}} - \sqrt{\frac{3}{2}}$$
       among its roots. What are the other roots of the equation?

     (ii) Prove that $4(x^2 + x + 1)^3 - 27x^2(x + 1)^2 = (x - 1)^2 (2x + 1)^2 (x + 2)^2.$

9. If $x + y + z = x^2 + y^2 + z^2 = 2$, show that
$$x(1 - x)^2 = y(1 - y)^2 = z(1 - z)^2.$$

10. If $a + b + c + d = 0$ and $x + y + z + t = 0$, prove that the results of rationalising the equations
$$\sqrt{ax} + \sqrt{by} + \sqrt{cz} + \sqrt{dt} = 0, \quad \sqrt{bx} + \sqrt{ay} + \sqrt{dz} + \sqrt{ct} = 0,$$
are equivalent. State the two other equations which lead to equivalent results.

11. Find the equation whose roots are given by
$$y_i = x_i - x_i^2 + (x_1^2 + x_2^2 + x_3^2), \quad (i = 1, 2, 3),$$
where x_1, x_2, x_3 are the roots of $x^3 - x^2 + 4 = 0$.

12. Solve the equation
$$6x^4 - 3x^3 + 8x^2 - x + 2 = 0,$$
being given that it has a pair of roots whose sum is zero.

13. If $\dfrac{a}{x - md} + \dfrac{b}{x - mc} + \dfrac{c}{x + mb} + \dfrac{d}{x + ma} = 0$ and $a + b + c + d = 0$, prove that the only finite value of x is $m(ac + bd)/(a + b)$.

14. If $(ca' - ac')^2 < 4(ab' - a'b)(bc' - b'c)$, prove that $b^2 > ac$ and $b'^2 > a'c'$.

15. If $a < b < c < d$, and $a - c \le b - d$, the roots of
$$(x - a)(x - c) = k(x - b)(x - d)$$
are real for all values of k.

16. If $l + m + n = 0$, show that the expression
$$a^2l^2 + b^2m^2 + c^2n^2 - 2bcmn - 2canl - 2ablm$$
is positive if $abc(a + b + c)$ is positive.

17. Show that the expression $(cx - az)^2 - (ay - bx)(bz - cy)$ is the product of two linear functions of x, y, z, and that, when a, b, c are real, the coefficients in the linear functions are real if $b^2 \ge 4ac$.

18. If the cubic derived from the equation
$$\frac{a}{x + a} + \frac{b}{x + b} = \frac{c}{x + c} + \frac{d}{x + d}$$
has a pair of equal roots, then either one of the numbers a, b is equal to one of the numbers c, d or
$$\frac{1}{a} + \frac{1}{b} = \frac{1}{c} + \frac{1}{d}.$$

19. Express in factors
$$(bc - a^2)(ca - b^2) + (ca - b^2)(ab - c^2) + (ab - c^2)(bc - a^2).$$

20. Show that if a, b, c are real, the roots of
$$1/(x + a) + 1/(x + b) + 1/(x + c) = 3/x$$
are real.

21. Eliminate x, y, z from the equations:
$$\frac{x}{y}+\frac{y}{z}+\frac{z}{x} = a, \qquad \frac{x}{z}+\frac{y}{x}+\frac{z}{y} = b,$$

$$\left(\frac{x}{y}+\frac{y}{z}\right)\left(\frac{y}{z}+\frac{z}{x}\right)\left(\frac{z}{x}+\frac{x}{y}\right) = c.$$

22. Prove that $10^n - (5 + \sqrt{17})^n - (5 - \sqrt{17})^n$ is divisible by 2^{n+1}.

23. (a) Solve $(x - 1)^3 + (x - 2)^3 + (x - 3)^3 + \ldots + (x - n)^3 = 0.$

(b) If x, y, z are unequal and
$$(x - 1)(y - 2) = (y - 1)(z - 2) = (z - 1)(x - 2).$$
show that each product is equal to -1 and that
$$yz + zx + xy - 3(x + y + z) + 9 = 0$$
and
$$xyz - 2(x + y + z) + 9 = 0.$$

24. By the method of partial fractions, prove that, if a, b, c, ... k are n numbers, then the sum of n fractions of the type
$$\frac{a^m}{(a-b)(a-c)(a-d)\ldots(a-k)}$$
is zero if m is an integer less than $n - 1$ and is unity if $m = n - 1$.

25. If n is a positive integer,
$$\frac{2}{\lfloor n+1\rfloor\lfloor n-1\rfloor}\cdot\frac{1}{a+1} - \frac{2}{\lfloor n+2\rfloor\lfloor n-2\rfloor}\cdot\frac{4}{a+4} + \frac{2}{\lfloor n+3\rfloor\lfloor n-3\rfloor}\cdot\frac{9}{a+9} - \cdots \text{ to } n \text{ terms}$$

$$= \frac{1}{(a+1)(a+4)(a+9)\ldots(a+n^2)}.$$

26. Under what conditions are the following equations consistent?
$$x + y + z = 1, \quad 2x - y + z = c_1, \quad (4a + b)y + az = c_2, \quad -bx + (a + 3b)y = c_3.$$

27. Show that, if the roots of $x^3 - ax^2 + bx - c = 0$ are not in A.P., then there are in general three transformations of the form $x = y + \lambda$, such that the transformed cubic in y has its roots in G.P.

28. (a) Prove algebraically that, if a, b, α, β are real numbers, then
$$\sqrt{(a+\alpha)^2 +(b+\beta)^2} \le \sqrt{a^2 +b^2} + \sqrt{\alpha^2 +\beta^2},$$
the positive root being indicated by the radical sign.

(b) If a and b are positive and $a + b = 1$, show that
$$\left(a+\frac{1}{a}\right)^2 + \left(b+\frac{1}{b}\right)^2 \ge 12\frac{1}{2}.$$

29. Examine, for different positive or negative values of x and p, the convergence or divergence of

(*i*) $\Sigma \dfrac{x^n}{n^p}$;

(*ii*) $\Sigma \dfrac{\sqrt{(n+1)} - \sqrt{n}}{n^p}$.

30. Show how to find the nth term and the sum of n terms of a recurring series of which the scale of relation is of the form $u_n + pu_{n-1} + qu_{n-2} = 0$ and of which the first four terms are known.

Apply the method to the series

(*i*) 1, 2, 5, 14, ... ,   ($p = -4, q = 3$)

(*ii*) 1, 2, 5, 12 ... ,   ($p = -2, q = -1$)

31. (*a*) Show that $2\Sigma b^2 c^2 (b^2 - c^2)(a^4 - b^2 c^2)$

$$= \Sigma (a^2 b^2 + a^2 c^2 - a^4)[(b^4 - c^2 a^2)(a^2 - b^2) + (c^4 - a^2 b^2)(c^2 - a^2)].$$

(*b*) Eliminate x, y, z from

$$x^2 y + x^2 z = a^3, \quad y^2 z + y^2 x = b^3, \quad z^2 x + z^2 y = c^3, \quad xyz = d^3.$$

32. (*a*) Show that any root of the number n can be expressed in the form

$$\sqrt[3]{a\sqrt[3]{b\sqrt[3]{c....}}},$$

where each of the quantities a, b, c, ... is n, 1 or n^{-1}.

(*b*) If x is less than $\sqrt{N}$, but greater than $\dfrac{1}{2}\sqrt{N}$, show that

$$\dfrac{3N + x^2}{N + 3x^2} \cdot x \text{ is nearer to } N^{\frac{1}{2}} \text{ than is either } \dfrac{N + x^2}{2x} \text{ or } x.$$

33. Show that

$$\begin{vmatrix} (a - a_1)^{-2} & (a - a_1)^{-1} & a_1^{-1} \\ (a - a_2)^{-2} & (a - a_2)^{-1} & a_2^{-1} \\ (a - a_3)^{-2} & (a - a_3)^{-1} & a_3^{-1} \end{vmatrix} = \pm \dfrac{a^2 \Pi (a_i - a_j)}{\Pi a_i \Pi (a - a_i)^2}.$$

Write out the terms of the product in the numerator, and give the resulting expression its correct sign.

34. Discuss the reality of the roots of

$$x^4 + 4x^3 - 2x^2 - 12x + a = 0,$$

for all real values of a.

[Denoting the equation by $f(x) = 0$, consider the roots of $f'(x) = 0$.]

35. Solve the equation $46x^3 + 72x^2 + 18x - 11 = 0$, by expressing the left-hand side as the sum of two cubes.

36. If $(1 + x)^n = 1 + c_1 x + c_2 x^2 + ... + c_n x^n$, show that

$$c_1(1 - x) - \dfrac{1}{2}c_2(1 - x)^2 + \dfrac{1}{3}c_3(1 - x)^3 - ... + (-1)^{n-1}\dfrac{1}{n}c_n(1 - x)^n$$

$$= 1 - x + \dfrac{1}{2}(1 - x^2) + \dfrac{1}{3}(1 - x^3) + ... + \dfrac{1}{n}(1 - x^n).$$

[Denoting the expressions by u and v, show that $\dfrac{du}{dx} = \dfrac{dv}{dx}$, etc.]

37. Prove that there is only one set of real values of x, y, z which satisfy the equation

$$(1 - x)^2 + (x - y)^2 + (y - z)^2 + z^2 = \frac{1}{4},$$

and find them.

[Writing a, b, c, d, for $(1 - x)$, $(x - y)$, $(y - z)$, z, we have $a + b + c + d = 1$ and $a^2 + b^2 + c^2 + d^2 = \dfrac{1}{4}$; hence $\Sigma(a - b)^2 = 0$.]

38. If ω is a fifth root of 2 and $x = \omega + \omega^2$, prove that
$$x^5 = 10x^2 + 10x + 6.$$

39. Prove that, if $r^2 = p^2 s$, the product of a pair of roots of the equation
$$x^4 + px^3 + qx^2 + rx + s = 0$$
is equal to the product of the other pair.

solve $\qquad\qquad x^4 - x^3 - 16x^2 - 2x + 4 = 0$

40. Obtain the roots of
$$x^4 - 6x^3 + 13x^2 - 6x + 1 = 0$$

in the form $(\sqrt{3} \pm \sqrt{2})\, e^{\pm\frac{\iota\pi}{6}}$.

41. Prove that $x^{4n+1} + \iota = (x + \iota).\ \Pi_{r=1}^{2n}\ (x^2 + 2\iota x \cos r\alpha - 1)$, where n is a positive integer and $(4n + 1)\alpha = 2\pi$.

42. If three real numbers satisfy the eqautions
$$x + y + z = 5 \quad \text{and} \quad yz + zx + xy = 8,$$

prove that none of them is less than 1 nor greater than $2\dfrac{1}{3}$.

43. Find a general formula for all the positive integers which, when divided by 5, 6, 7, will leave remainders 1, 2, 3 respectively, and show that 206 is the least of them.

44. Show that a determinant of order n can be expressed as one of order $n - 1$ as follows.

$$\text{If } \Delta = \begin{vmatrix} a_1b_1c_1...l_1 \\ a_2b_2c_2...l_2 \\ \cdots\cdots\cdots \\ a_nb_nc_n...l_n \end{vmatrix},\ \text{then } \Delta a_1{}^{n-2} = \begin{vmatrix} (a_1b_2)(a_1b_3)(a_1b_4)....(a_1b_n) \\ (a_1c_2)(a_1c_3)(a_1c_4)....(a_1c_n) \\ \cdots\cdots\cdots\cdots\cdots\cdots \\ (a_1l_2)(a_1l_3)(a_1l_4)....(a_1l_n) \end{vmatrix}$$

$$[\text{Multiply } \Delta \text{ by } \begin{vmatrix} a_1 & 0 & 0 & ... & 0 \\ -b_1 & a_1 & 0 & ... & 0 \\ -c_1 & 0 & a_1 & ... & 0 \\ \cdots & & & & \\ -l_1 & 0 & 0 & ... & a_1 \end{vmatrix}.]$$

45. Show that, if n is a positive integer, the number of solutions of the equation
$$n = 2n_1 + 3n_2,$$
where n_1 and n_2 are positive integers or zero, is eqaul to N or $N + 1$ according as r is or is not equal to 1, where N denotes the quotient and r the remainder when n is divided by 6.

46. Prove that $\dfrac{a^3 b^3}{(a+b)^4} \leq \dfrac{3^3 a^2}{4^4}$ or $\dfrac{3^3 b^2}{4^4}$, a, b being positive.

47. If n is a positive integer,

$$1 + 3n + \frac{3 \cdot 4}{1 \cdot 2} \cdot \frac{n(n-1)}{\underline{|2}} + \frac{4 \cdot 5}{1 \cdot 2} \cdot \frac{n(n-1)(n-2)}{\underline{|3}} + \dots + \frac{(n+1)(n+2)}{1 \cdot 2}$$

$$= (n^2 + 7n + 8)2^{n-3}.$$

48. If a, b, c, d are real and

$$\frac{b^2 - ac}{ad - bc} = \frac{ad - bc}{c^2 - bd},$$

show that each ratio $= \left(\dfrac{a}{d}\right)^{\frac{1}{3}}$ and that $a^{\frac{1}{3}}c + bd^{\frac{1}{3}} + a^{\frac{2}{3}}d^{\frac{2}{3}} = 0$, the real values of the roots being taken in all cases.

49. Prove that the condition that $(ax + by + cz)(bx + cy + az)(cx + ay + bz)$
can be expressed in the form $P^3 + Q^3$ where P and Q are linear functions of x, y, z is
$$a^3 + b^3 + c^3 = 3abc;$$
and find values of P and Q.

50. If $x_1, x_2, \dots x_n$ are positive numbers such that $\Sigma x = n$, prove that $\Sigma x^m \geq n$ if $m > 1$.

51. If r is an odd integer and $q_r = \dfrac{1 \cdot 3 \cdot 5 \dots (2r-1)}{2 \cdot 4 \cdot 6 \dots 2r}$, prove that
$$q_r + q_1 q_{r-1} + q_2 q_{r-2} + \dots$$
to $\dfrac{1}{2}(r+1)$ terms is equal to $\dfrac{1}{2}$.

52. If $\left(\dfrac{1+x}{1-x}\right)^n = 1 + a_1 x + a_2 x^2 + \dots + a_r x^r + \dots$, show that

$$(r + 1)a_{r+1} - 2na_r - (r - 1)a_{r-1} = 0.$$

Also show that

$$a_r = 2^n C_r^{n+r-1} - 2^{n-1} C_1^n C_r^{n+r-2} + 2^{n-2} C_2^n C_2^{n+r-3} - \dots + (-1)^{n-1} \cdot 2C_1^n.$$

53. Sum the series $\displaystyle\sum_{n=0}^{\infty} \frac{3n+2}{n^2 + 3n + 2} \cdot x^n.$

54. Examine the convergence of the series whose nth term is
$$x^n / (x^{2n} + x^n + 1)$$
for any value of x.

55. If m is a positive integer, show that

$$1 - m^2 + \frac{m^2(m^2 - 1)}{1^2 \cdot 2^2} + \dots + (-1)^m \frac{m^2(m^2 - 1^2) \dots \{(m^2 - (m-1)^2\}}{(\underline{|m})^2} = 0.$$

56. Prove that, as the real variable x changes steadily from $-\infty$ to $+\infty$, the function

$$y = \frac{(x-a)^2}{x-b},$$

where a and b are real and $a < b$, assumes twice overall values except those in an interval of length $4(b - a)$; and locate this range of values precisely.

57. Prove that, if
$$\begin{vmatrix} a & a^3 & a^4 & -1 \\ b & b^3 & b^4 & -1 \\ c & c^3 & c^4 & -1 \end{vmatrix} = 0,$$

and a, b, c are all different, then $abc\,(ab + bc + ca) = a + b + c$.

58. Prove that, in the expansion of $(1 + x)^n\,(1 - x)^{-2}$ in ascending powers of x, where n is a positive integer, the sum of the first $n + r$ coefficients is
$$2^{n-3}\{n^2 + n(4r + 3) + 4r(r + 1)\}.$$

59. The graph of $y = \dfrac{ax + b}{(x-1)(x-4)}$ has a turning point at $P(2, -1)$. Find the values of a and b, and show that y is a maximum at P. Sketch the curve.

60. Prove that, if $x \geq 100$, then $x^{\frac{1}{3}} - \dfrac{1}{2}\{(x+1)^{\frac{1}{3}} + (x-1)^{\frac{1}{3}}\} = \dfrac{1}{9}x^{\frac{-5}{3}}$ (nearly), the approximation being correct to at least eight decimal places.

61. If $f'(x)$ is positive, show that $f(x)$ is increasing.

Prove that $\quad 2x + x \cos x - 3 \sin x > 0$ if $0 < x < \dfrac{\pi}{2}$.

62. Prove that
$$\frac{1}{1\cdot 2\cdot 3} + \frac{1}{3\cdot 4\cdot 5} + \frac{1}{5\cdot 6\cdot 7} + \dots \text{ to infinity } = \log 2 - \frac{1}{2},$$

and that $\dfrac{1}{4} + \dfrac{1\cdot 3}{4\cdot 6} + \dfrac{1\cdot 3\cdot 5}{4\cdot 6\cdot 8} + \dots \text{ to infinity } = 1$.

63. Solve the equations:
$$\left. \begin{array}{l} y^2 + z^2 - x(y + z) = a \\ z^2 + x^2 - y(z + x) = b \\ x^2 + y^2 - z(x + y) = c \end{array} \right\}.$$
[Show that $b + c - a = 2\,(x^2 - yz)$, etc.]

64. By expanding $(1 - 2x - 3x^2)^{-1}$ in two different ways, prove that
$$2^n + (n - 1)\cdot 3\cdot 2^{n-2} + \frac{(n-2)(n-3)}{\lfloor 2} 3^2\cdot 2^{n-4}$$
$$+ \frac{(n-3)(n-4)(n-5)}{\lfloor 3} 3^3\cdot 2^{n-6} + \dots = \frac{1}{4}\{3^{n+1} + (-1)^n\}.$$

65. If $ax + by + cz = 1$ and a, b, c are positive, show that the values of x, y, z, for which
$$\frac{1}{x} + \frac{1}{y} + \frac{1}{z} \text{ is stationary, are given by}$$
$$ax^2 = by^2 = cz^2.$$
Show that this is a true maximum or minimum, if $xyz > 0$.

66. Show that the eqaution
$$\tan x = 1 + x$$

has an infinite number of real roots, and find graphically the approximate value of the smallest positive root.

67. Assuming that $x\{\log(1+x)\}^{-1}$ can be expanded in ascending powers of x, find the first four terms of the expansion.

By help of this result prove that a capital sum accumulating at compound interest at r per cent per annum will be increased ten-fold after

$$\left(\frac{230.26}{r}+1.15\right) \text{ years approximately.}$$

68. If $x = y - y^2 + 2y^3 - y^4$, and it is assumed that y can be expanded in ascending powers of x, show that

$$y = x + x^2 - 4x^4 + \ldots .$$

69. Draw graphs of

$$(i) \ \ y = \frac{2(x^2 - 6x)}{x^2 + x - 6};$$

$$(ii) \ \ y^2 = \frac{2(x^2 - 6x)}{x^2 + x - 6}.$$

70. By expanding

$$\left(2 + x + \frac{1}{x}\right)^{2m+n}$$

in two ways, prove that

$$\underline{|2m+n}\ \sum_{r=0}^{r=m} \frac{2^{2m-2r}}{\underline{|r}\,\underline{|2m-2r}\,\underline{|n+r}} \ = \ \frac{\underline{|4m+2n}}{\underline{|2m+2n}\,\underline{|2m}}.$$

71. Show that $\left(a - \dfrac{1}{a} - x\right)(4 - 3x^2)$, where a is a positive constant, has one maximum value and one minimum value. Find these values and show that the difference between them is $\dfrac{4}{9}\left(a + \dfrac{1}{a}\right)^3$. What is the least value of this difference for various values of a?

72. Prove that $$x^4 - 18x^2 + 4dx + 9 \ = \ 0$$
has four real roots, if $$d^4 \ \leq \ 1728.$$

73. If $$r^2x \ = \ aR + bR^3, \quad \text{where } R \geq r,$$
and the quantities denoted by all the letters are real and positive, prove that
$$x \ \geq \ 2\sqrt{ab}.$$

74. Prove that, if ε is small, one root of
$$x^3 + x^2 - (1 + 2\varepsilon^2)\, x - 1 \ = \ 0$$
is approximately $1 + \dfrac{1}{2}\varepsilon^2$, and find corresponding approximations to the other roots.

75. If α, β, γ are the roots of
$$x^3 - ax^2 + bx - c \ = \ 0,$$
the area of the triangle, of which the lengths of the sides are α, β, γ, is

$$\frac{1}{4}\{a(4ab - a^3 - 8c)\}^{\frac{1}{2}}.$$

If the triangle is right-angled, show that
$$a(4ab - a^3 - 8c)\,(a^2 - 2b) \ = \ 8c^2.$$

76. If n is a positive integer, prove that

(i) $\dfrac{2^n}{n} < \dfrac{n^n}{\lfloor n} < 3^{n-1}$ when $n > 3$;

(ii) $1^1 \cdot 2^2 \cdot 3^3 \ldots n^n < \left(\dfrac{2n+1}{3}\right)^{\frac{n(n+1)}{2}}$ when $n > 1$.

77. If y is positive, show that $\log y$ lies between $2\left(\dfrac{y-1}{y+1}\right)$ and $\dfrac{y^2-1}{y}$.

78. Prove that all the roots of the equation
$$x^4 - 14x^2 + 24x = k$$
are real if $8 < k < 11$.

79. Find a function $(ax^2 + 2bx + c)/(a'x^2 + 2b'x + c')$ which has turning values 3 and 4 when $x = 2$ and -2 respectively, and has the value 6 when $x = 0$.

80. If a, b are unequal positive quantities, show that
$$a^a b^b > \left(\dfrac{a+b}{2}\right)^{a+b}.$$

81. Show that two pencils of straight lines in one plane, containing p and q lines respectively, no two lines being coincident or parallel, divide the plane into $pq + 2p + 2q - 1$ parts.

82. Show that the number 30 may be divided into three unequal parts in 61 ways, while if equal parts are allowed, there are 75 ways.

83. In how many ways can a batsman make 14 runs in 6 balls, not scoring more than 4 off any ball?

84. Prove that the equation
$$x^4 + 4rx + 3s = 0$$
has no real roots if $r^4 < s^3$.

85. Find the minimum values of
$$\dfrac{a^m + x^m}{2} - \left(\dfrac{a+x}{2}\right)^m \quad \text{and} \quad \dfrac{a^m + b^m + x^m}{3} - \left(\dfrac{a+b+x}{3}\right)^m$$
for positive values of x, where a, b, m are positive and $m > 1$.

86. Find the condition that $ax + b/x$ can take any real value for real values of x. Express $\xi = (x - a)(x - b)/(x - c)(x - d)$ in terms of y where $y = (x - d)/(x - c)$; and hence, or otherwise, show that ξ can take all real values if
$$(c - a)(c - b)(d - a)\,(d - b)$$
is negative.

87. Find the maximum and minimum values of
$$y = (x + 1)^3 (x + 3)^2 (x + 2),$$
and draw a rough graph of the curve.

88. Prove that $7^{2n} - 48n - 1$ is divisible by 2304.

89. If
$$(x_1 - x_2)^2 + (y_1 - y_2)^2 = a^2,$$
$$(x_2 - x_3)^2 + (y_2 - y_3)^2 = b^2,$$
$$(x_3 - x_1)^2 + (y_3 - y_1)^2 = c^2;$$

$$\text{then } 4 \begin{vmatrix} x_1 & y_1 & 1 \\ x_2 & y_2 & 1 \\ x_3 & y_3 & 1 \end{vmatrix}^2 = (a+b+c)(b+c-a)(c+a-b)(a+b-c).$$

90. Show that an approximate solution of $x \log x + x - 1 = \varepsilon$, where ε is small, is
$$x = 1 + \varepsilon/2 - \varepsilon^2/16.$$

91. Find the limits of $\dfrac{x^3 + y^3}{x - y}$, as x, y tend to zero along the curves

    (i) $y = x - x^2$,   (ii) $y = x - x^3$,

    (iii) $y = x - x^6$.

92. If a, b, c are real, prove that square root of $a^2 + b^2 + c^2 - bc - ca - ab$ is greater than $\dfrac{1}{2}(2a - b - c)$ or than $\dfrac{\sqrt{3}}{2}(a - c)$, whre $a > b > c$.

93. Prove that if $\quad a_r = 1 + \dfrac{1}{\lfloor 1} + \dfrac{1}{\lfloor 2} + ... + \dfrac{1}{\lfloor r}$,

    then $\qquad 1 + \dfrac{1}{2}a_1 + \dfrac{1}{2^2}a_2 + ... \text{ to } \infty = 2\sqrt{e}.$

94. Show that a sum of n pence can be made up of pennies, halfpennies and farthings in $(n + 1)^2$ different ways.

95. If $x > 2$, prove that
$$1 + \frac{n}{x} + \frac{n(n+1)}{\lfloor 2x^2} + ... = 1 + \frac{n}{x-1} + \frac{n(n-1)}{\lfloor 2(x-1)^2} + ...$$

and deduce that
$$n + \frac{n(n-1)}{\lfloor 2}\frac{(r-1)}{\lfloor 1} + \frac{n(n-1)(n-2)}{\lfloor 3}\frac{(r-1)(r-2)}{\lfloor 2} + ... = \frac{n(n+1)...(n+r-1)}{\lfloor r}.$$

96. Find approximations to the roots of
$$(x-1)(x-2)(x-3) = \alpha,$$
where α is small.

97. If $3u_n - 7u_{n-1} + 5u_{n-2} - u_{n-3} = 0$ and $u_0 = 1$, $u_1 = 8$, $u_2 = 17$, prove that
$$2u_n = 20n - 7 + 3^{2-n}.$$

98. Expand $\left\{ 2x + \sqrt{(9 + 3x^2)} \right\}/(3 - x)$ in ascending powers of x, as far as the fifth power of x, and show that for small values of x it leads to a good approximation for e^x.

Deduce that $e^{\frac{1}{4}} = 1.2840 ...$.

99. If a, b are positive, and p and q are positive rationals such that $1/p + 1/q = 1$, prove that
$$ab \le a^p/p + b^q/q.$$

100. By means of the expansions of e^x and $\log(1 + x)$, prove that, when n is large,
$$\left(1 + \frac{1}{n}\right)^n = e\left(1 - \frac{1}{2n} + \frac{11}{24n^2} - \frac{7}{16n^3} + ...\right).$$

Show that e is given approximately by
$$2e = (1 \cdot 1)^{10} + (0.9)^{-10},$$
with an error of about $0 \cdot 46$ per cent.

101. If $r > p > q$, show that the number of combinations of $p + q$ things r at a time, p things being of one sort and q of another, is
$$p + q - r + 1.$$

102. Show that one root of the equation
$$x^3 = 100(x - 1)$$
is approximately 1.0103, and find the other roots correct to two places of decimals.

103. Prove that, if n is a positive integer, the coefficient of x^n in $(1 + x + x^2)^n$ is
$$1 + \frac{n(n-1)}{(\underline{|1})^2} + \frac{n(n-1)(n-2)(n-3)}{(\underline{|2})^2} + \dots .$$

104. The number of ways of selecting r pairs from n different things is
$$\frac{n(n-1)\dots(n-2r+1)}{2^r \underline{|r}} = \frac{\underline{|n}}{2^r \underline{|r}\,\underline{|n-2r}}.$$

105. Having given that
$$x + y + z = 1,$$
$$x^2 + y^2 + z^2 = 2,$$
$$x^3 + y^3 + z^3 = 3,$$
prove that $\quad x^4 + y^4 + z^4 = 4\dfrac{1}{6}.$

106. If a, b, c are real and
$$a^2 + b^2 + c^2 + 2abc = 1,$$
then a^2, b^2, c^2 are all greater than 1 or all less than 1.

107. If $x + y + z = xyz$ and $x^2 = yz$, then y and z may have any real values and $x^2 \geq 3$.

108. Solve $\quad x^{\frac{1}{4}} + (1552 - x)^{\frac{1}{4}} = 10.$

109. Show that the result of eliminating x and y from the three equations
$$\frac{1}{x-a} + \frac{1}{y-a} = \frac{1}{a}, \qquad \frac{1}{x-b} + \frac{1}{y-b} = \frac{1}{b},$$
and
$$x^2 + y^2 = 2(a^2 + b^2),$$
is
$$a^2 + b^2 - 6ab = 0.$$

110. Prove that $\log_e \{\log_e (1+x)^{\frac{1}{x}}\} = -\dfrac{1}{2}x + \dfrac{5}{24}x^2 - \dfrac{1}{8}x^3 - \dots .$

Find, without using tables, the value of $\log_e(\log_e 1.01)$ correct to five places of decimals, having given $\log_e 10 = 2.302585$.

111. Show that the coefficient of x^{3n+1} in the expansion of $\dfrac{8 - 2x}{(x+2)(x^3 + 8)}$ in a series of ascending powers of x is
$$(-1)^{n+r}\frac{3n+3}{2^{3n+3}}.$$

112. Find to five places of decimals the real root of $x^3 + x + 1 = 0$.

113. If $x^{3-\mu} = a^3$ where μ is small, show that an approximate root of the equation is given

by $x = a\left\{1 + \frac{1}{3}\mu\,\log_e a + \frac{1}{9}\mu^2\,\log_e a\left(1 + \frac{1}{2}\log_e a\right).\right\}$

114. Prove that $(1+x)^{\frac{1}{1-x}} = 1 + x + x^2 + \frac{3}{2}x^3 + \frac{11}{6}x^4 + \dots$.

115. The difference between the A.M. and the G.M. of a set of nearly equal numbers is to a first approximation equal to the quotient of half the difference between the square of their A.M. and the A.M. of their squares by any of the numbers.

116. Find the sum of the terms after the nth in the expansion of $(1 + x)/(1 - x)^2$ in ascending powers of x.

Prove that the ratio of this sum to the sum of the corresponding terms in the expansion of $(1 - x)^{-1}$ can be made equal to any given number λ, which is greater than $2n$, by suitable choice of x. Explain clearly why the restriction upon λ is necessary.

117. An examination consists of three papers, to each of which m marks are assigned as a maximum. One candidate obtains a total of $2m$ marks on the three papers. Show that

there are $\frac{1}{2}(m + 1)(m + 2)$ ways in which this may occur.

118. Show that the $2n$-th convergent of $\dfrac{3}{4-}\dfrac{4}{3+}\dfrac{3}{4-}\dfrac{4}{3+}\dots$ differs from the value of the continued fraction by $13/\{9 \cdot 12^n + 4\,(-1)^n\}$.

[Prove that

   (i) $p_{2n} = 11p_{2n-2} + 12p_{2n-4}$, with a like result for the q's ;

   (ii) $q_{2n} = p_{2n} + (-1)^n$ and $q_{2n-1}/4 = (q_{2n} - p_{2n-1})/5 = p_{2n}/9.$]

119. If the infinite product

$$(1 + x)^2\left(1 + \frac{x}{2}\right)^2\left(1 + \frac{x}{2^2}\right)^2\left(1 + \frac{x}{2^3}\right)^2 \dots$$

is expressed in the form $a_0 + a_1 x + a_2 x^2 + \dots + a_r x^r + \dots,$

then $\qquad (2^{r+1} - 1)\,a_{r+1} = 4\,(a_r + a_{r-1}).$

120. Find the coefficient of z^n in the expansion in ascending powers of z of $\dfrac{1 + xz}{1 - z - xz^2}$, and show that

$$\frac{1}{\sqrt{4x+1}}\left\{\left(\frac{1 + \sqrt{4x+1}}{2}\right)^{n+2} - \left(\frac{1 - \sqrt{4x+1}}{2}\right)^{n+2}\right\}$$

$$= 1 + nx + \frac{(n-1)(n-2)}{\lfloor 2}x^2 + \frac{(n-2)(n-3)(n-4)}{\lfloor 3}x^3 + \dots .$$

121. Prove that

$$mx^{m+n} - (m + n)x^m + n,$$

where m and n are positive integers, is divisible by $(x - 1)^2$, and that the quotient is

$$mx^{m-n-2} + \dots + p_r x^r + \dots + n$$

where $\quad p_r = m(m + n - r - 1)\ (r \geq m), \quad p_r = (r + 1)n\ (r < m).$

122. Prove that the number of ways in which n prizes may be distributed among q people so that everybody may have one *at least* is

$$q^n - q(q - 1)^n + \frac{q(q-1)}{\lfloor 2}(q - 2)^n - \dots .$$

123. Prove that when x is sufficiently small

$$\frac{1}{\log(1+x)} = \frac{1}{x} + \frac{1}{2} - \frac{1}{12}x + \frac{1}{24}x^2 - \dots .$$

Also, if $n > 1$,

$$\frac{1}{\log\left(1 + x + \dfrac{x^2}{\lfloor 2} + \dots + \dfrac{x^n}{\lfloor n}\right)} = \frac{1}{x} + \frac{x^{n-1}}{\lfloor n+1} - \frac{x^n}{(n+2)\lfloor n} + \dots .$$

124. A man owes a sum of £a, and repays £b at the end of each year, partly to pay the interest at $100n$ per cent, and partly to repay capital. Show that the sum will be repaid in λ years, where λ is the integer equal to or next greater than

$$-\log\left(1 - \frac{an}{b}\right)\Big/\log(1+n).$$

125. Show that the number of combinations taken n together of $3n$ letters of which n are a, n are b, and the rest unlike is

$$C_n^n + 2C_{n-1}^n + 3C_{n-2}^n + \dots + nC_1^n + n + 1,$$

and that this sum is equal to $2^{n-1}(n+2)$.

126. If a, b are positive integers, the probability that $\dfrac{1}{5}(a^2 + b^2)$ is a positive integer is $\dfrac{9}{25}$.

127. If
$$\begin{aligned} x + (b+c)y + bcz &= b - c, \\ x + (c+a)y + caz &= c - a, \\ x + (a+b)y + abz &= a - b, \end{aligned}$$

where $a \neq b \neq c$, prove that $zx - y^2 = 3$.

128. Ladders 15 ft. and 10 ft. long are placed in a passage, each with its top against a wall and its foot against the opposite wall. If the point where the ladders cross is 5 ft. from the ground, find the width of the passage.   (*Sunday Times*)

[If the width of the passage is x ft. and the tops of the ladders are a ft and b ft from the ground, then

$$\frac{1}{a} + \frac{1}{b} = \frac{1}{5}, \text{ and } a^2 - b^2 = 125.$$

Hence if $b = 5(y + 1)$, show that $0 < y < 1$, and
$$y^4 + 2y^3 + 5y^2 - 2y - 1 = 0,$$

giving $y = 0.5761287$, $b = 7.880644$, $x = \sqrt{(100 - b^2)} = 6.155929$.]

MISCELLANEOUS EXERCISES (B)

1. A party consisting of 7 robbers and a monkey had stolen a number N of nuts, which were to be divided equally among the robbers. In the night a robber woke and decided to take his share. He found that, if he gave one to the monkey, the remaining nuts could be divided into 7 equal lots. He gave one to the monkey and took his share. The other robbers woke in succession, found that the same thing was possible, and each gave one to the monkey and took one-seventh of the remaining nuts. In the morning they divided what were left, and they found that, if they gave one to the monkey, the nuts could be divided into 7 equal lots. Prove that the least value of N is $7^8 - 6 = 5764795$.

[The value -6 of N satisfies the numerical conditions; for $-6 -1 = -7$, $-7 \div 7 = -1$, $-7 - (-1) = -6$. Thus -6 is unaltered by the set of operations corresponding to what each robber did. Hence the values of N are given by $-6 + k.\,7^8$.]

2. Let a/b be a proper fraction and $c_1, c_2, c_3, \ldots$ an infinite sequence of positive integers. If $q_1, q_2, q_3, \ldots q_n$ are the quotients and $r_1, r_2, r_3, \ldots r_n$ the remainders when $ac_1, r_1c_2, r_2c_3, \ldots r_{n-1}c_n$ respectively, are divided by b, show that

(i)
$$\frac{a}{b} = \frac{q_1}{c_1} + \frac{q_2}{c_1c_2} + \frac{q_3}{c_1c_2c_3} + \ldots + \frac{q_n}{c_1c_2\ldots c_n} + R_n,$$

where
$$R_n = \frac{1}{c_1c_2\ldots c_n}\cdot\frac{r_n}{b};$$

(ii) $q_1 < c_1,\ q_2 < c_2,\ \ldots\ q_n < r_n.$

(iii) $r_n \equiv ac_1c_2 \ldots c_n \pmod{b}$;

(iv) If a is a prime to b and $a < b$, then

$$\frac{a}{b} = \frac{q_1}{c_1} + \frac{q_2}{c_1c_2} + \frac{q_3}{c_1c_2c_3} + \ldots,$$

where the series terminates if $c_1c_2 \ldots c_n$ is divisible by b, for some value of n: otherwise the series is convergent.

3. Express the following fractions in the form:

$$\frac{q_1}{3} + \frac{q_2}{3\cdot 5} + \frac{q_3}{3\cdot 5\cdot 7} + \ldots + \frac{q_n}{3\cdot 5 \ldots (2n+1)} + \ldots;$$

(i) $\dfrac{7}{9}$;   (ii) $\dfrac{1}{2}$;   (iii) $\dfrac{1}{4}$.

In the last two cases, find the value of q_n.

4. Show that $\dfrac{1}{8} = \dfrac{1}{3^2} + \dfrac{3}{3^2\cdot 5^2} + \dfrac{6}{3^2\cdot 5^2\cdot 7^2} + \ldots + \dfrac{\frac{1}{2}n(n+1)}{3^2\cdot 5^2(2n+1)^2} + \ldots.$

5. Use Ex. 2, (iv), to show that e is irrational.

6. Sum the series

$$n^2 + 2(n-1)^2 + 3(n-2)^2 + \ldots\,,$$

where n is a positive integer.

Prove that $S_1 + \dfrac{S_2}{\lfloor 2} + \ldots + \dfrac{S_n}{\lfloor n} + \ldots$ to infinity $= \dfrac{17e}{6}$,

where S_n is the sum of the squares of the first n positive integers.

7. Prove that, if x, y, z are positive and $x + y + z = 1$, then

$$\left(\frac{1}{x}-1\right)\left(\frac{1}{y}-1\right)\left(\frac{1}{z}-1\right) \geq 8.$$

8. From a cask containing x apples, a man sells half the contents of the cask and half an apple to one customer, half the remainder and half an apple to a second, and so on, always selling half of what are left and half an apple. In no case is an apple divided. After n sellings, a apples are left. Show that

$$x = 2^n(a + 1) - 1.$$

9. Show that the planes $ny - mz = \lambda$, $lz - nx = \mu$, $mx - ly = v$, have a common line if, and only if,

$$l\lambda + m\mu + nv = 0.$$

10. If
$$\beta = \frac{a\alpha + b}{c\alpha + d}, \quad \gamma = \frac{\alpha\beta + b}{c\beta + d}, \quad \text{where } ad - bc \neq 0,$$

and a, b, c, d are connected by the equation $a^2 + d^2 + ad + bc = 0$,

prove that
$$\alpha = \frac{a\gamma + b}{c\gamma + d}.$$

Show also that α, β, γ are the roots of

$$x.\frac{ax + b}{cx + d}.\frac{-dx + b}{cx - a} = \alpha.\frac{a\alpha + b}{c\alpha + d}.\frac{-d\alpha + b}{c\alpha - a}.$$

11. If $k_r = \dfrac{\lfloor 2r - 2}{\lfloor r \lfloor r - 1}$, show that $k_r = k_1 k_{r-1} + \ldots + k_p k_{r-p} + \ldots + k_{r-1}k_1$,

where r is a positive integer greater than unity, and $p < r$.

12. Show that $\dfrac{a^m + b^m + c^m}{3} > \dfrac{a^p + b^p + c^p}{3}.\dfrac{a^q + b^q + c^q}{3}.\dfrac{a^r + b^r + c^r}{3}$,

where p, q, r are positive quantities such that $m = p + q + r$.

13. If $x_r = x(x - 1)(x - 2) \ldots (x - r + 1)$, prove that
$$(2n - 1)_{n-1} . (2n - 3)_{n-3} . (2n - 5)_{n-5} . \ldots > n_n . n_{n-3} . n_{n-5} . n_{n-7} \ldots ,$$
each product being continued as long as the suffixes are greater than zero.

14. If l, m, l', m', l'' and m'' are integers, and if α/β is not rational, and if
$$l\alpha + m\beta = l'\alpha + m'\beta,$$
show that $l = l'$, and $m = m'$.

Also show that no two of the numbers
$$(2l + 5m)\alpha + l\beta, (2l' + 5m' + 1)\alpha + l'\beta, (2l'' + 5m'')\alpha + (l'' + 1)\beta$$
can be equal.

15. One root of the equation $x^6 - 9x^5 + 18x^4 + 9x^3 + 27x^2 - 54x - 36 = 0$ is $3 - 3^{\frac{1}{3}} - 3^{\frac{2}{3}}$; find all the roots.

16. If
$$a_1 yz + b_1(y + z) + c_1 = 0,$$
$$a_2 zx + b_2(z + x) + c_2 = 0,$$
$$a_3 xy + b_3(x + y) + c_3 = 0,$$
are true for an infinite series of values of x, y, z prove that, in general,
$$a_2 c_3 + a_3 c_2 = 2b_2 b_3, \quad a_3 c_1 + a_1 c_3 = 2b_1 b_3, \quad a_1 c_2 + a_2 c_1 = 2b_1 b_2.$$

17. If a, b, c are unequal, show that
$$ax^{b-c} + bx^{c-a} + cx^{a-b} \geq a + b + c.$$

18. If
$$a(x + y + b) + x^2 y^2 + bxy(x + y) = 0,$$
$$a(z + x + b) + z^2 x^2 + bzx(z + x) = 0,$$
where y and z are unequal, prove that
$$a(y + z + b) + y^2 z^2 + byz(y + z) = 0.$$

19. Prove that

$$\begin{vmatrix} 2 & \alpha + \beta + \gamma + \delta & \alpha\beta + \gamma\delta \\ \alpha + \beta + \gamma + \delta & 2(\alpha + \beta)(\gamma + \delta) & \alpha\beta(\gamma + \delta) + \gamma\delta(\alpha + \beta) \\ \alpha\beta + \gamma\delta & \alpha\beta(\gamma + \delta) + \gamma\delta(\alpha + \beta) & 2\alpha\beta\gamma\delta \end{vmatrix}$$

is equal to zero.

20. Prove that

$$\frac{nx}{(1+x)^n - (1-x)^n} = \Sigma \frac{A_r}{x^2 + \tan^2 \dfrac{r\pi}{n}},$$

where $A_r = (-1)^{r-1} \sin^2 \dfrac{r\pi}{n} \cos^{n-4} \dfrac{r\pi}{n}$ and $r = 1, 2, 3, \ldots \dfrac{1}{2}(n-1)$ or $\dfrac{1}{2}n - 1,$

according as n is odd or even.

21. Determine the condition that $(x^{n+1} - x^n + 1)$ shall be divisible by

$$(x^2 - x + 1).$$

22. Show that the first four terms in the expansions of

$$(1+x)^n + a\left(1 + \frac{x}{p}\right)^{pn} \quad \text{and} \quad b\left(1 + \frac{x}{q}\right)^{qn} + c\left(1 + \frac{x}{r}\right)^{rn}$$

are identical if

$$1 + a = b + c, \quad 1 + \frac{a}{p} = \frac{b}{q} + \frac{c}{r}, \quad 1 + \frac{a}{p^2} = \frac{b}{q^2} + \frac{c}{r^2},$$

and find the values of a, b, c in terms of $p, q, r.$

23. Find the limiting values of

$$(\cos x)^{\frac{1}{x}}, \quad (\cos x)^{\frac{1}{x^2}}, \quad (\cos x)^{\frac{1}{x^3}},$$

as x tends to zero through positive or negative values.

24. Show that, if $l_1, m_1, n_1; l_2, m_2, n_2; l_3, m_3, n_3$ are real quantities satisfying the six relations,

$$l_1^2 + m_1^2 + n_1^2 = l_2^2 + m_2^2 + n_2^2 = l_3^2 + m_3^2 + n_3^2 = 1,$$
$$l_2 l_3 + m_2 m_3 + n_2 n_3 = l_3 l_1 + m_3 m_1 + n_3 n_1 = l_1 l_2 + m_1 m_2 + n_1 n_2 = 0,$$

then

$$l_1^2 + l_2^2 + l_3^2 = m_1^2 + m_2^2 + m_3^2 = n_1^2 + n_2^2 + n_3^2 = 1,$$
$$m_1 n_1 + m_2 n_2 + m_3 n_3 = n_1 l_1 + n_2 l_2 + n_3 l_3 = l_1 m_1 + l_2 m_2 + l_3 m_3 = 0$$

and

$$\begin{vmatrix} l_1 & m_1 & n_1 \\ l_2 & m_2 & n_2 \\ l_3 & m_3 & n_3 \end{vmatrix} = \pm 1.$$

25. If $z^3 + 3hz + g = \dfrac{1}{\mu - v}\left[\mu(z + v)^3 - v(z + \mu)^3\right]$ for all values of z, show that

$$2\mu = \frac{g}{h} + \frac{\sqrt{\Delta}}{h} \quad \text{and} \quad 2v = \frac{g}{h} - \frac{\sqrt{\Delta}}{h},$$

where $\qquad\qquad\Delta = g^2 + 4h^3.$

Show that the cubic $4z^3 - 27a^2(z + a) = 0$ has two equal roots.

26. If $u = (b - c)^n + (c - a)^n + (a - b)^n$, where n is a positive integer, prove that

(*i*) If u is divisible by $\Sigma a^2 - \Sigma bc$, then n is of the form $3k \pm 1$, where k is an integer.

(*ii*) If u is divisible by $(\Sigma a^2 - \Sigma bc)^2$, then u is of the form $3k + 1$.

27. If all the roots of the equation

$$x^n + p_1 x^{n-1} + \ldots + p_n = 0$$

are rational and negative, show that

(*i*) $\left(\dfrac{p_1}{n}\right)^n \geq p_n,$ $\qquad\qquad$ (*ii*) $\left(\dfrac{p_1}{n}\right)^r \geq p_r \dfrac{\lfloor r \lfloor n - r}{\lfloor n}.$

28. Prove that, if n is a positive integer, and $\theta_r = (\alpha + 2r\pi)/n$, then

$$\frac{1}{x^{2n} - 2x^n \cos\alpha + 1} = \frac{1}{n\sin\alpha} \sum_{r=0}^{n-1} \frac{\sin\alpha - x\sin(\alpha - \theta_r)}{x^2 - 2x\cos\theta_r + 1}.$$

29. Investigate the maximum and minimum values of $(x + 1)^5/(x^5 + 1)$, and trace its graph. Prove that the equation $(x + 1)^5 = m(x^5 + 1)$ has three real roots if $0 < m < 16$ and only one real root (-1) if $m < 0$ or $m > 16$.

30. If n and $n + 2$ are both prime numbers, prove that $(n - 2) \lfloor n - 1 - 2$ is divisible by $n(n + 2)$.

31. Prove that, if ε is small, the equation

$$x^2 - 3x + 2 = \varepsilon \cdot x^3$$

has a root nearly equal to $1 - \varepsilon + 4\varepsilon^2$, and find approximations to the other two roots.

32. If $a, b, c, \ldots k, l$ are positive numbers arranged in descending order, and

$$a + b + c + \ldots + k + l = \frac{\mu a - l}{\mu - 1},$$

show that μ lies between the greatest and least of the numbers

$$a/b, \; b/c, \; c/d, \; \ldots, \; k/l.$$

33. Discuss the convergence of the series whose nth terms are

$$\frac{(\lfloor n)^2}{\lfloor 2n} x^n, \quad (-1)^n \frac{n}{n+1}.$$

34. Find the seventh roots of unity and show their positions in the z plane where $z = x + iy$. If z is any one of the imaginary roots, find the equation whose roots are

$$z + z^4 + z^2 \text{ and } z^6 + z^3 + z^5.$$

35. If d denotes the determinant

$$\begin{vmatrix} a_1 & a_2 & a_3 & a_4 & a_5 \\ a_5 & a_1 & a_2 & a_3 & a_4 \\ a_4 & a_5 & a_1 & a_2 & a_3 \\ a_3 & a_4 & a_5 & a_1 & a_2 \\ a_2 & a_3 & a_4 & a_5 & a_1 \end{vmatrix},$$

and D denotes the determinant obtained by changing a_r into A_r where

$$A_1 = a_1^2 + 2a_2a_5 + 2a_3a_4,$$
$$A_2 = a_4^2 + 2a_1a_2 + 2a_3a_5,$$
$$A_3 = a_2^2 + 2a_1a_3 + 2a_4a_5,$$
$$A_4 = a_5^2 + 2a_1a_4 + 2a_2a_3,$$
$$A_5 = a_3^2 + 2a_1a_5 + 2a_2a_4,$$

show that $d^2 = D$.

36. If m, n, p, q are positive integers and if x, y, z all tend to a, prove that

$$\lim \frac{x^m(y^n - z^n) + y^m(z^n - x^n) + z^m(x^n - y^n)}{x^p(y^q - z^q) + y^p(z^q - x^q) + z^p(x^q - y^q)} = \frac{mn(m-n)}{pq(p-q)}a^{m+n-p-q}.$$

37. Prove that, in the expansion of

$$(1 + x)^m + (1 - x)^m,$$

where $-1 < x < 1$, the terms are either all positive or, after a certain stage, become and remain negative, and that in the latter case the last positive term exceeds numerically the sum of the infinite series of negative terms.

Deduce or otherwise prove that

$$(1 + x)^m + (1 - x)^m - 2$$

vanishes with or has the same sign as $m(m - 1)x^2$.

38. Show that the number of distinct sets of three positive integers (none zero) whose sum is the odd integer $2n + 1$, is given by the least integer containing $\frac{1}{3}(n^2 + n)$.

39. Show that the number of distinct sets of three positive integers (none zero) whose sum is an even integer $2n$ is the integral part of $\frac{1}{3}n^2$.

40. Prove that

$$1 - \frac{1}{2}x^2 + \frac{1}{3}x^3 - (1 + x)e^{-x}$$

is positive for positive values of x, and increases as x increases.

41. Find to 6 places of decimals the root of

$$x^3 + 3x^2 - 42x - 41 = 0.$$

which lies between 5 and 6.

42. Prove that the coefficient of x^{3n} in the expansion of $\dfrac{1}{(1-x)(1-x^3)(1-x^6)}$ in powers of x is

$$\frac{1}{4}\left\{(n+2)^2 - \frac{1-(-1)^n}{2}\right\}.$$

43. If the volume of a right circular cone is 20 cubic feet, prove that, when the cone is such that the curved surface is a minimum, the radius of the base is approximately 2.381 feet.

44. Arrange the following numbers in order, so that as $x \to \infty$, the ratio of each number to the preceding may tend to infinity:

$$x^2,\ 2^x,\ x^x,\ e^x,\ x^{\log x},\ (\log x)^2,\ 2^{\log x}.$$

45. Prove that

$$\frac{(x-1)(x-2)\ldots(x-n)}{x(x+1)(x+2)\ldots(x+n)} = \sum_{r=0}^{n} (-1)^{n-r} \frac{\lfloor n+r}{(\lfloor r)^2 \lfloor n-r} \cdot \frac{1}{x+r};$$

and deduce, or otherwise prove, that

$$-\frac{\lfloor n+1}{\lfloor n-1} + \frac{\lfloor n+2}{\lfloor 1 \lfloor 2 \lfloor n-2} - \frac{\lfloor n+3}{\lfloor 2 \lfloor 3 \lfloor n-3} + \frac{\lfloor n+4}{\lfloor 3 \lfloor 4 \lfloor n-4} - \ldots + (-)^n \frac{\lfloor 2n}{\lfloor n-1 \lfloor n} = (-)^n n(n+1).$$

46. Prove that, if

$$\frac{a}{l^2} + \frac{b}{m^2} + \frac{c}{n^2} = \frac{a}{x^2} + \frac{b}{y^2} + \frac{c}{z^2} = \frac{al}{x^3} + \frac{bm}{y^3} + \frac{cn}{z^3} = 0,$$

where a, b, c are not zero and no two of $\dfrac{x}{l}, \dfrac{y}{m}, \dfrac{z}{n}$ are equal, then

$$\frac{x}{l} + \frac{y}{m} + \frac{z}{n} = 0.$$

47. If $a_1 a_2 a_3 \ldots a_n = \lambda^n$, prove that the sum of the products r together of the a's is greater than

$$\lfloor n \cdot \lambda^r / \{\lfloor r \lfloor n-r\}.$$

48. Solve the equations:

$$\begin{aligned}
x + y + z + w &= 1, \\
ax + by + cz + dw &= \lambda, \\
a^2 x + b^2 y + c^2 z + d^2 w &= \lambda^2, \\
a^3 x + b^3 y + c^3 z + d^3 w &= \lambda^3.
\end{aligned}$$

Prove that, if a, b, c, d, λ are all real and unequal, at least two and not more than three of x, y, z, w are positive.

49. Prove that if

$$\begin{aligned}
y^2 + yz + z^2 &= a^2, \\
z^2 + zx + x^2 &= b^2, \\
x^2 + xy + y^2 &= c^2, \\
yz + zx + xy &= 0;
\end{aligned}$$

then

$$a \pm b \pm c = 0.$$

50. Prove that $\dfrac{\lfloor 72}{(\lfloor 36)^2} - 1$ is divisible by 73.

51. Prove that if m is prime and $p < m$,

$$\lfloor p-1 \lfloor m-p + (-1)^{p-1} \equiv 0 \pmod{m}.$$

52. If $\alpha^2 + \beta^2 = \kappa\alpha\beta$, $\beta^2 + \gamma^2 = \kappa\beta\gamma$, $\gamma^2 + \delta^2 = \kappa\gamma\delta$, $\delta^2 + \varepsilon^2 = \kappa\delta\varepsilon$, and if α, β, γ, δ, ε are all different, then

$$\alpha^2 + \varepsilon^2 = (\kappa^4 - 4\kappa^2 + 2)\alpha\varepsilon.$$

53. Prove that

$$\frac{1}{\lfloor r} + \frac{1}{2^2} \cdot \frac{n(n-1)}{\lfloor 1 \lfloor r+1} + \frac{1}{2^4} \cdot \frac{n(n-1)(n-2)(n-3)}{\lfloor 2 \lfloor r+2} + \ldots = \frac{\lfloor 2n+2r}{2^n \lfloor n+r \lfloor n+2r}.$$

54. If
$$(1 + px + x^2)^n = 1 + a_1 x + a_2 x^2 + \ldots + a_{2n} x^{2n},$$

prove that
$$a_r = a_{2n-r};$$

and that $\ 1 + 3a_1 + 5a_2 + \ldots + (4n + 1)a_{2n} = (2n + 1)(2 + p)^n.$

55. If of $3n$ letters there are n a's, n b's and n c's, the number of combinations of these r together is equal to the number of combinations $3n - r$ together.

Also, if $n > r > 2n + 1$, the number of combinations is

$$\frac{1}{2}(n + 1)(n + 2) + (r - n)(2n - r)$$

and the maximum number of combinations is

$$\frac{1}{4}\{3(n + 1)^2 + 1\} \quad \text{or} \quad \frac{3}{4}(n + 1)^2,$$

according as n is even or odd.

56. Assuming that $1 + \dfrac{1}{2} + \dfrac{1}{3} + \ldots + \dfrac{1}{n} - \log n$ tends to a limit γ as $n \to \infty$, prove that

$$(i) \quad \lim_{n \to \infty} \left(1 + \frac{1}{2} + \ldots + \frac{1}{n} - \frac{1}{n+1} - \frac{1}{n+2} - \ldots - \frac{1}{n^2}\right) = \gamma;$$

$$(ii) \quad \lim_{n \to \infty} \frac{1}{n}\left(\frac{n}{1} + \frac{n-1}{2} + \frac{n-2}{3} + \ldots + \frac{1}{n} - \log\lfloor n\right) = \gamma.$$

57. Expand $\log(1 - x - x^2)$ as far as the term containing x^4, and if

$$\log(1 - x - x^2) = -u_1 x - \frac{1}{2}u_2 x^2 - \frac{1}{3}u_3 x^3 - \ldots ,$$

obtain an expression for u_n, and prove that
$$u_n = u_{n-1} + u_{n-2}.$$

58. Find $\ \displaystyle\lim_{x \to 1} \dfrac{1 - x + \log x}{1 - (2x - x^2)^{\frac{1}{2}}}.$

59. Find $\ \displaystyle\lim_{x \to 0} \dfrac{a^x - b^x}{c^x - d^x}$, where a, b, c, d are positive and $c \ne d$.

60. If
$$x^2 + y^2 + z^2 = \xi^2 + \eta^2 + \zeta^2 = 1,$$

and
$$x + y + z = \xi + \eta + \zeta = x\xi + y\eta + z\zeta = 0,$$

show that
$$3x^2 = (\eta - \zeta)^2 \quad \text{and} \quad x^2 + \xi^2 = \frac{2}{3}.$$

61. If n points are taken in space so that not more than 3 of them lie in any plane and not more than 2 in any straight line, show that the points define $\frac{1}{2}n(n - 1)$ straight lines

and $\frac{1}{2}\lfloor n - 1$ polygons of n sides.

62. Prove that the continued fraction $a - \dfrac{1}{a-}\dfrac{1}{a-}...\dfrac{1}{a-}\dfrac{1}{x}$ in which a is equal to -1 and is repeated any number of times, must have one of three values, and that if x satisfies the equation $2x^3 + 3x^2 - 3x - 2 = 0$, the fraction satisfies this equation.

63. If $(1 + x)^n = c_0 + c_1 x + ... c_n x^n$, where n is a positive integer, prove that

$$\frac{c_0}{1^2} - \frac{c_1}{2^2} + \frac{c_2}{3^2} ... + \frac{(-1)^n c_n}{(n+1)^2} = \frac{1}{n+1}\left(1 + \frac{1}{2} + \frac{1}{3} + ... + \frac{1}{n+1}\right).$$

64. If a, b, c are real and l, m, n are integers, find the values of x for which $(x - a)^l (x - b)^m (x - c)^n$ has maxima or minima.

Determine which of these values give maxima and which minima, (*i*) when l, m, n are all even, (*ii*) when l, m, n are all odd.

65. Solve the equations:

$$y + z = a(1 - yz), \quad z + x = b(1 - zx), \quad x + y + z - xyz = c(1 - xy - yz - zx).$$

66. Sum of the series $1 + \dfrac{2}{\lfloor 1} + \dfrac{3}{\lfloor 2} + \dfrac{8}{\lfloor 3} + \dfrac{13}{\lfloor 4} + \dfrac{30}{\lfloor 5} + \dfrac{55}{\lfloor 6} + ...$ to ∞.

67. Discuss the convergence of the series

$$\Sigma \frac{\lfloor 2n}{(\lfloor n)^2} z^n,$$

for all values of z, real or complex, distinguishing cases of absolute and conditional convergence.

68. The equations of four lines are

    (1) $y = a, z = a'$;   (2) $z = b, x = b'$;

    (3) $x = c, y = c'$;   (4) $x = y = z$.

Find the equations of the line drawn from the point (k, k, k) to cut the lines (1) and (2).

Prove that two lines can be drawn to cut all the four given lines, and that these lines cut the line (4) in two points determined by $x = y = z = k$, where

$$(k - a)(k - b)(k - c) = (k - a')(k - b')(k - c').$$

69. Discuss the reality and equality of the roots of

$$x^4 - 2\lambda(3x^2 - 4x + 3) + 3\lambda^2 = 0$$

for different real positive values of λ.

70. A bag contains 21 balls of 6 different colours, there being 6 of the first colour, 5 of the second, 4 of the third, and so on. Show that the number of different selections of 6 balls which can be made from the bag is 259.

71. If

$$\Delta f(a) = f(a + w) - f(a),$$

$$\Delta^2 f(a) = \Delta f(a + w) - \Delta f(a), \text{ etc.}$$

prove that

$$f(a + xw) = f(a) + x\Delta f(a) + \frac{x(x-1)}{\lfloor 2} \Delta^2 f(a) + ...$$

Use this to find log 7.063, given that

		Δ log	Δ^2 log	Δ^3 log
log 7	= 0.845098			
		12234		
log 7.2	= 0.857332		−334	
		11900		16
log 7.4	= 0.869232		−318	
		11582		
log 7.6	= 0.880814			

72. A series is such that the sum of the rth term and the $(r+1)$th is always r^4. Prove that

(*i*) the rth term is $\dfrac{1}{2}r(r-1)(r^2-r-1)-(-1)^r c$;

(*ii*) the sum of r terms is $\dfrac{r(r^2-1)(3r^2-7)}{30}+\dfrac{c}{2}\{1-(-1)^r\}$.

73. If n is a positive integer,

$$\frac{1}{2}<\frac{1}{3n+1}+\frac{1}{3n+2}+\ldots+\frac{1}{5n}+\frac{1}{5n+1}<\frac{2}{3}.$$

74. Prove that, if $\dfrac{p}{q},\dfrac{r}{s}$ are fractions such that $qr-ps=1$, then the denominator of any

fraction whose value lies between $\dfrac{p}{q}$ and $\dfrac{r}{s}$ is not less than $q+s$.

Prove that there are two and only two fractions with denominators less than 19 which

lie between $\dfrac{10}{13}$ and $\dfrac{4}{5}$.

75. Determine ranges of x for which the function $(\log x)/x$ (*i*) increases and (*ii*) decreases as x increases.

Hence prove the following theorems, wherein n is a given positive number and only positive values of x are considered:

(*i*) The equation $n^x = x$ has two roots, one root or none according to the value of n.

(*ii*) The inequality $n^x > x^n$ is a consequence of either $x > n$ or $x > n$, according to the value of n. State the critical values of n.

76. From a bag containing 9 red and 9 blue balls, 9 are drawn at random, the balls being replaced. Show that the probability that 4 balls of each colour will be included is a

little less than $\dfrac{1}{2}$.

77. For the continued fraction $\dfrac{1}{a_1+}\dfrac{1}{a_2+}\ldots$, prove that $p_{n-1}+q_n$ is not altered if

$a_1, a_2, \ldots a_n$ are permuted cyclically.

78. An approximate value of ϕ, measured in radians, is $3\sin\phi/(2+\cos\phi)$ provided that

$\phi<\dfrac{1}{2}\pi$. Establish this result when ϕ is small and show that the error is approximately

$\phi^5/180$.

Hence express approximately the acute angles of a right-angled triangle in terms of the sides and deduce that, if $c^2 = a^2 + b^2$, then

$$\frac{6(a+b)c + 3c^2}{ab + 2(a+b)c + 4c^2},$$

is nearly equal to $\dfrac{1}{2}\pi$ for all positive values of a and b.

79. Given a_1 things of one kind, a_2 things of another kind, etc., a_r things of an rth kind, show that the number of different groups which can be formed from one or more of the above things is

$$(a_1 + 1)(a_2 + 1) \dots (a_r + 1) - 1,$$

and show that if $r = 4$ and $a_1 < n < a_2 < a_3 < a_4$, the number of groups having n members is

$$\frac{1}{6}[(n + 1)(n + 2)(n + 3) - (n - a_1)(n - a_1 + 1)(n - a_1 + 2)].$$

80. When $n \geq 2$, if
$$x_n = a_n x_{n-1} + x_{n-2},$$
$$y_n = a_n y_{n-1} + y_{n-2},$$
and if
$$x_0 = 1, \quad x_1 = a_1, \quad y_0 = 0, \quad y_1 = 1,$$
verify by induction that $x_{n+1}y_n - x_n y_{n+1} = (-1)^{n+1}$, and deduce that

$$\frac{x_n}{y_n} = a_1 + \frac{1}{y_1 y_2} - \frac{1}{y_2 y_3} + \dots + (-1)^n \frac{1}{y_{n-1}y_n}.$$

81. Find the equation determining the values of x for which $\sin mx/\sin x$ is stationary. Hence, or otherwise, show that, if m is an integer, $\sin^2 mx/\sin^2 x$ never exceeds m^2.

82. A quadratic function of x takes the values y_1, y_2, y_3 corresponding to three equidistant values of x. Prove that if $y_1 + y_2 + 2y_2$, the minimum value of the function is

$$y_2 - \frac{(y_3 - y_1)^2}{8(y_1 + y_3 - 2y_2)}.$$

83. Show that $(n + 1)^{n-1}(n + 2)^n > 3^n(\lfloor n \rfloor)^2$.

84. The digits of a number are 1, 2, 3, 4, 5, 6, 7, 8, 9, written at random in any order. Show that the odds are 115 to 11 against the number being divisible by 11.

85. If m, n are positive, show that

$$\left(\frac{mn+1}{m+1}\right)^{m+1} > n^m.$$

86. The coefficient of x^{2n} in the expansion of $(1 + x^2)^n/(1 - x)^3$ is
$$2^{n-1}(n^2 + 4n + 2).$$

87. If
$$ax_1^2 + by_1^2 + cz_1^2 = ax_2^2 + by_2^2 + cz_2^2 = ax_3^2 + by_3^2 + cz_3^2 = d,$$
$$ax_2 x_3 + by_2 y_3 + cz_2 z_3 = ax_3 x_1 + by_3 y_1 + cz_3 z_1 = ax_1 x_2 + by_1 y_2 + cz_1 z_2 = f,$$

prove that
$$\begin{vmatrix} x_1 & y_1 & z_1 \\ x_2 & y_2 & z_2 \\ x_3 & y_3 & z_3 \end{vmatrix} = (d - f)\{(d + 2f)/abc\}^{1/2}.$$

88. Prove that

$$\begin{vmatrix} \sin\alpha & \sin\beta & \sin\gamma \\ \cos\alpha & \cos\beta & \cos\gamma \\ \sin 2\alpha & \sin 2\beta & \sin 2\gamma \end{vmatrix} = -4\sin\tfrac{1}{2}(\beta-\gamma)\sin\tfrac{1}{2}(\gamma-\alpha)\sin\tfrac{1}{2}(\alpha-\beta).\,\Sigma\sin(\alpha+\beta).$$

$$\begin{vmatrix} \sin^3\alpha & \sin^3\beta & \sin^3\gamma \\ \sin\alpha & \sin\beta & \sin\gamma \\ \cos\alpha & \cos\beta & \cos\gamma \end{vmatrix} = -\sin(\beta-\gamma)\sin(\gamma-\alpha)\sin(\alpha-\beta)\sin(\alpha+\beta+\gamma).$$

89. Denoting the number of combinations of n letters taken r together, all the letters being unlike, by $_nC_r$, show that if r of the letters are alike, the rest being unlike, the number of combinations is

$$_{n-r}C_r + \,_{n-r}C_{r-1} + \dots + \,_{n-r}C_1 + 1.\ (n \geq 2r)$$

Show that the number of combinations in groups of n which can be formed from $3n$ letters, of which n are a and n are b and the rest unlike, is

$$_nC_n + 2\,_nC_{n-1} + 3\,_nC_{n-2} + \dots + n\,_nC_1 + (n+1),$$

and show that this sum is $2^{n-1}(n+2)$.

90. Prove that if n is a positive integer,

$$1 - \frac{n}{1^2}x + \frac{n(n-1)}{1^2.2^2}x^2 - \frac{n(n-1)(n-2)}{1^2\cdot 2^2\cdot 3^2}x^3 + \dots = e^x\left\{1 - \frac{n+1}{1^2}x + \frac{(n+1)(n+2)}{1^2.2^2}x^2 - \dots\right\}.$$

91. Investigate the convergence of the series whose nth term is

$$e^{-2\left(1+\frac{1}{2}+\dots+\frac{1}{n}\right)}.$$

92. Prove that

$$\frac{1}{n(n+1)} - \frac{C_1^n}{(n+1)(n+2)} + \frac{C_2^n}{(n+2)(n+3)} - \dots + (-1)^n\frac{C_n^n}{2n(2n+1)} = \frac{\lfloor n-1 \rfloor \lfloor n+1 \rfloor}{\lfloor 2n+1 \rfloor}.$$

93. If $t^2 + at + b$ and $t^2 + at - b$ both have rational factors, then

$$a = \lambda(\mu^2 + v^2), \quad b = \lambda^2\mu v(\mu^2 - v^2),$$

where λ, μ, v are integers and μ is prime to v.

[For instance, $t^2 + 5t + 6$ and $t^2 + 5t - 6$.]

94. For the fraction $\dfrac{1^2}{1+}\dfrac{2^2}{1+}\dfrac{3^2}{1+}\dots$, prove that $p_n + q_n = \lfloor n+1 \rfloor$.

95. Discuss the reality of the roots of

$$16x^4 + 24x^2 + 16kx + 9 = 0,$$

for all real values of k, and solve the equation when $k = \sqrt{2}$.

96. If $ax^2 + 2hx + b > 0$ for all real values of x, show that

$$a(x^2 + y^2) + b(z^2 + 1) + h\{xz + y + \sqrt{3}(x - yz)\} > 0$$

for all real values of x, y, z, the coefficients being real.

97. 2^n players of equal skill enter for a tournament. They are drawn in pairs, and the winners of each round are drawn again for the next. Find the chance that two given competitors will play against each other in the course of the tournament. Also show that if $n = 7$ the chance that a given competitor either wins or is beaten by the actual winner is $\dfrac{1}{16}$.

98. If there are four relations,

$$A_i a_j + B_i b_j + C_i c_j + D_i d_j = 0 \quad \text{for} \quad i = 1, 2;\ j = 1, 2;$$

show that $\quad \dfrac{(B,C)}{(a,d)} = \dfrac{(C,A)}{(b,d)} = \dfrac{(A,B)}{(c,d)} = \dfrac{(A,D)}{(b,c)} = \dfrac{(B,D)}{(c,a)} = \dfrac{(C,D)}{(a,b)}.$

99. If $\sin x = xy^2$ wher x^2 and $y - 1$ are both small, show that

$\quad$ (i) $\ y = 1 - \dfrac{x^2}{12} + \dfrac{x^4}{1440}$ approx.;

$\quad$ (ii) $\ x^2 = -12(y - 1) + \dfrac{6}{5}(y - 1)^2$ approx.

100. In the game of *Craps*, each player in turn acts as 'banker'. The banker throws with two dice numbered 1 to 6. If he throws 7 or 11 he wins. If he throws 2, 3 or 12, he loses. If he throws any other number, he throws again and continues to throw until either the number he threw first or 7 turns up. In the first case he wins, in the second he loses.

Show that the odds *against* the bank are 251 to 244.

101. If $u_n u_{n-1} + au_n + bu_{n-1} + c = 0$, show that

$$\frac{u_n - \alpha}{u_n - \beta} = \frac{\beta + a}{\alpha + a} \cdot \frac{u_{n-1} - \alpha}{u_{n-1} - \beta},$$

where α, β are the roots of

$$x^2 + (a + b)x + c = 0,$$

unless $\alpha = \beta$, in which case

$$\frac{1}{u_n - \alpha} - \frac{1}{u_{n-1} - \alpha} = \frac{2}{a - b}.$$

Hence obtain the general solution of the first equation, and show that the results so obtained agree with those in Exercise XXXVIII, 15, p. 379.

102. If $\quad \begin{vmatrix} 1 & \cos x & 0 & 0 \\ \cos x & 1 & \cos\alpha & \cos\beta \\ 0 & \cos\alpha & 1 & \cos\gamma \\ 0 & \cos\beta & \cos\gamma & 1 \end{vmatrix} = 0,$

prove that $\sin^2\gamma \sin^2 x = \cos^2\alpha + \cos^2\beta - 2\cos\alpha \cos\beta \cos\gamma.$

103. If $\dfrac{ab}{b+x} - \dfrac{cd}{d+y} = \dfrac{bc}{x} - \dfrac{ad}{y} = z$, then either $\dfrac{x}{b} + \dfrac{y}{d} + 1 = 0$ or $z = a - c.$

104. If $\qquad a^2 x^2 + b^2 y^2 + c^2 z^2 = 0,$

$\qquad\qquad\quad a^2 x^3 + b^2 y^3 + c^2 z^3 = 0,$

and $\qquad \dfrac{1}{x} - a^2 = \dfrac{1}{y} - b^2 = \dfrac{1}{z} - c^2,$

prove that $\qquad a^4x^3 + b^4y^3 + c^4x^3 = 0,$

and $\qquad a^6x^3 + b^6y^3 + c^6z^3 = a^4x^2 + b^4y^2 + c^4z^2.$

105. X and Y are two places on a road, 4 miles apart, A man A, starting from X at any time between one and four o'clock, walks to Y at 4 miles an hour. A man B, starting from Y at any time between one and four o'clock, walks to X at 4 miles an hour. Show that the odds in favour of the men meeting on the way are as 5 : 4, all times of starting being equally likely.

[Suppose that A and B start at x hrs. and y hrs. past one, respectively. Taking axes Ox, Oy at right angles, set off $OL = 3$ units along Ox and draw a square on OL. The total number of cases is represented by the area of the square, and the number of favourable cases by the part of the square between the lines $x - y = \pm 1$.]

106. Show that, in a game of whist, the chance that a hand is void of a suit is about $\dfrac{1}{20}$.
Also the chance that in 25 deals a certain player has a hand void of a suit *at least twice* is about 0.358.

[Chance that a hand is void of a suit = 4. $C_{13}^{39} \div C_{13}^{52}$.]

107. If $u \equiv ax^4 + 4bx^3 + 6cx^2 + 4dx + e$ and t is any root of
$$4(x + H)^3 - a^2I(x + H) + a^2J = 0,$$
show that the roots of $u = 0$ are given by
$$(ax + b \pm \sqrt{t})^2 + t + 3H \mp G/\sqrt{t} = 0.$$

■ ■ ■

ANSWERS

EXERCISE I (*Page 8–9*)

7. $28^2 - 27^2$; $8^2 - 3^2$.　　**10.** 212, 213, ..., 219, 220.　　**22.** 20, 4836.

23. 6, 28, 496.　　**27.** 120.　　**28.** 360.　　**30.** $(i)\,\dfrac{(p^r-1)}{(p-1)}$; $(ii)\,\dfrac{(p^r-1)}{(p-1)} - r$

EXERCISE II (*Page 21*)

19. (i) 5.394, 0.006; (ii) 2.594; 0.002; (iii) 2.276; 0.002.

EXERCISE III (*Pages 29–31*)

1. $5x^3 - 20x^2 + 80x - 314$; 1258.　　**2.** $x^4 - 3x^3 - 6x^2 - 5x - 10$; -21.

3. $\dfrac{1}{3}x^3 - \dfrac{10}{9}x^2 + \dfrac{34}{27}x - \dfrac{40}{81}$; $\dfrac{1}{81}$.　　**4.** $\dfrac{1}{2}x^4 - \dfrac{1}{4}x^3 + \dfrac{1}{8}x^2 - \dfrac{1}{16}x + \dfrac{1}{32}$; $-\dfrac{33}{32}$;

5. $3x^2 + x - 3$; $-3x - 7$.　　**6.** $3x^3 + x - 3$; $5x + 1$.

7. 13.

8. (i) $(x+5)^3 - 13(x+5)^2 + 56(x+5)$;

　　(ii) $(x+1)^3 - (x+1) + 80$;

　　(iii) $(x+\dfrac{1}{3})^3 + (x+\dfrac{1}{3})^2 + \dfrac{2156}{27}$.

9. (i) $y^3 + 5y^2 - 9$;　(ii) $y^3 - 75y + 7$;　(iii) $y^3 - 15y^2 + 257$.

10. $-n + 7n(n+1) - 6n(n+1)(n+2) + n(n+1)(n+2)(n+3)$.

11. $-\dfrac{55}{4}$; 0, $\dfrac{11}{3}$.　　**12.** $(x+1)(x+2)(x-2)(x-3)$.

13. $(3x - 3y)(2x + y)(x^2 + xy + y^2)$.　　**14.** $3x^3 + 4x^2 - 5x - 2$.

15. $7x^2 - 11x - 6$.　　**16.** $3x^2 - 7xy + 4y^2$.

18. $(x+1)(x^2 - x + 1)(x^4 - x^3 + x^2 - x + 1)(x^8 + x^7 - x^5 - x^4 - x^3 + x + 1)$.

20. $a(ca' - c'a)^3 + b(ab' - a'b)^2(ca' - c'a) + c(ab' - a'b)^3 = 0$.

21. $a = 3, b = -3, c = 1$.

22. $p^2 - 3q^2 + 3r^2 - s^2 = 0$.　　**23.** $x^2 - 3y^2 + 3z^2 - w^2 = 0$.

24. $a = 5, b = -3, c = 1$.　　**25.** $a = 1, b = 2$.

26. $a = 8, b = 0$.　　**27.** $(3x + 4y - 1)(x - y + 3)$.

28. $(3x - 4y + 2)(x + 2y - 3)$.　　**29.** $x^2 - 2x + 4$.

30. $(x - 1)^2(x + 5)$; $(x + 3)^2(x - 3)$.　　**31.** (i) $3(p + a^2)x + q - 2a^3$.

33. $\lambda = 1$, $(2x + y - 2)(x - y + 1)$;　$\lambda = -7/8$, $(x - 7y + 4)(x - y - 4)/8$.

34. $\lambda = -1, (y-1)(y-2); \quad \lambda = -2, -(x+y)(x-y);$

$\lambda = -\dfrac{10}{9}, -(x+3y+4)(x-3y-4)/9.$

EXERCISE IV *(Pages 37–38)*

16. 120.   **17.** 60.   **18.** 15.   **20.** *(i)* 2^{2n}.

21. *(i)* $x^{\frac{1}{2}n(n+1)}$;   *(ii)* $\lfloor n+1$.

23. $(-1)^m C_m^n,$ if $n = 2m;$ $(-1)^m (n+1) C_m^n,$ if $n = 2m+1.$

EXERCISE V *(Page 42)*

1. $(x-1)^2.$ **2.** $2x+1.$ **3.** $x^2+x-3.$ **4.** $x^2+x-1.$

5. $(x-1)^2.$ **6.** $3x-4.$ **7.** $X = \dfrac{1}{4}(x+2),\ Y = \dfrac{1}{4}(x-2).$

8. $X = x^3+1,\ Y = x.$ **9.** $X = 1+x^2,\ Y = 1-x+x^2.$

10. $\lambda = -2\mu = 1;\ A = -\dfrac{1}{2},\ B = -1,\ C = \dfrac{3}{2},\ D = -\dfrac{1}{2}.$

EXERCISE VI *(Pages 48–50)*

3. $-2\Sigma x^3 + 3\Sigma x^2 y - 12xyz.$ **4.** $2\Sigma x^3 + 6\Sigma x^2 y - 12xyz.$

17. $(a-b)^2 + (b-c)^2 + (c-a)^2.$

18. $p = a+f-g-h,\ q = b+g-h-f,\ r = c+h-f-g.$

24. 6.   **25.** 1.

26. $-2.$ **27.** $\dfrac{(a-b)(b-c)(c-a)}{(1+ab)(1+bc)(1+ca)}.$

28. $-abc\,\dfrac{(b-c)(c-d)(a-b)}{(1+bc)(1+ca)(1+ab)}.$ **29.** $-(b+c-2a)(c+a-2b)(a+b-2c).$

30. $-9(b-c)(c-a)(a-b).$ **31.** $-4(b-c)(c-a)(a-b)(a+b+c).$

32. $3(b+c+2a)(c+a+2b)(a+b+2c).$

33. $24\,abc.$ **34.** $-(b-c)(c-a)(a-b)(\Sigma a^2 + 3\Sigma bc).$

35. $(1-abc)(1-a^2-b^2-c^2+2abc).$

36. (16) (25) (34);   (12) (24) (36) (65);   (16) (65) (53) (24).

EXERCISE VII *(Page 59)*

1. $2, \pi/6;\ 2, -\pi/6;\ 2, \dfrac{5\pi}{6};\ 2, -\dfrac{5\pi}{6}.$ **3.** $(18 + \iota)/25.$

4. *(i)* $(5 - \iota)/13;$ *(ii)* $(-2 + 2\iota)/8;$ *(iii)* $\dfrac{1}{2}\left(1 + \iota\cot\dfrac{\theta}{2}\right).$

5.   (*i*) $(x^2 + y^2)^{\frac{n}{2}}\left[\cos n\left(\tan^{-1}\dfrac{y}{x}\right) + \iota\sin n\left(\tan^{-1}\dfrac{y}{x}\right)\right]$;

    (*ii*) $(x - \iota y)\,(x^2 + y^2)$

    (*iii*) $(x^2 - y^2 - 1 - 2\iota xy)/(x^2 + y^2 - 2x + 1)$.

6. (*i*) 1625;   (*ii*) 13/125.   **7.** $-\pi/3$.

13. $1 - 2\iota$; $\dfrac{1}{2}(1 \pm \iota\sqrt{3}\,)$.

EXERCISE VIII (*Page 65*)

7. $2 - \iota, -4 - 3\iota$.

EXERCISE IX (*Page 75*)

3. If $X + \iota Y = (1 + x + \iota y)/(1 - x - \iota y)$, then $X = (1 - x^2 - y^2)/[(1 - x)^2 + y^2]$, and $Y = 2y/[(1 - x)^2 + y^2]$; $X = 0$, if $x^2 + y^2 = 1$. Geometrically, take any line *AOB,* so that $BO = OA$, and draw *AL*, *BM*, parallel to *Ox* to meet a line through 1, parallel to *BOA*, in *L* and *M* : construct *OMN* similar to *OL*1, with $\angle s$ 1*ON*, *LOM* described in opposite directions. If *A* represents *z*, then *L*, *M*, *N* represent $1 + z$, $1 - z$ and $(1 + z)/(1 - z)$. If $|z| = 1$, *OAL*1, *BO*1*M* are rhombi; hence $\angle OM$ is a right angle; and *N* lies on *Oy*.

5. If $a = p + \iota q$, $px + qy = p^2 + q^2$, which passes through (p, q) and is perpendicular to $qx = py$.

EXERCISE X (*Page 84*)

1. (*i*) $4x^3 + 2x^2 - 3x - 1 = 0$;   (*ii*) $2x^3 + 9x^2 + 11x + 3 = 0$;

    (*iii*) $4x^3 - 11x^2 + 9x - 2 = 0$;   (*iv*) $2x^3 - 9x^2 + 11x - 3 = 0$;

    (*v*) $4x^3 - 13x^2 + 17x + 1 = 0$.

2. (*i*) $4x^3 - 8x^2 + 8x - 3 = 0$;   (*ii*) $x^3 - 4x^2 + 8x - 6 = 0$.

3. $\dfrac{1}{2}(3 \pm \sqrt{5}\,)$; $\dfrac{1}{2}(-1 \pm \sqrt{5}\,)$.   **4.** $\pm 3/2$, $\dfrac{1}{2}(1 \pm \sqrt{5}\,)$.

5. 2, 9/2, –1/2.   **6.** 2, 1/3, –1/2.   **7.** 5, 3, 1, –1.   **8.** 4, 2, 1, $\dfrac{1}{2}$.

9. –2, 3, 6.   **10.** 3, –2, 1/2, –1/2.

11. (*i*) $x^3 - 8x - 15$;   (*ii*) $x^4 - 23x^2 + 59x - 52 = 0$.

12. (*i*) $x^3 - 14x^2 + 11x^3 - 75 = 0$;   (*ii*) $x^4 - 25x^3 + 375x^2 - 1260x - 11700 = 0$.

EXERCISE XI (*Pages 91–92*)

3. 4, –2.   **4.** 11, –2.   **5.** 4, –6.   **6.** 2, 0.

7. 7, –1.   **8.** 3.1, –3.   **9.** 7, –6.   **13.** 4.

14. 6.   **15.** 3, 6, –4.   **16.** 2/3.   **17.** 3/2, 2/3.

18. 2, 2, 2/3, –1/2. **19.** 3/2, 2/3.

EXERCISE XII (*Page 96*)

17. $-a, -b, -\dfrac{1}{2}(a+b) \pm \iota\sqrt{3a^2 + 2ab + 3b^2})$.

EXERCISE XIII   (*Pages 102–103*)

1. $\dfrac{1}{x-2} - \dfrac{1}{x-1} - \dfrac{1}{(x-1)^2}$.

2. $\dfrac{1}{x+3} - \dfrac{1}{x} + \dfrac{3}{x^2} - \dfrac{6}{x^3}$.

3. $\dfrac{10}{x-3} - \dfrac{10}{x-2} - \dfrac{9}{(x-2)^3} - \dfrac{5}{(x-2)^3}$.

4. $\dfrac{1}{x-1} + \dfrac{x}{x^2+x+1}$.

5. $\dfrac{2}{x-2} - \dfrac{x}{x^2+x+2}$

6. $\dfrac{3x+2}{x^2+3} - \dfrac{2x+1}{x^2+2}$.

7. $\dfrac{x+2}{x^2-x+1} + \dfrac{x-3}{(x^2-x+1)^2} - \dfrac{x-2}{(x^2-x+1)^3}$.

8. $\dfrac{2}{3(x-1)} + \dfrac{1}{(x-1)^2} + \dfrac{x+2}{3(x^2+x+1)}$.

9. $\dfrac{x-2}{x^2+1} - \dfrac{1}{x+2} + \dfrac{1}{(x+2)^2} + \dfrac{3}{(x+2)^3}$.

10. $\dfrac{4}{x^2+1} - \dfrac{3}{(x^2+1)^2} + \dfrac{1}{(x^2+1)^3} - \dfrac{4}{x^2+2}$.

11. $\dfrac{x-1}{x^2+1} + \dfrac{1}{(x^2+1)^2} - \dfrac{x-1}{x^2+2}$

12. $\left[\dfrac{4}{(x+1)^2} + \dfrac{5}{x^2+x+1} + \dfrac{3x}{(x^2+x+1)^2}\right]\Big/ 9$.

13. $\dfrac{1}{4(x-1)} - \dfrac{1}{4(x+1)} + \dfrac{1}{4(x-1)^2} + \dfrac{1}{4(x+1)^2} + \dfrac{1}{2(x-1)^3} - \dfrac{1}{2(x+1)^3}$.

14. $\dfrac{1}{2\sqrt{2}}\left[\dfrac{x+\sqrt{2}}{x^2+\sqrt{2}x+1} - \dfrac{x-\sqrt{2}}{x^2-\sqrt{2}x+1}\right]$.

15. $-1 + \dfrac{1}{1-ax} + \dfrac{1}{1-bx}$; $(a^{n+1} + b^{n+1})x^{n+1} - ab(a^n + b^n)x^{n+2}$.

16. $1 + \Sigma \dfrac{2a(a+b)(a+c)}{(a-b)(a-c)} \cdot \dfrac{1}{x-a}$.

17. $1 + \Sigma \dfrac{a^3}{(a-b)(a-c)} \cdot \dfrac{1}{x-a}$.

18. Put $x = d$ in Ex. 17.

19. $x + a + b + c + \Sigma \dfrac{a^4}{(a-b)(a-c)} \cdot \dfrac{1}{x-a}$; put $x = d$.

EXERCISE XIV (A) *(Pages 116–117)*

1. $\dfrac{1}{5}n\,(n+1)\,(n+2)\,(n+3)\,(n+4)$.

2. $\dfrac{1}{12}\,(3n-1)\,(3n+2)\,(3n+5)\,(3n+8)+\dfrac{20}{3}$.

3. $\dfrac{1}{3}n\,(2n+1)\,(2n-1)$.

4. $\dfrac{1}{6}n\,(n+1)\,(2n+13)$.

5. $\dfrac{1}{12}n\,(n+1)\,(n+2)\,(3n+1)$.

6. $\dfrac{1}{12}n\,(n+1)\,(n+2)\,(3n+5)$.

7. $\dfrac{1}{12}n\,(n+1)\,(n+2)\,(3n+13)$.

8. $\dfrac{3}{4}n\,(n+1)\,(3n+1)\,(3n-2)$.

9. $\dfrac{1}{2}n\,(4n^2+n-1)$.

10. 2926.

11. 286.

12. 50336.

13. 4300.

14. $\dfrac{1}{15}n\,(n+1)\,(n+2)\,(3n^2+6n+1)$.

15. $\dfrac{1}{6}n\,(n+1)\,(n+2)$.

16. $\dfrac{1}{12}n\,(n+1)^2\,(n+2)$.

19. (*i*) $\dfrac{1}{2}n\,(6n^2-3n-1)$; (*ii*) $\dfrac{1}{8}n\,(n-1)\,(9n^2-9n-2)$, where $n \geq 2$.

20. 220.

21. 344.

23. 715.

EXERCISE XIV (B) *(Pages 117–118)*

1. $\dfrac{n}{n+1}$.

2. $\dfrac{n}{2n+1}$.

3. $\dfrac{n(n+3)}{4(n+1)(n+2)}$.

4. $\dfrac{n(3n+7)}{20(3n+2)(3n+5)}$.

5. $\dfrac{n(n^2+6n+11)}{18(n+1)(n+2)(n+3)}$.

6. $\dfrac{n(3n+1)}{4(n+1)(n+2)}$.

7. $\dfrac{n(4n+5)}{3(2n+1)(2n+3)}$.

8. $\dfrac{n(29n^2+138n+157)}{36(n+1)(n+2)(n+3)}$.

9. $\dfrac{1}{2}-\dfrac{1\cdot3\cdot5...(2n+1)}{2\cdot4\cdot6...(2n+2)}$.

10. $\dfrac{4\cdot7....(3n+4)}{2\cdot5....(3n+2)}-2$.

11. $\dfrac{1-(n+1)x^n+nx^{n+1}}{(1-x)^2}$.

12. $6-\dfrac{2n+3}{2^{n-1}}$.

13. $\dfrac{a-\{a+(n-1)d\}r^n}{1-r}-\dfrac{dr(1-r^{n-1})}{(1-r)^2}.$ **14.** $34-(4n^2+12n+17)/2^{n-1}.$

15. $n/(1+x)\{1+(n+1)x\}.$ **16.** $\underline{|n+1}-1.$

17. $\left\{1-\dfrac{1}{1\cdot3\cdot5....(2n+1)}\right\}\Big/2.$ **18.** $\dfrac{n(n+2)}{(n+1)^2}.$

19. $\dfrac{1}{140}-\dfrac{1}{4(2n+7)(2n+9)}-\dfrac{2}{3(2n+5)(2n+7)(2n+9)}$

$$-\dfrac{1}{(2n+3)(2n+5)(2n+7)(2n+9)}$$

or $\quad \dfrac{1}{140}-\dfrac{1}{48}\left[\dfrac{1}{2n+3}+\dfrac{1}{2n+5}+\dfrac{1}{2n+7}-\dfrac{3}{2n+9}\right].$

20. $1-\dfrac{x^n}{(x+1)(x+2)...(x+n)}$ **21.** $\dfrac{1-x^n}{(1-x)^3}-\dfrac{nx^n}{(1-x)^2}-\dfrac{n(n+1)x^n}{2(1-x)}.$

22. $\dfrac{n(n+1)}{4(n+2)}.$ **23.** $\dfrac{2^n}{n+2}-\dfrac{1}{2}.$

24. $\dfrac{1}{2}-\dfrac{1}{n+2}\cdot\dfrac{1}{\underline{|n}}.$ **26.** $\dfrac{1}{x-1}-\dfrac{k}{x^k-1},$ where $k=2^n.$

27. $\dfrac{1}{1-x}-\dfrac{1}{1-x^k},$ where $k=2^n.$ **28.** $1-\dfrac{1}{(1+a_1)(1+a_2)...(1+a_n)}.$

EXERCISE XV (*Pages 130, 132*)

8. (*i*) $a^3+b^3;$ (*ii*) $-(a^3+b^3)^2;$
(*iii*) $36.$

20. (*i*) $(\beta-\gamma)(\gamma-\alpha)(\gamma-\beta);$ (*ii*) $(\beta-\gamma)(\gamma-\alpha)(\alpha-\beta)(\alpha+\beta+\gamma);$
(*iii*) $(\mu-v)(v-\lambda)(\lambda-\mu).$

EXERCISE XVI (*Pages 144, 147*)

1. $0,\pm\sqrt{\dfrac{3}{2}(a^2+b^2+c^2)}.$

19. $39^2=30^2+20^2+14^2+5^2.$

EXERCISE XVII (*Page 159*)

1. $\dfrac{11}{8},\dfrac{21}{16},-\dfrac{1}{16};$ (*ii*) $1,-2,-1,3.$ **2.** $3:-2:10.$

3. $-3/2.$

EXERCISE XVIII (*Pages 166–169*)

3. $(2, 0), (0, 2), \left(-\dfrac{1}{2}\pm\dfrac{11}{2}\iota,-\dfrac{1}{2}\mp\dfrac{1}{2}\iota\right), \left(-2\pm{}_\iota\sqrt{2}, -2\mp{}_\iota\sqrt{2}\right).$

4. $\left(1\pm\dfrac{1}{2}\sqrt{2}, 1\mp\dfrac{1}{2}\sqrt{2}\right), \left(-\dfrac{3}{4}\pm\dfrac{1}{4}\sqrt{57}, -\dfrac{3}{4}\mp\dfrac{1}{4}\sqrt{57}\right).$

5. $(0, 0), (\pm\sqrt{3}, \mp\sqrt{3}), (\pm\iota, \mp\iota).$ **6.** $2, 3, 4$, in any order.

7. $2, 1, \dfrac{1}{2}$, in any order.   **8.** $(3, 5), \left(-\dfrac{9}{2},\dfrac{5}{2}\right), \left(\dfrac{3}{4},-\dfrac{25}{4}\right).$

9. $(1, 1, 1), \left(-\dfrac{7}{5},-\dfrac{7}{3},7\right).$

10. $x, \omega x, \omega^2 x = 4$ or $-\sqrt[3]{36}$; $y, \omega^2 y \, \omega y = 2$ or $-\sqrt[3]{162}.$

11. $\left(\dfrac{1}{2}\sqrt{5}\pm\dfrac{1}{2}\iota\sqrt{3}, \dfrac{1}{2}\sqrt{5}\mp\dfrac{1}{2}\iota\sqrt{3}\right), \left(\dfrac{1}{2}\iota\sqrt{5}\pm\dfrac{1}{2}\sqrt{3},\dfrac{1}{2}\iota\sqrt{5}\mp\dfrac{1}{2}\sqrt{3}\right)$

or $x = -y = a$ fourth root of $\dfrac{3}{2}.$

[Use the identity, $x^4 + x^2 y^2 + y^4 = (x^2 + xy + y^2)(x^2 - xy + y^2).$]

12. The only solutions are $(0, a) (a, 0)$. [Remove fractions and show that $xy = 0$ or a^2, and that the latter alternative is impossible.]

13. $(1, 1), \left(\dfrac{a+b\pm\sqrt{(a-b)(a+3b)}}{2a},\dfrac{a+b\pm\sqrt{(b-a)(b+3a)}}{2b}\right)$ with any arrangement of signs.

14. $x = \dfrac{1}{2}\left\{1\pm\sqrt{(1+2b+2c-2a)}\right\}$, etc., with any arrangement of signs.

15. $x = \pm(a^2 b^2 + a^2 c^2 - b^2 c^2)/abc$, etc., the upper or lower sign to be taken throughout.

16. $kx = a^2(b^2 + c^2 - a^2)$, etc., where $k^2 = (b^2 + c^2 - a^2)(c^2 + a^2 - b^2)(a^2 + b^2 - c^2).$

17. (a, b, c) or $x = (ab - ab^2 c - 1)/(ab^2 c^2 - b^2 c + b)$, etc. [Put $x = a + \lambda, y = b + \mu, z = c + v.$]

18. $(1, 1, 1)$ or $x = (a + b + c)/(a - b - c)$, etc.

19. $(1, 1, 1), (-3, -3, -3), (\pm\sqrt{2}, \pm\sqrt{2}, 1\mp\sqrt{2})$, in any order.

20. $\left(\dfrac{1}{2},\dfrac{1}{6},-\dfrac{1}{6}\right).$

21. If $A = \pm\sqrt{(b^2 + c^2 - a^2)}, B = \pm\sqrt{(c^2 + a^2 - b^2)}, C = \pm\sqrt{(a^2 + b^2 - c^2)}$, then $2x = B + C, 2y = C + A, 2z = A + B$, giving eight solutions. [Show that $2x(y - z) = b^2 - c^2.$]

22. $(\pm\sqrt{ac},\pm\sqrt{bc}), \left(\dfrac{1}{2}(a - b + c)\pm\sqrt{Q}, \dfrac{1}{2}(b + c - a)\mp\sqrt{Q}\right)$ where

$$Q = \Sigma a^2 - 2\Sigma bc.$$

23. If $\Sigma bc \neq 0$, the only solutions are $(0, 0, 0)$, $(a, 0, 0)$, $(0, b, 0)$, $(0, 0, c)$. If $\Sigma bc = 0$, the equations are not independent. [Show that

$$x(y - z)/bc = \ldots = \ldots \, . \,]$$

24. $(0, 0, 0)$ or $x = \pm \dfrac{b^3 c^3 + c^3 a^3 + a^3 b^3 - 3a^2 b^2 c^2}{a(ca - b^2)(ab - c^2)}$, etc.

25. $(-1, -1, -1)$, $\left(\dfrac{1}{2}, \dfrac{1}{3}, \dfrac{1}{6}\right)$, $\left(\dfrac{4}{5}, \dfrac{3}{5}, 0\right)$.

26. The values of x, y, u, v are $\left(3, -2, \dfrac{3}{5}, \dfrac{7}{5}\right)$ or $\left(-2, 3, \dfrac{7}{5}, \dfrac{3}{5}\right)$.

27. $x = y = x' = y' = a.$

28. $\left(\dfrac{5}{4}, \dfrac{17}{4}\right)$.

29. $(2, 1)$, $(-1, -2)$, $(-5, 4)$.

31. $a^3 + b^3 + c^3 = a^2(b + c) + b^2(c + a) + c^2(a + b)$.

32. (*i*) Subtract the second equation from the first, and divide by $x - y$;

(*ii*) prove $(x^3 + y^3)(1 - xy) = 2a^4 - a^5$; then, from the fact that x, y are roots of $(2a - 1)t^2 + a(2a - 1)t + a^2(a - 1) = 0$, obtain $(2a - 1)(x^2 + y^2) = a^2$, $(2a - 1)(x^3 + y^3) = a^3(a - 2)$, and thus $xy = 2a^2 - a + 1 = a^2(a - 1) / (2a - 1)$

from (*i*); for (*iii*), (*iv*), (*v*), multiply the result (*ii*) by $a + 1$, and obtain $3a^4 = -2a - 1$, and the result follow.

34. $2x(b + c - a) = \ldots = \ldots -\Sigma a^2 + 2\Sigma bc$.

39. $(a^3 - 3ab^2 + 2c^3)(a^4 - 3a^2 b^2 + 3b^4 - 4ac^3) + 16d^3 (a^2 - b^2)^2 = 0$.

40. $a^4 + b^4 = c^2(a^2 + b^2)$.

42. $16 : 5 : -21$.

46. $(bc + ca + ab)(a^2 + b^2 + c^2 - bc - ca - ab) = 0$.

EXERCISE XIX (*Pages 179–180*)

1. $\dfrac{1}{2}(3 \pm \sqrt{5})$, $\dfrac{1}{2}(5 \pm \sqrt{21})$.   **2.** $(-7 \pm 3\sqrt{5})$, $\dfrac{1}{2}(1 \pm \iota\sqrt{3})$.

3. $\dfrac{1}{2}(-1 \pm \iota\sqrt{3})$, $\dfrac{1}{4}(-1 \pm \iota\sqrt{15})$.   **4.** $\dfrac{1}{2}(\pm\sqrt{5})$, $\dfrac{1}{4}(-7 \pm \sqrt{33})$.

5. 1, $\dfrac{1}{2}(-3 \pm \sqrt{15})$, $\dfrac{1}{4}(-1 \pm \iota\sqrt{15})$.   **6.** -1, $2 \pm \sqrt{3}$, $\dfrac{1}{2}(1 \pm \iota\sqrt{35})$.

7. -1, -1, $\pm \iota$, $\dfrac{1}{2}(3 + \iota\sqrt{7})$.

8. -1, $\pm \iota$, $\dfrac{1}{8}[3 + \sqrt{33} \pm \sqrt{(6\sqrt{33} - 42)}]$, $\dfrac{1}{8}[3 - \sqrt{33} \pm \iota\sqrt{(6\sqrt{33} + 42)}]$.

9. -1, -1, $\dfrac{1}{2}(-1 + \iota\sqrt{3})$.   **16.** $k = -3$.

EXERCISE XX (*Page 187*)

1. (*i*) 2.68061; (*ii*) 0.80125; (*iii*) 4.22524.
2. (*i*) 2 cos 40° or 1.53209, 0.34730, −1.87939.
    (*ii*) 0.69073, 0.11479, − 6.30553.

EXERCISE XXI (*Pages 201–202*)

9. $-1 \pm \sqrt{2},\ 1 \pm 2\iota.$

10. $\frac{1}{4}\left(-1+\sqrt{5}\pm\sqrt{-2+6\sqrt{5}}\right),\ \frac{1}{4}\left(-1+\sqrt{5}\pm\iota\sqrt{2+6\sqrt{5}}\right).$

11. $\frac{1}{2}\left(1\pm\sqrt{13}\right),\ \frac{1}{2}\left(1\pm\iota\sqrt{3}\right).$ 12. $\frac{1}{2}\left(1\pm\sqrt{5}\right),\ \frac{1}{2}\left(3\pm\sqrt{17}\right).$

13. $-2 \pm \sqrt{2},\ -4 \pm 2\iota.$

14. $(x^2 - x + 1)(x^2 - 3x + 3),\ \{x^2 - x(2 - \iota\sqrt{3}) - \iota\sqrt{3}\}\ \{x^2 - x(2 + \iota\sqrt{3}) + \iota\sqrt{3}\},$

 $\{x^2 - 2x + \frac{1}{2}(3 + \iota\sqrt{3})\}\ \{x^2 - 2x + \frac{1}{2}(3 - \iota\sqrt{3})\}.$

18. $\sqrt{5}\pm\sqrt{2+2\sqrt{5}},\ -\sqrt{5}\pm\sqrt{2-2\sqrt{5}}$; the other substitutions are given by $q^2 + 2q + 5 = 0.$

20. Roots as in Ex. 10; the other substitutions are given by
$$16t^2 + 4t - 3 = 0.$$

EXERCISE XXII (*Page 216*)

7. $\sqrt{2} = \frac{1}{2}(y^2 - 9y);\ \sqrt{3} = -\frac{1}{2}(y^3 - 11y);\ \sqrt{2} - \sqrt{6} + 3 = (y + 5)/(y + 1).$
8. $x^4 - 10x^2 + 1 = 0;\ y^3 - 18y - 110 = 0;\ z^9 - 15z^6 - 117z^3 - 125 = 0.$

EXERCISE XXIV (*Page 235*)

1. (*i*) Oscillates between 1 and 3; (*ii*) diverges; (*iii*) oscillates between −2 and +2; (*iv*) converges to 2; (*v*) diverges.

EXERCISE XXVI (*Pages 256–258*)

1. (*i*) Convergent; (*ii*) oscillates finitely.
7. Divergent.   8. Divergent.   9. Divergent.   10. Divergent.
11. Convergent.   12. Convergent.   13. Convergent.   14. Divergent.
15. Convergent.   16. Divergent.   17. Convergent.
18. Divergent if $p - q + 1 \geq 0$, convergent if $p - q + 1 < 0$.
19. Convergent.   20. Convergent.   21. Divergent.   22. Convergent.
23. $x < 1.$ 24. $x \leq 1.$ 25. $x \leq 1.$ 26. $x < 1.$
27. $x > 1.$ 28. $x < 1.$ 29. $x > a.$ 30. $x \neq 1.$

31. $x < 1.$ **38.** $\dfrac{1}{12}.$ **39.** $\dfrac{11}{18}.$ **40.** $\dfrac{49}{144}.$

41. $\dfrac{7}{24}.$

EXERCISE XXVII (*Page 267*)

10. (*i*) Between $\dfrac{1}{4}\left(5-\sqrt{41}\right)$ and $\dfrac{1}{2}$, or between 2 and $\dfrac{1}{4}\left(5+\sqrt{41}\right)$;

(*ii*) if $|x| < \dfrac{1}{4}\left(\sqrt{41}-5\right).$

13. $1 - \iota.$

EXERCISE XXVIII (*Pages 285–287*)

1. $\dfrac{1}{2}, \dfrac{3}{4}.$ **2.** $0, -\infty.$ **3.** $\dfrac{2}{3}, 0.$ **4.** 1 **5.** $7/3.$

6. $\infty.$ **7.** $\dfrac{3}{2}x^{\frac{1}{2}}.$ **8.** $1.$

15. (*i*) $2\{(ab' - ab')x^2 + (ac' - a'c)x + bc' - b'c\}/(a'x^2 + 2b'x + c')^2$;

(*ii*) $-(1-x)^{-\frac{1}{2}}(1+x)^{\frac{3}{2}}$;   (*iii*) $(a^2 - 2x^2)/\left(a^2 + x^2\right)^{\frac{5}{2}}$;

(*iv*) $n \sin^{n-1} x \cos x$;   (*v*) $-n \cos^{n-1} x \sin x$;

(*vi*) $n \tan^{n-1} x \sec^2 x.$

16. (*i*) $-(hx + by + f)/(ax + hy + g)$;   (*ii*) $(x^2 - ay)/(ax - y^2).$

21. $y = ax.$ **24.** $4/27$, when $x = 1/3.$

25. $1, 7/4$; the graph has a point of inflexion at $(1, 0)$, where the gradient is zero; another at $(3/2, -1/16)$, where the gradient is $-\dfrac{1}{4}$; a minimum at $(7/4, -27/256)$.

27. Between $x = -2$ and $x = 0$, and for $x > 3$; max. at $(0, -1)$; min. at $(-2, -65)$; min. at $(3, -190)$.

29. 18 cubic inches.

EXERCISE XXX (*Pages 306, 307*)

3. $a = 2, x = 1, 1, 5/2$; or $a = \dfrac{54}{25}, x = 3/5, 3/5, 5/6.$

18. Use $\Sigma x^9 = -27\Sigma(x + 3)^3$, and substitute the values of $\Sigma x^3, \Sigma x^2, \Sigma x.$

EXERCISE XXXI (*Pages 318, 319*)

2. $2e.$ **3.** $e.$ **4.** $e - 1.$ **5.** $3e/2.$

6. $1 - 2/e.$ **7.** $15e.$ **8.** $-1/e.$

9. $\dfrac{1}{x}[(x - 1) e^x + x + 1].$

EXERCISE XXXII (*Pages 326–328*)

5. (*ii*) 0.84510.

6. (*ii*) 1.04139.

7. (*ii*) 1.11394.

9. $-3/4m; + 1/(4m + 1)$.

10. $\log_e 2$.

11. $2 \log_e 2 - 1$.

12. $2 - \log_e 2$.

13. $\dfrac{1}{2\sqrt{x}} \log \dfrac{1+\sqrt{x}}{1-\sqrt{x}} + \dfrac{1}{2x} \log (1 - x)$.

14. $\dfrac{1}{4}\left[6+\left(x-\dfrac{1}{x}\right)\log\dfrac{1+x}{1-x}\right]$.

15. $c-\dfrac{b}{2}+\dfrac{1}{3}; \quad -\dfrac{c}{2}+\dfrac{b}{3}-\dfrac{1}{4}; \quad b = 1, c = \dfrac{1}{6}; a = \dfrac{1}{2}$.

17. $1.010299956\,\dfrac{6}{7}$; 0.3010299957.

19. $\dfrac{5}{4x}(x + 2) + \dfrac{x}{1-x} + \dfrac{1}{2x^2}(5 - x^2) \log (1 - x)$.

20. $\dfrac{4}{9x^2}(6 + 3x + 2x^2) + \dfrac{2x-x^2}{(1-x)^2} + \dfrac{1}{3x^3}(8 - x^3) \log (1 - x)$.

EXERCISE XXXIII (*Pages 343–344*)

10. Convergent for all values of θ (see Art. 11, Ex. 1).

13. (*i*) If $\gamma > \alpha + \beta + 1$. (*ii*) if $\gamma > \alpha + \beta$.

EXERCISE XXXIV (*Pages 354–356*)

1. 8th term. **2.** 11th term. **3.** 8th term. **4.** 5th term.

5. 6th and 7th terms. **6.** 5th and 6th terms.

7. $\dfrac{4}{5} < x < \dfrac{6}{7}$. **8.** 11th term.

21. $10.0033322284 ; 5.10^{-13}$.

22. $1.414214; 5.10^{-7}$.

23. $1.5860098; 5.10^{-8}$.

24. $1.70998 ; 3.10^{-6}$.

25. $1.2431626; 4.10^{-8}$.

26. $1.319508; 3.10^{-7}$

EXERCISE XXXV (*Page 363*)

10. $(1 + x)^{n-2} [1 + (3n + 2)x + (n + 1)^2x^2]$.

11. $\dfrac{(1+x)^n}{(n+1)(n+2)} \left[(n+1)^2 - \dfrac{n}{x} + \dfrac{1}{x^2}\right] - \dfrac{1}{(n+1)(n+2)x^2}$.

12. $P - Q$, where $P = \dfrac{(1+x)^n}{(n+1)(n+2)(n+3)}\left[n+2+\dfrac{2n+3}{x}+\dfrac{n}{x^2}-\dfrac{1}{x^3}\right]$ and

$$Q = \dfrac{1}{2(n+1)x} - \dfrac{1}{(n+1)(n+2)(n+3)x^3}.$$

13. $P - Q$, where $\quad Q = \dfrac{3}{2(n+1)x} - \dfrac{13}{(n+1)(n+2)(n+3)x^3}$ and

$$P = \frac{(1+x)^n}{(n+1)(n+2)(n+3)} \left[n^2 + n^2 - n + 2 - \frac{5n^2 - n - 9}{x} + \frac{13n}{x^2} - \frac{13}{x^3} \right].$$

EXERCISE XXXVI (*Page 366*)

1. (*i*) $1 + 4x + 11x^2 + 26x^3 + 57x^4 + 120x^5 \; ; \; 2^{n+2} - n - 3.$
   (*ii*) $1 + 2x + 2x^2 - 4x^3 - 34x^4 - 148x^5 \; ; \; -1 + 3 \cdot 2^n - 3^n.$

EXERCISE XXXVII (*Page 373*)

1. (*i*) $\dfrac{1}{2}(1 + 3^{n-1}), \; \dfrac{1}{4}(2n - 1 + 3^n);$

   (*ii*) $\dfrac{1}{2\sqrt{2}} [(1 + \sqrt{2})^n - (1 - \sqrt{2})^n], \; \dfrac{1}{4}[(1 + \sqrt{2})^{n+1} + (1 - \sqrt{2})^{n+1} - 2].$

4. $\dfrac{1}{2}(20n - 27 + 3^{3-n}), \; \dfrac{1}{4}(20n^2 - 34n + 27 - 3^{3-n}).$

5. $\dfrac{1}{9}[(23 - 3n)2^{n-2} + (-1)^n].$ **6.** $-\dfrac{1}{3} + \dfrac{4}{9e} + \dfrac{8e^2}{9}.$

EXERCISE XXXVIII (*Pages 378–379*)

1. $2^{n-2}.$

8. $u_n = \underline{|n+1} \left[u_0 + u_1 - 2u_0 \left\{ \dfrac{1}{\underline{|2}} - \dfrac{1}{\underline{|3}} + ... + (-1)^{n+1} \dfrac{1}{\underline{|n+1}} \right\} \right]$

10. $A \cdot 2^n + B \cdot 3^n + \dfrac{1}{4}(4n + 7).$ **11.** $A \cdot 2^n + B \cdot 3^n + \dfrac{1}{20} \cdot 7^{n+2}.$

12. $A\left(\dfrac{1+\sqrt{5}}{2}\right)^n + B\left(\dfrac{1-\sqrt{5}}{2}\right)^n - (n^2 + 6n + 13).$

14. $u_{2n} = \dfrac{2n-1}{2n} \cdot \dfrac{2n-3}{2n-2} \; ... \; \dfrac{1}{2} \cdot \dfrac{a}{u_1} \; ; \; u_{2n+1} = \dfrac{2n}{2n+1} \cdot \dfrac{2n-2}{2n-1} \; ... \; \dfrac{2}{3}u_1.$

EXERCISE XXXIX (*Page 392*)

1. $x^3 + 2x - 1.$

2. (*i*) $\dfrac{1}{3}n(4n^2 - 1);$ (*ii*) $\dfrac{n}{15}(3n^4 + 15n^3 + 25n^2 + 15n + 2).$

EXERCISE XL   (*Page 405*)

1. (*i*) $\dfrac{1}{2}, \dfrac{3}{8}, \dfrac{15}{38}, \dfrac{87}{222}$; (*ii*) $1, \dfrac{+2}{0}, \dfrac{+3}{-3}, \dfrac{+4}{-12}$.

3. $1 + \dfrac{1}{1+} \dfrac{1}{1+} \dfrac{1}{5+} \dfrac{1}{3+} \dfrac{1}{4}$.   **5.** $3 - \dfrac{1}{2-} \dfrac{1}{3-} \dfrac{1}{4-} \dfrac{1}{5}$.

6. $1 + \dfrac{1}{a-2+} \dfrac{1}{1+} \dfrac{1}{2(a-1)}$; $\dfrac{1}{1}, \dfrac{a-1}{a-2}, \dfrac{a}{a-1}, \dfrac{2a^2 - a - 1}{2a^2 - 3a}$.

EXERCISE XLI   (*Page 423*)

8. $\dfrac{86}{355} = 0.242253 \ldots$, $\dfrac{47}{194} = 0.242268 \ldots$.

10. $12\dfrac{1}{2}$ yards; 3 inches.

EXERCISE XLII   (*Pages 430–431*)

1. (*i*) 127, 55; (*ii*) 30, 13. **2.** $19 + 27t, 1 + 8t$.

3. $7 + 8t, 8 + 27t$. **4.** $9 + 13t, 20 + 29t$.

5. $29 + 41t, 12 + 17t$. **6.** $24 + 7t, 11 + 12t$.

7. $155 + 225t, 137 + 199t$. **8.** 6, 1.

9. 1, 20; 8, 9. **10.** 39, 8; 10, 20.

11. $\dfrac{3}{7} + \dfrac{5}{11}$. **12.** $\dfrac{4}{17} - \dfrac{3}{13}; \dfrac{10}{13} - \dfrac{13}{17}$.

13. 3, 4; 13, 10. **14.** 57, 43.

17. 67. **18.** $13, x = 205 - 17t; 12, x = 201 - 17t$.

20. 14, 2, 3; 1, 13, 8. **21.** 11, 3, 7.

22. 2, 1, 3; 3, 2, 2; 4, 3, 1. **23.** $2 \cdot 3 \cdot 7$.

24. 1149; 1149 + 1540t.

25.

2d.	3d.	9d.	1d.
3	2	1	2
3	3	2	1
3	6	1	1
6	4	1	1
6	1	2	1
9	2	1	1

26.

2s. 6d.	1s.	6d.	1d.
2	2	5	1
2	1	6	7
1	6	1	7
1	7	0	1

27. One; £12 12s. 8d.

EXERCISE XLIII　(*Page 438*)

12. 2, 9; 3, 6; 4, 13; 5, 7; 8, 15; 10, 12; 11, 14.

EXERCISE XLIV　(*Page 445*)

1. (*i*) $9 + 13t$;　(*ii*) $75 + 77t$;　(*iii*) $379 + 770t$.

2. (*i*) $149 + 179t$;　(*ii*) $30 + 179t$.

3. (*i*) $78 + 1217t$;　(*ii*) $1800 + 1861t$;　(*iii*) $540 + 1009t$;　(*iv*) $1683 + 1901t$.

8. $\dfrac{7}{32} + \dfrac{3}{5} + \dfrac{2}{11} - 1$.

9. $\dfrac{5}{7} + \dfrac{4}{11} + \dfrac{12}{13} - 2$.

10. $30k + 1$.

11. $\pm 17 + 77k$.

12. $3 + 23t$, $-7 + 23t$.

13. $\pm 13 + 66t$, $\pm 9 + 66t$.

14. $1735 + 3465t$, etc.

15. 2, 4; $x^2 + 1 \equiv$ (mod 7) has no solution.

16. $\pm 2, -3, -4$.

17. 3.

EXERCISE XLV　(*Pages 452, 454–455*)

1. 1, 10, 5, 12, etc., *i.e.* those of 19 in reverse order; 1, 7, 11, a period of 3, since 7 is the 6th number after 1, and 6 is a factor of $(19 - 1)$; 2^1, 2^5, 2^7, 2^{11}, 2^{13}, 2^{17}, or 2, 13, 14, 15, 3, 10.

2. 1, 3, 9, 10, 13, 5, 15, 11, 16, 14, 8, 7, 4, 12, 2, 6; 3^{2n+1}, *i.e.* every other number in the preceding, since $2n + 1$ is prime to $17 - 1$, or 3, 10, 5, 11, 14, 7, 12, 6.

12. 99999 is divisible by 41, hence recurring periods contain 5 digits; 2, 20, 36, 32, 37, the remainders in the division of 2 by 41.

14. If $n = 41$, $n' = 271$; $n = 123$, $n' = 903342366757$.

16. Because $p - 1$ occurs as a remainder.

EXERCISE XLVI　(*Page 456*)

1. 1, 7, 11.　　**2.** 15.　　**3.** 5.　　**4.** 25.

5. 8, 33　　**6.** 2, 10, 19.　　**7.** $19k \pm 9$, $19k^2 \pm 18k + 4$.

8. (25, 214), (54, 2157), (67, 4120).

EXERCISE XLVII　(*Page 458*)

1. 15.　　**2.** 10.　　**3.** 4.　　**4.** 4.

5. 5.

EXERCISE XLVIII　(*Page 467*)

2. $(-2, -1), (1, 2), (3, 4)$.　　**3.** $(-8, -9), (13, 14), (14, 15)$.

4. $(-6, -7)$ and two in $(0, 1)$.　　**5.** $(-9, -10)$ and two in $(3, 4)$.

6. $(-3, -4)$ and two in $(5, 6)$.　　**7.** $(2, 3), (3, 4)$.

8. Two in (1, 2).   **9.** (−2, −1), (0, 1) and two in (1, 2).
10. (−6, −5), (−1, 0), (4, 5).   **11.** One root in (1, 2).

EXERCISE XLIX (*Page 477*)

1. 1.3569, 1.6920.   **2.** 1.7838.
3. 2.6306166.   **4.** 16.0428539.
5. 0.3472964, 1.5320862, − 1.8793826. **6.** 0.1147994, 0.6907309.
7. 13.8440609, 14.2895592.   **8.** 5.2100150, 5.2973245.
9. 17.7459667, −1.9216606.   **10.** 1.2783089, 1.5511616.
11. 1.4007219, 1.5823564.   **12.** 1.0156820, 1.52858586.
13. 3.4334634, 3.6617081.   **14.** 1.7320508, 1.8666161.
15. 4.4707625, 4.7814825.   **16.** 2.59861938, 2.76190631.
17. 2.1010205, 2.1084952.   **18.** 7.3355540.
19. (*i*) 3.3548487;   (*ii*) 4.4641016;   (*iii*) 0.6386058.

EXERCISE LIII (*Pages 515–517*)

1. (*i*) 6720; (*ii*) 480.   **2.** $\lfloor n-1 / \lfloor r$.

7. $C_a^{m+a-1}.C_b^{m+b-1}.C_c^{m+c-1}$.   **9.** 1024.

10. 116280.
13. (*i*) 44; (*ii*) 20.   **17.** (*i*) 1895040; (*ii*) 145680.

20. $\lfloor n \cdot \lfloor\lfloor n$.
22. (*i*) 1771; (*ii*) 969; (*iii*) 885; (*iv*) 165; (*v*) 552.
27. (*i*) 57; (*ii*) 47; (*iii*) 15.   **28.** 23.
29. 42.   **30.** 15.

EXERCISE LIV (*Page 518*)

1. $C_3^n - C_3^p$.

EXERCISE LV (*Page 536–539*)

1. (*i*) 6/55, (*ii*) 4/11.   **3.** 14/33.
4. 5/33.   **6.** (*i*) 1/36; (*ii*) 5/12; (*iii*) 5/9.
7. 1/8.   **8.** (*i*) 1/12; (*ii*) 125/1296; (*iii*) 155/648.
9. 25.   **10.** 236/270725.
12. (*i*) 1/7; (*ii*) 2/7.
13. (*i*) $2/n$; (*ii*) $2(n - m - 1)/n(n - 1)$; (*iii*) $(m + 1)(2n - m - 2)/n(n - 1)$.
15. $1/2n$.   **18.** (*i*) 20/27, (*ii*) 496/729.
19. (*i*) $1/n$; (*ii*) $(n - 1)/2n$; (*iii*) $(n - 1)2n$.
21. $(n + 1)(3n + 2)$ pence.   **22.** (*i*) 2162/54145; (*ii*) 2257/54145.

23. (*i*) $\Sigma p_1 p_2 (1 - p_3)(1 - p_4) = \Sigma p_1 p_2 - 3\Sigma p_1 p_2 p_3 + 6 p_1 p_2 p_3 p_4$;

  (*ii*) $\Sigma p_1 p_2 - 2\Sigma p_1 p_2 p_3 + 3 p_1 p_2 p_3 p_4$.

24. 73/648.

25. $(0.55)^7$.

26. (*i*) $p = \lfloor 3 \cdot \dfrac{13}{51} \cdot \dfrac{13}{50} \cdot \dfrac{13}{49} = \dfrac{2197}{20825}$; (*ii*) $p' = \lfloor 4 \cdot \dfrac{12}{51} \cdot \dfrac{13}{50} \cdot \dfrac{13}{49} \cdot \dfrac{13}{48} = p$;

  (*iii*) $\dfrac{4}{13} p + \dfrac{9}{13} p' = p$.

27. (*i*) $p = 3 \cdot \dfrac{25}{51} \cdot \dfrac{26}{50} \cdot \dfrac{25}{49} = \dfrac{325}{833}$; (*ii*) $p' = 6 \cdot \dfrac{25}{51} \cdot \dfrac{24}{50} \cdot \dfrac{26}{49} \cdot \dfrac{25}{48} = p$;

  (*iii*) $\dfrac{4}{13} p + \dfrac{9}{13} p' = p$.

29. (*i*) $6/n(n-1)$; (*ii*) $6(n-3)/n(n-1)$;

  (*iii*) $(n-3)(n-4)/n(n-1)$.

33. (*i*) 1/6; (*ii*) 7/9.

34. (*i*) 1/4; (*ii*) 2/3.

35. 1/2.

36. (*i*) $pp'p''$ to $(1-p)(1-p')(1-p'')$; (*ii*) $pp'(1-p'')$ to $(1-p)(1-p')p''$.

EXERCISE LVI (*Page 555*)

5. (*i*) (2, 1), (50, 31); (*ii*) (3, 2), (29, 18);

  (*iii*) (5, 2), (12, 5).

6. (*i*) (25, 1) (623, 25), ... ; (7775, 312), ...

8. (*i*) $\left[\dfrac{1}{2} a(a-1) + 1, \dfrac{1}{2}(a-1) \right]$; (*ii*) $\left[\dfrac{1}{2} a(a+1) + 1, \dfrac{1}{2}(a+1) \right]$.

MISCELLANEOUS EXERCISES (A) (*Pages 556–568*)

2. n^2 or $n^2 - 1$ and $n^2 + 1$.

4. (*iii*) $\dfrac{1}{1} - \dfrac{1}{2} + \dfrac{1}{6} - \dfrac{1}{24} + \dfrac{1}{408}$; $\dfrac{1}{2} - \dfrac{1}{34} + \dfrac{1}{884} - \dfrac{1}{46852}$.

5. $a + c - b - d,\ a + d - b - c$. **6.** $x = -\dfrac{1}{2}$.

8. $36x^4 - 156x^2 + 25 = 0,\ \sqrt{\dfrac{2}{3}} \pm \sqrt{\dfrac{3}{2}}$.

10. $\sqrt{cx} + \sqrt{dy} + \sqrt{az} + \sqrt{bt} = 0,\ \sqrt{dx} + \sqrt{cy} + \sqrt{bz} + \sqrt{at} = 0$.

11. $y^3 - 3y^2 - y + 19 = 0$. **12.** $\pm \dfrac{1}{\sqrt{3}}, \dfrac{1 \pm \sqrt{-15}}{4}$

19. $\Sigma ab(\Sigma ab - \Sigma a^2)$.

21. $ab = c + 1$.

23. $x = \dfrac{1}{2}(n + 1)$ or $\dfrac{1}{2}\left\{n + 1 \pm \sqrt{(1 - n^2)}\right\}$.

26. $a + b \neq 0$ and $c_1(a - b) + c_2 - c_3 = 2a - b$.

29. (*i*) Convergent if $|x| < 1$ or if $x = -1$. Divergent if $|x| > 1$. When $x = 1$ the series is convergent if $p > 1$, divergent if $p \leq 1$.

    (*ii*) Convergent if $p > \dfrac{1}{2}$, divergent if $p \leq \dfrac{1}{2}$.

30. (*i*) $\dfrac{1}{2}(1 + 3^{n-1})$;

    (*ii*) $\dfrac{\sqrt{2}}{4}\{(1 + \sqrt{2})^n - (1 - \sqrt{2})^n\}$.

31. $d^4(a^3 + b^3 + c^3 + 2d^3) = a^3b^3c^3$.

33. $-\dfrac{a^2(a_2 - a_3)(a_3 - a_1)(a_1 - a_2)}{\Pi a_i \Pi(a - a_i)^2}$.

34. If $a > 9$, none real, $-7 \leq a \leq 9$, four real, $a \leq -7$ two real.

35. $46x = -24 + 15 \cdot 4^{\frac{1}{3}}k + 10 \cdot 2^{\frac{1}{3}}k^2$ where $k = 1$, ω or ω^2.

37. $x = \dfrac{3}{4}$, $y = \dfrac{1}{2}$, $z = \dfrac{1}{4}$.

39. $x = -2 \pm \sqrt{2}$ or $\dfrac{1}{2}(5 \pm \sqrt{17})$.

43. $206 + 210n$.

49. Equate the factors to $P + Q$, $P + \omega Q$, $P + \omega^2 Q$.

53. $\dfrac{x - 4}{x^2}\log(1 - x) - \dfrac{4}{x}$.

54. Convergent unless $x = \pm 1$.

56. $0 < y < 4(b - a)$.

59. $a = 1$, $b = 0$. [$y = 0$, $x = 1$, $x = 4$, are asymptotes; y is a minimum at $\left(-2, -\dfrac{1}{9}\right)$.]

63. $x = \dfrac{b^2 + c^2 - ab - ac}{2\sqrt{(\Sigma a^3 - 3abc)}}$, etc.

66. $64° \, 50'$ nearly.

67. $1 + \dfrac{x}{2} - \dfrac{x^2}{12} + \dfrac{x^3}{24}$.

71. $32/9$.

74. $-1 + \varepsilon - \dfrac{\varepsilon^2}{4}$, $-1 - \varepsilon - \dfrac{\varepsilon^2}{4}$.

76. (*i*) Since $\left(1 + \dfrac{1}{m}\right)^m < e < 3$, $\therefore \dfrac{2}{1} \cdot \left(\dfrac{3}{2}\right)^2 \cdot \left(\dfrac{4}{3}\right)^3 \cdots \left(\dfrac{n}{n-1}\right)^{n-1} < 3^{n-1}$.

    (*ii*) Consider the arithmetic and geometric means of the factors of
$$1^1 \cdot 2^2 \cdot 3^3 \cdot \ldots \cdot n^n$$
    when written at full length thus $1 \cdot 2 \cdot 2 \cdot 3 \cdot 3 \cdot 3. \ldots$.

79. $\dfrac{6(x^2 - 12x + 4)}{x^2 - 20x + 4}.$ **83.** 31.

85. Minimum values when (i) $x = a$, (ii) $x = \dfrac{1}{2}(a + b)$.

87. $\dfrac{27}{64}$ is a maximum value; 0 and $-\dfrac{128}{729}$ are minimum values.

91. $0,\ 2,\ \infty$.   **96.** $1 + \dfrac{\alpha}{2} + \dfrac{3\alpha^2}{8},\ 2 - \alpha,\ 3 + \dfrac{\alpha}{2} - \dfrac{3\alpha^2}{8}.$

99. Put $p = 1/m$, $q = 1/n$.   **102.** $9.46,\ -10.47$.

108. $4^4,\ 6^4,\ \left(5 \pm \sqrt{-151}\right)^4.$ **110.** -4.61015.

112. -0.68233.

123. Take $1 + x + \dfrac{x^2}{\lfloor 2} + \ldots + \dfrac{x^n}{\lfloor n} = e^x - \left(\dfrac{x^{n+1}}{\lfloor n+1} + \ldots\right).$

MISCELLANEOUS EXERCISES (B)   (*Pages 569–579*)

3. (i) $\dfrac{2}{3} + \dfrac{1}{3 \cdot 5} + \dfrac{4}{3 \cdot 5 \cdot 7} + \dfrac{6}{3 \cdot 5 \cdot 7 \cdot 9};$ (ii) $q_n = n;$

    (iii) $q_1 = 0,\quad q_{4n+2} = 3 + 6n,\quad q_{4n+3} = 5 + 6n,$
        $q_{4n+4} = q_{4n+5} = 2 + 2n,\quad$ for $n = 0, 1, 2, \ldots$.

6. $\dfrac{1}{12} n\,(n+1)^2(n+2).$

15. $3 - 3^{\frac{1}{3}} k - 3^{\frac{2}{3}} k^2$ and $3^{\frac{1}{3}} k$, where k has the values $1,\ \omega,\ \omega^2$.

21. n is of the form $6k + 1$.

22. The values of $-a$, b, c are $p^2\,\dfrac{(q-1)(r-1)}{(p-q)(p-r)}$ and similar expressions.

23. $1,\ e^{-\frac{1}{2}},\ 0.$ **29.** 16 and 0.

31. $2 + 8\varepsilon + 32\varepsilon^2$ and $\dfrac{1}{\varepsilon} - 3 - 7\varepsilon.$

33. (i) Convergent if $|x| < 4$, divergent if $x \geq 4$, oscillates if $x = -4$.
    (ii) Oscillates between finite limits.

35. $D = \pi\,(a_1 + a_2\omega + a_3\omega^2 + a_4\omega^3 + a_5\omega^4)^2$, where $\omega^5 = 1$.

36. Put $x = a + h$, $y = a + k$, $z = a + l$, where h, k and $l \to 0$.

41. 5.674619.   **44.** $(\log x)^2,\ 2^{\log x},\ x^2,\ x^{\log x},\ 2^x,\ e^x,\ x^x.$

48. $x = \dfrac{(\lambda - b)(\lambda - c)(\lambda - d)}{(a - b)(a - c)(a - d)},$ etc.

58. -1.   **59.** $(\log a - \log b)/(\log c - \log d)$.

64. Let $\alpha,\ \beta,\ (\alpha < \beta)$ be the roots of $\Sigma\ l(x - b)\ (x - c) = 0$.

    (*i*) If $l,\ m,\ n$ are odd, $x = \alpha$ gives a max. and $x = \beta$ a min. value.

    (*ii*) If $l,\ m,\ n$ are even, $x = \alpha$ and $x = \beta$ give max. value, and $x = a, b, c$ give min. values, each zero.

65. $x = \dfrac{c - a}{1 + ca},\quad y = \dfrac{c - b}{1 + cb},\quad z = \dfrac{a + b - c + abc}{1 + bc + ca - ab}.$

    Put $x = \tan X,\quad y = \tan Y, z = \tan Z$.

66. $\dfrac{1}{9}(8e^2 + 4e^{-1})$. The sum $= \displaystyle\sum_{1}^{\infty} \dfrac{2^{n+2} + (-1)^n(3n - 4)}{9\lfloor n - 1}.$

67. Converges absolutely if $|z| < \dfrac{1}{4}$, diverges if $|z| > \dfrac{1}{4}$. If $|z| = \dfrac{1}{4}$, it converges, but not absolutely.

68. $\dfrac{x - k}{(a' - k)(b' - k)} = \dfrac{y - k}{(a - k)(b - k)} = \dfrac{z - k}{(a' - k)(b - k)}.$

69. If $\lambda > 2$, the roots are all real; if $0 < \lambda < 2$, two are real and two imaginary; if $\lambda < 0$, all are imaginary; equal roots if $\lambda = 1$ or 2.

71. 0.8489892.

73. For the first part, use Chap. 14, 15, p. 224, putting $m = -1$. For the second part, show that the sum of the series $< \dfrac{n}{3n + 1} + \dfrac{n + 1}{4n + 1} < \dfrac{2}{3}.$

75. $(\log x)/x$ increases as x increases from 0 to e and decreases as x increases from e to ∞.

    (*i*) One root if $0 < n < 1$; two roots if $1 < n < e^{\frac{1}{e}}$; no root if $n > e^{\frac{1}{e}}$.

    (*ii*) $n^x > x^n$ if $x < n < e$, or if $e < n < x$.

    A rough graph of $(\log x)/x$ should be drawn.

84. Consider the number of ways in which 4 digits can be chosen so that their sum is 17 or 28.

85. This is so if $n\left(1 - \dfrac{n - 1}{n(m + 1)}\right)^{m+1} > 1$; then use Chap. 14, 11.

91. Convergent; use Euler's constant, p. 315, and comparison test (*iii*), p. 251.

95. If $k^2 < 4$, all the roots are imaginary; if $k^2 \geq 4$ two are real and two imaginary.

    If $\qquad\qquad k = \sqrt{2},\quad x = -\dfrac{\iota}{2} \pm \dfrac{3^{\frac{1}{4}}}{\sqrt{2}}\left(\iota\cos\dfrac{\theta}{2} - \sin\dfrac{\theta}{2}\right)$ or

$$\dfrac{\iota}{2} \pm \dfrac{3^{\frac{1}{4}}}{\sqrt{2}}\left(\iota\cos\dfrac{\theta}{2} + \sin\dfrac{\theta}{2}\right) \text{ where } \tan\theta = \sqrt{2}.$$

◼ ◼ ◼

■ ■ ■